MW00563140

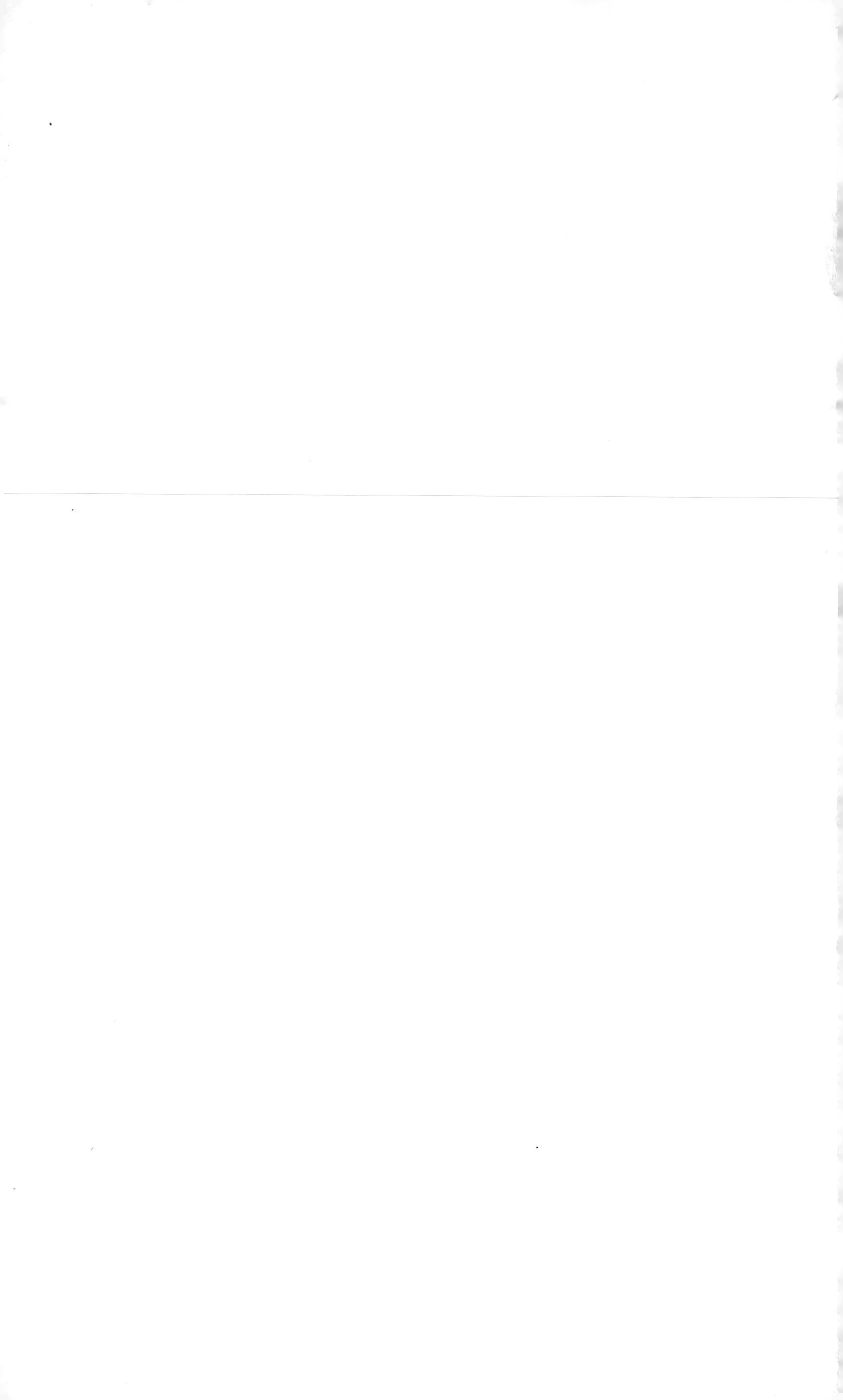

IMAGE PROCESSING AND PATTERN RECOGNITION

IEEE Press
445 Hoes Lane
Piscataway, NJ 08854

IMAGE PROCESSING AND PATTERN RECOGNITION

Fundamentals and Techniques

FRANK Y. SHIH

Published by John Wiley & Sons, Inc., Hoboken, New Jersey.
Published simultaneously in Canada.

Library of Congress Cataloging-in-Publication Data is Available

Shih, Frank Y.
Image processing and pattern recognition : fundamentals and techniques / Frank Shih.
p. cm.
ISBN 978-0-470-40461-4 (cloth)
1. Image processing. 2. Signal processing. 3. Pattern recognition systems.
I. Title.
TA1637.S4744 2010
621.36′7–dc22 2009035856

Printed in the United States of America

10 9 8 7 6 5 4 3 2 1

CONTENTS

PART I
FUNDAMENTALS

PART I

FUNDAMENTALS

CHAPTER 1

INTRODUCTION

An image is a subset of a signal. A *signal* is a function that conveys information generally about the behavior of a physical system or attributes of some phenomenon. A simple example is the traffic signal that uses three universal color codes (red, yellow, and green) signaling the moment to stop, drive, or walk. Although signals can be represented in many ways, in all cases the information is contained in a pattern of variations of some form, and with that information is typically transmitted and received over a medium. Electrical quantities such as current and voltage are called *electrical signals*, which are often used in radio, radar, sonar, telephone, television, and many other areas. An acoustic wave signal can convey speech or music information, in which people often speak of a strong or weak signal when the sound is referred to its clarity and audibility. A thermocouple can convey temperature, and a pH meter can convey the acidity of a solution.

A signal may take a form of time variations or a spatially varying pattern. Mathematically speaking, signals are represented as functions of one or more independent variables that can be either continuous or discrete. *Continuous-time* signals are defined at a continuum of the time variable. *Discrete-time* signals are defined at discrete instants of time. *Digital* signals are those for which both time and amplitude are discrete. The continuous-time and continuous-amplitude signals are called *analog* signals. Analog signals that have been converted to digital forms can be processed by a computer or other digital devices.

Signal processing is the process of extracting information from the signal. *Digital signal processing* (DSP) is concerned with the representation of signals by sequences of numbers or symbols and processing of these sequences. It was initiated in the seventeenth century and has become an important modern tool in the tremendously diverse fields of science and technology. The purpose of such processing is to estimate characteristic parameters of a signal or to transform a signal into a form that is more sensible to human beings. DSP includes subfields such as digital image processing, video processing, statistical signal processing, signal processing for communications, biomedical signal processing, audio and speech signal processing, sonar and radar signal processing, sensor array processing, spectral estimation, and so on.

Human beings possess a natural signal processing system. "Seeing" takes place in the visual system and "hearing" takes place in the auditory system. Human visual system (HVS) plays an important role in navigation, identification, verification, gait, gesture, posture, communication, psychological interpretation, and so on. Human

Image Processing and Pattern Recognition by Frank Shih

auditory system converts sound waves into nerve impulses, to analyze auditory events, remember and recognize sound sources, and perceive acoustic sequences. As the speed, capability, and economic advantages of modern signal processing devices continue to increase, there is simultaneously an increase in efforts aimed at developing sophisticated, real-time automatic systems capable of emulating human abilities. Because of digital revolution, digital signals have been increasingly used. Most household electronic devices are based entirely or almost entirely upon digital signals. The entire Internet is a network of digital signals, as is modern mobile phone communication.

1.1 THE WORLD OF SIGNALS

The world is filled with many kinds of signals; each has its own physical meaning. Sometimes the human body is incapable of receiving a special signal or interpreting (decoding) a signal, so the information that the signal intends to convey cannot be captured. Those signals are not to be said nonsense or insignificant, but conversely they are exactly what people are working very hard to understand. The more we learn from the world's signals, the better living environment we can provide. Furthermore, some disaster or damage can be avoided if a warning signal can be sensed in advance. For example, it was recorded historically that animals, including rats, snakes, and weasels, deserted the Greek city of Helice in droves just days before a quake devastated the city in 373 B.C. Numerous claims have been made that dogs and cats usually behave strangely before earthquake by barking, whining, or showing signs of nervousness and restlessness.

The characteristics of a signal may be one of a broad range of shapes, amplitudes, time durations, and perhaps other physical properties. Based on the sampling of time axis, signals can be divided into continuous-time and discrete-time signals. Based on the sampling of time and amplitude axes, signals can be divided into analog and digital signals. If signals repeat in some period, they are called periodic signals; otherwise, aperiodic or nonperiodic signals. If each value of a signal is fixed by a mathematical function, it is called a deterministic signal; otherwise, a random signal that has uncertainty about its behavior. In the category of dimensionality, signals are divided into one-dimensional (1D), two-dimensional (2D), three-dimensional (3D), and multidimensional signals, which are further explained below.

1.1.1 One-Dimensional Signals

A 1D signal is usually modeled as an ensemble of time waveforms, for example, $x(t)$ or $f(t)$. One-dimensional signal processing has a rich history, and its importance is evident in such diverse fields as biomedical engineering, acoustics (Beranek, 2007), sonar (Sun et al., 2004), radar (Gini et al., 2001), seismology (Al-Alaoui, 2001), speech communication, and many others. When we use a telephone, our voice is converted to an electrical signal and through telecommunication systems circulates around the Earth. The radio signals, which are propagated through free space and by radio receivers, are converted into sound. In speech transmission and recognition, one may wish to extract some characteristic parameters of the linguistic messages, representing the temporal and spectral behavior of acoustical speech input. Alternatively, one may wish to remove interference, such as noise, from the signal or to modify the signal to present it in a form more easily interpreted by an expert.

1.1.2 Two-Dimensional Signals

Signal processing problems are not confined to 1D signals. A 2D signal is a function of two independent variables, for example, $f(x, y)$. In particular, one is concerned with the functional behavior in the form of an intensity variation over the (x, y)-plane. Everyday scenes viewed by a human observer can be considered to be composed of illuminated objects. The light energy reflected from these objects can be considered to form a 2D intensity function, which is commonly referred to as an *image*.

As a result of numerous applications, not least as a consequence of cheap computer technology, image processing now influences almost all areas of our daily life: automated acquisition, processing and production of documents, industrial process automation, acquisition and automated analysis of medical images, enhancement and analysis of aerial photographs for detection of forest fires or crop damage, analysis of satellite weather photos, and enhancement of television transmission from lunar and deep-space probes.

1.1.3 Three-Dimensional Signals

Photographs of a still scene are the images that are functions of the (x, y)-plane. By adding a time variable, the 3D signals represent image sequences of a dynamic scene that are called *video signals*. Computer analysis of image sequences requires the development of internal representations for the entities in a depicted scene as well as for discernible changes in appearance and configuration of such entities. More fundamental approaches result from efforts to improve application-oriented solutions. Some illustrative examples are given as follows.

Image sequences obtained from satellite sensors are routinely analyzed to detect and monitor changes. Evaluation of image series recorded throughout the growth and harvest periods can result in more reliable cover type mapping as well as improved estimates of crop field. Very important is the determination of cloud displacement vector fields. These are used to estimate wind velocity distributions that in turn are employed for weather prediction and meteorological modeling (Desportes et al., 2007).

Biomedical applications are concerned with the study of growth, transformation, and transport phenomena. Angiocardiography, blood circulation, and studies of metabolism are the primary areas of medical interest for the evaluation of temporal image sequences (Charalampidis et al., 2006). Architects who have to design pedestrian circulation areas would appreciate quantitative data about how pedestrians walk in halls and corridors. Efforts to extract such data from TV-frame sequences could be considered as behavioral studies. They might as well be assigned to a separate topic such as object tracking (Qu and Schonfeld, 2007), which is of special concern in cases of traffic monitoring (Zhou et al., 2007), target tracking, and visual feedback for automated navigation (Negahdaripour and Xun, 2002; Xu and Tso, 1999).

1.1.4 Multidimensional Signals

When a signal is represented in more than one dimension, it is often called a multidimensional signal. As discussed in previous sections, an image is a two-dimensional signal, and a video is a three-dimensional signal. A multidimensional signal is vector valued and may be a function of multiple relevant independent variables. One chooses the variable domain in which to process a signal by making an informed guess as to which domain best represents the essential characteristics of the signal. Multidimensional signal processing is an innovative field interested in developing technology that can capture and analyze information in more than one dimension. Some of its applications include 3D face modeling (Roy-Chowdhury et al., 2004), 3D object tracking (Wiles et al., 2001), and multidimensional signal filtering.

The need for a generally applicable artificial intelligence approach for optimal dimensionality selection in high-dimensional signal spaces is evident in problems involving vision since the dimensionality of the input data often exceeds 10^6. It is likely to fail if vision problems are handled by reducing the dimensionality by means of throwing away almost certain available information in a basically ad hoc manner. Therefore, designing a system capable of learning the relevant information extraction mechanisms is critical.

1.2 DIGITAL IMAGE PROCESSING

Images are produced by a variety of physical devices, including still and video cameras, scanners, X-ray devices, electron microscopes, radar, and ultrasound, and are used for a variety of purposes, including entertainment, medical, business, industrial, military, civil, security, and scientific. The interests in digital image processing stem from the improvement of pictorial information for human interpretation and the processing of scene data for autonomous machine perception.

Webster's Dictionary defines an image as: "An image is a representation, likeness, or imitation of an object or thing, a vivid or graphic description, something introduced to represent something else." The word "picture" is a restricted type of image. Webster's Dictionary defines a picture as: "A representation made by painting, drawing, or photography; a vivid, graphic, accurate description of an object or thing so as to suggest a mental image or give an accurate idea of the thing itself." In image processing, the word "picture" is sometimes equivalent to "image."

Digital image processing starts with one image and produces a modified version of that image. Webster's Dictionary defines digital as: "The calculation by numerical methods or discrete units," defines a digital image as: "A numerical representation of an object," defines processing as: "The act of subjecting something to a process," and

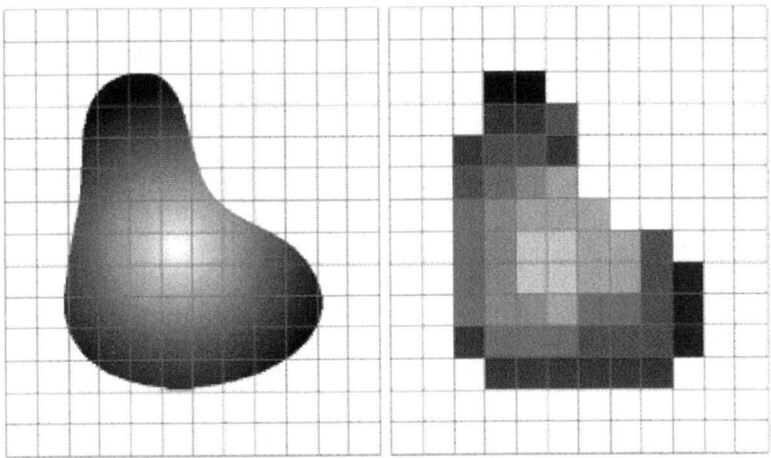

Figure 1.1 Image digitization. (Courtesy of Gonzalez and Woods, 2008)

defines a process as: "A series of actions or operations leading to a desired result." An example of a process is car wash that changes an automobile from dirty to clean.

Digital image analysis is a process that converts a digital image into something other than a digital image, such as a set of measurement data or a decision. Image digitization is a process that converts a pictorial form to numerical data. A digital image is an image $f(x, y)$ that has been discretized in both spatial coordinates and brightness (intensity). The image is divided into small regions called *picture elements* or *pixels* (see Fig. 1.1).

Image digitization includes image sampling (i.e., digitization of spatial coordinates (x, y)) and gray-level quantization (i.e., brightness amplitude digitization). An image is represented by a rectangular array of integers. The image sizes and the number of gray levels are usually integer powers of 2. The number at each pixel represents the brightness or darkness (generally called the intensity) of the image at that point. For example, Figure 1.2 shows a digital image of size 8×8 with 1 byte (i.e., 8 bits = 256 gray levels) per pixel.

1	8	219	51	69	171	81	41
94	108	20	121	17	214	15	74
233	93	197	83	177	215	183	78
41	84	118	62	210	71	122	38
222	73	197	248	125	226	210	5
35	36	127	5	151	2	197	165
196	180	142	52	173	151	243	164
254	62	172	75	21	196	126	224

Figure 1.2 A digital image and its numerical representation.

The quality of an image strongly depends upon the number of samples and gray levels; the more are these two, the better would be the quality of an image. But, this will result in a large amount of storage space as well because the storage space for an image is the product of dimensions of an image and the number of bits required to store gray levels. At lower resolution, an image can result in checkerboard effect or graininess. When an image of size 1024×1024 is reduced to 512×512, it may not show much deterioration, but when reduced to 256×256 and then rescaled back to 1024×1024 by duplication, it might show discernible graininess.

The visual quality of an image required depends upon its applications. To achieve the highest visual quality and at the same time the lowest memory requirement, we can perform fine sampling of an image in the neighborhood of sharp gray-level transitions and coarse sampling in the smooth areas of an image. This is known as sampling based on the characteristics of an image (Damera-Venkata et al., 2000). Another method, known as tapered quantization, can be used for the distribution of gray levels by computing the occurrence frequency of all allowed levels. Quantization level is finely spaced in the regions where gray levels occur frequently, but when gray levels occur rarely in other regions, the quantization level can be coarsely spaced. Images with large amounts of details can sometimes still enjoy a satisfactory appearance despite possessing a relatively small number of gray levels. This can be seen by examining isopreference curves using a set of subjective tests for images in the Nk-plane, where N is the number of samples and k is the number of gray levels (Huang, 1965).

In general, image processing operations can be categorized into four types:

1. *Pixel operations:* The output at a pixel depends only on the input at that pixel, independent of all other pixels in that image. Thresholding, a process of making the corresponding input pixels above a certain threshold level white and others black, is simply a pixel operation. Other examples include brightness addition/subtraction, contrast stretching, image inverting, log, and power law.
2. *Local (neighborhood) operations:* The output at a pixel depends on the input values in a neighborhood of that pixel. Some examples are edge detection, smoothing filters (e.g., the averaging filter and the median filter), and sharpening filters (e.g., the Laplacian filter and the gradient filter). This operation can be adaptive because results depend on the particular pixel values encountered in each image region.
3. *Geometric operations:* The output at a pixel depends only on the input levels at some other pixels defined by geometric transformations. Geometric operations are different from global operations, such that the input is only from some specific pixels based on geometric transformation. They do not require the input from all the pixels to make its transformation.
4. *Global operations:* The output at a pixel depends on all the pixels in an image. It may be independent of the pixel values in an image, or it may reflect statistics calculated for all the pixels, but not a local subset of pixels. A popular distance transformation of an image, which assigns to each object pixel the minimum distance from it to all the background pixels, belongs to a global operation.

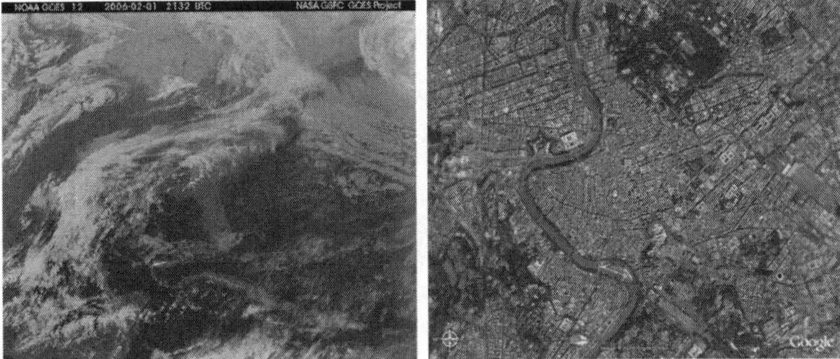

Figure 1.3 Remote sensing images for tracking Earth's climate and resources.

Other examples include histogram equalization/specification, image warping, Hough transform, and connected components.

Nowadays, there is almost no area that is not impacted in some way by digital image processing. Its applications include

1. *Remote sensing:* Images acquired by satellites and other spacecrafts are useful in tracking Earth's resources, solar features, geographical mapping (Fig. 1.3), and space image applications (Fig. 1.4).
2. *Image transmission and storage for business:* Its applications include broadcast television, teleconferencing, transmission of facsimile images for office automation, communication over computer networks, security monitoring systems, and military communications.
3. *Medical processing:* Its applications include X-ray, cineangiogram, transaxial tomography, and nuclear magnetic resonance (Fig. 1.5). These images may be

Figure 1.4 Space image applications.

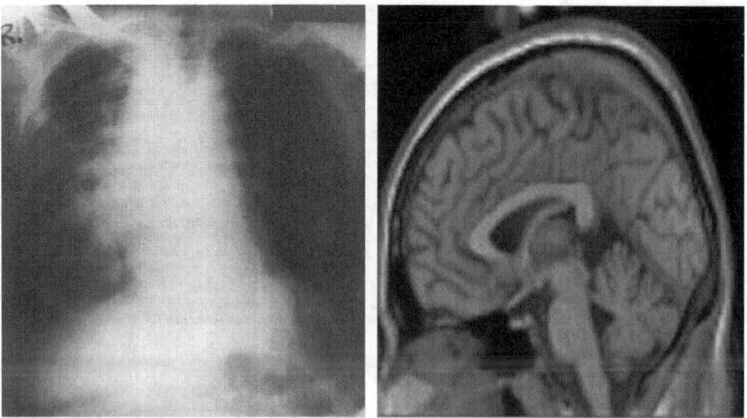

Figure 1.5 Medical imaging applications.

used for patient screening and monitoring or for detection of tumors or other diseases in patients.

4. *Radar, sonar, and acoustic image processing:* For example, the detection and recognition of various types of targets and the maneuvering of aircraft (Fig. 1.6).
5. *Robot/machine vision:* Its applications include the identification or description of objects or industrial parts in 3D scenes (Fig. 1.7).

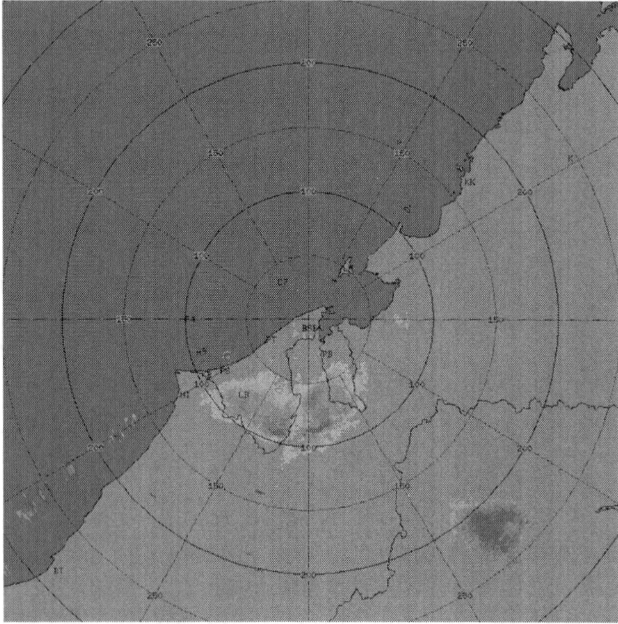

Figure 1.6 Radar imaging.

Figure 1.7 Robot and machine vision applications.

1.3 ELEMENTS OF AN IMAGE PROCESSING SYSTEM

Elements of an image processing system include

1. *Image acquisition:* A physical device that is sensitive to a band in the electromagnetic energy spectrum can produce an electrical signal output. A digitizer is used for converting the electrical signal output into a digital form. Digital images can be obtained by conversion of the analog images (such as 35 mm prints, slides, transparencies, or reflective art) into digital images with a scanner, or else by directly capturing the object or scene into digital forms by means of a digital camera or video-capturing device.
2. *Storage:*
   - (**a**) Short-term storage for use during processes. One of the means of providing short-term storage is computer memory. Another is a specialized board, called a *frame buffer*.
   - (**b**) Online storage for relatively fast recall.
   - (**c**) Archival storage characterized by infrequent access. The term "archival quality" is used to designate materials that are permanent and durable, and therefore can be safely used for preservation purposes. The objective of archival storage is the protection against tampering, deletion, viruses, and disaster.
3. *Processing:* Most image processing functions can be implemented in software, running on a host computer.
4. *Communication:* Cable modem Internet services on average promise higher levels of bandwidth than DSL Internet services, and this bandwidth roughly translates to communication speed. Cable Internet theoretically runs faster than DSL. Cable technology can support approximately 30 Mbps (megabits per second) of bandwidth, whereas most forms of DSL cannot reach 10 Mbps.
5. *Display:* An image display device may take the form of an illuminating picture lamp providing means by which images may be illuminated by a light source on a selectable and removably attachable basis. Monochrome and color monitors are the principal display devices used. Other display media include random access cathode ray tubes (CRTs) and printing devices.

To illustrate the systematic procedures of an image processing system, we give an example of human face identification (Adler and Schuckers, 2007). The problem domain is the faces of people. The objective is to associate the face with the identity of the person. The output is a person's unique identifier (e.g., social security number). The necessary procedures to achieve the goal of face identification could include

1. *Image acquisition:* The face image could be acquired through a high-resolution still digital camera and compressed to an image file.
2. *Preprocessing:* The acquired image may be enhanced by improving contrast, sharpness, color, and so on.
3. *Segmentation*: The image may first be cropped to only the facial area. Then, the face may be segmented into eyes, mouth, nose, chin, and so on.
4. *Representation and description:* In this step, each of the segmented areas may be characterized by statistical data, for example, principal components analysis, texture, aspect ratios of eyes and nose, or the color of eyes.
5. *Matching recognition and interpretation:* This step may involve using the characteristics derived in the previous step to match each individually segmented area based on specific recognition algorithms. For example, eyes may be processed to determine, based on its features, what class of eye it is. Then, all of these interpretations are used to create a composite description of the "ensemble," perhaps in the form of a feature vector for the subject.
6. *Knowledge base:* Finally, the feature vector above may be fed to a knowledge base of all known subjects to associate it with one of the subjects in the database, thus returning perhaps the individual's social security number or perhaps a confidence score of the match.

APPENDIX 1.A SELECTED LIST OF BOOKS ON IMAGE PROCESSING AND COMPUTER VISION FROM YEAR 2000

Acharya, T. and Ray, A. K., *Image Processing: Principles and Applications*, Wiley, Sept. 2005.

Barner, K. E. and Arce, G. R., *Nonlinear Signal and Image Processing: Theory, Methods, and Applications*, CRC Press, 2004.

Bhanu, B. and Pavlidis, I., *Computer Vision Beyond the Visible Spectrum*, Springer, 2004.

Bose, T., *Digital Signal and Image Processing*, Wiley, Nov. 2003.

Burger, W. and Burge, M. J., *Digital Image Processing: An Algorithmic Introduction Using Java*, Springer, 2008.

Chan, T. F. and Shen, J., *Image Processing and Analysis: Variational, PDE, Wavelet, and Stochastic Methods*, SIAM, 2005.

Chen, C. H., *Image Processing for Remote Sensing*, CRC Press, 2008.

Cyganek, B. and Siebert, J. P., *An Introduction to 3-D Computer Vision Techniques and Algorithms*, Wiley, Feb. 2009.

Dhawan, A., *Medical Image Analysis*, Wiley, July 2003.

Fisher, R. B., *Dictionary of Computer Vision and Image Processing*, Wiley, 2005.

Forsyth, D. A. and Ponce, J., *Computer Vision: A Modern Approach*, Prentice Hall, 2003.

Gonzalez, R. C. and Woods, R. E., *Digital Image Processing*, 3rd edition, Prentice Hall, 2008.

Gonzalez, R. C., Woods, R. E., and Eddins, S. L., *Digital Image Processing Using MATLAB*, Prentice Hall, 2003.

Hartley, R. and Zisserman, A., *Multiple View Geometry in Computer Vision*, Cambridge University Press, 2003.

Hornberg, A., *Handbook of Machine Vision*, Wiley, 2006.

Jähne, B., *Digital Image Processing*, Springer, 2005.

Koschan, A. and Abidi M., *Digital Color Image Processing*, Wiley, Apr. 2008.

Kulkarni, A. D., *Computer Vision and Fuzzy-Neural Systems*, Prentice Hall, 2001.

Lukac, R. and Plataniotis, K. N., *Color Image Processing: Methods and Applications*, CRC Press, 2007.

Maître, H., *Image Processing*, Wiley, Aug. 2008.

Medioni, G. and Kang, S. B., *Emerging Topics in Computer Vision*, Prentice Hall, 2005.

McAndrew, A., *Introduction to Digital Image Processing with Matlab*, Thomson Course Technology, 2004.

Mitra, S. K. and Sicuranza, G. L., *Nonlinear Image Processing*, Academic Press, 2001.

Morris, T., *Computer Vision and Image Processing*, Palgrave Macmillan, 2003.

Nixon, M. S. and Aguado, A. S., *Feature Extraction and Image Processing*, Academic Press, 2008.

Paragios, N., Chen, Y., and Faugeras, O., *Handbook of Mathematical Models in Computer Vision*, Springer, 2006.

Petrou, M. and Sevilla, P. G., *Image Processing: Dealing With Texture*, Wiley, Mar. 2006.

Pitas, I., *Digital Image Processing Algorithms and Applications*, Wiley–IEEE, 2000.

Pratt, W. K., *Digital Image Processing: PIKS Scientific Inside*, 4th edition, Wiley, 2007.

Ritter, G. X. and Wilson, J. N., *Handbook of Computer Vision Algorithms in Image Algebra*, CRC Press, 2001.

Russ, J. C., *The Image Processing Handbook*, CRC Press, 2006.

Sebe, N., *Machine Learning in Computer Vision*, Springer, 2005.

Shapiro, L. G. and Stockman, G. C., *Computer Vision*, Prentice Hall, 2001.

Shih, F.Y., Image Processing and Mathematical Morphology: Fundamentals and Applications, CRC Press, 2009.

Snyder, W. E. and Qi, H., *Machine Vision*, Cambridge University Press, 2004.

Sonka, M., Hlavac, V., and Boyle, R., *Image Processing, Analysis, and Machine Vision*, Thomson Wadsworth, 2007.

Tuzlukov, V. P., *Signal and Image Processing in Navigational Systems*, CRC Press, 2005.

Umbaugh, S. E., *Computer Imaging: Digital Image Analysis and Processing*, CRC Press, 2005.

Whelan, P. F. and Molloy, D., *Machine Vision Algorithms in Java: Techniques and Implementation*, Springer, 2001.

Woods, J. W., *Multidimensional Signal, Image, and Video Processing and Coding*, Academic Press, 2006.

Zuech, N., *Understanding and Applying Machine Vision*, CRC Press, 2000.

1.A.1 Selected List of Books on Signal Processing from Year 2000

Blackledge, J. M., *Digital Signal Processing: Mathematical and Computational Methods, Software Development, and Applications*, Horwood Publishing, 2006.

Blanchet, G. and Charbit, M., *Digital Signal and Image Processing Using MATLAB*, Wiley, May 2006.

Garello, R., *Two-Dimensional Signal Analysis*, Wiley, Apr. 2008.

Gaydecki, P., *Foundations of Digital Signal Processing: Theory, Algorithms, and Hardware Design*, IET, 2004.

Gray, R. M. and Davisson, L. D., *An Introduction to Statistical Signal Processing*, Cambridge University Press, 2004.

Ifeachor, E. C. and Jervis, B. W., *Digital Signal Processing: A Practical Approach*, Prentice Hall, 2002.

Ingle, V. K. and Proakis, J. G., *Digital Signal Processing Using MATLAB*, Brooks/Cole, 2000.

Khan, A. A., *Digital Signal Processing Fundamentals*, Da Vinci Engineering Press, 2005.

Mitra, S. K., *Digital Signal Processing: A Computer-Based Approach*, McGraw-Hill, 2001.

Narasimhan, S. V. and Veena, S., *Signal Processing: Principles and Implementation*, Alpha Science Int'l Ltd., 2005.

Proakis, J. G. and Manolakis, D. G., *Digital Signal Processing*, Pearson Prentice Hall, 2007.

Quinquis, A., *Digital Signal Processing Using Matlab*, Wiley, Apr. 2008.

Rockmore, D. N. and Healy, D. M., *Modern Signal Processing*, Cambridge University Press, 2004.

Sundararajan, D., *Digital Signal Processing: Theory and Practice*, World Scientific, 2003.

Wang, B.-C., *Digital Signal Processing Techniques and Applications in Radar Image Processing*, Wiley, Aug. 2008.

Wysocki, T., Honary, B., and Wysocki, B. J., *Signal Processing for Telecommunications and Multimedia*, Springer, 2005.

1.A.2 Selected List of Books on Pattern Recognition from Year 2000

Abe, S., *Support Vector Machines for Pattern Classification*, Springer, 2005.

Bhagat, P., *Pattern Recognition in Industry*, Elsevier, 2005.

Bishop, C. M., *Pattern Recognition and Machine Learning*, Springer, 2006.

Bunke, H., Kandel, A., and Last, M., *Applied Pattern Recognition*, Springer, 2008

Chen, D. and Cheng, X., *Pattern Recognition and String Matching*, Springer, 2002.

Corrochano, E. B., *Handbook of Geometric Computing: Applications in Pattern Recognition, Computer Vision, Neuralcomputing, and Robotics*, Springer, 2005.

Duda, R. O., Hart, P. E., and Stork, D. G., *Pattern Classification*, 2nd edition, Wiley, 2003.

Dunne, R. A., *A Statistical Approach to Neural Networks for Pattern Recognition*, Wiley, July 2007

Gibson, W., *Pattern Recognition*, Penguin Group, 2005.

Hand, D. J., Mannila, H., and Smyth, P., *Principles of Data Mining*, MIT Press, Aug. 2001.

Hastie, T., Tibshirani, R., and Fridman, J., *The Elements of Statistical Learning: Data Mining, Inference, and Prediction*, Springer, 2001.

Hyvärinen, A., Karhunen, J., and Oja, E., *Independent Component Analysis*, Wiley, 2001.

Kuncheva, L. I., *Combining Pattern Classifiers: Methods and Algorithms*, Wiley–IEEE, 2004.

McLachlan, G. J. and Peel, D., *Finite Mixture Models*, Wiley, New York, 2000.

Pal, S. K. and Pal, A., *Pattern Recognition: From Classical to Modern Approaches*, World Scientific, 2001.

Raudys, S., *Statistical and Neural Classifiers*, Springer, 2001.

Schlesinger, M. I. and Hlavác, V., *Ten Lectures on Statistical and Structural Pattern Recognition*, Kluwer Academic Publishers, 2002.

Suykens, A. K., Horvath, G., Basu, S., Micchelli, C., and Vandewalle, J., *Advances in Learning Theory: Methods, Models and Applications, NATO Science Series III: Computer & Systems Sciences*, vol. 190, IOS Press, Amsterdam, 2003.

Theodoridis, S. and Koutroumbas, K., *Pattern Recognition*, Academic Press, 2003.

Webb, A. R., *Statistical Pattern Recognition*, 2nd edition, Wiley, 2002.

REFERENCES

Adler, A. and Schuckers, M. E., "Comparing human and automatic face recognition performance," *IEEE Trans. Syst. Man Cybernet. B*, vol. 37, no. 5, pp. 1248–1255, Oct. 2007.

Al-Alaoui, M. A., "Low-frequency differentiators and integrators for biomedical and seismic signals," *IEEE Trans. Circuits Syst. I: Fundam. Theory Appl.*, vol. 48, no. 8, pp. 1006–1011, Aug. 2001.

Beranek, L. L., "Seeking concert hall acoustics," *IEEE Signal Process. Mag.*, vol. 24, no. 5, pp. 126–130, Sept. 2007.

Charalampidis, D., Pascotto, M., Kerut, E. K., and Lindner, J. R., "Anatomy and flow in normal and ischemic microvasculature based on a novel temporal fractal dimension analysis algorithm using contrast enhanced ultrasound," *IEEE Trans. Med. Imaging*, vol. 25, no. 8, pp. 1079–1086, Aug. 2006.

Damera-Venkata, N., Kite, T. D., Geisler, W. S., Evans, B. L., and Bovik, A. C., "Image quality assessment based on a degradation model," *IEEE Trans. Image Process.*, vol. 9, no. 4, pp. 636–650, Apr. 2000.

Desportes, C., Obligis, E., and Eymard, L., "On the wet tropospheric correction for altimetry in coastal regions," *IEEE Trans. Geosci. Remote Sens.*, vol. 45, no. 7, pp. 2139–2149, July 2007.

Gini, F., Farina, A., and Greco, M., "Selected list of references on radar signal processing," *IEEE Trans. Aerospace Electron. Syst.*, vol. 37, no. 1, pp. 329–359, Jan. 2001.

Gonzalez, R. C. and Woods, R. E., Digital Image Processing, 3rd edition, Prentice Hall, 2008.

Huang, T. S., "PCM picture transmission," *IEEE Spectrum*, vol. 2, no. 12, pp. 57–63, Dec. 1965.

Negahdaripour, S. and Xun X., "Mosaic-based positioning and improved motion-estimation methods for automatic navigation of submersible vehicles," *IEEE J. Oceanic Eng.*, vol. 27, no. 1, pp. 79–99, Jan. 2002.

Qu, W. and Schonfeld, D., "Real-time decentralized articulated motion analysis and object tracking from videos," *IEEE Trans. Image Process.*, vol. 16, no. 8, pp. 2129–2138, Aug. 2007.

Roy-Chowdhury, A. K., Chellappa, R., and Keaton, T., "Wide baseline image registration with application to 3-D face modeling," *IEEE Trans. Multimedia*, vol. 6, no. 3, pp. 423–434, June 2004.

Sun, Y., Willett, P., and Lynch, R., "Waveform fusion in sonar signal processing," IEEE Trans. Aerospace Electron. Syst., vol. 40, no. 2, pp. 462–477, Apr. 2004.

Wiles, C. S., Maki, A., and Matsuda, N., "Hyperpatches for 3-D model acquisition and tracking," *IEEE Trans. Pattern Anal. Mach. Intell.*, vol. 23, no. 12, pp. 1391–1403, Dec. 2001.

Xu, W. L. and Tso, S. K., "Sensor-based fuzzy reactive navigation of a mobile robot through local target switching," *IEEE Trans. Syst. Man Cybernet. C*, vol. 29, no. 3, pp. 451–459, Aug. 1999.

Zhou, J., Gao, D., and Zhang, D., "Moving vehicle detection for automatic traffic monitoring," *IEEE Trans. Veh. Technol.*, vol. 56, no. 1, pp. 51–59, Jan. 2007.

CHAPTER 2

MATHEMATICAL PRELIMINARIES

Most images are recorded and processed in the time domain or spatial domain. The spatial domain refers to the aggregate of pixels composing an image, and the spatial-domain processing involves operations that apply directly on these pixels. However, it is sometimes convenient and efficient to process images in the frequency domain because edge pixels generally correspond to high-frequency components and interior pixels of an object region correspond to low-frequency components. In this chapter, the mathematical preliminaries that are often used in image processing for converting an image from spatial domain to frequency domain are introduced. These include Laplace transform, Fourier transform, Z-transform, cosine transform, and wavelet transform.

2.1 LAPLACE TRANSFORM

Laplace transform is named after Pierre Simon Laplace (1749–1827), a French mathematician and astronomer. Given a function $x(t)$ of the continuous-time variable t, the Laplace transform, denoted by $X(s)$, is a function of the complex variable $s=\sigma + j\omega$ defined by

$$X(s) = \int_0^\infty x(t)e^{-st}\, dt \tag{2.1}$$

Note that time-domain signals are denoted by lowercase letters and the Laplace transforms (i.e., the frequency-domain signals) by uppercase letters.

Contrarily, given a Laplace transform $X(s)$, one can compute $x(t)$ from $X(s)$ by taking the inverse Laplace transform of $X(s)$ as

$$x(t) = \frac{1}{2\pi j} \lim_{\omega \to \infty} \int_{\sigma - j\omega}^{\sigma + j\omega} X(s)e^{st}\, ds \tag{2.2}$$

The integral is evaluated along the path $s = \sigma + j\omega$ in the complex plane from $\sigma - j\infty$ to $\sigma + j\infty$, where σ is any fixed real number for which $s = \sigma$ is a point in the region of absolute convergence of $X(s)$. Equation (2.1) is called the *one-sided Laplace transform*,

Image Processing and Pattern Recognition by Frank Shih

which depends only on the values of the signal $x(t)$ for $t \geq 0$ and must satisfy the following constraint:

$$\int_0^\infty |x(t)|e^{-\sigma t}\, dt < \infty \tag{2.3}$$

where σ is any fixed real number. If there is no such σ existence, the function $x(t)$ does not have a Laplace transform.

Example 2.1 Suppose that $x(t)$ is the unit-step function $u(t)$. Then, the Laplace transform of $u(t)$ is given by

$$L[u(t)] = \int_0^\infty u(t)e^{-st}\, dt = \int_0^\infty e^{-st}\, dt = \frac{-1}{s}\, e^{-st}\Big|_0^\infty = \frac{1}{s}$$

Example 2.2 Let $x(t)$ be the unit impulse $\delta(t)$. The Laplace transform of $\delta(t)$ is given by

$$L[\delta(t)] = \int_{0_-}^\infty \delta(t)e^{-st}\, dt \tag{2.4}$$

The lower limit of the integral in equation (2.4) must be taken from 0_-, since the impulse $\delta(t)$ is not defined at $t=0$. Since $\delta(t)=0$ for all $t \neq 0$,

$$L[\delta(t)] = \int_{0_-}^{0_+} \delta(t)e^{0}\, dt = 1$$

Example 2.3 Let $x(t) = e^{ct}$, where c is an arbitrary real number. The Laplace transform is given by

$$L[e^{ct}] = \int_0^\infty e^{ct}e^{-st}\, dt = \int_0^\infty e^{-(s-c)t}\, dt = \frac{-1}{s-c}\, e^{-(s-c)t}\Big|_0^\infty \tag{2.5}$$

By setting $s = \sigma + j\omega$ in equation (2.5) at $t = \infty$,

$$e^{-(s-c)\infty} = \lim_{t\to\infty} e^{-(\sigma-c)t}e^{-j\omega t} \tag{2.6}$$

The limit in equation (2.6) exists if and only if $\sigma - c > 0$, in which case the limit is zero, and the Laplace transform is

$$L[e^{ct}] = \frac{1}{s-c}$$

Example 2.4 Let $x(t)$ be a sinusoidal function, that is, $x(t) = \sin(\omega_0 t)$. It can be written as the sum of two exponential functions:

$$\sin(\omega_0 t) = \frac{e^{j\omega_0 t} - e^{-j\omega_0 t}}{2j}$$

The Laplace transform of $\sin(\omega_0 t)$ is given by

$$L[\sin(\omega_0 t)] = \int_0^\infty \frac{e^{j\omega_0 t} - e^{-j\omega_0 t}}{2j} e^{-st}\, dt = \int_0^\infty \frac{e^{-(s-j\omega_0)t} - e^{-(s+j\omega_0)t}}{2j}\, dt$$

$$= \frac{1}{2j}\left(\frac{1}{s-j\omega_0} - \frac{1}{s+j\omega_0}\right) = \frac{\omega_0}{s^2+\omega_0^2}$$

The Laplace transform changes a signal in the time domain into a signal in the s-domain. It has many applications in mathematics, physics, engineering, and signal processing. It has been used for solving differential and integral equations and analysis of electrical circuits, harmonic oscillators, optical devices, and mechanical systems. The Laplace transform can generally analyze any system governed by differential equations. The most powerful application of the Laplace transform is the design of systems directly in the s-domain. The Laplace transform can be used for designing a low-pass filter (Oliaei, 2003).

2.1.1 Properties of Laplace Transform

Laplace transform satisfies a number of properties that are useful in the transformation between frequency domain and time domain (Schiff, 1999; Graf, 2004). By using these properties, it is helpful in deriving many new transformation pairs from a basic set of pairs. Therefore, $x(t)$ and $X(s)$ are called the *Laplace transform pair*. A number of often-used transform pairs are given as follows:

1. *Unit-step function*: $u(t) \leftrightarrow \frac{1}{s}$
2. *Constant function*: $c \leftrightarrow \frac{c}{s}$
3. *Unit-impulse function*: $\delta(t) \leftrightarrow 1$
4. *Linear function*: $t \leftrightarrow \frac{1}{s^2}$
5. *Power function*: $t^n \leftrightarrow \frac{n!}{s^n}$
6. *Exponential function*: $e^{ct} \leftrightarrow \frac{1}{s-c}$
7. *Sinusoidal function*: $\sin(\omega_0 t) \leftrightarrow \frac{\omega_0}{s^2+\omega_0^2}$ and $\cos(\omega_0 t) \leftrightarrow \frac{s}{s^2+\omega_0^2}$

Several fundamental properties of Laplace transform are given below.

1. *Linearity*: $ax(t)+by(t) \leftrightarrow aX(s)+bY(s)$
2. *Shift in time domain for $c>0$*: $x(t-c)u(t-c) \leftrightarrow e^{-cs}X(s)$
   Except time shift, one can also perform frequency shift, which corresponds to the multiplication by an exponential in time domain, as follows:
3. *Shift in frequency domain for any real or complex number c*: $e^{ct}x(t) \leftrightarrow X(s-c)$

4. *Time scale*: $x(ct) \leftrightarrow \frac{1}{c}X\left(\frac{s}{c}\right)$
5. *Differentiation in time domain*: $X'(t) \leftrightarrow sX(s) - x(0)$
   Laplace transform of second-order or high-order derivatives of $x(t)$ can also be expressed in terms of $X(s)$ and the initial conditions. Let n be an arbitrary positive integer.

$$X^{(n)}(t) \leftrightarrow s^n X(s) - \sum_{k=1}^{n} s^{n-k} x^{(k-1)}(0)$$

6. *Integration in time domain*: Given a function $x(t)$ with $x(t) = 0$ for all $t < 0$.

$$X^{(-1)}(t) = \int_0^t x(\tau)d\tau \leftrightarrow \frac{X(s)}{s}$$

   The above two properties are differentiation and integration with respect to time domain. These operators are also performed in frequency domain, and their time-domain responses are shown below.
7. *Differentiation in frequency domain*: $t^n x(t) \leftrightarrow (-1)^n \frac{d^n X(s)}{ds^n}$
8. *Integration in frequency domain*: $\frac{x(t)}{t} \leftrightarrow \int_0^\infty X(s)ds$
9. *Convolution*: Given two functions $x(t)$ and $h(t)$ whose values are equal to zero when $t < 0$. The convolution in time domain corresponds to a product in frequency domain by Laplace transform. Conversely, the convolution in frequency domain corresponds to a product in time domain.

$$x(t) * h(t) \leftrightarrow X(s)H(s)$$

$$x(t)h(t) \leftrightarrow X(s) * H(s)$$

The convolution in the 2D case can be represented as

$$g(x, y) = f * h = \int_{-\infty}^{\infty}\int_{-\infty}^{\infty} f(\alpha, \beta)h(x-\alpha, y-\beta)d\alpha\, d\beta \tag{2.7}$$

The integrand, called *convolution*, is a product of two functions, $f(\alpha, \beta)$ and $h(\alpha, \beta)$, with the latter rotated by 180° and shifted by x and y along the x- and y-directions, respectively. A simple change of variables produces

$$g(x, y) = \int_{-\infty}^{\infty}\int_{-\infty}^{\infty} f(x-\alpha, y-\beta)h(\alpha, \beta)d\alpha\, d\beta = h * f \tag{2.8}$$

Therefore, $f * h = h * f$.

Another operation similar to convolution, called *correlation*, can be represented as

$$g(x,y) = f \circ h = \int_{-\infty}^{\infty}\int_{-\infty}^{\infty} f^*(\alpha,\beta)h(x+\alpha,y+\beta)d\alpha\, d\beta, \qquad (2.9)$$

where f^* is the complex conjugate of f. In image processing, the difference of correlation as compared to convolution is that the function h is not folded about the origin.

Example 2.5 Compute the convolution $g(x) = f(x) * h(x)$, where

$$f(x) = \begin{cases} x & \text{if } 1 \le x \le 2 \\ 0 & \text{elsewhere} \end{cases} \quad \text{and} \quad h(x) = \begin{cases} 1 & \text{if } 3 \le x \le 4 \\ 0 & \text{elsewhere} \end{cases}$$

Answer: The convolution is calculated as follows. It can be separated into four cases, as illustrated in Figure 2.1, where (a) is $f(x)$, (b) is $h(x)$, (c) is the case when $x < 4$, (d) is the case when $4 \le x < 5$, (e) is the case when $5 \le x < 6$, and (f) is the case when $x \ge 6$.

$$g(x) = \int_{-\infty}^{\infty} f(t)h(x-t)dt$$

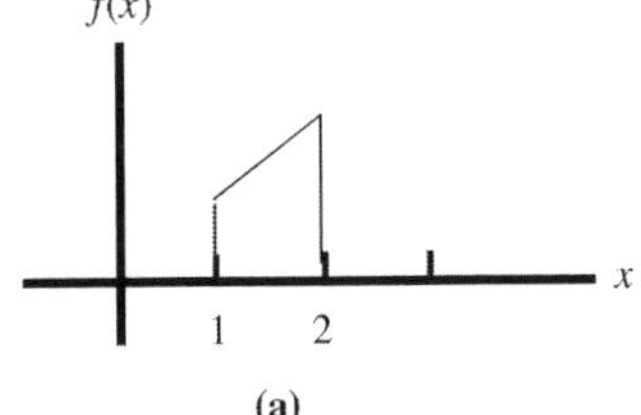

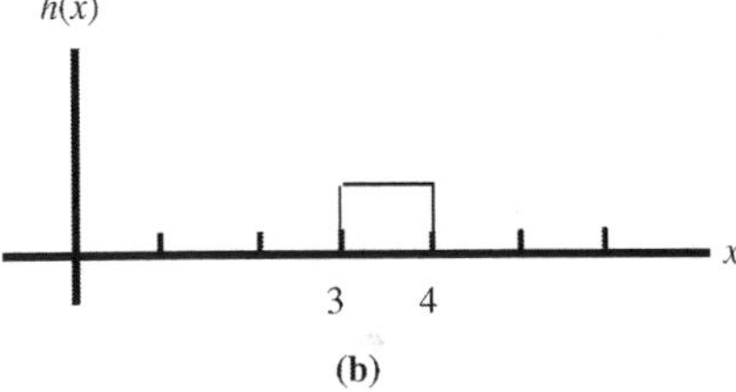

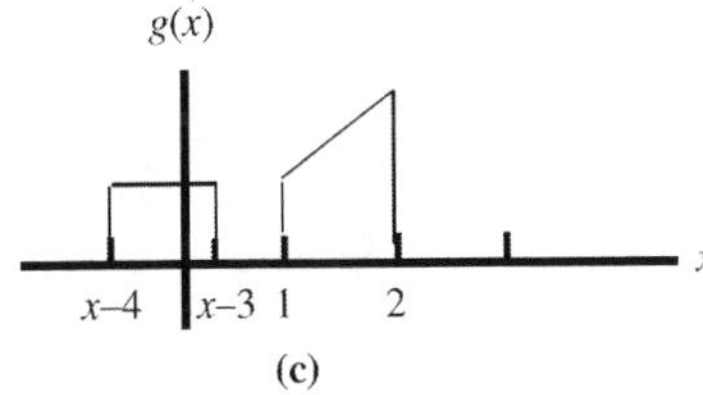

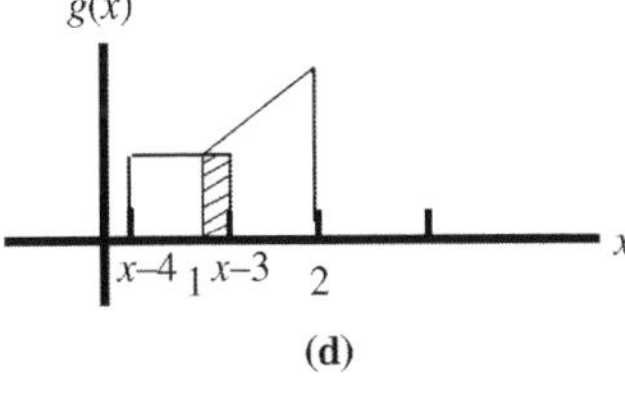

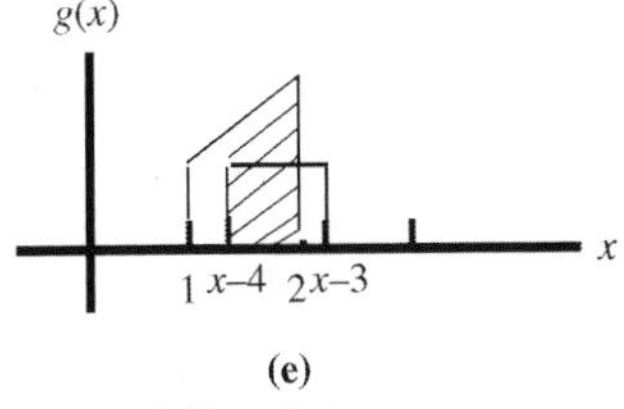

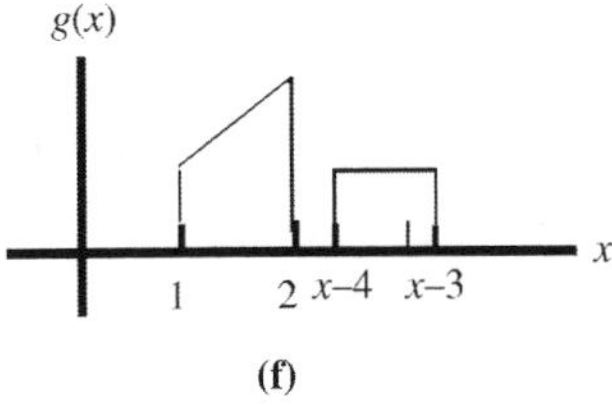

Figure 2.1 The illustration of convolution (a) is $f(x)$, (b) is $h(x)$, (c) is the case when $x < 4$, (d) is the case when $4 \le x < 5$, (e) is the case when $5 \le x < 6$, and (f) is the case when $x \ge 6$.

1. When $x < 4$, $g(x) = 0$.
2. When $4 \le x < 5$, $g(x) = \int_1^{-3+x} t \cdot 1\, dt = \frac{x^2}{2} - 3x + 4$.
3. When $5 \le x < 6$, $g(x) = \int_{-4+x}^{2} t \cdot 1\, dt = \frac{-x^2}{2} + 4x - 6$.
4. When $x \ge 6$, $g(x) = 0$.

Example 2.6 Let two images f and h are represented in the matrix form as below. Determine the convolution and correlation of f by h. Use the lower-left pixel of h (i.e., the underlined pixel) as its origin and deal with the pixels outside the boundary of image f to be zeros.

$$f: \begin{bmatrix} 3 & 5 & 1 \\ 2 & 6 & 2 \end{bmatrix}, \qquad h: \begin{bmatrix} 2 & 1 \\ \underline{-1} & 3 \end{bmatrix}$$

Answer: The convolution $f * h$ equates to flipping the latter image about the origin and then overlaying the result. The flipped h is

$$\begin{bmatrix} 3 & \underline{-1} \\ 1 & 2 \end{bmatrix}$$

where the underline indicates the origin. The result of convolution is

$$\begin{bmatrix} -1(3)+2(2) & 3(3)-1(5)+1(2)+2(6) & 3(5)-1(1)+1(6)+2(2) \\ -1(2) & 2(3)-1(6) & 3(6)-1(2) \end{bmatrix} \text{ or}$$

$$\begin{bmatrix} 1 & 18 & 24 \\ -2 & 0 & 16 \end{bmatrix}$$

The correlation $f \circ h$ does not flip the latter image. The result of correlation is

$$\begin{bmatrix} -1(3)+3(5) & -1(5)+3(1) & -1(1) \\ -2(3)+1(5)-1(2)+3(6) & 2(5)+1(1)-1(6)+3(2) & -1(2)+1(2) \end{bmatrix} \text{ or}$$

$$\begin{bmatrix} 12 & -2 & -1 \\ 27 & 11 & 0 \end{bmatrix}$$

1. *Separation of real and imaginary numbers*:

$$x(t) = \mathrm{Re}[x(t)] + j\,\mathrm{Im}[x(t)] \leftrightarrow X(s) = \mathrm{Re}[X(s)] + j\,\mathrm{Im}[X(s)]$$

2. *Initial value theorem*: $x(0) \leftrightarrow \lim_{s \to \infty} sX(s)$
3. *Final value theorem*: $\lim_{t \to \infty} x(t) \leftrightarrow \lim_{s \to 0} sX(s)$

2.2 FOURIER TRANSFORM

In the early 1800s, French mathematician Joseph Fourier, with his studies of the problem of heat flow, introduced Fourier series for the representation of continuous-time periodic signals. He made a serious attempt to prove that any piecewise smooth function can be expressed into a trigonometric sum. Fourier's theory has significant applications in signal processing. The frequency spectrum of a signal can be generated by representing the signal as a sum of sinusoids (or complex exponentials), called a *Fourier series*. This weighted sum represents the frequency content of a signal called *spectrum*.

When a signal becomes nonperiodic, its period becomes infinite and its spectrum becomes continuous. An image is considered as a spatially varying function. The Fourier transform decomposes such an image function into a set of orthogonal functions and converts the spatial intensity image into its frequency domain. From the continuous form, one can digitize it for discrete-time images. Signal processing in the frequency domain simplifies computational complexity in filtering analysis; thus, the Fourier transform has played the leading role in signal processing and engineering control for a long time (Sneddon, 1995; Bracewell, 2000).

Let T be a fixed positive real number. A continuous-time signal $x(t)$ is said to be *periodic* with a period T if

$$x(t) = x(t+T) \text{ for all } t$$

That is, a periodic signal repeats every T seconds, the frequency of the signal is $u = 1/T$, and the angular frequency is $\omega = 2\pi u$. In image processing, the frequency parameter is often used. According to Fourier's theorem, $x(t)$ can be expressed as a linear sum of complex exponentials:

$$x(t) = \sum_{n=-\infty}^{\infty} \alpha_n e^{jn\omega t} \tag{2.10}$$

where the constants α_n are given by

$$\alpha_n = \frac{1}{T} \int_{-T/2}^{T/2} x(t) e^{-jn\omega t}\, dt \tag{2.11}$$

The trigonometric identity is applied as

$$A\cos(n\omega t) - B\sin(n\omega t) = \sqrt{A^2+B^2}\cos(n\omega t+\theta) \tag{2.12}$$

where

$$\theta = \begin{cases} \tan^{-1}(B/A) & \text{when } A > 0 \\ 180^\circ + \tan^{-1}(B/A) & \text{when } A < 0 \end{cases}$$

Therefore, the following is obtained:

$$x(t) = \alpha_0 + \sum_{n=1}^{\infty} [a_n \cos(n\omega t) + b_n \sin(n\omega t)] \tag{2.13}$$

where $a_n = 2\mathrm{Re}[\alpha_n]$ and $b_n = -2\mathrm{Im}[\alpha_n]$.

2.2.1 Basic Theorems

Given a continuous-time signal $x(t)$, its *Fourier integral* or *Fourier transform* $X(\omega)$ is formed as

$$X(\omega) = \int_{-\infty}^{\infty} x(t) e^{-j\omega t}\, dt \tag{2.14}$$

When $X(\omega)$ is given, the *inverse Fourier transform* can be used to obtain $x(t)$ as

$$x(t) = \frac{1}{2\pi} \int_{-\infty}^{\infty} X(\omega) e^{j\omega t}\, d\omega \tag{2.15}$$

Equations (2.14) and (2.15) are called the *Fourier transform pair* and can be shown to exist if $x(t)$ is continuous and a valid integral and $X(\omega)$ is also a valid integral. These conditions are almost satisfied in practice. The Fourier transform is generally complex:

$$X(\omega) = R(\omega) + jI(\omega) = |X(\omega)| e^{j\phi(\omega)}$$

where $R(\omega)$ and $I(\omega)$ are the real and imaginary components of $X(\omega)$, respectively, and

$$|X(\omega)| = \sqrt{R^2(\omega) + I^2(\omega)}$$

and

$$\phi(\omega) = \tan^{-1}(I(\omega)/R(\omega))$$

$|X(\omega)|$ is called the *Fourier spectrum* of $x(t)$, $X^2(\omega)$ is the *energy* (or *power*) *spectrum*, and $\phi(\omega)$ is the *phase angle*.

Example 2.7 Consider a rectangular pulse $x(t)$ defined by

$$x(t) = \begin{cases} 1 & \text{if } |t| \leq T \\ 0 & \text{otherwise} \end{cases}$$

Its Fourier transform is

$$\begin{aligned} X(\omega) &= \int_{-\infty}^{\infty} x(t) e^{-j\omega t}\, dt = \int_{-T}^{T} e^{-j\omega t}\, dt \\ &= \frac{-1}{j\omega} [e^{-j\omega t}]_{-T}^{T} = \frac{-1}{j\omega} (e^{-j\omega T} - e^{j\omega T}) = \frac{2\sin(\omega T)}{\omega} = 2T\,\mathrm{sinc}(T) \end{aligned}$$

Let $\mathrm{sinc}(T) = \sin(\omega T)/\omega T$. The Fourier spectrum is $|2T\,\mathrm{sinc}(T)|$.

Example 2.8 Derive the Fourier transform of $f(x)$ that is defined in the following figure:

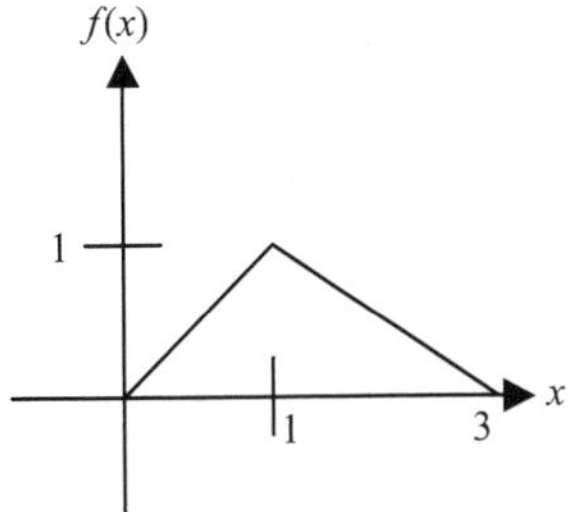

Answer: The function $f(x)$ is formulated as

$$f(x) = \begin{cases} x & \text{if } 0 \le x \le 1 \\ -\frac{1}{2}(x-3) & \text{if } 1 < x \le 3 \\ 0 & \text{otherwise} \end{cases}$$

Its Fourier transform $F(\omega)$ is computed as follows:

$$\begin{aligned} F(\omega) &= \int_{-\infty}^{+\infty} f(x)e^{-j\omega x}\,dx = \int_0^1 xe^{-j\omega x}\,dx + \int_1^3 \left[-\frac{1}{2}(x-3)e^{-j\omega x}\right]dx \\ &= \left[\left(\frac{1}{\omega^2}+\frac{j}{\omega}\right)e^{-j\omega} - \frac{1}{\omega^2}\right] + \left[\left(\frac{1}{2\omega^2}-\frac{j}{\omega}\right)e^{-j\omega} - \frac{1}{2\omega^2}e^{-j3\omega}\right] \\ &= \frac{3}{2\omega^2}e^{-j\omega} - \frac{1}{2\omega^2}e^{-j3\omega} - \frac{1}{\omega^2} \end{aligned}$$

Example 2.9 Compute the Fourier transform $F(\omega)$ of $f(x) = |x|$.

Answer:

$$F(\omega) = \int_{-\infty}^{\infty} |x|e^{-j\omega x}\,dx = \int_{-\infty}^{0} (-x)e^{-j\omega x}\,dx + \int_0^{\infty} xe^{-j\omega x}\,dx = \frac{-2}{\omega^2}$$

The basic theorems for Fourier transform are given below. All functions under consideration are assumed to have Fourier integrals.

1. *Linearity:* If $X_1(\omega)$ and $X_2(\omega)$ are Fourier transforms of $x_1(t)$ and $x_2(t)$, respectively, and a and b are two arbitrary constants, then

$$ax_1(t) + bx_2(t) \leftrightarrow aX_1(\omega) + bX_2(\omega)$$

That is, Fourier transform possesses the linear combination property, which is called *superposition theory*.

2. *Symmetry:* If the parameters t and ω are exchanged,

$$x(t) \leftrightarrow X(\omega)$$

$$2\pi x(-\omega) \leftrightarrow X(t)$$

3. *Time shifting:* For constant a, $x(t-a) \leftrightarrow X(\omega)e^{-j\omega a}$

   That is, the Fourier spectrum through a time shift remains the same, but the term $-\omega a$ is added to its phase angle.
4. *Frequency shifting:* $x(t)e^{jat} \leftrightarrow X(\omega-a)$
5. *Time scaling:* If "a" denotes a real constant, $x(at) \leftrightarrow \frac{1}{|a|}X\left(\frac{\omega}{a}\right)$
6. *Time differentiation:* $\frac{d^n x(t)}{dt^n} \leftrightarrow (j\omega)^n X(\omega)$
7. *Frequency differentiation:* $(-jt)^n x(t) \leftrightarrow \frac{d^n X(\omega)}{d\omega^n}$
8. *Conjugate function:* Let a complex function $x(t) = a(t) + jb(t)$. The *complex conjugate* of $x(t)$ is $\text{conj}[x(t)] = a(t) - jb(t)$ and its Fourier transform is $\text{conj}[x(t)] \leftrightarrow \text{conj}[X(-\omega)]$.
9. *Parseval's relation:* $\int_{-\infty}^{\infty} |x(t)|^2 dt = \frac{1}{2\pi} |X(\omega)|^2 d\omega$. There is a physical interpretation of Parseval's relation. The energy (density) of a wave is related to the square of its amplitude. Therefore, the energy density of a periodic function is the sum of energy densities of its Fourier components.

2.2.2 Discrete Fourier Transform

In the previous section, the complex Fourier series representation is an infinite sum of products of Fourier coefficients and exponentials. In this section, the *discrete Fourier transform* (DFT) is described and evaluated in practical applications. The input to the DFT is a sequence of numbers rather than a continuous function of time $x(t)$. The sequence of numbers usually results from periodically sampling a continuous signal $x(t)$ at an interval of T.

The DFT is derived from a continuous-time function $x(t)$ using N samples taken at times $t = 0, T, 2T, \ldots, (N-1)T$, where T is the sampling interval. These N samples of $x(t)$ form a data sequence: $\{x(t_0), x(t_0+T), x(t_0+2T), \ldots, x(t_0+(N-1)T)\}$.

It will be convenient to use t as either a continuous or a discrete variable, depending on the context of discussion. Therefore, $x(t) = x(t_0 + tT)$, where t now assumes the discrete values 0, 1, 2, ..., $N-1$. The simplified notations $\{x(0), x(1), x(2), \ldots, x(N-1)\}$ mean N uniformly spaced samples of a continuous function.

Let $\omega = 2\pi u$, where $u = 0, 1, 2, \ldots, N-1$. The DFT equation for evaluating $X(u)$ is given by

$$X(u) = \frac{1}{N}\sum_{t=0}^{N-1} x(t)e^{-j2\pi ut/N} \tag{2.16}$$

The multiplying constant 1/N is included for convenience and, of course, has no important effect on the nature of the representation. The integer values of u are referred to as the *transform sequence number* or *frequency bin number*. The inverse DFT is given by

$$x(t) = \sum_{u=0}^{N-1} X(u) e^{j2\pi ut/N}, \tag{2.17}$$

where $t = 0, 1, 2, \ldots, N-1$. In the 2D image case, let $f(x, y)$ and $F(u, v)$, respectively, denote spatial domain and frequency domain. The DFT pair is

$$F(u, v) = \frac{1}{MN} \sum_{x=0}^{M-1} \sum_{y=0}^{N-1} f(x, y) e^{-j2\pi(ux/M + vy/N)} \tag{2.18}$$

for $u = 0, 1, 2, \ldots, M-1$ and $v = 0, 1, 2, \ldots, N-1$, and

$$f(x, y) = \sum_{u=0}^{M-1} \sum_{v=0}^{N-1} F(u, v) e^{j2\pi(ux/M + vy/N)} \tag{2.19}$$

for $x = 0, 1, 2, \ldots, M-1$ and $y = 0, 1, 2, \ldots, N-1$.

If an image is a square array, that is, $M = N$, the DFT pair is rearranged as

$$F(u, v) = \frac{1}{N} \sum_{x=0}^{N-1} \sum_{y=0}^{N-1} f(x, y) e^{-j2\pi(ux + vy)/N} \tag{2.20}$$

for $u, v = 0, 1, 2, \ldots, N-1$, and

$$f(x, y) = \frac{1}{N} \sum_{u=0}^{N-1} \sum_{v=0}^{N-1} F(u, v) e^{j2\pi(ux + vy)/N} \tag{2.21}$$

for $x, y = 0, 1, 2, \ldots, N-1$. Note that both equations include a constant factor of 1/N. The repetition of coefficients at intervals of T is the periodic property of the DFT.

$$X(u + kN) = \frac{1}{N} \sum_{t=0}^{N-1} x(t) e^{-j2\pi(u + kN)t/N} = X(u) \tag{2.22}$$

for any integer k. The DFT coefficients separated by N frequency bins are equal because the sampled sinusoids for frequency bin number $u + kN$ complete k cycles between sampling times and take the same value as the sinusoid for the frequency bin number u.

The folding property of DFT states that coefficient $X(N-u)$ is the complex conjugate of coefficient $X(u)$.

$$X(N-u) = \frac{1}{N} \sum_{t=0}^{N-1} x(t) e^{-j2\pi(N-u)t/N} = \left[\frac{1}{N} \sum_{t=0}^{N-1} x(t) e^{-j2\pi ut/N} \right]^* = X^*(u) \tag{2.23}$$

Therefore, an output in the DFT frequency bin $N-u$ is obtained even though the only input is in bin u.

The folding property is closely related to the time-domain sampling theorem: If a real signal is sampled at a rate at least twice the frequency of the highest frequency sinusoid in the signal, then the signal can be completely reconstructed from these samples. The minimum sampling frequency f_s for which the time-domain sampling theorem is satisfied is called the *Nyquist sampling rate*. In this case, the frequency $f_s/2$ about which the spectrum of the real signal folds is called the *Nyquist frequency*. The number of complex multiplications and additions required to implement the DFT is proportional to N^2. Its calculation is usually performed using a method known as the *fast Fourier transform* (FFT) (Brigham, 1988), whose decomposition can make the number of multiplication and addition operations proportional to $N \log_2 N$. Cheng and Parhi (2007) developed a low-cost fast VLSI algorithm for DFT.

Example 2.10 Let a function be sampled as $\{f(x_0) = 4, f(x_1) = 3, f(x_2) = 1, f(x_3) = 1\}$. Calculate the Fourier spectrum of $F(2)$ using the discrete Fourier transform.

Answer: Applying equation (2.16),

$$F(u) = \frac{1}{N}\sum_{x=0}^{N-1} f(x)e^{-j2\pi ux/N}$$

In this case, $N=4$. When $u=2$,

$$F(2) = \frac{1}{4}\sum_{x=0}^{3} f(x)e^{-j4\pi x/4} = \frac{1}{4}\left[4e^0 + 3e^{-j\pi} + e^{-j2\pi} + e^{-j3\pi}\right] = \frac{1}{4}[1 + j0]$$

Therefore, the Fourier spectrum of $F(2)$ is 0.25.

2.2.3 Fast Fourier Transform

The direct calculation of equation (2.16) requires the computational complexity proportional to N^2. That is, for each of N values of ω, expansion of the summation requires N complex multiplications of $x(t)$ by $\exp(j\omega t/N)$ and $N-1$ additions of the results. For a large value of N, this requires extremely lengthy calculation time on a computer.

An alternative method of obtaining the DFT is the *fast Fourier transform* (Cooley and Tukey, 1965; Brigham, 1988). This is a recursive algorithm based on the factorization of matrix multiplication and the removal of the redundancy in the calculation of repeated products. The reduction in complexity from N^2 to $N \log_2 N$ represents a significant saving in computational effort. Sheridan (2007) proposed a methodology to perform FFT based on considerations of the natural physical constraints imposed by image capture devices.

Let $W(N) = \exp(-j2\pi/N)$. By removing the constant term $1/N$, the DFT equation (2.16) can be expressed as

$$X(u) = \sum_{t=0}^{N-1} x(t)W(N)^{ut} \tag{2.24}$$

and N is assumed to be of the form $N = 2^n$, where n is a positive integer. Since N is a power of 2, $N/2$ is an integer, and the samples separated by $N/2$ in the data sequence can be combined to yield

$$X(u) = \sum_{t=0}^{(N/2)-1} x(2t)W(N)^{2ut} + \sum_{t=0}^{(N/2)-1} x(2t+1)W(N)^{(2t+1)u} \tag{2.25}$$

Because $W(N)^{2ut} = W(N/2)^{ut}$,

$$X(u) = \sum_{t=0}^{(N/2)-1} x(2t)W(N/2)^{ut} + \sum_{t=0}^{(N/2)-1} x(2t+1)W(N/2)^{ut}\,W(N)^{u} \tag{2.26}$$

Define

$$X_{\text{even}}(u) = \sum_{t=0}^{(N/2)-1} x(2t)W(N/2)^{ut} \tag{2.27}$$

and

$$X_{\text{odd}}(u) = \sum_{t=0}^{(N/2)-1} x(2t+1)W(N/2)^{ut} \tag{2.28}$$

for $u = 0, 1, 2, \ldots, (N/2)-1$. Therefore,

$$X(u) = X_{\text{even}}(u) + X_{\text{odd}}(u)W(N)^{u} \tag{2.29}$$

Because

$$W(N/2)^{u+(N/2)} = W(N/2)^{u} \quad \text{and} \quad W(N)^{u+(N/2)} = -W(N)^{u} \tag{2.30}$$

the following can be obtained:

$$X\left(u + \frac{N}{2}\right) = X_{\text{even}}(u) - X_{\text{odd}}(u)W(N)^{u} \tag{2.31}$$

Note that the DFT is decomposed from N-points to ($N/2$)-points, then to ($N/4$)-points, and so on, until a two-point output is obtained. Also note that the Fourier transforms for the indices u and $u+N/2$ only differ by the sign of the second term. Thus, for the composition of two terms, only one complex multiplication is needed. The strategy aimed to be applied recursively for a more efficient solution to a search problem is called *divide and conquer*.

The Fourier transform has many applications in any field of solving physical problems, such as engineering, physics, applied mathematics, and chemistry. In electromagnetic theory, the intensity of light is proportional to the square of the

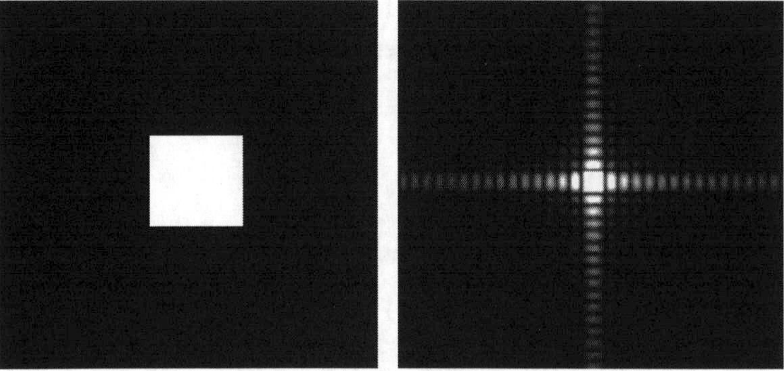

Figure 2.2 A rectangular function and its Fourier transform, a sinc function.

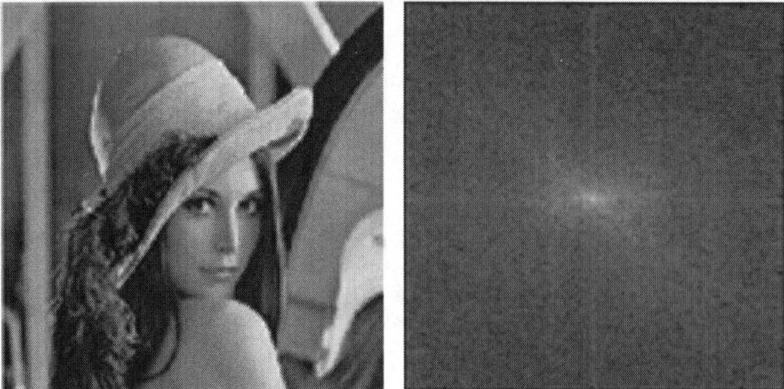

Figure 2.3 Lena image and its Fourier transform.

oscillating electric field that exists at any point in space. The Fourier transform of this signal is equivalent to breaking the light into its component parts of the spectrum, a mathematical spectrometer. An example of applying the Fourier transform in optics is the diffraction of light when it passes through narrow slits. The ideas are similarly applied to acoustic, X-ray, and microwave diffraction, or any other form of wave diffraction. In signal and image processing, the Fourier transform is used for frequency-domain transformation, representation, encoding, smoothing and sharpening, restoration, blur removal, noise estimation, and Wiener high-pass, low-pass, and band-pass filters (Yazici and Xie, 2006). The Fourier transform of a rectangular function is a 2D sinc function as shown in Figure 2.2. The Fourier transform of the Lena image is shown in Figure 2.3.

2.3 *Z*-TRANSFORM

In the study of time-invariant discrete-time systems, the *Z*-transform plays a role similar to that of the Laplace transform. By applying the *Z*-transform, a difference equation in the discrete variable can be transformed into an algebraic equation in

z (Jury, 1964). Usually, the algebraic equations are simpler to solve and provide more insight into the behavior of the system.

2.3.1 Definition of Z-Transform

Given a sequence of scalars, $\{x(t)|t = 0, 1, 2, \ldots\}$, the Z-transform of this sequence is the function of the complex variable z defined by

$$X(z) = \sum_{t=0}^{\infty} x(t)z^{-t} \tag{2.32}$$

In a similar but not identical way, there are a number of important relationships between Z-transform and Fourier transform. To explore these relationships, the complex variable z is represented in the form

$$z = e^{j2\pi\omega/N} \tag{2.33}$$

Therefore, in this case, Z-transform becomes equivalent to Fourier transform. The *inverse Z-transform* is defined by

$$x(t) = \frac{1}{j2\pi} \int_T X(z)z^t \, dz \tag{2.34}$$

where T is any counterclockwise closed circular contour centered at the origin and with radius r.

The value of r can be chosen as any value for which $X(z)$ converges.

Example 2.11 Let $x(t) = a^t u(t)$. The Z-transform is given by

$$X(z) = \sum_{t=0}^{\infty} a^t u(t)z^{-t} = \sum_{t=0}^{\infty} (az^{-1})^t$$

For convergence of $X(z)$, the following is required:

$$\sum_{t=0}^{\infty} \left|az^{-1}\right|^t < \infty$$

Thus, the region of convergence is the range of values of z for which $\left|az^{-1}\right| < 1$, or equivalently $|z| > |a|$. Therefore,

$$X(z) = \frac{1}{1-az^{-1}} = \frac{z}{z-a}, \quad \text{for} \quad |z| > |a|$$

Consequently, the Z-transform converges for any finite value of a. Readers should note that the Fourier transform of $x(t)$ only converges if $|a| < 1$. For $a = 1$, $x(t)$ is the unit-step function with its Z-transform

$$Z[u(t)] = \frac{z}{z-1} \tag{2.35}$$

By the definition of Z-transform, a list of properties can be obtained.

2.3.2 Properties of Z-Transform

In this section, some important properties of Z-transform are introduced.

1. *Linearity*: Suppose that $x_1(t)$ and $x_2(t)$ have their corresponding Z-transforms $X_1(z)$ and $X_2(z)$. For any constants a and b,

$$ax_1(t)+bx_2(t) \leftrightarrow aX_1(z)+bX_2(z)$$

Example 2.12 Let $x(t) = \cos(t\theta)$. The Z-transform is given by

$$Z[\cos(t\theta)] = Z\left[\frac{e^{jt\theta}}{2}+\frac{e^{-jt\theta}}{2}\right] = \frac{1}{2}\left[\sum_{t=0}^{\infty} e^{jt\theta}z^{-t} + \sum_{t=0}^{\infty} e^{-jt\theta}z^{-t}\right]$$

Use the result of the previous example by replacing $a = e^{j\theta}$ and $a = e^{-j\theta}$, respectively, for the above first and second terms. Thus,

$$Z[\cos(t\theta)] = \frac{1}{2}\left[\frac{z}{z-e^{j\theta}}+\frac{z}{z-e^{-j\theta}}\right]$$

This can be simplified as

$$Z[\cos(t\theta)] = \left[\frac{2z(z-\cos\theta)}{z^2-2z\cos\theta+1}\right], \quad \text{for} \quad |z|>1$$

2. *Shift of a sequence for* $c>0$: $x(t-c)u(t-c) \leftrightarrow z^{-c}X(z)$
3. *Multiplication by an exponential sequence* a^t, *where "a" may be complex*: $a^t x(t) \leftrightarrow X\left(\frac{z}{a}\right)$
4. *Differentiation in z-domain*: $tx(t) \leftrightarrow z\frac{dX(z)}{dz}$
5. *Conjugate of a complex sequence*: $x^*(t) \leftrightarrow X^*(z^*)$
6. *The initial value theorem*: $x(0) = \lim_{z\to\infty} X(z)$
7. *The final value theorem*: $\lim_{t\to\infty} x(t) = \lim_{z\to 1}(1-z^{-1})X(z)$
8. *Convolution of sequences*: $x(t)*h(t) \leftrightarrow X(z)H(z)$

The Z-transform is useful for discrete signal and image processing and plays an important role in the formulation and analysis of discrete-time systems (Psenicka et al., 2002). It has been used extensively in the areas of digital signal processing, control theory, applied mathematics, population science, and economics. Similar to the Laplace transform, the Z-transform is used in the solution of differential equations.

2.4 COSINE TRANSFORM

Fourier series was originally motivated by the problem of heat conduction, and later found a vast number of applications as well as provided a basis for other transforms, such as cosine transform. Many video and image compression algorithms apply the discrete cosine transform (DCT) to transform an image to frequency domain and perform quantization for data compression. One of the advantages of the DCT is its energy compaction property; that is, the signal energy is concentrated on a few components while most other components are zero or negligibly small. This helps separate an image into parts (or spectral subbands) of hierarchical importance (with respect to the image's visual quality). A well-known JPEG compression technology uses the DCT to compress an image.

The Fourier transform kernel is complex valued. The DCT is obtained by using only the real part of Fourier complex kernel. Let $f(x, y)$ denote an image in spatial domain, and let $F(u, v)$ denote an image in frequency domain. The general equation for a 2D DCT is defined as

$$F(u, v) = C(u)C(v)\sum_{x=0}^{N-1}\sum_{y=0}^{N-1} f(x, y)\cos\left(\frac{(2x+1)u\pi}{2N}\right)\cos\left(\frac{(2y+1)v\pi}{2N}\right) \quad (2.36)$$

where if $u = v = 0$, then $C(u) = C(v) = \sqrt{1/N}$; otherwise, $C(u) = C(v) = \sqrt{2/N}$.

The inverse DCT can be represented as

$$f(x, y) = \sum_{u=0}^{N-1}\sum_{v=0}^{N-1} C(u)C(v)F(u, v)\cos\left(\frac{(2x+1)u\pi}{2N}\right)\cos\left(\frac{(2y+1)v\pi}{2N}\right). \quad (2.37)$$

A more convenient method for expressing the 2D DCT is using matrix products as $F = MfM^T$, and its inverse DCT is $f = M^T FM$, where F and f are 8×8 data matrices, and M is the matrix as

$$M = \begin{bmatrix}
\frac{1}{\sqrt{8}} & \frac{1}{\sqrt{8}} & \frac{1}{\sqrt{8}} & \frac{1}{\sqrt{8}} & \frac{1}{\sqrt{8}} & \frac{1}{\sqrt{8}} & \frac{1}{\sqrt{8}} & \frac{1}{\sqrt{8}} \\
\frac{1}{2}\cos\frac{1}{16}\pi & \frac{1}{2}\cos\frac{3}{16}\pi & \frac{1}{2}\cos\frac{5}{16}\pi & \frac{1}{2}\cos\frac{7}{16}\pi & \frac{1}{2}\cos\frac{9}{16}\pi & \frac{1}{2}\cos\frac{11}{16}\pi & \frac{1}{2}\cos\frac{13}{16}\pi & \frac{1}{2}\cos\frac{15}{16}\pi \\
\frac{1}{2}\cos\frac{2}{16}\pi & \frac{1}{2}\cos\frac{6}{16}\pi & \frac{1}{2}\cos\frac{10}{16}\pi & \frac{1}{2}\cos\frac{14}{16}\pi & \frac{1}{2}\cos\frac{18}{16}\pi & \frac{1}{2}\cos\frac{22}{16}\pi & \frac{1}{2}\cos\frac{26}{16}\pi & \frac{1}{2}\cos\frac{30}{16}\pi \\
\frac{1}{2}\cos\frac{3}{16}\pi & \frac{1}{2}\cos\frac{9}{16}\pi & \frac{1}{2}\cos\frac{15}{16}\pi & \frac{1}{2}\cos\frac{21}{16}\pi & \frac{1}{2}\cos\frac{27}{16}\pi & \frac{1}{2}\cos\frac{33}{16}\pi & \frac{1}{2}\cos\frac{39}{16}\pi & \frac{1}{2}\cos\frac{45}{16}\pi \\
\frac{1}{2}\cos\frac{4}{16}\pi & \frac{1}{2}\cos\frac{12}{16}\pi & \frac{1}{2}\cos\frac{20}{16}\pi & \frac{1}{2}\cos\frac{28}{16}\pi & \frac{1}{2}\cos\frac{36}{16}\pi & \frac{1}{2}\cos\frac{44}{16}\pi & \frac{1}{2}\cos\frac{52}{16}\pi & \frac{1}{2}\cos\frac{60}{16}\pi \\
\frac{1}{2}\cos\frac{5}{16}\pi & \frac{1}{2}\cos\frac{15}{16}\pi & \frac{1}{2}\cos\frac{25}{16}\pi & \frac{1}{2}\cos\frac{35}{16}\pi & \frac{1}{2}\cos\frac{45}{16}\pi & \frac{1}{2}\cos\frac{55}{16}\pi & \frac{1}{2}\cos\frac{65}{16}\pi & \frac{1}{2}\cos\frac{75}{16}\pi \\
\frac{1}{2}\cos\frac{6}{16}\pi & \frac{1}{2}\cos\frac{18}{16}\pi & \frac{1}{2}\cos\frac{30}{16}\pi & \frac{1}{2}\cos\frac{42}{16}\pi & \frac{1}{2}\cos\frac{54}{16}\pi & \frac{1}{2}\cos\frac{66}{16}\pi & \frac{1}{2}\cos\frac{78}{16}\pi & \frac{1}{2}\cos\frac{90}{16}\pi \\
\frac{1}{2}\cos\frac{7}{16}\pi & \frac{1}{2}\cos\frac{21}{16}\pi & \frac{1}{2}\cos\frac{35}{16}\pi & \frac{1}{2}\cos\frac{49}{16}\pi & \frac{1}{2}\cos\frac{63}{16}\pi & \frac{1}{2}\cos\frac{77}{16}\pi & \frac{1}{2}\cos\frac{91}{16}\pi & \frac{1}{2}\cos\frac{105}{16}\pi
\end{bmatrix} \quad (2.38)$$

The obtained DCT coefficients indicate the correlation between the original 8×8 block and the respective DCT basis image. These coefficients represent the amplitudes of all cosine waves that are used to synthesize the original signal in the

Figure 2.4 Lena image and its discrete cosine transform.

inverse process. DCT inherits many properties from Fourier transform and provides a wide range of applications such as image coding, data compression, feature extraction, multiframe detection, and filter banks (Yip and Rao, 1990; Siu and Siu, 1997; Lin and Chen, 2000; Zeng and Fu, 2008). Zeng et al. (2001) developed integer DCTs and fast algorithms.

Discrete cosine transform has emerged as the de facto image transformation in most visual systems. It has been widely deployed by modern video coding standards, for example, MPEG, Joint Video Team (JVT), and so on, for image compression (Chen and Pratt, 1984). It is also widely employed in solving partial differential equations by spectral methods, where the different variants of the DCT correspond to slightly different even/odd boundary conditions at the two ends of the array. Figure 2.4 show the Lena image and its DCT image.

2.5 WAVELET TRANSFORM

Industrial standards for compressing still images (e.g., JPEG) and motion pictures (e.g., MPEG) have been based on the DCT. Both standards have produced good results, but have limitations at high compression ratios. At low data rates, the DCT-based transforms suffer from a "blocking effect" due to the unnatural block partition that is required in the computation. Other drawbacks include mosquito noise (i.e., a distortion that appears as random aliasing occurs close to object's edges) and aliasing distortions. Furthermore, the DCT does not improve the performance as well as the complexities of motion compensation and estimation in video coding.

Due to the shortcomings of DCT, discrete wavelet transform (DWT) has become increasingly important. The main advantage of DWT is that it provides space–frequency decomposition of images, overcoming the DCT and Fourier transform that only provide frequency decomposition (Addison, 2002; Jensen and Cour-Harbo, 2001). By providing space–frequency decomposition, the DWT allows energy compaction at the low-frequency subbands and the space localization of edges at the high-frequency subbands. Furthermore, the DWT does not present a blocking effect at the low data rates.

Wavelets are functions that integrate to zero waving above and below the x-axis. Like sines and cosines in Fourier transform, wavelets are used as the basis functions for signal and image representation. Such basis functions are obtained by dilating and translating a *mother wavelet* $\psi(x)$ by amounts s and τ, respectively:

$$\Psi_{\tau,s}(x) = \left\{\psi\left(\frac{x-\tau}{s}\right), (\tau, s) \in R \times R^{+}\right\}. \tag{2.39}$$

The translation τ and dilation s allow the wavelet transform to be localized in time and frequency. Also, wavelet basis functions can represent functions with discontinuities and spikes in a more compact way than sines and cosines.

The continuous wavelet transform (CWT) can be defined as

$$\text{cwt}_{\psi}(\tau, s) = \frac{1}{\sqrt{|s|}} \int x(t)\Psi^{*}_{\tau,s}(t)dt \tag{2.40}$$

where $\Psi^{*}_{\tau,s}$ is the complex conjugate of $\Psi_{\tau,s}$ and $x(t)$ is the input signal defined in the time domain.

The inverse CWT can be obtained as

$$x(t) = \frac{1}{C^{2}_{\psi}} \int_{s}\int_{\tau} \text{cwt}_{\psi}(\tau, s)\frac{1}{s^{2}}\Psi_{\tau,s}(t)d\tau\, ds \tag{2.41}$$

where C_{ψ} is a constant and depends on the wavelet used.

The Haar DWT (Falkowski, 1998; Grochenig and Madych, 1992) is the simplest of all wavelets. Haar wavelets are orthogonal and symmetric. The minimum support property allows arbitrary grid intervals. Boundary conditions are easier than other wavelet-based methods. Haar DWT works very well to detect the characteristics such as edges and corners. Figure 2.5 shows an example of DWT of size 4×4.

To discretize the CWT, the simplest case is the uniform sampling of the time–frequency plane. However, the sampling could be more efficient by using

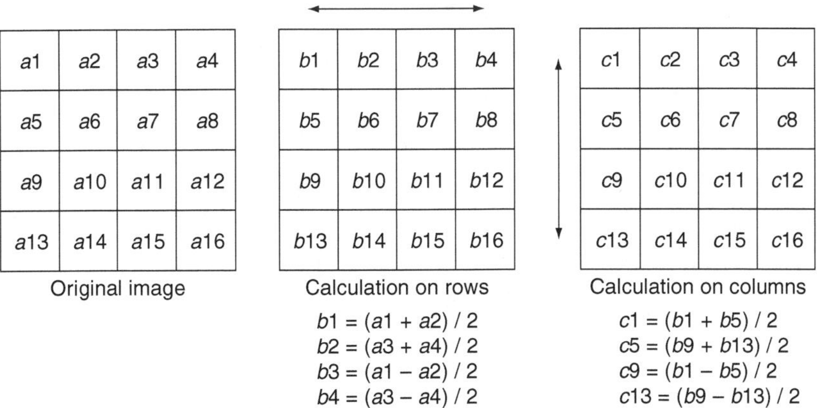

Figure 2.5 An example of Haar DWT.

the Nyquist's rule:

$$N_2 = \frac{s_1}{s_2} N_1 \tag{2.42}$$

where N_1 and N_2 denote the number of samples at scales s_1 and s_2, respectively, and $s_2 > s_1$. This rule means that at higher scales (lower frequencies), the number of samples can be decreased. The sampling rate obtained is the minimum rate that allows the original signal to be reconstructed from a discrete set of samples. A dyadic scale satisfies the Nyquist's rule by discretizing the scale parameter into a logarithmic series, and then the time parameter is discretized by applying equation (2.42) with respect to the corresponding scale parameters. The following equations set the translation and dilation to the dyadic scale with logarithmic series of base 2 for $\psi_{k,j}$: $\tau = k \cdot 2^j$, and $s = 2^j$. These coefficients can be viewed as filters that are classified into two types. One set, H, works as a low-pass filter, and the other set, G, as a high-pass filter. These two types of coefficients are called *quadrature mirror filters* used in the pyramidal algorithms.

For a 2D signal, it is not necessary, although straightforward, to extend the 1D wavelet transform to its 2D one. The strategy is described as follows. The 1D transform can be applied individually to each of the dimensions of the image. By using the quadrature mirror filters, an $n \times n$ image I can be decomposed into the wavelet coefficients as below. Filters H and G are applied on the rows of an image, splitting the image into two subimages of dimensions $n/2 \times n$ (half the columns) each. One of these subimages $H_r I$ (where the subscript r denotes row) contains the low-pass information, and the other, $G_r I$, contains the high-pass information. Next, the filters H and G are

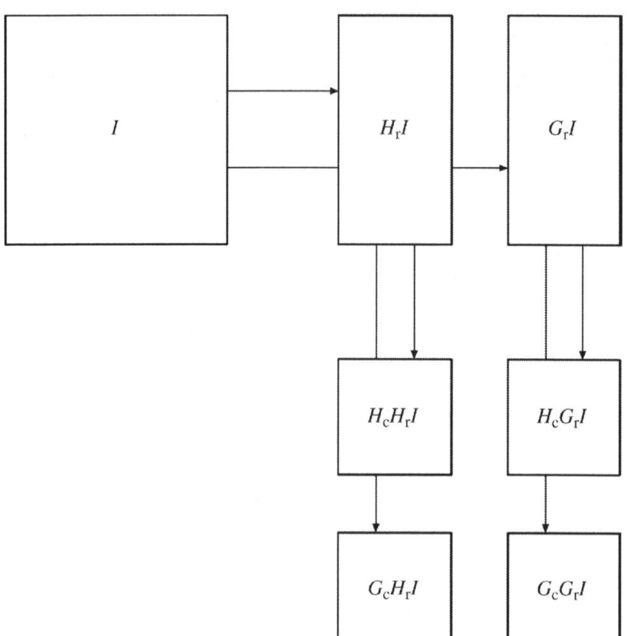

Figure 2.6 Wavelet decomposition of an image.

<table>
<tr><td>L_3</td><td>H_3</td><td rowspan="2">H_2</td><td rowspan="3">H_1</td></tr>
<tr><td>V_3</td><td>D_3</td></tr>
<tr><td colspan="2">V_2</td><td>D_2</td></tr>
<tr><td colspan="3">V_1</td><td>D_1</td></tr>
</table>

Figure 2.7 Three-level resolutions in a 2D wavelet transform.

applied to the columns of both subimages. Finally, four subimages with dimensions $n/2 \times n/2$ are obtained. Subimages H_cH_rI, G_cH_rI, H_cG_rI, and G_cG_rI (where the subscript c denotes column) contain the low–low, high–low, low–high, and high–high passes, respectively. Figure 2.6 illustrates this decomposition. The same procedures are applied iteratively to the subimage containing the most low band information until the subimage's size reaches 1×1. Therefore, the initial dimensions of the image are required to be powers of 2.

Figure 2.8 Wavelet representation on three-level resolutions of the Lena image.

In practice, it is not necessary to carry out all the possible decompositions until the size of 1×1 is reached. Usually, just a few levels are sufficient. Figure 2.7 shows three-level resolutions in a 2D wavelet transform, where L_i, V_i, H_i, and D_i denote the ith level subimages containing the low–low, high–low, low–high, and high–high passes, respectively. Figure 2.8 shows the Lena image decomposed in three levels. Each of the resulting subimages is known as a *subband.*

Since the wavelet functions are compact, the wavelet coefficients only measure the variations around a small region of the data array. This property makes wavelet analysis very useful for signal or image processing. The localized nature of the wavelet transform allows one to easily pick out features such as spikes, objects, and edges. Wavelets are also extensively used for the purposes of filtration and preprocessing data, analysis and prediction of stock markets situations, image recognition, as well as for processing and synthesizing various signals, such as speech or medical signals, image compressing and processing (Morales and Shih, 2000; Ding et al., 2007; Liu et al., 2007; Liu and Ngan, 2008), training neural networks, and so on (Broughton and Bryan, 2008).

REFERENCES

Addison, P. S., *The Illustrated Wavelet Transform Handbook: Introductory Theory and Applications in Science, Engineering, Medicine and Finance*, CRC Press, 2002.

Bracewell, R. N., *The Fourier Transform and Its Applications*, McGraw-Hill, 2000.

Brigham, E., *The Fast Fourier Transform and Its Applications*, Prentice-Hall, 1988.

Broughton, S. A. and Bryan, K. M., *Discrete Fourier Analysis and Wavelets: Applications to Signal and Image Processing*, Wiley, Nov. 2008.

Chen, W. H. and Pratt, W. K., "Scene adaptive coder," *IEEE Trans. Commun.*, vol. 32, no. 2, pp. 225–232, Mar. 1984.

Cheng, C. and Parhi, K. K., "Low-cost fast VLSI algorithm for discrete Fourier transform," *IEEE Trans. Circuits Syst. I*, vol. 54, no. 4, pp. 791–806, Apr. 2007.

Cooley, J. W. and Tukey, J. W., "An algorithm for the machine calculation of complex Fourier series," *Math. Comput.*, vol. 19, no. 90, pp. 297–301, Apr. 1965.

Ding, W., Wu, F., Wu, X., Li, S., and Li, H., "Adaptive directional lifting-based wavelet transform for image coding," *IEEE Trans. Image Process.*, vol. 16, no. 2, pp. 416–427, Feb. 2007.

Falkowski, B. J., "Forward and inverse transformations between Haar wavelet and arithmetic functions," *Electron. Lett.*, vol. 34, no. 11, pp. 1084–1085, May 1998.

Graf, U., *Applied Laplace Transforms and Z*-Transforms for Scientists and Engineers: A Computational Approach Using a Mathematica Package, Springer, 2004.

Grochenig, K. and Madych, W. R., "Multiresolution analysis, Haar bases, and self-similar tilings of R^n," *IEEE Trans. Inform. Theory*, vol. 38, no. 2, pp. 556–569, Mar. 1992.

Jensen, A. and Cour-Harbo, A., *Ripples in Mathematics: The Discrete Wavelet Transform*, Springer, 2001.

Jury, E. I., *Theory and Application of the Z*-Transform Method, Wiley, 1964.

Lin, S. and Chen, C., "A robust DCT-based watermarking for copyright protection," *IEEE Trans. Consum. Electron.*, vol. 46, no. 3, pp. 415–421, Aug. 2000.

Liu, Y. and Ngan, K. N., "Weighted adaptive lifting-based wavelet transform for image coding," *IEEE Trans. Image Process.*, vol. 17, no. 4, pp. 500–511, Apr. 2008.

Liu, Y., Wu, F., and Ngan, K. N., "3-D object-based scalable wavelet video coding with boundary effect suppression," *IEEE Trans. Circuits Syst. Video Technol.*, vol. 17, no. 5, pp. 639–644, May 2007.

Morales, E. and Shih, F. Y., "Wavelet coefficients clustering using morphological operations and pruned quadtrees," *Pattern Recogn.*, vol. 33, no. 10, pp. 1611–1620, Oct. 2000.

Oliaei, O., "Laplace domain analysis of periodic noise modulation," *IEEE Trans. Circuits Syst. I*, vol. 50, no. 4, pp. 584–588, Apr. 2003.

Psenicka, B., Garcia-Ugalde, F., and Herrera-Camacho, A., "The bilinear *Z* transform by Pascal matrix and its application in the design of digital filters," *IEEE Signal Process. Lett.*, vol. 9, no. 11, pp. 368–370, Nov. 2002.

Schiff, J. L., *The Laplace Transform: Theory and Applications*, Springer, 1999.

Sheridan, P., "A method to perform a fast Fourier transform with primitive image transformations," *IEEE Trans. Image Process.*, vol. 16, no. 5, pp. 1355–1369, May 2007.

Siu, Y. L. and Siu, W. C., "Variable temporal-length 3-D discrete cosine transform coding," *IEEE Trans. Image Process.*, vol. 6, no. 5, pp. 758–763, May 1997.

Sneddon, I. N., *Fourier Transforms*, Courier Dover, 1995.

Yazici, B. and Xie, G., "Wideband extended range-Doppler imaging and waveform design in the presence of clutter and noise," *IEEE Trans. Inform. Theory*, vol. 52, no. 10, pp. 4563–4580, Oct. 2006.

Yip, P. and Rao, K., *Discrete Cosine Transform: Algorithms, Advantages, and Applications*, Academic Press, 1990.

Zeng, B. and Fu, J., "Directional discrete cosine transforms—a new framework for image coding," *IEEE Trans. Circuits Syst. Video Technol.*, vol. 18, no. 3, pp. 305–313, Mar. 2008.

Zeng, Y., Cheng, L., Bi, G., and Kot, A. C., "Integer DCTs and fast algorithms," *IEEE Trans. Signal Process.*, vol. 49, no. 11, pp. 2774–2782, Nov. 2001.

CHAPTER 3

IMAGE ENHANCEMENT

Whenever a picture is converted from one form to another, for example, imaged, copied, scanned, transmitted, or displayed, the "quality" of the output picture is lower than that of the input. In the absence of knowledge about how the given picture was actually degraded, it is difficult to predict in advance how effective a particular enhancement method will be. Image enhancement aims to improve human perception and interpretability of information in images or to provide more useful input for other automated image processing techniques (Munteanu and Rosa, 2004; Kober, 2006; Panetta et al., 2008). In general, image enhancement techniques can be divided into three categories:

1. Spatial-domain methods that directly manipulate pixels in an image.
2. Frequency-domain methods that operate on the Fourier transform or other frequency domains of an image.
3. Combinational methods that process an image in both spatial and frequency domains.

When conducting image enhancement, we need to keep in mind that there is no general rule for determining what the best image enhancement technique is. Consequently, the enhancement methods are application specific and often developed empirically. When image enhancement techniques are used as preprocessing tools for other image processing techniques, some quantitative measures can be used to determine which techniques are most appropriate.

In general, the process of image enhancement involves three types of processes: point process, mask process, and global process. In a point process, each pixel is modified according to a particular equation depending on the input only at the same pixel, which is independent of other pixel values. The input may be one or more images. For example, the difference or product of two images can be taken point by point. In a mask process, each pixel is modified according to the values of the pixel's neighbors using convolution masks. For example, an average of the pixels can be taken in the neighborhood as a low-pass filter. In a global process, all the pixel values in the image (or subimage) are taken into consideration. For example, histogram equalization remaps the histogram of the entire input pixels to a uniformly distributed histogram. Spatial-domain processing methods include all three types, but frequency-domain methods, by the nature of frequency transforms, are global processes. Of course, frequency-domain operations can become mask operations based only on a

Image Processing and Pattern Recognition by Frank Shih

local neighborhood by performing the transform on small image blocks instead of the entire image.

In practice, it is very difficult to enhance an image that has been recorded in the presence of one or more sources of degradation in a single-pass filtering. Noise must be first eliminated and edges have to be preserved. Several passes may be needed to clarify the perceived subpatterns. It is similar to the activation of a human's visual system. When we see a picture, our brain will request the visual system to focus on certain subpatterns that are the most interesting. Humans can identify object edges and regions with exceptional accuracy seemingly instantaneously, while conventional computers are not always nearly as accurate or as fast.

In this chapter we introduce the commonly used image enhancement techniques, including grayscale transformation, piecewise linear transformation, bit plane slicing, histogram equalization, histogram specification, enhancement by arithmetic operations, smoothing filter, and sharpening filter. We also introduce image blur types and image quality measures.

3.1 GRAYSCALE TRANSFORMATION

Grayscale transformation aims to change the gray levels of an entire image in a uniform way or intends to modify the gray levels within a defined window by a mapping function. This transformation is usually expected to enhance the image contrast, so the details of an image can be more visible. The value of a pixel with coordinates (x, y) in the enhanced image $\hat{f}$ is the result of performing some operation on the value of f at (x, y). Thresholding is the simplest case to replace the intensity profile by a step function, jumping at a chosen threshold value. In this case, any pixel with a gray level below the threshold in the input image receives 0 in the output image, and above or equal to the threshold receives 255.

Another simple operation, image negative, reverses the order of pixel intensities from black to white, so the intensity of output decreases as the intensity of input increases. It is a reversed image where the image that is usually black on a white background is reversed to be white on a black background. Let the input gray level r and the output gray level s be in the range $[0, L-1]$. The relationship between the input and output gray levels is given by $s = L-1-r$. Figure 3.1 shows the Lena image and its image negative. This image negative operation is equivalent to a photographic negative. It is quite useful to show the details of small white or gray regions when they appear in a large dark background.

Some examples of mapping from the input gray level r to the output gray level s are illustrated in Figure 3.2, including identity, negative, and thresholding. The mapping function can use mathematical operations (e.g., logarithm, exponential, root, power, etc.) and any degree of polynomial functions. For example, the log mapping is given by $s = c\log(1+r)$, where c is a constant. A useful application of log mapping is to compress the large dynamic range of gray-level values (e.g., Fourier spectrum), so the brightest pixels will not dominate the display and the darker pixels will still be visible. Grayscale transformation can be performed using lookup tables and is often used for human perception purposes.

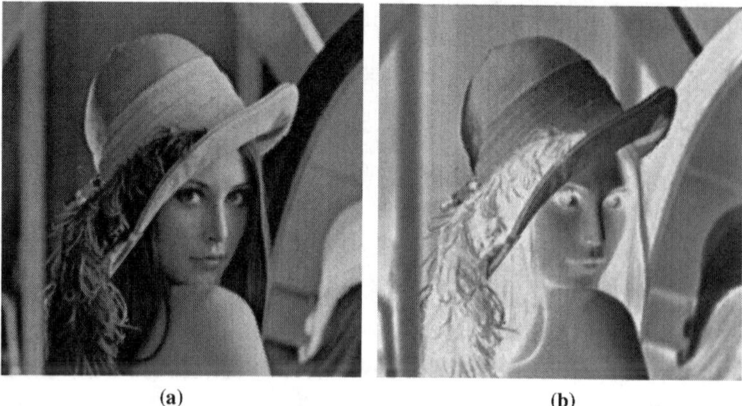

(a) (b)

Figure 3.1 (a) The Lena image and (b) its image negative.

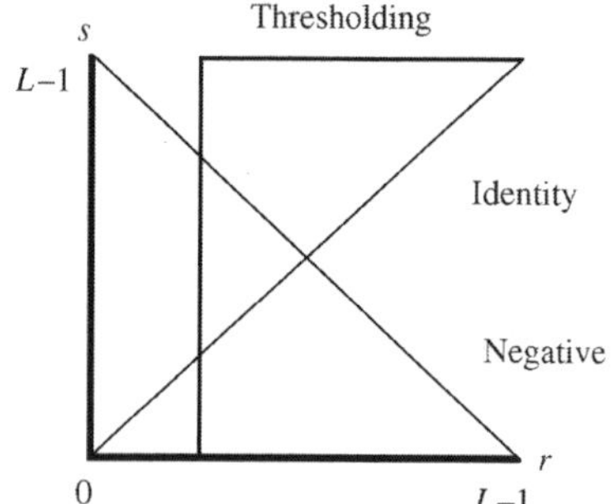

Figure 3.2 The mapping from an input image to an output image.

3.2 PIECEWISE LINEAR TRANSFORMATION

The piecewise linear transformation uses different linear functions to map different input gray-level intervals. A low-contrast image could occur due to poor illumination, lack of dynamic range in the sensor, or a wrong setting in a lens. Contrast stretching (or sometimes called *image normalization*) aims to improve the contrast in an image by stretching the narrow range of input intensity values to span a desired range of intensity values (usually the full range of gray values). It only applies a linear scaling function to the input. A general function for contrast stretching is illustrated in Figure 3.3. The locations of (r_1, s_1) and (r_2, s_2) control the shape of the transformation function. The constraints for this function are $r_1 \leq r_2$ and $s_1 \leq s_2$. The function is single valued and monotonically increasing, so the order of gray levels in the output is preserved.

If $r_1 = s_1$ and $r_2 = s_2$, it is a linear function indicating no changes in the output gray levels. If $r_1 = r_2$, $s_1 = 0$, and $s_2 = L-1$, it is a thresholding function and the output is a binary image. If $r_1 < r_2$, $s_1 = 0$, and $s_2 = L-1$, it is a linear scaling. Let the input gray level r be mapped to the output gray level s in the full range of $[0, L-1]$. Let the minimum and maximum values of the input image be denoted as "min" and "max," respectively. The following equation is used to perform the linear

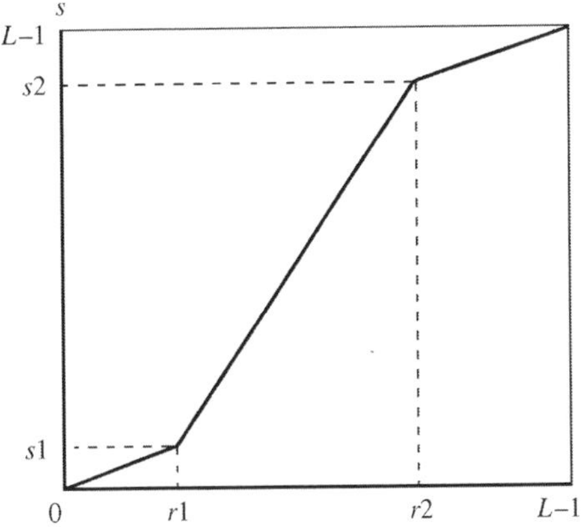

Figure 3.3 The mapping function of contrast stretching.

scaling and round the floating point value to the closest integer for the output image to display as

$$s = \frac{L-1}{\max - \min}(r - \min) \tag{3.1}$$

Figure 3.4 shows an example of contrast stretching.

Example 3.1 Given $r_1 = 2$, $s_1 = 1$, $r_2 = 5$, and $s_2 = 7$. Show the contrast-stretched output of the following image in the range [0, 7].

2	2	3	3	3	3	3	2
2	2	2	2	3	2	3	2
2	0	4	4	5	5	3	3
3	0	4	4	5	5	5	5
2	4	4	4	4	5	7	3
2	2	4	5	6	5	6	2
2	3	4	4	2	5	6	2
2	3	2	2	2	3	5	3

(a)

(b)

Figure 3.4 (a) A low-contrast image and (b) the image after contrast stretching.

Answer: The transform function is depicted by the following graph using the given (2, 1) and (5, 7) coordinates.

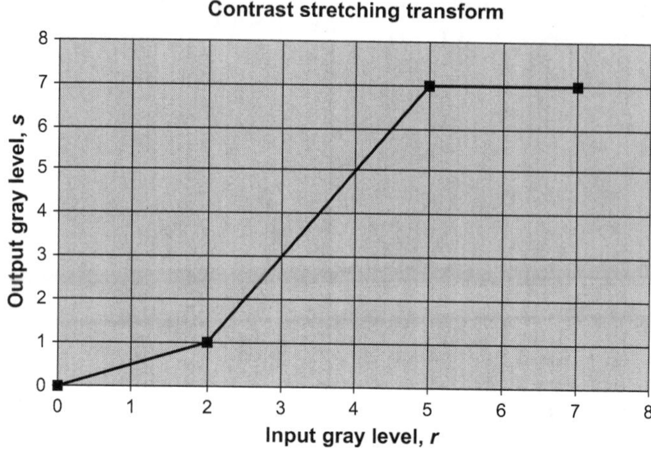

This results in the following intensity transformations from a given input original value to the contrast-stretched value.

Original value	Contrast-stretched value
0	0
1	0
2	1
3	3
4	5
5	7
6	7
7	7

The following contrast-stretched image is obtained by making the above replacements in the original image.

1	1	3	3	3	3	3	1
1	1	1	1	3	1	3	1
1	0	5	5	7	7	3	3
3	0	5	5	7	7	7	7
1	5	5	5	5	7	7	3
1	1	5	7	7	7	7	1
1	3	5	5	1	7	7	1
1	3	1	1	1	3	7	3

Another piecewise linear transformation uses the histogram of an input image to find the cluster of pixels that contain the most relevant information. The histogram is a diagram that shows the distribution of gray levels. In a digital image, the histogram shows at each gray level the number of pixels in the image having that gray level. It has a particular inclination for most images to be multimodal. The determination of this cluster can become a nontrivial process. This application searches for the lowest valley between the background and the start of the cluster. For example, if a histogram is trimodal, the two valley points A and B around the clusters are selected. Let the output gray levels be in the range [0, 255]. This grayscale transformation maps the input gray level r into the output gray level s and is given as

$$s = \begin{cases} \dfrac{255-A+B}{2(B-A)}(r-A) + \dfrac{A}{2} & \text{if } A \leq r \leq B \\ \dfrac{r}{2} & \text{if } r < A \\ \dfrac{r+255}{2} & \text{if } r > B \end{cases} \tag{3.2}$$

This transformation compresses the grayscale by a factor of 2 in the ranges $[0, A]$ and $[B, 255]$. The cluster is expanded by a factor of 2 over the range $[A, B]$. Note that the linear functions used can be replaced by any desired mathematical function, such as quadratic and high-degree polynomials.

3.3 BIT PLANE SLICING

Since a pixel contains 8 bits in a 256-grayscale image, the bit plane slicing collects all the bits in the specific 8-bit plane of the image at hand. An 8-bit per pixel image can be sliced into eight bit planes. Zero is the least significant bit (LSB) and 7 is the most significant bit (MSB). Each plane constitutes a collection of zeros and ones from that plane where zero translates to gray level 0 and one translates to gray level 255. The higher order bits contain more visually significant data, and the lower order bits contribute the more detailed information about the image. Figure 3.5 shows the bit plane slicing of the Lena image. The images show that most image information is contained in the higher (i.e., more significant) bits, whereas the less significant bits contain some of the finer details and noise.

An application of bit plane slicing is for data compression in image processing. Bit plane combining is the reverse process of the slicing. The planes are recombined in order to reconstruct the image. But, it is not needed to take into consideration all the slice contributions. Especially, in the case where the data rate is important, some planes can be ignored until the changes in gray level have an acceptable impact on the image.

3.4 HISTOGRAM EQUALIZATION

Histogram equalization employs a monotonic nonlinear mapping that reassigns the intensity values of pixels in an input image, such that the output image contains a uniform distribution of intensities (i.e., a histogram that is constant for all brightness values). This corresponds to a brightness distribution where all values are equally probable. Unfortunately, we can only achieve the approximation of this uniform distribution for

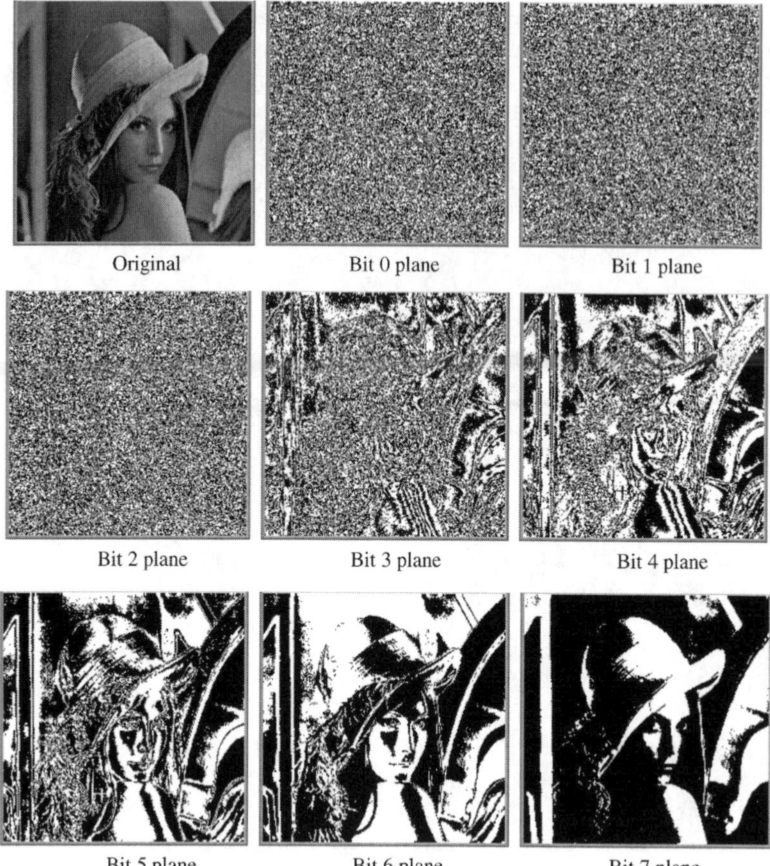

Figure 3.5 The Lena image and its bit planes 0 (least significant bit) through 7 (most significant bit).

a digital image. This technique is often used in image comparison because it is effective in detail enhancement and in the correction of nonlinear effects introduced by a digitizer or a display system. In general, since the histogram equalization causes the dynamic range of an image to be stretched, the density distribution of the resultant image is made flat, so that the contrast of the image is enhanced. However, the histogram equalization method causes some problems. Since the contrast is enhanced by stretching the dynamic range, background noise is simultaneously increased by the equalization, and the image quality in a near-constant region may be degraded.

Consider an image pixel value $r \geq 0$ to be an element of random variable R with a continuous probability density function $p_R(r)$ and cumulative probability distribution $F_R(r) = P[R \leq r]$. Let the mapping function be $s = f(r)$ between the input and output images. To equalize the histogram of the output image, let $p_S(s)$ be a constant. In particular, if the gray levels are assumed to be in the range between 0 and 1, then $p_S(s) = 1$ forms a uniform random variable S. The mapping function for histogram equalization

$$s = F_R(r) = \int_0^r p_R(r)dr \tag{3.3}$$

will be uniformly distributed over (0, 1).

To implement this transformation on digital images, let n denote the total number of pixels, n_G the total number of gray levels, and n_{r_j} the number of pixels in the input image with intensity value r_j. Let the input and output gray values be in the range of $[0, 1, \ldots, n_G-1]$. Then, the histogram equalization transformation maps the input value r_k (where $k = 0, 1, \ldots, n_G-1$) to the output value s_k as

$$s_k = T(r_k) = (n_G-1)\sum_{j=0}^{k}\frac{n_{r_j}}{n} = \frac{(n_G-1)}{n}\sum_{j=0}^{k} n_{r_j} \tag{3.4}$$

Note that the resulting floating point value will be rounded to its closest integer as the output value. Figure 3.6 shows a low-contrast input image and its resulting image after histogram equalization.

Example 3.2 Suppose that a digital image is subjected to histogram equalization. Show that a second pass of histogram equalization will produce exactly the same result as the first pass.

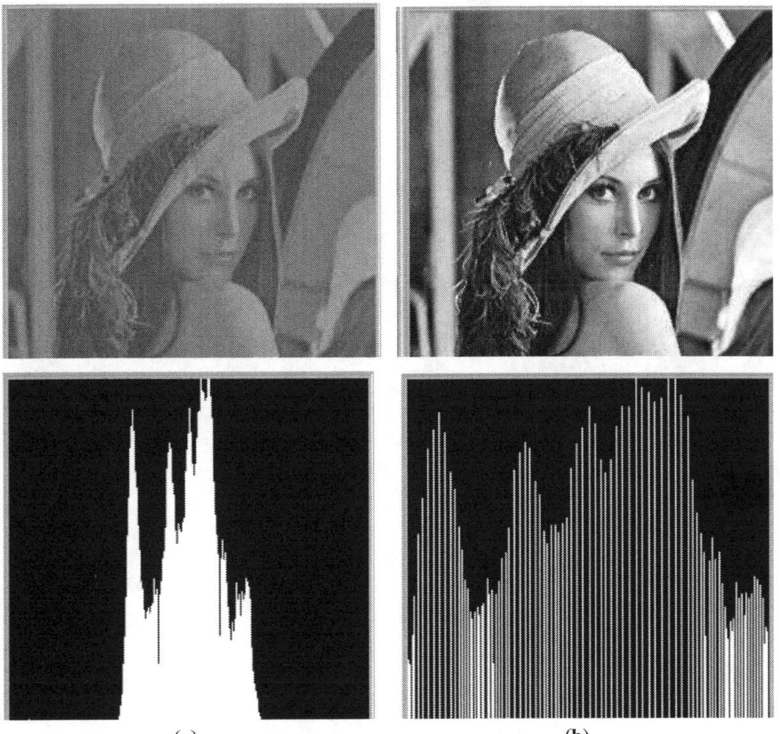

(a) (b)

Figure 3.6 Column (a) shows a low-contrast image and its histogram, and column (b) shows the equalized image and its histogram.

Answer: Assume that a first pass of histogram equalization transforms the input value r_k to s_k and then a second pass transforms s_k to v_k. The second pass is conducted according to the following transformation:

$$v_k = T(s_k) = \frac{(n_{\mathrm{G}}-1)}{n}\sum_{j=0}^{k} n_{s_j}$$

Since every pixel (and no others) with value r_k is mapped to value s_k, it follows $n_{s_k} = n_{r_k}$. Therefore,

$$v_k = T(s_k) = \frac{(n_{\mathrm{G}}-1)}{n}\sum_{j=0}^{k} n_{r_j} = s_k$$

which shows that a second pass of histogram equalization would yield the same result as the first pass. Note that negligible roundoff errors are assumed in this case.

Example 3.3 Apply histogram equalization on the input image of 8 × 8 below. Let the input and output gray levels be in the range of [0, 7].

1	1	5	5	0	0	1	0
1	1	2	2	0	1	0	1
1	7	6	6	5	5	0	0
0	7	6	7	5	5	5	5
4	7	6	7	3	5	7	0
1	1	4	1	6	5	6	1
2	2	4	1	1	5	1	1
1	2	2	0	0	0	0	5

Answer: In this case, $n_G = 8$ and $n = 64$. Using equation (3.4), it yields

$$s_k = \frac{7}{64}\sum_{j=0}^{k} n_{r_j}$$

r_k	n_{r_j}	$\sum_{j=0}^{k} n_{r_j}$	s_k
0	13	13	1
1	17	30	3
2	6	36	4
3	1	37	4
4	3	40	4
5	12	52	6
6	6	58	6
7	6	64	7

Therefore, the output image is

3	3	6	6	1	1	3	1
3	3	4	4	1	3	1	3
3	7	6	6	6	6	1	1
1	7	6	7	6	6	6	6
4	7	6	7	4	6	7	1
3	3	4	3	6	6	6	3
4	4	4	3	3	6	3	3
3	4	4	1	1	1	1	6

Histogram equalization has been extensively applied. Menotti et al. (2007) developed multi-histogram equalization methods for contrast enhancement and brightness preserving. Ibrahim and Kong (2007) proposed the method of brightness preserving dynamic histogram equalization (BPDHE), which can produce the output image with the mean intensity almost equal to the mean intensity of the input. Agaian et al. (2007) presented transform coefficient histogram-based image enhancement algorithms using contrast entropy. Wang and Ward (2007) developed fast image/video contrast enhancement based on weighted thresholded histogram equalization.

3.5 HISTOGRAM SPECIFICATION

Histogram equalization intends to map any input image into an output image with the uniformly distributed histogram. Sometimes, a particularly distributed histogram of output images is desired in specific applications. Therefore, histogram specification is used to convert an image, so that it has a particular histogram of output images as specified (Sun et al., 2005; Coltuc et al., 2006; Wan and Shi, 2007). Let x and y denote the gray levels of the input and output images, respectively. The probability $p_X(x)$ from the input image can be computed, and the specified probability $p_z(z)$, which the output image is desired to have, is designed for a particular application. In this method, the histogram equalization is first conducted as

$$y = f(x) = \int_0^x p_X(u)du \tag{3.5}$$

When the gray levels of the desired image z are available, it can also be equalized as

$$y' = g(z) = \int_0^z p_z(u)du \tag{3.6}$$

The inverse of the above transform is $z = g^{-1}(y')$. Since the images y and y' have the same equalized histogram, they are the same image and the overall transform from the given image x to the desired image z can be obtained by

$z = g^{-1}(y) = g^{-1}(f(x)) = h(x)$, where $h(x) = g^{-1}(f(x))$ is the overall transform, and both f and g can be found from the corresponding histograms of the given image x and the desired image z, respectively.

Example 3.4 Apply histogram specification on the following image. Let the input and output gray levels be in the range of [0, 7]. Assume that the expected grayscale specification is {0: 5%, 1: 5%, 2: 10%, 3: 10%, 4: 25%, 5: 5%, 6: 25%, 7: 15%}. Show the output image.

1	1	0	0	0	0	0	1
1	1	1	1	0	1	0	1
1	3	4	4	5	5	0	0
0	3	4	4	5	5	5	5
2	4	4	4	3	5	7	0
1	1	4	5	6	5	6	1
1	0	4	4	1	5	6	1
1	0	1	0	0	0	5	0

Answer: From the input image,

x	p_x	$\int_0^x p_x(u)du$
0	17/64	17/64 = 0.2656
1	18/64	35/64 = 0.5469
2	1/64	36/64 = 0.5625
3	3/64	39/64 = 0.6094
4	10/64	49/64 = 0.7656
5	11/64	60/64 = 0.9375
6	3/64	63/64 = 0.9844
7	1/64	64/64 = 1.0000

From the expected specification,

z	p_z	$\int_0^z p_z(u)du$
0	0.05	0.05
1	0.05	0.10
2	0.10	0.20
3	0.10	0.30
4	0.25	0.55
5	0.05	0.60
6	0.25	0.85
7	0.15	1.00

Mapping from the input gray level to the output gray level of the expected specification yields

x	Mapping	z
0	$0.2656 \approx 0.30$	3
1	$0.5469 \approx 0.55$	4
2	$0.5625 \approx 0.55$	4
3	$0.6094 \approx 0.60$	5
4	$0.7656 \approx 0.85$	6
5	$0.9375 \approx 1.00$	7
6	$0.9844 \approx 1.00$	7
7	$1.0000 \approx 1.00$	7

Therefore, the output image after histogram specification is

4	4	3	3	3	3	3	4
4	4	4	4	3	4	3	4
4	5	6	6	7	7	3	3
3	5	6	6	7	7	7	7
4	6	6	6	5	7	7	3
4	4	6	7	7	7	7	4
4	3	6	6	4	7	7	4
4	3	4	3	3	3	7	3

3.6 ENHANCEMENT BY ARITHMETIC OPERATIONS

Arithmetic operations (e.g., subtraction, addition, multiplication, division, and mean) are often used to combine and transform two or more images into a new image that can better display or highlight certain features in the scene. It is also possible to just use a single image as input and perform arithmetic operations on all the pixels to modify brightness and enhance contrast.

The image subtraction operator takes the difference of two input images. It usually uses the absolute difference between pixel values, rather than the straightforward signed output. Image subtraction can be used to detect changes in a series of images of the same scene or recognize a moving object.

Image averaging works if the noise in the image pixels and the associated noise are not correlated and the noise has a zero averaging value. These conditions are necessary because the image averaging method relies on the summing of N different noisy images. If the noise did not average out to zero, then artifacts of the noise would appear in the averaged image. The mathematical representation of this method is described below.

Let the captured K images be $g_i(x, y)$, $i = 1, 2, \ldots, K$. The averaged image, denoted as $\bar{g}(x, y)$, can be computed by

$$\bar{g}(x, y) = \frac{1}{K}\sum_{i=1}^{K} g_i(x, y) \tag{3.7}$$

The captured image, $g_i(x, y)$, is assumed to be formed by the addition of noise $\eta_i(x, y)$ to an original image$f(x, y)$ as

$$g_i(x, y) = f(x, y) + \eta_i(x, y) \tag{3.8}$$

At every pair of coordinates (x, y), the noise is uncorrelated and has zero average value. If an image $\bar{g}(x, y)$ is formed by averaging K different noisy image as in equation (3.7), then it follows that

$$E\{\bar{g}(x, y)\} = f(x, y) \quad \text{and} \quad \sigma^2_{\bar{g}(x,y)} = \frac{1}{K}\sigma^2_{\eta(x,y)} \tag{3.9}$$

where $E\{\bar{g}(x, y)\}$ is the expected value of $\bar{g}$, and $\sigma^2_{\bar{g}(x,y)}$ and $\sigma^2_{\eta(x,y)}$ are the variances of $\bar{g}$ and η, all at coordinates (x, y). The standard deviation at any point in the average image is

$$\sigma_{\bar{g}(x,y)} = \frac{1}{\sqrt{K}}\sigma_{\eta(x,y)} \tag{3.10}$$

As K increases, the variability of the pixel values at each location (x, y) decreases. Because $E\{\bar{g}(x, y)\} = f(x, y)$, this means that $\bar{g}(x, y)$ approaches $f(x, y)$ as the number of noisy images used in the averaging process increases. This method can result in improved image quality because the noise is effectively canceled if enough images are averaged together since a largely distributed set of images should bring the noise average to zero if it is truly uncorrelated.

Similarly, frame averaging provides a way to average multiple video frames to create a more stable image. This module can be used to eliminate pixel vibrations or high-frequency image changes. It works by adding each frame into a moving average of frames. This effectively creates the same effect of averaging many frames without the significant memory and time that averaging hundreds of frames would take.

3.7 SMOOTHING FILTER

A simple mean smoothing filter or operation intends to replace each pixel value in an input image by the mean (or average) value of its neighbors, including itself. This has an effect of eliminating pixel values that are unrepresentative of their surroundings. Like other convolution filters, it is based around a kernel, which represents the shape and size of the neighborhood to be sampled in calculation. Often a 3×3 square kernel is used, as shown below, although larger kernels (e.g., a 5×5 square) can be used for more severe

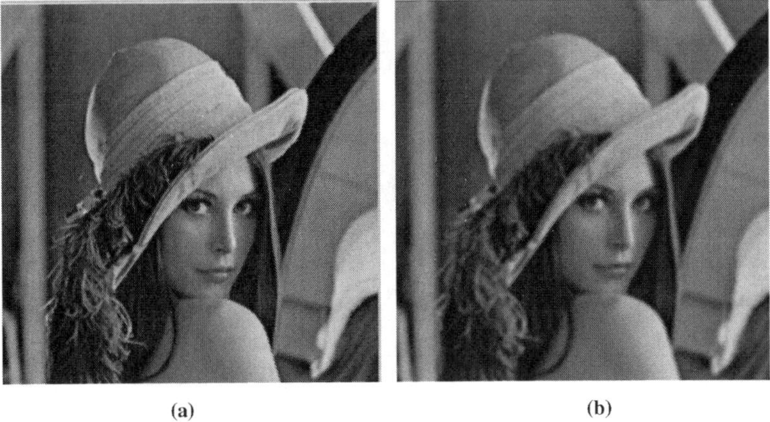

(a) (b)

Figure 3.7 (a) The Lena image and (b) the resulting image after a 3×3 averaging filter.

smoothing. Figure 3.7 shows an example of applying the 3×3 averaging kernel.

$$\frac{1}{9}\begin{bmatrix} 1 & 1 & 1 \\ 1 & 1 & 1 \\ 1 & 1 & 1 \end{bmatrix}$$

Another method of image smoothing convolves an input image by the Gaussian filter. The Gaussian filter, representing the shape of a Gaussian (bell-shaped) hump, will screen noise with the high spatial frequencies and produce a smoothing effect. The 1D Gaussian filter function is

$$G(x) = \frac{1}{\sqrt{2\pi}\sigma} e^{-x^2/2\sigma^2} \tag{3.11}$$

where σ is the standard deviation of the distribution. The distribution of mean zero and $\sigma = 1$ is shown in Figure 3.8. In two dimensions, an isotropic (i.e., circularly symmetric) Gaussian filter function is

$$G(x, y) = \frac{1}{\sqrt{2\pi\sigma^2}} e^{-(x^2+y^2)/2\sigma^2} \tag{3.12}$$

The distribution of mean (0, 0) and $\sigma = 1$ is shown in Figure 3.9.

In image processing, a discrete representation of the Gaussian function is required for conducting the convolution. In theory, the Gaussian distribution is nonzero everywhere, meaning an infinitely large convolution kernel. In practice, it is effectively zero more than about three standard deviations from the mean, so the kernels beyond this point can be truncated. Figure 3.10 shows integer-valued convolution kernels of 3×3 and 5×5 that approximate a Gaussian filter function with $\sigma = 1$. Note that a constant scaling factor is multiplied to ensure that the output gray levels are in the same range as the input gray levels.

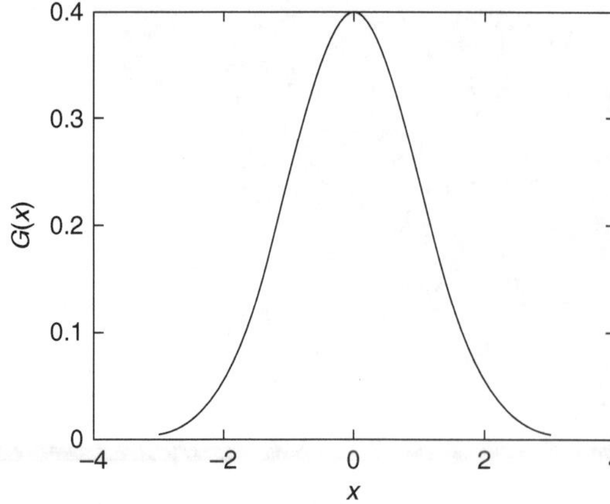

Figure 3.8 The 1D Gaussian filter function with mean zero and $\sigma = 1$.

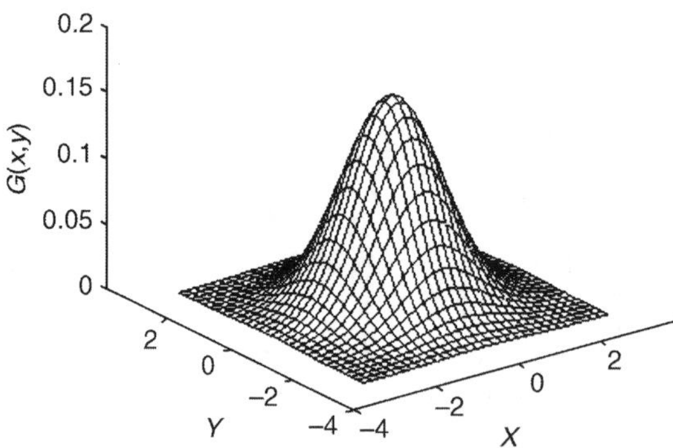

Figure 3.9 The 2D Gaussian filter function with mean (0,0) and $\sigma = 1$.

Example 3.5 Suppose a low-pass spatial filter is formed by averaging the four immediate neighbors of a point (x, y) but excluding the point itself. Find the equivalent filter $H(u, v)$ in the frequency domain.

Answer: The spatial average is

$$g(x,y) = \frac{1}{4}[f(x,y+1) + f(x+1,y) + f(x-1,y) + f(x,y-1)]$$

Perform Fourier transform to obtain

$$G(u,v) = \frac{1}{4}\left[e^{j2\pi v/N} + e^{j2\pi u/M} + e^{-j2\pi u/M} + e^{-j2\pi v/N}\right]F(u,v) = H(u,v)F(u,v)$$

$$\frac{1}{16}\begin{bmatrix}1 & 2 & 1\\ 2 & 4 & 2\\ 1 & 2 & 1\end{bmatrix}$$

(a)

$$\frac{1}{273}\begin{bmatrix}1 & 4 & 7 & 4 & 1\\ 4 & 16 & 26 & 16 & 4\\ 7 & 26 & 41 & 26 & 7\\ 4 & 16 & 26 & 16 & 4\\ 1 & 4 & 7 & 4 & 1\end{bmatrix}$$

(b)

Figure 3.10 Discrete Gaussian low-pass convolution filters (a) 3×3 and (b) 5×5.

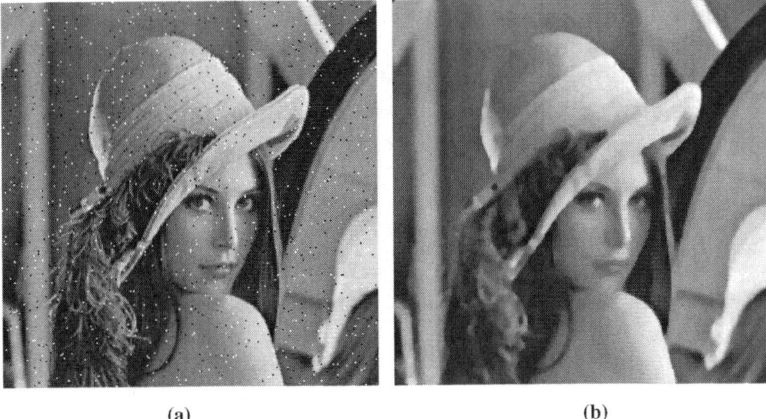

(a) (b)

Figure 3.11 (a) A noisy image and (b) the resulting image after median filtering.

where $H(u, v) = \frac{1}{2}[\cos(2\pi u/M) + \cos(2\pi v/N)]$ is the filter transfer function in the frequency domain.

Another smoothing filter, called a *median filter*, is used to reduce noise in an image, somewhat like a mean filter. However, it performs better than a mean filter in the sense of preserving useful details in the image. It is especially effective for removing impulse noise, which is characterized by bright and/or dark high-frequency features appearing randomly over the image. Statistically, impulse noise falls well outside the peak of the distribution of any given pixel neighborhood, so the median is well suited to learn where impulse noise is not present, and hence to remove it by exclusion. The median of a distribution is the value for which larger and smaller values are equally probable. To calculate the median of a list of sample values, we sort them in a descending or ascending order, and then select the central value. An example of applying a median filter on an added salt-and-pepper noisy Lena image is shown in Figure 3.11.

A median filter is a kind of nonlinear filter and a subset of so-called *order-statistics filters*. A maximum, minimum, or any rank in the descending or ascending order, instead of the median value, can be picked up in the neighborhood around a pixel. Note that the maximum filter corresponds to the morphological dilation and the minimum filter corresponds to the morphological erosion, which will be discussed in Chapter 4.

3.8 SHARPENING FILTER

Sharpening filter is used to enhance the edges of objects and adjust the contrast of object and background transitions. They are sometimes used as edge detectors by combining with thresholding. Sharpening or high-pass filter allows high-frequency components to pass and delete the low-frequency components. For a kernel to be a high-pass filter, the coefficients near the center must be set positive and in the outer periphery must be set negative. Sharpening filter can be categorized into four types: high-pass filter, Laplacian of Gaussian filter, high-boost filter, and derivative filter.

A high-pass filter, opposite of a low-pass filter, is a filter that passes high-frequency components, but attenuates (or delete) the components whose frequency is lower than the cutoff frequency. A simple 3×3 high-pass filter is given as

$$\begin{bmatrix} -1 & -1 & -1 \\ -1 & 8 & -1 \\ -1 & -1 & -1 \end{bmatrix}$$

An example of applying the high-pass filter on the Lena image is shown in Figure 3.12.

Another sharpening filter is the derivative (or gradient) operator. For an image function $f(x, y)$, the gradient vector is defined as $\nabla f \equiv (\partial f/\partial x)\vec{i} + (\partial f/\partial y)\vec{j}$, where $\vec{i}$ and $\vec{j}$ are, respectively, the unit vectors along x- and y-axes. Let $f_x \equiv \partial f/\partial x$ and $f_y \equiv \partial f/\partial y$. The magnitude of ∇f is $\sqrt{f_x^2 + f_y^2}$, and its direction is $\theta = \tan^{-1}(f_y/f_x)$. For digital images, the derivatives of the gradient operator are the two differences: $f_x = f(x+1, y) - f(x, y)$ and $f_y = f(x, y+1) - f(x, y)$. Both can be obtained by convolution with the following kernels:

$$\begin{bmatrix} -1 & 1 \\ 0 & 0 \end{bmatrix}, \quad \begin{bmatrix} -1 & 0 \\ 1 & 0 \end{bmatrix}$$

If we consider the diagonal directions, the two kernels are

$$\begin{bmatrix} 0 & 1 \\ -1 & 0 \end{bmatrix}, \quad \begin{bmatrix} 1 & 0 \\ 0 & -1 \end{bmatrix}$$

which are called *Roberts cross-gradient operators*. If considering kernels of size 3×3 and first finding the averages of one direction and then finding the difference of

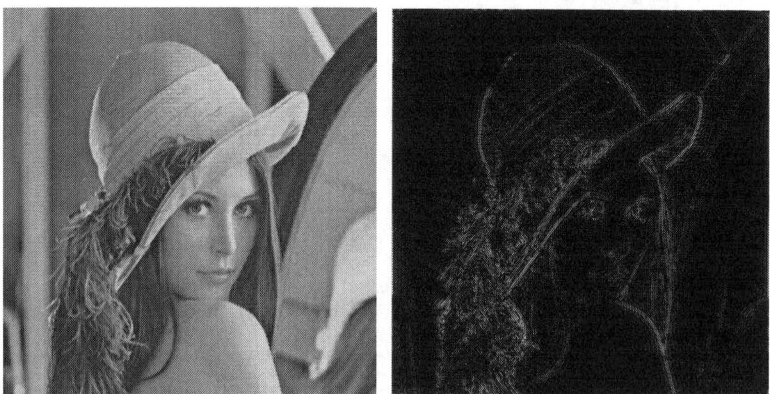

Figure 3.12 Applying the high-pass filter on the Lena image.

these averages in the other direction, we have

$$\begin{bmatrix} -1 & 0 & 1 \\ -1 & 0 & 1 \\ -1 & 0 & 1 \end{bmatrix}, \quad \begin{bmatrix} -1 & -1 & -1 \\ 0 & 0 & 0 \\ 1 & 1 & 1 \end{bmatrix}$$

which are called *Prewitt operators*. If considering a double weight in the horizontal and vertical neighbors, we have

$$\begin{bmatrix} -1 & 0 & 1 \\ -2 & 0 & 2 \\ -1 & 0 & 1 \end{bmatrix}, \quad \begin{bmatrix} -1 & -2 & -1 \\ 0 & 0 & 0 \\ 1 & 2 & 1 \end{bmatrix}$$

which are called *Sobel operators*.

The Laplacian of Gaussian filter (LoG) is the combination of the Laplacian and Gaussian filters where its characteristic is determined by the σ parameter and the kernel size, as shown in the mathematical expression of the kernel:

$$\text{LoG}(i,j) = -\frac{1}{\pi\sigma^4}\left(1-\frac{i^2+j^2}{2\sigma^2}\right)e^{-(i^2+j^2)/2\sigma^2} \tag{3.13}$$

The 3D graphics of the LoG filter is shown in Figure 3.13.

A discrete 9×9 kernel that approximates this function (for a Gaussian $\sigma = 1.4$) is shown in Figure 3.14. An example of applying this kernel on the Lena image is shown in Figure 3.15.

The high-boost filter is used to emphasize high-frequency components representing the image details without eliminating low-frequency components. It multiplies the original image by an amplification factor A as

$$\begin{aligned} \text{High boost} &= A \times \text{original} - \text{low pass} \\ &= (A-1) \times \text{original} + (\text{original} - \text{low pass}) \\ &= (A-1) \times \text{original} + \text{high pass} \end{aligned} \tag{3.14}$$

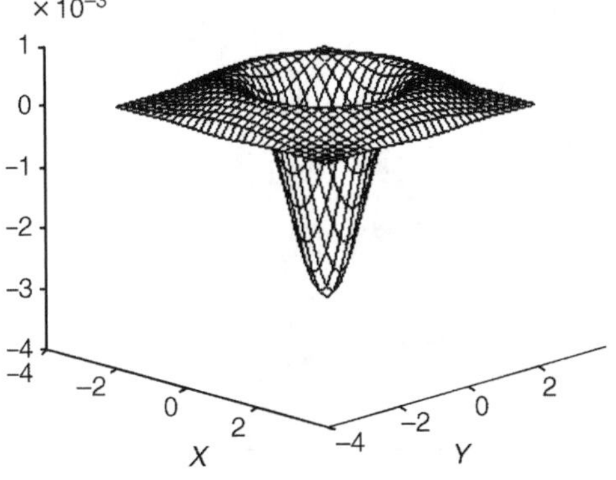

Figure 3.13 The Laplacian of Gaussian filter.

0	1	1	2	2	2	1	1	0
1	2	4	5	5	5	4	2	1
1	4	5	3	0	3	5	4	1
2	5	3	-12	-24	-12	3	5	2
2	5	0	-24	-40	-24	0	5	2
2	5	3	-12	-24	-12	3	5	2
1	4	5	3	0	3	5	4	1
1	2	4	5	5	5	4	2	1
0	1	1	2	2	2	1	1	0

Figure 3.14 The discrete approximation to LoG function with Gaussian $\sigma = 1.4$.

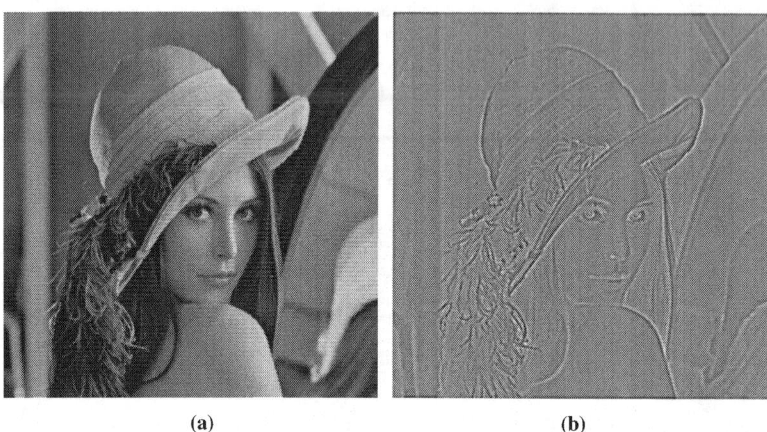

(a) (b)

Figure 3.15 (a) The Lena image and (b) the result after applying a discrete 9×9 LoG filter.

Note that if $A = 1$, it becomes the standard high-pass filter. For a 3×3 kernel,

$$\begin{bmatrix} 0 & 0 & 0 \\ 0 & A & 0 \\ 0 & 0 & 0 \end{bmatrix} - \frac{1}{9}\begin{bmatrix} 1 & 1 & 1 \\ 1 & 1 & 1 \\ 1 & 1 & 1 \end{bmatrix} = \frac{1}{9}\begin{bmatrix} -1 & -1 & -1 \\ -1 & 9A-1 & -1 \\ -1 & -1 & -1 \end{bmatrix} \tag{3.15}$$

An example of applying the high-boost filter with $A = 1.5$ on the Lena image is shown in Figure 3.16.

Figure 3.16 Applying the high-boost filter with $A = 1.5$ on the Lena image.

3.9 IMAGE BLUR TYPES AND QUALITY MEASURES

Three types of blur usually occurred in image processing are motion, Gaussian, and compression blurs (Molina et al., 2006). Motion blur may be due to object movement when a camera shutter remains open for an extended period of time and the object motion within this interval is visible in a single snapshot. Motion blur may also be caused by camera movement, in which the cameraman shifts the camera along objects. A 2D motion blur is shown in Figure 3.17(b). Gaussian blur is made by a soft lens to spread out the light on the focal plane, rather than all going toward a spot. It produces a smoothing effect by removing image details or noises. It can be simulated by mixing each pixel with a Gaussian density function, as shown in Figure 3.17(c). Compression blur is triggered by the loss of high-frequency components in JPEG compression. An example of JPEG compression ratio 1:160 to simulate the compression blur is shown in Figure 3.17(d).

Image quality is characterized by a close interaction between the display and human visual perception. There are many *image quality measures* (IQMs) proposed to evaluate imaging systems and image coding/processing techniques (Sanchez-Marin et al., 2001; Shnayderman et al., 2006). With the understanding of these measures, it is hoped that subjective decisions can be kept to a minimum. Therefore, computational image quality analysis aims at finding an accurate and meaningful

Figure 3.17 (a) Original image, (b) motion blur at orientation 45° and magnitude 40, (c) Gaussian blur with a 17 × 17 window, and (d) compression blur with compression ratio of 1:160.

model to evaluate image quality. The computational model takes an image as the input and returns a quality number as output. If the input takes one parameter, it is the blurred image. If the input takes two parameters, they are the original image and the blurred image.

Let $f(x_i)$denote the IQM score of an image under the degree of blur x_i. The IQMs used for measuring image blurs must satisfy the monotonically increasing or decreasing property. That is, if $x_{i+1} > x_i$, then$f(x_{i+1})-f(x_i) > 0$ for monotonically increasing or $f(x_{i+1})-f(x_i) < 0$ for monotonically decreasing property. The sensitivity of IQMs is defined as the score of the aggregate relative distance:

$$\sum_i \frac{f(x_{i+1})-f(x_i)}{f(x_i)} \tag{3.16}$$

Nine IQMs, which are grouped into three categories based on pixel distance, correlation, and mean square error, are introduced here. Let $F(j,k)$ denote the pixel value of row j and column k in a reference image of size $M \times N$, and $\hat{F}(j,k)$ denote the pixel value in a testing image.

Category I: IQMs based on pixel distance

1. AD (average distance)

$$\text{AD} = \sum_{j=1}^{M}\sum_{k=1}^{N} |F(j,k)-\hat{F}(j,k)|/MN \tag{3.17}$$

2. L2D (L2 Euclidean distance)

$$\text{L2D} = \frac{1}{MN}\left(\sum_{j=1}^{M}\sum_{k=1}^{N} (F(j,k)-\hat{F}(j,k))^2\right)^{1/2} \tag{3.18}$$

Category II: IQMs based on correlation

3. SC (structure content)

$$\text{SC} = \sum_{j=1}^{M}\sum_{k=1}^{N} F(j,k)^2 / \sum_{j=1}^{M}\sum_{k=1}^{N} \hat{F}(j,k)^2 \tag{3.19}$$

4. IF (image fidelity)

$$\text{IF} = 1-\left(\sum_{j=1}^{M}\sum_{k=1}^{N} (F(j,k)-\hat{F}(j,k))^2 / \sum_{j=1}^{M}\sum_{k=1}^{N} F(j,k)^2\right) \tag{3.20}$$

5. NK (N cross-correlation)

$$\text{NK} = \sum_{j=1}^{M}\sum_{k=1}^{N} F(j,k)\hat{F}(j,k) / \sum_{j=1}^{M}\sum_{k=1}^{N} F(j,k)^2 \tag{3.21}$$

Category III: IQMs based on mean square error

6. NMSE (normal mean square error)

$$\text{NMSE} = \sum_{j=1}^{M}\sum_{k=1}^{N}(F(j,k)-\hat{F}(j,k))^2 / \sum_{j=1}^{M}\sum_{k=1}^{N}F(j,k)^2 \tag{3.22}$$

7. LMSE (least mean square error)

$$\text{LMSE} = \sum_{j=1}^{M}\sum_{k=1}^{N}(F(j,k)-\hat{F}(j,k))^2 / \sum_{j=1}^{M}\sum_{k=1}^{N}O(F(j,k))^2 \tag{3.23}$$

where $O(F(j,k)) = F(j+1,k)+F(j-1,k)+F(j,k+1)+F(j,k-1)-4F(j,k)$.

8. PMSE (peak mean square error)

$$\text{PMSE} = \frac{1}{MN}\sum_{j=1}^{M}\sum_{k=1}^{N}[F(j,k)-\hat{F}(j,k)]^2 / \{\max_{j,k}[F(j,k)]\}^2 \tag{3.24}$$

9. PSNR (peak signal to noise ratio)

$$\text{PSNR} = 20 \times \log_{10}\left\{255 / \left\{\sum_{j=1}^{M}\sum_{k=1}^{N}[F(j,k)-\hat{F}(j,k)]^2\right\}^{1/2}\right\} \tag{3.25}$$

An automatic method to measure the degree of similarity with respect to the human vision system (HVS) is described as follows:

1. Perform *discrete cosine transform* (DCT) on the original image $C(x, y)$ and the blurred image $\hat{C}(x,y)$ to obtain $D(u, v)$ and $\hat{D}(u,v)$, respectively.
2. Convolve $D(u, v)$ and $\hat{D}(u,v)$ with the following band-pass filter H to obtain $E(u, v)$ and $\hat{E}(u,v)$, respectively.

$$H(p) = \begin{cases} 0.05e^{p^{0.554}} & \text{if } p < 7 \\ e^{-9|\log_{10}p-\log_{10}9|^{2.3}} & \text{if } p \geq 7 \end{cases} \quad \text{where } p = \sqrt{u^2+v^2} \tag{3.26}$$

3. Perform inverse DCT of $E(u, v)$ and $\hat{E}(u,v)$, respectively, to obtain $F(x, y)$ and $\hat{F}(x,y)$.
4. Calculate the Euclidean distance between $F(x,y)$ and $\hat{F}(x,y)$ using equation (3.18).

REFERENCES

Agaian, S. S., Silver, B., and Panetta, K. A., "Transform coefficient histogram-based image enhancement algorithms using contrast entropy," *IEEE Trans. Image Process.*, vol. 16, no. 3, pp. 741–758, Mar. 2007.

Coltuc, D., Bolon, P., and Chassery, J.-M., "Exact histogram specification," *IEEE Trans. Image Process.*, vol. 15, no. 5, pp. 1143–1152, May 2006.

Ibrahim, H. and Kong, N. S. P., "Brightness preserving dynamic histogram equalization for image contrast enhancement," *IEEE Trans. Consumer Electron.*, vol. 53, no. 4, pp. 1752–1758, Nov. 2007.

Kober, V., "Robust and efficient algorithm of image enhancement," *IEEE Trans. Consumer Electron.*, vol. 52, no. 2, pp. 655–659, May 2006.

Menotti, D., Najman, L., Facon, J., and de Araujo, A. A., "Multi-histogram equalization methods for contrast enhancement and brightness preserving," *IEEE Trans. Consumer Electron.*, vol. 53, no. 3, pp. 1186–1194, Aug. 2007.

Molina, R., Mateos, J., and Katsaggelos, A. K., "Blind deconvolution using a variational approach to parameter, image, and blur estimation," *IEEE Trans. Image Process.*, vol. 15, no. 12, pp. 3715–3727, Dec. 2006.

Munteanu, C. and Rosa, A., "Gray-scale image enhancement as an automatic process driven by evolution," *IEEE Trans. Syst. Man Cybernet. B*, vol. 34, no. 2, pp. 1292–1298, Apr. 2004.

Panetta, K. A., Wharton, E. J., and Agaian, S. S., "Human visual system-based image enhancement and logarithmic contrast measure," *IEEE Trans. Syst. Man Cybernet. B*, vol. 38, no. 1, pp. 174–188, Feb. 2008.

Sanchez-Marin, F. J., Srinivas, Y., Jabri, K. N., and Wilson, D. L., "Quantitative image quality analysis of a nonlinear spatio-temporal filter," *IEEE Trans. Image Process.*, vol. 10, no. 2, pp. 288–295, Feb. 2001.

Shnayderman, A., Gusev, A., and Eskicioglu, A. M., "An SVD-based grayscale image quality measure for local and global assessment," *IEEE Trans. Image Process.*, vol. 15, no. 2, pp. 422–429, Feb. 2006.

Sun, C., Ruan, S., Shie, M., and Pai, T., "Dynamic contrast enhancement based on histogram specification," *IEEE Trans. Consumer Electron.*, vol. 51, no. 4, pp. 1300–1305, Nov. 2005.

Wan, Y. and Shi, D., "Joint exact histogram specification and image enhancement through the wavelet transform," *IEEE Trans. Image Process.*, vol. 16, no. 9, pp. 2245–2250, Sept. 2007.

Wang, Q. and Ward, R. K., "Fast image/video contrast enhancement based on weighted thresholded histogram equalization," *IEEE Trans. Consumer Electron.*, vol. 53, no. 2, pp. 757–764, May 2007.

CHAPTER 4

MATHEMATICAL MORPHOLOGY

Image processing techniques have been tremendously developed during the past five decades, and among them, mathematical morphology has been continuously receiving a great deal of attention. It is because mathematical morphology provides quantitative description of geometric structure and shape, as well as mathematical description of algebra, topology, probability, and integral geometry. Mathematical morphology has been proved to be extremely useful in many image processing and analysis applications.

From Webster's dictionary, the word *morphology* refers to any scientific study of form and structure. This term has been widely used in biology, linguistics, and geography. In image processing, a well-known general approach is provided by *mathematical morphology*, where the images being analyzed are considered as sets of points and the set theory is applied on the morphological operations. This approach is based upon logical relations between pixels, rather than arithmetic relations, and can extract geometric features by choosing a suitable structuring shape as a probe.

Quoted from Haralick et al. (1987) is "As the identification and decomposition of objects, object features, object surface defects, and assembly defects correlate directly with shape, mathematical morphology clearly has an essential structural role to play in machine vision." Also, quoted from Shih and Mitchell (1989) is "Mathematical morphology has been becoming increasingly important in image processing and analysis applications for object recognition and defect inspection." Numerous morphological architectures and algorithms have been developed during last decades. The history of morphological image processing has followed a series of developments in the areas of mathematics and computer architecture.

From the mathematics aspect, the early work includes Minkowski (1903), Dineen (1955), Kirsch (1957), Preston (1961), Moore (1968), and Golay (1969). It was formalized at the Centre de Morphologie Mathematique on the campus of Paris School of Mines at Fontainebleau, France, in the mid-1970s for studying geometric and milling properties of ores. Matheron (1975) wrote a book on mathematical morphology, entitled "*Random Sets and Integral Geometry*." Two volumes containing sophisticated mathematical theory were written by Serra 1982, 1988. Other books describing fundamental applications can be referred in Giardina and Doughherty (1988), Dougherty (1993), Goutsias and Heijmans (2000) and Shih (2009). Haralick et al. (1987) presented a tutorial providing a quick understanding for a beginner. Shih

Image Processing and Pattern Recognition by Frank Shih

and Mitchell 1989, 1991 created threshold decomposition of grayscale morphology into binary morphology that provides a new insight into exploring grayscale morphological processing.

From the computer architecture aspect, morphological operations are the essence of the cellular logic machines, such as Golay logic processor (Golay, 1969), Diff3 (Graham and Norgren, 1980), PICAP (Kruse, 1977), Leitz texture analysis system (TAS) (Klein and Serra, 1977), CLIP processor arrays (Duff, 1979), cytocomputers (Lougheed and McCubbrey, 1980; Sternberg, 1980), and Delft image processor (DIP) (Gerritsen and Aardema, 1981).

Mathematical morphology can extract image shape features, such as edges, fillets, holes, corners, wedges, and cracks, by operating with various shaped structuring elements (Maragos and Schafer, 1987b). In industrial vision applications, mathematical morphology can be used to implement fast object recognition, image enhancement, segmentation, and defect inspection.

In this chapter, we will introduce binary morphology, opening and closing, hit-or-miss transform (HMT), grayscale morphology, basic morphological algorithms, and several variations of morphological filters, including alternating sequential filters (ASFs), recursive morphological filters, soft morphological filters, order-statistic soft morphological (OSSM) filters, recursive soft morphological filters, recursive order-statistic soft morphological filters, regulated morphological filters, and fuzzy morphological filters.

4.1 BINARY MORPHOLOGY

Mathematical morphology involves geometric analysis of shapes and textures in images. An image can be represented by a set of pixels. Morphological operators work with two images. The image being processed is referred to as the *active image*, and the other image, being a kernel, is referred to as the *structuring element*. Each structuring element has a designed shape, which can be thought of as a probe or a filter of the active image. The active image can be modified by probing it with various structuring elements. The elementary operations in mathematical morphology are dilation and erosion, which can be combined in sequence to produce other operations, such as opening and closing.

Let E^N denote the set of all points $p = (x_1, x_2, \ldots, x_N)$ in N-dimensional Euclidean space. A binary image in E^2 is a silhouette, a set representing foreground regions. A binary image in E^3 is a solid, a set representing the surface and interior of objects. The definitions and properties of dilation and erosion from a tutorial on mathematical morphology by Haralick et al. (1987) are adopted, which are slightly different from Matheron's (1975) and Serra's (1982) definitions.

4.1.1 Binary Dilation

Binary dilation combines two sets using vector addition of set elements. It was first introduced by Minkowski and is named *Minkowski addition*. Let A and B denote two sets in E^N with elements a and b, respectively, where $a = (a_1, a_2, \ldots, a_N)$ and

$b = (b_1, b_2, \ldots, b_N)$ being N-tuples of element coordinates. The binary dilation of A by B is the set of all possible vector sums of pairs of elements, one coming from A and the other from B. Since A and B are both binary, the morphological operators applied on the two sets are called *binary morphology*.

Definition 4.1. Let a set $A \subset E^N$ and an element $b \in E^N$. The *translation* of A by b, denoted by $(A)_b$, is defined as

$$(A)_b = \{c \in E^N | c = a + b \text{ for some } a \in A\} \tag{4.1}$$

Example 4.1 Let the coordinates (r, c) in A denote (row number, column number).

$$\begin{aligned} A &= \{(0,2),(1,1),(1,2),(2,0),(2,2),(3,1)\} \\ b &= (0,1) \\ (A)_b &= \{(0,3),(1,2),(1,3),(2,1),(2,3),(3,2)\} \end{aligned}$$

	0	1	2	3
0	0	0	1	0
1	0	1	1	0
2	1	0	1	0
3	0	1	0	0

A

	0	1	2	3
0	0	0	0	1
1	0	0	1	1
2	0	1	0	1
3	0	0	1	0

$(A)_{(0,1)}$

Definition 4.2. Let $A, B \subset E^N$. The *binary dilation* of A by B, denoted by $A \oplus_b B$, is defined as

$$A \oplus_b B = \{c \in E^N | c = a + b \text{ for some } a \in A \text{ and } b \in B\} \tag{4.2}$$

The subscript of dilation b indicates binary. Equivalently, we may write

$$A \oplus_b B = \bigcup_{b \in B} (A)_b = \bigcup_{a \in A} (B)_a \tag{4.3}$$

The representation, $A \oplus_b B = \cup_{b \in B} (A)_b$, states that the dilation of A by B can be implemented by delaying the raster scan of A by the amounts corresponding to the points in B and then ORing the delayed raster scans. That is,

$$(A \oplus_b B)_{(i,j)} = \underset{m,n}{\text{OR}}[\text{AND}(B_{(m,n)}, A_{(i-m,j-n)})] \tag{4.4}$$

Example 4.2

$$\begin{aligned} &A = \{(0,1),(1,1),(2,1),(3,1)\} \\ &B = \{(0,0),(0,2)\} \\ &A \oplus_b B = \{(0,1),(1,1),(2,1),(3,1),(0,3),(1,3),(2,3),(3,3)\} \end{aligned}$$

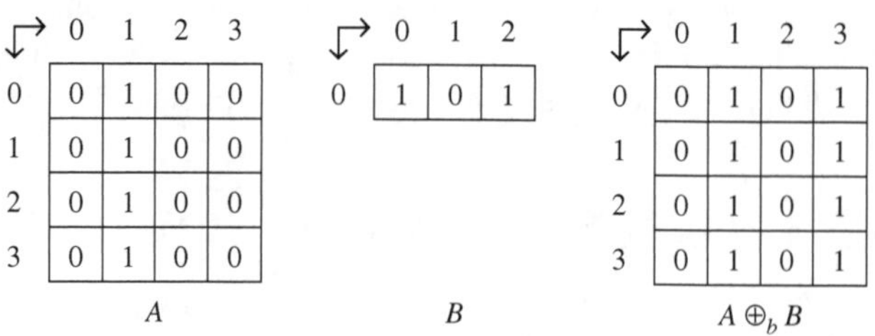

In dilation, the roles of sets A and B are symmetric. Dilation has a local interpretation. The dilation $A \oplus_b B$ is the locus of all centers c, such that the translation $(B)_c$ (by placing the origin of B at c) hits the set A. That is, if we think of each point, $a \in A$, as a seed from which grows a flower $(B)_a$, then the union of all the flowers is the dilation of A by B. Note that the dilation by a disk structuring element corresponds to the isotropic expansion algorithm popular in image processing. Dilation by a small square (3×3) is an 8-neighborhood operation easily implemented by adjacently connected array architectures and is the one known by the name "fill," "expand," or "grow." Other properties of dilation are

Properties of dilation:

1. If B contains the origin, that is, $0 \in B$, then $A \oplus_b B \supseteq A$
2. $A \oplus_b B = B \oplus_b A$ (commutative)
3. $(A \oplus_b B) \oplus_b C = A \oplus_b (B \oplus_b C)$ (associative)
4. $(A)_x \oplus_b B = (A \oplus_b B)_x$ (translation invariance)
5. $(A)_x \oplus_b (B)_{-x} = A \oplus_b B$
6. If $A \subseteq B$, then $A \oplus_b C \subseteq B \oplus_b C$ (increasing)
7. $(A \cap B) \oplus_b C \subseteq (A \oplus_b C) \cap (B \oplus_b C)$
8. $(A \cup B) \oplus_b C = (A \oplus_b C) \cup (B \oplus_b C)$(distributive)

4.1.2 Binary Erosion

Erosion is the morphological dual to dilation. It combines two sets using vector subtraction of set elements. If A and B denote two sets in E^N with elements a and b, respectively, then the *binary erosion* of A by B is the set of all elements x, for which $x + b \in A$ for every $b \in B$.

Definition 4.3. The *binary erosion* of A by B, denoted by A $A \ominus_b B$, is defined as

$$A \ominus_b B = \{x \in E^N | x + b \in A \text{ for } every\ b \in B\} \tag{4.5}$$

Equivalently, we may write

$$A \ominus_b B = \bigcap_{b \in B} (A)_{-b} \tag{4.6}$$

Note that the binary erosion defined here is slightly different from *Minkowski subtraction*, which is $\cap_{b \in B} (A)_b$. The above equation indicates that the implementation of binary erosion is similar to dilation except changing the OR function to an AND function and using the image translated by the negated points of B. That is,

$$A \ominus_b B_{(ij)} = \underset{m,n}{\text{AND}}\{\text{OR}[\text{AND}(B_{(m,n)}, A_{(i+m,j+n)}), \overline{B}_{(m,n)}]\} \tag{4.7}$$

where "$\overline{B}$" denotes an inverse function. Equivalently, we may simplify

$$A \ominus_b B_{(ij)} = \underset{m,n}{\text{AND}}[\text{OR}(A_{(i+m,j+n)}, \bar{B}_{(m,n)})] \tag{4.8}$$

Erosion $A \ominus_b B$ can be interpreted as the locus of all centers c, such that the translation $(B)_c$ is entirely contained within the set A.

$$A \ominus_b B = \{x \in E^N | (B)_x \subseteq A\} \tag{4.9}$$

Example 4.3

$$A = \{(0,1), (1,1), (2,1), (2,2), (3,0)\}$$
$$B = \{(0,0), (0,1)\}$$
$$A \ominus_b B = \{(2,1)\}$$

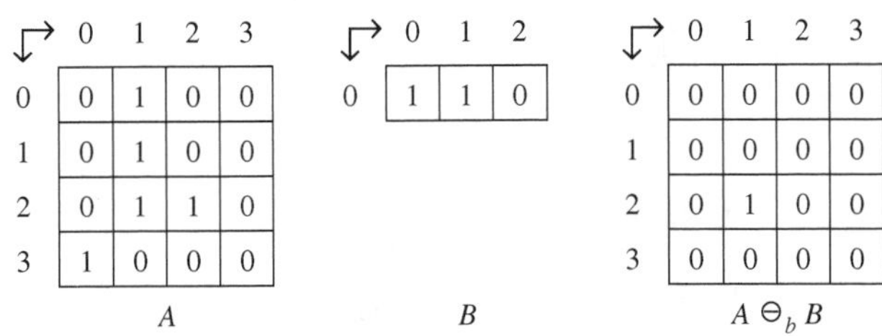

Erosion does not possess the commutative property. Some erosion equivalent terms are "shrink" and "reduce." It should be noted that the erosion by B given here is defined as in Haralick et al. (1987). In Serra (1982), this definition is the erosion by $\hat{B}$ (the dual structuring element). Other properties of erosion are

Properties of erosion:

1. If B contains the origin, that is, $0 \in B$, then $A \ominus_b B \subseteq A$
2. $(A \ominus_b B) \ominus_b C = A \ominus_b (B \oplus_b C)$
3. $A \oplus_b (B \ominus_b C) \subseteq (A \oplus_b B) \ominus_b C$
4. $A_x \ominus_b B = (A \ominus_b B)_x$ (translation invariance)
5. If $A \subseteq B$, then $A \ominus_b C \subseteq B \ominus_b C$ (increasing)
6. $(A \cap B) \ominus_b C = (A \ominus_b C) \cap (B \ominus_b C)$ (distributive)
7. $(A \cup B) \ominus_b C \supseteq (A \ominus_b C) \cup (B \ominus_b C)$
8. $A \ominus_b (B \cup C) = (A \ominus_b B) \cap (A \ominus_b C)$

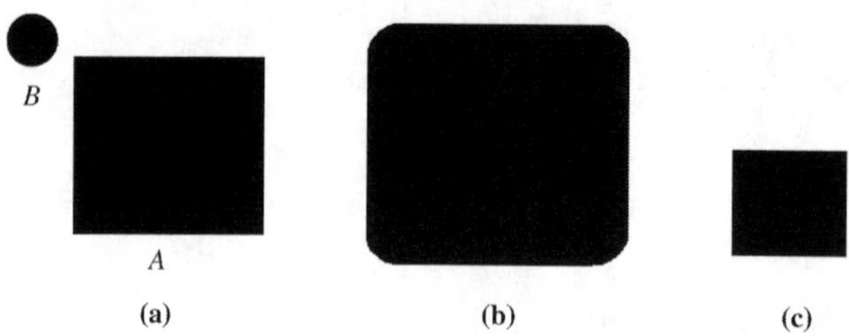

Figure 4.1 An example of binary morphological operations. (a) A and B, (b) $A \oplus_b B$, and (c) $A \ominus_b B$.

9. $A \ominus_b (B \cap C) \supseteq (A \ominus_b B) \cup (A \ominus_b C)$
10. *Erosion dilation duality*: Let $A^c = \{x \in E^N | x \notin A\}$ and $\hat{B} = \{x |$ for some $b \in B,\ x = -b\}$. Then $(A \ominus_b B)^c = A^c \oplus_b \hat{B}$.

It should be noted that although dilation and erosion are dual, this does not imply that the equality can hold for any morphological cancellation. For example, if $C = A \oplus_b B$, then eroding both sides of the expression by B results in $C \ominus_b B = (A \oplus_b B) \ominus_b B \neq A$. Instead of equality, a containment relationship does hold: $(A \oplus B) \ominus B \supseteq$ A. Note that if the dilation and erosion are clearly used in the binary image A and binary structuring element B, then the binary "b" subscript will be skipped. An example of binary dilation and erosion is shown in Figure 4.1.

4.2 OPENING AND CLOSING

In practical applications, dilation and erosion pairs are combined in sequence, either the dilation of an image followed by the erosion of the dilated result, or vice versa. In either case, the result of iteratively applying dilations and erosions is an elimination of specific image details whose sizes are smaller than the structuring element without the global geometric distortion of unsuppressed features. The properties were first explored by Matheron (1975) and Serra (1982). Both of their definitions for opening and closing are identical to the ones given here, but their formulas appear different because they use the symbol $\ominus$ to mean Minkowski subtraction rather than erosion.

Definition 4.4. The *opening* of image A by structuring element B, denoted by $A \circ B$, is defined as

$$A \circ B = (A \ominus B) \oplus B \tag{4.10}$$

Definition 4.5. The *closing* of image A by structuring element B, denoted by $A \bullet B$, is defined as

$$A \bullet B = (A \oplus B) \ominus B \tag{4.11}$$

Note that the symbols $\oplus$ and $\ominus$ can represent either binary or grayscale dilation and erosion. Equivalently, we may write

$$A \bullet B = \bigcup_{B_y \subseteq A} B_y \tag{4.12}$$

The opening and closing can be interpreted as follows: The opening will remove all of the pixels in the regions that are too small to contain the probe. The opposite sequence, closing, will fill in holes and concavities smaller than the probe. Such filters can be used to suppress object features or discriminate against objects based on their shape or size distribution. As an example, if a disk-shaped structuring element of diameter h is used, the opening of an image is equivalent to a *low-pass filter*. The opening residue is a *high-pass filter*. The difference of the openings of an image with two nonequal diameters is a *band-pass filter*. The symbolic expressions are

$$\begin{aligned} \text{Low pass} &= A \circ B^h \\ \text{High pass} &= A - (A \circ B^h) \\ \text{Band pass} &= (A \circ B^{h_1}) - (A \circ B^{h_2}) \end{aligned}$$

where diameters of B, $h_1 < h_2$

Properties of opening and closing:

1. $(A \circ B) \subseteq A \subseteq (A \bullet B)$
2. $A \oplus B = (A \oplus B) \circ B = (A \bullet B) \oplus B$
3. $A \ominus B = (A \ominus B) \bullet B = (A \circ B) \ominus B$
4. If $X \subseteq$ Y, then $X \circ B \subseteq Y \circ B$ (increasing)
5. If $X \subseteq$ Y, then $X \bullet B \subseteq Y \bullet B$ (increasing)
6. $(A \circ B) \circ B = A \circ B$ (idempotent)
7. $(A \bullet B) \bullet B = A \bullet B$ (idempotent)
8. *Opening closing duality*: $(A \bullet B)^c = A^c \circ \hat{B}$

Example 4.4 Let A and B be shown below. Draw the diagrams of $A \circ B$ and $A \bullet B$.

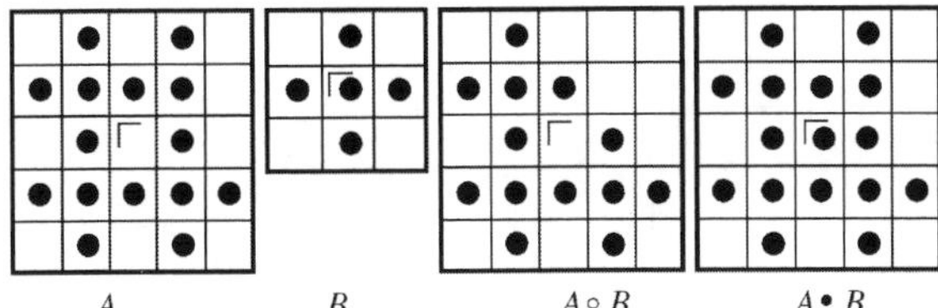

4.3 HIT-OR-MISS TRANSFORM

The hit-or-miss transformation on a binary image A is defined as follows: The structuring element B is a pair of binary images B_1 and B_2, where $B_1 \subseteq A$ and $B_2 \subseteq \bar{A}$

(complement of A). The *hit-or-miss transformation* of A by (B_1, B_2), denoted by $A \circledast (B_1, B_2)$, is defined as

$$A \circledast (B_1, B_2) = (A \ominus B_1) \cap (\bar{A} \ominus B_2) \tag{4.13}$$

Note that erosion is a special case of hit-or-miss transform, where B_2 is an empty set. Because $\bar{A} \ominus B_2 = \overline{A \oplus \hat{B}_2}$, we have

$$A \circledast (B_1, B_2) = (A \ominus B_1) \cap (\overline{A \oplus \hat{B}_2}) \tag{4.14}$$

It is equivalent to set difference

$$A \circledast (B_1, B_2) = (A \ominus B_1) - (A \oplus \hat{B}_2) \tag{4.15}$$

The morphological hit-or-miss transform is a natural tool for shape recognition. The hit-or-miss transform is a powerful morphological tool for the processing of binary images. There have been several attempts to generalize it to grayscale images, based on the grayscale erosion (Bloomberg and Maragos, 1990; Naegel et al., 2007a, 2007b). Khosravi and Schafer (1996) proposed a class of rank-order-based template matching criteria using grayscale hit-or-miss transform that are multiplier-free and independent of the dc variations of the image. Raducanu and Grana (2000) proposed the hit-or-miss transform based on level sets, called the level set hit-or-miss transform (LSHMT), to obtain a translation invariant recognition tool, with some robustness regarding small deformations and variations of illumination. It is applied to face localization on grayscale images based on a set of face patterns.

Example 4.5 Let A be

0	0	0	0	0	0	0	0
0	0	1	0	1	0	0	0
0	1	0	1	1	1	1	0
0	0	1	0	1	0	0	0
0	0	0	0	1	1	1	0
0	1	1	0	1	0	1	0
0	1	1	0	0	0	1	0
0	0	0	0	0	0	0	0

Let B_1 and B_2 be

B_1:

0	1	0
0	1	1
0	1	0

B_2:

1	0	1
1	0	0
1	0	1

The $A \circledast (B_1, B_2)$ is

0	0	0	0	0	0	0	0
0	0	0	0	0	0	0	0
0	0	0	0	0	0	0	0
0	0	0	0	0	0	0	0
0	0	0	0	1	0	0	0
0	0	0	0	0	0	0	0
0	0	0	0	0	0	0	0
0	0	0	0	0	0	0	0

4.4 GRAYSCALE MORPHOLOGY

Mathematical morphology represents image objects as sets in a Euclidean space. In morphological analysis, the set is the primary notion and a function is viewed as a particular case of a set (e.g., an N-dimensional, multivalued function can be viewed as a set in $(N+1)$-dimensional space). Then in this viewpoint, any function- or set-processing system is viewed as a set mapping (transformation) from one class of sets into another class of sets. The extensions of the morphological transformations from binary to grayscale processing by Serra (1982) and Sternberg (1986) introduce a natural morphological generalization of the dilation and erosion operations.

4.4.1 Grayscale Dilation and Erosion

We begin with the definitions of the top surface of a set and the umbra of a surface. Given a set A in E^N, the top surface of A is a function defined on the projection of A onto its first $(N-1)$ coordinates, and the highest value of the N-tuple is the function value (or gray value). We assume that for every $x \in F$, $\{y|(x,y) \in A\}$ is topologically closed, and also the sets E and K are finite.

Definition 4.6. Let $A \subseteq E^N$ and $F = \{x \in E^{N-1}|\text{ for some } y \in E, (x,y) \in A\}$. The *top* or *top surface* of A, denoted by $T[A]$: $F \to E$, is defined as

$$T[A](x) = \max\{y|(x,y) \in A\} \tag{4.16}$$

Definition 4.7. Let $F \subseteq E^{N-1}$ and f: $F \to E$. The *umbra* of f, denoted by $U[f]$, $U[f] \subseteq F \times E$, is defined as

$$U[f] = \{(x,y) \in F \times E|y \leq f(x)\} \tag{4.17}$$

Example 4.6 Let f be $\{2, 1, 0, 2\}$. Show the umbra $U[f]$.

	0	1	2	3
0	2	1	0	2

f

	0	1	2	3
3	0	0	0	0
2	1	0	0	1
1	1	1	0	1
0	1	1	1	1
.	1	1	1	1
.	1	1	1	1
.	1	1	1	1
$-\infty$	1	1	1	1

$U[f]$

The rows below the zeroth row will not be shown in the following examples whenever they are all 1's. A function is the top of its umbra. The umbra can also be illustrated by a graphic in R^2, as done by Serra (1982 ch. XII) (e.g., (1982, Fig. XII.1)).

Definition 4.8. Let $F, K \subseteq E^{N-1}$ and $f: F \rightarrow E$ and $k: K \rightarrow E$. The *grayscale dilation* of f by k, denoted by $f \oplus_g k$, is defined as

$$f \oplus_g k = T[U[f] \oplus_b U[k]] \tag{4.18}$$

An alternate definition for $f \oplus_g k$ is

$$U[f] \oplus_b U[k] = U[f \oplus_g k] \tag{4.19}$$

Example 4.7 Let f and k be {2, 1, 0, 2} and {2, 1, 0}, respectively. Show $U[f] \oplus_b U[k]$.

	0	1	2
0	2	1	0

k

	0	1	2
4	0	0	0
3	0	0	0
2	1	0	0
1	1	1	0
0	1	1	1

$U[k]$

	0	1	2	3
4	1	0	0	1
3	1	1	0	1
2	1	1	1	1
1	1	1	1	1
0	1	1	1	1

$U[f] \oplus_b U[k]$

Definition 4.9. Let $F, K \subseteq E^{N-1}$ and $f: F \rightarrow E$ and $k: K \rightarrow E$. The *grayscale erosion* of f by k, denoted by $f \ominus_g k$, is defined as

$$f \ominus_g k = T[U[f] \ominus_b U[k]] \tag{4.20}$$

An alternate definition for $f \ominus_g k$ is

$$U[f] \ominus_b U[k] = U[f \ominus_g k] \tag{4.21}$$

Example 4.8 Let f and k be {2, 1, 0, 2} and {2, 1, 0}, respectively. Show $U[f] \ominus_b U[k]$.

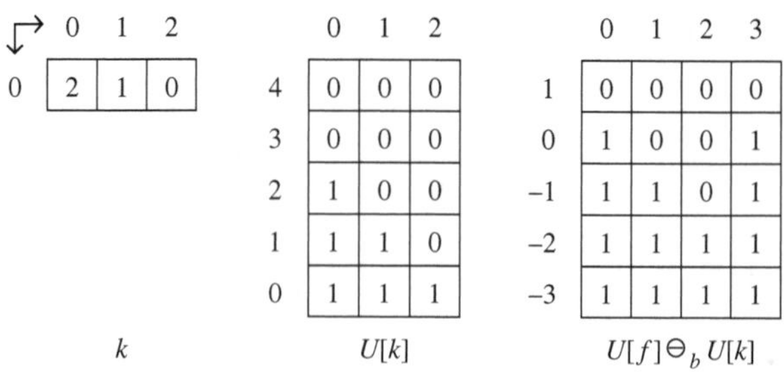

	0	1	2
0	2	1	0

k

	0	1	2
4	0	0	0
3	0	0	0
2	1	0	0
1	1	1	0
0	1	1	1

$U[k]$

	0	1	2	3
1	0	0	0	0
0	1	0	0	1
–1	1	1	0	1
–2	1	1	1	1
–3	1	1	1	1

$U[f] \ominus_b U[k]$

The following propositions give an alternative method of computing grayscale dilation and erosion (for a proof, see Haralick et al. (1987)). Let F and K be the domains of the grayscale image $f(x, y)$ and the grayscale structuring element $k(m, n)$, respectively. When f or k is grayscale, the morphological operators applied on the two functions are called *grayscale morphology*.

Proposition 4.1. Let $f(x, y)$: $F \rightarrow E$ and $k(m, n)$: $K \rightarrow E$. Then $(f \oplus_g k)(x, y)$: $F \oplus_g K \rightarrow E$ can be computed by

$$(f \oplus_g k)(x,y) = \max\{f(x-m, y-n) + k(m,n)\} \tag{4.22}$$

for all $(m,n) \in K$ and $(x-m, y-n) \in F$.

Example 4.9 Compute $f \oplus_g k$.

f (origin at 0; columns 0–3):

	0	1	2	3
0	2	1	0	2

k (origin at 0; columns 0–2):

	0	1	2
0	2	1	0

$f \oplus_g k$ (origin at 0; columns 0–3):

	0	1	2	3
0	4	3	2	4

Example 4.10 Perform the morphological dilation of f and k as shown below. Use the upper-left pixel of k as the origin. Note: do not consider the computation to the pixels which are located outside the boundary of f.

f:

2	3	0	4
1	3	5	7

k:

−1	1
2	3

Answer: The grayscale dilation is equivalent to reflecting k about the origin, pointwise adding it to f, and taking the maximum. The reflected k is, therefore,

$\hat{k}$:

3	2
1	-1

where the origin is now located at the lower right pixel, and the values produced can be represented by

$2+(-1)$	$2+1,\ 3+(-1)$	$3+1,\ 0+(-1)$	$0+1,\ 4+(-1)$
$2+2,\ 1+(-1)$	$2+3,\ 3+2,$ $1+1,\ 3+(-1)$	$3+3,\ 0+2,$ $3+1,\ 5+(-1)$	$0+3,\ 4+2,$ $5+1,\ 7+(-1)$

where the maximum is selected among those values. The result of grayscale dilation is

1	3	4	3
4	5	6	6

Proposition 4.2. Let $f(x, y)$: $F \rightarrow E$ and $k(m, n)$: $K \rightarrow E$. Then $(f \ominus_g k)(x, y)$: $F \ominus_g K \rightarrow E$ can be computed by

$$(f \ominus_g k)(x,y) = \min\{f(x+m, y+n)\} - k(m,n)\} \tag{4.23}$$

for all $(m,n) \in K$ and $(x+m, y+n) \in F$.

Example 4.11 Compute $f \ominus_g k$.

f (origin at 0; columns 0–3):

	0	1	2	3
0	2	1	0	2

k (origin at 0; columns 0–2):

	0	1	2
0	2	1	0

$f \ominus_g k$ (origin at 0; columns 0–3):

	0	1	2	3
0	0	-1	-2	0

Example 4.12 Perform the grayscale erosion of f and k that are shown below. Use the upper left pixel of k as the origin (as circled) and do not consider the computation on the pixels that are outside the boundary of f.

f:

2	3	0	4
1	3	5	7

k:

(−1)	1
2	3

Answer: The grayscale erosion is equivalent to pointwise subtracting k from f and taking the minimum. In this case, the values can be represented by

$2-(-1), 3-1, 1-2, 3-3$	$3-(-1), 0-1, 3-2, 5-3$	$0-(-1), 4-1, 5-2, 7-3$	$4-(-1), 7-2$
$1-(-1), 3-1$	$3-(-1), 5-1$	$5-(-1), 7-1$	$7-(-1)$

Taking the minimum gives the result of grayscale erosion as

−1	−1	1	5
2	4	6	8

Properties of umbra and top surface:

1. $T[U[f]] = f$
2. $U[T[U[f]]] = U[f]$
3. $A \subseteq U[T[A]]$
4. $U[f \oplus_g k] = U[f] \oplus_b U[k]$
5. $U[f \ominus_g k] = U[f] \ominus_b U[k]$

Properties of dilation:

1. $f \oplus_g k = k \oplus_g f$
2. $(f \oplus_g k_1) \oplus_g k_2 = f \oplus_g (k_1 \oplus_g k_2)$
3. If $\vec{P}_c$ is a constant vector, then $(f + \vec{P}_c) \oplus_g k = (f \oplus_g k) + \vec{P}_c$
4. $(f + \vec{P}_c) \oplus_g (k - \vec{P}_c) = f \oplus_g k$
5. If $f_1 \leq f_2$, then $f_1 \oplus_g k \leq f_2 \oplus_g k$

Properties of erosion:

1. $(f \ominus_g k_1) \ominus_g k_2 = f \ominus_g (k_1 \oplus_g k_2)$
2. If $\vec{P}_c$ is a constant vector, then $(f + \vec{p}_c) \ominus_g k = (f \ominus_g k) + \vec{p}_c$
3. $(f + \vec{P}_c) \ominus_g (k - \vec{P}_c) = f \ominus_g k$
4. If $f_1 \leq f_2$, then $f_1 \ominus_g k \leq f_2 \ominus_g k$

Note that the grayscale dilation and erosion are similar to the convolution operator, except that addition/subtraction is substituted for multiplication and maximum/

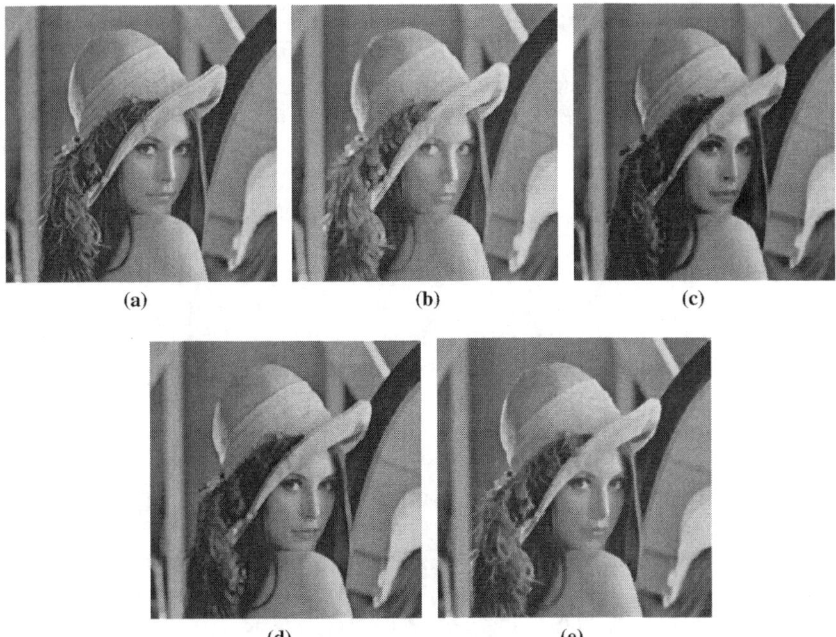

Figure 4.2 An example of grayscale morphological operations by a 3×3 square with all values 0. (a) The Lena image, (b) the grayscale dilation, (c) the grayscale erosion, (d) the grayscale opening, and (e) the grayscale closing.

minimum is substituted for summation. Unlike convolution, morphological operations are, however, highly nonlinear. Let a structuring element be a 3×3 square with all values 0. An example of grayscale dilation, erosion, opening, and closing is shown in Figure 4.2.

4.4.2 Grayscale Dilation Erosion Duality Theorem

Let f: $F \to E$ and k: $K \to E$. Let $x \in (F \oplus K) \cap (F \ominus \hat{K})$ be given. Then

$$-(f \oplus_g k) = (-f) \ominus_g \hat{k}, \text{ where } \hat{k}(x) = k(-x) \text{ denotes the reflection of } k \quad (4.24)$$

Hence, the grayscale erosion can be obtained by computing a grayscale dilation as follows:

$$f \ominus_g k = -((-f) \oplus_g \hat{k}) \quad (4.25)$$

Proof: According to equation (4.24), $(-f') \ominus_g \widehat{k'} = -(f' \oplus_g k')$

Let $f' = -f$. We get $f \ominus_g \widehat{k'} = -((-f) \oplus_g k')$.

Let $k' = \hat{k}$. Because $\hat{\hat{k}}(x) = \hat{k}(-x) = k(x)$, we get $f \ominus_g k = -((-f) \oplus_g \hat{k})$.

4.5 BASIC MORPHOLOGICAL ALGORITHMS

In image processing and analysis, it is important to extract object features, describe shapes, and recognize patterns. Such tasks often refer to geometric concept, such as size, shape, and orientation. Mathematical morphology takes the concept from set theory, geometry, and topology to analyze geometric structures in an image. Most essential image processing algorithms can be represented in the form of morphological operations.

The operations desired for image processing, described in morphological notations, can be easily implemented in a parallel processor designed for cellular transformations. Each pixel can be thought of as a cell in a given state. If we define a neighborhood relationship and a pixel transition function, the application of the transition function in parallel across the space will cause the configuration of pixel states forming the image to be modified or transformed into new configurations. The implementation of these transforms by neighborhood processing stages allows the parallel processing of images in real time with conceptually high-level operations. This allows the development of extremely powerful image processing algorithms in a short time.

In this section, we introduce several basic morphological algorithms, including boundary extraction, region filling, connected component extraction, convex hull, thinning, thickening, skeletonization, pruning, and morphological edge operator.

4.5.1 Boundary Extraction

When an input image is grayscale, one can perform the segmentation process by *thresholding* the image. This involves the selection of a gray-level threshold between 0 and 255 via a histogram. Any pixels greater than this threshold are assigned 255 and otherwise 0. This produces a simple binary image in preparation for boundary extraction, in which 255 is replaced by 1 as the foreground pixels.

In a binary image, an object set is a connected component having the pixels of value 1. The boundary pixels are those object pixels whose 8-neighbors have at least one with value 0. Boundary extraction of a set A is first eroding A by a structuring element B and then taking the set difference between A and its erosion. The structuring element must be isotropic; ideally, it is a circle, but in digital image processing, a 3×3 matrix of 1's is often used. That is, the boundary of a set A is obtained by

$$\partial A = A - (A \ominus B) \tag{4.26}$$

For example, let A be

0	0	0	0	0	0	0	0	0
0	1	1	1	0	0	0	0	0
0	1	1	1	0	0	0	0	0
0	1	1	1	0	1	1	1	0
0	1	1	1	0	1	1	1	0
0	1	1	1	1	1	1	1	0
0	1	1	1	1	1	1	1	0
0	1	1	1	1	1	1	1	0
0	0	0	0	0	0	0	0	0

The boundary of A is extracted as

0	0	0	0	0	0	0	0	0
0	**1**	**1**	**1**	0	0	0	0	0
0	**1**	0	**1**	0	0	0	0	0
0	**1**	0	**1**	0	**1**	**1**	**1**	0
0	**1**	0	**1**	0	**1**	0	**1**	0
0	**1**	0	**1**	**1**	**1**	0	**1**	0
0	**1**	0	0	0	0	0	**1**	0
0	**1**	**1**	**1**	**1**	**1**	**1**	**1**	0
0	0	0	0	0	0	0	0	0

Note that the size of structuring element determines the thickness of object contour. For instances, a 3×3 structuring element will generate the thickness of 1 and a 5×5 structuring element will generate the thickness of 3.

4.5.2 Region Filling

Region filling intends to fill value 1 into the entire object region. It is given a binary image set A that contains all boundary pixels labeled 1 and nonboundary pixels labeled 0. Region filling starts by assigning 1 to a pixel p inside the object boundary and grows the seed by performing iterative dilations under a limited condition restricted by A^c. If the restriction is not placed, the growing process would flood the entire image area. Let the initial step $X_0 = p$ and B be a cross structuring element as shown in Figure 4.3. The iterative process is conducted as in the kth step (Gonzalez and Woods, 2007):

$$X_k = (X_{k-1} \oplus B) \cap A^c, \quad k = 1, 2, 3, \ldots \tag{4.27}$$

The iteration terminates at step k if $X_k = X_{k-1}$. Therefore, X_{k-1} is the interior region of the set A. The union of X_{k-1} and A will yield the filled region and the boundary. An example of region filling is shown in Figure 4.3, where the region is entirely painted after seven iterations.

4.5.3 Extraction of Connected Components

Similar to region filling, the extraction of connected components applies the iterative process. Assume that a binary image set A contains several connected components in which all object pixels are labeled 1 and background pixels are labeled 0. One can extract a desired connected component by picking up a pixel in the component and growing the seed by performing iterative dilations under a limited condition restricted by A. Let the initial step $X_0 = p$ and B be a 3×3 structuring element of 1's. The iterative process is conducted as in the kth step

$$X_k = (X_{k-1} \oplus B) \cap A, \quad k = 1, 2, 3, \ldots \tag{4.28}$$

The iteration terminates at step k if $X_k = X_{k-1}$. Therefore, X_{k-1} is the extracted connected component. An example of connected component extraction is shown in Figure 4.4.

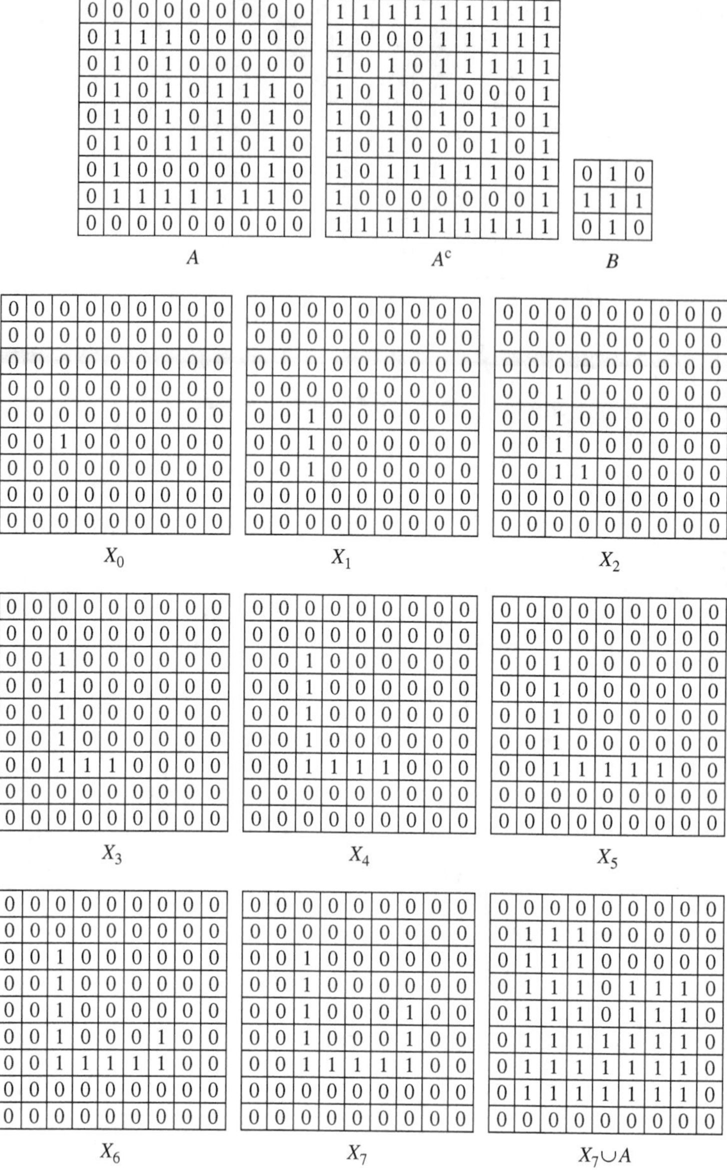

Figure 4.3 An example of region filling using mathematical morphology.

4.5.4 Convex Hull

A *convex* set is the one for which the straight line joining any two points in the set consists of points that are also in the set. The *convex hull* of a set A is defined as the smallest convex set that contains A. One can visualize the convex hull of an object as using an elastic band surrounding the object, so that the elastic band crosses over

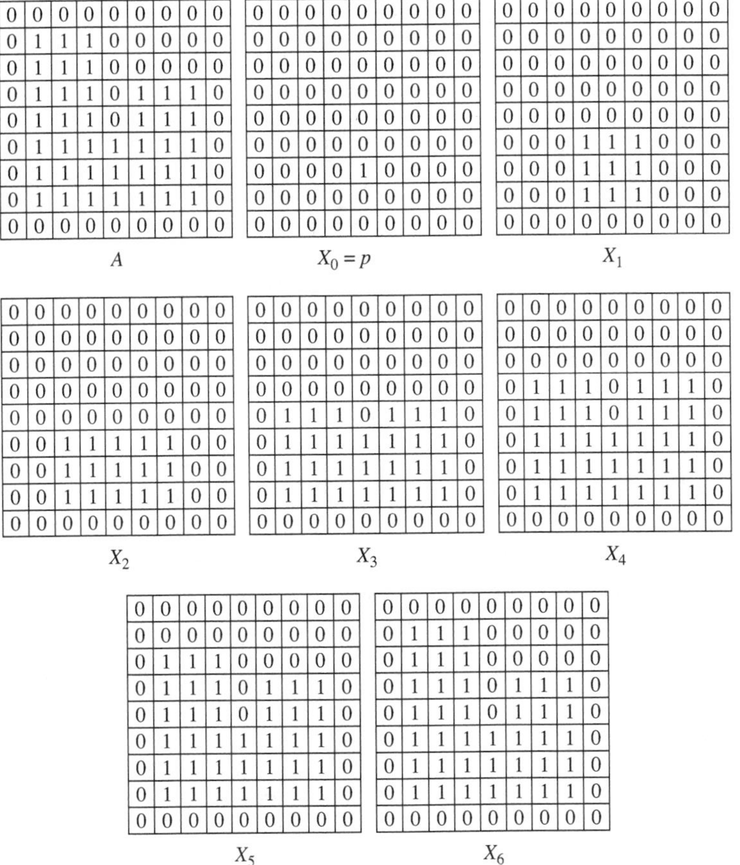

Figure 4.4 An example of connected component extraction by mathematical morphology.

concavity parts and follows the convex contours of the object. It results in a convex set informing the object's shape without concavities. In digital images, we use an approximated 45° convex hull set, which considers its contours in orientations of multiples of 45°.

The convex hull can be easily obtained by hit-or-miss operations with four 45°-rotated structuring elements. Let A be a binary image set and let B^i, $i=1, 2, 3, 4$, represent four structuring elements as

$$B^1 = \begin{bmatrix} 1 & \times & \times \\ 1 & 0 & \times \\ 1 & \times & \times \end{bmatrix}, \quad B^2 = \begin{bmatrix} 1 & 1 & 1 \\ \times & 0 & \times \\ \times & \times & \times \end{bmatrix},$$

$$B^3 = \begin{bmatrix} \times & \times & 1 \\ \times & 0 & 1 \\ \times & \times & 1 \end{bmatrix}, \quad B^4 = \begin{bmatrix} \times & \times & \times \\ \times & 0 & \times \\ 1 & 1 & 1 \end{bmatrix} \tag{4.29}$$

0	0	0	0	0	0	0	0	0
0	0	0	0	0	0	0	0	0
0	0	1	1	0	0	0	0	0
0	1	1	1	0	1	0	0	0
0	1	1	1	0	1	1	0	0
0	1	1	1	1	1	1	1	0
0	0	1	1	1	1	1	0	0
0	0	0	0	1	1	0	0	0
0	0	0	0	1	0	0	0	0

$X_0^i = A$

0	0	0	0	0	0	0	0	0
0	0	0	0	0	0	0	0	0
0	0	1	1	0	0	0	0	0
0	1	1	1	1	1	0	0	0
0	1	1	1	1	1	1	0	0
0	1	1	1	1	1	1	1	0
0	0	1	1	1	1	1	0	0
0	0	0	0	1	1	0	0	0
0	0	0	0	1	0	0	0	0

X_1^1

0	0	0	0	0	0	0	0	0
0	0	0	0	0	0	0	0	0
0	0	1	1	0	0	0	0	0
0	1	1	1	0	1	0	0	0
0	1	1	1	0	1	1	0	0
0	1	1	1	1	1	1	1	0
0	0	1	1	1	1	1	0	0
0	0	0	1	1	1	0	0	0
0	0	0	0	1	0	0	0	0

X_1^2

0	0	0	0	0	0	0	0	0
0	0	0	0	0	0	0	0	0
0	0	1	1	0	0	0	0	0
0	1	1	1	0	1	0	0	0
1	1	1	1	1	1	1	0	0
0	1	1	1	1	1	1	1	0
0	0	1	1	1	1	1	0	0
0	0	0	1	1	1	0	0	0
0	0	0	0	1	0	0	0	0

X_1^3

0	0	0	0	0	0	0	0	0
0	0	0	1	0	0	0	0	0
0	0	1	1	1	0	0	0	0
0	1	1	1	1	1	0	0	0
0	1	1	1	1	1	1	0	0
0	1	1	1	1	1	1	1	0
0	0	1	1	1	1	1	0	0
0	0	0	0	1	1	0	0	0
0	0	0	0	1	0	0	0	0

X_4^4

0	0	0	0	0	0	0	0	0
0	0	0	1	0	0	0	0	0
0	0	1	1	1	0	0	0	0
0	1	1	1	1	1	0	0	0
1	1	1	1	1	1	1	0	0
0	1	1	1	1	1	1	1	0
0	0	1	1	1	1	1	0	0
0	0	0	1	1	1	0	0	0
0	0	0	0	1	0	0	0	0

$C(A)$

Figure 4.5 An example of convex hull by mathematical morphology.

where the symbol "×" denotes don't care. Let the initial step be $X_0^i = A$. The iterative process is conducted as in the kth step

$$X_k^i = (X \circledast B^i) \cup A, \quad i = 1, 2, 3, 4 \text{ and } k = 1, 2, 3, \ldots \tag{4.30}$$

If $X_k^i = X_{k-1}^i$, then it converges. Let $D^i = X_{\text{conv}}^i$. Finally, the convex hull of A is

$$C(A) = \bigcup_{i=1}^{4} D^i \tag{4.31}$$

An example of convex hull by mathematical morphology is shown in Figure 4.5. Note that in this example, we obtain $D^1 = X_1^1$, $D^2 = X_1^2$, $D^3 = X_1^3$, and $D^4 = X_4^4$.

4.5.5 Thinning

Thinning is similar to erosion, but it does not cause disappearance of object components. It intends to reduce objects to the thickness of 1 pixel, generating a minimally connected axis that is equidistant from the object edges. Thinning is a fundamental early processing step in representing the structural shape of a pattern as a graph. It can be applied in industrial part inspection, fingerprint recognition, optical character recognition, and biomedical diagnosis.

Digital skeleton, generated by thinning algorithms, is often used to represent objects in a binary image for shape analysis and classification. Jang and Chin (1990) presented a precise definition of digital skeletons and a mathematical framework for the analysis of a class of thinning algorithms based on morphological set transformation. The thinning of a binary image set can be represented in terms of hit-or-miss operation as in Serra (1982) and Meyer (1988):

$$A \otimes B = A - (A \circledast B) = A \cap (A \circledast B)^c \tag{4.32}$$

This can be thought of as a search-and-delete process. The operation $A \circledast B$ locates all occurrences of B in A, and the set subtraction removes those pixels that have been located from A. For thinning A symmetrically, we use a sequence of eight structuring elements:

$$\{B\} = \{B^1, B^2, \ldots, B^8\} \tag{4.33}$$

where B^i is a rotated version of B^{i-1}. In digital images, this sequence contains eight structuring elements as

$$B^1 = \begin{bmatrix} 0 & 0 & 0 \\ \times & 1 & \times \\ 1 & 1 & 1 \end{bmatrix}, \quad B^2 = \begin{bmatrix} \times & 0 & 0 \\ 1 & 1 & 0 \\ 1 & 1 & \times \end{bmatrix}, \quad B^3 = \begin{bmatrix} 1 & \times & 0 \\ 1 & 1 & 0 \\ 1 & \times & 0 \end{bmatrix}, \quad B^4 = \begin{bmatrix} 1 & 1 & \times \\ 1 & 1 & 0 \\ \times & 0 & 0 \end{bmatrix}$$
$$B^5 = \begin{bmatrix} 1 & 1 & 1 \\ \times & 1 & \times \\ 0 & 0 & 0 \end{bmatrix}, \quad B^6 = \begin{bmatrix} \times & 1 & 1 \\ 0 & 1 & 1 \\ 0 & 0 & \times \end{bmatrix}, \quad B^7 = \begin{bmatrix} 0 & \times & 1 \\ 0 & 1 & 1 \\ 0 & \times & 1 \end{bmatrix}, \quad B^8 = \begin{bmatrix} 0 & 0 & \times \\ 0 & 1 & 1 \\ \times & 1 & 1 \end{bmatrix} \tag{4.34}$$

Therefore, thinning a sequence of eight structuring elements is represented as

$$A \otimes \{B\} = (\cdots((A \otimes B^1) \otimes B^2) \cdots) \otimes B^8 \tag{4.35}$$

The thinning process removes pixels from the outside edges of an object. The structuring elements are designed to find those edge pixels whose removal will not change the object's connectivity. After thinning for the first pass with these eight structuring elements is completed, the entire process for the second pass is repeated until no further changes occur. An example of thinning is shown in Figure 4.6, where the convergence is achieved in two passes. Note that the skeleton after this thinning process often contains undesirable short spurs as small irregularities in the object boundary. These *spurs* can be removed by a process called *pruning*, which will be introduced later.

4.5.6 Thickening

Thickening is similar to dilation, but it does not cause merging of disconnected objects. Thickening is the morphological dual to thinning. It is used to grow some concavities in an object and can be represented in terms of hit-or-miss operation and union as

$$A \odot B = A \cup (A \circledast B) \tag{4.36}$$

The thickened image contains the original set plus some filled-in pixels determined by the hit-or-miss operation. The sequence of structuring elements used in thickening is the same as that in thinning. The sequence of thickening operations can be represented as

$$A \odot \{B\} = (\cdots((A \odot B^1) \odot B^2) \cdots) \odot B^n \tag{4.37}$$

0	1	1	1	1	1	1	1	0
0	1	1	1	1	1	1	1	0
0	1	1	1	1	1	1	1	0
0	0	0	1	1	1	0	0	0
0	0	0	1	1	1	0	0	0
0	0	0	1	1	1	0	0	0
0	0	0	1	1	1	0	0	0
0	0	0	1	1	1	0	0	0
0	0	0	1	1	1	0	0	0

A

0	1	0	0	0	0	0	1	0
0	1	1	1	1	1	1	1	0
0	1	1	1	1	1	1	1	0
0	0	0	1	1	1	0	0	0
0	0	0	1	1	1	0	0	0
0	0	0	1	1	1	0	0	0
0	0	0	1	1	1	0	0	0
0	0	0	1	1	1	0	0	0
0	0	0	1	1	1	0	0	0

$A \otimes B^1$

0	1	0	0	0	0	0	1	0
0	1	1	1	1	1	1	1	0
0	1	1	1	1	1	1	1	0
0	0	0	1	1	1	0	0	0
0	0	0	1	1	1	0	0	0
0	0	0	1	1	1	0	0	0
0	0	0	1	1	1	0	0	0
0	0	0	1	1	1	0	0	0
0	0	0	1	1	1	0	0	0

$(A \otimes B^1) \otimes B^2$

0	1	0	0	0	0	0	1	0
0	1	1	1	1	1	1	1	0
0	1	1	1	1	1	1	1	0
0	0	0	1	1	1	0	0	0
0	0	0	1	1	0	0	0	0
0	0	0	1	1	0	0	0	0
0	0	0	1	1	0	0	0	0
0	0	0	1	1	0	0	0	0
0	0	0	1	1	1	0	0	0

$((A \otimes B^1) \otimes B^2) \otimes B^3$

0	1	0	0	0	0	0	1	0
0	1	1	1	1	1	1	1	0
0	1	1	1	1	1	1	0	0
0	0	0	1	1	0	0	0	0
0	0	0	1	1	0	0	0	0
0	0	0	1	1	0	0	0	0
0	0	0	1	1	0	0	0	0
0	0	0	1	1	0	0	0	0
0	0	0	1	1	1	0	0	0

$(((A \otimes B^1) \otimes B^2) \otimes B^3) \otimes B^4$

0	1	0	0	0	0	0	1	0
0	1	1	1	1	1	1	1	0
0	1	1	1	1	1	1	0	0
0	0	0	1	1	0	0	0	0
0	0	0	1	1	0	0	0	0
0	0	0	1	1	0	0	0	0
0	0	0	1	1	0	0	0	0
0	0	0	1	1	0	0	0	0
0	0	0	1	1	1	0	0	0

$(((A \otimes B^1) \otimes B^2) \otimes ...) \otimes B^5$

0	1	0	0	0	0	0	1	0
0	1	1	1	1	1	1	1	0
0	0	1	1	1	1	1	0	0
0	0	0	1	1	0	0	0	0
0	0	0	1	1	0	0	0	0
0	0	0	1	1	0	0	0	0
0	0	0	1	1	0	0	0	0
0	0	0	1	1	0	0	0	0
0	0	0	0	1	1	0	0	0

$(((A \otimes B^1) \otimes B^2) \otimes ...) \otimes B^6$

0	1	0	0	0	0	0	1	0
0	1	1	1	1	1	1	1	0
0	0	1	1	1	1	1	0	0
0	0	0	1	1	0	0	0	0
0	0	0	0	1	0	0	0	0
0	0	0	0	1	0	0	0	0
0	0	0	0	1	0	0	0	0
0	0	0	0	1	0	0	0	0
0	0	0	0	1	1	0	0	0

$(((A \otimes B^1) \otimes B^2) \otimes ...) \otimes B^7$

0	1	0	0	0	0	0	1	0
0	1	1	1	1	1	1	1	0
0	0	1	1	1	1	1	0	0
0	0	0	1	1	0	0	0	0
0	0	0	0	1	0	0	0	0
0	0	0	0	1	0	0	0	0
0	0	0	0	1	0	0	0	0
0	0	0	0	1	0	0	0	0
0	0	0	0	1	1	0	0	0

First pass of thinning

0	1	0	0	0	0	0	1	0
0	1	1	0	0	0	1	1	0
0	0	1	1	1	1	1	0	0
0	0	0	0	1	0	0	0	0
0	0	0	0	1	0	0	0	0
0	0	0	0	1	0	0	0	0
0	0	0	0	1	0	0	0	0
0	0	0	0	1	0	0	0	0
0	0	0	0	1	1	0	0	0

Second pass of thinning

Figure 4.6 An example of morphological thinning.

4.5.7 Skeletonization

Skeletonization is similar to thinning, but it explores more structure information of an object. The skeleton emphasizes certain properties of images; for instance, curvatures of the contour correspond to topological properties of the skeleton. The concept of skeleton was first proposed by Blum (1967). *Skeleton*, *medial axis*, or *symmetrical axis* has been extensively used for characterizing objects satisfactorily using the structures that are composed of line or arc patterns. This has the advantages of reducing the memory space required for storage of the essential structural information and simplifying the data structure required in pattern analysis. Applications include the representation and recognition of handwritten characters, fingerprint ridge patterns, biological cell structures, circuit diagrams, engineering drawings, robot path planning, and the like. Note that the original shape can be reconstructed using the skeleton points and their distances to shape boundary. On the contrary, one cannot reconstruct the original shape from a thinned image obtained by a thinning algorithm.

Let us visualize a connected object region as a field of grass and place a fire starting from its contour. Assume that this fire burning spreads uniformly in all directions. The skeleton is where waves collide with each other in a frontal or circular manner. It receives the name of medial axis because the pixels are located at midpoints or along local symmetrical axes of the region. The skeleton of an object pattern is a line representation of the object. It must preserve the topology of the object, is 1-pixel thick, and locate in the middle of the object. However, in digital images some objects are not always realizable for skeletonization, as shown in Figure 4.7. In case (a), it is impossible to generate 1-pixel thick skeleton from a 2-pixel thick line. In case (b), it is impossible to produce the skeleton from an alternative pixel pattern in order to preserve the topology.

In this section, we introduce the skeletonization of an object by morphological erosion and opening (Serra, 1982). A study on the use of morphological set operations to represent and encode a discrete binary image by parts of its skeleton can be found in Maragos and Schafer (1986). The morphological skeleton was shown to unify many skeletonization algorithms, and fast algorithms for skeleton decomposition and reconstruction were developed. Fast homotopy-preserving skeletons using mathematical morphology was proposed to preserve homotopy (Ji and Piper, 1992). Another approach to obtaining the skeleton by Euclidean distance transformation and maximum value tracking can be referred in Shih and Pu (1995a) and Shih and Wu (2004).

Let A denote a binary image set containing object pixels of 1's and background pixels of 0's. Let B be a 3×3 structuring element of 1's. The skeleton of A is obtained by

$$S(A) = \bigcup_{k=0}^{K} S_k(A) \tag{4.38}$$

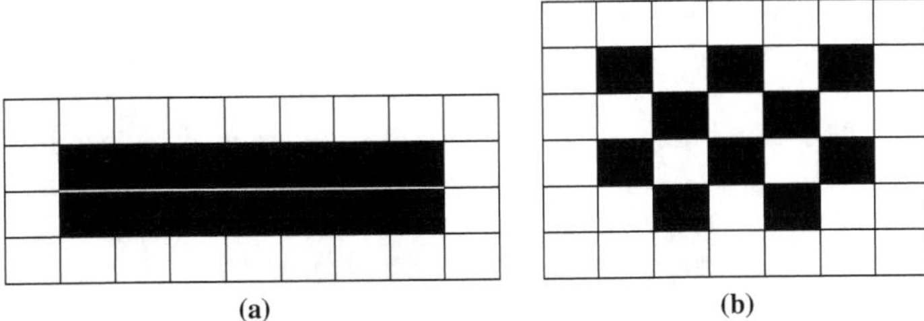

Figure 4.7 Two cases of nonrealizable skeletonization.

0	1	1	1	1	1	1	1	0
0	1	1	1	1	1	1	1	0
0	1	1	1	1	1	1	1	0
0	0	0	1	1	1	0	0	0
0	0	0	1	1	1	0	0	0
0	0	0	1	1	1	0	0	0
0	0	0	1	1	1	0	0	0
0	0	0	1	1	1	0	0	0
0	0	0	1	1	1	0	0	0

A

0	0	0	0	0	0	0	0	0
0	0	1	1	1	1	1	0	0
0	0	0	0	1	0	0	0	0
0	0	0	0	1	0	0	0	0
0	0	0	0	1	0	0	0	0
0	0	0	0	1	0	0	0	0
0	0	0	0	1	0	0	0	0
0	0	0	0	1	0	0	0	0
0	0	0	0	0	0	0	0	0

$A \ominus B$

0	0	0	0	0	0	0	0	0
0	0	0	0	0	0	0	0	0
0	0	0	0	0	0	0	0	0
0	0	0	0	0	0	0	0	0
0	0	0	0	0	0	0	0	0
0	0	0	0	0	0	0	0	0
0	0	0	0	0	0	0	0	0
0	0	0	0	0	0	0	0	0
0	0	0	0	0	0	0	0	0

$(A \ominus B) \circ B$

0	0	0	0	0	0	0	0	0
0	0	1	1	1	1	1	0	0
0	0	0	0	1	0	0	0	0
0	0	0	0	1	0	0	0	0
0	0	0	0	1	0	0	0	0
0	0	0	0	1	0	0	0	0
0	0	0	0	1	0	0	0	0
0	0	0	0	1	0	0	0	0
0	0	0	0	0	0	0	0	0

$S_1(A) = (A \ominus B) - (A \ominus B) \circ B$

0	0	0	0	0	0	0	0	0
0	0	0	0	0	0	0	0	0
0	0	0	0	0	0	0	0	0
0	0	0	0	0	0	0	0	0
0	0	0	0	0	0	0	0	0
0	0	0	0	0	0	0	0	0
0	0	0	0	0	0	0	0	0
0	0	0	0	0	0	0	0	0
0	0	0	0	0	0	0	0	0

$S_2(A)$

0	0	0	0	0	0	0	0	0
0	0	1	1	1	1	1	0	0
0	0	0	0	1	0	0	0	0
0	0	0	0	1	0	0	0	0
0	0	0	0	1	0	0	0	0
0	0	0	0	1	0	0	0	0
0	0	0	0	1	0	0	0	0
0	0	0	0	1	0	0	0	0
0	0	0	0	0	0	0	0	0

$S(A)$

Figure 4.8 An example of morphological skeleton.

where $S_k(A) = \bigcup_{k=0}^{K}\{(A \ominus kB) - [(A \ominus kB) \circ B]\}$. Note that $(A \ominus kB)$ denotes k successive erosions of $(A \ominus B)$, and K is the final step before A is eroded to be an empty set. An example of morphological skeleton is shown in Figure 4.8. After obtaining the skeleton, one can easily reconstruct the original set A from these skeleton subsets using successive dilations as

$$A = \bigcup_{k=0}^{K} (S_k(A) \oplus kB) \tag{4.39}$$

4.5.8 Pruning

The skeleton of a pattern after thinning usually appears extra short noisy branches that need to be cleaned up in the postprocessing. This process is called *pruning*. For instance, in automated recognition of handprinted characters, the skeleton is often characterized by "spurs" caused during erosion by nonuniformities in the strokes composing the characters. The short noisy branch can be located by finding end points, which have only one neighboring pixel in the set, and moving along the path until touching another branch within a very limited number of pixels. The length of a noisy branch is related to object characteristics and image size; it is often given as at most three pixels. A branch junction is considered as the pixel having at least three neighboring pixels. The pruning process could be complicated if an object has a complex shape structure. A detailed pruning process can be referred in (Serra, 1982, chapter 11).

0	1	0	0	1	1	0	0	0
0	0	1	1	0	0	0	0	0
0	0	0	1	0	0	0	0	0
0	0	0	1	0	0	0	0	0
0	0	0	1	0	1	1	0	0
0	0	0	1	1	0	0	1	0
0	0	0	1	0	0	0	1	0
0	0	0	1	1	0	0	1	0
0	0	0	1	0	1	1	0	1

A

0	0	0	0	0	0	0	0	0
0	0	0	1	0	0	0	0	0
0	0	0	1	0	0	0	0	0
0	0	0	1	0	0	0	0	0
0	0	0	1	0	1	1	0	0
0	0	0	1	1	0	0	1	0
0	0	0	1	0	0	0	1	0
0	0	0	1	1	0	0	1	0
0	0	0	1	0	1	1	0	0

X

Figure 4.9 An example of pruning a character "b" by mathematical morphology.

The pruning is represented in terms of thinning as $X = A \otimes \{B\}$, where $\{B\} = \{B^1, B^2, \ldots, B^8\}$ is given below.

$$B^1 = \begin{bmatrix} \times & 0 & 0 \\ 1 & 1 & 0 \\ \times & 0 & 0 \end{bmatrix}, \quad B^2 = \begin{bmatrix} \times & 1 & \times \\ 0 & 1 & 0 \\ 0 & 0 & 0 \end{bmatrix}, \quad B^3 = \begin{bmatrix} 0 & 0 & \times \\ 0 & 1 & 1 \\ 0 & 0 & \times \end{bmatrix}, \quad B^4 = \begin{bmatrix} 0 & 0 & 0 \\ 0 & 1 & 0 \\ \times & 1 & \times \end{bmatrix}$$
$$B^5 = \begin{bmatrix} 1 & 0 & 0 \\ 0 & 1 & 0 \\ 0 & 0 & 0 \end{bmatrix}, \quad B^6 = \begin{bmatrix} 0 & 0 & 1 \\ 0 & 1 & 0 \\ 0 & 0 & 0 \end{bmatrix}, \quad B^7 = \begin{bmatrix} 0 & 0 & 0 \\ 0 & 1 & 0 \\ 0 & 0 & 1 \end{bmatrix}, \quad B^8 = \begin{bmatrix} 0 & 0 & 0 \\ 0 & 1 & 0 \\ 1 & 0 & 0 \end{bmatrix} \tag{4.40}$$

An example of pruning a character "b" after two passes is shown in Figure 4.9.

4.5.9 Morphological Edge Operator

Edge operators based on grayscale morphological operations were introduced in Lee et al. (1987). These operators can be efficiently implemented in real-time machine vision systems that have special hardware architecture for grayscale morphological operations. The simplest morphological edge detectors are the dilation residue and erosion residue operators. Another *grayscale gradient* is defined as the difference between the dilated result and the eroded result. Figure 4.10 shows an example of grayscale gradient. A blur-minimum morphological edge operator is also introduced.

4.5.9.1 The Simple Morphological Edge Operators A simple method of performing grayscale edge detection in a morphology-based vision system is to take the difference between an image and its erosion by a small structuring element. The difference image is the image of edge strength. We can then select an appropriate threshold value to threshold the edge strength image into a binary edge image.

Let f denote a grayscale image and let k denote a structuring element as below, where "×" means don't care.

$k =$

×	0	×
0	0	0
×	0	×

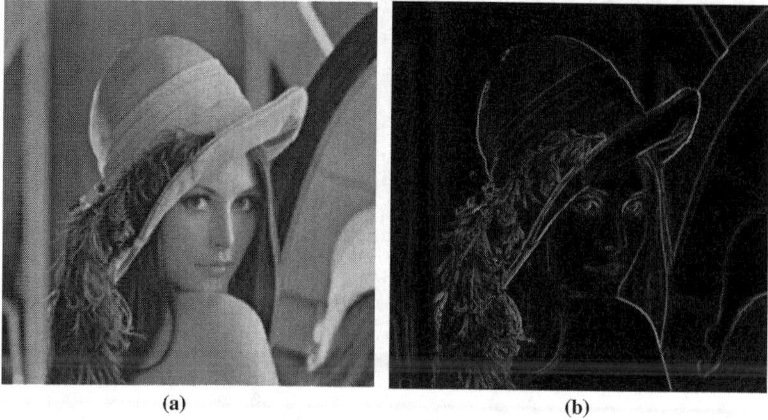

Figure 4.10 (a) Lena image. (b) The grayscale gradient.

The erosion of f by k is given by $e(x, y) = \min[f(x+i, y+j) - k(x, y)]$, for $(i,j) \in N_4(x, y)$. Since k contains four-connected neighbors of position (0, 0), the erosion residue edge operator produces the edge strength image G_e as

$$G_e(x, y) = f(x, y) - e(x, y) = \max[f(x, y) - f(i,j)], \quad \text{for} \quad (i,j) \in N_4(x, y) \tag{4.41}$$

where $N_4(x, y)$ is the set of four-connected neighbors of pixel (x, y).

A natural nonmorphological variation of this operator takes the summation instead of maximization. This is the familiar *linear digital Laplacian operator*, $\nabla^2 f(x, y)$, which is the digital convolution of $f(x, y)$ with the kernel

0	−1	0
−1	4	−1
0	−1	0

It is also possible to increase the neighborhood size of the morphological edge operator by increasing the size of the structuring element used for the erosion. For example, we can have an 8-connected neighborhood edge operator by changing the structuring element to the following:

$k =$

0	0	0
0	0	0
0	0	0

The erosion residue edge operator produces the edge strength image G_e defined by

$$G_e(x, y) = \max[f(x, y) - f(i,j)], \quad \text{for} \quad (i,j) \in N_8(x, y) \tag{4.42}$$

where $N_8(x, y)$ is the set of 8-connected neighbors of pixel (x, y).

The corresponding linear Laplacian operator, which has 8-connected neighborhood support, can be implemented as the digital convolution of $f(x, y)$ with

−1	−1	−1
−1	8	−1
−1	−1	−1

The performance of the edge operator can be easily evaluated. This operator performs perfectly on ideal step edge patterns. However, it is noise sensitive and position biased. To correct this bias and give both inside and outside boundaries of the checkers their corresponding edge strengths, the erosion residue morphological edge detector can be used in conjunction with dilation residue operator. The dilation residue operator takes the difference between a dilated image and its original image. The dilation residue edge strength image is

$$G_{\mathrm{d}}(x,y) = \max[f(i,j) - f(x,y)], \quad \text{for} \quad (i,j) \in N_8(x,y) \tag{4.43}$$

A position unbiased edge operator can be obtained by a combination of the operators $G_{\mathrm{e}}(x,y)$ and $G_{\mathrm{d}}(x,y)$ using the pixelwise minimum, maximum, or summation (Lee et al., 1987). For example, considering summation we have the edge operator strength $E(x,y) = G_{\mathrm{e}}(x,y) + G_{\mathrm{d}}(x,y)$. The summation and maximum versions perform perfectly on ideal step edge patterns, but they are also noise sensitive. The minimum version is noise insensitive but has no response when applied to a single noise point. Unfortunately, it is unable to detect ideal step edge patterns. This motivates a new edge operator that first performs a blur operation to convert all the ideal step edges into ideal ramp edges and then applies the minimum version of edge operator on them.

4.5.9.2 Blur-Minimum Morphological Edge Operator

The blur-minimum morphological edge operator is defined by

$$I_{\text{edge-strength}} = \min\{\tilde{I} - \text{erosion}(\tilde{I}), \text{dilation}(\tilde{I}) - \tilde{I}\} \tag{4.44}$$

where $\tilde{I} = \text{blur}(I)$, which is the blurred image after a blurring operation (Lee et al., 1987). An example of the blur-minimum morphological edge operator is shown in Figure 4.11. The blur-minimum morphological edge operator is noise insensitive. For the ideal step edge, it produces a result that has nonzero edge strength

(a)

(b)

Figure 4.11 (a) The original image. (b) The edge image by the blur-minimum morphological edge operator.

on both the edge pixels. However, due to the effect of blurring, the edge strength assigned to the edge pixels is one-third of the edge contrast. For ideal ramp edges of larger spatial extent, it will assign nonzero edge strength to more than 1 pixel. However, the true edge pixel is usually given edge strength higher than its neighbors.

The performance of the blur-minimum morphological edge operator may be poor in the presence of noise. To overcome this problem, Feehs and Arce (1987) proposed a twofold modification that uses the so-called *alpha-trimmed mean* (ATM) filter to smooth noise models other than Gaussian distribution. They defined the morphological edge operator as

$$I_{\text{atm}} = \min\{[(f_{\text{atm}} \circ k)-(f_{\text{atm}} \ominus k)], [(f_{\text{atm}} \oplus k)-(f_{\text{atm}} \bullet k)]\} \tag{4.45}$$

where f_{atm} is the image f smoothed by an alpha-trimmed mean filter. Since opening and closing are smoothing and noise-suppressing operations, this result will be less sensitive to noise. Furthermore, Moran (1990) developed a local neighborhood operator to sharpen edges using a selective erosion followed by a selective dilation.

4.6 MORPHOLOGICAL FILTERS

Image filters are specifically designed to remove unwanted image components and/or enhance wanted ones. Mathematical morphology, applying set-theoretic concept, provides an approach to digital image processing and analysis based on the geometric shape of objects. Using an *a priori* determined structuring element, features in an image can be extracted, suppressed, or preserved by applying morphological operators. *Morphological filters*, using the basic operations such as dilation, erosion, opening, and closing, are suited for many purposes including digitization, enhancement, compression, restoration, segmentation, and description.

Linear filtering techniques have serious limitations in dealing with the images that have been created or processed by a system exhibiting some degree of nonlinearity. A nonlinear filter produces the output that is not a linear function of its input. Nonlinear filtering techniques have shown clear superiority over linear filters in image processing. For example, an image often contains meaningful high-frequency components, such as edges and fine details. A linear low-pass filter would blur sharp edges, thus producing unacceptable results, and so nonlinear filters must be used.

Morphological filters belong to nonlinear operations. Morphological dilation is used to smooth small dark regions. If all the values in the structuring element are positive, the output image tends to be brighter than the input. Dark elements are reduced or eliminated depending on how their shapes and sizes relate to the structuring element used. Morphological erosion is used to smooth small light regions as opposed to dilation. Morphological opening removes bright objects that are small in size and breaks narrow connections between two bright objects. Morphological closing preserves small objects that are brighter than background and connects bright objects with small gaps in between. A morphological top-hat operation, that is, the difference between the original image and the opened image, is used to enhance low-contrast, high-frequency details within an image.

In this section, we will introduce alternating sequential filters, recursive morphological filters, soft morphological filters, order-statistic soft morphological filters, recursive soft morphological filters, recursive order-statistic soft morphological filters, regulated morphological filters, and fuzzy morphological filters. Another type of morphological filters called *general sweep morphology*, which allows simultaneous rotation and scaling of the structuring element when operating with an image, was proposed by Shih and Gaddipati (2003). Its geometric modeling and representation can be referred in Shih and Gaddipati (2005).

4.6.1 Alternating Sequential Filters

Alternating sequential filters in mathematical morphology are a combination of iterative morphological filters with increasing sizes of structuring elements. They offer a hierarchical structure for extracting geometric characteristics of objects. In addition, they provide less distortion in feature extraction than those filters that directly process the images with the largest structuring element (Serra, 1982). The alternating filter (AF) is composed of morphological openings and closings whose primitive morphological operations are dilation and erosion.

The class of AF has been demonstrated to be useful in image analysis applications. Sternberg (1986) introduced a new class of morphological filters called ASFs that consist of iterative operations of openings and closings with structuring elements of increasing sizes. Schonfeld and Goutsias (1991) showed that ASFs are the best in preserving crucial structures of binary images in the least differentiation sense. Pei et al. (1997) proposed an efficient class of alternating sequential filters in mathematical morphology to reduce the computational complexity in the conventional ASFs about a half.

The ASFs have been successfully used in a variety of applications, such as remote sensing and medical imaging (Destival, 1986; Preteux et al., 1985). Morales et al. (1995) presented a relationship between alternating sequential filters and the morphological sampling theorem (MST) developed by Haralick et al. (1989). The motivation is to take advantage of the computational efficiency offered by the MST to implement morphological operations.

Let X denote a binary image and let B denote a binary structuring element. The AF is defined as an opening followed by a closing or a closing followed by an opening, and is represented as

$$AF_B(X) = (X \circ B) \bullet B \tag{4.46}$$

or

$$AF_B(X) = (X \bullet B) \circ B \tag{4.47}$$

Another type of AF is defined as

$$AF_B(X) = ((X \circ B) \bullet B) \circ B \tag{4.48}$$

or

$$AF_B(X) = ((X \bullet B) \circ B) \bullet B \tag{4.49}$$

An ASF is an iterative application of $AF_B(X)$ with increasing sizes of structuring elements, denoted as

$$ASF(X) = AF_{B_N}(X)AF_{B_{N-1}}(X) \cdots AF_{B_1}(X) \tag{4.50}$$

where N is an integer and $B_N, B_{N-1}, \ldots, B_1$ are structuring elements with decreasing sizes. The B_N is constructed by

$$B_N = B_{N-1} \oplus B_1, \quad \text{for } N \geq 2 \tag{4.51}$$

The ASF offers a method of extracting image features hierarchically. The features can be divided into different layers according to their corresponding structuring element sizes. The features, such as size distribution (Giardina and Doughherty, 1988), of each layer can be used in many applications, for example, feature classification and recognition.

4.6.2 Recursive Morphological Filters

Recursive morphological filters can be used to avoid the iterative processing in the distance transform and skeletonization (Shih and Mitchell, 1992; Shih et al., 1995). The intent of the recursive morphological operations is to feed back the output at the current scanning pixel to overwrite its corresponding input pixel to be considered into computation at the following scanning pixels. Therefore, each output value obtained depends on all the updated input values at positions preceding the current location. The resulting output image in recursive mathematical morphology inherently varies with different image scanning sequences.

Let $N(O)$ be the set of the pixel O plus the neighbors prior to it in a scanning sequence of the picture within the domain of a defined structuring element. Suppose that a set A in the Euclidean N-space (E^N) is given. Let F and K be $\{x \in E^{N-1} | \text{ for some } y \in E, (x, y) \in A\}$, and let the domain of the grayscale image be f and of the grayscale structuring element be k. The *recursive dilation* of f by k, which is denoted by $f \circledcirc k$, is defined as follows:

$$(f \circledcirc k)(x, y) = \max\{[(f \circledcirc k)(x - m, y - n)] + k(m, n)\} \tag{4.52}$$

for all $(m, n) \in K$ and $(x-m, y-n) \in (N \cap F)$. The *recursive erosion* of f by k, which is denoted by $f \circledcirc k$, is defined as follows:

$$(f \circledcirc k)(x, y) = \min\{[(f \circledcirc k)(x + m, y + n)] - k(m, n)\} \tag{4.53}$$

for all $(m, n) \in K$ and $(x + m, y + n) \in (N \cap F)$

The scanning order will affect the output image in recursive morphology. Starting with a corner pixel, the scanning sequences in a 2D digital image are, in general, divided into

(i) "LT" denotes left-to-right and top-to-bottom. (Note that this is a usual television raster scan.)

(ii) "RB" denotes right-to-left and bottom-to-top.

(iii) "LB" denotes left-to-right and bottom-to-top.

(iv) "RT" denotes right-to-left and top-to-bottom.

Assume that a 3×3 structuring element k is denoted by

$$k = \begin{bmatrix} A_1 & A_2 & A_3 \\ A_4 & A_5 & A_6 \\ A_7 & A_8 & A_9 \end{bmatrix}$$

Because they are associated with different scanning directions, not all of the nine elements $A_1, \ldots, A_9$ are used in computation. The redefined structuring element in recursive mathematical morphology must satisfy the following criterion: Wherever a pixel is being dealt with, all its neighbors defined within the structuring element k must have already been visited before by the specified scanning sequence. Thus, the k is redefined as

$$k_{\mathrm{LT}} = \begin{bmatrix} A_1 & A_2 & A_3 \\ A_4 & A_5 & \times \\ \times & \times & \times \end{bmatrix}, \quad k_{\mathrm{RB}} = \begin{bmatrix} \times & \times & \times \\ \times & A_5 & A_6 \\ A_7 & A_8 & A_9 \end{bmatrix},$$

$$k_{\mathrm{LB}} = \begin{bmatrix} \times & \times & \times \\ A_4 & A_5 & \times \\ A_7 & A_8 & A_9 \end{bmatrix}, \quad k_{\mathrm{RT}} = \begin{bmatrix} A_1 & A_2 & A_3 \\ \times & A_5 & A_6 \\ \times & \times & \times \end{bmatrix} \tag{4.54}$$

where "×" means don't care. A complete scanning in the image space must contain the dual directions, that is, the first one being any of the four scans LT, RB, LB, RT working with its associated structuring element and the second one being the opposite scanning sequence of the first scan and its associated structuring element. For example, the opposite scanning of LT is RB.

The advantage of recursive mathematical morphology can be illustrated by applying distance transform. The distance transform is to convert a binary image that consists of object (foreground) and nonobject (background) pixels into an image in which every object pixel has a value corresponding to the minimum distance from the background. Three types of distance measures in digital image processing are usually used: Euclidean, city-block, and chessboard. The city-block and chessboard distances are our concern since their structuring elements can be decomposed into iterative dilations of small structuring components. The distance computation is, in principle, a global operation. By applying the well-developed decomposition properties of mathematical morphology, we can significantly reduce the tremendous cost of global operations to that of small neighborhood operations, which is suited for parallel-pipelined computers.

The iteration in the distance transform results in a bottleneck in real-time implementation. To avoid iteration, the recursive morphological operation is used. Only two scannings of the image are required: one is left-to-right top-to-bottom, and the other is right-to-left bottom-to-top. Thus, the distance transform becomes independent of the object size.

By setting the origin point as P, we can represent all of the points by their distances from P (so that P is represented by 0). The values of the structuring element are selected to be negative of the associated distance measures because the erosion is the minimum selection after the subtraction is applied. Let f denote a binary image that consists of two classes, object pixels (foreground) and nonobject pixels (background). Let the object pixels be the value "$+\infty$" (or any number larger than a half of the object's diameter) and the nonobject pixels be the value "0." Let k_1 and k_2 be the structuring elements associated with the LT and RB scannings, respectively; that is, in the chessboard distance

$$k_1 = \begin{bmatrix} -1 & -1 & -1 \\ -1 & 0 & \times \\ \times & \times & \times \end{bmatrix}, \quad k_2 = \begin{bmatrix} \times & \times & \times \\ \times & 0 & -1 \\ -1 & -1 & -1 \end{bmatrix} \tag{4.55}$$

In the city-block distance

$$k_1 = \begin{bmatrix} -2 & -1 & -2 \\ -1 & 0 & \times \\ \times & \times & \times \end{bmatrix}, \quad k_2 = \begin{bmatrix} \times & \times & \times \\ \times & 0 & -1 \\ -2 & -1 & -2 \end{bmatrix} \tag{4.56}$$

Distance transform using recursive morphology:

1. $d_1 = f \circledcirc k_1$.
2. $d_2 = d_1 \circledcirc k_2$.

The distance transform algorithm has widely used applications in image analysis. One of the applications is to compute a shape factor that is a measure of the compactness of the shape based on the ratio of the total number of object pixels to the summation of all distance measures (Danielsson, 1978). Another is to obtain the medial axis (or skeleton) that can be used for features extraction.

Example 4.13 Apply the two-step distance transformation using recursive morphology to calculate the chessboard distance transform on the following image:

0	0	0	0	0	0	0	0	0	0	0	0	0
0	7	7	7	7	7	0	0	0	7	7	7	0
0	7	7	7	7	7	0	0	0	7	7	7	0
0	7	7	7	7	7	0	0	0	7	7	7	0
0	7	7	7	7	7	7	7	7	7	7	7	0
0	7	7	7	7	7	7	7	7	7	7	7	0
0	7	7	7	7	7	7	7	7	7	7	7	0
0	7	7	7	7	7	7	7	7	7	7	7	0
0	7	7	7	7	7	7	7	7	7	7	7	0
0	0	0	0	0	0	0	0	0	0	0	0	0

Answer: First, we apply $d_1 = f \circledcirc k_1$, where

$$k_1 = \begin{bmatrix} -1 & -1 & -1 \\ -1 & 0 & \times \\ \times & \times & \times \end{bmatrix}$$

and obtain the following:

0	0	0	0	0	0	0	0	0	0	0	0	0
0	1	1	1	1	1	0	0	0	1	1	1	0
0	1	2	2	2	1	0	0	0	1	2	1	0
0	1	2	3	2	1	0	0	0	1	2	1	0
0	1	2	3	2	1	1	1	1	1	2	1	0
0	1	2	3	2	2	2	2	2	2	2	1	0
0	1	2	3	3	3	3	3	3	3	2	1	0
0	1	2	3	4	4	4	4	4	3	2	1	0
0	1	2	3	4	5	5	5	4	3	2	1	0
0	0	0	0	0	0	0	0	0	0	0	0	0

Second, we apply $d_2 = d_1 \circleddash k_2$, where

$$k_2 = \begin{bmatrix} \times & \times & \times \\ \times & 0 & -1 \\ -1 & -1 & -1 \end{bmatrix}$$

and obtain the chessboard distance transform as

0	0	0	0	0	0	0	0	0	0	0	0	0
0	1	1	1	1	1	0	0	0	1	1	1	0
0	1	2	2	2	1	0	0	0	1	2	1	0
0	1	2	3	2	1	0	0	0	1	2	1	0
0	1	2	3	2	1	1	1	1	1	2	1	0
0	1	2	3	2	2	2	2	2	2	2	1	0
0	1	2	3	3	3	3	3	3	3	2	1	0
0	1	2	2	2	2	2	2	2	2	2	1	0
0	1	1	1	1	1	1	1	1	1	1	1	0
0	0	0	0	0	0	0	0	0	0	0	0	0

Figure 4.12 shows a binary image of human and bird figures and its chessboard and city-block distance transformations. Note that after distance transformations, the distances are linearly rescaled to be in the range of [0, 255] for clear display.

Note that the recursive morphological filters defined here are different from the commonly used recursive filters in signal processing, which reuse one or more of the outputs as an input and can efficiently achieve a long impulse response, without having to perform a long convolution (Smith, 1997). Other recursive algorithms may use different definitions. For example, the recursive algorithm based on observation of the basis matrix and block basis matrix of grayscale structuring elements was used to avoid redundant steps in computing overlapping local maximum or minimum operations (Ko et al., 1996). Another recursive two-pass algorithm that runs at a

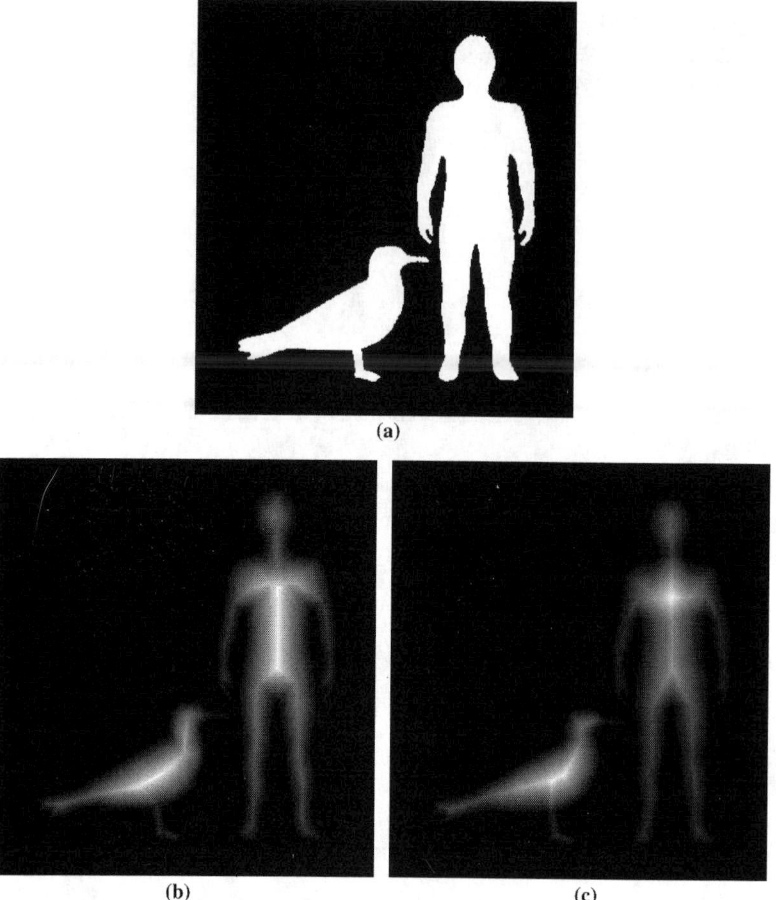

Figure 4.12 (a) A binary image of human and bird figures, (b) its chessboard distance transformation, and (c) its city-block distance transformation.

constant time for simultaneously obtaining binary dilations and erosions with all possible length line structuring elements can be referred in Nadadur and Haralick (2000).

4.6.3 Soft Morphological Filters

Soft morphological filters adopt the concept of weighted order-statistic filters and morphological filters. The primary difference from traditional morphological filters is that the maximum and minimum operations are replaced by general weighted order statistics and the "soft" boundary is added to the structuring element. In general, soft morphological filters are less sensitive to additive noise and to small variations in object shape, and can preserve most of the desirable properties of traditional morphological filters (Koskinen et al., 1991; Kuosmanen and Astola, 1995; Shih and Pu, 1995b; Shih and Puttagunta, 1995). Hamid et al. (2003) presented a technique for the optimization of multidimensional grayscale soft morphological filters for applications in automatic film archive restoration, specific to the problem of film dirt removal.

The basic idea of soft morphological operations is that the structuring element B is split into two subsets: the *core* subset A (i.e., $A \subseteq B$) and the *soft boundary* subset $B \backslash A$, where "\" denotes the *set difference*. Soft morphological dilation (erosion) of a function with respect to these two finite sets A and B using the order index k is a function whose value at location x is obtained by sorting in a descendant (ascendant) order the total Card($B \backslash A$) + $k \times$ Card(A) values of the input function, including the elements within $(B \backslash A)_x$ and repetition k times of the elements within $(A)_x$, and then selecting the kth order from the sorted list. Let $\{k \diamond f(a)\}$ denote the repetition k times of $f(a)$, which means $\{k \diamond f(a)\} = \{f(a), f(a), \ldots, f(a)\}$ (k times). The soft morphological operations of function-by-set, where f is a grayscale image and A and B are flat structuring elements, are defined as follows:

Definition 4.10. The soft morphological dilation of f by $[B, A, k]$ is defined as

$$(f \oplus [B, A, k])(x) = k\text{th largest of}(\{k \diamond f(a) | a \in A_x\} \cup \{f(b) | b \in (B \backslash A)_x\}) \tag{4.57}$$

Definition 4.11. The soft morphological erosion of f by $[B, A, k]$ is defined as

$$(f \ominus [B, A, k])(x) = k\text{th smallest of } (\{k \diamond f(a) | a \in A_x\} \cup \{f(b) | b \in (B \backslash A)_x\}) \tag{4.58}$$

The soft morphological operations of set-by-set can be simplified as in the definition below, where $n = \text{Card}(A)$ and $N = \text{Card}(B)$.

Definition 4.12. The soft morphological dilation of a set X by $[B, A, k]$ is defined as

$$X \oplus [B, A, k] = \{x | k \times \text{Card}(X \cap A_x) + \text{Card}(X \cap (B \backslash A)_x) \geq k\} \tag{4.59}$$

Definition 4.13. The soft morphological erosion of a set X by $[B, A, k]$ is defined as

$$X \ominus [B, A, k] = \{x | k \times \text{Card}(X \cap A_x) + \text{Card}(X \cap (B \backslash A)_x) \geq N + (k-1) \times n - k + 1\} \tag{4.60}$$

The logic-gate implementation of soft morphological dilation and erosion is illustrated in Figure 4.13, where the parallel counter counts the number of 1's of the input signal and the comparator outputs 1 if the input is greater than or equal to the index k.

The cross-section $X_t(f)$ of f at level t is the set obtained by thresholding f at level t:

$$X_t(f) = \{x | f(x) \geq t\}, \text{ where } -\infty < t < \infty. \tag{4.61}$$

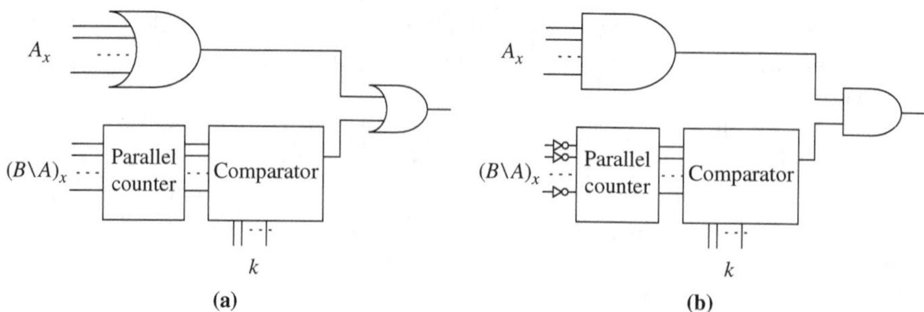

Figure 4.13 The logic-gate implementation of soft morphological operations (a) dilation and (b) erosion.

Theorem 4.1. The soft morphological operations of function-by-set commute with thresholding. That is, for any t, we have

$$X_t(f \oplus [B,A,k]) = X_t(f) \oplus [B,A,k] \tag{4.62}$$

This theorem can be easily derived from Shih and Mitchell (1989) and Shih and Pu (1995b). The implementation and analysis of soft morphological operations of function-by-set can be achieved by using the set-by-set operations, which are much easier to deal with because they are involved in only counting the number of pixels instead of sorting numbers. Consider the thresholding of a binary image:

$$f_a(x) = \begin{cases} 1 & \text{if } f(x) \geq a \\ 0 & \text{otherwise} \end{cases} \tag{4.63}$$

where $0 \leq a \leq L$ and L is the largest value in f. It is simple to show that f can be reconstructed from its thresholded binary images as

$$f(x) = \sum_{a=1}^{L} f_a(x) = \max\{a | f_a(x) = 1\} \tag{4.64}$$

A transformation Ψ is said to possess *threshold-linear superposition* if it satisfies

$$\Psi(f) = \sum_{a=1}^{L} \Psi(f_a) \tag{4.65}$$

Such a transformation Ψ can be realized by decomposing f into all its binary images f_a's, processing each thresholded images by Ψ, and creating the output $\Psi(f)$ by summing up the processed f_a's. The soft morphological operations of function-by-set can be easily shown to obey the threshold-linear superposition.

Theorem 4.2. The soft morphological operations of functions by sets obey the threshold-linear superposition. That is,

$$f \oplus [B,A,k] = \sum_{a=1}^{L} (f_a \oplus [B,A,k]) \tag{4.66}$$

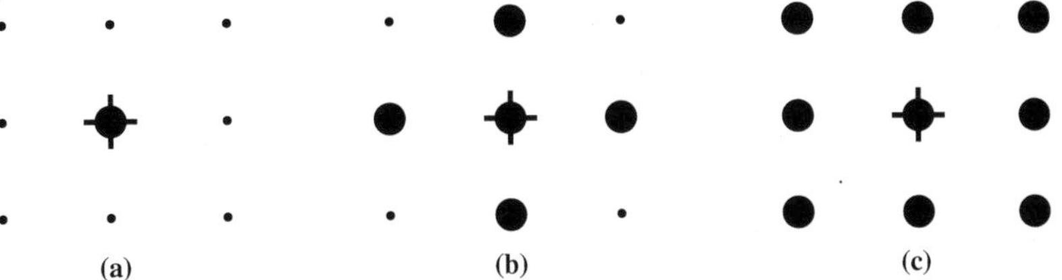

Figure 4.14 Two core sets (a) A_1, (b) A_2, and (c) the structuring element B.

This theorem is very useful in the sense that the grayscale input image is thresholded at each gray level a to get f_a, followed by the soft morphological operations of sets by sets for f_a, and then summing up all f_a's, which is exactly the same as the soft morphological operations of functions by sets. Considering A_1, A_2, and B in Figure 4.14, an example of illustrating the threshold decomposition is given in Figure 4.15.

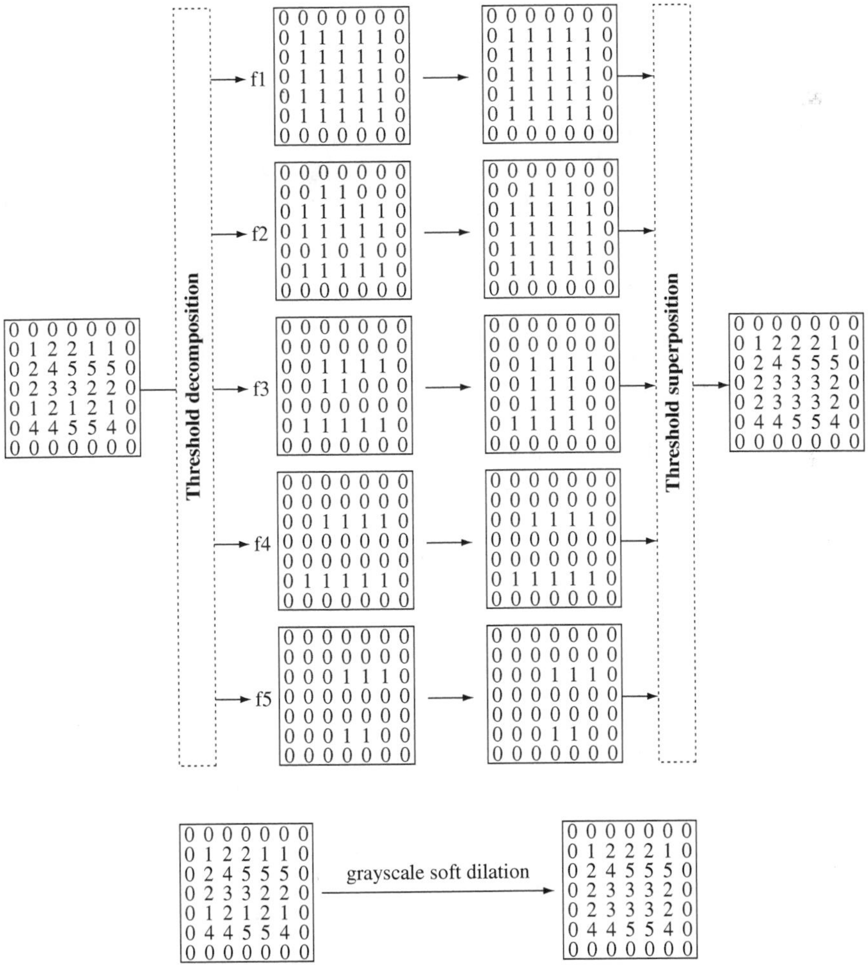

Input (Threshold decomposition):

```
0 0 0 0 0 0 0
0 1 2 2 1 1 0
0 2 4 5 5 5 0
0 2 3 3 2 2 0
0 1 2 1 2 1 0
0 4 4 5 5 4 0
0 0 0 0 0 0 0
```

f1:

```
0 0 0 0 0 0 0        0 0 0 0 0 0 0
0 1 1 1 1 1 0        0 1 1 1 1 1 0
0 1 1 1 1 1 0        0 1 1 1 1 1 0
0 1 1 1 1 1 0   →    0 1 1 1 1 1 0
0 1 1 1 1 1 0        0 1 1 1 1 1 0
0 1 1 1 1 1 0        0 1 1 1 1 1 0
0 0 0 0 0 0 0        0 0 0 0 0 0 0
```

f2:

```
0 0 0 0 0 0 0        0 0 0 0 0 0 0
0 0 1 1 0 0 0        0 0 1 1 1 0 0
0 1 1 1 1 1 0        0 1 1 1 1 1 0
0 1 1 1 1 1 0   →    0 1 1 1 1 1 0
0 0 1 0 1 0 0        0 1 1 1 1 1 0
0 1 1 1 1 1 0        0 1 1 1 1 1 0
0 0 0 0 0 0 0        0 0 0 0 0 0 0
```

f3:

```
0 0 0 0 0 0 0        0 0 0 0 0 0 0
0 0 0 0 0 0 0        0 0 0 0 0 0 0
0 0 1 1 1 1 0        0 0 1 1 1 1 0
0 0 1 1 0 0 0   →    0 0 1 1 1 0 0
0 0 0 0 0 0 0        0 0 1 1 1 0 0
0 1 1 1 1 1 0        0 1 1 1 1 1 0
0 0 0 0 0 0 0        0 0 0 0 0 0 0
```

f4:

```
0 0 0 0 0 0 0        0 0 0 0 0 0 0
0 0 0 0 0 0 0        0 0 0 0 0 0 0
0 0 1 1 1 1 0        0 0 1 1 1 1 0
0 0 0 0 0 0 0   →    0 0 0 0 0 0 0
0 0 0 0 0 0 0        0 0 0 0 0 0 0
0 1 1 1 1 1 0        0 1 1 1 1 1 0
0 0 0 0 0 0 0        0 0 0 0 0 0 0
```

f5:

```
0 0 0 0 0 0 0        0 0 0 0 0 0 0
0 0 0 0 0 0 0        0 0 0 0 0 0 0
0 0 0 1 1 1 0        0 0 0 1 1 1 0
0 0 0 0 0 0 0   →    0 0 0 0 0 0 0
0 0 0 0 0 0 0        0 0 0 0 0 0 0
0 0 0 1 1 0 0        0 0 0 1 1 0 0
0 0 0 0 0 0 0        0 0 0 0 0 0 0
```

Threshold superposition → output:

```
0 0 0 0 0 0 0
0 1 2 2 2 1 0
0 2 4 5 5 5 0
0 2 3 3 3 2 0
0 2 3 3 3 2 0
0 4 4 5 5 4 0
0 0 0 0 0 0 0
```

grayscale soft dilation:

```
0 0 0 0 0 0 0                              0 0 0 0 0 0 0
0 1 2 2 1 1 0                              0 1 2 2 2 1 0
0 2 4 5 5 5 0   grayscale soft dilation    0 2 4 5 5 5 0
0 2 3 3 2 2 0   ----------------------→    0 2 3 3 3 2 0
0 1 2 1 2 1 0                              0 2 3 3 3 2 0
0 4 4 5 5 4 0                              0 4 4 5 5 4 0
0 0 0 0 0 0 0                              0 0 0 0 0 0 0
```

Figure 4.15 The soft morphological dilation obeys threshold superposition with A_1 and B as shown in Figure 4.14, and k is equal to 4.

The grayscale soft morphological operation can be decomposed into binary soft morphological operation (Pu and Shih, 1995). A new hardware structure for implementation of soft morphological filters can be referred in Gasteratos et al. (1997). Let the reflected sets of A and B be $\hat{A}(x) = A(-x)$ and $\hat{B}(x) = B(-x)$, respectively.

Definition 4.14. The soft morphological closing of f by $[B, A, k]$ is defined as

$$f \bullet [B, A, k] = (f \oplus [B, A, k]) \ominus [\hat{B}, \hat{A}, k] \tag{4.67}$$

Definition 4.15. The soft morphological opening of f by $[B, A, k]$ is defined as

$$f \circ [B, A, k] = (f \ominus [B, A, k]) \oplus [\hat{B}, \hat{A}, k] \tag{4.68}$$

It is interesting to note that if $k = 1$, the soft morphological operations are equivalent to standard morphological operations. If $k > \text{Card}(B \backslash A)$, the soft morphological operations are reduced to consider only the hard set A (i.e., the soft boundary set $(B \backslash A)$ will actually not affect the result). Therefore, in order to preserve the nature of soft morphological operations, the constraint that $k \leq \min\{\text{Card}(B)/2, \text{Card}(B \backslash A)\}$ is used.

In Figure 4.16, (a) shows an original image, and (b) shows the same image but with added Gaussian noise being zero mean and standard deviation 20. Figure 4.16c shows the result of applying a standard morphological closing with a structuring element of size 3×3, and (d) shows the result of applying a soft morphological closing with $B = \langle(0, -1), (-1, 0), (0, 0), (1, 0), (0, 1)\rangle$, $A = \langle(0, 0)\rangle$, and $k = 3$. The mean-square signal-to-noise ratios (SNR) for Figure 4.16c and d are 39.78 and 55.42, respectively. One can observe that the result of soft morphological filtering produces a smoother image and a higher signal-to-noise ratio.

Example 4.14 Perform the soft morphological dilation $f \oplus [B, A, 3]$ and the soft morphological erosion $f \ominus [B, A, 3]$. Use the center (underlined) of A and B as the origin.

$$f: \begin{bmatrix} 0 & 0 & 0 & 1 & 0 \\ 0 & 5 & 5 & 5 & 1 \\ 4 & 5 & 1 & 5 & 0 \\ 3 & 6 & 5 & 6 & 0 \\ 0 & 0 & 0 & 0 & 1 \end{bmatrix}, \quad A = \begin{bmatrix} 0 & 0 & 0 \\ 0 & \underline{1} & 0 \\ 0 & 0 & 0 \end{bmatrix}, \quad B = \begin{bmatrix} 1 & 1 & 1 \\ 1 & \underline{1} & 1 \\ 1 & 1 & 1 \end{bmatrix}$$

Answer: The soft morphological dilation $f \oplus [B, A, 3]$ repeats the central element of the 3×3 window three times and others once, and then arranges them in an ascending order. Taking out the third largest element, we obtain the following result:

$$\begin{matrix} 0 & 0 & 5 & 1 & 1 \\ 4 & 5 & 5 & 5 & 1 \\ 5 & 5 & 5 & 5 & 5 \\ 4 & 6 & 5 & 6 & 1 \\ 0 & 3 & 5 & 1 & 1 \end{matrix}$$

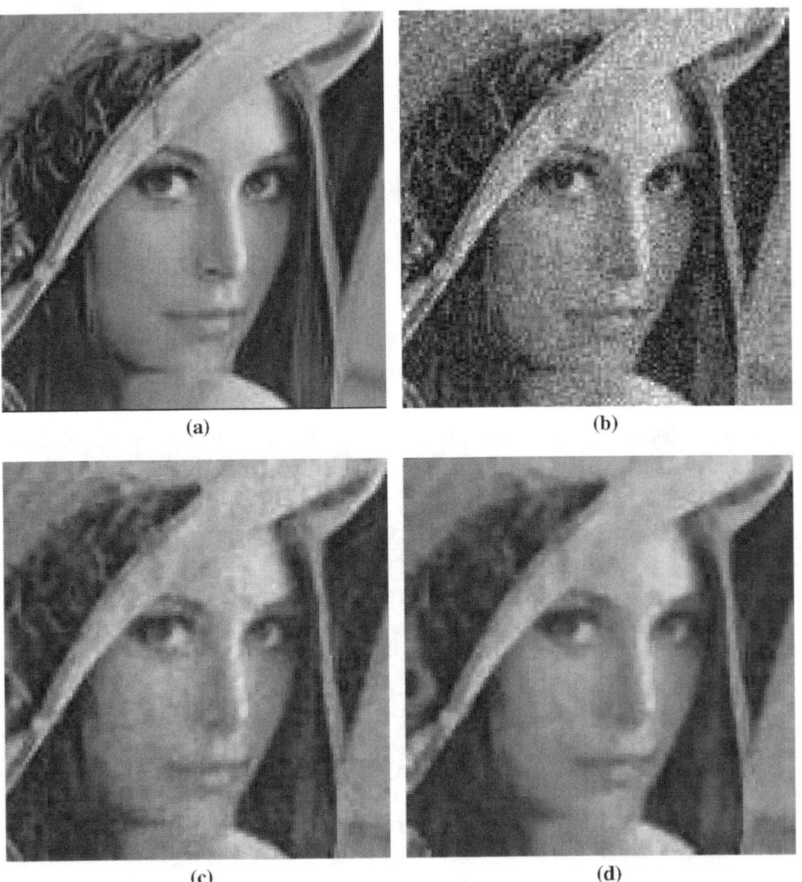

(a) (b)

(c) (d)

Figure 4.16 (a) An original image, (b) image corrupted by Gaussian noise, (c) the result of applying a standard morphological closing, and (d) the result of applying a soft morphological closing.

The soft morphological erosion $f \ominus [B, A, 3]$ repeats the central element of the 3×3 window three times and others once, and then arranges in an ascending order. Taking out the third smallest element we obtain the following result:

$$\begin{matrix} 0 & 0 & 0 & 1 & 0 \\ 0 & 0 & 1 & 0 & 1 \\ 4 & 3 & 1 & 1 & 0 \\ 3 & 0 & 0 & 0 & 0 \\ 0 & 0 & 0 & 0 & 1 \end{matrix}$$

4.6.4 Order-Statistic Soft Morphological (OSSM) Filters

The disadvantage of soft morphological filters is that the repetition number is set up to be the rank order. This constraint limits the flexibility and applicability of the filters in noise removal and edge preservation. In this section, we present the OSSM filters that provide a free order selection parameter. We also develop the properties and explore the restoration performance when a signal is corrupted by variant noises.

Many nonlinear image processing techniques have their root in statistics. Their applications in images, image sequences, and color images with order statistics (David, 1981) and their closely related morphological filters have by far been very successful (Maragos and Schafer, 1987a, 1987b; Stevenson and Arce, 1987; Soille, 2002). The median filter was first introduced into statistics for smoothing economic time series and was then used in image processing applications to remove noises with retaining sharp edges. The stacking property unifies all subclasses of stack filters, such as ranked order, median, weighted median, and weighted order-statistic filters.

Let f be a function defined on m-dimensional discrete Euclidean space Z^m. Let W be a window of a finite subset in Z^m that contains N points, where $N = \mathrm{Card}(W)$, the *cardinality* of W. The kth order statistics (OS^k) of a function $f(x)$ with respect to the window W is a function whose value at location x is obtained by sorting in a descendent order the N values of $f(x)$ inside the window W whose origin is shifted to location x and picking up the kth number from the sorted list, where k ranges from 1 to N. Let W_x denote the translation of the origin of the set W to location x. The kth OS is represented by

$$OS^k(f, W)(x) = k\text{th largest of } \{f(a) | a \in W_x\} \tag{4.69}$$

The OSSM operations, like in the soft morphology, adopt order statistics and use the split subsets, core set A and boundary set $B \backslash A$, as the structuring element. The obvious difference is that a free order selection parameter is introduced to increase flexibility and efficiency of the operations. That is, we select the lth order, rather than the kth order, from the sorted list of the total "Card($B \backslash A$) $+ k \times$ Card(A)" values of the input function. The definitions of function by sets are given as follows, where f is a grayscale image and B is a flat structuring element.

Definition 4.16. The order-statistic soft morphological dilation of f by $[B, A, k, l]$ is defined as

$$(f \oplus [B, A, k, l])(x) = l\text{th largest of}(\{k \diamond f(a) | a \in A_x\} \cup \{f(b) | b \in (B \backslash A)_x\}) \tag{4.70}$$

Definition 4.17. The order-statistic soft morphological erosion of f by $[B, A, k, l]$ is defined as

$$(f \ominus [B, A, k, l])(x) = l\text{th smallest } of\ (\{k \diamond f(a) | a \in A_x\} \cup \{f(b) | b \in (B \backslash A)_x\}) \tag{4.71}$$

One of the advantages of using the OSSM operators over standard morphological operators can be observed in the following example. Assume that we have a flat signal f corrupted by impulsive noise as shown in Figure 4.17a. In order to remove

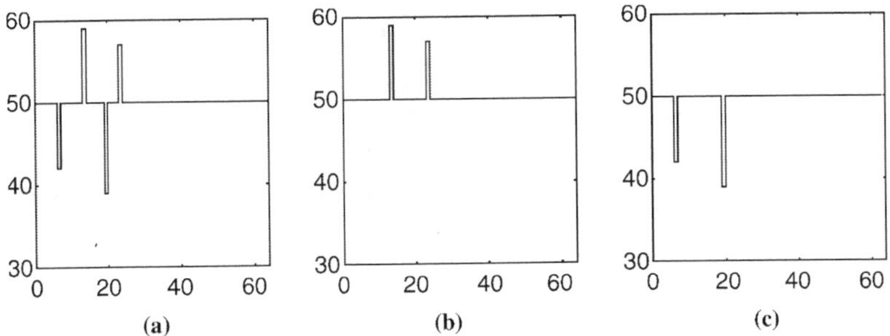

Figure 4.17 (a) A flat signal corrupted by positive and negative impulsive noises, and the filtered results by using (b) a soft dilation filter and (c) a soft erosion filter.

the impulsive noise, a standard morphological opening and a closing are applied to the distorted signal. Note that the positive impulse is removed by the opening operation and the negative impulse by closing. On the other hand, by applying a single soft opening and closing, the impulse noise can be removed. It is due to that the soft dilation can remove one part (positive or negative) of the noise and the soft erosion can remove the other. Figure 4.17b and c, respectively, shows the results of a soft dilation and a soft erosion by $[(1, 1, \underline{1}, 1, 1), (\underline{1}), 3]$, where $B = (1, 1, \underline{1}, 1, 1)$ with underlined one indicating the origin, $A = (\underline{1})$, and $k = 3$. Therefore, four standard morphological operations or two soft morphological operations are needed to eliminate the impulse noise. The same goal of filtering can be achieved by using an OSSM dilation by $[(1, 1, \underline{1}, 1, 1), (\underline{1}), 3, 4]$. This leads to a significant reduction in computational complexity near a half amount.

Another important characteristic of the OSSM filters is the edge preserving ability. Figure 4.18a and b, respectively, shows a signal with upward and downward step edges. The edges of applying a standard morphological dilation or an erosion to the signal are shifted and are preserved if an opening or a closing is used. However, if an OSSM dilation by $[(1, 1, \underline{1}, 1, 1), (\underline{1}), 2, 3]$ is used, the location of edges is preserved.

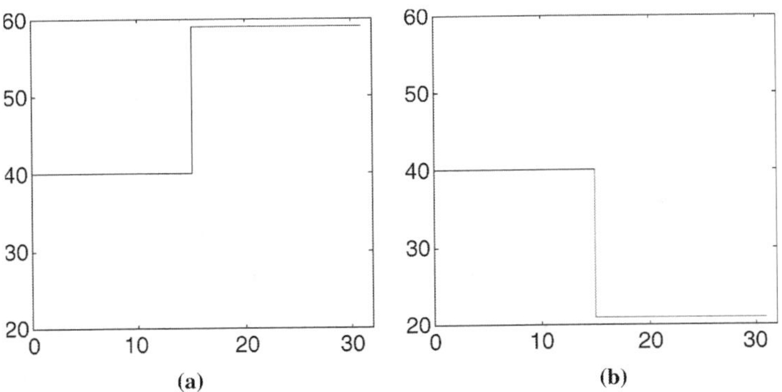

Figure 4.18 A signal with (a) upward step edge and (b) downward step edge.

4.6.5 Recursive Soft Morphological Filters

The structuring element in morphological filters can be regarded as a template that is translated to each pixel location in an image. These filters can be implemented in parallel because each pixel's value in the transformed image is only a function of its neighboring pixels in the given image. Also, the sequence in which the pixels are processed is completely irrelevant. Thus, these parallel image operations can be applied to each pixel simultaneously if a suitable parallel architecture is available. Parallel image transformations are also referred to as *nonrecursive transformations*.

In contrast to nonrecursive transformations, a class of *recursive transformations* is also widely used in signal and image processing, for example, sequential block labeling, predictive coding, and adaptive dithering (Serra, 1982). The main distinction between these two classes of transformations is that in the recursive transformations, the pixel's value of the transformed image depends upon the pixel's values of both the input image and the transformed image itself. Due to this reason, some partial order has to be imposed on the underlying image domain, so that the transformed image can be computed recursively according to this imposed partial order. In other words, a pixel's value of the transformed image may not be processed until all the pixels preceding it have been processed.

Recursive filters are the filters that use previously filtered outputs as their inputs. Let x_i and y_i denote the input and output values at location i, respectively, where $i = \{0, 1, \ldots, N-1\}$. Let the domain of the structuring element be $\langle -L, \ldots, -1, 0, 1, \ldots, R \rangle$, where L is the left margin and R is the right margin. Hence, the structuring element has the size of $L + R + 1$. Start-up and end effects are accounted for by appending L samples to the beginning and R samples to the end of the signal sequence. The L appended samples are given the value of the first signal sample; similarly, the R appended samples receive the value of the last sample of the signal.

Definition 4.18. The recursive counterpart of a nonrecursive filter Ψ given by

$$y_i = \Psi(x_{i-L}, \ldots, x_{i-1}, x_i, x_{i+1}, \ldots, x_{i+R}) \tag{4.72}$$

is defined as

$$y_i = \Psi(y_{i-L}, \ldots, y_{i-1}, x_i, x_{i+1}, \ldots, x_{i+R}) \tag{4.73}$$

by assuming that the values of $y_{i-L}, \ldots, y_{i-1}$ are already given.

Example 4.15 Let $B = \langle -1, 0, 1 \rangle$, $A = \langle 0 \rangle$, and $k = 2$. Let the input signal $f = \{4\,7\,2\,9\,6\,8\,5\,4\,7\}$. We have

$$f \oplus_r [B, A, 2] = \{4\,7\,7\,9\,8\,8\,5\,5\,7\}$$
$$f \ominus_r [B, A, 2] = \{4\,4\,2\,6\,6\,6\,5\,4\,7\}$$

where "$\oplus_r$" and "$\ominus_r$" denote the recursive soft morphological dilation and erosion, respectively.

In Figure 4.19, (a) shows a 125×125 Lena image and (b) shows the same image but with added Gaussian noise having zero mean and standard deviation 20. The mean-square signal-to-noise ratio is used. Figure 4.19b has SNR = 40.83. Figure 4.19c

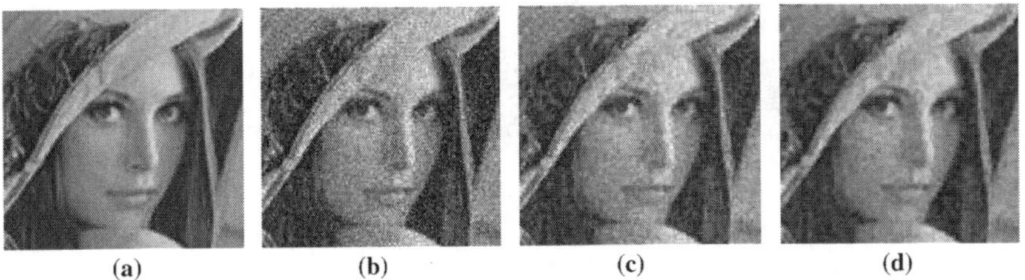

Figure 4.19 (a) An original image, (b) image corrupted by Gaussian noise, (c) result of a recursive soft morphological dilation, (d) followed by a recursive soft morphological erosion (i.e., closing).

shows the result of applying a recursive soft morphological dilation with $B = \langle(0,-1),(-1,0),(0,0),(1,0),(0,1)\rangle$, $A = \langle(0,0)\rangle$, and $k = 3$, where SNR = 64.65. Figure 4.19d shows the result of applying a recursive soft morphological erosion on Figure 4.19c (i.e., closing) with the same structuring elements and rank order, where SNR = 101.45.

In Figure 4.20, (a) shows a 512×432 grayscale image, and (b) shows the same image but with added Gaussian noise having zero mean and standard deviation 20.

(a) (b)

(c) (d)

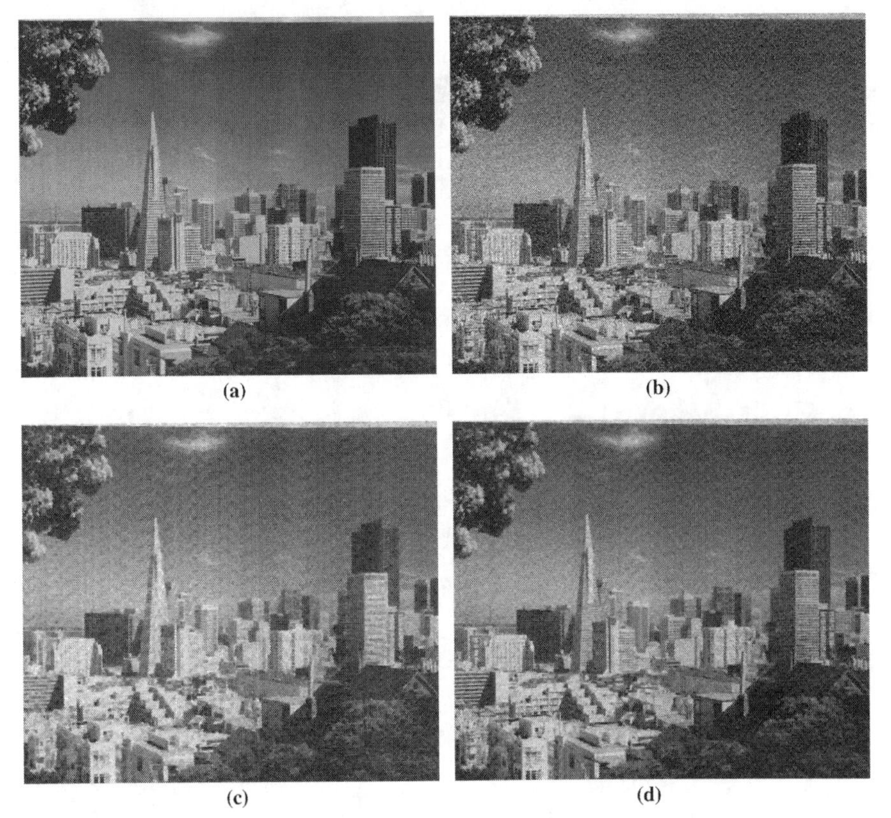

Figure 4.20 (a) An original image, (b) image corrupted by Gaussian noise, (c) result of a standard morphological closing, and (d) result of a recursive soft morphological closing.

Figure 4.20b has SNR = 35.54. Figure 4.20c shows the result of applying a standard morphological closing with a structuring element of size 3 × 3, where SNR = 25.15. Figure 4.20d shows the result of applying a recursive soft morphological closing with $B = \langle(0,-1),(-1,0),(0,0),(1,0),(0,1)\rangle$, $A = \langle(0,0)\rangle$, and $k = 3$, where SNR = 69.41.

4.6.6 Recursive Order-Statistic Soft Morphological Filters

A class of *recursive order-statistic soft morphological* (ROSSM) filters is introduced and their properties are described in this section. It has been shown through experimental results that the ROSSM filters, compared to the order-statistic soft morphological filters or other nonlinear filters, have better outcomes in signal reconstruction (Pei et al., 1998).

To define the recursive OSSM filters, let Φ denote an ROSSM operation and *mBnA* be the symmetric structuring element sets with m pixels in B and n pixels in A. For example, the structuring element [–2, –1, $\underline{0}$, 1, 2] is denoted as 5*B*1*A*, while [–2, $\underline{-1, 0, 1}$, 2] is denoted as 5*B*3*A*, where the center of the underlined denotes the center of symmetry. Then, applying a recursive order-statistic soft morphological filter on input f at index i can be written as

$$y_i = \Phi(y_{i-L}, \ldots, y_{i-1}, y_i, y_{i+1}, \ldots, y_{i+R}) \tag{4.74}$$

where $y_{i-L}, \ldots, y_{i-1}$ are already obtained by $\Phi(f_{i-L}), \ldots, \Phi(f_{i-1})$.

The two-dimensional ROSSM operation on an image F with a structuring element S is represented as

$$Y_{ij} = \Phi(Y_p, F_{ij}, F_q) \tag{4.75}$$

where $S = (S_B, S_A)$ is a two-dimensional structuring element with core set S_A and boundary set $S_B \backslash S_A$ (note that $S_A \subseteq S_B$), and p and q are domain index sets that precede and succeed the index (i, j) in a scanning order (see Fig. 4.21).

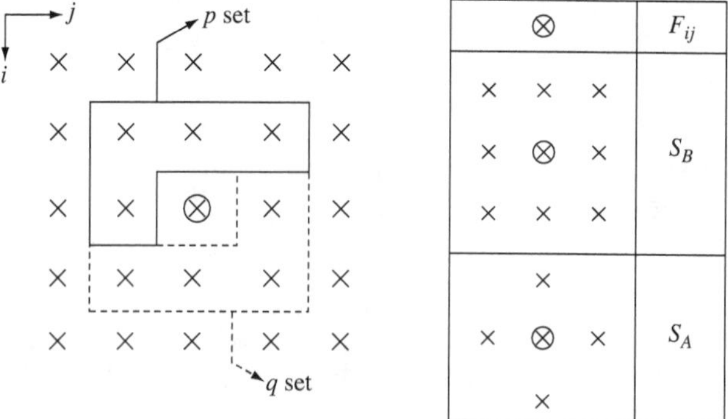

Figure 4.21 Two-dimensional structuring element and the preceding (succeeding) domain index sets when using the raster scan order.

An example is given below to show the difference between the OSSM and ROSSM filters.

Example 4.16 Let the structuring element be $5B1A$. Let $k=2$ and $l=3$. Assume that the input sequence be $f=\{74, 14, 28, 65, 55, 8, 61, 20, 17, 65, 26, 94\}$ and the values that are outside the region of f are set to be boundary values of f. Then the output sequences obtained from the OSSM and ROSSM dilations are

$$\text{OSSM}\,(f, B, A, k, l) = \{74, 65, 55, 55, 55, 55, 55, 20, 26, 65, 65, 94\}$$

and

$$\text{ROSSM}\,(f, B, A, k, l) = \{74, 65, 65, 65, 61, 61, 61, 61, 61, 65, 65, 94\}$$

respectively. Figure 4.22a and b shows a 256×256 image and its disturbed version with additive Gaussian noise having zero mean and standard deviation 20. The reconstructed images of applying the OSSM and ROSSM filters with $S_B=[(-1,0),$

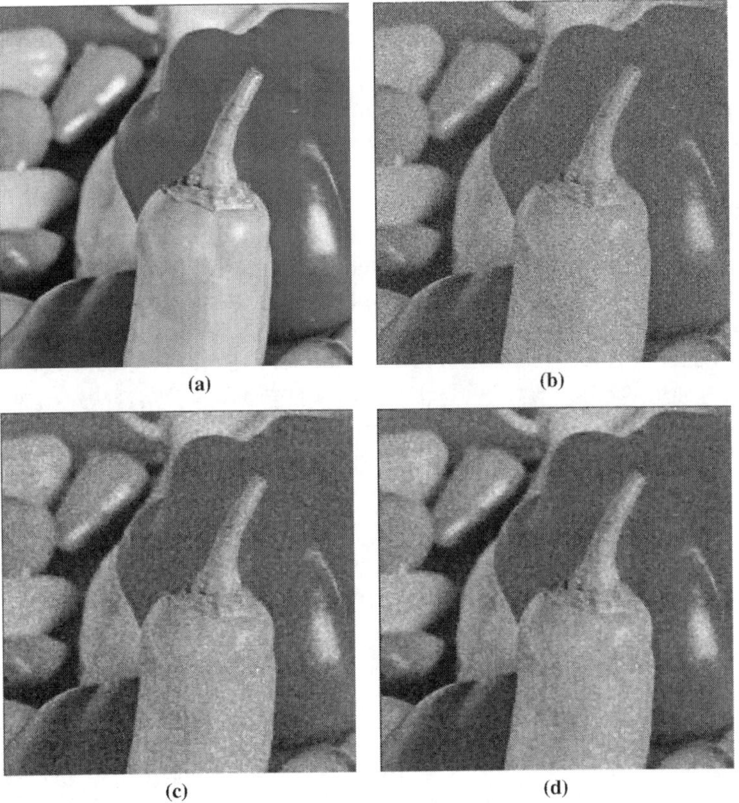

Figure 4.22 (a) Original image, (b) noisy image disturbed by zero mean Gaussian noise ($\sigma = 20$), (c) reconstructed image obtained from the OSSMF, and (d) reconstructed image obtained from the ROSSMF.

(0,−1), (0,0), (1,0), (0,1)], $S_A = [(0,0)]$, $k=3$, and $l=4$ are shown in Figure 4.22c and d. The mean square signal to noise ratio, defined as

$$\mathrm{SNR_{ms}} = \frac{\sum (y'_i)^2}{\sum (y'_i - y_i)^2} \tag{4.76}$$

with y and y' denoting the reconstructed and original signal, respectively, is used to measure the fidelity of the reconstructed signals. The $\mathrm{SNR_{ms}}$ values of Figure 4.22c and d are 84.3 and 91.6, respectively. It is observed that the visual perception effect of Figure 4.22d is slightly better than Figure 4.22c.

An important feature of the OSSM filter is the duality property. This relationship property also holds on the ROSSM operators. That is, choosing l in an ROSSM dilation is the same as choosing $T - l + 1$ in an ROSSM erosion, where T represents the total number in the order list. If $\tilde{\oplus}_R$ and $\tilde{\ominus}_R$ denote the ROSSM dilation and erosion, respectively, the duality relationship can be represented as $f\tilde{\oplus}_R[B, A, k, l] = f\tilde{\ominus}_R[B, A, k, T-l+1]$. However, if the structuring element is obtained by taking the intersection (union) of two or more structuring elements, this duality property may be unavailable.

4.6.7 Regulated Morphological Filters

Agam and Dinstein (1999) developed regulated morphological operators and showed how the fitting property can be adapted for analyzing the map and line drawing images. Since regulated morphology inherits many properties of the traditional standard morphology, it is feasible to apply in image processing and optimize its strictness parameters. Tian et al. (2002) extended the regulated morphological operators by adjusting the weights in the structuring element.

Regulated morphological operations adopt a strictness parameter to control their sensitivity with respect to noise and small intrusions or protrusions on object boundary and thereby prevent excessive dilation or erosion. The *regulated dilation* of a set A by a structuring element set B with a strictness parameter s is defined by (Agam and Dinstein, 1999):

$$A\oplus^s B = \{x | \#(A \cap (\hat{B})_x) \geq s, s \in [1, \min(\#A, \#B)]\} \tag{4.77}$$

where the symbol "#" denotes the cardinality of a set.

The properties of regulated morphological dilation are introduced below. Their proofs can be referred in Agam and Dinstein (1999).

Property 4.1. The regulated dilation is decreasing with respect to the strictness s.

$$A\oplus^{s1} B \subseteq A\oplus^{s2} B \Longleftrightarrow s1 \geq s2 \tag{4.78}$$

Property 4.2. A regulated dilation generates a subset of a standard dilation.

$$A\oplus^s B \subseteq A \oplus B \tag{4.79}$$

Property 4.3. The regulated dilation is commutative, increasing with respect to the first and second arguments, and translation invariant.

$$A \oplus^s B = B \oplus^s A \tag{4.80}$$

$$A \subseteq B \Rightarrow A \oplus^s D \subseteq B \oplus^s D \tag{4.81}$$

$$B \subseteq C \Rightarrow A \oplus^s B \subseteq A \oplus^s C \tag{4.82}$$

$$(A)_x \oplus^s B = (A \oplus^s B)_x \tag{4.83}$$

$$A \oplus^s (B)_x = (A \oplus^s B)_x \tag{4.84}$$

Property 4.4. The regulated dilation of a union (intersection) of sets is bigger (smaller than) or equal to the union (intersection) of the regulated dilation of the individual sets.

$$(A \cup B) \oplus^s D \supseteq (A \oplus^s D) \cup (B \oplus^s D) \tag{4.85}$$

$$(A \cap B) \oplus^s D \subseteq (A \oplus^s D) \cap (B \oplus^s D) \tag{4.86}$$

The *regulated erosion* of a set A by a structuring element set B with a strictness parameter s is defined by Agam and Dinstein (1999)

$$A \ominus^s B = \{x | \#(A^c \cap (B)_x)\} < s, s \in [1, \#B]\} \tag{4.87}$$

The properties of regulated morphological erosion are introduced below. Their proofs can be referred in Agam and Dinstein (1999).

Property 4.5. The regulated erosion is increasing with respect to the strictness s.

$$A \ominus^{s1} B \subseteq A \ominus^{s2} B \Longleftrightarrow s1 \leq s2 \tag{4.88}$$

Property 4.6. A regulated erosion results in a superset of a standard dilation.

$$A \ominus^s B \supseteq A \ominus B \tag{4.89}$$

Property 4.7. The regulated erosion is increasing with respect to the first argument, decreasing with respect to the second argument, and translation invariant.

$$A \subseteq B \Rightarrow A \ominus^s D \subseteq B \ominus^s D \tag{4.90}$$

$$B \subseteq C \Rightarrow A \ominus^s B \supseteq A \ominus^s C \tag{4.91}$$

$$(A)_x \ominus^s (B) = (A \ominus^s B)_x \tag{4.92}$$

$$A \ominus^s (B)_x = (A \ominus^s B)_{-x} \tag{4.93}$$

Property 4.8. The regulated erosion of a union (intersection) of sets is bigger (smaller) or equal to the union (intersection) of the regulated dilation of the individual sets.

$$(A \cup B) \ominus^s D \supseteq (A \ominus^s D) \cup (B \ominus^s D) \tag{4.94}$$

$$(A \cap B) \ominus^s D \subseteq (A \ominus^s D) \cap (B \ominus^s D) \tag{4.95}$$

Property 4.9. The regulated dilation and erosion are dual in the same sense that exists for the standard dilation and erosion.

$$A \ominus^s B = (A^c \oplus^s \hat{B})^c \tag{4.96}$$

In order to maintain the same properties as standard morphological operators, the *regulated closing* of a set A by a structuring element set B with a strictness parameter s is defined by

$$A \bullet^s B = ((A \underline{\oplus}^s B) \ominus B) \cup A \tag{4.97}$$

where $A \underline{\oplus}^s B = (A \oplus^s B) \cup A$ is defined as the *extensive regulated dilation.* The *regulated opening* of a set A by a structuring element set B with a strictness parameter s is defined by

$$A \circ^s B = ((A \underline{\ominus}^s B) \oplus B) \cap A \tag{4.98}$$

where $A \underline{\ominus}^s B = (A \ominus^s B) \cap A$ is defined as the *antiextensive regulated erosion.*

By adjusting the strictness parameter, we can alleviate the noise sensitivity problem and small intrusions or protrusions on the object boundary. An application of using regulated morphological filters is for corner detection. The regulated morphological corner detector is described as follows (Shih et al., 2005):

Step 1: $A_1 = (A \oplus^s B) \ominus^s B$. Corner strength: $C_1 = |A - A_1|$.

Step 2: $A_2 = (A \ominus^s B) \oplus^s B$. Corner strength: $C_2 = |A - A_2|$.

Step 3: Corner detector $= C_1 \cup C_2$.

The original image is performed using a regulated dilation by a 5×5 circular structuring element with a strictness s, and then followed by a regulated erosion by the same structuring element with the same strictness. The corner strength C_1 is computed by finding the absolute value of the difference between the original and resulting images. This step is to extract concave corners. By reversing the order such that the regulated erosion is applied first and then followed by the regulated dilation, the convex corners can be extracted. Finally, both types of corners are combined. Figure 4.23 shows the corner detection for an airplane image using the regulated morphological corner detector at strictness $= 2$, 3, and 4.

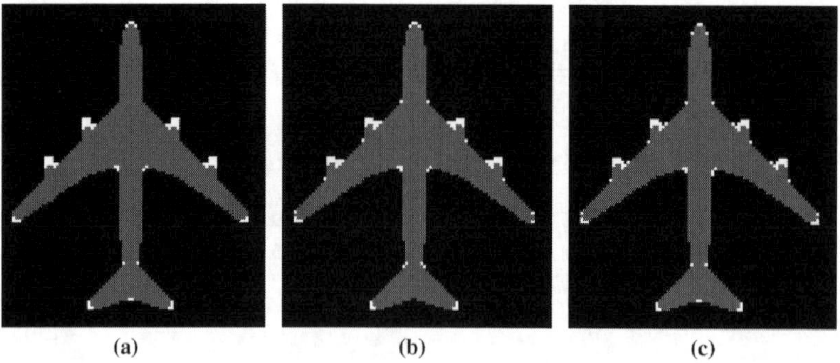

(a) (b) (c)

Figure 4.23 (a–c) Corner detection for an airplane image using the regulated morphological corner detector at strictness $= 2$, 3, and 4, respectively.

4.6.8 Fuzzy Morphological Filters

Fuzzy set theory has found a promising field of applications in the domain of digital image processing since fuzziness is an intrinsic property of image. The fuzzy logic has been developed in order to capture the uncertainties associated with human cognitive processes such as in thinking, reasoning, perception, and so on. Images are modeled as fuzzy subsets of the Euclidean plane or Cartesian grid, and the morphological operators are defined in terms of a fuzzy index function. Zadeh (1965) introduced fuzzy sets that contain elementary operations such as intersection, union, complementation, and inclusion. The fuzzy set theoretical operations are defined as follows. Let the characteristic function of a crisp set A be denoted as $\mu_A : U \rightarrow \{0, 1\}$, and it is defined as

$$\mu_A(x) = \begin{cases} 1 & \text{if } x \in A \\ 0 & \text{otherwise} \end{cases} \tag{4.99}$$

where A is any finite set. However, the object classes generally encountered in the real world are not so "precisely" or "crisply" defined. In most cases, several ambiguities arise in the determination of whether a particular element belongs to a set or not. A good example mentioned by Zadeh is a class of animals. This class clearly includes dogs, cats, tigers, and so on and excludes rocks, plants, houses, and so on. However, an ambiguity arises in the context of objects such as bacteria and starfish with respect to the class of animals.

The membership function of a fuzzy set A, denoted as $\mu_A : U \rightarrow [0, 1]$, is defined in such a way that $\mu_A(x)$ denotes the degree to which x belongs to A. Note that in the crisp set, the membership function has the value either 1 or 0, but in the fuzzy set, the member function has the value in the range of 0–1. The higher the value of $\mu_A(x)$, the more x belongs to A, and conversely, the smaller the value of $\mu_A(x)$, the less likelihood of x being in the set A. For further details, see Zadeh (1965, 1977).

The *union*, *intersection*, *difference*, and *complement* operations on crisp as well as fuzzy sets can be defined in terms of their characteristic/membership functions as

$$\mu_{A \cup B}(x) = \max[\mu_A(x), \mu_B(x)] \tag{4.100}$$

$$\mu_{A \cap B}(x) = \min[\mu_A(x), \mu_B(x)] \tag{4.101}$$

$$\mu_{A \backslash B}(x) = \min[\mu_A(x),\ 1 - \mu_B(x)] \tag{4.102}$$

$$\mu_A^c(x) = 1 - \mu_A(x) \tag{4.103}$$

The subset relation can be expressed as

$$A \subseteq B \Longleftrightarrow \mu_B = \max[\mu_A, \mu_B] \Longleftrightarrow \mu_A = \min[\mu_A, \mu_B] \tag{4.104}$$

The *support* of a set A, denoted as $S(A)$, is a crisp set of those elements of U that belong to A with some certainty:

$$S(A) = \{x | \mu_A(x) > 0\} \tag{4.105}$$

The *translation* of a set A by a vector $v \in U$, denoted by $\Im(A; v)$, is defined as

$$\mu_{\Im(A;v)}(x) = \mu_A(x - v) \tag{4.106}$$

The *reflection* of a set A, denoted by $\hat{A}$, is defined as

$$\mu_{\hat{A}}(x) = \mu_A(-x) \tag{4.107}$$

The *scalar addition* of a fuzzy set A and a constant α, denoted as $A \dagger \alpha$, is defined as

$$\mu_{A\dagger\alpha}(x) = \min(1, \max[0, \mu_A(x) + \alpha]) \tag{4.108}$$

Giles (1976) proposed fuzzy operations, *bold union*, $X\Delta Y$ of two sets X and Y as

$$\mu_{X\Delta Y}(z) = \min[1, \mu_X(z), \mu_Y(z)] \tag{4.109}$$

and *bold intersection*$X\nabla Y$ as

$$\mu_{X\nabla Y}(z) = \max[0, \mu_X(z), \mu_Y(z) - 1] \tag{4.110}$$

If X and Y are crisp sets, then $X\Delta Y \equiv X \cup Y$ and $X\nabla Y \equiv X \cap Y$.
An *index* function I: $2^U \times 2^U \rightarrow \{0, 1\}$ is defined as

$$I(A,B) = \begin{cases} 1 & \text{if } A \subseteq B \\ 0 & \text{otherwise} \end{cases} \tag{4.111}$$

The above equation can be rewritten so as to express the index function directly in terms of characteristic functions as

$$\begin{aligned} I(A,B) &= \inf_{x\in A}\mu_B(x) \\ &= \min[\inf_{x\in A}\mu_B(x), \inf_{x\notin A} 1] \\ &= \inf_{x\in U}\mu_{A^c\Delta B}(x) \end{aligned} \tag{4.112}$$

The last relation follows because for crisp sets the bold union has the following properties:

$$x \in A \Rightarrow \mu_{A^c\Delta B}(x) = \mu_B(x)$$

$$x \notin A \Rightarrow \mu_{A^c\Delta B}(x) = 1$$

The index function can be generalized, so that $I(A, B)$ gives the degree to which A is a subset of B. The formulation in equation (4.112) suffices for this purpose. The properties of morphological operations will be induced by the properties of the index function.

Consider any two fuzzy subsets $A, B \subset U$; index function $I(A, B)$ for different values of set B and set A essentially fixed and $A \neq \phi$. The properties (axioms) for the index function are as follows:

$$I(A,B) \in [0,1]$$

If A and B are crisp sets, then $I(A,B) \in \{0,1\}$.

$$A \subseteq B \Longleftrightarrow I(A,B) = 1$$

If $B \subseteq C$, then $I(A,B) \leq I(A,C)$.
If $B \subseteq C$, then $I(C,A) \leq I(B,A)$.

Invariant under translation, complement, and reflection:

(a) $I(A, B) = I(\Im(A; v), \Im(B; v))$.

(b) $I(A, B) = I(A^c, B^c)$.

(c) $I(A, B) = I(\hat{A}, \hat{B})$.

If B and C are subsets of A, then so is $B \cup C$. The converse also holds:

$(B \subseteq A) \wedge (C \subseteq A) \Longleftrightarrow (B \cup C) \subseteq A$; that is, $I(B \cup C, A) = \min[I(B, A), I(C, A)]$

Similarly, $(A \subseteq B) \wedge (A \subseteq C) \Longleftrightarrow A \subseteq (B \cup C)$; that is, $I(A, B \cup C) = \min[I(A, B), I(A, C)]$.

Gasteratos et al. (1998) proposed a framework that extends the concepts of soft mathematical morphology into fuzzy sets. They studied its compatibility with binary soft mathematical morphology as well as the algebraic properties of fuzzy soft operations. The fuzzy morphological operations, dilation, erosion, opening, and closing (Sinha and Dougherty, 1992, 1995) are defined in terms of index function as follows. The fuzzy erosion of a set A by another set B, denoted as $A \ominus B$, is defined by

$$\mu_{A \ominus B}(x) = I(\Im(B; x), A) \tag{4.113}$$

The fuzzy dilation of a set A by another set B, denoted as $A \oplus B$, is defined by

$$\begin{aligned} \mu_{A \oplus B}(x) &= \mu_{(A^c \ominus \hat{B})^c}(x) = 1 - \mu_{A^c \ominus \hat{B}}(x) \\ &= I^c(\Im(\hat{B}; x), A^c) = \sup_{z \in U} \max[0, \mu_{\Im(\hat{B}; x)}(z) + \mu_A(z) - 1] \end{aligned} \tag{4.114}$$

The fuzzy opening of a set A by another set B, denoted as $A \circ B$, is defined by

$$A \circ B = (A \ominus B) \oplus B \tag{4.115}$$

The fuzzy closing of a set A by another set B, denoted as $A \bullet B$, is defined by

$$A \bullet B = (A \oplus B) \ominus B \tag{4.116}$$

Sinha and Dougherty (1992) did not just introduce the modeling of grayscale images and the fuzzy morphological operations by simply replacing ordinary set theoretic operations by their fuzzy counterparts. Instead, they developed an intrinsically fuzzy approach to mathematical morphology. A new fitting paradigm is used by employing an index for set inclusion to measure the degree to which one image is beneath another image for erosion. Fuzzy morphological operations are defined based on this approach.

In binary morphology, erosion of image A by structuring element B is defined by

$$A \ominus B = \bigcap_{x \in -B} T(A; x) \tag{4.117}$$

where $T(A; x)$ is translation of A by a vector. Dilation of image A by structuring element B can be defined by

$$A \oplus B = \bigcup_{x \in B} T(A; x) \tag{4.118}$$

The definition of fuzzy morphology requires an index function for set inclusion:

$$\begin{aligned} I(A,B) &= \inf_{x\in A} \mu_B(x)A \\ &= \min[\inf_{x\in A} \mu_B(x), \inf_{x\notin A} 1] = \inf_{x\in U} \mu_{A^c\Delta B}(x) \end{aligned} \tag{4.119}$$

The *fuzzy complement* of the set inclusion indicator is then defined as

$$I^c(A,B) = 1-I(A,B) = \sup_{x\in U} \mu_{A\nabla B^c}(x) \tag{4.120}$$

Based on equations (4.119) and (4.120), the *fuzzy erosion* is defined as

$$\begin{aligned} \mu_{E(A,B)}(x) &= I(T(B;x),A) \\ &= \inf_{x\in U} \min[1, 1+\mu_A(x)-\mu_T(B;x)(x)] \end{aligned} \tag{4.121}$$

and the *fuzzy dilation* is defined as

$$\begin{aligned} \mu_{D(A,B)}(x) &= 1-\mu_{E(A,B)}(x) \\ &= \sup_{x\in U} \max[0,\ \mu_T(-B;x)(x)+\mu_A(x)-1] \end{aligned} \tag{4.122}$$

The property of being dual operations between fuzzy erosion and fuzzy dilation is preserved.

$$\mu_{D(A,B)}(x) = \mu_{E(A^c,-B)^c}(x) \tag{4.123}$$

The fuzzy morphological operations are illustrated in the following examples. The area of interest in a given image is represented by a rectangle, and the membership values within this rectangular region are specified in a matrix format. The membership values outside this region are assumed to be fixed as a constant. This value is specified as the superscript of the matrix, and the coordinates of the upper left element of the matrix are indicated by the subscript.

Example 4.17

$$A = \begin{bmatrix} 0.2 & 1.0 & 0.8 & 0.1 \\ 0.3 & 0.9 & 0.9 & 0.2 \\ 0.1 & 0.9 & 1.0 & 0.3 \end{bmatrix}^{0.0}_{(0,0)} \quad \text{and} \quad B = [0.8 \quad 0.9]^{0.0}_{(0,0)}$$

Since the structuring element fits entirely under the image when it translated by $(1,-1), A \ominus B\,(1,-1) = 1.0$. Similarly, $A \ominus B\,(1,-2) = 1.0$. For vector $(1,0)$, even though $\Im(B;1,0) \not\subset A$, the subset relationship almost holds:

$$A \cap \Im(B;1,0) = (\,0.8 \quad 0.8\,)_{(1,0)} \approx \Im(B;1,0)$$

Therefore, a relatively high value of $A \ominus B(1,0)$ is expected.

$$I[\Im(B;1,0),A] = \min\{\min[1, 1+1-0.8], \min[1, 1+0.8-0.9]\} = 0.9$$

Thus, $A \ominus B\,(1,0) = 0.9$. Proceeding along this way, the eroded image is obtained as

$$A \ominus B = \begin{bmatrix} 0.2 & 0.4 & 0.9 & 0.2 \\ 0.2 & 0.5 & 1.0 & 0.3 \\ 0.2 & 0.3 & 1.0 & 0.4 \end{bmatrix}^{0.1}_{(-1,0)}$$

Note that the coordinates of the upper left element are now $(-1, 0)$ instead of $(0, 0)$. By appropriately thresholding the image at a value between 0.5 and 0.9, we obtain

$$A \ominus B = \begin{bmatrix} 0.9 \\ 1.0 \\ 1.0 \end{bmatrix}^{0.0}_{(1,0)}$$

Example 4.18

$$A = \begin{bmatrix} 0.7 \\ 0.9 \\ 0.8 \end{bmatrix}^{0.0}_{(0,0)} \quad \text{and} \quad B = \begin{bmatrix} 0.8 & 0.9 \end{bmatrix}^{0.0}_{(0,0)}$$

We obtain the reflected set: $\hat{B} = \begin{bmatrix} 0.9 & 0.8 \end{bmatrix}^{0.0}_{(-1,0)}$.

$$I^{c}(\Im(\hat{B}; x), A^{c}) = \begin{cases} 0.5 & \text{if } x = (0,0) \\ 0.6 & \text{if } x = (1,0), (0,-2) \\ 0.7 & \text{if } x = (0,-1), (1,-2) \\ 0.8 & \text{if } x = (1,-1) \\ 0.0 & \text{otherwise} \end{cases}$$

Hence, the fuzzy dilation is

$$A \oplus B = \begin{bmatrix} 0.5 & 0.6 \\ 0.7 & 0.8 \\ 0.6 & 0.7 \end{bmatrix}^{0.0}_{(0,0)}$$

Example 4.19 Consider the image from Example 4.17.

$$P \equiv A \ominus B = \begin{bmatrix} 0.2 & 0.4 & 0.9 & 0.2 \\ 0.2 & 0.5 & 1.0 & 0.3 \\ 0.2 & 0.3 & 1.0 & 0.4 \end{bmatrix}^{0.1}_{(-1,0)}$$

Therefore,

$$P^c \ominus \hat{B} = \begin{bmatrix} 0.8 & 0.3 & 0.2 & 0.9 \\ 0.7 & 0.2 & 0.1 & 0.8 \\ 0.9 & 0.2 & 0.1 & 0.7 \end{bmatrix}_{(-1,0)}^{1.0}$$

Hence,

$$A \circ B = (P^c \ominus \hat{B})^c = \begin{bmatrix} 0.2 & 0.7 & 0.8 & 0.1 \\ 0.3 & 0.8 & 0.9 & 0.2 \\ 0.1 & 0.8 & 0.9 & 0.3 \end{bmatrix}_{(-1,0)}^{0.0}$$

By appropriately thresholding the image at a value between 0.3 and 0.7, we obtain

$$A \circ B = \begin{bmatrix} 0.7 & 0.8 \\ 0.8 & 0.9 \\ 0.8 & 0.9 \end{bmatrix}_{(0,0)}^{0.0}$$

Some applications of fuzzy morphological filters to image processing are given below. Grobert et al. (1996) applied fuzzy morphology to detect infected regions of a leaf. It was shown that the fuzzy morphological operation achieves a compromise between the deletion of isolated bright pixels and the connection of scattered regions of bright pixels. Wirth and Nikitenko (2005) applied fuzzy morphology to contrast enhancement. Yang et al. (2006) proposed an edge detection algorithm based on adaptive fuzzy morphological neural network. The gradient of the fuzzy morphology utilizes a set of structuring elements to detect the edge strength with a view to decrease the spurious edge and suppress the noise.

Chatzis and Pitas (2000) proposed the generalized fuzzy mathematical morphology (GFMM) based on a novel definition of the fuzzy inclusion indicator (FII). FII is a fuzzy set used as a measure of the inclusion of a fuzzy set into another. The GFMM provides a very powerful and flexible tool for morphological operations. It can be applied on the skeletonization and shape decomposition of two-dimensional and three-dimensional objects. Strauss and Comby (2007) presented variable structuring element-based fuzzy morphological operations for single viewpoint omnidirectional images.

REFERENCES

Agam, G. and Dinstein, I., "Regulated morphological operations," *Pattern Recognit.*, vol. 32, no. 6, pp. 947–971, June 1999.

Bloomberg, D. and Maragos, P., "Generalized hit-miss operations," in *Image Algebra and Morphological Image Processing 1998, Proceedings* (P. Gader, ed.), vol. 1350, SPIE, San Diego, CA, pp. 116–128, 1990.

Blum, H., "A transformation for extracting new descriptors of shape," *Models for the Perception of Speech and Visual Forms* in (W. Wathen-Dunn,ed.), Proceeding of Meeting, Nov. 1964, MIT Press, Cambridge, MA, pp. 362–380, 1967.

Chatzis, V. and Pitas, I., "A generalized fuzzy mathematical morphology and its application in robust 2-D and 3-D object representation," *IEEE Trans. Image Process.*, vol. 9, no. 10, pp. 1798–1810, Oct. 2000.

Danielsson, P. E., "A new shape factor," *Comput. Graph. Image Process.*, vol. 7, no. 2, pp. 292–299, Apr. 1978.

David, H., *Order Statistics*, Wiley, New York, 1981.

Destival, I., "Mathematical morphology applied to remote sensing," *Acta Astronaut.*, vol. 13, no. 6–7, pp. 371–385, June–July, 1986.

Dineen, G. P., "Programming pattern recognition," Proceedings of the Western Joint Computer Conference, Los Angeles, CA, pp. 94–100, Mar. 1955.

Dougherty, E. R., *Mathematical Morphology in Image Processing*, CRC Press, 1993.

Duff, M. J. B., "Parallel processors for digital image processing," in *Advances in Digital Image Processing* (P. StuckiEd.), Plenum, New York, pp. 265–276, 1979.

Feehs, R. J. and Arce, G. R., "Multidimensional morphological edge detection," *Proc. SPIE Visual Commun. Image Process. II*, vol. 845, pp. 285–292, 1987.

Gasteratos, A., Andreadis, I., and Tsalides, P., A new hardware structure for implementation of soft morphological filters, *Computer Analysis of Images and Patterns, Lecture Notes in Computer Science*, Springer, vol. 1296, pp. 488–494, 1997.

Gasteratos, A., Andreadis, I., and Tsalides, P., "Fuzzy soft mathematical morphology," *IEE Proc. Vis. Image Signal Process.*, vol. 145, no. 1, pp. 41–49, Feb. 1998.

Gerritsen, F. A. and Aardema, L. G., "Design and use of DIP-1: a fast flexible and dynamically microprogrammable image processor," *Pattern Recognit.*, vol. 14, no. 1–6, pp. 319–330, 1981.

Giardina, C. and Doughherty, E., *Morphological Methods in Image and Signal Processing*, Prentice-Hall, Englewood Cliffs, NJ, 1988.

Giles, R., "Lukasiewicz logic and fuzzy theory," *Int. J. Man Mach. Stud.*, vol. 8, no. 3, pp. 313–327, 1976.

Golay, M. J. E., "Hexagonal parallel pattern transformations," *IEEE Trans. Comput.*, vol. 18, no. 8, pp. 733–740, Aug. 1969.

Gonzalez, R. and Woods, R., *Digital Image Processing*, third edition, Prentice Hall, 2007.

Goutsias, J. and Heijmans, H. J., *Mathematical Morphology*, IOS Press, 2000.

Graham, D. and Norgren, P. E., "The diff3 analyzer: a parallel/serial Golay image processor," in *Real-Time Medical Image Processing* (D. Onoe, K. Preston, Jr. and A. Rosenfeld, eds.), Plenum, London, pp. 163–182, 1980.

Grobert, S., Koppen, M., and Nickolay, B. A., "New approach to fuzzy morphology based on fuzzy integral and its application in image processing," Proceedings of the International Conference on Pattern Recognition, vol. 2, pp. 625–630, Aug. 1996.

Hamid, M. S., Harvey, N. R., and Marshall, S., "Genetic algorithm optimization of multidimensional grayscale soft morphological filters with applications in film archive restoration," *IEEE Trans. Circuits Syst. Video Technol.*, vol. 13, no. 5, pp. 406–416, May 2003.

Haralick, R. M., Sternberg, S. R., and Zhuang, X., "Image analysis using mathematical morphology," *IEEE Trans. Pattern Anal. Mach. Intell.*, vol. 9, no. 7, pp. 532–550, July 1987.

Haralick, R., Zhuang, X., Lin, C., and Lee, J., "The digital morphological sampling theorem," *IEEE Trans. Acoust. Speech Signal Process.*, vol. 37, no. 12, pp. 2067–2090, Dec. 1989.

Jang, B. and Chin, R. T., "Analysis of thinning algorithms using mathematical morphology," *IEEE Trans. Pattern. Anal. Mach. Intell.*, vol. 12, no. 6, pp. 541–551, June 1990.

Ji, L. and Piper, J., "Fast homotopy—preserving skeletons using mathematical morphology," *IEEE Trans. Pattern Anal. Mach. Intell.*, vol. 14, no. 6, pp. 653–664, June 1992.

Khosravi, M. and Schafer, R. W., "Template matching based on a grayscale hit-or-miss transform," *IEEE Trans. Image Process.*, vol. 5, no. 6, pp. 1060–1066, June 1996.

Kirsch, R. A., "Experiments in processing life motion with a digital computer," Proceedings of the Eastern Joint Computer Conference, pp. 221–229, 1957.

Klein, J. C. and Serra, J., "The texture analyzer," *J. Microsc.*, vol. 95, no. 2, pp. 349–356, Apr. 1977.

Ko, S.-J., Morales, A., and Lee, K.-H., "Fast recursive algorithms for morphological operators based on the basis matrix representation," *IEEE Trans. Image Process.*, vol. 5, no. 6, pp. 1073–1077, June 1996.

Koskinen, L., Astola, J. T., and Neuvo, Y. A., "Soft morphological filters," in *Image Algebra and Morphological Image Processing II, Proceedings*, (P. D. Gader and E. R. Dougherty, eds.), vol. 1568, SPIE, pp. 262–270, July 1991.

Kruse, B., Design and implementation of a picture processor, Science and Technology Dissertations, no. 13, University of Linkoeping, Linkoeping, Sweden 1977.

Kuosmanen, P. and Astola, J., "Soft morphological filtering," *J. Math. Imaging Vis.*, vol. 5, no. 3, pp. 231–262, Sep. 1995.

Lee, J. S., Haralick, R. M., and Shapiro, L. G., "Morphologic edge detection," *IEEE Trans. Robot. Autom.*, vol. 3, no. 4, pp. 142–156, Apr. 1987.

Lougheed, R. M. and McCubbrey, D. L., "The cytocomputer: a practical pipelined image processor," Proceedings of the IEEE Annual Symposium on Computer Architecture, La Baule, France, pp. 271–278, 1980.

Maragos, P. A. and Schafer, R. W., "Morphological skeleton representation and coding of binary images," *IEEE Trans. Acoust. Speech Signal Process.*, vol. 34, no. 5, pp. 1228–1244, Oct. 1986.

Maragos, P., and Schafer, R. "Morphological filters. Part I: their set-theoretic analysis and relations to linear shift-invariant filters," *IEEE Trans. Acoust. Speech Signal Process.*, vol. 35, no. 8, pp. 1153–1169, Aug. 1987a.

Maragos, P. and Schafer, R., "Morphological filters. Part II: their relations to median, order-statistics, and stack filters," *IEEE Trans. Acoust. Speech Signal Process.*, vol. 35, no. 8, pp. 1170–1184, Aug. 1987b.

Matheron, G., *Random Sets and Integral Geometry*, J. Wiley & Sons, Inc., 1975.

Meyer, F., "Skeletons in digital spaces," in *Image Analysis and Mathematical Morphology* (J. Serra, ed.), vol. 2, Academic Press, New York, pp. 257–296, 1988.

Minkowski, H., "Volumen und oberflache," *Math. Ann.*, vol. 57, pp. 447–495, 1903.

Moore, G. A., "Automatic sensing and computer process for the quantitative analysis of micrographs and equivalued subjects," *Picturial Pattern Recognition*, Thompson Book Co., pp. 275–326, 1968.

Morales, A., Acharya, R., and Ko, S.-J., "Morphological pyramids with alternating sequential filters," *IEEE Trans. Image Process.* vol. 4, no. 7, pp. 965–977, July 1995.

Moran, C. J., "A morphological transformation for sharpening edges of features before segmentation," *Comput. Vis. Graph. Image Process.*, vol. 49, no. 1, pp. 85–94, Jan. 1990.

Nadadur, D. and Haralick, R. M., "Recursive binary dilation and erosion using digital line structuring elements in arbitrary orientations," *IEEE Trans. Image Process.*, vol. 9, no. 5, pp. 749–759, May 2000.

Naegel, B., Passat, N., and Ronse, C., "Grey-level hit-or-miss transforms. Part I: unified theory," *Pattern Recognit.*, vol. 40, no. 2, pp. 635–647, Feb. 2007a.

Naegel, B., Passat, N., and Ronse, C., "Grey-level hit-or-miss transforms. Part II: application to angiographic image processing," *Pattern Recognit.*, vol. 40, no. 2, pp. 648–658, Feb. 2007b.

Pei, S., Lai, C., and Shih, F. Y., "An efficient class of alternating sequential filters in morphology," *Graph. Models Image Process.*, vol. 59, no. 2, pp. 109–116, Mar. 1997.

Pei, S., Lai, C., and Shih, F. Y., "Recursive order-statistic soft morphological filters," *IEE Proc. Vis. Image Signal Process.*, vol. 145, no. 5, pp. 333–342, Oct. 1998.

Preston, K., Jr., "Machine techniques for automatic identification of binucleate lymphocyte," Proceedings of the International Conference on Medical Electronics, Washington, DC, 1961.

Preteux, F., Laval-Jeantet, A. M., Roger, B., and Laval-Jeantet, M. H., "New prospects in C.T. image processing via mathematical morphology," *Eur. J. Radiol.*, vol. 5, no. 4, pp. 313–317, Nov. 1985.

Pu, C. C. and Shih, F. Y., "Threshold decomposition of gray-scale soft morphology into binary soft morphology," *Graph. Model Image Process.*, vol. 57, no. 6, pp. 522–526, Nov. 1995.

Raducanu, B. and Grana, M., "A grayscale hit-or-miss transform based on level sets," Proceedings of the IEEE International Conference on Image Processing, vol. 2, Vancouver, BC, Canada, pp. 931–933, Sep. 2000.

Schonfeld, D. and Goutsias, J., "Optimal morphological pattern restoration from noisy binary image," *IEEE Trans. Pattern Anal. Mach. Intell.*, vol. 13, no. 1, pp. 14–29, Jan. 1991.

Serra, J., *Image Analysis and Mathematical Morphology*, Academic Press, New York, 1982.

Serra, J., *Image Analysis and Mathematical Morphology: Theoretical Advances*, vol. 2, Academic Press, New York, 1988.

Shih, F. Y., *Image Processing and Mathematical Morphology: Fundamentals and Applications*, CRC Press, 2009.

Shih, F. Y. and Gaddipati, V., "General sweep mathematical morphology," *Pattern Recognit.*, vol. 36, no. 7, pp. 1489–1500, July 2003.

Shih, F. Y. and Gaddipati, V., "Geometric modeling and representation based on sweep mathematical morphology," *Inform. Sci.*, vol. 171, no. 3, pp. 213–231, Mar. 2005.

Shih, F. Y. and Mitchell, O. R., "Threshold decomposition of grayscale morphology into binary morphology," *IEEE Trans. Pattern Anal. Mach. Intell.*, vol. 11, no. 1, pp. 31–42, Jan. 1989.

Shih, F. Y. and Mitchell, O. R., "Decomposition of grayscale morphological structuring elements," *Pattern Recognit.*, vol. 24, no. 3, pp. 195–203, Mar. 1991.

Shih, F. Y. and Mitchell, O. R., "A mathematical morphology approach to Euclidean distance transformation," *IEEE Trans. Image Process.*, vol. 1, no. 2, pp. 197–204, Apr. 1992.

Shih, F. Y. and Pu, C. C., "A skeletonization algorithm by maxima tracking on Euclidean distance transform," *Pattern Recognit.*, vol. 28, no. 3, pp. 331–341, Mar. 1995a.

Shih, F. Y. and Pu, C. C., "Analysis of the properties of soft morphological filtering using threshold decomposition," *IEEE Trans. Signal Process.*, vol. 43, no. 2, pp. 539–544, Feb. 1995b.

Shih, F. Y. and Puttagunta, P., "Recursive soft morphological filters," *IEEE Trans. Image Process.*, vol. 4, no. 7, pp. 1027–1032, July 1995.

Shih, F. Y. and Wu, Y., "The efficient algorithms for achieving Euclidean distance transformation," *IEEE Trans. Image Process.*, vol. 13, no. 8, pp. 1078–1091, Aug. 2004.

Shih, F. Y., King, C. T., and Pu, C. C., "Pipeline architectures for recursive morphological operations," *IEEE Trans. Image Process.*, vol. 4, no. 1, pp. 11–18, Jan. 1995.

Shih, F. Y., Chuang, C., and Gaddipati, V., "A modified regulated morphological corner detector," *Pattern Recognit. Lett.*, vol. 26, no. 7, pp. 931–937, June 2005.

Sinha, D. and Dougherty, E. R., "Fuzzy mathematical morphology," *J. Vis. Commun. Image Represent.*, vol. 3, no. 3, pp. 286–302, Sep. 1992.

Sinha, D. and Dougherty, E. R., "A general axiomatic theory of intrinsically fuzzy mathematical morphologies," *IEEE Trans. Fuzzy Syst.*, vol. 3, no. 4, pp. 389–403, Nov. 1995.

Smith, S. W., *The Scientist and Engineer's Guide to Digital Signal Processing*, California Technical Publishing, 1997.

Soille, P., "On morphological operators based on rank filters," *Pattern Recognit.*, vol. 35, no. 2, pp. 527–535, Feb. 2002.

Sternberg, S. R., "Language and architecture for parallel image processing," in *Proceedings of the International Conference on Pattern Recognition in Practice* (E. S., Gelsema and L. N., Kanal, eds.), North-Holland, Amsterdam, p. 35, 1980.

Sternberg, S. R., "Grayscale morphology," *Comput. Vis. Graph. Image Process.*, vol. 35, no. 3, pp. 333–355, Sep. 1986.

Stevenson, R. L. and Arce, G. R., "Morphological filters: statistics and further syntactic properties," *IEEE Trans. Circuits Syst.*, vol. 34, no. 11, pp. 1292–1305, Nov. 1987.

Strauss, O. and Comby, F., "Variable structuring element based fuzzy morphological operations for single viewpoint omnidirectional images," *Pattern Recognit.*, vol. 40, no. 12, pp. 3578–3596, Dec. 2007.

Tian, X.-H., Li, Q.-H., and Yan, S.-W., "Regulated morphological operations with weighted structuring element," Proceedings of the International Conference on Machine Learning and Cybernetics, Beijing, China, pp. 768–771, Nov. 2002.

Wirth, M. A. and Nikitenko, D., "Applications of fuzzy morphology to contrast enhancement," Proceedings of the Annual Meeting of the North American Fuzzy Information Processing Society, Ann Arbor, MI, pp. 355–360, June 2005.

Yang, G.-Q., Guo, Y.-Y., and Jiang, L.-H., "Edge detection based on adaptive fuzzy morphological neural network," Proceedings of the International Conference on Machine Learning and Cybernetics, Heibei, China, pp. 3725–2728, Aug. 2006.

Zadeh, L. A., "Fuzzy sets," *Inform. Control*, vol. 8, no. 3, pp. 338–353, June 1965.

Zadeh, L. A., "Theory of fuzzy sets," in *Encyclopedia of Computer Science and Technology* (J. Belzer, A. Holzman, and A. Kent, eds.), Dekker, New cYork, 1977.

CHAPTER 5

IMAGE SEGMENTATION

Image segmentation, a process of pixel classification, aims to extract or segment objects or regions from the background. It is a critical preprocessing step to the success of image recognition (Pachowicz, 1994), image compression (Belloulata and Konrad, 2002), image visualization (Hartmann and Galloway, 2000), and image retrieval (Chen and Wang, 2002). Pal and Pal (1993) provided a review on various segmentation techniques. It should be noted that there is no single standard approach to segmentation. Many different types of scene parts can serve as the segments on which descriptions are based, and there are many different ways in which one can attempt to extract these parts from the image. Selection of an appropriate segmentation technique depends on the type of images and applications.

The level of segmentation or subdivision relies on the problem domain being dealt with. For example, in the optical character recognition (OCR), the text is separated from the document image, and further partitioned into columns, lines, words, and connected components. In building character subimages, one is often confronted with touching or broken characters that occur in degraded documents (such as fax, scan, photocopy, etc.). It is still challenging to develop techniques for properly segmenting words into their characters.

There are primarily four types of segmentation techniques: thresholding, boundary-based, region-based, and hybrid techniques. Thresholding is based on the assumption that clusters in the histogram correspond to either background or objects of interest that can be extracted by separating these histogram clusters. In addition to thresholding, many image segmentation algorithms are based on two basic properties of the pixel intensities in relation to their local neighborhood: discontinuity and similarity. Methods based on pixel discontinuity are called boundary-based or edge extraction methods, whereas methods based on pixel similarity are called region-based methods. Boundary-based methods assume that the pixel properties, such as intensity, color, and texture (Khotanzad and Chen, 1989; Panjwani and Healey, 1995), should change abruptly between different regions. Region-based methods assume that neighboring pixels within the same region should have similar values (e.g., intensity, color, and texture).

It is well known that such segmentation techniques—based on boundary or region information alone—often fail to produce accurate segmentation results. Hence, there has been a tendency toward hybrid segmentation algorithms that take advantage of the complementary nature of such information. Hybrid methods combine boundary detection and region growing together to achieve better segmentation (Pavlidis and Liow, 1990; Haris et al., 1998; Fan et al., 2001). Note that both results should achieve the foreground and background segmentation coherently.

The evaluation of segmentation algorithms has been mostly subjective; therefore, people usually judge an algorithm's effectiveness based on intuition and the results from several segmented images. Cardoso and Corte-Real (2005) proposed a generic framework for segmentation evaluation using symmetric and asymmetric distance metric alternatives to meet the specificities of a wide class of applications. Unnikrishnan et al. (2007) demonstrated how a proposed measure of similarity, the normalized probabilistic rand (NPR) index, can be used to perform a quantitative comparison between image segmentation algorithms using a hand-labeled set of ground truth segmentations.

In this chapter, we discuss a number of image segmentation techniques, including thresholding, component labeling, locating object contours by the snake model, edge detection, linking edges by adaptive mathematical morphology, automatic seeded region growing (SRG), and top–down region dividing (TDRD). Other variations of image segmentation algorithms can be referred to as hidden Markov model (Pyun et al., 2007; Wu and Chung, 2007; Rivera et al., 2007), Bayesian methods (Wong and Chung, 2005), genetic algorithm and artificial neural network (Awad et al., 2007), and multicue partial differential equation (Sofou and Maragos, 2008).

5.1 THRESHOLDING

Thresholding provides an easy and convenient way to perform image segmentation based on different intensities or colors in the foreground and background regions of an image (Saha and Udupa, 2001). The automatic selection of optimum thresholds has remained a challenge over decades. Not all images can be segmented successfully into foreground and background using simple thresholding. Its validity relies on the distribution of the intensity histogram. If the intensity distribution of foreground objects is quite distinct from the intensity distribution of background, it will be clear to apply thresholding for image segmentation. In this case, we expect to see distinct peaks in the histogram corresponding to foreground objects, so that threshold values can be picked to isolate those peaks accordingly. If such a peak does not exist, it is unlikely that simple thresholding can achieve a good segmentation.

There are several methods of choosing the threshold value λ. For example, the universal thresholding by Donoho et al. (1996) sets

$$\lambda = \frac{\sigma\sqrt{2\log n}}{\sqrt{n}} \tag{5.1}$$

where σ is the standard deviation of the wavelet coefficients and n is the total size of samples. Other possibility is quantile thresholding, where λ is statistically set to replace a percentage of the coefficients with the smallest magnitude to zero. Another adaptive method of automatically choosing the best threshold λ consists of four steps: (1) choose an initial estimate λ; (2) calculate the two mean values μ_1 and μ_2 within the two groups of pixels after thresholding at λ; (3) calculate the new threshold value $\lambda = (1/2)(\mu_1 + \mu_2)$; and (4) if the new threshold value has a little change (i.e., smaller than a predefined constant), then the threshold selection is done; otherwise, go back to step 2.

Example 5.1 Apply the adaptive threshold selection method on the following image.

1	1	1	1	0	0	1	0
1	1	0	1	1	1	1	1
1	2	8	8	8	9	1	0
0	1	8	9	9	9	1	0
0	1	8	8	9	8	1	0
1	1	8	9	9	8	1	1
0	1	1	1	1	1	0	0
1	0	0	0	0	0	0	0

Answer: We first compute the average gray level of the image, which is 2.5781, as the initial threshold λ_0. We segment the image by λ_0 into two groups of pixels and calculate the mean values μ_1 and μ_2. The new threshold value $\lambda_1 = 0.5(\mu_1 + \mu_2) = 4.5312$. Then the image is segmented by λ_1 into two groups of pixels and the new mean values μ_1 and μ_2 are calculated. The new threshold value $\lambda_2 = 0.5(\mu_1 + \mu_2) = 4.5312$. Since $\lambda_1 = \lambda_2$, the best threshold value is 4.5312. Note that if a random number is chosen as the initial threshold, it would take more number of iterations to converge to the best threshold value.

An image can be regarded as a fuzzy subset of a plane. A fuzzy entropy measuring the blur in an image is a functional that increases when the sharpness of its argument image decreases. Zenzo et al. (1998) generalized and extended the relation "sharper than" between fuzzy sets in view of implementing the properties of a relation "sharper than" between images. They showed that there are infinitely many implementations of this relation into an ordering between fuzzy sets (equivalently, images). Relying upon these orderings, classes of fuzzy entropies are constructed that are useful for image thresholding by cost minimization.

Saha and Udupa (2001) took advantage of both intensity-based class uncertainty (a histogram-based property) and region homogeneity (an image morphology-based property) to achieve optimal thresholding. A scale-based formulation is used for region homogeneity computation. At any threshold, intensity-based class uncertainty is computed by fitting a Gaussian to the intensity distribution of each of the two regions segmented at that threshold. The theory of the optimum thresholding method is based on the postulate that objects manifest themselves with fuzzy boundaries in any digital image acquired by an imaging device. The optimal threshold value is selected at which pixels having high class uncertainty accumulate mostly around object boundaries.

Tao et al. (2008) used a normalized graph-cut measure as thresholding principle to distinguish an object from the background. The weight matrices used in evaluating the graph cuts are based on the gray levels of the image, rather than the commonly used image pixels. For most images, the number of gray levels is much smaller than the number of pixels. Therefore, their algorithm requires much smaller storage space and lower computational complexity than other image segmentation algorithms based on graph cuts.

5.2 OBJECT (COMPONENT) LABELING

It is very possible to have more than one object in a scene. All the objects must be individually extracted for the purpose of establishing the object model base. The object-labeling technique is used, so that the array representation of these objects is a multivalued picture, in which the points of each component all have a unique nonzero label and the points of background are all zeros. This technique only requires two raster scans. To label 4-connected components, only the upper and left neighbors are checked. If the 8-connectedness is used, the upper two diagonal neighbors are also included.

Athanasiadis et al. (2007) presented a framework for simultaneous image segmentation and object labeling leading to automatic image annotation. Xu et al. (2003) proposed object segmentation and labeling by learning from examples. They used color and two-dimensional shape matching to automatically group those low-level segments into objects based on their similarity to a set of example object templates presented by the user. A hierarchical content tree data structure is used for each database image to store matching combinations of low-level regions as objects. The learning phase refers to labeling of combinations of low-level regions that have resulted in successful color and/or 2D shape matches with the example templates. Moga and Gabbouj (1997) developed parallel image component labeling using watershed transformation.

Let the value of object points be "1" and of background points be "0". Assume that the 8-connectedness is adopted. If all the four previously scanned neighbors of the point *P* are zeros, then *P* is assigned a new label. If one of the four neighbors is 1, then *P* gets the same label as that neighbor. If two or more of them are 1's, then *P* gets any one of their labels, and the equivalence table is established by marking the different labels together for later adjustment. Equivalence processing consists in merging the equivalent pair into the same class; that is, a unique label is assigned to each class. Finally, a second scan is performed to replace each label by the representative of its class. Each component has now been uniquely labeled. After these processes, each individual model object can be fetched by its label.

Example 5.2 Perform object labeling on the following figure using 4-connectedness for foreground.

```
            1  1  1        1  1     1  1
      1  1  1     1  1  1  1              1
1  1        1  1  1     1  1  1  1  1
```

Answer: The two-step procedure is conducted as follows:

```
                   a  a  a        b  b     c  c
(1)         d  d   a     a  a  a  a           c  (Mark d = a, b = a)
      e  e         a  a  a     a  a  a  a  a

                   a  a  a        a  a     c  c
(2)         a  a   a     a  a  a  a           c
      e  e         a  a  a     a  a  a  a  a
```

He et al. (2008) presented an efficient run-based two-scan algorithm for labeling connected components in a binary image. Unlike the label equivalence-based algorithm using label equivalences between provisional labels, their algorithm resolves label equivalences between provisional label sets. At any time, all provisional labels that are assigned to a connected component are combined in a set, and the smallest label is used as the representative label. The corresponding relation of a provisional label and its representative label is recorded in a table. Whenever different connected components are found to be connected, all provisional label sets concerned with these connected components are merged together, and the smallest provisional label is taken as the representative label. When the first scan is finished, all the provisional labels that were assigned to each connected component in the given image will have a unique representative label. During the second scan, each provisional label is replaced by its representative label.

5.3 LOCATING OBJECT CONTOURS BY THE SNAKE MODEL

In the original snake formulation of Kass et al. (1987), the best snake position was defined as the solution of a variational problem requiring the minimization of the sum of internal and external energies integrated along the length of the snake. The corresponding Euler equations, which give the necessary conditions for this minimizer, comprise a force balance equation. The snake model has provided a number of applications in object segmentation, stereo matching, motion tracking, and so on. In image processing, the snake model defines a snake as an energy-minimizing spline guided by external constraint forces and influenced by such forces that pull it toward image features such as lines and edges. It is a kind of the active contour model in the way that it locks on nearby edges, localizing them accurately.

There are two parts in the energy function of the snake model. The first part reflects geometric properties of the contour, and the second part utilizes the external force field to drive the snake. The first part imposes a piecewise smoothness constraint, and the second part is responsible for putting the snake near the local minimum of energy. The traditional snakes suffer from a great disadvantage that when an object resides in a complex background, the strong edges may not be the object edges of interest. Therefore, researchers have proposed statistical and variational methods to enrich the energy function and extend its flexibility. They calculated the difference between the target object and the background using statistical analysis, but these limitations come from the a priori knowledge requirements, such as independent probability models and template models. Unfortunately, such a priori knowledge is usually unavailable unless the captured images are under very constrained settings. A problem with the snake model is that a user needs to place the initial snake points sufficiently close to the feature of interest.

Li et al. (2005) described a variational formulation for geometric active contours that forces the level set function to be close to a signed distance function. They used the level set function to allow flexible initialization. Shen and Davatzikos (2000) proposed an adaptive-focus deformable model using statistical and geometric information. Michailovich et al. (2007) proposed a generalization of the segmentation methods, in which active contours maximize the difference between a finite number of empirical moments. Melonakos et al. (2008) proposed an image segmentation technique based on augmenting the conformal (or geodesic) active contour framework with directional information.

Most snake models do not discern between positive and negative step edges. Cheng and Foo (2006) proposed dynamic directional gradient vector flow to use this information for better performance. Recently, the proposed Sobolev active contours (Sundaramoorthi et al., 2007) introduced a new paradigm for minimizing energies defined on curves by changing the traditional cost of perturbing a curve and thereby redefining gradients associated with these energies. Sobolev active contours evolve more globally and are less attracted to certain intermediate local minima than traditional active contours, and they are based on a well-structured Riemannian metric, which is important for shape analysis and shape priors. Sundaramoorthi et al. (2008) analyzed Sobolev active contours using scale-space analysis to understand their evolution across different scales. Chen et al. (2006) presented an integrated approach to extend the geodesic and edgeless active contour approaches for texture segmentation. Xie and Mirmehdi (2008) proposed an active contour model using an external force field that is based on magnetostatics and hypothesized magnetic interactions between the active contour and object boundaries.

5.3.1 The Traditional Snake Model

A snake is a controlled continuity spline that moves and localizes onto a specified contour under the influence of the objective function. Let a snake be a parametric curve: $v(s) = [x(s), y(s)]$, where parameter $s \in [0, 1]$. It moves around the image spatial domain to minimize the objective energy function as defined by

$$E_{\text{snake}}(v) = \sum_{i=1}^{n} [\alpha \times E_{\text{cont}}(v_i) + \beta \times E_{\text{curv}}(v_i) + \gamma \times E_{\text{image}}(v_i)] \tag{5.2}$$

where α, β, and γ are weighting coefficients that control the snake's tension, rigidity, and attraction, respectively. The first and second terms are the first- and second-order continuity constraints, respectively. The third term measures the edge strength (i.e., the image force).

The continuity force E_{cont} encouraging even spacing of points can be calculated as

$$E_{\text{cont}}[v_i] = \frac{|\bar{d}-|v_i-v_{i-1}||}{\max_j\{|\bar{d}-|v_i(j)-v_{i-1}||\}} \tag{5.3}$$

where $\{v_i(j)|j = 1, 2, \ldots, m\}$ denotes the snake point v_i's m neighbors and $\bar{d}$ denotes the average length of all the pairs of adjacent points on the snake contour as given by

$$\bar{d} = \frac{\sum_{i=1}^{n} |v_i - v_{i-1}|}{n} \tag{5.4}$$

where $v_0 = v_n$. This term tends to keep the distances between each pair of adjacent vertices equal.

The energy of the second-order continuity E_{curv} is represented by

$$E_{\text{curv}}[v_i] = \frac{|v_{i-1}-2v_i+v_{i+1}|}{\max\{|v_{i-1}-2v_i+v_{i+1}|\}} \tag{5.5}$$

The numerator can be rearranged as

$$v_{i-1}-2v_i+v_{i+1} = (v_{i+1}-v_i)-(v_i-v_{i-1}) \tag{5.6}$$

If the ith vertex is pushed toward the midpoint of two adjacent vertices, E_{curv} is minimized; that is, the shape of the contour will remain C^2 continuity.

The third term, image energy E_{image}, is derived from the image so that it takes on smaller values at the features of interest, such as boundaries. It considers the gradient (denoted as grad) magnitude, leading the active contour toward step edges. It is normalized to measure the relative magnitude as

$$E_{\text{image}}[v_i] = \frac{\min\{|\text{grad}|\}-|\text{grad}_{v_i}|}{\max\{|\text{grad}|\}-\min\{|\text{grad}|\}} \tag{5.7}$$

where min and max denote the minimum and maximum gradients in the v_i's local m-neighborhood, respectively. Note that because the numerator in equation (5.7) is always negative, E_{image} can be minimized for locating the largest gradient, which is the edge. In general, the traditional snake model can locate object contours in a simple background. If the background becomes complex, it will fail since the complex background will generate noisy edges to compete with the object edges for attracting the snake.

5.3.2 The Improved Snake Model

The following two issues are observed when the snake model fails to locate object contours in complex backgrounds. One is the gray-level sensitivity; that is, the more abrupt change the gray levels have (e.g., noises), the larger impact on the energy function the snake makes. The other is that the snake mistakenly locates the edges belonging to the background details due to their closeness to the snake point. Figure 5.1 shows a disk object in a complex background. If the snake points are initialized outside the disk, the snake cannot locate the disk contour accurately due to the disturbances from the background grids. The idea is to push the mean intensity of the polygon enclosed by the snake contour to be as close as possible to the mean intensity of the target object. The smaller the intensity difference between the polygon and the object, the closer the snake approaches the object contour. Therefore, a new energy term, called the *regional similarity energy* (RSE), is established for calculating the gray-level differences to be added into the overall energy (Shih and Zhang, 2004).

Supposedly, the intensities of the target disk are homogeneous while its background is complex. As illustrated in Figure 5.2, let the snake contour in the current iteration be C, which contains a point i. An area of neighborhood is defined and the neighbors of i are searched for the new point in the next iteration, say i', which produces the minimum area compared to the other neighbors. Such a contour C' is regarded as the base area and the mean intensity of its enclosed area is regarded as the base mean intensity M_i.

Let p denote a point in the neighborhood of i. The RSE of point p is calculated as

$$E_{\mathrm{RSE}}(i,p) = |m(p) - M_i| \tag{5.8}$$

where $m(p)$ denotes the mean intensity in the neighborhood of point p. Therefore, the RSE of the snake contour is

$$E_{\mathrm{RSE}}[v(s)] = \sum_i E_{\mathrm{RSE}}(i,p) \tag{5.9}$$

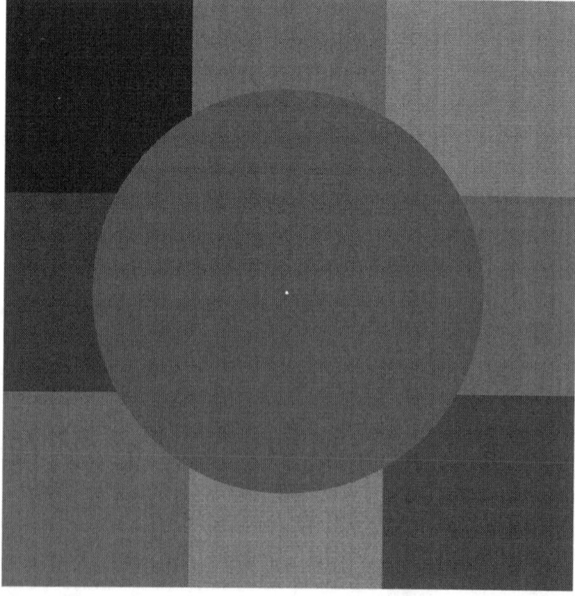

Figure 5.1 A disk object resides in a complex background.

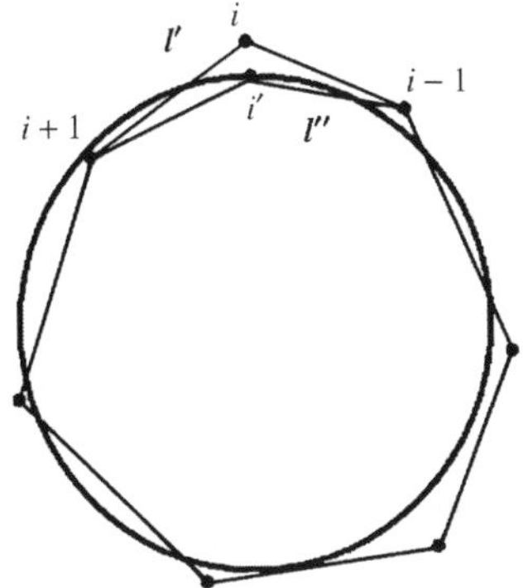

Figure 5.2 The regional similarity energy. See context for explanation.

For convex and concave regions, the snake point that could be located outside, on, or inside the object contour is analyzed. As shown in Figure 5.3, there are six cases and the left-hand side image in each case displays a convex or concave region being overlaid with a snake contour. The right-hand side image shows the RSE distribution, where the brightness indicates the energy value. The square box implies the region of neighborhood. Since the mean intensity of the snake contour is desired to be the closest to that of the object contour, a minimum RSE is favored. Cases (a)–(c) are for convex regions. In case (a), the RSE will pull the snake point toward the boundary. In case (b), part of the square box inside the object boundary (i.e., black region) will keep

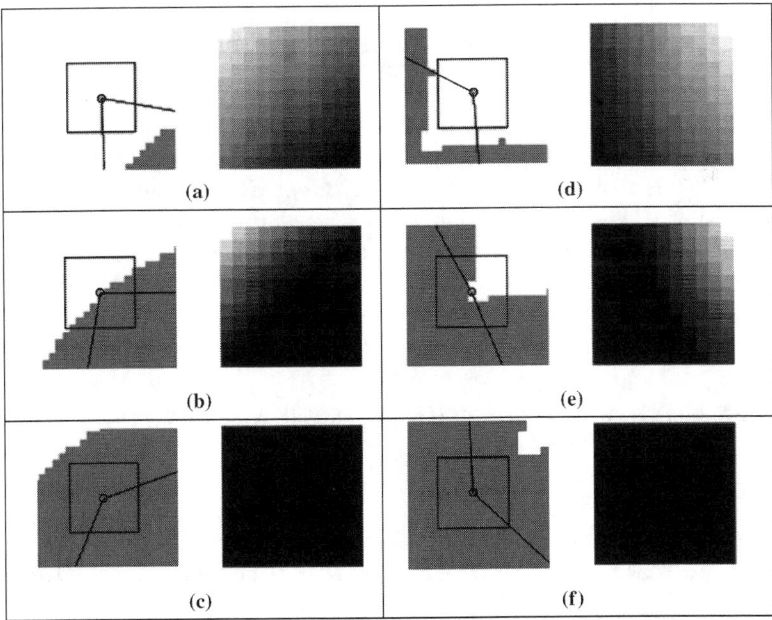

Figure 5.3 Cases (a)–(c) show snake points located outside, on, and inside the boundary for convex regions; cases (d)–(f) show snake points located outside, on, and inside the boundary for concave regions, respectively.

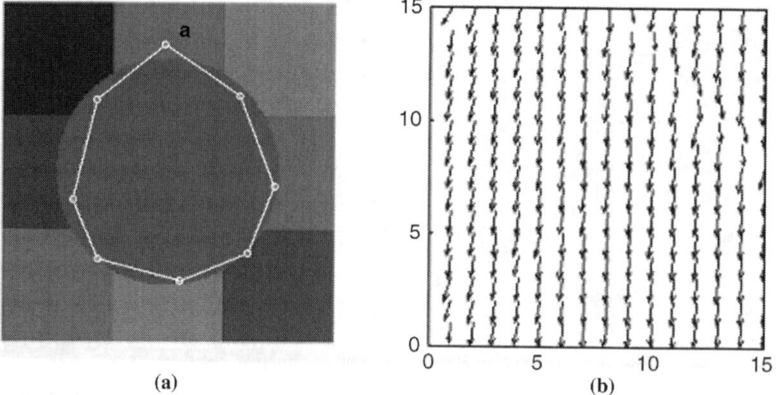

Figure 5.4 The energy flow of the snake point "a" within a 15×15 area.

the snake point the same; however, the other part outside the object boundary (i.e., white region) will pull the snake point toward the boundary. In case (c), the RSE will have no effect since the square box is entirely inside the object boundary. Similarly, cases (d)–(f) are for concave regions.

Figure 5.4(a) shows an image of 300×300 pixels, and Figure 5.4(b) illustrates the RSE for point "a" within a 15×15 area. It is observed that the energy flow pushes the snake point "a" downward to the disk contour.

5.3.3 The Gravitation External Force Field and the Greedy Algorithm

In this section, the gravitation external force field is introduced and the greedy algorithm (Ji and Yan, 2002) is used for the active contour. The concept of gravitation external force field is taken from physics. Two objects attract each other by a force, which is proportional to their mass product and inversely proportional to the distance between their mass centers. Based on this concept, an external energy field, called the *gravitation energy field* (GEF), is developed as given by

$$E_{\text{gravitation}} = \int \frac{g(\vec{r})}{\|\vec{r}\|} \vec{r}\, d\vec{r} \tag{5.10}$$

where $\vec{r}$ is a position vector and $g(\vec{r})$ is a first-order derivative. Note that the edge pixels have the local maxima in the first-order derivative. The attractive force enables the snake points to move toward the object. Figure 5.5 illustrates the effect of the gravitation force field on Figure 5.1. Note that the brightness represents the energy value. With the gravitation energy field, the active contour can be dragged toward the object even if the snake points are far away. Therefore, the total energy function becomes

$$E_{\text{snake}} = \alpha E_{\text{cont}}[v(s)] + \beta E_{\text{curv}}[v(s)] + \gamma E_{\text{image}}[v(s)] + \mu E_{\text{gravitation}}[v(s)] + \delta E_{\text{RSE}}[v(s)] \tag{5.11}$$

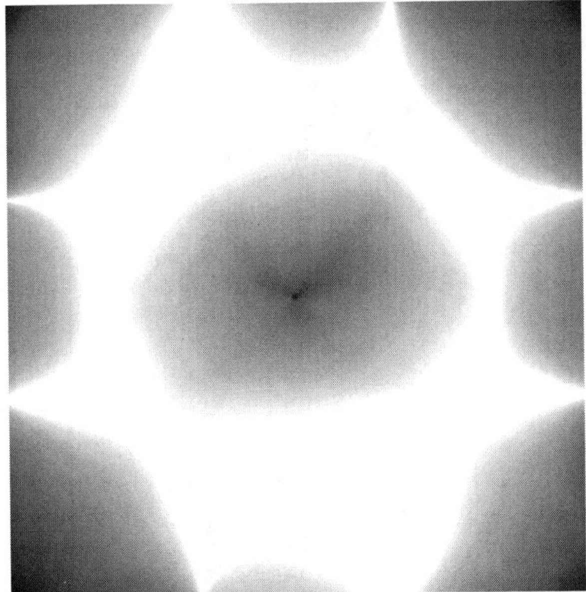

Figure 5.5 The effect of the gravitation force field on Figure 5.1.

These weighting coefficients can be adjusted according to the user's application. For simplicity, $\alpha = \beta = \gamma = \mu = \delta = 1$.

A set of snake points are initialized, and then a loop is entered. At each loop, every snake point is temporally moved to its neighbor within a defined neighborhood, and the energy is calculated. The one with the least energy is chosen to be the new location. When all the snake points are processed, the next loop is repeated until all the snake points are stabilized or a given number of iterations is reached. The greedy algorithm is described as follows:

```
while n < iterations
  for j = 1 to numPoints /* Check every snake point. */
    for k = 1 to numNeighbors
  E_snake = α*E_cont[v_k(s)] + β*E_curv[v_k(s)] + γ*E_image[v_k(s)];
    end;
    Move this point to the new location that has the least energy;
  end; /* main loop*/
end; /* iteration loop*/
```

5.3.4 Experimental Results

Both the improved and the traditional snake models are applied on the added salt-and-pepper noise of Figure 5.1. The results are shown in Figure 5.6. It is observed that the improved model can locate the disk contour, but the traditional model fails. The improved model is suitable for random noise or fixed pattern noise. The banding noise is highly camera-dependent and is the one that is introduced by the camera when it

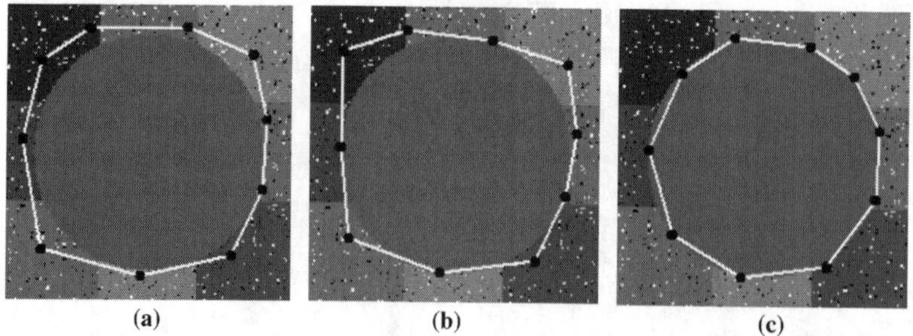

Figure 5.6 (a) The initial snake points, (b) the result by the traditional snake model, and (c) the result by the improved snake model.

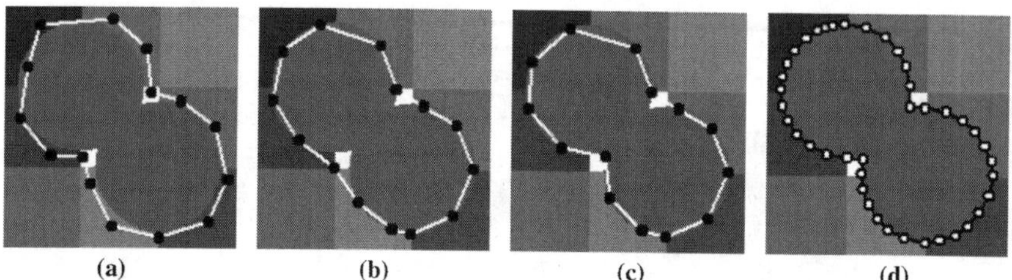

Figure 5.7 (a) The initial snake points, (b) the result by the traditional snake model, (c) the result by the improved snake model, and (d) the interpolation appearance

reads data from the digital sensor. The improved model may not perform well for an image with such noise.

Figure 5.7 illustrates that the proposed model can work successfully for a concave object, but the traditional model fails. Figure 5.7(d) shows that an interpolation between adjacent snake points can be applied for smooth appearance of the object contour. Figure 5.8 shows more results for convex and concave objects in a variety of complex backgrounds using the proposed model. It is observed that some snake point is attracted by the background grid using the traditional model; however, it can locate the object boundary using the proposed model.

The snake model expects the resulting contour to be stabilized on the boundary. If a snake point is inside the object, the force field can pull it out toward the boundary. In Figure 5.8(c2), due to the display limitation and in order to clearly augment the snake points, the snake contour points may look deviating from the true boundary, but in fact they are close to the boundary. The size of neighborhood affects the smoothness of snake contour. In the experiments, a 7×7 neighborhood and 50 as the maximum number of iterations are used. The improved snake model works more effectively in an image with convex or concave objects under a variety of complex backgrounds.

5.4 EDGE OPERATORS

Image segmentation is the fundamental step in image analysis to partition an image into individual regions of interest. Gray-level edge detection is one of the image

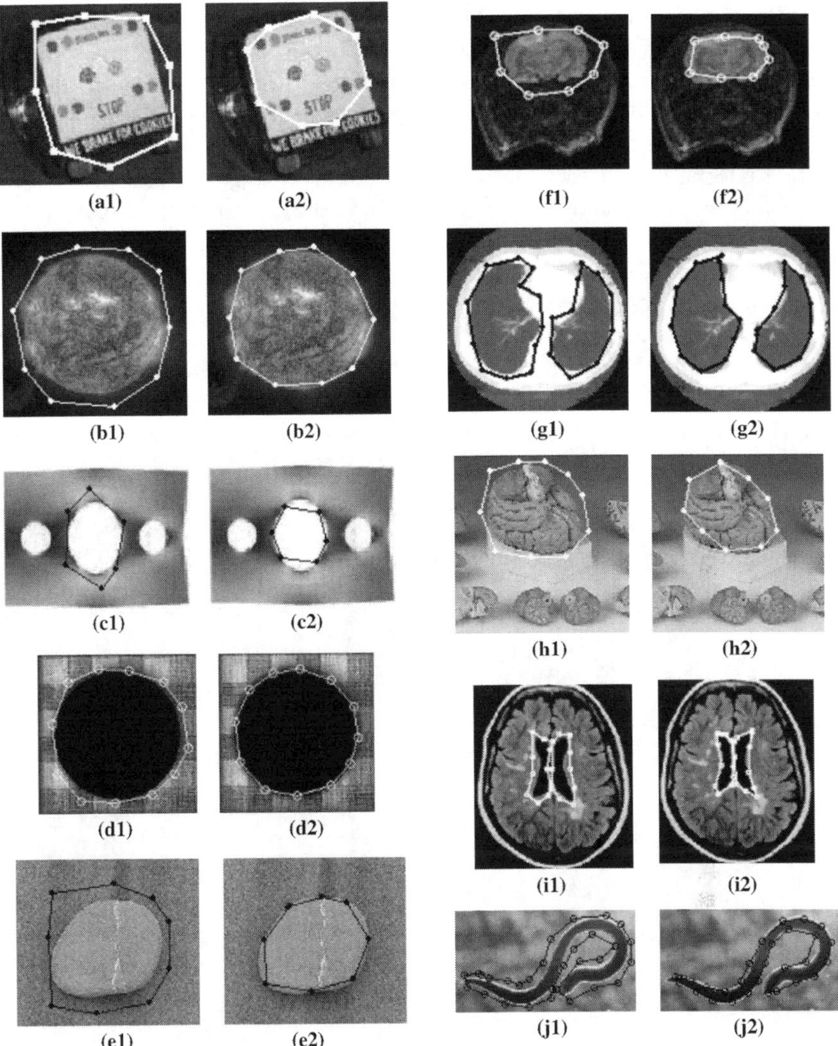

Figure 5.8 More examples for convex and concave objects in complex backgrounds. The first column shows the initial snake points and the second column shows the resulting snake points using the improved model.

segmentation techniques. The output of edge operators should form continuous boundaries of the regions of interest to be further processed. A large number of edge operators have already been proposed (Davis, 1974).

Law et al. (1996) characterized the problem of detecting edges in images as a fuzzy reasoning problem. The edge detection problem is divided into three stages: filtering, detection, and tracing. Images are filtered by applying fuzzy reasoning based on local pixel characteristics to control the degree of Gaussian smoothing. Ahmad and Choi (1999) developed an edge detection method based on local threshold and Boolean function. Meer and Georgescu (2001) proposed edge detection with embedded confidence. Konishi et al. (2003) formulated edge detection as statistical inference. Bao et al. (2005) analyzed the technique of scale multiplication in the framework of Canny edge detection (Canny, 1986). A scale multiplication function is defined as the product of the responses of the detection filter at two scales. Edge maps are constructed as the local maxima by thresholding the scale multiplication results. The detection and localization criteria of the scale multiplication are derived. Wu et al. (2007a) proposed a fast multilevel fuzzy edge detection (FMFED) algorithm to realize the fast and accurate detection of the edges from the blurry images.

The derivative filter emphasizes the edges (i.e., high-frequency components) of objects. As an image is a 2D function, the direction in which the derivative is taken needs to be defined. Edges could be in the horizontal, vertical, or arbitrary direction. Let h_x, h_y, and h_θ, respectively, denote horizontal, vertical, and arbitrary angle derivative filters. Their relationship can be represented as

$$[h_\theta] = \cos\theta \cdot [h_x] + \sin\theta \cdot [h_y] \tag{5.12}$$

For a function $f(x, y)$, the gradient filter generates a vector derivative $\nabla \mathbf{f}$ as the *gradient*:

$$\nabla \mathbf{f} = \begin{bmatrix} \partial f/\partial x \\ \partial f/\partial y \end{bmatrix} = \frac{\partial f}{\partial x}\mathbf{u}_x + \frac{\partial f}{\partial y}\mathbf{u}_y = (h_x * f)\mathbf{u}_x + (h_y * f)\mathbf{u}_y \tag{5.13}$$

where "$*$" denotes the convolution operation, and $\mathbf{u}_x$ and $\mathbf{u}_y$ are unit vectors in the horizontal and vertical directions, respectively. The magnitude of the gradient vector is

$$|\nabla \mathbf{f}| = \left[\left(\frac{\partial f}{\partial x}\right)^2 + \left(\frac{\partial f}{\partial y}\right)^2\right]^{1/2} = \sqrt{(h_x * f)^2 + (h_y * f)^2} \tag{5.14}$$

and the direction of the gradient vector is

$$\theta(\nabla \mathbf{f}) = \tan^{-1}\left(\frac{\partial f/\partial y}{\partial f/\partial x}\right) = \tan^{-1}\left(\frac{h_y * f}{h_x * f}\right) \tag{5.15}$$

For computational simplicity, the gradient magnitude is approximately taken as

$$|\nabla \mathbf{f}| = |h_x * f| + |h_y * f| \tag{5.16}$$

Basic derivative filters are $h_x = [-1 \quad 1]$ and $h_y = \begin{bmatrix} 1 \\ -1 \end{bmatrix}$. The center of the filter is at the middle (i.e., the grid) of two horizontal or vertical neighboring pixels. In order to use a pixel (instead of a grid) as the center, the filter is extended to three pixels as

$$h_x = [-1 \quad 0 \quad 1] \quad \text{and} \quad h_y = \begin{bmatrix} 1 \\ 0 \\ -1 \end{bmatrix}$$

Roberts cross-gradient operator uses the 2D case of two pixels as

$$\text{Roberts}: \quad h_x = \begin{bmatrix} 1 & 0 \\ 0 & -1 \end{bmatrix} \quad \text{and} \quad h_y = \begin{bmatrix} 0 & 1 \\ -1 & 0 \end{bmatrix}$$

Prewitt gradient operator uses the 2D case of three pixels as

$$\text{Prewitt}: \quad h_x = \begin{bmatrix} -1 & 0 & 1 \\ -1 & 0 & 1 \\ -1 & 0 & 1 \end{bmatrix} \quad \text{and} \quad h_y = \begin{bmatrix} 1 & 1 & 1 \\ 0 & 0 & 0 \\ -1 & -1 & -1 \end{bmatrix}$$

Sobel gradient operator slightly modifies the Prewitt gradient operator by putting more weight on the pixels closer to the center as

$$\text{Sobel}: \quad h_x = \begin{bmatrix} -1 & 0 & 1 \\ -2 & 0 & 2 \\ -1 & 0 & 1 \end{bmatrix} \quad \text{and} \quad h_y = \begin{bmatrix} 1 & 2 & 1 \\ 0 & 0 & 0 \\ -1 & -2 & -1 \end{bmatrix}$$

Another popularly used edge operation, called Canny edge operator (Canny, 1986), used the calculus of variations to find the function that optimizes a given functional, described by the sum of four exponential terms in horizontal, vertical, and diagonal edges.

Example 5.3 Apply two masks of Sobel operator on the image below and combine the two results using the so-called "squares and square roots." Show the output edge image after rounding real numbers to integers. After that, how can we separate the edge pixels and nonedge pixels? What is the rough shape of the object in this image?

1	1	1	1	0	0	1	0
1	1	0	1	1	1	1	1
1	2	8	8	8	9	1	0
0	1	8	9	9	9	1	0
0	1	8	8	9	8	1	0
1	1	8	9	9	8	1	1
0	1	1	1	1	1	0	0
1	0	0	0	0	0	0	0

Answer: Let the parts of the masks falling outside the image are zeroed out. After taking the square root of the sum of the squares of the two results convolving with

h_x and h_y, we obtain the output image as

4	3	2	3	4	4	4	4
5	10	23	29	32	27	13	4
6	22	31	32	32	33	28	4
6	31	29	3	1	31	35	4
4	31	30	4	1	32	32	4
4	24	31	29	30	34	25	3
3	13	27	35	35	28	14	3
1	3	4	4	4	3	1	0

We can separate edge pixels from nonedge pixels by thresholding. Using histogram thresholding by selecting a valley between two significant peaks, we can obtain the binary edge image as follows. The rough shape of the object in this image is a square.

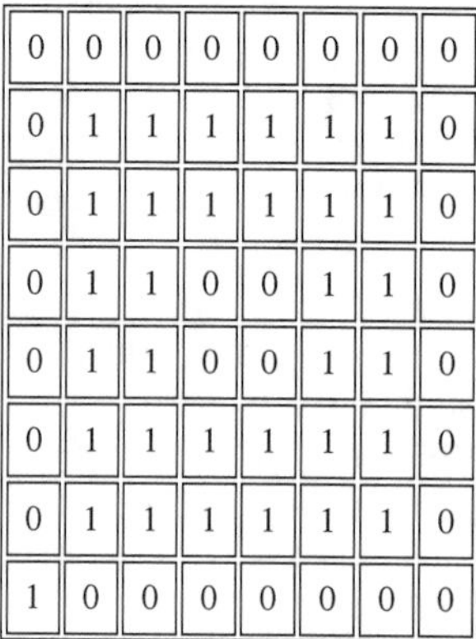

0	0	0	0	0	0	0	0
0	1	1	1	1	1	1	0
0	1	1	1	1	1	1	0
0	1	1	0	0	1	1	0
0	1	1	0	0	1	1	0
0	1	1	1	1	1	1	0
0	1	1	1	1	1	1	0
1	0	0	0	0	0	0	0

Figure 5.9 shows the Lena image and the resulting images after applying the Roberts, Prewitt, and Sobel operators, respectively. It should be noted that both h_x and h_y are separable. That is, they can be decomposed into 1D cases. For example, in the Prewitt gradient operator,

$$\text{Prewitt}: \quad h_x = \begin{bmatrix} 1 \\ 1 \\ 1 \end{bmatrix} \cdot \begin{bmatrix} -1 & 0 & 1 \end{bmatrix} \quad \text{and} \quad h_y = \begin{bmatrix} 1 \\ 0 \\ -1 \end{bmatrix} \cdot \begin{bmatrix} 1 & 1 & 1 \end{bmatrix}$$

Higher order derivatives of functions of two variables can also be adopted in image derivative filters. Laplacian second-derivative filter plays an important role. Unlike the first-derivative filters, the Laplacian is isotropic; that is, it gives the same response to edges in any direction. However, the disadvantages are the loss of edge direction (which may be useful in some applications) and the edge responses being not closely related to the edge magnitudes.

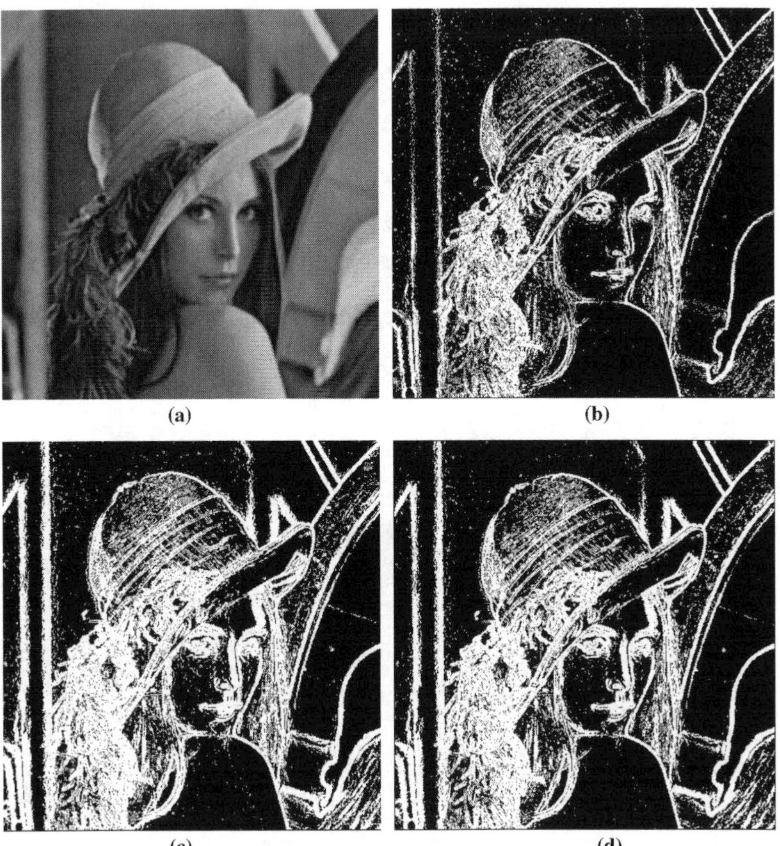

Figure 5.9 (a) The Lena image and the resulting images after applying (b) Roberts, (c) Prewitt, and (d) Sobel filters.

Let a 3 × 3 window of an input image f be represented as

$f(x-1, y+1)$	$f(x, y+1)$	$f(x+1, y+1)$
$f(x-1, y)$	$f(x, y)$	$f(x+1, y)$
$f(x-1, y-1)$	$f(x, y-1)$	$f(x+1, y-1)$

The Laplacian is defined as

$$\nabla^2 f = \frac{\partial^2 f}{\partial x^2} + \frac{\partial^2 f}{\partial y^2} \tag{5.17}$$

Since

$$\begin{aligned} \frac{\partial^2 f}{\partial x^2} &= [f(x+1, y) - f(x, y)] - [f(x, y) - f(x-1, y)] \\ &= f(x+1, y) + f(x-1, y) - 2f(x, y) \end{aligned} \tag{5.18}$$

Similarly,

$$\frac{\partial^2 f}{\partial y^2} = f(x, y+1) + f(x, y-1) - 2f(x, y) \tag{5.19}$$

Therefore,

$$\nabla^2 f = [f(x+1, y) + f(x-1, y) + f(x, y+1) + f(x, y-1)] - 4f(x, y) \tag{5.20}$$

The kernel of the Laplacian second-derivative filter h is denoted as

$$h = \begin{bmatrix} 0 & 1 & 0 \\ 1 & -4 & 1 \\ 0 & 1 & 0 \end{bmatrix}$$

Example 5.4 Apply the Laplacian second-derivative filter on the image below. Show the output edge image after rounding real numbers to integers. After that, how can the edge pixels and nonedge pixels be separated? What is the rough shape of the object in this image?

1	1	1	1	0	0	1	0
1	1	0	1	1	1	1	1
1	2	8	8	8	9	1	0
0	1	8	9	9	9	1	0
0	1	8	8	9	8	1	0
1	1	8	9	9	8	1	1
0	1	1	1	1	1	0	0
1	0	0	0	0	0	0	0

Answer: The Laplacian second-derivative filter mask used is as follows:

0	1	0
1	-4	1
0	1	0

The output is

−2	−1	−2	−2	2	2	−3	2
−1	0	11	6	6	7	0	−3
−1	3	−14	−6	−5	−17	7	2
2	7	−6	−3	−1	−9	7	1
2	6	−7	3	2	−5	6	2
−3	7	−13	−10	−9	−13	6	−3
3	−2	6	7	7	5	2	1
−4	2	1	1	1	1	0	0

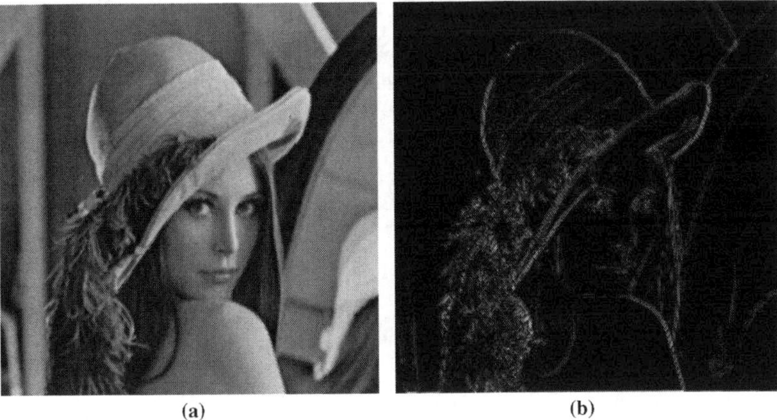

(a) (b)

Figure 5.10 (a) The Lena image and (b) the resulting image after applying the Laplacian filter.

The negative and positive values are rescaled to be from 0 to 255. Sometimes one encounters the absolute value being used for this purpose, but this really is not correct because it produces double lines of nearly equal magnitude, which can be confusing. After performing thresholding on the above image to obtain the edge image, it is observed that the object is a square.

Figure 5.10 shows an example of applying the Laplacian second-derivative filter. Sometimes, the kernel of the Laplacian second-derivative filter h is extended to include diagonal neighbors as

$$h = \begin{bmatrix} 1 & 1 & 1 \\ 1 & -8 & 1 \\ 1 & 1 & 1 \end{bmatrix}$$

5.5 EDGE LINKING BY ADAPTIVE MATHEMATICAL MORPHOLOGY

In edge or line linking, two assumptions are usually made: (1) True edge and line points in a scene follow some continuity patterns, whereas the noisy pixels do not follow any such continuity. (2) The strengths of the true edge or line pixels are greater than those of the noisy pixels. The "strength" of pixels is defined differently for different applications.

Nevatia (1976) presented an algorithm to link edge points by fitting straight lines. Linking direction is based on the direction of edge elements within a defined angular interval. However, portions of small curved edge segments may be neglected. Nalwa and Pauchon (1987) presented an approach based upon local information. The defined *edgels* (i.e., short, linear edge elements, each characterized by a direction and a position) as well as curved segments are linked based solely on proximity and relative orientation. No global factor has been taken into consideration, and there is little concern about noise reduction.

Liu et al. (1988) proposed an edge-linking algorithm to fill gaps between edge segments. The filling operation is performed in an iterative manner rather than a single step. They first connect so-called *tip ends* (i.e., the set of edge pixels that are very likely to be the knots of some contour containing the one being considered) of two segments with a line segment and then try to modify the resulting segment by straight-line fitting. In each iteration, they define the dynamic threshold, and noises are removed gradually. However, it is difficult to get accurate tip ends. Another method (Russ, 1992) locates all end points of broken edges and uses a relaxation method to link them up so that the line direction is maintained. Lines are not allowed to cross, and the points that are closer are matched first. However, this method may fail if unmatched end points or noises are present.

Hajjar and Chen (1999) presented a real-time algorithm and its VLSI implementation for edge linking. The linking process is based on the break points' directions and the weak level points. Stahl and Wang (2007, 2008) proposed an edge-grouping method to detect perceptually salient structures in noisy images by combining boundary and region information. Papari and Petkov (2008) introduced the adaptive pseudodilation that uses context-dependent structuring elements to identify long curvilinear structure in the edge map.

5.5.1 The Adaptive Mathematical Morphology

Traditional morphological operators process an image using a structuring element of fixed shape and size over the entire image pixels. In adaptive morphology, rotation and scaling factors are incorporated. The structuring elements are adjusted according to local properties of an image. In order to introduce adaptive mathematical morphology, the following terminologies need to be first introduced.

Assume that the sets under consideration are always connected and bounded. Let the boundary ∂B of a set B be the set of points, all of whose neighborhoods intersect both B and its complement B^c. If a set B is connected and has no holes, it is called *simply connected*. If it is connected but has holes, it is called *multiply connected*. The concept of ∂B is referred to as the continuous boundary of a structuring element in Euclidean plane and only the simply connected structuring element is considered. Therefore, the transformations performed on the structuring element B are replaced by the transformations on its boundary ∂B in Euclidean plane and followed by a positive filling operator. The positive filling $[\partial B]_+$ of a set B is defined as the set of points that are inside ∂B. The adaptive morphological dilation and erosion can be defined by slightly modifying the traditional morphological operations (Shih and Cheng, 2004).

Definition 5.1. Let A and B be subsets of E^N. The *adaptive morphological dilation* of A by a structuring element B is defined by

$$A\hat{\oplus} B = \{c \in E^N | c = a + \hat{b}, \text{ for some } a \in A \text{ and } \hat{b} \in [(R(a)S(a)\partial B]_+\} \quad (5.21)$$

where S and R are the scaling and rotation matrices, respectively.

Definition 5.2. Let A and B be subsets of E^N. The *adaptive morphological erosion* of A by a structuring element B is defined by

$$A\hat{\Theta} B = \{c \in E^N | c + \hat{b} \in A, \text{ for some } a \in A \text{ and every } \hat{b} \in [(R(a)S(a)\partial B]_+\} \quad (5.22)$$

The broken edge segments can be linked gradually using the adaptive morphological dilation as shown in Figure 5.11. Figure 5.11(a) shows the input signal with gaps, and Figure 5.11(b) shows an adaptive dilation at the end points.

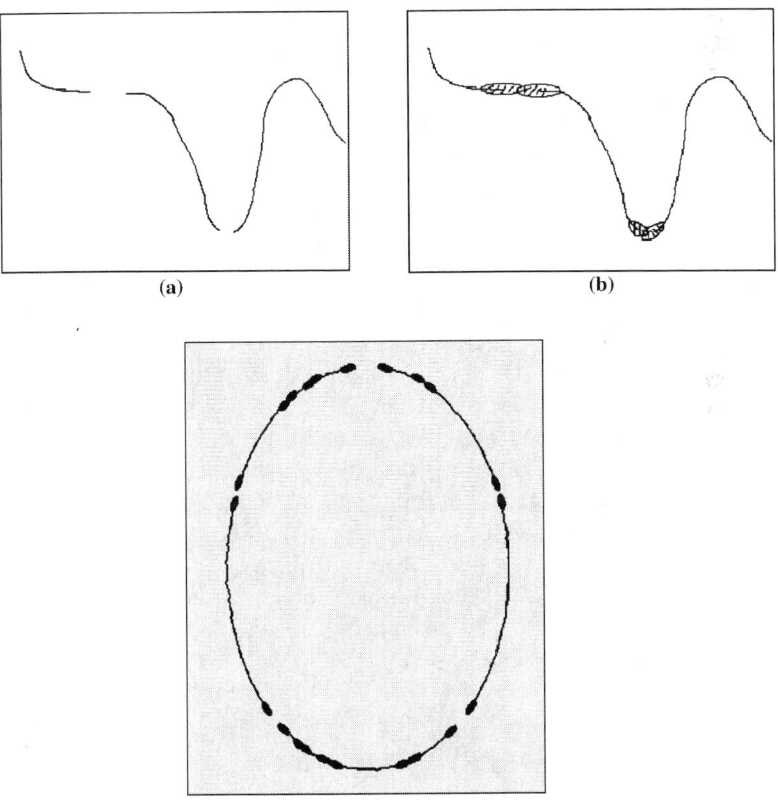

Figure 5.11 (a) Input signal with gaps, (b) adaptive dilation using elliptical structuring elements, and (c) the result of an adaptive dilation operation. Note that the white intensity of edge pixels and the black intensity of background pixels are inverted for clear display purposes.

The reason for choosing the elliptical structuring element is that by using appropriate major and minor axes, all kinds of curves can be linked smoothly. Figure 5.11(c) shows the result of an adaptive dilation operation. Note that the white intensity of edge pixels and the black intensity of background pixels are inverted for clear display purposes.

5.5.2 The Adaptive Morphological Edge-Linking Algorithm

Removing noisy edge segments, checking end points, applying adaptive dilation, thinning, and pruning constitute a complete edge-linking algorithm. If large gaps occur, this algorithm needs to be applied several times until no more gap exists or a predefined number of iterations has been reached.

Step 1: *Removing noisy edge segments*
If the length of one edge segment is shorter than a threshold value, then this segment is removed. A reasonable threshold value 3 is used.

Step 2: *Detecting all the end points*
An *end point* is defined as any pixel having only one 8-connected neighbor. All the edge points are checked to extract the entire end point set.

Step 3: *Applying adaptive dilation at each end point*
From each end point, a range of pixels along the edge is chosen. From the properties of this set of pixels, the rotation angle and size of the elliptical structuring element are obtained. In the program, an increasingly sized range of pixels for each iteration is used. It is referred to as an adjustable parameter s denoting the size of the range of pixels from the end point. Two pixels are taken from this range of pixels, the leftmost side $p_1(x_1, y_1)$ and the rightmost side $p_2(x_2, y_2)$. The following equation is used to compute the slope: $\text{slope} = (y_2 - y_1)/(x_2 - x_1)$. The rotation angle of the elliptical structuring element can be obtained by using the equation $\theta = \tan^{-1}(\text{slope})$. The size of the elliptical structuring element is adjusted according to the number of the pixels in the set, the rotation angle θ, and the distance between p_1 and p_2. In the program, the elliptical structuring element with fixed $b = 3$ and a changing from 5 to 7 is used. These values are based on the experimental results. For a big gap, by using $a = 7$ the gap can be linked gradually in several iterations. For a small gap, using $a = 5$ is better than using 7. If $a = 3$ or 4 is used for a small gap, the structuring element will be almost a circle, and the thinning algorithm does not work well.

An ellipse can be represented as

$$\frac{x^2}{a^2} + \frac{y^2}{b^2} = 1 \tag{5.23}$$

where a and b denote the major and minor axes, respectively. If the center of the ellipse is shifted from the origin to (x_0, y_0), the equation becomes

$$\frac{(x - x_0)^2}{a^2} + \frac{(y - y_0)^2}{b^2} = 1 \tag{5.24}$$

If the center is at (x_0, y_0) and the rotation angle is θ $(-\pi/2 \leq \theta \leq \pi/2)$, the equation becomes

$$\frac{((x-x_0)\cos\theta-(y-y_0)\sin\theta)^2}{a^2}+\frac{((x-x_0)\sin\theta-(y-y_0)\cos\theta)^2}{b^2}=1 \quad (5.25)$$

Because the image is discrete, the rounded values are used in defining the elliptical structuring element. At each end point, an adaptive dilation is performed using the elliptical structuring element. Therefore, the broken edge segment will be extended along the slope direction by the shape of ellipse.

Step 4: *Thinning*

After applying the adaptive dilation at each end point, the edge segments are extended in the direction of local slope. Because the elliptical structuring element is used, the edge segments grow a little fat. Morphological thinning is used to obtain the edge with one pixel of width.

Step 5: *Branch pruning*

The adaptively dilated edge segments after thinning may contain noisy short branches. These short branches must be pruned away. The resulting skeletons after thinning have a width of one pixel. A *root point* is defined as a pixel having at least three pixels in 8-connected neighbors. From each end point, it is traced back along the existing edge. If the length of this branch is shorter than a threshold value after it reaches a root point, the branch is pruned.

Step 6: *Decision*

The program terminates when no end point exists or when a predefined number of iterations has been reached. In the program, eight iterations are used as the limit.

5.5.3 Experimental Results

Figure 5.12(a) shows an original elliptical edge, and Figure 5.12(b) shows its randomly discontinuous edge. The edge-linking algorithm is experimented on Figure 5.12(b).

1. *Using circular structuring elements*
   Figure 5.13 shows the results of using circular structuring elements with $r = 3$ and $r = 5$, respectively, in five iterations. Compared to the original ellipse in Figure 5.12(a), if the gap is larger than the radius of the structuring element, it is difficult to link smoothly. However, if a very big circular structuring element is used, the edge will look hollow and protuberant. In addition, it can obscure the details of the edge.
2. *Using a fixed sized elliptical structuring element and a fixed range of pixels to measure the slope*
   In this experiment, an elliptical structuring element with a fixed minor axis $b = 3$ and a fixed major axis $a = 5$, 7, and 9, respectively, is used. A fixed range of pixels is used to measure the local slope for each end point. That is, the same range of pixels counting from the end point in each iteration is used. The parameter s denotes the range of pixels from end point.

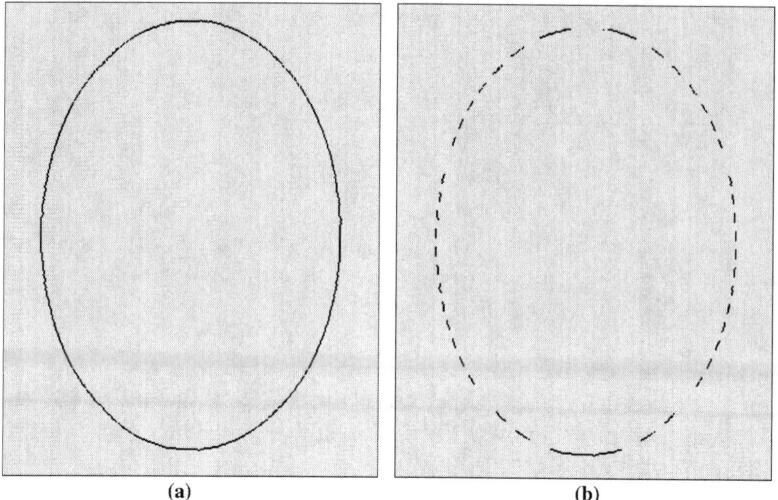

Figure 5.12 (a) Original elliptical edge and (b) its randomly discontinuous edge.

Figure 5.14 shows the result of using an elliptical structuring element with $a=5$, $b=3$, and $s=7$. The result has some big gaps because $a=5$ is too small. Figure 5.15 shows the result of using $a=7$, $b=3$, and $s=9$. Figure 5.16 shows the result of using $a=7$, $b=3$, and $s=11$. There is not much difference between Figures 5.15 and 5.16. Compared to the original ellipse in Figure 5.12 (a), Figures 5.15 and 5.16 have a few shortcomings, but they are much better than Figure 5.14. Figure 5.17 shows the result of using $a=9$, $b=3$, and $s=11$. Therefore, using $a=7$ or $a=9$, a reasonable good result is obtained except for a few shifted edge pixels.

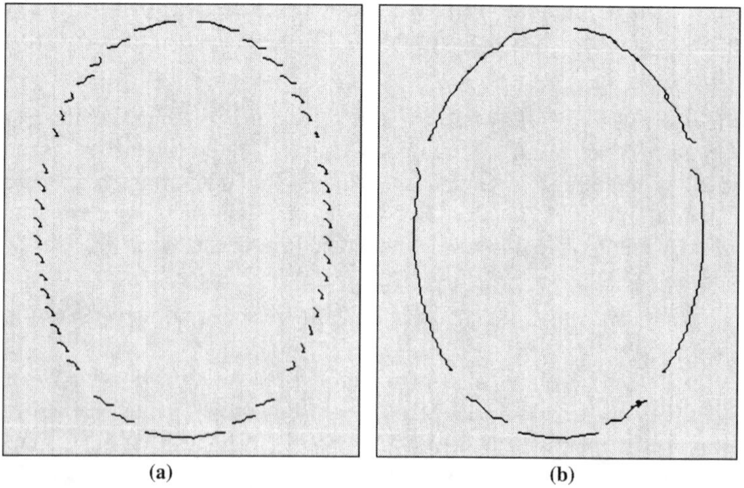

Figure 5.13 Circular structuring elements in five iterations with (a) $r=3$ and (b) $r=5$.

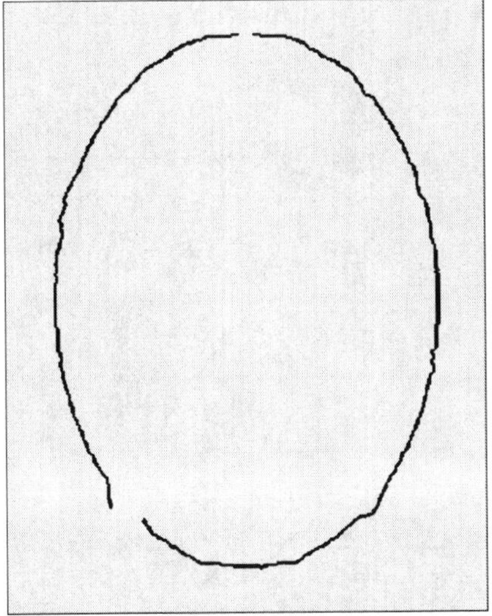

Figure 5.14 An elliptical structuring element with $a = 5$, $b = 3$, and $s = 7$.

3. *Using a fixed sized elliptical structuring element for every end point but using adjustable sized range of pixels to measure local slope in each iteration* Figure 5.18 shows the result of using $a = 5$, $b = 3$, and an adjustable s, and Figure 5.19 shows the result of using $a = 7$, $b = 3$, and an adjustable s. Compared to Figures 5.16 and 5.17, Figures 5.18 and 5.19 are better in terms of elliptical smoothness. This is true because after each iteration the range of pixels used to measure local slope is increased and more information from the original edge is taken into account.

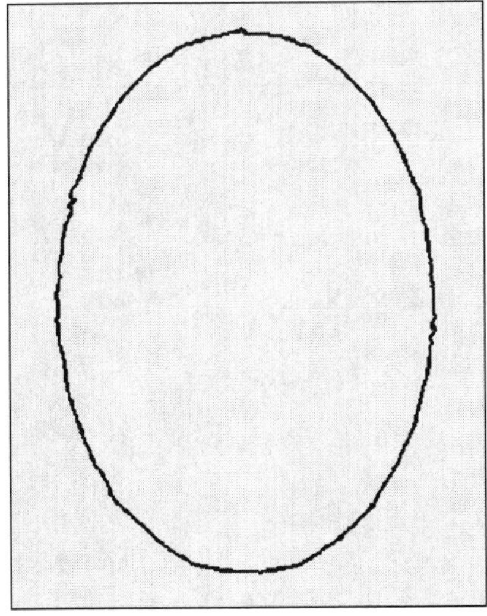

Figure 5.15 An elliptical structuring element with $a = 7$, $b = 3$, and $s = 9$.

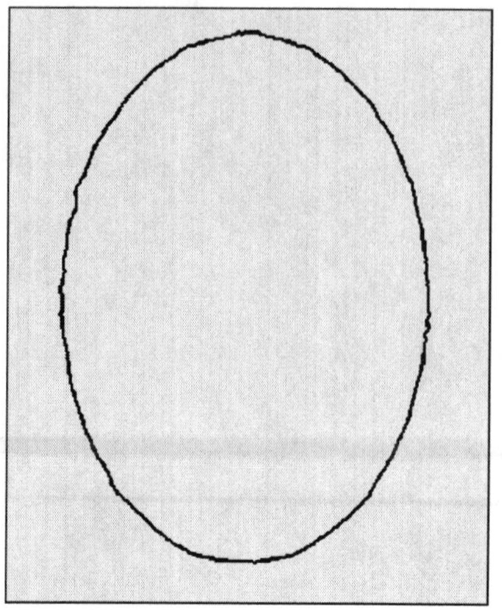

Figure 5.16 An elliptical structuring element with $a = 7$, $b = 3$, and $s = 11$.

4. *Using adjustable sized elliptical structuring elements for every end point and using adjustable sized range of pixels to measure local slope in each iteration (the adaptive morphological edge-linking algorithm)*
   In this experiment, the adjustable sized elliptical structuring elements with a changing from 5 to 7, $b = 3$, and an adjustable s are used.

   Figure 5.20 shows the result of using the adaptive morphological edge-linking algorithm. Compared to Figures 5.16 and 5.19, Figure 5.20 seems about the same. However, the adaptive method has the advantage of

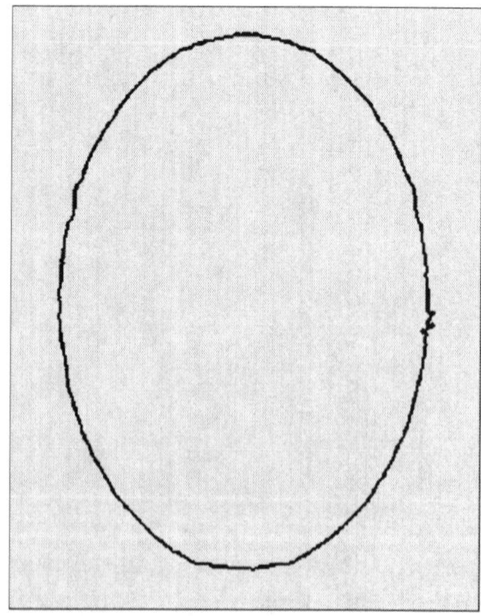

Figure 5.17 An elliptical structuring element with $a = 9$, $b = 3$, and $s = 11$.

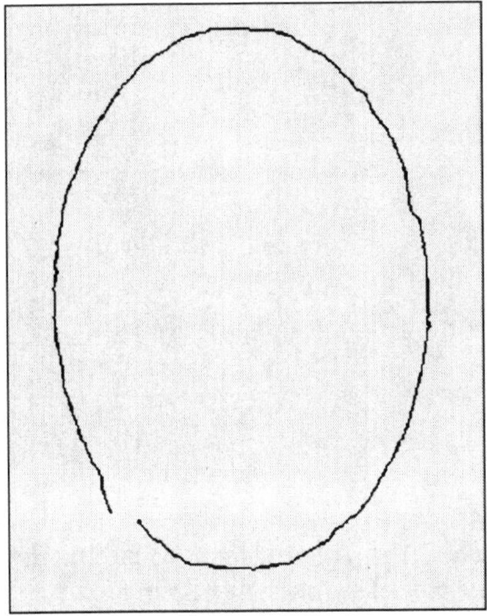

Figure 5.18 An elliptical structuring element with $a = 5$, $b = 3$, and an adjustable s.

adjusting a and s automatically, while the parameters in Figures 5.16 and 5.19 are fixed; that is, they work well in a certain case, but may not work well in other cases.

Figure 5.21(a) shows the elliptical shape with added uniform noise. Figure 5.21(b) shows the shape after removing the noise. Figure 5.21(c) shows the result of using the adaptive morphological edge-linking algorithm. Compared to Figure 5.20, Figure 5.21(b) has several shortcomings. This is because after removing the added noise, the shape is changed at some places.

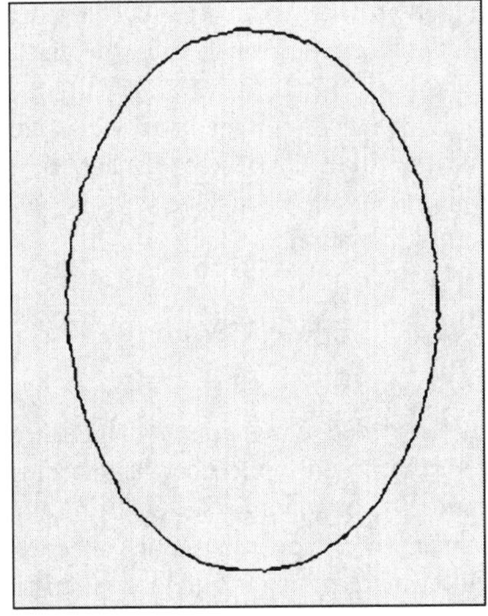

Figure 5.19 An elliptical structuring element with $a = 7$, $b = 3$, and an adjustable s.

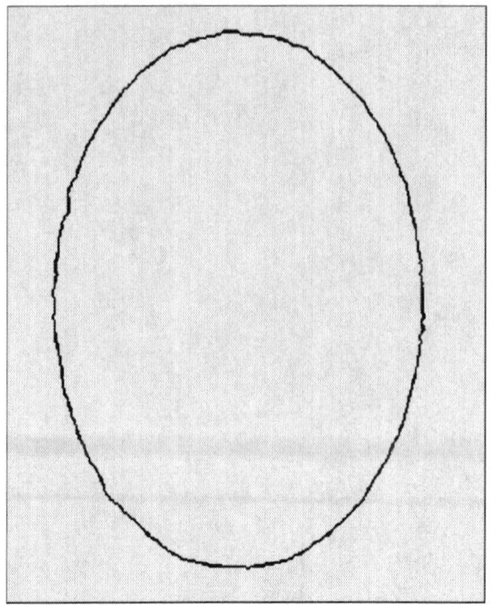

Figure 5.20 The result of using the adaptive morphological edge-linking algorithm.

5.6 AUTOMATIC SEEDED REGION GROWING

Seeded region growing is one of the hybrid methods proposed by Adams and Bischof (1994). It starts with assigned seeds and grows regions by merging a pixel into its nearest neighboring seed region. Mehnert and Jackway (1997) pointed out that SRG has two inherent pixel order dependencies that cause different resulting segments. The first order dependency occurs whenever several pixels have the same difference measure to their neighboring regions. The second order dependency occurs when one pixel has the same difference measure to several regions. They used parallel processing and reexamination to eliminate the order dependencies.

Fan et al. (2001) presented an automatic color image segmentation algorithm by integrating color-edge extraction and seeded region growing on the *YUV* color space. Edges in *Y*, *U*, and *V* are detected by an isotropic edge detector, and the three components are combined to obtain edges. The centroids between adjacent edge regions are taken as the initial seeds. The disadvantage is that their seeds are overgenerated. Tao et al. (2007) incorporated the advantages of the mean shift segmentation and the normalized cut partitioning methods to achieve low computational complexity and real-time color image segmentation.

5.6.1 Overview of the Automatic Seeded Region Growing Algorithm

Figure 5.22 presents the overview of the automatic seeded region growing algorithm (Shih and Cheng, 2005). First, the color image is converted from *RGB* to YC_bC_r color space. Second, automatic seed selection is applied to obtain initial seeds. Third, the seeded region growing algorithm is used to segment the image into regions, where a region corresponds to a seed. Finally, the region-merging algorithm is applied to

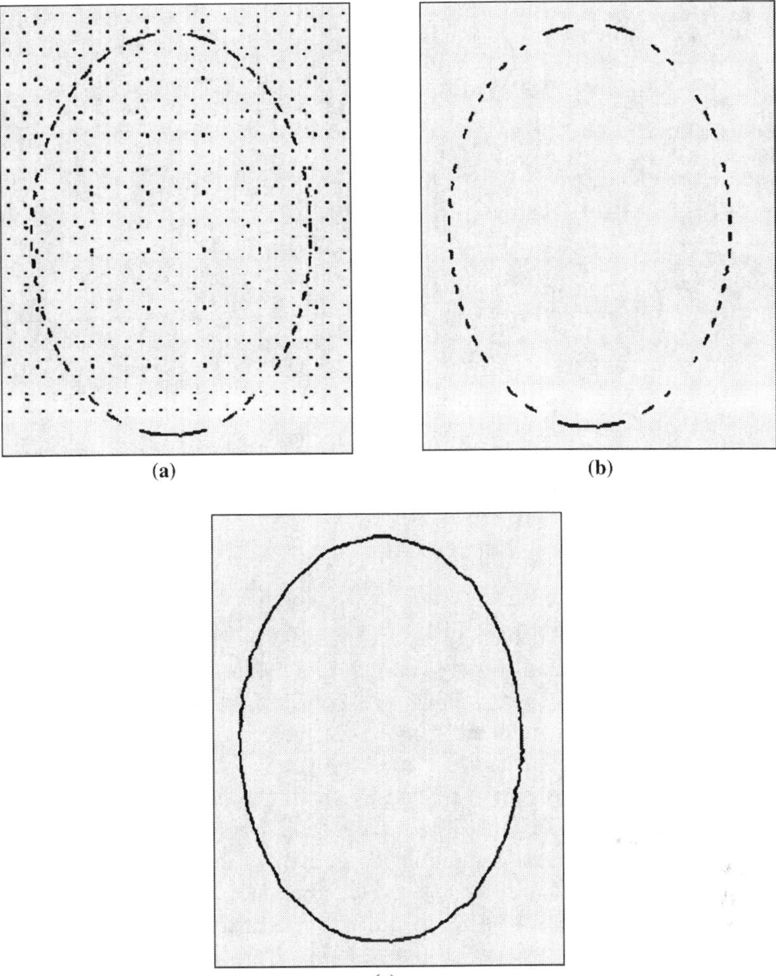

Figure 5.21 (a) The elliptical shape with added uniform noise, (b) the shape after removing noise, and (c) the result of using the adaptive morphological edge-linking algorithm.

merge similar regions, and small regions are merged into their nearest neighboring regions.

A captured color image is stored in *RGB* components. The *RGB* model is suitable for color display, but is not good for color analysis because of its high correlation among *R*, *G*, and *B* components. Also, the distance in *RGB* color space does not represent the perceptual difference in a uniform scale. In image processing and analysis, the *RGB* color space is often converted into other color spaces. Cheng et al. (2001) compared several color spaces including *RGB*, *YIQ*, *YUV*, normalized *RGB*, *HIS*, *CIE* $L^*a^*b^*$, and *CIE* $L^*u^*v^*$ for color image segmentation purposes. Every color space has its advantages and disadvantages. Garcia and Tziritas (1999) noticed that the intensity value *Y* has little influence on the distribution in the C_bC_r plane and

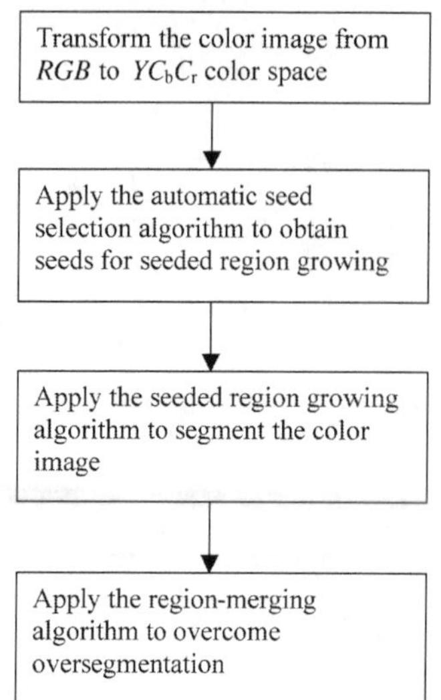

Figure 5.22 Outline of the automatic seeded region growing algorithm.

the sample skin colors form a small and very compact cluster in the C_bC_r plane. The YC_bC_r color space has been widely chosen for skin color segmentation (Hsu et al., 2002). The reasons are as follows: (1) YC_bC_r color space is used in video compression standards (e.g., MPEG and JPEG); (2) the color difference of human perception is related to the Euclidean distance of the YC_bC_r color space; and (3) the intensity and chromatic components can be easily and independently controlled. The YC_bC_r color space is a scaled and offset version of YUV color space, where Y, U, and V denote luminance, color, and saturation, respectively. The C_b (C_r, respectively) is the difference between the blue (red, respectively) component and a reference value (Chai and Bouzerdoum, 2000). The transformation from RGB to YC_bC_r can be performed using the following equation, where R, G, and B are in the range [0, 1], Y is [16, 235], and C_b and C_r are [16, 240]:

$$\begin{bmatrix} Y \\ C_b \\ C_r \end{bmatrix} = \begin{bmatrix} 65.481 & 128.553 & 24.966 \\ -39.797 & -74.203 & 112 \\ 112 & -93.786 & -18.214 \end{bmatrix} \times \begin{bmatrix} R \\ G \\ B \end{bmatrix} + \begin{bmatrix} 16 \\ 128 \\ 128 \end{bmatrix} \tag{5.26}$$

5.6.2 The Method for Automatic Seed Selection

For automatic seed selection, the following three criteria must be satisfied. First, the seed pixel must have high similarity to its neighbors. Second, for an expected region, at least one seed must be generated to produce this region. Third, seeds for different regions must be disconnected.

The similarity of a pixel to its neighbors can be computed as follows. Considering a 3×3 neighborhood, the standard deviations of Y, C_b, and C_r components are calculated using

$$\sigma_x = \sqrt{\frac{1}{9}\sum_{i=1}^{9}(x_i - \bar{x})^2} \tag{5.27}$$

where x can be Y, C_b, or C_r, and the mean value $\bar{x} = \frac{1}{9}\sum_{i=1}^{9} x_i$. The total standard deviation is

$$\sigma = \sigma_Y + \sigma_{C_b} + \sigma_{C_r} \tag{5.28}$$

The standard deviation is normalized to [0, 1] by

$$\sigma_N = \sigma/\sigma_{\max} \tag{5.29}$$

where $\sigma_{\max}$ is the maximum of the standard deviation in the image. The similarity of a pixel to its neighbors is defined as

$$H = 1 - \sigma_N \tag{5.30}$$

From the similarity, the first condition for the seed pixel candidate is defined as follows:

Condition 1: A seed pixel must have the similarity higher than a threshold value.

Second, the relative Euclidean distance (in terms of YC_bC_r) of a pixel to its eight neighbors is calculated as

$$d_i = \frac{\sqrt{(Y - Y_i)^2 + (C_b - C_{b_i})^2 + (C_r - C_{r_i})^2}}{\sqrt{Y^2 + C_b^2 + C_r^2}}, \quad i = 1, 2, \ldots, 8. \tag{5.31}$$

From experiments, the performance using relative Euclidean distance is better than that using normal Euclidean distance. For each pixel, the maximum distance to its neighbors is calculated as

$$d_{\max} = \max_{i=1}^{8}(d_i) \tag{5.32}$$

From the maximum distance, the second condition for the seed pixel candidate is defined below.

Condition 2: A seed pixel must have the maximum relative Euclidean distance to its eight neighbors, which is less than a threshold value.

A pixel is classified as a seed pixel if it satisfies the above two conditions. In order to choose the threshold automatically in condition 1, Otsu's method (Otsu, 1979) is used. The threshold is determined by choosing the value that maximizes the discrimination criterion σ_B^2/σ_w^2, where σ_B^2 is the between-class variance and σ_w^2 is the within-class variance. In condition 2, the value 0.05 is selected as the threshold based on our experiments.

Each connected component of seed pixels is taken as one seed. Therefore, the seeds generated can be one pixel or one region with several pixels. Condition 1 checks whether the seed pixel has high similarity to its neighbors. Condition 2 makes sure that the seed pixel is not on the boundary of two regions. It is possible that for one desired region, several seeds are detected to split it into several regions. The oversegmented

Figure 5.23 (a) Original color image, (b) the detected seeds are shown in red color, (c) seeded region growing result, (d) the result of merging adjacent regions with relative Euclidean distance less than 0.1, (e) the result of merging small regions with size less than 1/150 of the image, and (f) final segmented result. (See Shih and Cheng, 2005 for color images)

regions can be merged later in the region-merging step. Figure 5.23(a) shows a color image, and Figure 5.23(b) shows the detected seeds marked in red color. Note that the connected seed pixels are considered as one seed.

Regarding the threshold value for the relative Euclidean distance, if a higher value is used, a smaller number of pixels will be classified as seeds and some objects may be missed; conversely, a higher number of pixels will be classified as seeds and different regions may be connected. For example, in Figure 5.24, if the threshold is 0.04, one flower is missed in (c) because there is no seed pixel on the flower as shown in (b). On the other hand, if the threshold is 0.08, then Figure 5.24(f) shows that some part of the boat is merged with the water since these seed pixels are connected as shown in (e). Note that due to variant background colors appearing in the figure, different colors are used to display object boundaries.

5.6.3 The Segmentation Algorithm

Let $A_1, A_2, \ldots, A_i$ denote initial seeds and S_i denote the region corresponding to A_i. The mean of all seed pixels in S_i in terms of Y, C_b, and C_r components is denoted as $(\bar{Y}, \bar{C}_b, \bar{C}_r)$. The segmentation algorithm is described as follows:

1. Perform automatic seed selection.
2. Assign a label to each seed region.

3. Record neighbors of all regions in a sorted list T in a decreasing order of distances.
4. While T is not empty, remove the first point p and check its 4-neighbors. If all labeled neighbors of p have a same label, set p to this label. If the labeled neighbors of p have different labels, calculate the distances between p and all neighboring regions and classify p to the nearest region. Then, update the mean of this region and add 4-neighbors of p, which are neither classified yet nor in T, to T in a decreasing order of distances.
5. Perform region merging.

Note that in step 3 T denotes the set of pixels that are unclassified and are neighbors of at least one of the regions S_i. The relative Euclidean distance d_i between the pixel i and its adjacent region is calculated by

$$d_i = \frac{\sqrt{(Y_i - \overline{Y})^2 + (C_{\mathrm{b}_i} - \overline{C}_b)^2 + (C_{\mathrm{r}_i} - \overline{C}_r)^2}}{\sqrt{Y_i^2 + C_{\mathrm{b}_i}^2 + C_{\mathrm{r}_i}^2}} \tag{5.33}$$

where $(\overline{Y}, \overline{C}_{\mathrm{b}}, \overline{C}_{\mathrm{r}})$ are the mean values of Y, C_{b}, and C_{r} components in that region. In step 4, the pixel p with the minimum distance value is extracted. If several pixels have the same minimum value, the pixel corresponding to the neighboring region having the largest size is chosen. If p has the same distance to several neighboring regions, it is classified as the largest region. Figure 5.23(c) shows the result of the algorithm, where boundaries of regions are marked in white color.

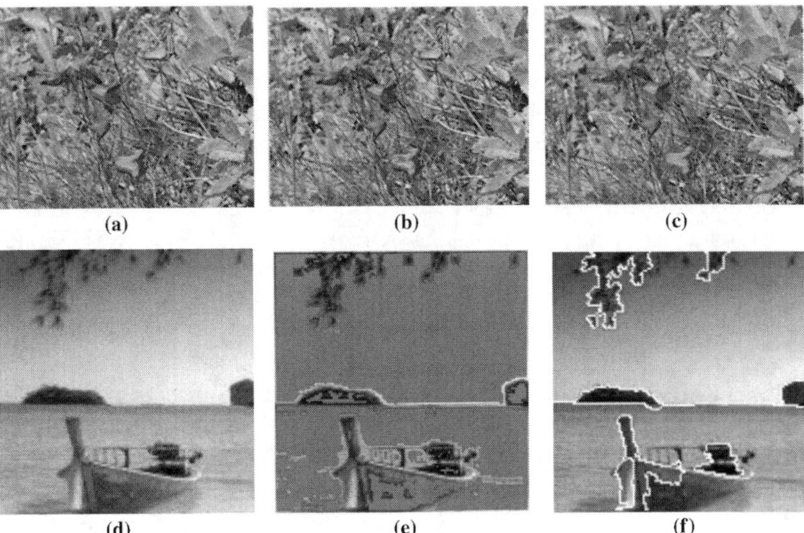

Figure 5.24 Examples of setting threshold for relative Euclidean distance. (a) A color image, (b) the red pixels are seed pixels generated using threshold 0.04, (c) the segmented result, (d) a color image with boat, water, and sky, (e) the red pixels are the seed pixels generated using threshold 0.08, and (f) the segmented result. (See Shih and Cheng, 2005 for color images)

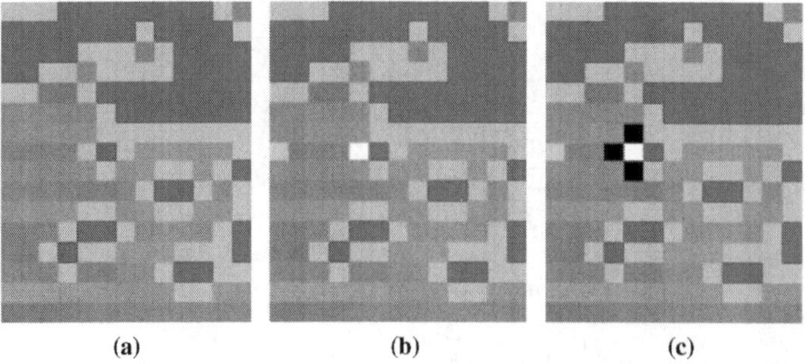

Figure 5.25 Step 4 of region growing. (a) A small window of Figure 5.23(b), where the red pixels are the seeds and the green pixels are the pixels in the sorted list T, (b) the white pixel is the pixel with the minimum distance to the seed regions, and (c) the white pixel is connected to the neighboring red region and the black pixels are added to T. (See Shih and Cheng, 2005 for color images)

The above step 4 is explained graphically in Figure 5.25. Figure 5.25(a) is a small window of Figure 5.23(b), where the red pixels are the seed pixels and the green pixels are the 4-neighbor pixels of the seed regions. Therefore, the green pixels are actually the pixels stored in the sorted list T. In Figure 5.25(b), the white pixel is the one among all the green pixels that has the minimum distance to its adjacent regions. Therefore, the white pixel will be connected to its neighboring red seed region. In Figure 5.25(c), the black pixels are the three 4-neigbor pixels of the white pixel that are added to T. Note that these three black pixels are neither classified nor previously stored in T.

It is possible that several seeds are generated to split a region into several small ones. To overcome the oversegmentation problem, region merging is applied. Two criteria are used: one is the similarity and the other is the size. If the mean color difference between two neighboring regions is less than a threshold value, the two regions are merged. Unfortunately, the result of region merging depends on the order in which regions are examined. The two regions having the smallest distance value among others are first checked. In each iteration, if this value is less than a threshold, the two regions are merged, and the mean of the new region and the distances between the new region and its neighboring regions are recalculated. The process is repeated until no region has the distance less than the threshold. The color difference between two adjacent regions R_i and R_j is defined as the relative Euclidean distance

$$d(R_i, R_j) = \frac{\sqrt{(\overline{Y}_i - \overline{Y}_j)^2 + (\overline{C}_{b_i} - \overline{C}_{b_j})^2 + (\overline{C}_{r_i} - \overline{C}_{r_j})^2}}{\min\left(\sqrt{\overline{Y}_i^2 + \overline{C}_{b_i}^2 + \overline{C}_{r_i}^2}, \sqrt{\overline{Y}_j^2 + \overline{C}_{b_j}^2 + \overline{C}_{r_j}^2}\right)} \tag{5.34}$$

Based on the experiments, 0.1 is selected as the threshold for color similarity measurement.

Next, the size of regions is checked. If the number of pixels in a region is smaller than a threshold, the region is merged into its neighboring region with the smallest color difference. This procedure is repeated until no region has size less than the threshold. Based on the experiments, 1/150 of the total number of pixels is selected in an image as the threshold. Figure 5.23(d) shows the result of merging all similar neighboring regions, and Figure 5.23(e) shows the result of merging small regions.

Since variations of image complexity could be large, some have a high number of regions and some have small sized regions. In order to achieve better segmentation results, the merging by controlling the size of regions and the color difference between regions is conducted. In the region merging, the relative Euclidean difference threshold is 0.1 for initial merging. If this threshold is too high, some desired regions may be merged with other regions; conversely, there will be too many regions. The size of region threshold is set as 1/150 of the image size. By using too high threshold, some important objects may be merged to other regions, while too low threshold may lead to oversegmentation. In the further merging step, the threshold value 0.2 is chosen for final merging because all the regions with large difference from their surroundings and with the size larger than 1/150 of the image must be kept. It is observed that two regions are obviously different if the relative Euclidean distance is over 0.2. The size threshold is set as 1/10 of the image. All the regions with difference higher than 0.1 and size larger than 1/10 of the image must be kept. If this is not used, the sky and water will be merged as shown in Figure 5.23(f). If the number of regions in an image is known, this information will help in improving the segmentation result; however, it is usually unknown. Therefore, in the improved algorithm the final result is obtained by controlling the size of regions and the difference between regions.

The relative Euclidean distance is used as the merging condition because it outperforms the fixed Euclidean distance. Figure 5.23(d) shows that using 0.1 as the threshold value for initial merging can merge the most similar regions, while the water and sky can be separated. If 0.15 is used, then the sky and water are merged. Figure 5.26(b) shows the segmented result of using 0.1 as the distance threshold and 1/150 of the image size as the size threshold. Figure 5.26(c) shows the result of using threshold 0.2 for further merging, which produces a reasonably good result. Figure 5.26(d) shows the result of using threshold 0.15 for further merging. The two small regions on the right should be merged. Figure 5.26(e) shows the result of using threshold 0.25 for further merging. It is observed that the two different regions at the bottom are merged.

5.6.4 Experimental Results and Discussions

The proposed segmentation algorithm is conducted on 150 color nature scene images randomly collected from the Internet. Figure 5.27 illustrates some of the proposed results. In Figure 5.28, the comparison with the JSEG algorithm developed by Deng and Manjunath (2001) is shown. Figure 5.28(a3), (b3), and (c3) shows their segmented results on Figure 5.28(a1), (b1), and (c1), and Figure 5.28(a2), (b2), and (c2) shows the proposed results, respectively. Compared to Figure 5.28(a2), baboon's nose in Figure 5.25(a3) is separated into several regions and its background is oversegmented. Their JSEG algorithm is downloaded from the web site and Figure 5.28(d3) and (e3) shows their segmented results.

The results of the segmentation algorithm are less noisy compared to the results in Deng and Manjunath (2001) and Fan et al. (2001). This is due to the fact that the merging algorithm uses the dynamic control over the size of regions and the difference

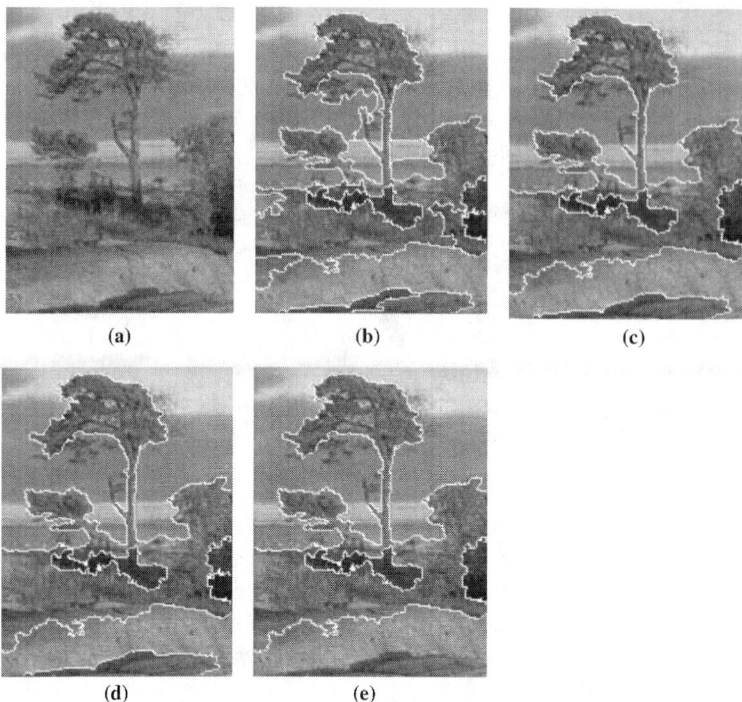

(a) (b) (c)

(d) (e)

Figure 5.26 Example of choosing the relative Euclidean distance threshold for further region merging. (a) Original image, (b) the result after merging using 0.1 as the distance threshold value and 1/150 of the image size as the size threshold, (c) further merging using the relative Euclidean distance threshold 0.2, (d) further merging using threshold 0.15, and (e) further merging using threshold 0.25.

between regions. However, the merging step in Deng and Manjunath (2001) and Fan et al. (2001) uses a fixed threshold value. In Figure 5.28(d3), two objects are missed (Fan et al., 2001). The reason is that either there is no seed generated in these objects or the merging threshold is too high, so they are merged to the surroundings. Compared to Figure 5.28(e3), Figure 5.28(e2) produces more detailed and accurate boundaries. The reason is that in Fan et al. (2001) color quantization is applied to form the class map, so that some details of color information may be lost.

The comparison with the segmentation algorithm by Fan et al. (2001) is shown in Figure 5.29. Figure 5.29(a3), (b3), (c3), and (d3) shows their segmented results on Figure 5.29(a1), (b1), (c1), and (d1), and Figure 5.29(a2), (b2), (c2), and (d2) shows the proposed results, respectively. By observing Figure 5.29(a3), their results are oversegmented and the hair is inappropriately merged to background. By comparing Figure 5.29(b2), (c2), and (d2) with (b3), (c3), and (d3), respectively, their results are oversegmented.

There are mainly two disadvantages of the proposed algorithm. First, although using the fixed threshold values can produce reasonably good results as shown in Figures 5.27–30, it may not generate the best results for all the images. From Figures 5.28(b2), (c2), (d2), and (5.29)(a2), some small objects are missed. Some

Figure 5.27 Some results of the color image segmentation algorithm.

applications require more detailed information. Users can specify these threshold values according to their applications. For example, if the threshold values are adjusted for the size of objects and color differences, more detailed information can be segmented as shown in Figure 5.30. Second, when an image is highly color textured (i.e., there are numerous tiny objects with mixed colors), the proposed algorithm may fail to obtain satisfactory results because the mean value could not represent the property of the region well. Figure 5.31 shows an example that the proposed algorithm performs worse than the method in Fan et al. (2001). Figure 5.31(b) shows that the tree is merged to the flower.

The time complexity for the segmentation algorithm consists of three components: seed selection, region growing, and region merging. In automatic seed selection, complexity of calculating the standard deviation and maximum distance for each pixel is $O(n)$, where n is the total number of pixels in an image. In region growing, each unclassified pixel is inserted into the sorted list exactly once.

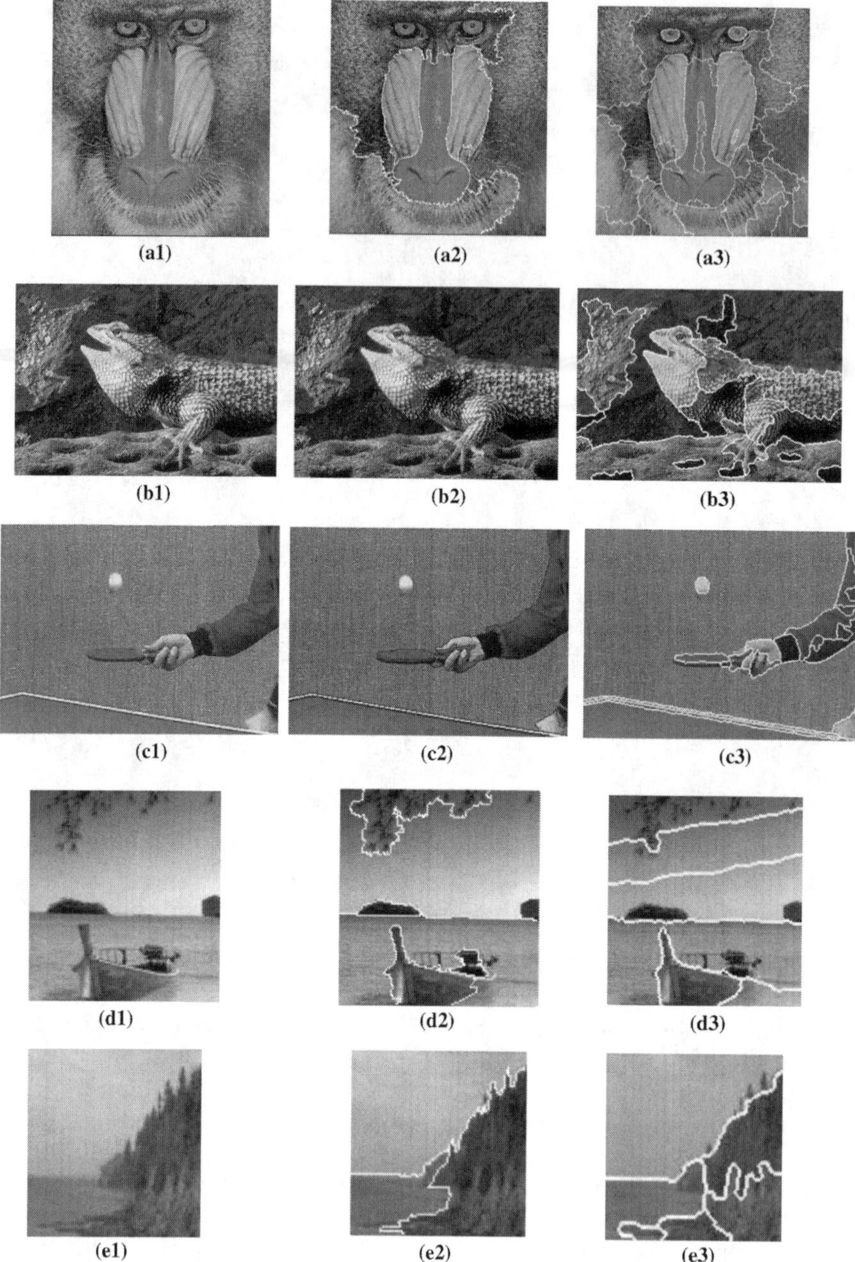

Figure 5.28 The first column shows the original images, the second column shows our segmented results, and the third column shows the results of the JSEG algorithm.

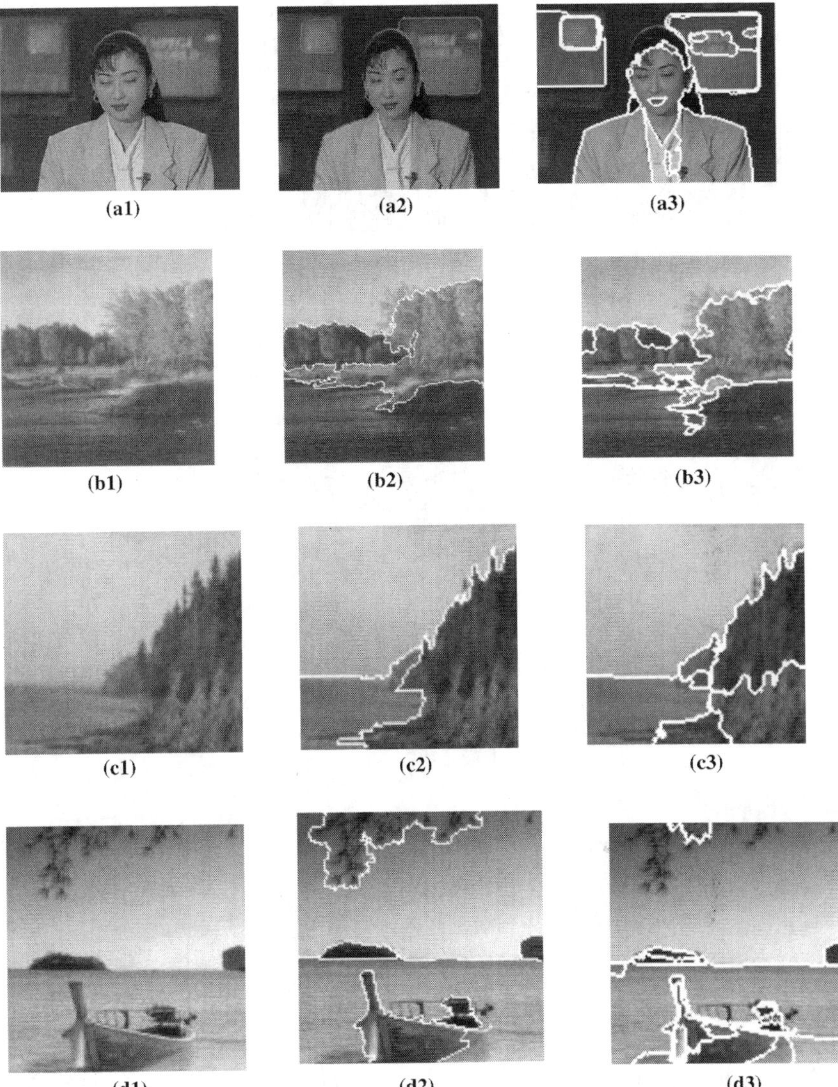

Figure 5.29 The first column shows the original images, the second column shows our segmented results, and the third column shows the results of Fan's algorithm.

Neighboring regions can be checked and distances can be calculated in constant time. Putting a pixel into the sorted list requires $\log(n)$. Therefore, complexity is $O(n \log(n))$ for region growing.

In region merging, complexity of calculating the differences between regions is $O(m^2)$, where m is the number of regions. To calculate sizes for all the regions, complexity is $O(n)$. Usually, m is much less than n. To merge two regions, we need to label the pixels in the two regions, calculate the mean for the new merged region, and calculate the distances between this and other regions. Therefore, complexity is

(a) (b) (c) (d)

Figure 5.30 The segmented results with more details by adjusting the size of the minimum objects and the threshold value for the relative Euclidean distance.

$O(n + m) \approx O(n)$ to merge two regions. The complexity for region merging is $O(m^2 + n + mn) \approx O(mn)$. Therefore, the total time complexity for the proposed algorithm is $O(n + n \log(n) + mn) \approx O((m + \log(n))n)$.

5.7 A TOP–DOWN REGION DIVIDING APPROACH

Histogram-based and region-based segmentation approaches have been widely used in image segmentation. Difficulties arise when we use these techniques, such as the selection of a proper threshold value for the histogram-based technique and the oversegmentation followed by the time-consuming merge processing for the region-based technique. To provide efficient algorithms that not only produce better segmentation results but also maintain low computational complexity, a novel top–down region dividing based approach is developed for image segmentation,

(a) (b) (c)

Figure 5.31 (a) An image with highly colored texture, (b) our segmented result showing the tree is merged to the flower region, and (c) the results of Fan et al. (2001).

which combines the advantages of both histogram-based and region-based approaches.

5.7.1 Introduction

The histogram-based (or feature-based) segmentation technique produces a binary image based on the threshold value (Tobias and Seara, 2002). The intensities of object and background pixels tend to cluster into two sets in the histogram. The histogram will be bimodal by a threshold value, which is selected from the valley between the two sets. The optimal segmentation intends to find the threshold value that minimizes the misclassification. If the threshold is too high, many object pixels will be lost and object contour will be severely destroyed. Although the complexity of histogram-based technique is low, the threshold selection is difficult, especially when the histogram is multimodal. Furthermore, the histogram-based technique considers only the feature image (histogram) without checking the spatial relationship among connected pixels.

Watershed-based segmentation, a region-based approach, uses a bottom–up strategy that segments an image into several small regions, followed by a merging procedure. The immersion-based system (Vincent and Soille, 1991) and the drainage rainfall system (Chibanga et al., 2001) are two approaches for performing watershed transformation. It considers an image as a topographic surface and the image intensity as the altitude. The drop of water will progressively fill up the ascending catchment basins from the minima of lowest altitude (lowest intensity) of the surface. Each pixel will flow along a descending path to a local minimum. When the altitude of water gradually increases, two catchment basins will reach at some points, called *watershed points*. A collection of watershed pixels on the contour is defined as the *watershed line*. Since the watershed algorithm is highly sensitive to the local minimum, it usually results in oversegmentation. In other words, there are overcrowded regions segmented in an image. Furthermore, in order to merge similar smaller connected regions, the *region adjacency graph* (RAG) is used for region growing. Although the watershed-based image segmentation provides better results than the histogram-based approach, its computational complexity is high.

In order to provide more efficient algorithms that not only obtain better results but also maintain low complexity, a novel TDRD-based approach is presented to iteratively divide subregions if the size of a subregion is larger than a predefined threshold or the homogeneity of a subregion is larger than a predefined threshold.

5.7.2 Overview of the TDRD-Based Image Segmentation

5.7.2.1 Problem Motivation The histogram-based image segmentation method, although its complexity is low, does not consider the spatial relationship of neighborhood and may fail in some cases. Figure 5.32(a) shows an image containing dark and bright parts on the left and right areas, respectively. In each part, gray values gradually increase from left to right. For example, the pixels on the leftmost column of left part are "32" and are increased by 1 on each column to the right, until the pixels on the rightmost column of left part are "96". Similarly, the

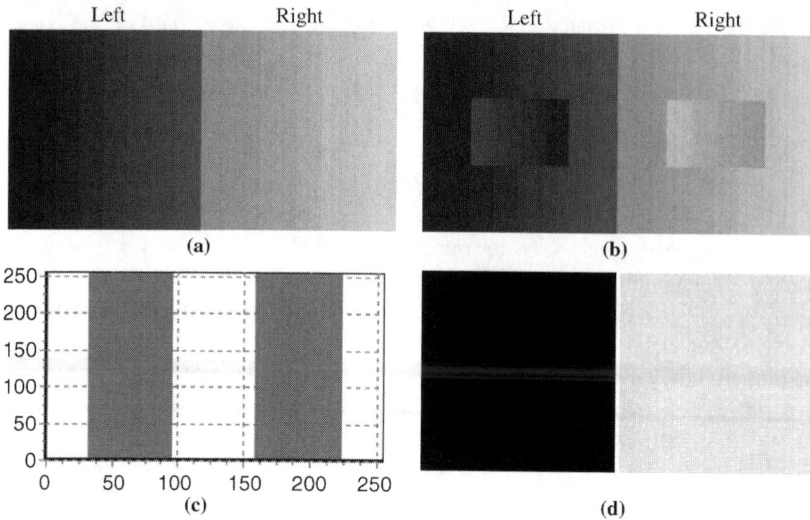

Figure 5.32 An example showing the problem of histogram-based image segmentation.

pixels on the leftmost and the rightmost columns of right part are "160" and "224", respectively. In Figure 5.32(b), we reverse a small rectangular area on each part by swapping the pixels on the corresponding right and left columns within the rectangle. Figure 5.32(c) shows the histogram of the original image. Since we only exchange the positions of pixels, the histogram remains the same. After thresholding, the same result as shown in Figure 5.32(d) is obtained for Figure 5.32(a) and (b).

The watershed-based image segmentation method, although offering better performance than the histogram-based method in general, may fail in some cases. Figure 5.33(a) shows an image composed of five rectangular layers with different gray values. From the property of watershed, the central lowest intensity box is considered as the catchment basin, and after watershed-based image segmentation, we obtain Figure 5.33(b), which shows only one object in existence. It is because the watershed approach could destroy the spatial structure of an image.

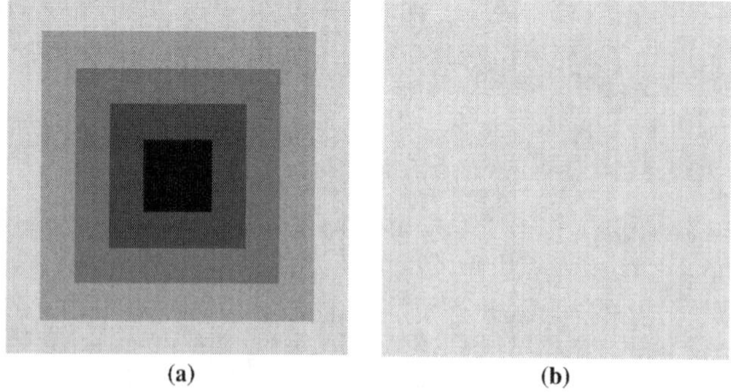

Figure 5.33 A counterexample showing the problem of watershed-based image segmentation.

5.7.2.2 The TDRD-Based Image Segmentation In order to perform image segmentation in arbitrary cases successfully and reduce computational complexity, the TDRD-based method is proposed. Its concept is opposite to the watershed algorithm. The watershed algorithm finds local minimum areas as the catchment basins and segments each catchment basin. Then, the region growing procedure is repeated to merge connected similar catchment basins. It is a bottom–up approach that first segments an image as detailed as possible and then merges the segments from small to large regions. The TDRD-based segmentation consists of two main steps, region dividing and subregion evaluation procedures, to iteratively segment an input region into several subregions by evaluating both feature and spatial properties. In the region dividing procedure, the previously developed image simplification algorithm (Wu et al., 2006a) is used to segment an input region. First, two histograms of each input region are used to determine suspicious intensities. Second, suspicious pixels are obtained using the suspicious intensities. A pixel will be considered as suspicious if its intensity is suspicious. Third, the final intensity of a suspicious pixel can be derived by analyzing the fuzziness of the suspicious pixel and its 8-connected neighbors. In the subregion evaluation procedure, the size and homogeneity of each divided subregion will be evaluated to determine whether the subregion has to be continuously divided. The TDRD-based image segmentation algorithm is presented below and its flowchart is shown in Figure 5.34.

TDRD-Based Image Segmentation Algorithm

1. Consider an input image as a big region and add it into a dividing list.
2. *Region dividing procedure*: Segment each region in the dividing list into several subregions.
3. *Subregion evaluation procedure*: On each divided obtained subregion, determine whether a subregion needs to be continuously divided. If it does, add the subregion into the dividing list.

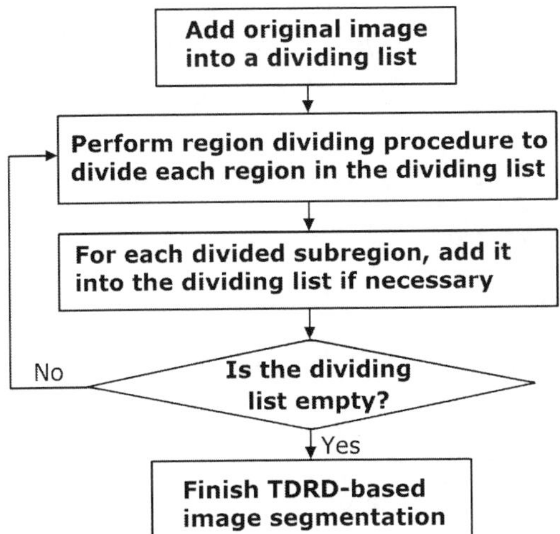

Figure 5.34 The flowchart of our TDRD-based image segmentation.

4. Repeat steps 2 and 3 until the dividing list is empty.
5. Output the segmentation result by collecting all the subregions.

5.7.3 The Region Dividing and Subregion Examination Strategies

5.7.3.1 Region Dividing Procedure The region dividing procedure combines the advantages from histogram-based and region-based segmentation methods. It includes three steps: (a) suspicious intensities determination, (b) suspicious pixels determination, and (c) final intensity determination (FID).

Suspicious Intensities Determination The suspicious intensities are determined by comparing the histograms of two transformed images using histogram equalization. As mentioned earlier, the selection of an appropriate threshold value is challenging in the histogram-based approach. Many researchers have tried to solve this problem. However, it is really difficult to find a general rule for all cases to determine the threshold value for optimal segmentation because this method lacks the relationship between a pixel and its neighbors. To solve this problem, we propose a new idea by avoiding the selection of such a threshold value on the histogram. That is, we do not intend to find a general rule for determining the threshold value; instead, we determine the suspicious intensities by comparing histograms of two transformed image using histogram equalization (HE) as shown in Figure 5.35 and determine the final intensity of suspicious pixels by checking their local spatial information.

Figure 5.35(a) shows the thresholding of the histogram after HE into binary. The input grayscale image of 256 colors will become a binary image of 2 colors. Let $\mathrm{Num}(m)$ denote the total number of pixels of gray level m in an image. In an ideal case, we obtain the intensity, say i, such that $S_1 = S_2$, where $S_1 = \sum_{m=0}^{i} \mathrm{Num}(m)$ and $S_2 = \sum_{m=i+1}^{255} \mathrm{Num}(m)$. Similarly, we perform another histogram graph of a translated image after HE, in which the intensity range is changed from [0, 255] to [0, 2] (i.e., three colors) as shown in Figure 5.35(b). Also, we can find two intensities, say j and k, such that $W_1 = W_2 = W_3$, where $W_1 = \sum_{m=0}^{j} \mathrm{Num}(m)$, $W_2 = \sum_{m=j+1}^{k} \mathrm{Num}(m) \sum$, and $W_1 = \sum_{m=k+1}^{255} \mathrm{Num}(m)$.

Let OriGray and NewGray indicate the original and the translated intensities, respectively. In Figure 5.35(a), NewGray = 0 (black) if OriGray $\leq i$, and NewGray = 1 (white) for the others. Similarly, in Figure 5.35(b), NewGray = 0 (black) if

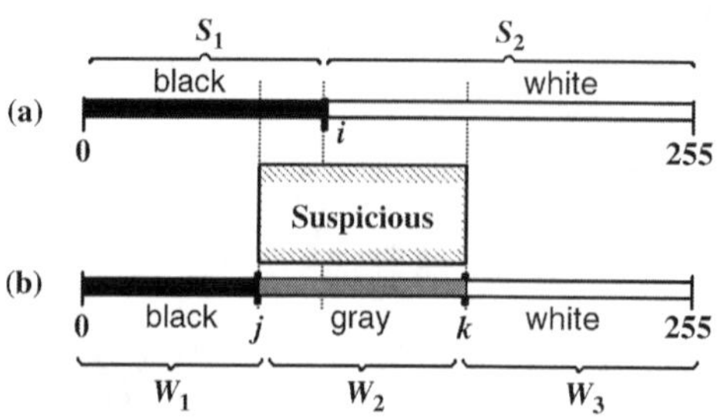

Figure 5.35 The idea of finding suspicious intensities by the DHE.

OriGray $\leq j$, NewGray = 1 (gray) if $j <$ OriGray $\leq k$, and NewGray = 2 (white) if OriGray $> k$. The *suspicious intensity* is defined as the one whose new intensities are different between two histograms of an image. For example, the suspicious intensities in Figure 5.35 possess the following two properties: NewGray is black in Figure 5.35 (a) but gray in Figure 5.35(b), and NewGray is white in Figure 5.35(a) but gray in Figure 5.35(b).

Suspicious Pixels Determination A pixel is considered as *a suspicious pixel* if its intensity is one of the suspicious intensities. Figure 5.36 shows an example of determining suspicious pixels based on the suspicious intensities. Figure 5.36(a) is the original image. Let HE_i indicate a transformed image after HE, meaning that there are only i colors remaining. Figure 5.36(b) and (c) is obtained by translating Figure 5.36 (a) into HE_2 and HE_3, respectively, after performing histogram equalization. Figure 5.36(d) is obtained by checking the pixel intensity in the corresponding location between Figure 5.36(b) and (c), where white and gray pixels are used to indicate suspicious pixels. Note that the white pixels in Figure 5.36(d) mean that the intensity is white in Figure 5.36(b) but gray in Figure 5.36(c), and the gray pixels in Figure 5.36(d) mean that the intensity is black in Figure 5.36(b) but gray in Figure 5.36(c).

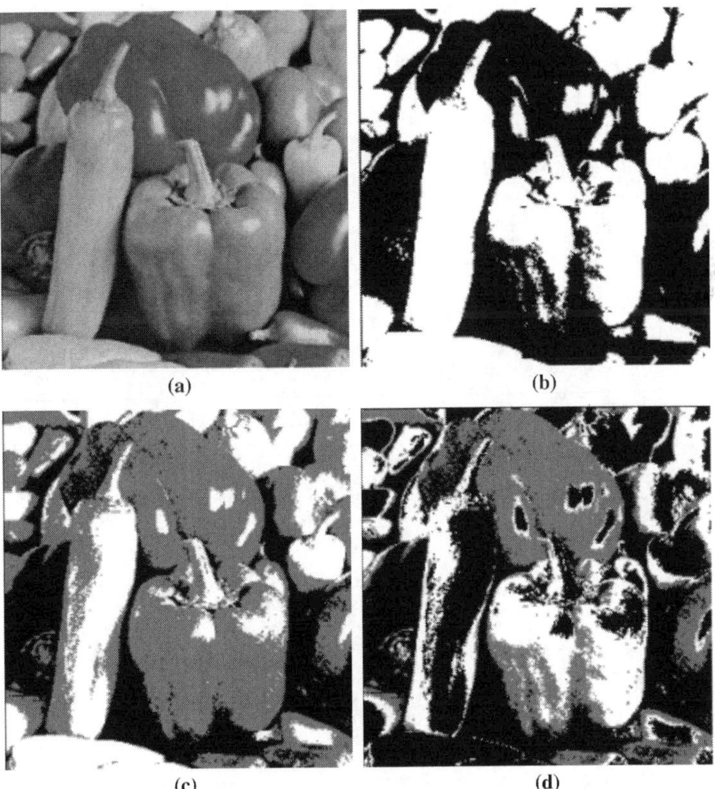

Figure 5.36 An example of determining the suspicious pixels.

Final Intensity Determination As mentioned earlier, it is difficult to determine the proper threshold in the histogram-based approach. In order to avoid the difficulty, a range of suspicious intensities is first determined based on the histogram analysis. Once the suspicious pixels are derived, the local spatial information is used to determine the final intensity of suspicious pixels. We analyze the pixels and their neighbors to determine the final intensity. The weighted fuzzy mean (WFM) filter introduced by Lee et al. (1997) works well in image enhancement, especially for removing additive impulse noise. Therefore, we develop our FID based on the WFM filter as follows.

In the first step, three intensities i, j, and k are derived. For each of them, a membership function is generated as shown in Figure 5.37. Let FM_i, FM_j, and FM_k denote the fuzzy subsets for each intensity, respectively. The fuzzy subsets used are the L-R type fuzzy member (Zimmermann, 1991) as

$$f(x) = \begin{cases} L\left(\dfrac{m-x}{\alpha}\right) & \text{for } x \leq m \\ R\left(\dfrac{m-x}{\beta}\right) & \text{for } x \geq m \end{cases} \tag{5.35}$$

where the functions $L(y) = R(y) = \max(0, 1-y)$, $f(x)$ is represented by a triple $[m, \alpha, \beta]$, and α and β are used to determine the slopes for the values smaller than and larger than m, respectively. Note that, once i, j, and k are obtained, the three fuzzy subsets can be derived by the three triples $[m_i, \alpha_i, \beta_i]$, $[m_j, \alpha_j, \beta_j]$, and $[m_k, \alpha_k, \beta_k]$ using the following formulas, where $0.5 \leq c_1 \leq 1$:

$$m_i = i, \quad \alpha_i = c_1(i-j), \quad \beta_i = c_1(k-i)$$

$$m_j = j, \quad \alpha_j = c_1(i-j), \quad \beta_j = c_1(i-j)$$

$$m_k = k, \quad \alpha_k = c_1(k-i), \quad \beta_k = c_1(k-i)$$

Let $p(n, m)$ denote a suspicious pixel located at (n, m) of an image. Three fuzziness evaluations, $\text{Fuzzy}_i(n, m)$, $\text{Fuzzy}_j(n, m)$, and $\text{Fuzzy}_k(n, m)$, for their corresponding intensities, i, j, and k, are calculated for determining the proper intensity.

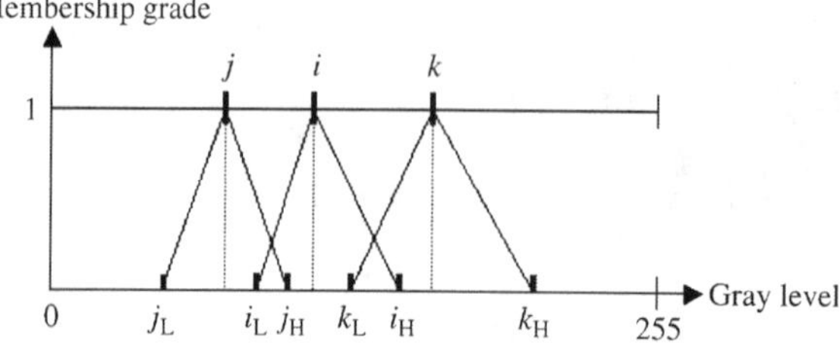

Figure 5.37 The membership functions of intensities i, j, and k.

Fuzziness Evaluation for Intensity *i*

If

$$\sum_{u=-1}^{1}\sum_{v=-1}^{1} \mathrm{FM}_i(\mathrm{Ori}(n+u,m+v)) \neq 0$$

then

$$\mathrm{Fuzzy}_i(n,m) = \frac{\sum_{u=-1}^{1}\sum_{v=-1}^{1} \mathrm{FM}_i(\mathrm{Ori}(n+u,m+v))^*\mathrm{Ori}(n+u,m+v)}{\sum_{u=-1}^{1}\sum_{v=-1}^{1} \mathrm{FM}_i(\mathrm{Ori}(n+u,m+v))}$$

else

$$\mathrm{Fuzzy}_i(n,m) = 0$$

where $\mathrm{Ori}(n,m)$ is the intensity of pixel p in the original image. We obtain the other two fuzziness evaluations, $\mathrm{Fuzzy}_j(n,m)$ and $\mathrm{Fuzzy}_k(n,m)$, similarly.

In order to determine the proper intensity of a suspicious pixel, we use *fuzzy estimator* (FE) (Lee et al., 1997) to evaluate the maximum likelihood. Let $\mathrm{FM}_{\mathrm{FE}}$ be the fuzzy set of fuzzy estimator, which is a fuzzy interval I of L-R type as shown in Figure 5.38 and denoted by $I = [\mathrm{FE_L}, \mathrm{FE_H}, \alpha, \beta]_{LR}$, where $\mathrm{FE_L}$ and $\mathrm{FE_H}$ are the low and high gray levels, and α and β are used to determine the slopes for the values smaller than $\mathrm{FE_L}$ and larger than $\mathrm{FE_H}$, respectively.

$$f_{LR_I}(x) = \begin{cases} L\left(\dfrac{\mathrm{FE_L}-x}{\alpha}\right) & \text{for } x \leq \mathrm{FE_L} \\ 1 & \text{for } \mathrm{FE_L} \leq x \leq \mathrm{FE_H} \\ R\left(\dfrac{x-\mathrm{FE_H}}{\beta}\right) & \text{for } x \geq \mathrm{FE_H} \end{cases} \tag{5.36}$$

Note that $\mathrm{FE_L}$ and $\mathrm{FE_H}$ in Figure 5.38 can be calculated by $\mathrm{FE_L} = j - c_2(i-j)$ and $\mathrm{FE_H} = k + c_2(k-i)$, respectively, where $0 \leq c_2 \leq 0.5$.

The fuzzy estimator of a suspicious pixel can be calculated by

$$\mathrm{Fuzzy_{FE}}(n,m) = \frac{\sum_{u=-1}^{1}\sum_{v=-1}^{1} f_{LR_I}(\mathrm{Ori}(n+u,m+v)) \times \mathrm{Ori}(n+u,m+v)}{\sum_{u=-1}^{1}\sum_{v=-1}^{1} f_{LR_I}(\mathrm{Ori}(n+u,m+v))}$$

Let "New" be the adjusted image after performing the FID on an original region. The FID algorithm is presented below, and Figure 5.39 shows its result where the whole

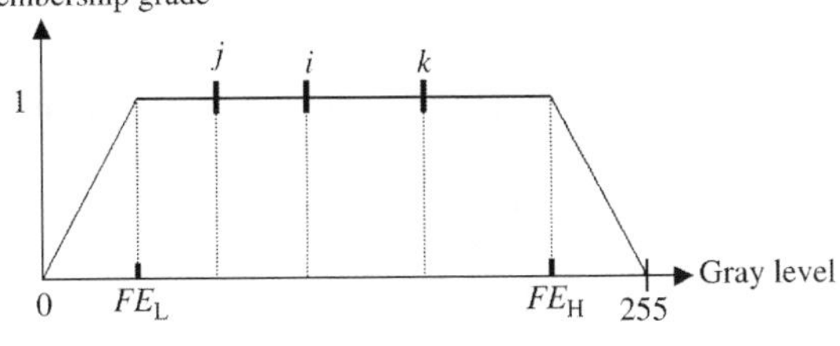

Figure 5.38 The membership functions of fuzzy interval.

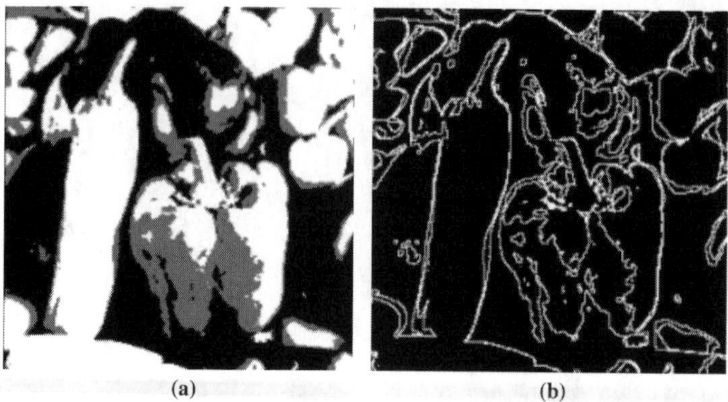

Figure 5.39 (a) The result of performing FID on Figure 5.36(d) and (b) the result after edge detection.

image is considered as an original region. Figure 5.39(a) is obtained by performing FID on Figure 5.36(d), and Figure 5.39(b) shows the result of Figure 5.39(a) after edge detection. Note that Figure 5.39(a) also shows the result after performing the TDRD-based image segmentation, in which all subregions derived require only at most one dividing procedure.

The Final Intensity Determination Algorithm

1. Set up *L-R* type fuzzy subsets FM_i, FM_j, FM_k, and $\mathrm{FM}_{\mathrm{FE}}$ using arrays. For example, $\mathrm{FM}_i = \text{array}\ \mathrm{fm}_i[h]$, where $0 \le h \le 255$ and $0 \le \text{array}\ \mathrm{fm}_i[h] \le 1$.
2. For each suspicious pixel (n, m), calculate $\mathrm{Fuzzy}_i(n,m)$, $\mathrm{Fuzzy}_j(n,m)$, $\mathrm{Fuzzy}_k(n,m)$, and $\mathrm{Fuzzy}_{\mathrm{FE}}(n,m)$.
3. If $(|\mathrm{Fuzzy}_j(n,m) - \mathrm{Fuzzy}_{\mathrm{FM}}(n,m)| < |\mathrm{Fuzzy}_i(n,m) - \mathrm{Fuzzy}_{\mathrm{FM}}(n,m)|)$ and $(|\mathrm{Fuzzy}_j(n,m) - \mathrm{Fuzzy}_{\mathrm{FM}}(n,m)| < |\mathrm{Fuzzy}_k(n,m) - \mathrm{Fuzzy}_{\mathrm{FM}}(n,m)|)$ then $\mathrm{Fin}(n, m) = 0$
   else If $(|\mathrm{Fuzzy}_i(n,m) - \mathrm{Fuzzy}_{\mathrm{FM}}(n,m)| \le |\mathrm{Fuzzy}_j(n,m) - \mathrm{Fuzzy}_{\mathrm{FM}}(n,m)|)$ and $(|\mathrm{Fuzzy}_i(n,m) - \mathrm{Fuzzy}_{\mathrm{FM}}(n,m)| \le |\mathrm{Fuzzy}_k(n,m) - \mathrm{Fuzzy}_{\mathrm{FM}}(n,m)|)$ then $\mathrm{Fin}(n, m) = 2$
   else $\mathrm{New}(n, m) = 1$.
4. Repeat steps 1 to 3 to determine the final intensity for all suspicious pixels.

5.7.3.2 Subregion Examination Strategy In the top–down dividing image segmentation, two factors of size and homogeneity are used to determine whether a subregion needs to be further segmented. The size factor is the total number of pixels in the subregion, and the homogeneity factor is its standard deviation. The subregion is continuously segmented if its size is larger than a predefined size threshold and its homogeneity value is larger than a predefined standard deviation threshold. Otherwise, the subregion is not considered for further segmentation. Note that the criteria are varied with the purpose of using the TDRD-based segmentation algorithm. For example, for an application to the breast boundary segmentation, only one decomposing iteration is enough since we do not need to consider the detailed patterns of the breast. If the TDRD-based image segmentation is merely used to reduce the gray levels of an image such as the texture feature extraction in the application of mass detection, more decomposing iterations would be required.

5.7.4 Experimental Results

We perform our algorithm using two dividing iterations on an input image in Figure 5.36 (a). After obtaining Figure 5.39(a), performing one dividing iteration, we continuously segment all the subregions to achieve two dividing iterations for an input image. The results of performing the second iteration of our TDRD-based image simplification are shown in Figure 5.40. Since there are three classes in white, gray, and black in Figure 5.39(a), the results of the second iteration are shown below for each class.

1. Figure 5.40(a1)–(a4): The processing on the "black" class in Figure 5.39(a).
2. Figure 5.40(b1)–(b4): The processing on the "gray" class in Figure 5.39(a).
3. Figure 5.40(c1)–(c4): The processing on the "white" class in Figure 5.39(a).

Figure 5.40(a1) and (a2) shows the HE_2 and HE_3 of the "black." Figure 5.40(a3) shows the suspicious pixels represented as "light gray" and "dark gray." Note that the white pixels in Figure 5.40 indicate "ignored pixels," which are not considered during the second iteration of the TDRD-based image segmentation. Figure 5.40(a4) shows the result after performing the FID procedure on the suspicious pixels of Figure 5.40 (a3). Similarly, Figure 5.40(b1), (b2), (b3), and (b4) and (c1), (c2), (c3), and (c4) corresponds to HE_2,HE_3, the suspicious-pixel image, and the final result after performing the second iteration of the TDRD-based segmentation, respectively, on the "gray" and "white" areas of Figure 5.39(a). Figure 5.41 shows the final result of collecting all subregions in Figure 5.40(a4), (b4), and (c4).

Figure 5.42 shows the results when we fix the homogeneity threshold (Th_H) but change the size threshold (Th_S). Figure 5.42(a) shows four original images: vegetable, Lena, cameraman, and baboon. Figure 5.42(b), (c), and (d) shows the results after applying the TDRD-based image segmentation with Th_S = 500, 250, and 50, respectively. Th_H is fixed to be 10. The number of connected components (NCC) and the within-class standard deviation (WCSD) in each condition are listed in Table 5.1. Note that the WCSD is defined below and is used to represent the homogeneity of all the segmented subregions.

$$\mathrm{WCSD} = \sqrt{\sum_{i=1}^{m} \left(\frac{N_i \sigma_i^2}{N_{\mathrm{Total}}} \right)} \tag{5.37}$$

where m is the total number of subregions in the segmented image, N_i is the number of pixels in the ith subregion, and $\sigma_i^2 = \sum_{k=1}^{N_i} (I_i(k) - \overline{I_i})$ in which $\overline{I_i}$ and $I_i(k)$ are the average intensity and the intensity of the kth pixels in ith subregion, respectively. The WCSD measures the average homogeneity of all subregions in a given image. The ideal value of WCSD is 0, which indicates that each subregion has the same pixel intensity. In practice, the WCSD is subject to the range of pixel intensities. In this chapter, all images are normalized by the pixel intensities ranging from 0 to 255. We estimate that the good WCSD range is from 0 to 12, and it would be bad if the WCSD value is larger than 24.

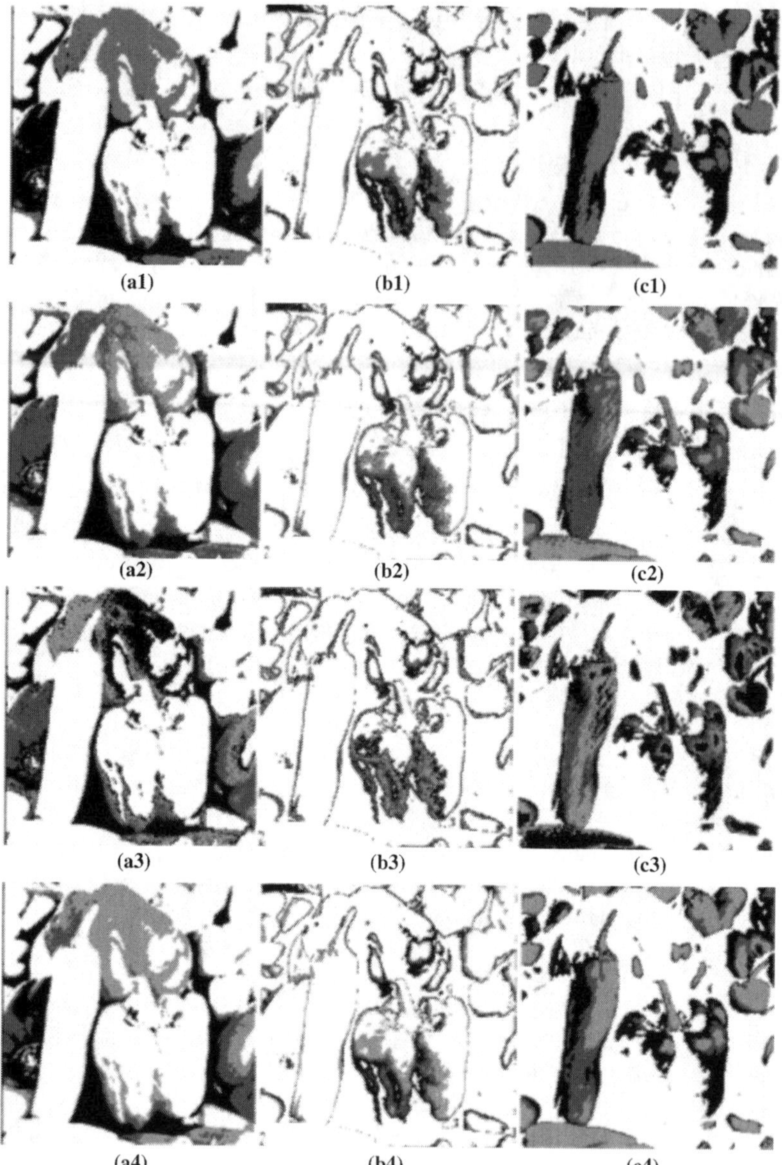

Figure 5.40 An example of performing the second iteration of the TDRD-based image segmentation on each class of Figure 5.39(a).

Figure 5.43 shows the results when we fix the size threshold (Th_S) but change the homogeneity threshold (Th_H). Figure 5.43(a), (b), and (c) shows the results after applying the TDRD-based image segmentation with Th_H = 40, 25, and 10, respectively. Th_S is fixed to be 100. The NCC and the WCSD values in each condition are presented in Table 5.2.

Figure 5.41 The final result of performing the second iteration of the TDRD-based image segmentation.

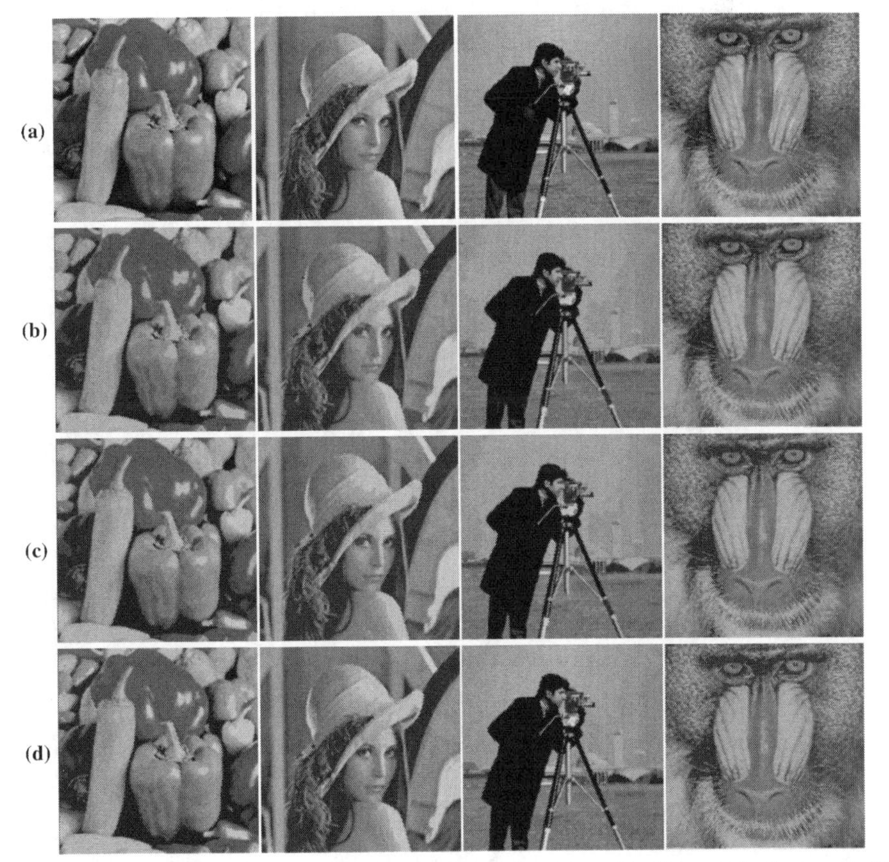

Figure 5.42 An example of performing the TDRD algorithm with different Th_S when Th_H is fixed.

TABLE 5.1 The NCC and WCSD Values When Performing the TDRD Algorithm with (I) Th_H = 10, Th_S = 500, (II) Th_H = 10, Th_S = 250, and (III) Th_H = 10, Th_S = 50

		Vegetable	Lena	Cameraman	Baboon
(I)	NCC	1556	2294	1407	8876
	WCSD	11.84	8.82	11.69	12.57
(II)	NCC	2055	2226	1519	9296
	WCSD	10.92	8.62	11.47	12.34
(III)	NCC	3555	3401	2396	11476
	WCSD	9.14	7.91	10.10	11.38

We compare the performance of the TDRD-based image segmentation against the watershed segmentation provided in Matlab and the histogram-based image segmentation presented by Tobias and Seara (2002). Since the watershed algorithm contains the problem of oversegmentation, the thresholds (Th_S and Th_H) are intentionally set as low as possible, so that the NCC derived in the TDRD approach is close to the one derived in watershed. Figure 5.44 shows the segmentation results of the watershed and our TDRD-based approaches. Figure 5.44(b) and (a) shows the results of watershed segmentation with and without displaying the edges, respectively. Similarly, Figure 5.44(d) and (c) shows the results of our TDRD-based segmentation with and without displaying the edges, respectively. The thresholds (Th_S, Th_H) are (100, 8), (100, 8), (80, 2), and (30, 16) for images "vegetable," "Lena," "cameraman," and "baboon," respectively. The detailed results are listed in

Figure 5.43 An example of performing the TDRD algorithm with different Th_H when Th_S is fixed.

TABLE 5.2 The NCC and WCSD Values When Performing the TDRD Algorithm with (I) Th_S = 100, Th_H = 40, (II) Th_S = 100, Th_H = 25, and (III) Th_S = 100, Th_H = 10

		Vegetable	Lena	Cameraman	Baboon
(I)	NCC	230	267	365	1586
	WCSD	22.95	19.12	16.97	21.17
(II)	NCC	753	267	694	3305
	WCSD	15.59	19.12	13.84	17.64
(III)	NCC	3198	2958	2213	10290
	WCSD	8.89	7.50	10.44	11.85

Table 5.3. Note that the watershed segmentation will somehow destroy the spatial structure of an image since the strong edges will be ignored if there is only one catchment basin in the area as shown in Figure 5.33. Figure 5.44(a) again shows the drawback of watershed segmentation. Our TDRD-based segmentation will avoid the spatial-structure destruction problem, so the WCSD derived in our algorithm is much lower than the one from the watershed segmentation.

Figure 5.44 The results of the watershed and the TDRD-based image segmentations.

TABLE 5.3 The NCC and WCSD Values When Performing the Watershed and the TDRD Segmentations in Four Images

		Vegetable	Lena	Cameraman	Baboon
Watershed	NCC	3686	3744	4287	5663
	WCSD	28.51	24.80	21.70	23.39
TDRD	NCC	3782	3443	4341	5625
	WCSD	8.54	6.98	9.66	15.60
Original image	NCC	1	1	1	1
	SD	54.06	47.95	62.35	42.57

TABLE 5.4 The NCC and WCSD Values When Performing the Histogram and the TDRD Segmentations in Four Images

		Vegetable	Lena	Camera man	Baboon
Histogram	NCC	130	210	213	850
	WCSD	27.05	25.12	29.40	23.34
TDRD	NCC	228	300	307	1217
	WCSD	22.92	19.12	16.97	21.17

(Tobias and Seara, 2002)

The histogram-based image segmentation by Tobias and Seara (2002) segments a given image into a binary image. To compare our TDRD-based and their histogram-based algorithm, only one decomposing iteration is performed. Table 5.4 shows the comparison.

In order to conduct numerical analysis, the 1000 images used in the SIMPLIcity paper (Wang et al., 2001) were downloaded for testing our TDRD-based algorithm. Tables 5.5 and 5.6 show the numerical results when performing the watershed and the TDRD-based algorithm, in which two parameter sets, parameter_1 (Th_S = 150, Th_H = 15) and parameter_2 (Th_S = 300, Th_H = 30), are applied. There are 81%, 40%, and 20% of images having NCC values that are smaller than or equal to 2000 for the watershed, TDRD-parameter_1, and TDRD-parameter_2, respectively, and 55%, 91%, and 67% of images having WCSD values that are smaller than or equal to 15 for the watershed, TDRD-parameter_1, and TDRD-parameter_2, respectively. The

TABLE 5.5 The NCC Values When Performing the Watershed and the TDRD-Based Algorithm in Two Parameter Sets to the 1000 Images Used in the SIMPLIcity Paper

	<1000	1000	2000	3000	4000	5000	6000	>6000
Watershed	0	65	135	185	203	174	150	88
TDRD(150, 15)	32	212	155	120	192	190	46	53
TDRD(300, 30)	191	425	189	112	56	17	5	5

(Wang et al., 2001)

TABLE 5.6 The WCSD Values When Performing the Watershed and the TDRD-Based Algorithm in Two Parameter Sets to the 1000 Images Used in the SIMPLIcity Paper

	5	10	15	20	25	>25
Watershed	4	72	473	326	82	43
TDRD(150, 15)	26	690	196	48	30	10
TDRD(300, 30)	10	244	416	280	40	10

(Wang et al., 2001)

results show that both the NCC and WCSD values in the TDRD-based algorithm are smaller than those in the watershed approach.

The complexity of the TDRD-based algorithm relies on the total number of pixels (Num_Image), the number of decomposing iterations (Num_Iteration), and the required calculations (e.g., the important intensities and their corresponding fuzzy subsets) in each iteration. By using the array structure, the required calculation of each connected component can be completed within a certain number of image scans. The complexity of calculating the required information of all connected components and determining the final intensity of suspicious pixels is $O(\text{Num\_Image})$. Therefore, the overall computation complexity of the TDRD algorithm is $O(\text{Num\_Image} \times \text{Num\_Iteration})$.

5.7.5 Potential Applications in Medical Image Analysis

Medical image analysis is important since it provides assistance for medical doctors to find out the diseases inside the body without the surgery procedure (Duncan and Ayache, 2000). The TDRD-based image segmentation provides useful applications due to the properties of medical images. Generally, medical images contain three major regions: background, soft tissue, and object. We present the ideas of two potential applications, breast boundary segmentation and lung segmentation. Further detailed implementation is still under investigation.

5.7.5.1 Breast Boundary Segmentation Mammography is a useful tool for detecting breast cancer (Zuckerman, 1987). It has been shown that CAD can provide assistance to radiologists to increase the cancer detection rate (Chan et al., 1990). Automated detection of breast boundary is the fundamental step in the mammographic CAD system (Wu et al., 2006b). A breast boundary detection algorithm was presented in Wu et al. (2007b) based on multiple thresholding. However, it can be improved by incorporating more related information. Figure 5.45 shows an example of applying the TDRD approach in breast boundary segmentation. Figure 5.45(a) shows a mammogram image obtained from the mammography database at http://marathon.csee.usf.edu/Mammography/Database.html. Figure 5.45(b) shows the segmented result after the first iteration of the TDRD-based approach. The image is segmented into three regions marked in dark gray, gray, and light gray indicating background, soft tissue, and breast areas, respectively. Most breast boundary detection algorithms (Wu et al., 2006b) obtain

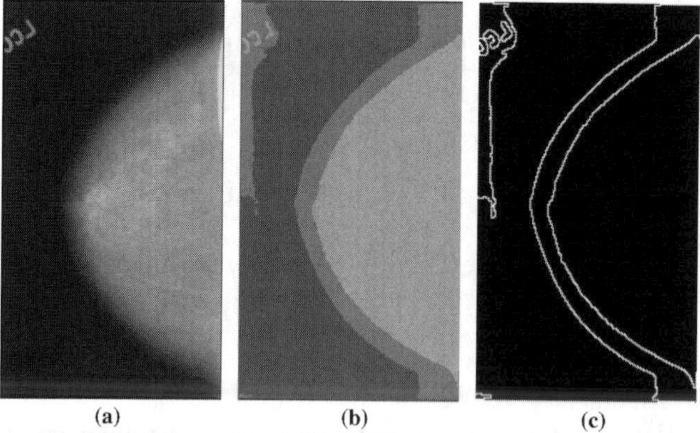

Figure 5.45 An example of applying the TDRD approach in breast boundary segmentation.

an initial breast area by analyzing the histogram followed by a refinement procedure to determine the final breast boundary. However, they may fail if the initial breast boundary obtained from the thresholding approach is incorrect, such that the refinement procedure is led to the wrong direction. After edge detection, we obtain two outer and inner breast boundaries, as shown in Figure 5.45(c), in which the outer is close to the true boundary and the inner can evaluate the correctness of the outer boundary using the slopes of two boundaries in the corresponding positions.

5.7.5.2 Lung Segmentation Lung segmentation plays an important role in the CAD systems. Appropriate segmentation helps lung nodule detection (Ge et al., 2005) and alignment of lung nodules from multiple chest CT scans (Shi et al., 2007). One lung cancer CAD system uses the k-means clustering as the initial segmentation. Its disadvantage is the high computational cost. Since the proposed segmentation method is computationally efficient, it can be used to substitute for the k-means clustering. Figure 5.46(a) shows a typical chest CT image from the LIDC

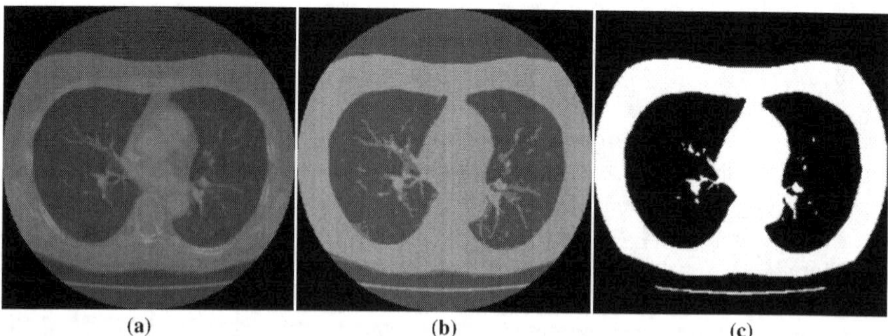

Figure 5.46 An example of applying the TDRD approach in lung segmentation.

database (see http://imaging.cancer.gov/programsandresources/InformationSystems/LIDC). The left and right lungs with low intensities are well separated by the mediastinum in the middle. Using the new method with one dividing iteration yields Figure 5.46(b), where the three major regions are marked in dark gray, gray, and light gray indicating the lung, vessel, and body areas, respectively. The profiles of two lungs are delineated roughly as shown in Figure 5.46(c) by a simple thresholding to keep the light gray area. Furthermore, it is possible to build a 3D vessel tree by considering the gray area in Figure 5.46(b), so that the false positive can be reduced.

REFERENCES

Adams, R. and Bischof, L., "Seeded region growing," *IEEE Trans. Pattern Anal. Mach. Intell.*, vol. 16, no. 6, pp. 641–647, June 1994.

Ahmad, M. B. and Choi, T., "Local threshold and Boolean function based edge detection," *IEEE Trans. Consumer Electron.*, vol. 45, no. 3, pp. 674–679, Aug. 1999.

Athanasiadis, T., Mylonas, P., Avrithis, Y., and Kollias, S., "Semantic image segmentation and object labeling," *IEEE Trans. Circuits Syst. Video Technol.*, vol. 17, no. 3, pp. 298–312, Mar. 2007.

Awad, M., Chehdi, K., and Nasri, A., "Multicomponent image segmentation using a genetic algorithm and artificial neural network," *IEEE Trans. Geosci. Remote Sens. Lett.*, vol. 4, no. 4, pp. 571–575, Oct. 2007.

Bao, P., Zhang, L., and Wu, X., "Canny edge detection enhancement by scale multiplication," *IEEE Trans. Pattern Anal. Mach. Intell.*, vol. 27, no. 9, pp. 1485–1490, Sept. 2005.

Belloulata, K. and Konrad, J., "Fractal image compression with region-based functionality," *IEEE Trans. Image Process.*, vol. 11, no. 4, pp. 351–362, Apr. 2002.

Canny, J., "A computational approach to edge detection," *IEEE Trans. Pattern Anal. Mach. Intell.*, vol. 8, no. 6, pp. 679–698, Nov. 1986.

Cardoso, J. S. and Corte-Real, L., "Toward a generic evaluation of image segmentation," *IEEE Trans. Image Process.*, vol. 14, no. 11, pp. 1773–1782, Nov. 2005.

Chai, D. and Bouzerdoum, A., "A Bayesian approach to skin color classification in YCbCr color space," Proceedings of IEEE TENCON 2000, Kuala Lumpur, Malaysia, pp. 421–424, 2000.

Chan, H. P., Doi, K., Vyborny, C. J., Schmidt, R. A., Metz, C. E., Lam, K. L., Ogura, T., Wu, Y., and Macmahon, H., "Improvement in radiologists' detection of clustered microcalcifications on mammograms: the potential of computer-aided diagnosis," *Invest. Radiol.*, 25, pp. 1102–1110, 1990.

Chen, Y. and Wang, J. Z., "A region-based fuzzy feature matching approach to content-based image retrieval," *IEEE Trans. Pattern Anal. Mach. Intell.*, vol. 24, no. 9, pp. 1252–1267, Sept. 2002.

Chen, S., Sochen, N. A., and Zeevi, Y. Y., "Integrated active contours for texture segmentation," *IEEE Trans. Image Process.*, vol. 15, no. 6, pp. 1633–1646, June 2006.

Cheng, J. and Foo, S., "Dynamic directional gradient vector flow for snakes," *IEEE Trans. Image Process.*, vol. 15, no. 6, pp. 1563–1571, June 2006.

Cheng, H. D., Jiang, X. H., Sun, Y., and Wang, J., "Color image segmentation: advance and prospects," *Pattern Recogn.*, vol. 34, no. 12, pp. 2259–2281, Dec. 2001.

Chibanga, R., Berlamont, J., and Vandewalle, J., "Artificial neural networks in hydrological watershed modeling: surface flow contribution from the ungauged parts of a catchment," Proceedings of the 13th IEEE International Conference on Tools with Artificial Intelligence, Dallas, TX, pp. 367–374, Nov. 2001.

Davis, L. S., "A survey of edge detection technique," *Comput. Graph. Image Process.*, 4, pp. 248–270, 1974.

Deng, Y. and Manjunath, B. S., "Unsupervised segmentation of color–texture regions in images and video," *IEEE Trans. Pattern Anal. Mach. Intell.*, vol. 23, no. 8, pp. 800–810, Aug. 2001.

Donoho, D., Johnstone, I., Kerkyacharian, G., and Picard, D., "Density estimation by wavelet thresholding," *Ann. Stat.*, vol. 24, no. 2, pp. 508–539, 1996.

Duncan, J. S. and Ayache, N., "Medical image analysis: progress over two decades and the challenges ahead," *IEEE Trans. Pattern Anal. Mach. Intell.*, vol. 22, no. 1, pp. 85–106, Jan. 2000.

Fan, J., Yau, D. K., Elmagarmid, A. K., and Aref, W. G., "Automatic image segmentation by integrating color-edge extraction and seeded region growing," *IEEE Trans. Image Process.*, vol. 10, no. 10, pp. 1454–1466, Oct. 2001.

Garcia, C. and Tziritas, G., "Face detection using quantized skin color regions merging and wavelet packet analysis," *IEEE Trans. Multimedia*, vol. 1, no. 3, pp. 264–277, Sept. 1999.

Ge, Z., Sahiner, B., Chan, H. P., Hadjiiski, L. M., Cascade, P. N., Bogot, N., Kazerooni, E. A., Wei, J., and Zhou, C., "Computer aided detection of lung nodules: false positive reduction using a 3D gradient field method and 3D ellipsoid fitting," *Med. Phys.*, vol. 32, no. 8, pp. 2443–2454, Aug. 2005.

Hajjar, A. and Chen, T., "A VLSI architecture for real-time edge linking," *IEEE Trans. Pattern Anal. Mach. Intell.*, vol. 21, no. 1, pp. 89–94, Jan. 1999.

Haris, K., Efstratiadis, S. N., Maglaveras, N., and Katsaggelos, A. K., "Hybrid image segmentation using watersheds and fast region merging," *IEEE Trans. Image Process.*, vol. 7, no. 12, pp. 1684–1699, Dec. 1998.

Hartmann, S. L. and Galloway, R. L., "Depth-buffer targeting for spatially accurate 3-D visualization of medical images," *IEEE Trans. Med. Imaging*, vol. 19, no. 10, pp. 1024–1031, Oct. 2000.

He, L., Chao, Y., and Suzuki, K., "A run-based two-scan labeling algorithm," *IEEE Trans. Image Process.*, vol. 17, no. 5, pp. 749–756, May 2008.

Hsu, R.-L., Abdel-Mottaleb, M., and Jain, A. K., "Face detection in color image," *IEEE Trans. Pattern Anal. Mach. Intell.*, vol. 24, no. 5, pp. 696–706, May 2002.

Ji, L. and Yan, H., "Attractable snakes based on the greedy algorithm for contour extraction," *Pattern Recogn.*, vol. 35, no. 4, pp. 791–806, Apr. 2002.

Kass, M., Witkin, A., and Terzopoulos, D., "Snakes: active contour models," *Int. J. Comput. Vision*, vol. 1, no. 4, pp. 321–331, 1987.

Khotanzad, A. and Chen, J. Y., "Unsupervised segmentation of textured images by edge detection in multidimensional feature," *IEEE Trans. Pattern Anal. Mach. Intell.*, vol. 11, no. 4, pp. 414–421, Apr. 1989.

Konishi, S., Yuille, A. L., Coughlan, J. M., and Zhu, S., "Statistical edge detection: learning and evaluating edge cues," *IEEE Trans. Pattern Anal. Mach. Intell.*, vol. 25, no. 1, pp. 57–74, Jan. 2003.

Law, T., Itoh, H., and Seki, H., "Image filtering, edge detection, and edge tracing using fuzzy reasoning," *IEEE Trans. Pattern Anal. Mach. Intell.*, vol. 18, no. 5, pp. 481–491, May 1996.

Lee, C.-S., Kuo, Y.-H., and Yu, P.-T., "Weighted fuzzy mean filters for image processing," *Fuzzy Sets Syst.*, vol. 89, no. 2, pp. 157–180, 1997.

Li, C., Xu, C., Gui, C., and Fox, M. D., "Level set evolution without re-initialization: a new variational formulation," Proceedings of the IEEE Conference on Computer Vision and Pattern Recognition, San Diego, CA, 1, pp. 430–436, June 2005.

Liu, S.-M., Lin, W.-C., and Liang, C.-C., "An iterative edge linking algorithm with noise removal capability," Proceedings of the 9th International Conference on Pattern Recognition, Rome, Italy, pp. 1120–1122, Nov. 1988.

Meer, P. and Georgescu, B., "Edge detection with embedded confidence," *IEEE Trans. Pattern Anal. Mach. Intell.*, vol. 23, no. 12, pp. 1351–1365, Dec. 2001.

Mehnert, A. and Jackway, P., "An improved seeded region growing algorithm," *Pattern Recogn. Lett.*, vol. 18, no. 10, pp. 1065–1071, Oct. 1997.

Melonakos, J., Pichon, E., Angenent, S., and Tannenbaum, A., "Finsler active contours," *IEEE Trans. Pattern Anal. Mach. Intell.*, vol. 30, no. 3, pp. 412–423, Mar. 2008.

Michailovich, O., Rathi, Y., and Tannenbaum, A., "Image segmentation using active contours driven by the Bhattacharyya gradient flow," *IEEE Trans. Image Process.*, vol. 16, no. 11, pp. 2787–2801, Nov. 2007.

Moga, A. N. and Gabbouj, M., "Parallel image component labelling with watershed transformation," *IEEE Trans. Pattern Anal. Mach. Intell.*, vol. 19, no. 5, pp. 441–450, May 1997.

Nalwa, V. S. and Pauchon, E., "Edgel aggregation and edge description," *Comput. Vision Graph. Image Process.*, vol. 40, no. 1, pp. 79–94, Oct. 1987.

Nevatia, R., "Locating objects boundaries in textured environments," *IEEE Trans. Comput.*, vol. 25, no. 11, pp. 1170–1175, Nov. 1976.

Otsu, N., "A threshold selection method from gray-level histogram," *IEEE Trans. Syst. Man Cybernet.*, vol. 9, no. 1, pp. 62–66, Jan. 1979.

Pachowicz, P. W., "Semi-autonomous evolution of object models for adaptive object recognition," *IEEE Trans. Syst. Man Cybernet.*, vol. 24, no. 8, pp. 1191–1207, Aug. 1994.

Pal, N. R. and Pal, S. K., "A review on image segmentation techniques," *Pattern Recogn.*, vol. 26, no. 9, pp. 1277–1294, Sept. 1993.

Panjwani, D. K. and Healey, G., "Markov random field models for unsupervised segmentation of textured color images," *IEEE Trans. Pattern Anal. Mach. Intell.*, vol. 17, no. 10, pp. 939–954, Oct. 1995.

Papari, G. and Petkov, N., "Adaptive pseudo dilation for Gestalt edge grouping and contour detection," *IEEE Trans. Image Process.*, vol. 17, no. 10, pp. 1950–1962, Oct. 2008.

Pavlidis, T. and Liow, Y. T., "Integrating region growing and edge detection," *IEEE Trans. Pattern Anal. Mach. Intell.*, vol. 12, no. 3, pp. 225–233, Mar. 1990.

Pyun, K. P., Lim, J., Won, C., and Gray, R. M., "Image segmentation using hidden Markov Gauss mixture models," *IEEE Trans. Image Process.*, vol. 16, no. 7, pp. 1902–1911, July 2007.

Rivera, M., Ocegueda, O., and Marroquin, J. L., "Entropy-controlled quadratic Markov measure field models for efficient image segmentation," *IEEE Trans. Image Process.*, vol. 16, no. 12, pp. 3047–3057, Dec. 2007.

Russ, J. C., *The Image Processing Handbook*, CRC Press, 1992.

Saha, P. K. and Udupa, J. K., "Optimum image thresholding via class uncertainty and region homogeneity," *IEEE Trans. Pattern Anal. Mach. Intell.*, vol. 23, no. 7, pp. 689–706, July 2001.

Shen, D. and Davatzikos, C., "An adaptive-focus deformable model using statistical and geometric information," *IEEE Trans. Pattern Anal. Mach. Intell.*, vol. 22, no. 8, pp. 906–913, Aug. 2000.

Shi, J., Sahiner, B., Chan, H.-P., Hadjiiski, L., Zhou, C., Cascade, P. N., Bogot, N., Kazerooni, E. A., Wu, Y.-T., and Wei, J., "Pulmonary nodule registration in serial CT scans based on rib anatomy and nodule template matching," *Med. Phys.*, vol. 34, no. 4, pp. 1336–1347, Apr. 2007.

Shih, F. Y. and Cheng, S., "Adaptive mathematical morphology for edge linking," *Inform. Sci.*, vol. 167, no. 1–4, pp. 9–21, Dec. 2004.

Shih, F. Y. and Cheng, S., "Automatic seeded region growing for color image segmentation," *Image Vision Comput.*, vol. 23, no. 10, pp. 877–886, Sept. 2005.

Shih, F. Y. and Zhang, K., "Efficient contour detection based on improved snake model," *Pattern Recogn. Artif. Intell.*, vol. 18, no. 2, pp. 197–209, Mar. 2004.

Sofou, A. and Maragos, P., "Generalized flooding and multicue PDE-based image segmentation," *IEEE Trans. Image Process.*, vol. 17, no. 3, pp. 364–376, Mar. 2008.

Stahl, J. S. and Wang, S., "Edge grouping combining boundary and region information," *IEEE Trans. Image Process.*, vol. 16, no. 10, pp. 2590–2606, Oct. 2007.

Stahl, J. S. and Wang, S., "Globally optimal grouping for symmetric closed boundaries by combining boundary and region information," *IEEE Trans. Pattern Anal. Mach. Intell.*, vol. 30, no. 3, pp. 395–411, Mar. 2008.

Sundaramoorthi, G., Yezzi, A., and Mennucci, A., "Sobolev active contours," *Int. J. Comput. Vision*, vol. 73, no. 3, pp. 345–366, July 2007.

Sundaramoorthi, G., Yezzi, A., and Mennucci, A. C., "Coarse-to-fine segmentation and tracking using Sobolev active contours," *IEEE Trans. Pattern Anal. Mach. Intell.*, vol. 30, no. 5, pp. 851–864, May 2008.

Tao, W., Jin, H., and Zhang, Y., "Color image segmentation based on mean shift and normalized cuts," *IEEE Trans. Syst. Man Cybernet. B*, vol. 37, no. 5, pp. 1382–1389, Oct. 2007.

Tao, W., Jin, H., Zhang, Y., Liu, L., and Wang, D., "Image thresholding using graph cuts," *IEEE Trans. Syst. Man Cybernet. A*, vol. 38, no. 5, pp. 1181–1195, Sept. 2008.

Tobias, O. J. and Seara, R., "Image segmentation by histogram thresholding using fuzzy sets," *IEEE Trans. Image Process.*, vol. 11, no. 12, pp. 1457–1465, Dec. 2002.

Unnikrishnan, R., Pantofaru, C., and Hebert, M., "Toward objective evaluation of image segmentation algorithms," *IEEE Trans. Pattern Anal. Mach. Intell.*, vol. 29, no. 6, pp. 929–944, June 2007.

Vincent, L. and Soille, P., "Watersheds in digital spaces: an efficient algorithm based on immersion simulations," *IEEE Trans. Pattern Anal. Mach. Intell.*, vol. 13, no. 6, pp. 583–198, June 1991.

Wang, J. Z., Li, J., and Wiederhold, G., "SIMPLIcity: semantics-sensitive integrated matching for picture libraries," *IEEE Trans. Pattern Anal. Mach. Intell.*, vol. 23, no. 9, pp. 947–963, Sept. 2001.

Wong, W. and Chung, A., "Bayesian image segmentation using local iso-intensity structural orientation," *IEEE Trans. Image Process.*, vol. 14, no. 10, pp. 1512–1523, Oct. 2005.

Wu, J. and Chung, A., "A segmentation model using compound Markov random fields based on a boundary model," *IEEE Trans. Image Process.*, vol. 16, no. 1, pp. 241–252, Jan. 2007.

Wu, Y.-T., Shih, F. Y., and Wu, Y.-T., "Differential histogram equalization based image simplification," Proceedings of the Sixth IEEE International Conference on Electro/Information Technology, East Lansing, MI, USA, pp. 344–349, May 2006a.

Wu, Y.-T., Hadjiiski, L. M., Wei, J., Zhou, C., Sahiner, B., and Chan, H.-P., "Computer-aided detection of breast masses on mammograms: bilateral analysis for false positive reduction," Proceedings of the 2006 SPIE on Medical Imaging, San Diego, CA, Feb. 2006b.

Wu, J., Yin, Z. P., and Xiong, Y., "The fast multilevel fuzzy edge detection of blurry images," *IEEE Trans. Signal Process. Lett.*, vol. 14, no. 5, pp. 344–347, May 2007a.

Wu, Y.-T., Zhou, C., Hadjiiski, L. M., Shi, J., Paramagul, C., Sahiner, B., and Chan, H.-P., "A dynamic multiple thresholding method for automated breast boundary detection in digitized mammograms," Proceedings of the 2007 SPIE Medical Imaging, San Diego, CA, Feb. 2007b.

Xie, X. and Mirmehdi, M., "MAC: magnetostatic active contour model," *IEEE Trans. Pattern Anal. Mach. Intell.*, vol. 30, no. 4, pp. 632–646, Apr. 2008.

Xu, Y., Saber, E., and Tekalp, A. M., "Object segmentation and labeling by learning from examples," *IEEE Trans. Image Process.*, vol. 12, no. 6, pp. 627–638, June 2003.

Zenzo, D. S., Cinque, L., and Levialdi, S., "Image thresholding using fuzzy entropies," *IEEE Trans. Syst. Man Cybernet. B*, vol. 28, no. 1, pp. 15–23, Feb. 1998.

Zimmermann, H. J., *Fuzzy Set Theory and Its Applications*, Kluwer Academic Publishers, Boston, MA, 1991.

Zuckerman, H. C., "The role of mammography in the diagnosis of breast cancer," *Breast Cancer: Diagnosis and Treatment* (I. M. Ariel and J. B. Cleary, eds.), pp. 152–172, McGraw-Hill, 1987.

CHAPTER 6

DISTANCE TRANSFORMATION AND SHORTEST PATH PLANNING

A grayscale image after image segmentation is changed into a binary image, which contains object (foreground) and nonobject (background) pixels. Distance transformation (DT) is used to convert a binary image into another image where each object pixel has a value corresponding to the minimum distance from the background by a distance function (Vincent, 1991). The distance transformation has wide applications in image analysis. One of the applications is computing a shape factor, which measures shape compactness based on the ratio of total number of object pixels and the summation of all distance measures (Danielsson, 1978). Another is obtaining the medial axis (or skeleton), which is used for feature extractions (Blum, 1967; Hilditch, 1969; Lantuejoul, 1980). Computing the distance from an object pixel to background is a global operation, which is often prohibitively costly; therefore, a decomposition strategy using local neighborhood operations is needed to speed up the computation (Shih and Mitchell, 1992).

Advances in robotics, artificial intelligence, and computer vision have stimulated considerable interest in the problems of robot motion and shortest path planning (Latombe, 1991). The path planning problem is concerned with finding paths connecting different locations in an environment (e.g., a network, a graph, or a geometric space). Depending on specific applications, the desired paths often need to satisfy some constraints (e.g., obstacle avoiding) and optimize certain criteria (e.g., variant distance metrics and cost functions). Two approaches in shortest path planning are often used: one is via configuration space where the geometry of a robot is added to the obstacle to create a new free space (Lozano-Perez, 1983; Pei et al., 1998), and the other is via computational geometry where the space is searched directly without transforming the workspace to the configuration space (Brooks, 1983; Kambhampati and Davis, 1994).

This chapter is organized as follows. The general concept of distance transformation is introduced in Section 6.1. Distance transformation by mathematical morphology is outlined in Section 6.2. Approximation of Euclidean distance is described in Section 6.3. The decomposition of distance structuring element is presented in Section 6.4. The 3D Euclidean distance is discussed in Section 6.5. Distance transformations by acquiring and deriving approaches are described in Sections 6.6 and 6.7, respectively. The shortest path planning is presented in

Image Processing and Pattern Recognition by Frank Shih

Section 6.8. A new way of forward and backward chain codes for motion planning is introduced in Section 6.9. A few examples are given in Section 6.10.

6.1 GENERAL CONCEPT

The interior of a closed object boundary is considered as object pixels and the exterior as background pixels. Let S denote a set of object pixels in an image. The function d mapping a binary image containing S to a matrix of nonnegative integers is called *a distance function*, if it satisfies the following three criteria:

(a) *Positive definite:* That is, $d(p,q) \geq 0$, if and only if $p=q$, for all $p,q \in S$.

(b) *Symmetric:* That is, $d(p,q) = d(q,p)$, for all $p,q \in S$.

(c) *Triangular:* That is, $d(p,r) \leq d(p,q)+d(q,r)$, for all $p,q,r \in S$.

The approaches to achieve distance transformation do not directly adopt the definition of the minimum distance from an object pixel to all background border pixels, since their computations are extremely time-consuming. Previous researches by Shih and Mitchell (1992), Saito and Toriwaki (1994), Eggers (1998), Cuisenaire and Macq (1999) and Datta and Soundaralakshmi (2001) represent a sampled set of successful efforts in improving speed efficiency. In general, the distance transformation algorithms can be categorized into two classes: one is the *iterative* method that is efficient in a cellular array computer since all the pixels at each iteration can be processed in parallel, and the other is the *sequential* (or *recursive*) method that is suited for a conventional computer by avoiding iterations with the efficiency independent of object size. Using the general machines that most people have access to, sequential algorithms are often much more efficient than iterative ones.

Three types of distance measures in digital image processing are often used: Euclidean, city block, and chessboard. City-block and chessboard distances are easy to compute and can be recursively accumulated by considering only a small neighborhood at a time. The algorithms for city block, chessboard, or a combination of both (called *octagon*) have been widely developed (Rosenfeld and Pfaltz, 1966; Shih and Mitchell, 1987; Toriwaki and Yokoi, 1981). One distance transformation algorithm uses an iterative operation by peeling off border pixels and summing the distances layer by layer (Rosenfeld and Kak, 1982). This is quite efficient on a cellular array computer.

The *city-block distance* between two points $P=(x,y)$ and $Q=(u,v)$ is defined as

$$d_4(P,Q) = |x-u| + |y-v| \tag{6.1}$$

The *chessboard distance* between P and Q is defined as

$$d_8(P,Q) = \max(|x-u|, |y-v|) \tag{6.2}$$

The *Euclidean distance* between P and Q is defined as

$$d_e(P,Q) = \sqrt{(x-u)^2 + (y-v)^2} \tag{6.3}$$

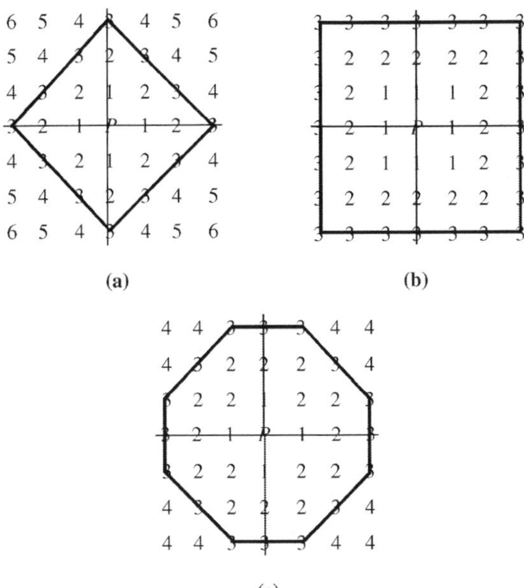

Figure 6.1 (a) City-block, (b) chessboard, and (c) octagon distances.

Note that the subscripts "4" and "8" indicate the 4-neighbor and 8-neighbor considerations in calculating the respective distances. Figure 6.1 shows (a) city-block (b) chessboard, and (c) octagon distances from a center point P, where the octagon distance applies the city-block and chessboard distances alternatively. From the shape of distance value "3", we observe that city block is like a diamond, chessboard is like a square, and the combination of both is like an octagon.

The city-block and chessboard distances satisfy the following property: For all P, Q such that $d(P, Q) \geq 2$, there exists a point R, different from P and Q, such that $d(P,Q) = d(P,R) + d(R,Q)$. Given χ_S, which is 1 at point S and 0 elsewhere, we define $\chi_S^{(m)}$ inductively for $m = 1, 2, \ldots$ as follows (Rosenfeld and Kak, 1982):

$$\chi_S^{(m)}(P) = \chi_S^{(0)}(P) + \min_{d(Q,P) \leq 1} \chi_S^{(m-1)}(Q) \tag{6.4}$$

where $\chi_S^{(0)} = \chi_S$. Note that equation (6.4) requires iterative operations, which are time-consuming when an image size is large.

Example 6.1 Apply city-block and chessboard distance transformations on the following two images:

$$\text{(a)} \quad \begin{matrix} 1 & 1 & 1 & 1 & 1 \\ 1 & 1 & 1 & 1 & 1 \\ 1 & 1 & 1 & 1 & 1 \\ 1 & 1 & 1 & 1 & 1 \\ 1 & 1 & 1 & 1 & 1 \end{matrix}, \qquad \text{(b)} \quad \begin{matrix} 0 & 0 & 1 & 0 & 0 \\ 0 & 1 & 1 & 1 & 0 \\ 1 & 1 & 1 & 1 & 1 \\ 0 & 1 & 1 & 1 & 0 \\ 0 & 0 & 1 & 0 & 0 \end{matrix}$$

Answer: The city-block and chessboard distance transformations of (a) are obtained as follows (note that both are same in this case):

$$\text{City block}: \begin{matrix} 1 & 1 & 1 & 1 & 1 \\ 1 & 2 & 2 & 2 & 1 \\ 1 & 2 & 3 & 2 & 1 \\ 1 & 2 & 2 & 2 & 1 \\ 1 & 1 & 1 & 1 & 1 \end{matrix}, \quad \text{Chessboard}: \begin{matrix} 1 & 1 & 1 & 1 & 1 \\ 1 & 2 & 2 & 2 & 1 \\ 1 & 2 & 3 & 2 & 1 \\ 1 & 2 & 2 & 2 & 1 \\ 1 & 1 & 1 & 1 & 1 \end{matrix}$$

The city-block and chessboard distance transformations of (b) are obtained as follows:

$$\text{City block}: \begin{matrix} 0 & 0 & 1 & 0 & 0 \\ 0 & 1 & 2 & 1 & 0 \\ 1 & 2 & 3 & 2 & 1 \\ 0 & 1 & 2 & 1 & 0 \\ 0 & 0 & 1 & 0 & 0 \end{matrix}, \quad \text{Chessboard}: \begin{matrix} 0 & 0 & 1 & 0 & 0 \\ 0 & 1 & 1 & 1 & 0 \\ 1 & 1 & 2 & 1 & 1 \\ 0 & 1 & 1 & 1 & 0 \\ 0 & 0 & 1 & 0 & 0 \end{matrix}$$

Another algorithm needs only two steps in different directional scans of an image: one scan is in the left-to-right, top-to-bottom direction, and the other is in the right-to-left, bottom-to-top direction. This algorithm does not require a large number of iterations and can be implemented efficiently using a conventional computer. Let $N_1(P)$ be the set of (4- or 8-) neighbors that precede P in a row-by-row (left-to-right, top-to-bottom) scan of the picture, and let $N_2(P)$ be the remaining (4- or 8-) neighbors of P. The two-step city-block and chessboard distance transformations are given as follows (Rosenfeld and Kak, 1982):

$$\chi_S'(P) = \begin{cases} 0 & \text{if} \quad P \in \bar{S} \\ \min\limits_{Q \in N_1} \quad \chi_S'(Q) + 1 & \text{if} \quad P \notin \bar{S} \end{cases} \tag{6.5}$$

$$\chi_S''(P) = \min_{Q \in N_2} \left[\chi_S'(P), \chi_S''(Q) + 1\right] \tag{6.6}$$

Thus, we can compute χ_S' in a single left-to-right, top-to-bottom scan of the image, since for each P, χ_S' has already been computed for the Q's in N_1. Similarly, we can compute χ_S'' in a single reverse scan (right-to-left, bottom-to-tp).

Example 6.2 Perform the two-step city-block and chessboard distance transformations on the following image:

$$\begin{matrix} 1 & 1 & 1 & 1 & 1 & 1 & 0 & 0 \\ 1 & 1 & 1 & 1 & 1 & 1 & 0 & 0 \\ 1 & 1 & 1 & 1 & 1 & 1 & 1 & 0 \\ 1 & 1 & 1 & 0 & 0 & 1 & 1 & 1 \\ 1 & 1 & 1 & 0 & 0 & 1 & 1 & 1 \\ 1 & 1 & 1 & 1 & 1 & 1 & 1 & 0 \end{matrix}$$

Answer: (a) Applying city-block distance transformation, the result of the first step is

$$\begin{matrix} 1&1&1&1&1&1&0&0\\ 1&2&2&2&2&2&0&0\\ 1&2&3&3&3&3&1&0\\ 1&2&3&0&0&1&2&1\\ 1&2&3&0&0&1&2&2\\ 1&2&3&1&1&2&3&0 \end{matrix}$$

and the result of the second step is

$$\begin{matrix} 1&1&1&1&1&1&0&0\\ 1&2&2&2&2&1&0&0\\ 1&2&2&1&1&2&1&0\\ 1&2&1&0&0&1&2&1\\ 1&2&1&0&0&1&2&1\\ 1&1&1&1&1&1&1&0 \end{matrix}$$

(b) Applying chessboard distance transformation, the result of the first step is

$$\begin{matrix} 1&1&1&1&1&1&0&0\\ 1&2&2&2&2&1&0&0\\ 1&2&3&3&2&1&1&0\\ 1&2&3&0&0&1&1&1\\ 1&2&1&0&0&1&2&1\\ 1&2&1&1&1&1&2&0 \end{matrix}$$

and the result of the second step is

$$\begin{matrix} 1&1&1&1&1&1&0&0\\ 1&2&2&2&2&1&0&0\\ 1&2&1&1&1&1&1&0\\ 1&2&1&0&0&1&1&1\\ 1&2&1&0&0&1&1&1\\ 1&1&1&1&1&1&1&0 \end{matrix}$$

Another algorithm applies binary morphological erosions recursively and accumulates the distances step by step (Serra, 1982). Note that all the above algorithms can only be applied to city-block and chessboard distance measures, which are very sensitive to the orientation of the object. In other words, if an object is rotated, its distance measures would be different. The Euclidean distance measurement is rotation invariant. However, the square-root operation is costly and the global operation is difficult to decompose into small neighborhood operations because of its nonlinearity. Hence, the algorithms dealing with the approximation of Euclidean distance transformation (EDT) have been extensively discussed (Borgefors, 1984; Danielsson, 1978; Vossepoel, 1988). In the next few sections, we will introduce a mathematical morphology approach to construct the distance transformation and apply its theorems to accomplish the decomposition of the global operation into local operations with special emphasis on the Euclidean distance computation.

6.2 DISTANCE TRANSFORMATION BY MATHEMATICAL MORPHOLOGY

The city-block and chessboard distance transformations can be simply obtained using the aforementioned iterative or the two-step algorithm. The Euclidean distance transformation is often used because of its rotational invariance property, but it involves time-consuming calculations such as square, square-root, and the minimum over a set of floating point numbers. Although many techniques have been presented to obtain EDT, most of them are either inefficient or complex to implement and understand. Furthermore, they require extra cost such as special structure or storage for recording information. Cuisenaire and Macq (1999) proposed a fast EDT by propagation using multiple neighborhoods, but they need bucket sorting for calculating the EDT. Datta and Soundaralakshmi (2001) proposed a constant-time algorithm, but their algorithm is based on a special hardware structure, the reconfigurable mesh. Eggers (1998) proposed an algorithm by avoiding unnecessary calculations, but some data must be recorded in the lists. Saito and Toriwaki (1994) proposed an algorithm to compute the EDT in an n-dimensional domain, but the time complexity is high in their algorithm.

In Shih and Mitchell (1992), Shih and Wu (1992), and Huang and Mitchell (1994), a *mathematical morphology* (MM) approach was proposed to realize the EDT using grayscale erosions with successive small distance *structuring elements* (SEs) by decomposition. Furthermore, a *squared Euclidean distance structuring element* (SEDSE) was used to perform the *squared Euclidean distance transform* (SEDT). Shih and Wu (1992) decomposed the SEDSE into successive dilations of a set of 3×3 structuring components. Hence, the SEDT is equivalent to the successive erosions of the results at each preceding stage by each structuring component. The EDT can be finally obtained by simply a square-root operation over the entire image. In fact, the task of image analysis can directly take the result of SEDT as an input for object feature extraction and recognition.

There are six types of distance measures discussed in Borgefors (1986): city block, chessboard, Euclidean, octagonal, Chamfer 3-4, and Chamfer 5-7-11. The morphological distance transformation algorithm presented here is a general method, which uses a predefined structuring element to obtain the type of distance measure desired. The Euclidean distance between two points $P = (x, y)$ and $Q = (u, v)$ is defined as

$$d_e(P, Q) = \sqrt{(x-u)^2 + (y-v)^2} \tag{6.7}$$

By placing the center of the structuring element at the origin P, all the points can be represented by their distances from P (so that P is represented by 0). The weights of the structuring element are selected to be negative of related distance measures because grayscale erosion is the minimum selection after subtraction. This gives a

(1) Euclidean distance:

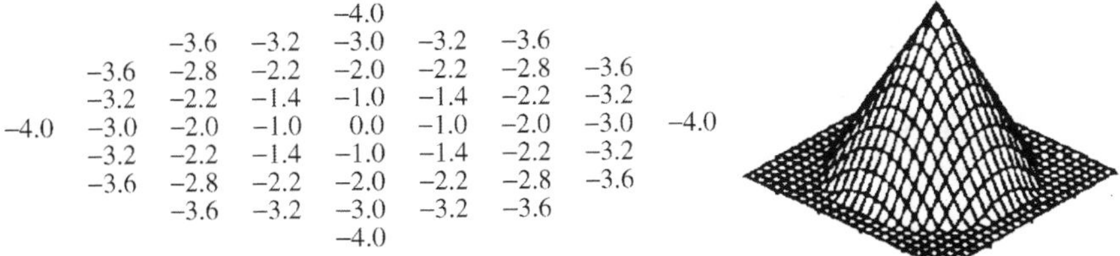

				−4.0				
		−3.6	−3.2	−3.0	−3.2	−3.6		
	−3.6	−2.8	−2.2	−2.0	−2.2	−2.8	−3.6	
	−3.2	−2.2	−1.4	−1.0	−1.4	−2.2	−3.2	
−4.0	−3.0	−2.0	−1.0	0.0	−1.0	−2.0	−3.0	−4.0
	−3.2	−2.2	−1.4	−1.0	−1.4	−2.2	−3.2	
	−3.6	−2.8	−2.2	−2.0	−2.2	−2.8	−3.6	
		−3.6	−3.2	−3.0	−3.2	−3.6		
				−4.0				

(2) City-block distance:

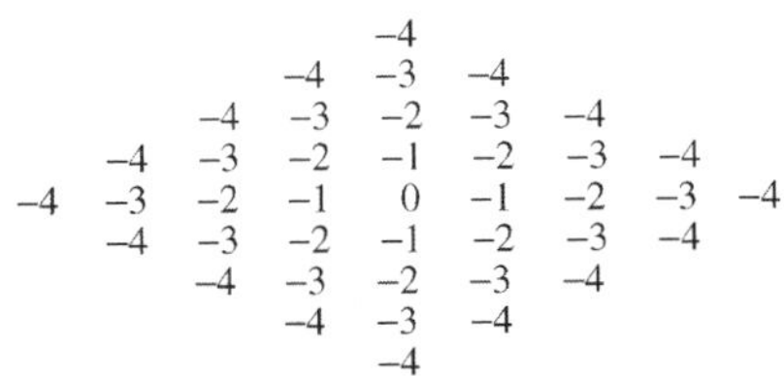

				−4				
			−4	−3	−4			
		−4	−3	−2	−3	−4		
	−4	−3	−2	−1	−2	−3	−4	
−4	−3	−2	−1	0	−1	−2	−3	−4
	−4	−3	−2	−1	−2	−3	−4	
		−4	−3	−2	−3	−4		
			−4	−3	−4			
				−4				

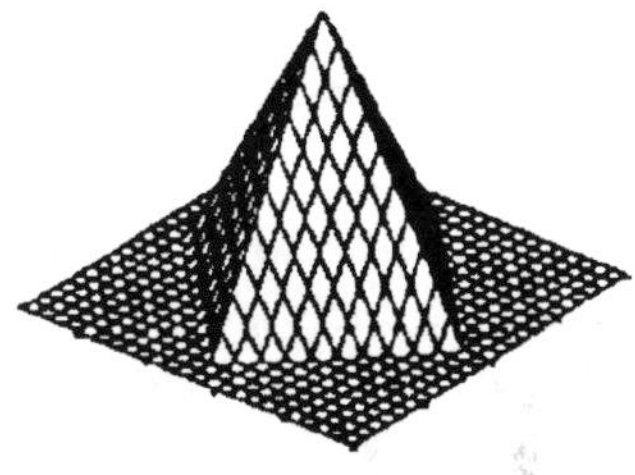

(3) Chessboard distance:

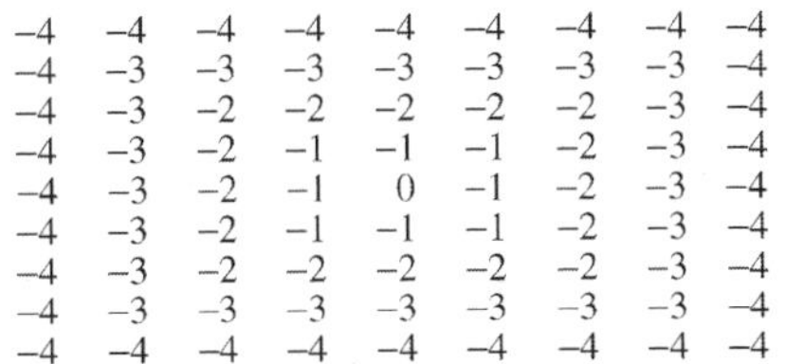

−4	−4	−4	−4	−4	−4	−4	−4	−4
−4	−3	−3	−3	−3	−3	−3	−3	−4
−4	−3	−2	−2	−2	−2	−2	−3	−4
−4	−3	−2	−1	−1	−1	−2	−3	−4
−4	−3	−2	−1	0	−1	−2	−3	−4
−4	−3	−2	−1	−1	−1	−2	−3	−4
−4	−3	−2	−2	−2	−2	−2	−3	−4
−4	−3	−3	−3	−3	−3	−3	−3	−4
−4	−4	−4	−4	−4	−4	−4	−4	−4

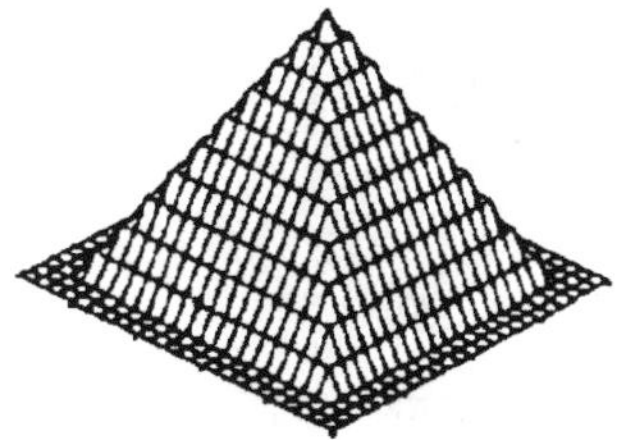

Figure 6.2 Three distance structuring elements. Only distances of 4 or smaller are shown.

positive distance measure. Three types of distance structuring elements, Euclidean, city block, and chessboard, are illustrated in Figure 6.2 in both tabular and graphical forms. Note that only distances of 4 or smaller are shown.

A binary image f consists of two classes, object pixels (foreground) and nonobject pixels (background). Let the object pixels have the value "$+\infty$" (or any number larger than the object's greatest distance) and the nonobject pixels have the value "0". Let k be the selected distance structuring element. The distance transformation algorithm is described as follows:

Distance Transformation Algorithm

A binary image is eroded by a distance structuring element; that is, $g = f \ominus_g k$.

The reason for assigning a large number to all the object pixels is to avoid interfering with the minimum selection of grayscale erosion that is designed for replacing the object value by a smaller distance value. Note that the size of the distance structuring element should be at least as large as the largest expected object size in the image. Otherwise, the central portion of the object pixels will not be reached by the minimum selection. The big size of the neighborhood computation is global and prohibitively costly. Most morphological hardware implementations are limited to a fixed size of the structuring element. The decomposition technique of a big structuring element into the combination of small ones can be referred in Shih and Mitchell (1991, 1992) and Shih and Wu (2004, 2005).

Algorithm verification: According to the erosion definition and the fact that the pixels of the binary image f are only represented by two values "$+\infty$" and "0", we have

$$f \ominus_g k(x) = \min\{\infty - k(z), 0 - k(z')\}, \quad \text{where} \quad z, z' \in K \tag{6.8}$$

Because the differences of all $k(z)$ and $k(z')$ are finite, "$\infty - k(z)$" is always greater than "$0 - k(z')$". Hence,

$$f \ominus_g k(x) = \min\{-k(z')\} \tag{6.9}$$

where for any $x \in F$, all z' satisfy that the pixel of $x + z'$ is located in the background (since a "0" in f corresponds to the background). Because the weights of the structuring element are replaced by the negative of the related distance measures from the center of the structuring element, the binary image f eroded by the distance structuring element k is exactly equal to the minimum distance from the object point to the outer boundary.

6.3 APPROXIMATION OF EUCLIDEAN DISTANCE

The distance transformation is essentially a global computation. All distance measures are positive definite, symmetrical, and satisfy the triangle inequality. Hence, global distances in the image can be approximated by propagating the local distance, that is, distances between neighboring pixels. The city-block distance is based on 4-neighborhood computation and the chessboard distance is based on 8-neighborhood computation. Both distance measures are linearly additive and can be easily decomposed into the accumulation of neighboring distances. The Euclidean distance is a nonlinear measure. However, we can still make a reasonable approximation.

The octagonal distance transformation is identical to applying the 3×3 city-block and the 3×3 chessboard distance transformations alternatively. Borgefors (1986) optimized the local distances in 3×3, 5×5, and 7×7 neighborhoods by minimizing the maximum absolute value of the difference between the distance transformation proposed by him and the Euclidean distance transformation. Vossepoel (1988) determined the coefficients of a distance transform by minimizing the above maximum absolute difference and minimizing the root-mean-square difference between the small neighborhood distance transformation and the Euclidean distance transformation.

In the following discussion, we present three different approximation methods. The first method combines the two outputs of relative x- and y-coordinates, where the minimum city-block and chessboard distances accumulate individually from each object point, and maps into the Euclidean distance by a lookup table. During city-block and chessboard distance transformations, we also accumulate the relative x- and y-coordinates with respect to the current point and the iteration will stop when there is no distance change. After the iteration stops, we obtain the relative x- and

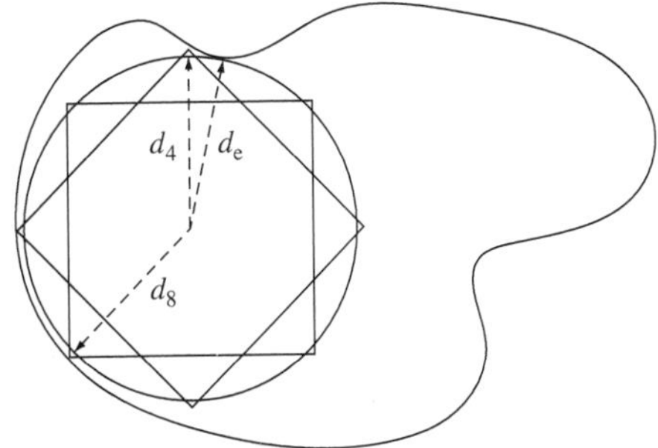

Figure 6.3 Special case when the closest boundary point for Euclidean distance (d_e) does not occur at the same position as that for city-block (d_4) or for chessboard (d_8).

y-coordinates of the closest point on the boundary, which has the shortest path in the image. Let (dx_1, dy_1) and (dx_2, dy_2) denote the relative coordinates of the closest points on the boundary for city block and chessboard, respectively. The quasi-Euclidean distance is computed by the following formula to construct the mapping lookup table with 5-tuples $(dx_1, dy_1, dx_2, dy_2, d_e)$:

$$d_e = \min\left(\sqrt{dx_1^2 + dy_1^2}, \sqrt{dx_2^2 + dy_2^2}\right) \tag{6.10}$$

It can be easily proved from the definitions of distance measure that the relationship among city block (d_4), chessboard (d_8), and Euclidean (d_e) is $d_4 \geq d_e \geq d_8$. In most of the situations, the Euclidean distance can be exactly obtained by equation (6.10) except when the closest point on the boundary for the Euclidean distance is not located at the same position as that for the city-block or for the chessboard distances (e.g., see Fig. 6.3).

The second method combines city-block and chessboard distances to map into the Euclidean distance. The city-block and chessboard distance structuring elements are expressed as follows.

$$k_4 = \begin{bmatrix} -2 & -1 & -2 \\ -1 & 0 & -1 \\ -2 & -1 & -2 \end{bmatrix}, \quad k_8 = \begin{bmatrix} -1 & -1 & -1 \\ -1 & 0 & -1 \\ -1 & -1 & -1 \end{bmatrix} \tag{6.11}$$

We can calculate the Euclidean distance by the following formula:

$$d_e = \sqrt{(d_4 - d_8)^2 + (d_8)^2} \tag{6.12}$$

Because the city-block distance is the summation of two absolute values in x- and y-coordinates, and the chessboard distance is the maximum of them, the subtraction $d_4 - d_8$ is equal to the minimum of two absolute values. Equation (6.12) is exactly equivalent to the definition of Euclidean distance. In determining a minimum distance, this equality does not hold when the city-block and chessboard distances are measured from the different closest boundary points.

The third method improves upon the second method by separating the neighborhood into four quadrants. We use four city-block and another four chessboard distance structuring elements as

$$(k_4)_1 = \begin{bmatrix} -1 & -2 \\ \underline{0} & -1 \end{bmatrix}, \quad (k_4)_2 = \begin{bmatrix} -2 & -1 \\ -1 & \underline{0} \end{bmatrix}, \quad (k_4)_3 = \begin{bmatrix} -1 & \underline{0} \\ -2 & -1 \end{bmatrix}, \quad (k_4)_4 = \begin{bmatrix} \underline{0} & -1 \\ -1 & -2 \end{bmatrix} \tag{6.13}$$

$$(k_8)_1 = \begin{bmatrix} -1 & -1 \\ \underline{0} & -1 \end{bmatrix}, \quad (k_8)_2 = \begin{bmatrix} -1 & -1 \\ -1 & \underline{0} \end{bmatrix}, \quad (k_8)_3 = \begin{bmatrix} -1 & \underline{0} \\ -1 & -1 \end{bmatrix}, \quad (k_8)_4 = \begin{bmatrix} \underline{0} & -1 \\ -1 & -1 \end{bmatrix} \tag{6.14}$$

Note that the origin of each structuring element is underlined. The origins of the first pair, $(k_4)_1$ and $(k_8)_1$, are located at the bottom-left corner. The origins of the second, third, and fourth pairs are located at the bottom-right, upper-right, and upper-left corners, respectively. The Euclidean distance can be obtained by computing four pairs of city-block and chessboard distances according to equation (6.12) and then selecting the minimum value of the four. This method has the advantage of confining the location of the closest boundary points to be within the same quadrant. But, the aforementioned special case in Figure 6.2 still remains a problem.

In summary, the first method provides the most precise Euclidean distance measure but requires extra computation. The second method is the simplest for implementation but has the least precision, and its accuracy is improved by the third method that requires some additional computation.

Different approximations to Euclidean distance measure, such as octagonal using the alternative city block and chessboard (Rosenfeld and Kak, 1982), Chamfer 3-4 using a 3×3 kernel, and Chamfer 5-7-11 using a 5×5 kernel (Borgefors, 1984), can also be adapted to the morphological approach. The structuring elements of the Chamfer 3-4 and 5-7-11 distance measures are constructed as follows:

$$\text{Chamfer 3-4}: \quad \begin{bmatrix} -4 & -3 & -4 \\ -3 & 0 & -3 \\ -4 & -3 & -4 \end{bmatrix}$$

$$\text{Chamfer 5-7-11}: \quad \begin{bmatrix} -14 & -11 & -10 & -11 & -14 \\ -11 & -7 & -5 & -7 & -11 \\ -10 & -5 & 0 & -5 & -10 \\ -11 & -7 & -5 & -7 & -11 \\ -14 & -11 & -10 & -11 & -14 \end{bmatrix}$$

6.4 DECOMPOSITION OF DISTANCE STRUCTURING ELEMENT

To achieve distance transformation by grayscale erosion, the size of the distance structuring element must be sufficiently large. When an image is operated using a large-size structuring element, implementation difficulties arise. To solve this problem, the decomposition of large-size structuring elements into recursive operations with small-size structuring components is quite critical for computational efficiency.

6.4.1 Decomposition of City-Block and Chessboard Distance Structuring Elements

According to the properties of grayscale morphological operations, when an image f is eroded by a large-size structuring element k that can be decomposed into dilations of several small structuring components k_i, we may obtain the same result by sequential erosions with these small structuring components (Serra, 1982; Shih and Mitchell, 1989). This can be expressed as follows:

$$k = k_1 \oplus k_2 \oplus \cdots \oplus k_n \tag{6.15}$$

$$f \ominus k = [\cdots [[f \ominus k_1] \ominus k_2] \cdots] \ominus k_n \tag{6.16}$$

City-block and chessboard distance structuring elements have a linear separable slope (Shih and Mitchell, 1991) in any direction from the center. k_i used for the decomposition is identical to k_4 and k_8 in equation (6.11). That is, if the dimension of the city-block or chessboard distance structuring element is $(2n+3) \times (2n+3)$, it can be decomposed into n successive morphological dilations using the 3×3 components.

Example 6.3 Let $k(x, y) = -(|x| + |y|)$, which is a city-block distance structuring element. For simplicity, only the structuring element of size 9×9 is shown.

$$k_{(9\times 9)} = \begin{bmatrix} -8 & -7 & -6 & -5 & -4 & -5 & -6 & -7 & -8 \\ -7 & -6 & -5 & -4 & -3 & -4 & -5 & -6 & -7 \\ -6 & -5 & -4 & -3 & -2 & -3 & -4 & -5 & -6 \\ -5 & -4 & -3 & -2 & -1 & -2 & -3 & -4 & -5 \\ -4 & -3 & -2 & -1 & 0 & -1 & -2 & -3 & -4 \\ -5 & -4 & -3 & -2 & -1 & -2 & -3 & -4 & -5 \\ -6 & -5 & -4 & -3 & -2 & -3 & -4 & -5 & -6 \\ -7 & -6 & -5 & -4 & -3 & -4 & -5 & -6 & -7 \\ -8 & -7 & -6 & -4 & -4 & -5 & -6 & -7 & -8 \end{bmatrix}$$

We select a 3×3 structuring element as

$$k_{(3\times 3)} = \begin{bmatrix} -2 & -1 & -2 \\ -1 & 0 & -1 \\ -2 & -1 & -2 \end{bmatrix}$$

Hence,

$$f \ominus k_{9\times 9} = (((f \ominus k_{3\times 3}) \ominus k_{3\times 3}) \ominus k_{3\times 3}) \ominus k_{3\times 3}$$

Example 6.4 Let $k(x,y) = -\max(|x|,|y|)$, which is a chessboard distance structuring element. The structuring element of size 9×9 is shown.

$$k_{(9\times 9)} = \begin{bmatrix} -4 & -4 & -4 & -4 & -4 & -4 & -4 & -4 & -4 \\ -4 & -3 & -3 & -3 & -3 & -3 & -3 & -3 & -4 \\ -4 & -3 & -2 & -2 & -2 & -2 & -2 & -3 & -4 \\ -4 & -3 & -2 & -1 & -1 & -1 & -2 & -3 & -4 \\ -4 & -3 & -2 & -1 & 0 & -1 & -2 & -3 & -4 \\ -4 & -3 & -2 & -1 & -1 & -1 & -2 & -3 & -4 \\ -4 & -3 & -2 & -2 & -2 & -2 & -2 & -3 & -4 \\ -4 & -3 & -3 & -3 & -3 & -3 & -3 & -3 & -4 \\ -4 & -4 & -4 & -4 & -4 & -4 & -4 & -4 & -4 \end{bmatrix}$$

We select a 3×3 structuring element as

$$k_{(3\times 3)} = \begin{bmatrix} -1 & -1 & -1 \\ -1 & 0 & -1 \\ -1 & -1 & -1 \end{bmatrix}$$

Hence,

$$f \ominus k_{9\times 9} = (((f \ominus k_{3\times 3}) \ominus k_{3\times 3}) \ominus k_{3\times 3}) \ominus k_{3\times 3}$$

6.4.2 Decomposition of the Euclidean Distance Structuring Element

The decomposition of the Euclidean distance structuring element cannot be treated as dilations of small structuring components, since it poses an additional problem in the off-axis and off-diagonal directions. However, if a Euclidean distance structuring element k can be segmented into the pointwise maximum selection of multiple linearly sloped structuring components k_i (see equation (6.17)), the grayscale erosion of an image with a Euclidean structuring element is equivalent to the minimum of the outputs (see equation (6.18)) when the image is individually eroded with these structuring components. This can be expressed as

$$k = \max(k_1, k_2, \ldots, k_n) \tag{6.17}$$

$$f \ominus k = \min(f \ominus k_1, f \ominus k_2, \ldots, f \ominus k_n) \tag{6.18}$$

Since each structuring component k_i has a linear slope, it can be further decomposed by equation (6.15) into the dilation of its structuring subcomponents k_{ij}. The procedure to construct these structuring components and subcomponents is described and illustrated as follows.

6.4.2.1 ***Construction Procedure*** Here we describe the construction procedure by illustrating a 9×9 Euclidean distance structuring element. Any $N \times N$ Euclidean distance structuring element can be decomposed into the operations of 3×3 ones by following a similar procedure. Detailed generalization and mathematical proof of the decomposition technique can be found in Shih and Mitchell (1991). A 9×9 Euclidean distance structuring element can be represented as

$$k_{e(9\times9)} = \begin{bmatrix} -d_4 & -d_3 & -d_2 & -d_1 & -d_0 & -d_1 & -d_2 & -d_3 & -d_4 \\ -d_3 & -c_3 & -c_2 & -c_1 & -c_0 & -c_1 & -c_2 & -c_3 & -d_3 \\ -d_2 & -c_2 & -b_2 & -b_1 & -b_0 & -b_1 & -b_2 & -c_2 & -d_2 \\ -d_1 & -c_1 & -b_1 & -a_1 & -a_0 & -a_1 & -b_1 & -c_1 & -d_1 \\ -d_0 & -c_0 & -b_0 & -a_0 & 0 & -a_0 & -b_0 & -c_0 & -d_0 \\ -d_1 & -c_1 & -b_1 & -a_1 & -a_0 & -a_1 & -b_1 & -c_1 & -d_1 \\ -d_2 & -c_2 & -b_2 & -b_1 & -b_0 & -b_1 & -b_2 & -c_2 & -d_2 \\ -d_3 & -c_3 & -c_2 & -c_1 & -c_0 & -c_1 & -c_2 & -c_3 & -d_3 \\ -d_4 & -d_3 & -d_2 & -d_1 & -d_0 & -d_1 & -d_2 & -d_3 & -d_4 \end{bmatrix}$$

where $a_0 = 1$, $a_1 = \sqrt{2}$, $b_0 = 2$, $b_1 = \sqrt{5}$, $b_2 = 2\sqrt{2}$, $c_0 = 3$, $c_1 = \sqrt{10}$, $c_2 = \sqrt{13}$, $c_3 = 3\sqrt{2}$, $d_0 = 4$, $d_1 = \sqrt{17}$, $d_2 = 2\sqrt{5}$, $d_3 = 5$, and $d_4 = 4\sqrt{2}$.

1. Select the central 3×3 window to be k_{11}:

$$k_{11} = \begin{bmatrix} -a_1 & -a_0 & -a_1 \\ -a_0 & 0 & -a_0 \\ -a_1 & -a_0 & -a_1 \end{bmatrix}$$

The value k_{11} is the structuring component k_1 in equations (6.17) and (6.18).

2. Because b_0, b_1, and b_2 do not represent equal increments, we need to account for the increasing difference offset. k_{21} consists of $-b_1$ and $-b_2$. k_{22} consisting of $b_1 - b_0$ and 0 is used as the offset. These matrices are shown as follows:

$$k_{21} = \begin{bmatrix} -b_2 & -b_1 & -b_2 \\ -b_1 & \times & -b_1 \\ -b_2 & -b_1 & -b_2 \end{bmatrix}$$

$$k_{22} = \begin{bmatrix} 0 & b_1 - b_0 & 0 \\ b_1 - b_0 & \times & b_1 - b_0 \\ 0 & b_1 - b_0 & 0 \end{bmatrix}$$

where "×" means a don't care term. The dilation by 3×3 k_{21} and 3×3 k_{22} will construct 5×5 k_2 in equations (6.17) and (6.18) because the extended values outside the window are regarded as "$-\infty$". k_2, which is shown below, has the same values on the boundary as k_e and has smaller values in the interior than k_e. The maximum of k_1 and k_2 will generate exactly the same values as the 5×5 Euclidean distance structuring element.

$$k_2 = \begin{bmatrix} -b_2 & -b_1 & -b_0 & -b_1 & -b_2 \\ -b_1 & -b_0 & -b_1 & -b_0 & -b_1 \\ -b_0 & -b_1 & -b_0 & -b_1 & -b_0 \\ -b_1 & -b_0 & -b_1 & -b_0 & -b_1 \\ -b_2 & -b_1 & -b_0 & -b_1 & -b_2 \end{bmatrix}$$

3. Repeat the selection procedure in step 2 until the total size of the Euclidean distance structuring element is reached. For example, k_{31}, k_{32}, and k_{33} are chosen as follows:

$$k_{31} = \begin{bmatrix} -c_3 & -c_2 & -c_3 \\ -c_2 & \times & -c_2 \\ -c_3 & -c_2 & -c_3 \end{bmatrix}$$

$$k_{32} = \begin{bmatrix} 0 & c_2-c_1 & 0 \\ c_2-c_1 & \times & c_2-c_1 \\ 0 & c_2-c_1 & 0 \end{bmatrix}$$

$$k_{33} = \begin{bmatrix} 0 & c_1-c_0 & 0 \\ c_1-c_0 & \times & c_1-c_0 \\ 0 & c_1-c_0 & 0 \end{bmatrix}$$

Hence, we have

$$\begin{aligned} k_{e(9\times9)} &= \max\left(k_{1(3\times3)}, k_{2(5\times5)}, k_{3(7\times7)}\right) \\ &= \max\left(k_{1(3\times3)}, k_{21(3\times3)} \oplus k_{22(3\times3)}, k_{31(3\times3)} \oplus k_{32(3\times3)} \oplus k_{33(3\times3)}\right) \end{aligned}$$

An application of the Euclidean distance structuring element may be found in Borgefors (1984).

A decomposition diagram for a 9×9 Euclidean distance structuring element is shown in Figure 6.4, in which a binary image f is simultaneously fed to four levels and sequentially processed with the grayscale erosion. Finally, all the outputs are combined by a minimum operator.

6.4.2.2 *Computational Complexity* In general, supposing that the image size is $(2n + 1) \times (2n + 1)$, the Euclidean distance computation needs $n(n + 1)/2$ grayscale erosions. For the 9×9 example, $n = 4$. The decomposition can be implemented quite efficiently on a parallel pipelined computer. The input goes to all n levels simultaneously and is sequentially pipelined to the next operator at each level. After appropriate delays for synchronization, all outputs are fed to the minimum operator. Alternatively, we can compare the minimum of first- and second-level outputs, and again compare the result with the third-level output, and so forth. Hence, the time complexity of this decomposition algorithm is of order n. The real-valued distance transformations are considered during the process, and the result is approximated to the closest integer in the final output image.

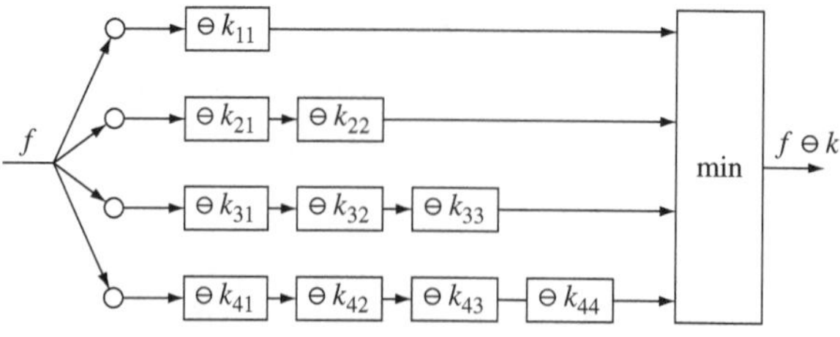

Figure 6.4 The decomposition diagram of a 9×9 Euclidean distance structuring element.

6.5 THE 3D EUCLIDEAN DISTANCE

6.5.1 The 3D Volumetric Data Representation

The 3D digitized volumetric data can be represented as a 3D matrix $T^3[x][y][z]$ (or $T^3[x, y, z]$) of dimensions $X \times Y \times Z$, where x, y, and z, respectively, denote the row, column, and height coordinates, as shown in Figure 6.5. Each voxel has the physical size of $D_x \times D_y \times D_z$ in physical units (e.g., mm or μm). For a binary image, the values of voxels are either 0 or 1. For a grayscale image, the voxels are composed of 256 values (0, 1, 2, ..., 255). Note that the representation can be extended to an arbitrarily dimensional domain as shown in Figure 6.6, where an nD digitized domain can be represented as an nD matrix $T^n[t_1][t_2] \cdots [t_n]$ of dimensions $t_1 \times t_2 \times \cdots \times t_n$.

6.5.2 Distance Functions in the 3D Domain

Let $p = (p_1, p_2, p_3)$ and $q = (q_1, q_2, q_3)$ denote two points in a 3D digital volumetric data, where p_1, p_2, p_3, q_1, q_2, and q_3 are integers. Three frequently used distance functions are defined as follows:

1. *City-block distance*: $d_{ci}(p, q) = |p_1 - q_1| + |p_2 - q_2| + |p_3 - q_3|$.
2. *Chessboard distance*: $d_{ch}(p, q) = \max(|p_1 - q_1|, |p_2 - q_2|, |p_3 - q_3|)$.
3. *Euclidean distance*: $d_e(p, q) = \sqrt{(p_1 - q_1)^2 + (p_2 - q_2)^2 + (p_3 - q_3)^2}$.

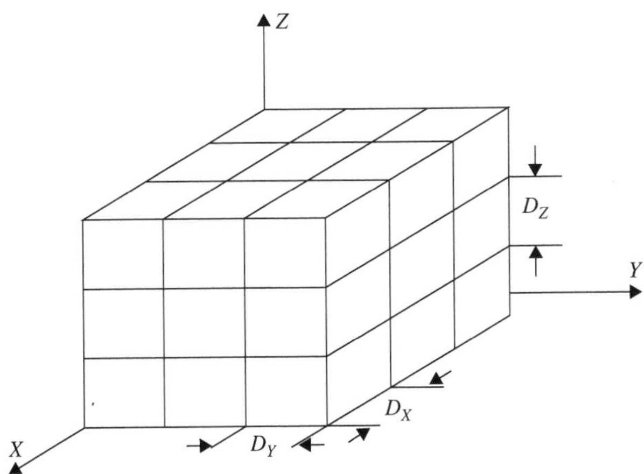

Figure 6.5 The 3D volumetric data representation.

$$
\begin{array}{l}
t_1 = \{\ -\infty\ \ldots\ 0\ \ldots\ +\infty\ \} \\
t_2 = \{\ -\infty\ \ldots\ 0\ \ldots\ +\infty\ \} \\
t_3 = \{\ -\infty\ \ldots\ 0\ \ldots\ +\infty\ \} \\
\ \vdots\quad \vdots\quad \vdots\quad \vdots\quad \vdots\quad \vdots\quad \vdots\quad \vdots \\
t_n = \{\ -\infty\ \ldots\ 0\ \ldots\ +\infty\ \}
\end{array}
$$

Figure 6.6 The representation of n-dimensional space.

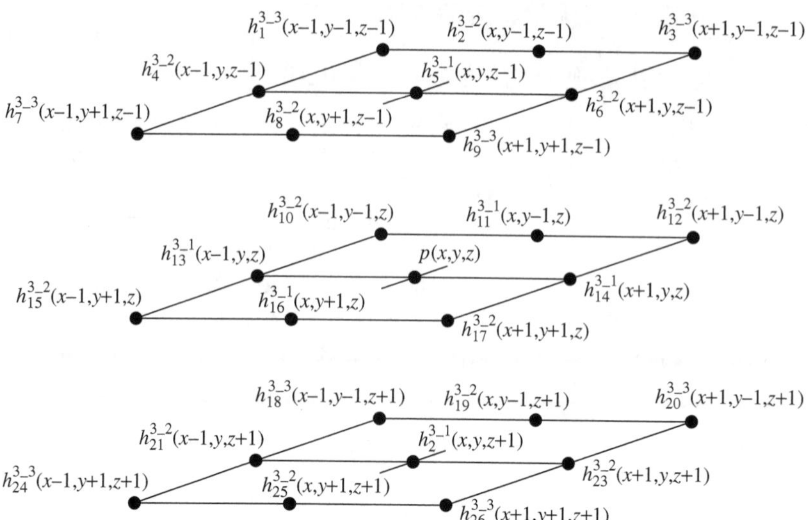

Figure 6.7 The 26 neighbors of pixel p.

Note that the voxels with *city-block distance* 1 counting from p correspond to 6-neighbors of p and those with *chessboard distance* 1 correspond to 26-neighbors. d_{ci} and d_{ch} are integer-valued; however, d_e is not. The *squared Euclidean distance* (SED) is defined as

$$d_s(p, q) = [d_e(p, q)]^2 \tag{6.19}$$

Note that the distance functions above can be similarly extended to an nD domain.

6.5.3 The 3D Neighborhood in the EDT

In 3D space, we denote the 26 neighbors of pixel p by $N^3(p) = \{h_1^3, h_2^3, \ldots, h_{26}^3\}$, as illustrated in Figure 6.7. They can be categorized into three groups, $N^{3\text{-}1}(p)$, $N^{3\text{-}2}(p)$, and $N^{3\text{-}3}(p)$, in which the corresponding $h_{\text{index}}^{3\_d}$ adopts the notation index and d, representing the ordering number in the neighborhood and the number of moves between p and its neighbor, respectively. Generally, in the nD domain the number of neighbors of p is $3^n - 1$, and they can be categorized into n groups.

6.6 THE ACQUIRING APPROACHES

Basically, the approaches for computing distance transformation can be categorized into acquiring and deriving approaches. The acquiring approach is to acquire the minimum distance of each pixel from its neighborhood as illustrated in Figure 6.8(a). The deriving approach uses the strategy that each pixel broadcasts relative distances to its children (i.e., the pixels in its neighborhood) as illustrated in Figure 6.8(b). The

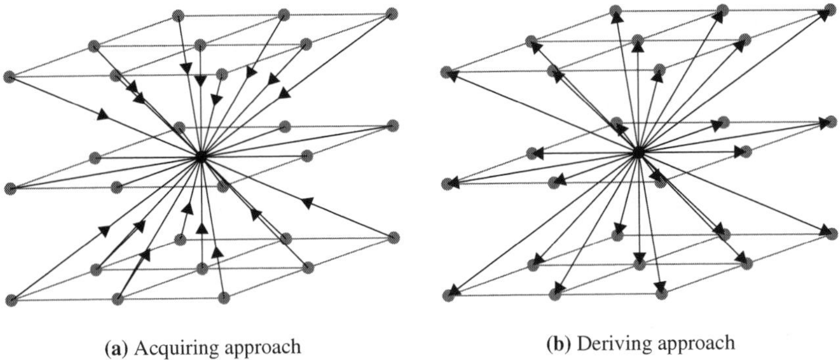

(a) Acquiring approach (b) Deriving approach

Figure 6.8 The two approaches for achieving the EDT.

acquiring approaches for city-block, chessboard, and Euclidean distances are presented in this section. The deriving approach for Euclidean distance is described in the next section.

6.6.1 Acquiring Approaches for City-Block and Chessboard Distance Transformations

3D *city-block* and *chessboard* distances are very easy to compute since they can be recursively accumulated by considering only 6- and 26-neighbors, respectively, at one time. The iterative algorithm for *city-block* and *chessboard* distances is described below. Given an X_S, which is 1 at the pixels of foreground S and 0 elsewhere, X_S^m is defined recursively for $m = 1, 2, \ldots$ as

$$X_S^m(p) = X_S^0 + \min_{d(q,p)\leq 1} X_S^{m-1}(q) \tag{6.20}$$

where $X_S^0 = X_S$, the initial binary image. Thus, X_S^m can be computed by performing a local operation on the pair of arrays X_S^0 and X_S^{m-1} at each point.

The iterative algorithm is very efficient on a cellular array computer since each iteration can be performed at all points in parallel, and the number of iterations is at most the radius of the image. However, in a conventional computer, each iteration requires the processing of the entire image that presents inefficiency. A two-scan algorithm is therefore developed below.

Let $N_1(p)$ be the set of (6- or 26-) neighbors that precede p in the left-to-right, top-to-bottom, near-to-far scan of the picture, and let $N_2(p)$ be the remaining neighbors of p. The algorithm is computed by

$$X'_S(p) = \begin{cases} 0 & \text{if } p \in \overline{S} \\ \min_{q\in N_1(p)} X'_S(q) + 1 & \text{otherwise} \end{cases} \tag{6.21}$$

$$X''_S(p) = \min_{q\in N_2(p)} [X'_S(p), X''_S(q) + 1] \tag{6.22}$$

$X'_S(p)$ is computed in a left-to-right, top-to-bottom, near-to-far scan, since for each p, X'_S has already been computed for the q's in N_1. Similarly, X''_S is computed in a reverse scan.

6.6.2 Acquiring Approach for Euclidean Distance Transformation

In this section, a morphology-based approach to compute the 3D EDT is introduced. The grayscale dilation and erosion have been defined in Section 4.4.1. The 2D *Euclidean distance structuring element* (EDSE) is adopted for computing the EDT, and it can be decomposed using the concave decomposition strategy (see Shih and Mitchell (1992)). Shih and Wu (1992) developed a morphology-based algorithm by decomposing the big 2D SEDSE into successive dilations of a set of 3×3 structuring components. Hence, the SEDT is equivalent to the successive erosions of the results at each preceding stage by the iteration-related structuring components, which are represented as

$$g(l) = \begin{bmatrix} -(4l-2) & -(2l-1) & -(4l-2) \\ -(2l-1) & 0 & -(2l-1) \\ -(4l-2) & -(2l-1) & -(4l-2) \end{bmatrix} \tag{6.23}$$

where l indicates the number of iterations. Note that the input binary image is represented by "$+\infty$" (or a large number) for foreground and "0" for background. In distance transformation, the 3D images and structuring elements are not represented in a temporal stack. By extending the structuring components into three dimensions, $g^3(l)$ is obtained as

$$g^3(l) = \begin{bmatrix} g_a^3(l) \\ g_b^3(l) \\ g_c^3(l) \end{bmatrix} \tag{6.24}$$

$$\text{where } g_a^3(l) = \begin{bmatrix} -(6l-3) & -(4l-2) & -(6l-3) \\ -(4l-2) & -(2l-1) & -(4l-2) \\ -(6l-3) & -(4l-2) & -(6l-3) \end{bmatrix},$$

$$g_b^3(l) = \begin{bmatrix} -(4l-2) & -(2l-1) & -(4l-2) \\ -(2l-1) & 0 & -(2l-1) \\ -(4l-2) & -(2l-1) & -(4l-2) \end{bmatrix}, \text{ and } \quad g_c^3(l) = g_a^3(l)$$

Figure 6.9 illustrates the 3D structuring components. The SEDT is the iterative erosions by a set of small structuring components and a square-root operation. The following is the 3D EDT algorithm:

1. Let the values of foreground be a large number, and let values of background be "0".
2. Initialize $l = 1$.
3. $d = f \ominus g(l)$.
4. If $d \neq f$, let $f = d$ and $l{+}{+}$; repeat steps 3 and 4 until $d = f$.
5. EDT $= \sqrt{d}$.

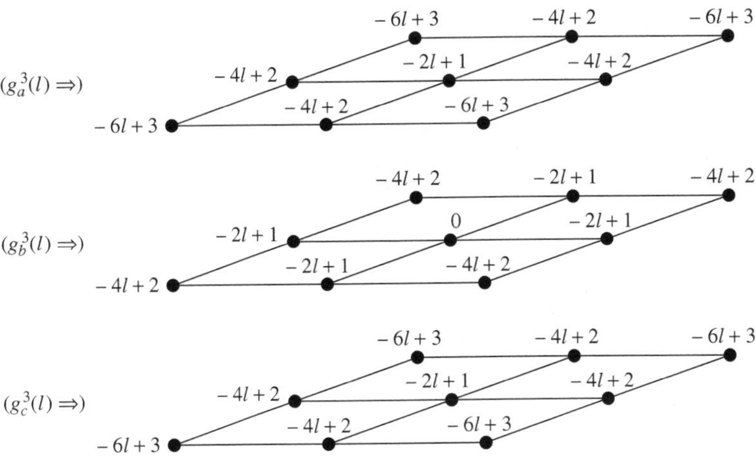

Figure 6.9 The extended structuring components in the 3D space.

Figure 6.10 shows the resulting image of SEDT using the 3D iterative algorithm with step 5 skipped when the input contains only one background pixel at the center of layer 4. Note that due to the symmetric property, layer 3 = layer 5, layer 2 = layer 6, and layer 1 = layer 6.

27	22	19	18	19	22	27
22	17	14	13	14	17	22
19	14	11	10	11	14	19
18	13	10	9	10	13	18
19	14	11	10	11	14	19
22	17	14	13	14	17	22
27	22	19	18	19	22	27

Layer 1

22	17	14	13	14	17	22
17	12	9	8	9	12	17
14	9	6	5	6	9	14
13	8	5	4	5	8	13
14	9	6	5	6	9	14
17	12	9	8	9	12	17
22	17	14	13	14	17	22

Layer 2

19	14	11	10	11	14	19
14	9	6	5	6	9	14
11	6	3	2	3	6	11
10	5	2	1	2	5	10
11	6	3	2	3	6	11
14	9	6	5	6	9	14
19	14	11	10	11	14	19

Layer 3

18	13	10	9	10	13	18
13	8	5	4	5	8	13
10	5	2	1	2	5	10
9	4	1	0	1	4	9
10	5	2	1	2	5	10
13	8	5	4	5	8	13
18	13	10	9	10	13	18

Layer 4

27	22	19	18	19	22	27
22	17	14	13	14	17	22
19	14	11	10	11	14	19
18	13	10	9	10	13	18
19	14	11	10	11	14	19
22	17	14	13	14	17	22
27	22	19	18	19	22	27

Layer 7

22	17	14	13	14	17	22
17	12	9	8	9	12	17
14	9	6	5	6	9	14
13	8	5	4	5	8	13
14	9	6	5	6	9	14
17	12	9	8	9	12	17
22	17	14	13	14	17	22

Layer 6

19	14	11	10	11	14	19
14	9	6	5	6	9	14
11	6	3	2	3	6	11
10	5	2	1	2	5	10
11	6	3	2	3	6	11
14	9	6	5	6	9	14
19	14	11	10	11	14	19

Layer 5

Figure 6.10 An SEDT example.

6.7 THE DERIVING APPROACHES

Let o denote a background pixel. The method of calculating $d_s(N^{3\text{-}1}(p), o)$, $d_s(N^{3\text{-}2}(p), o)$, and $d_s(N^{3\text{-}3}(p), o)$ is described below. The children of pixel p in forward and backward scans, each containing 13 pixels, are illustrated in Figures 6.11 and 6.12, respectively. For notation simplicity, $d_s(p, q)$ is substituted by $d_s(p)$ and $d_s(h^{3\text{-}d}_{\text{index}}, q)$ is replaced by $d_s(h^{3}_{\text{index}})$ if q is a background pixel.

6.7.1 The Fundamental Lemmas

Let "q" denote a background pixel. The following three lemmas describe the relationship between an object pixel p and its neighbors.

Lemma 6.1. The SEDT of each pixel can be divided into the summation of three squared integers.

Note that Lemma 6.1 can be easily proved from the definition of SED. Therefore, let $d_s(p, q) = (I_{\max})^2 + (I_{\text{mid}})^2 + (I_{\min})^2$, where $I_{\max} \geq I_{\text{mid}} \geq I_{\min}$ and $I_{\max}, I_{\text{mid}}$, and $I_{\min}$ are integers.

Lemma 6.2. The SEDT of the further children of p can be obtained by equations (6.25)–(6.27) depending on its group.

$$d_s(N^{3\text{-}1}(p), q) = \begin{cases} (I_{\max}+1)^2 + (I_{\text{mid}})^2 + (I_{\min})^2, & \text{or} \\ (I_{\max})^2 + (I_{\text{mid}}+1)^2 + (I_{\min})^2, & \text{or} \\ (I_{\max})^2 + (I_{\text{mid}})^2 + (I_{\min}+1)^2 \end{cases} \tag{6.25}$$

$d_s(p) = x^2 + y^2 + z^2$

$d_s(h^3_{14}) = (x+1)^2 + y^2 + z^2$

$d_s(h^3_{15}) = (x+1)^2 + (y+1)^2 + z^2$

$d_s(h^3_{16}) = x^2 + (y+1)^2 + z^2$

$d_s(h^3_{17}) = (x+1)^2 + (y+1)^2 + z^2$

$d_s(h^3_{18}) = (x+1)^2 + (y+1)^2 + (z+1)^2$

$d_s(h^3_{19}) = x^2 + (y+1)^2 + (z+1)^2$

$d_s(h^3_{20}) = (x+1)^2 + (y+1)^2 + (z+1)^2$

$d_s(h^3_{21}) = (x+1)^2 + y^2 + (z+1)^2$

$d_s(h^3_{22}) = x^2 + y^2 + (z+1)^2$

$d_s(h^3_{23}) = (x+1)^2 + y^2 + (z+1)^2$

$d_s(h^3_{24}) = (x+1)^2 + (y+1)^2 + (z+1)^2$

$d_s(h^3_{25}) = x^2 + (y+1)^2 + (z+1)^2$

$d_s(h^3_{26}) = (x+1)^2 + (y+1)^2 + (z+1)^2$

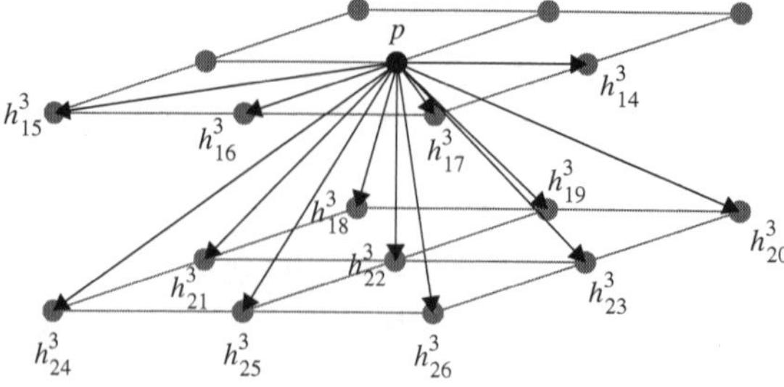

Figure 6.11 The 13 children in the forward scan.

$d_s(p) = x^2 + y^2 + z^2$

$d_s(h_1^3) = (x+1)^2 + (y+1)^2 + (z+1)^2$

$d_s(h_2^3) = x^2 + (y+1)^2 + (z+1)^2$

$d_s(h_3^3) = (x+1)^2 + (y+1)^2 + (z+1)^2$

$d_s(h_4^3) = (x+1)^2 + y^2 + (z+1)^2$

$d_s(h_5^3) = x^2 + y^2 + (z+1)^2$

$d_s(h_6^3) = (x+1)^2 + y^2 + (z+1)^2$

$d_s(h_7^3) = (x+1)^2 + (y+1)^2 + (z+1)^2$

$d_s(h_8^3) = x^2 + (y+1)^2 + (z+1)^2$

$d_s(h_9^3) = (x+1)^2 + y^2 + z^2$

$d_s(h_{10}^3) = (x+1)^2 + (y+1)^2 + z^2$

$d_s(h_{11}^3) = x^2 + (y+1)^2 + z^2$

$d_s(h_{12}^3) = (x+1)^2 + (y+1)^2 + z^2$

$d_s(h_{13}^3) = (x+1)^2 + y^2 + z^2$

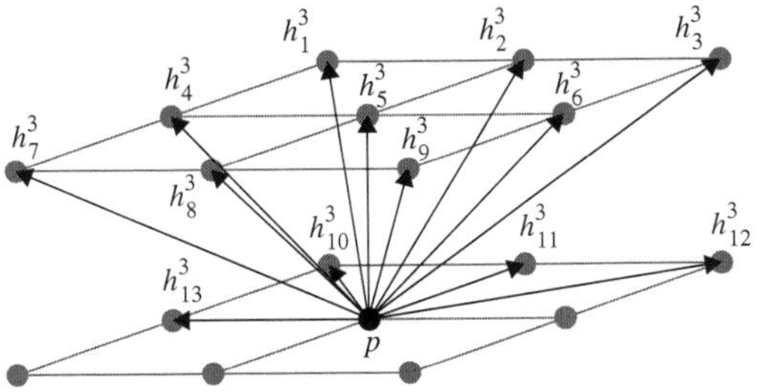

Figure 6.12 The 13 children in the backward scan.

$$d_s(N^{3\text{-}2}(p), q) = \begin{cases} (I_{\max}+1)^2 + (I_{\text{mid}}+1)^2 + (I_{\min})^2, & \text{or} \\ (I_{\max}+1)^2 + (I_{\text{mid}})^2 + (I_{\min}+1)^2, & \text{or} \\ (I_{\max})^2 + (I_{\text{mid}}+1)^2 + (I_{\min}+1)^2 \end{cases} \tag{6.26}$$

$$d_s(N^{3\text{-}3}(p), q) = (I_{\max}+1)^2 + (I_{\text{mid}}+1)^2 + (I_{\min}+1)^2 \tag{6.27}$$

Proof. From $d_s(p,q) = (I_{\max})^2 + (I_{\text{mid}})^2 + (I_{\min})^2$, we know that the relative distance between p and q in each of the three dimensions is, respectively, $I_{\max}$, I_{mid}, and $I_{\min}$, as shown in Figure 6.13. Note that the relative distance between $N^{3\text{-}1}(p)$ and p is one pixel in one dimension. Therefore, the SEDT of $N^{3\text{-}1}(p)$ can be obtained by equation (6.25). Similarly, the relative distances between $(N^{3\text{-}2}(p), p)$ and $(N^{3\text{-}3}(p), p)$ are one pixel in each of the two and three dimensions, respectively. Therefore, equations (6.26) and (6.27) are used to obtain the SEDT of $N^{3\text{-}2}(p)$ and $N^{3\text{-}3}(p)$, respectively.

Lemma 6.3. If $d_s(p,q)$ can be decomposed into more than one set of the summation of three squared integers, that is,

$$d_s(p,q) = \begin{cases} (I_{\max\_1})^2 + (I_{\text{mid}\_1})^2 + (I_{\min\_1})^2 \\ (I_{\max\_2})^2 + (I_{\text{mid}\_2})^2 + (I_{\min\_2})^2 \\ \vdots \\ (I_{\max\_n})^2 + (I_{\text{mid}\_n})^2 + (I_{\min\_n})^2 \end{cases}$$

we can check the parents of p to determine the correctly decomposed set.

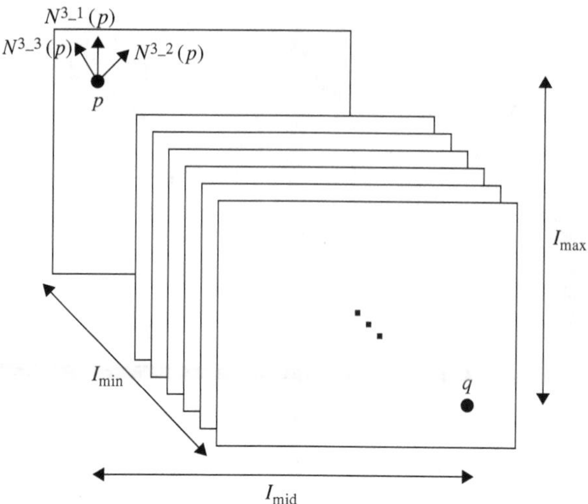

Figure 6.13 The $d_s(N^{3\text{-}1}(p), q)$, $d_s(N^{3\text{-}2}(p), q)$, and $d_s(N^{3\text{-}3}(p), q)$.

Proof. According to Lemma 6.2, the value of SEDT of pixel p can be obtained from its parents. Suppose that there are n possibilities when $d_s(p, q)$ is decomposed into the summation of three squared integers. If p is derived from H_1, $d_s(H_1, q)$ should be equal to $(I_{max}-1)^2 + (I_{mid})^2 + (I_{min})^2$, $(I_{max})^2 + (I_{mid}-1)^2 + (I_{min})^2$, or $(I_{max})^2 + (I_{mid})^2 + (I_{min}-1)^2$. Note that H_1 is the parent of p and has only 1D difference to p. Similarly, if p is derived from H_2, which has 2D difference to p, then $d_s(H_2, q)$ should be equal to $(I_{max}-1)^2 + (I_{mid}-1)^2 + (I_{min})^2$,$(I_{max})^2 + (I_{mid}-1)^2 + (I_{min}-1)^2$, or $(I_{max}-1)^2 + (I_{mid})^2 + (I_{min}-1)^2$. If p is derived from H_3, which has the 3D difference to p, then $d_s(H_3, q)$ should be equal to $(I_{max}-1)^2 + (I_{mid}-1)^2 + (I_{min}-1)^2$. Therefore, by checking the parents of p, we can find out the exact decomposition when dividing $d_s(p, q)$.

6.7.2 The Two-Scan Algorithm for EDT

In fact, the concepts in the acquiring and deriving approaches are similar. In the acquiring approach, it is obvious that the minimum EDT of a pixel can be obtained by one of its neighbors with the minimum EDT value. In the deriving approach, a pixel can derive the minimum EDT value to its neighbors based on the aforementioned lemmas. To correctly compute the EDT in two scans to overcome the time-consuming iterative operations, the deriving approach is adopted to substitute for the acquiring approach. Let the object pixels be a large number (i.e., larger than the square of a half of the image size) and the background pixels be 0. The algorithm is described below, and its flowchart is shown in Figure 6.14.

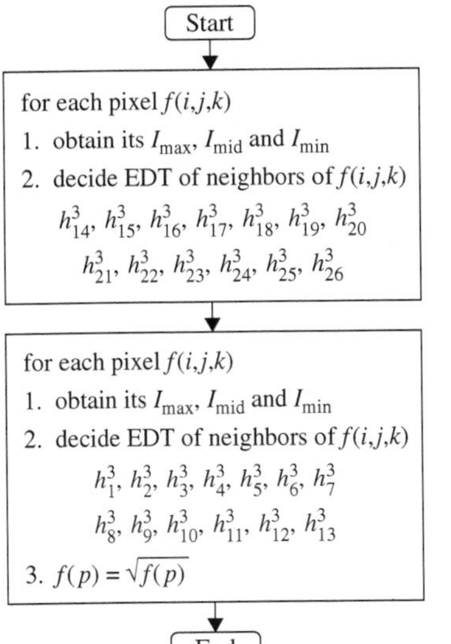

Figure 6.14 The flowchart of the two-scan algorithm.

A. *Forward scan*: left-to-right, top-to-bottom, and near-to-far

```
for ( i = 0; i ≤ length; i++ )
        for ( j = 0; j ≤ length; j++ )
                for ( k = 0; k ≤ length; k++ )
{
Decompose f(i,j,k) into I_max, I_mid and I_min, where
f(i,j) = (I_max)^2 + (I_mid)^2 + (I_min)^2;
If there exists more than one decomposition, check the parents of
f(i,j,k) to choose the exact I_max, I_mid and I_min;
For the following three types of neighborhoods, perform:
f(N^{3-1}(p)) = min(f(N^{3-1}(p)), (I_max + 1)^2 + (I_mid)^2 + (I_min)^2);
f(N^{3-2}(p)) = min(f(N^{3-2}(p)), (I_max + 1)^2 + (I_mid + 1)^2 + (I_min)^2);
f(N^{3-3}(p)) = min(f(N^{3-3}(p)), (I_max + 1)^2 + (I_mid + 1)^2 + (I_min + 1)^2);
}
```

B. *Backward scan*: far-to-near, bottom-to-top, and right-to-left

```
for ( i = length; i ≥ 0; i-- )
        for ( j = length; j ≥ 0; j-- )
                for ( k = length; k ≥ 0; k-- )
{
Decompose f(i,j,k)into I_max, I_mid and I_min, where
f(i,j) = (I_max)^2 + (I_mid)^2 + (I_min)^2;
If there exists more than one decomposition, check the parents of
f(i,j,k) to choose the exact I_max, I_mid and I_min;
For the following three types of neighborhoods, perform:
f(N^{3-1}(p)) = min(f(N^{3-1}(p)), (I_max + 1)^2 + (I_mid)^2 + (I_min)^2);
```

1000	1000	1000	1000	1000	1000	1000
1000	1000	1000	1000	1000	1000	1000
1000	1000	1000	1000	1000	1000	1000
1000	1000	1000	1000	1000	1000	1000
1000	1000	1000	1000	1000	1000	1000
1000	1000	1000	1000	1000	1000	1000
1000	1000	1000	1000	1000	1000	1000

Layer 1

1000	1000	1000	1000	1000	1000	1000
1000	1000	1000	1000	1000	1000	1000
1000	1000	1000	1000	1000	1000	1000
1000	1000	1000	0	1	4	9
1000	1000	2	1	2	5	10
1000	8	5	4	5	8	13
18	13	10	9	10	13	18

Layer 2

1000	1000	1000	1000	1000	1000	1000
1000	1000	1000	1000	1000	1000	1000
1000	1000	3	2	3	6	11
1000	9	2	1	2	5	10
19	8	3	2	3	6	11
18	9	6	5	6	9	14
19	14	11	10	11	14	19

Layer 3

1000	1000	1000	1000	1000	1000	1000
1000	12	9	8	9	12	17
22	9	6	5	6	9	14
19	8	5	0	1	4	9
18	9	2	1	2	5	10
19	8	5	4	5	8	13
18	13	10	9	10	13	18

Layer 4

27	22	19	18	19	22	27
22	17	14	13	14	17	22
19	14	3	2	3	6	11
18	9	2	1	2	5	10
19	8	3	2	3	6	11
18	9	6	5	6	9	14
19	14	11	10	11	14	19

Layer 7

30	29	26	25	26	29	30
29	12	9	8	9	12	17
22	9	6	5	6	9	14
19	8	5	0	1	4	9
18	9	2	1	2	5	10
19	8	5	4	5	8	13
18	13	10	9	10	13	18

Layer 6

27	22	19	18	19	22	27
22	17	14	13	14	17	22
19	14	3	2	3	6	11
18	9	2	1	2	5	10
19	8	3	2	3	6	11
18	9	6	5	6	9	14
19	14	11	10	11	14	19

Layer 5

Figure 6.15 The forward scan of the two-scan algorithm.

$f(N^{3\text{-}2}(p)) = \min(f(N^{3\text{-}2}(p)), (I_{\max}+1)^2 + (I_{\text{mid}}+1)^2 + (I_{\min}{}^2);$
$f(N^{3\text{-}3}(p)) = \min(f(N^{3\text{-}3}(p)), (I_{\max}+1)^2 + (I_{\text{mid}}+1)^2 + (I_{\min}+1)^2);$
$E(f(i,j,k)) = \sqrt{f(i,j,k)};$
}

Figures 6.15 and 6.16 show the forward and backward scanning results of the two-scan algorithm on an image of $7 \times 7 \times 7$. Note that there are only three background

19	14	11	10	11	14	19
14	9	6	5	6	9	14
11	6	3	2	3	6	11
10	5	2	1	2	5	10
11	6	3	2	3	6	11
14	9	6	5	6	9	14
19	14	11	10	11	14	19

Layer 1

18	13	10	9	10	13	18
13	8	5	4	5	8	13
10	5	2	1	2	5	10
9	4	1	0	1	4	9
10	5	2	1	2	5	10
13	8	5	4	5	8	13
18	13	10	9	10	13	18

Layer 2

19	14	11	10	11	14	19
14	9	6	5	6	9	14
11	6	3	2	3	6	11
10	5	2	1	2	5	10
11	6	3	2	3	6	11
14	9	6	5	6	9	14
19	14	11	10	11	14	19

Layer 3

18	13	10	9	10	13	18
13	8	5	4	5	8	13
10	5	2	1	2	5	10
9	4	1	0	1	4	9
10	5	2	1	2	5	10
13	8	5	4	5	8	13
18	13	10	9	10	13	18

Layer 4

19	14	11	10	11	14	19
14	9	6	5	6	9	14
11	6	3	2	3	6	11
10	5	2	1	2	5	10
11	6	3	2	3	6	11
14	9	6	5	6	9	14
19	14	11	10	11	14	19

Layer 7

18	13	10	9	10	13	18
13	8	5	4	5	8	13
10	5	2	1	2	5	10
9	4	1	0	1	4	9
10	5	2	1	2	5	10
13	8	5	4	5	8	13
18	13	10	9	10	13	18

Layer 6

19	14	11	10	11	14	19
14	9	6	5	6	9	14
11	6	3	2	3	6	11
10	5	2	1	2	5	10
11	6	3	2	3	6	11
14	9	6	5	6	9	14
19	14	11	10	11	14	19

Layer 5

Figure 6.16 The backward scan of the two-scan algorithm.

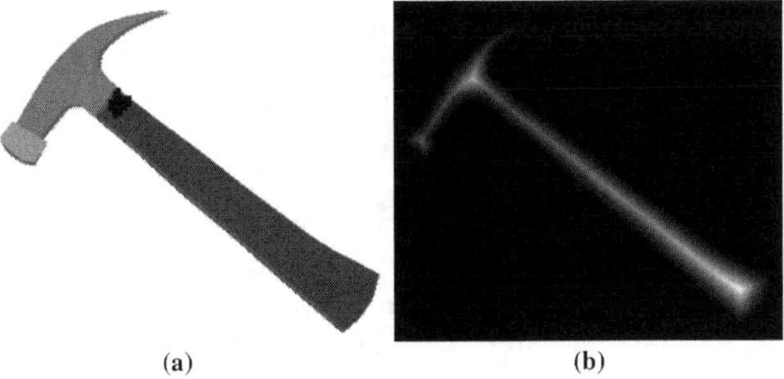

(a) (b)

Figure 6.17 The EDT example of an image.

pixels in the image, which are located at the centers of layers 2, 4, and 6. Also note that, for some image with obstacle, we need to repeat the above algorithm several times to achieve the exact EDT. Let P be the SEDT image that stores the previous processing image. The algorithm is

1. Set $P = f$.
2. Obtain the SEDT of f by the above two-scan algorithm with the last step being excluded.
3. Repeat steps 1 to 2 until $P = f$.
4. Take a square root of P (the distance image). That is, EDT $= \sqrt{P}$.

Figure 6.17 shows an example of an image with size 300×350. Figure 6.17(a) and (b) shows the original image and its EDT result, respectively. Note that, in Figure 6.17(a), the pixels within the hammer are set as the object pixels. In Figure 6.17(b), the brighter pixels indicate the larger EDT values, and we rescale the EDT values to the range 0–255 for display purpose.

6.7.3 The Complexity of the Two-Scan Algorithm

Given an image of size $n \times n \times n$, the computational complexity of the approximation algorithm by Danielsson (1980) is $O(n^3)$. The exact EDT algorithm proposed by Eggers (1998) strongly depends on the image content, and its complexity is varied between $O(n^3)$ and $O(n^4)$. The complexity of the proposed two-scan algorithm is $O(n^3)$.

Coeurjolly and Montanvert (2007) presented time optimal algorithms to solve the reverse Euclidean distance transformation and the reversible medial axis extraction problems for d-dimensional images. Maurer et al. (2003) proposed a sequential algorithm for computing the Euclidean distance transform of a k-dimensional binary image in time linear in the total number of voxels N.

6.8 THE SHORTEST PATH PLANNING

The shortest path planning problem is a classical topic in computational geometry and motion planning. The shortest path planning is finding a path between two locations such that some constraints can be satisfied and certain criteria can be optimized. Besides the problem of finding a path with minimal length according to the Euclidean metric, different optimization criteria are studied in the literature, for example, the sum of all the turning angles or the number of line segments (Mitchell et al., 1990; Arkin et al., 1991).

Pei and Horng (1998) proposed an algorithm to solve the path planning problem in an automated vehicle guidance system using the modified constrained distance transformation. Diaz de Leon and Sossa (1998) presented a fast path planning method for a mobile robot (MR) among objects of arbitrary shape. It comprises two phases. During the first phase, the graph including all possible collision-free paths from a top view of the environment is obtained. During the second phase, the optimal path for the MR is selected.

Mezouar and Chaumette (2002) developed an efficient approach to couple path planning in image space and image-based control. Sun and Reif (2007) proposed robotic optimal path planning in polygonal regions with pseudo-Euclidean metrics. Jan et al. (2008) presented some optimal path planning algorithms for navigating mobile rectangular robot among obstacles and weighted regions. Willms and Yang (2008) proposed an efficient grid-based distance-propagating dynamic system for real-time robot path planning in dynamic environments, which incorporates safety margins around obstacles using local penalty functions.

Since we set the value zero at the destination point in the distance map, we can always find a neighbor with a smaller SED value from any other pixel. Therefore, the shortest path from an arbitrary source to the destination point can be definitely achieved. Assume that an arbitrarily shaped object moves from a starting point to a destination point in a finite space with arbitrarily shaped obstacles in it. We relate this model to mathematical morphology using the following strategy:

1. The finite space consists of free regions as foreground and obstacles as "forbidden" areas with value "-1", and the moving object is modeled as the structuring element.
2. Set the destination point to be "0".
3. Apply a morphological erosion to the free regions, followed by a distance transformation on the region with the grown obstacles excluded, and then trace the distance map from any starting point to its neighbor of the minimum distance until the destination point is reached.

6.8.1 A Problematic Case of Using the Acquiring Approaches

When dealing with SPP using the SEDT, we cannot obtain the correct SEDT by the acquiring approach. Figure 6.18 shows the problematic case in a 2D domain where several obstacles denoted by "-1" exist. Note that the SED at the location enclosed by a rectangular box should be actually "29" instead of "35" obtained by the iterative algorithm. The reason why the iterative algorithm fails is that the structure component, $g(l)$, only records the relations between a pixel and its neighborhood. For example, in Figure 6.18, when the pixel enclosed by the rectangular box is processed, the adopted structure component is

2	1	2	5	10	17	26	37	50
1	0	1	4	9	16	25	36	49
2	1	2	5	10	17	26	37	50
5	4	5	8	13	20	29	40	53
10	9	10	13	18	25	34	45	58
17	16	17	20	25	32	41	52	65
26	25	26	29	34	41	50	61	74
37	36	37	40	45	52	61	72	85
50	49	50	53	58	65	74	85	98

(a) Without obstacles

2	1	2	-1	-1	-1	194	211	230
1	0	1	-1	-1	-1	169	186	205
2	1	2	-1	-1	-1	146	163	182
5	4	5	-1	-1	-1	125	142	161
10	9	10	-1	-1	-1	106	123	142
17	16	17	-1	-1	-1	89	106	125
26	25	26	35	46	59	74	91	110
37	36	37	46	57	70	85	102	121
50	49	50	59	70	83	98	115	134

(b) With obstacles

2	1	2	-1	-1	-1	178	185	194
1	0	1	-1	-1	-1	153	160	169
2	1	2	-1	-1	-1	130	137	146
5	4	5	-1	-1	-1	109	116	125
10	9	10	-1	-1	-1	90	97	116
17	16	17	-1	-1	-1	73	90	109
26	25	26	29	40	53	68	85	104
37	36	37	40	45	58	73	90	109
50	49	50	53	58	65	80	97	116

(c) Correct SEDT

Figure 6.18 The problematic case.

$$g(5) = \begin{bmatrix} -18 & -9 & -18 \\ -9 & 0 & -9 \\ -18 & -9 & -18 \end{bmatrix}$$

After an erosion is applied, the resulting SEDT is "29" from its neighbor as circled in Figure 6.18(a). When we add obstacles in Figure 6.18(b), the pixel enclosed by the rectangle is "35", which is apparently an error. This error could be corrected by the proposed two-scan algorithm to generate Figure 6.18(c). As we can see, the actual SED of this pixel relative to the background pixel "0" is correctly obtained as "29".

6.8.2 Dynamically Rotational Mathematical Morphology

Rotational morphology is considered in a digital image to equally divide the full 360° neighborhood into eight parts. Let A and B denote a binary image and a binary structuring element, respectively. The *rotational dilation* of A by B is defined as

$$\begin{aligned} A \tilde{\oplus} B &= [A \oplus B_7, A \oplus B_6, \ldots, A \oplus B_1, A \oplus B_0] \\ &= [P_7, P_6, \ldots, P_1, P_0] = P \end{aligned} \tag{6.28}$$

Similarly, the *rotational erosion* of A by B is defined as

$$A \tilde{\ominus} B = [A \ominus B_7, A \ominus B_6, \ldots, A \ominus B_1, A \ominus B_0] \tag{6.29}$$

Let P denote an 8-bit codeword with a base 2. The value can be interpreted as the degree of the structuring element's orientation that can be fitted into the object region. All the directional erosion can be used to indicate free positions in the path finding, so the object can be in the free space for at least one orientation under collision-free conditions.

In Pei et al. (1998), although rotational morphology is used for collision-free paths, the additional space and time complexity in their algorithm are not optimal. The additional record is an 8-bit codeword with a base 2, and the processing of mathematical morphology is eight times for each pixel in the 2D domain. Moreover, the cost increases dramatically with the increasing dimensions. For example, in the 3D space, the additional record is a 26-bit codeword and the morphological processing is 26 times for each pixel.

To reduce computational complexity and additional space, we develop the *dynamically rotational mathematical morphology* (DRMM). We do not calculate the morphological erosion for each pixel in the original image until the pixel is selected in the SPP. Figure 6.19 shows an example of deciding the next step by the DRMM.

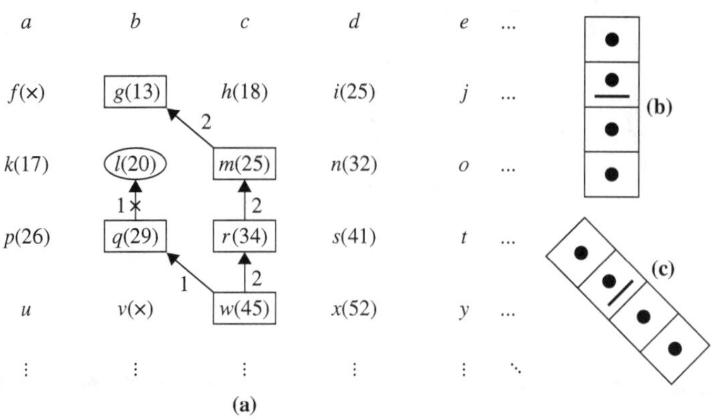

Figure 6.19 An example of DRMM.

Figure 6.19(a) and (b) shows the original image and the structure element, respectively. The underlined pixel indicates the origin of the structuring element. Figure 6.19(c) shows the structuring element with a rotation of 45°. Note that a counterclockwise rotation is counted as a positive angle. Let pixel w be the starting point. When deciding the next neighboring pixel in the SPP, we use the following two criteria:

1. The neighbor that has the minimum SED is selected.
2. The neighbor remaining as an object pixel after a rotational erosion (i.e., without collision) is selected.

When we decide the next step of pixel w, pixel q is selected because of its minimum SED and the satisfaction of morphological erosion. Note that the adopted structuring element is rotated by 45° because pixel q is in the direction of 45° of pixel w. Assume that pixels v and f are obstacles. Continually, when we decide the next step of pixel q, pixel l is not selected because the condition of applying the erosion on pixel l by the structuring element in Figure 6.18(b) is not satisfied (i.e., collision with the obstacle pixel v). Therefore, we go back to w and reconsider the next step of w. The pixel r is selected since it is satisfied by the above two criteria. Later on, m and g are selected.

6.8.3 The Algorithm for Shortest Path Planning

Let W, O, E, T, D, and S denote working domain, obstacle set, SE set, stack, destination point, and starting point, respectively. The algorithm for shortest path planning is described below, and its flowchart is shown in Figure 6.20.

1. Set the values of the obstacle pixels O and the destination pixel D to be "−1" and "0", respectively.
2. Obtain the SEDT by the two-scan algorithm.
3. Select a starting point S on the SEDT image.
4. Choose the pixel m that has the minimum SED from the neighborhood of S, $N(S)$, and decide the direction R_1 that is a vector from S to m to be recorded as a chain code.

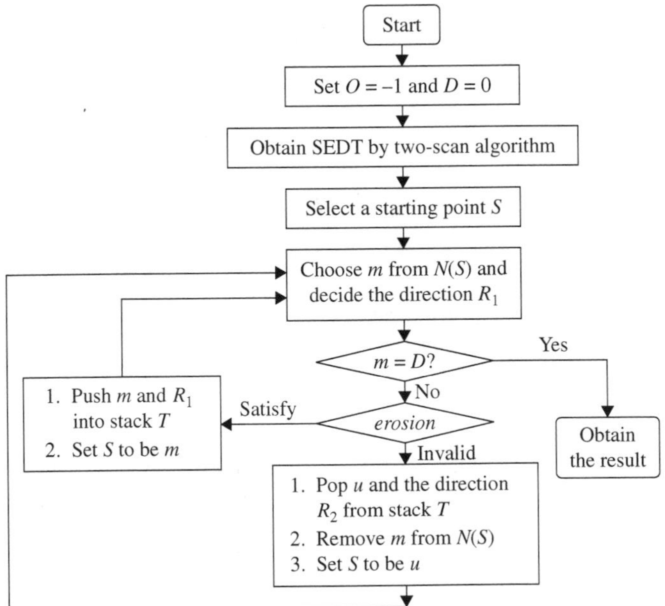

Figure 6.20 The flowchart for achieving the SPP based on DRMM.

5. Apply a morphological erosion on pixel m by the structuring element that is rotated based on the direction obtained from the previous step. If the result is satisfied, both m and R_1 are pushed into the stack T and set S to be m; otherwise, we pop the preceding pixel u from the stack, remove pixel m from $N(S)$, and set S to be u.
6. Obtain the shortest path by repeating steps 4 and 5 until the destination point D is reached.

6.8.4 Some Examples

Figure 6.21 illustrates an example of performing the SPP based on the DRMM, in which the two pixels circled with values "0" and "2377" indicate the destination and the starting points, respectively. The dark pixels with value "−1" are obstacles and the pixels enclosed by a rectangle constitute the shortest path. Figure 6.22(a) shows the result of applying our algorithm on the 3D space with convex obstacles, and Figure 6.22(b) shows the result when there is a concave obstacle. The starting and destination points are located at (3, 3, 1) and (27, 27, 3), respectively.

An example is also shown in Figure 6.23, where (a) is the original image with the hammer obstacle and (b) is the distance map obtained by the two-scan EDT. Note that Figure 6.23(b) is different from Figure 6.17(b) since different settings in the object and background are used. In Figure 6.23(a), the pixels within the hammer are set to "−1" to represent the obstacle, the white pixels are set to "1" to represent the object pixels, and only one pixel is set to "0" as shown by the sign "×" to represent the destination point. We obtain the shortest path as shown in Figure 6.23(c).

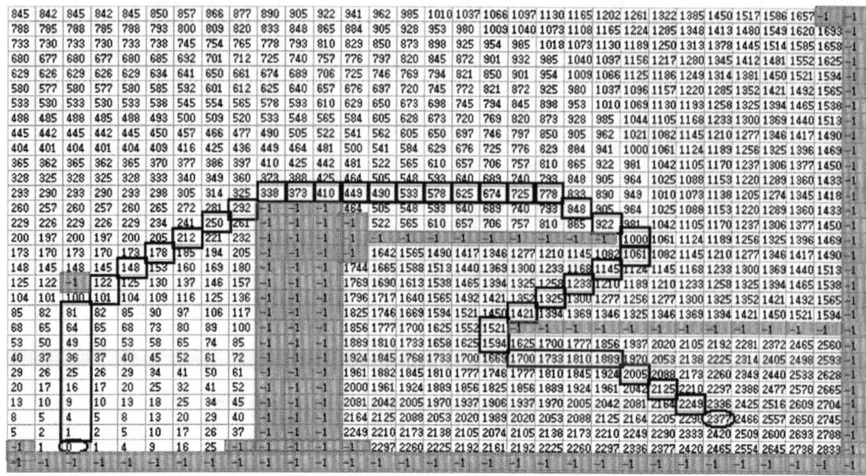

Figure 6.21 The experimental results for achieving SPP based on DRMM.

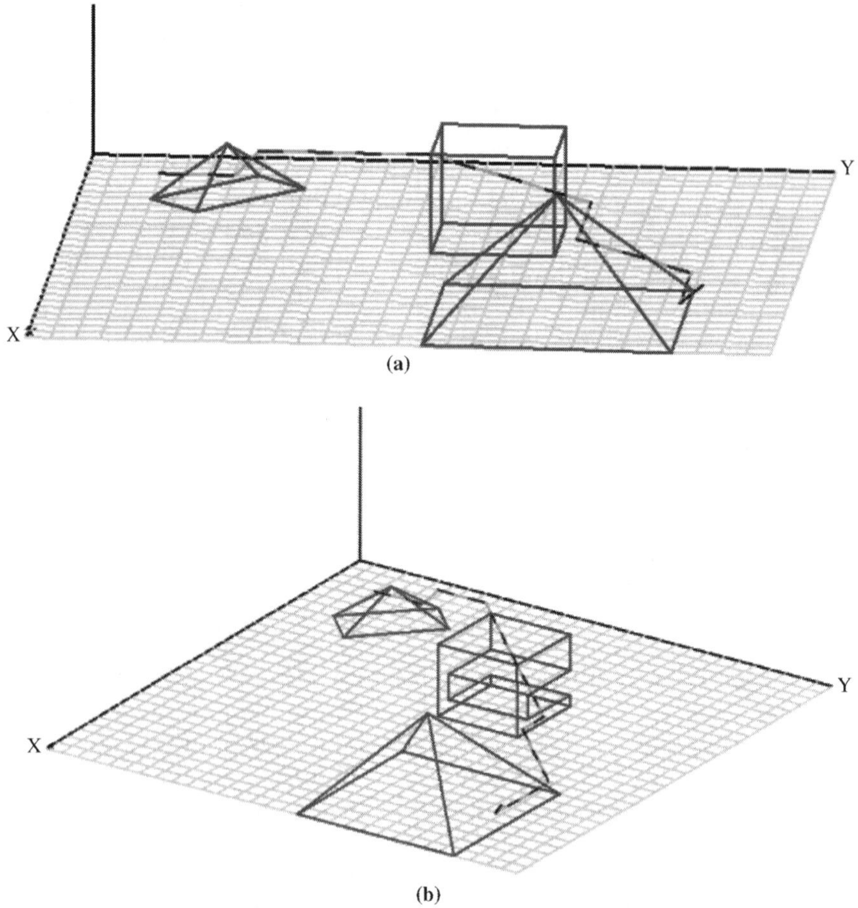

(a)

(b)

Figure 6.22 The results of applying our SPP algorithm on the 3D space (a) with convex obstacles and (b) with convex and concave obstacles.

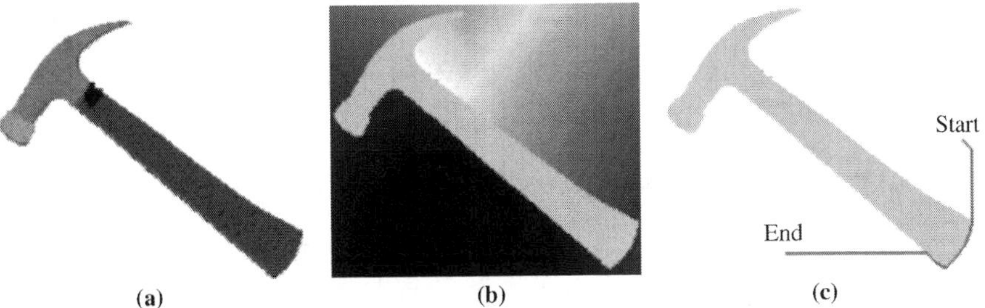

Figure 6.23 The SPP example of an image.

6.9 FORWARD AND BACKWARD CHAIN CODES FOR MOTION PLANNING

The chain-code representation (Shih and Wong, 1994, 2001) is efficient in contour tracing. It is based on 4- or 8-connectivity of the segments as shown in Figure 6.24 (Shih et al., 2004), where each direction is represented by a number i and $i = 0, 1, 2, 3$ indicating $90i°$ or $i = 0, 1, \ldots, 7$ indicating $45i°$, respectively. The chain code will be mentioned again in Section 7 for contour representation. Here we introduce its new variant that is used for motion planning.

Since the backward movement is allowed in car motion, the traditional chain-code representation cannot satisfy this requirement. Therefore, we add one additional set of chain codes prefixed by "*" to denote "backward," as shown in Figure 6.25. For

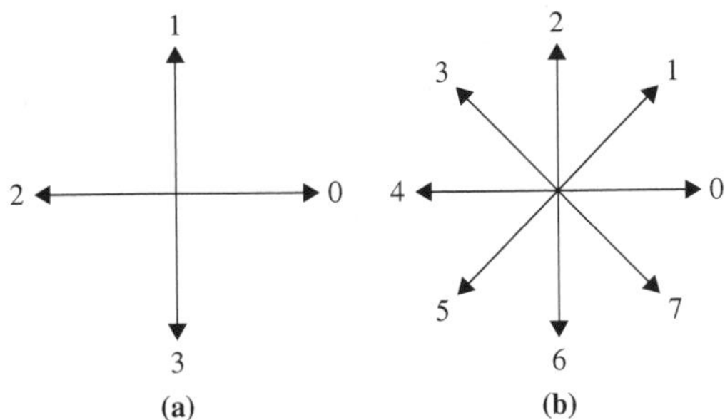

Figure 6.24 The representation of 4- and 8-connectivity chain codes.

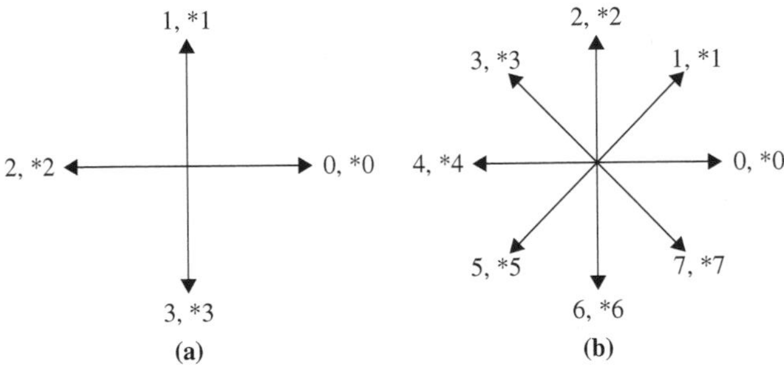

Figure 6.25 The new chain-code representation.

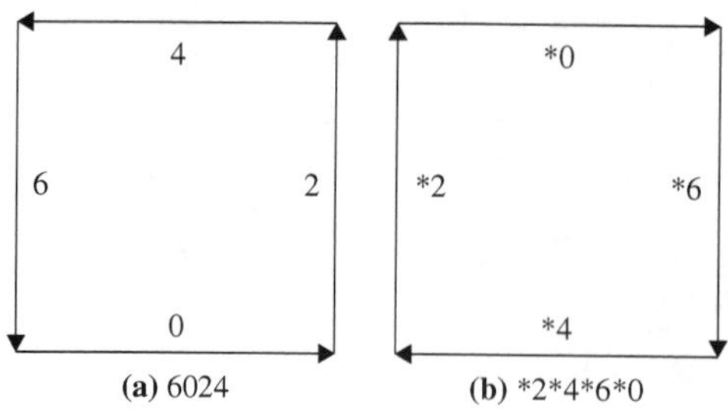

Figure 6.26 The examples of (a) the traditional and (b) the new chain codes.

example, "0" indicates the forward movement from the center to the east with the car oriented toward east, and "*0" indicates the backward movement from the east to the center with the car oriented toward east.

We adopt the 8-connectivity chain codes. Figure 6.26(a) and (b) shows the examples of the traditional (considering only forward movement) and the new chain codes, respectively. Note that the starting point is located at the top-left corner of the square and the tracing direction is counterclockwise. For the illustration purpose, we allow 90° turns in these examples. Therefore, we obtain "6024" and "*2*4*6*0" by using the traditional and the new chain codes, respectively.

Three distance metrics, Euclidean, city-block (4-connected), and chessboard (8-connected) distances, are often used in distance transformation. Note that, as long as the distance metric between a pixel p and its horizontal and vertical neighbors is smaller than the distance between p and its diagonal neighbors, it can be applied to generate the distance map for the shortest path planning. We choose the city-block distance transformation because of its simplicity in computation.

The propagated distance from a point p to a point q, denoted as $d(p, q)$, is defined as the length of the shortest path between p and q. Let eight neighbors of p be denoted by $q_1, q_2, \ldots, q_8$ as illustrated in Figure 6.27, and the distances between p and its neighbors are $d(p, q_1) = d(p, q_3) = d(p, q_5) = d(p, q_7) = 1$ and $d(p, q_2) = d(p, q_4) = d(p, q_6) = d(p, q_8) = 2$.

Let O, Q, F, B, and $N(p)$ denote the obstacle set, queue, the set of foreground pixels, the set of background pixels, and the 8-neighbors of p, respectively. The algorithm of *recursive propagation distance transformation* (RPDT) is described below.

1. Set $d[p] = \infty$ for all $p \in F$, $d[p] = 0$ for all $p \in B$, and $d[p] = -1$ for all $p \in O$.
2. Push all p into queue Q for all $p \in B$.
3. Choose one pixel r from Q. Modify the value of $d[e]$ if $d[e] \neq -1$ and $d[e] > d[r]$, where $e \in N(r)$. Place the modified e into Q.
4. Obtain the city-block distance map by repeating steps 2 and 3 until Q is empty.

q_2	q_3	q_4
q_1	p	q_5
q_8	q_7	q_6

Figure 6.27 The pixel p and its eight neighbors.

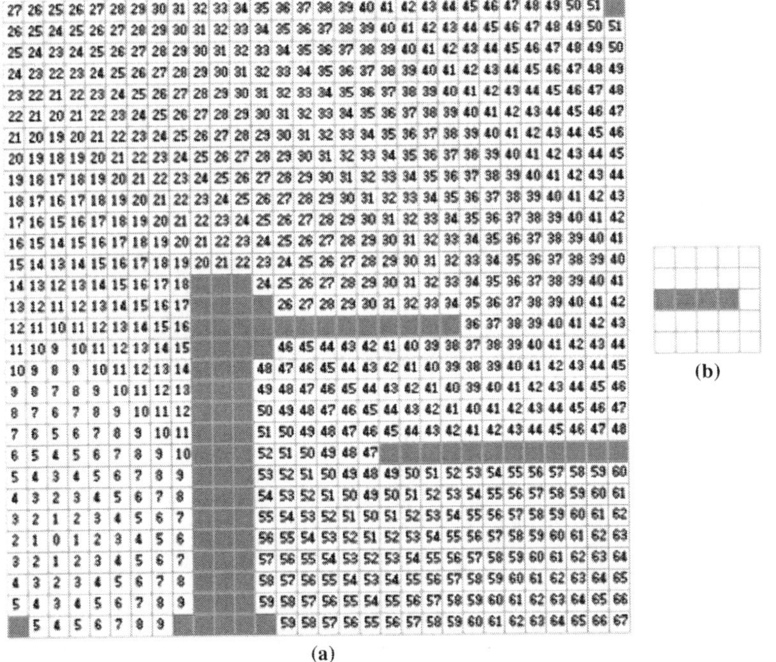

Figure 6.28 The distance map and the model of a four-pixel line vehicle.

An example of using a four-pixel line vehicle in a field of size 30×30 is shown in Figure 6.28. Figure 6.28(a) is the city-block distance map with a destination point "0" located at (3, 26), where (1, 1) is placed at the top-left pixel. We use the dark pixels to denote the obstacles. Figure 6.28(b) shows the model of a four-pixel line vehicle with its orientation toward east.

Let W, T, C, R, G, D, and S denote the working domain, stack, car orientation, moving direction, rotation angle, destination point, and starting point, respectively. We present the shortest path planning algorithm considering forward and backward movements below, and its flowchart is shown in Figure 6.29.

Shortest Path Planning Algorithm

1. Set D to be 0.
2. Obtain the distance map by the recursive propagation algorithm.
3. Randomly select a starting point S and a beginning car orientation C on the distance map.
4. Choose a pixel m from $N(S)$ that has the minimum city-block distance value and the collision-free code "1", and record the corresponding R and G values.
5. If $m = D$, the destination point is reached.
6. If m is a previously selected pixel m', pop a record in which we can obtain a new set of data S, m', R, and G from the stack T. Set $C = R$ and go to step 4.

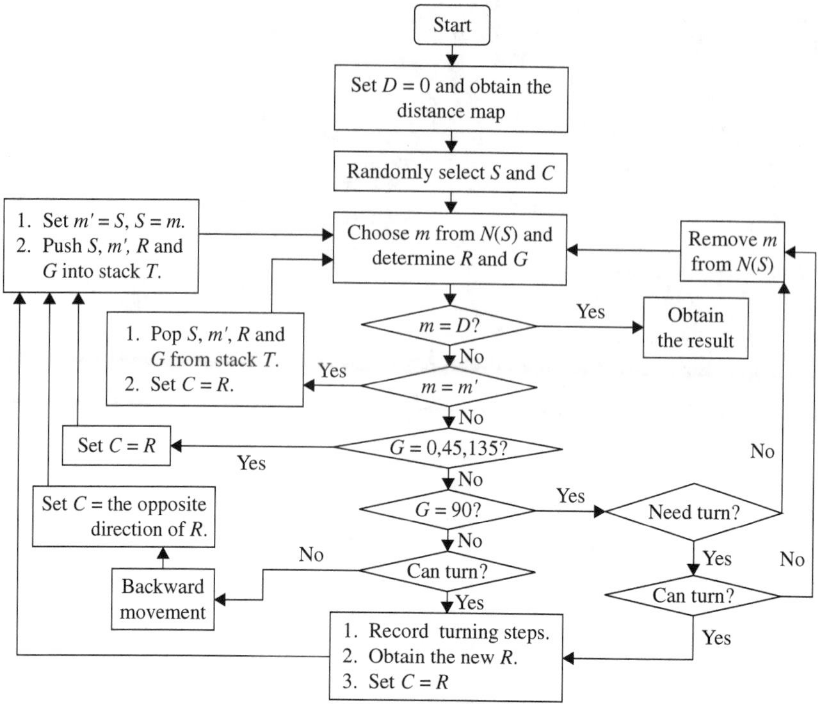

Figure 6.29 The flowchart of the shortest path planning.

7. If $G = 0°$, $45°$, or $135°$, it indicates that the car can move to the location. Therefore, we set $C = R$, $m' = S$, $S = m$, and push S, m', R, and G into the stack T, and then go to step 4.
8. If $G = 90°$, we need to check whether the car is required to turn. If the car orientation is different from the next moving direction, it is required to turn. However, if the turn causes collision, we remove the pixel m from $N(S)$ and go to step 4. Otherwise, continue with step 9.
9. If the car can be turned at the location m, record the turning steps and obtain the new car orientation R. Set $C = R$, $m' = S$, $S = m$, and push S, m', R, and G into the stack T. Otherwise, the car should take a backward movement at the location m. Note that if the car moves backward, its orientation is equal to the opposite direction of R. Go to step 4 for continuing to seek the next pixel until the destination point is reached.

The turning cases in step 9 can be categorized into three classes, 90°, 135°, and 180°, depending on the inner angle between the car orientation and the moving direction. Figure 6.30(a) shows an example of a backward movement followed by a forward movement in order to turn 45°. In other words, two movements are required to turn 45° at the same location. Similarly, we need three and four steps for turning the car if the angles between the car orientation and the moving direction are, respectively, 135° and 180°, as shown in Figure 6.30(b) and (c).

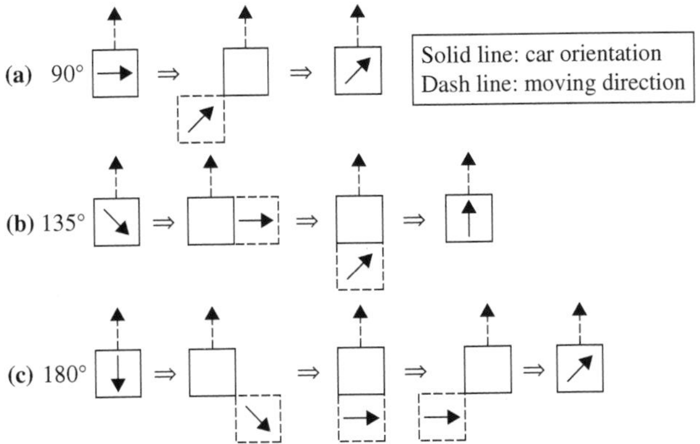

Figure 6.30 The turning cases for different degrees.

6.10 A FEW EXAMPLES

A few examples of autonomous vehicles moving in a factory space of size 30 × 30 are presented in this section. By placing the starting point at (25, 27), Figure 6.31(a) and (b) shows two examples when the car is oriented in the north and east directions, respectively. In Figure 6.31(a), since the angle between the car orientation and the moving direction is 45°, we do not need to turn the car orientation. Therefore, the chain-code representation is "33344332111112334444444444445555555666666". In Figure 6.31(b), since the angle between the car orientation and the moving direction is 135°, we need three additional steps to rotate the car orientation. Therefore, the chain-code representation is "0*1233344332111112334444444444445555555666666".

The proposed shortest path planning algorithm is experimented on a computer with Pentium III 600 MHz and 256 MB DRAM under Windows 2000 environment. The programs were written in Delphi 5. In Figure 6.31(a), the execution time for establishing the city-block distance map and collision-free codes is 0.56 s.

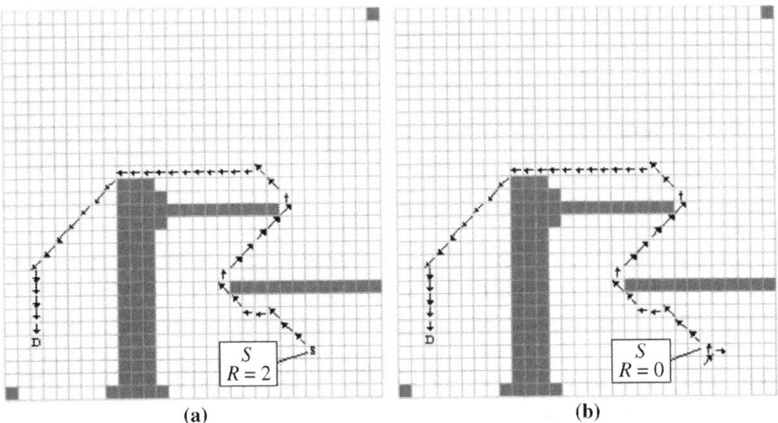

Figure 6.31 The example of the shortest path planning in the different car orientations.

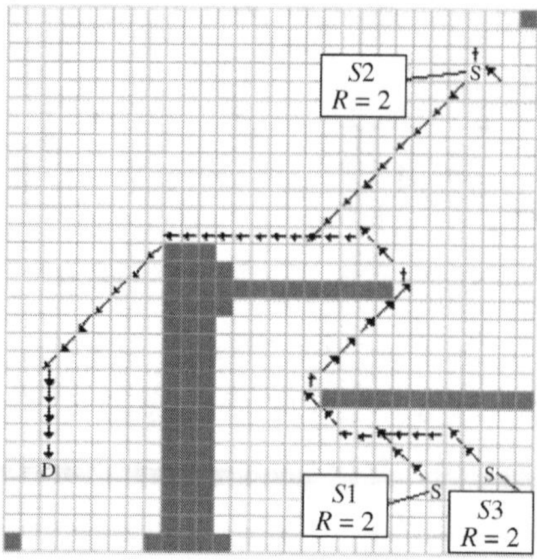

Figure 6.32 The shortest path planning with different starting points.

Once the required data are established, the time for obtaining the shortest path is as fast as 0.12 s.

As soon as the distance map and the collision-free code have been established offline in the configuration space, the shortest paths of cars starting from any location toward the destination can be promptly obtained online. In Figure 6.32, the corresponding chain codes for the three starting points are "333443321111123344 4444444445555555666666" for *S*1, "2*34555555555444444444555555566666" for *S*2, and "3344444433211111233444444444445555555 666666" for *S*3.

The allowance for the backward motion can sometimes provide a feasible path by using the complicated geometric methods (Reeds and Shepp, 1990; Laumond et al., 1994). When the forward movement is solely used in finding a path in a narrow space and no solution can be found, the backward movement will be considered. Figure 6.33 shows the car that cannot pass through the narrow space because of the collision with obstacles. By adding the backward movement, we can obtain the path as "44444444445*6*6*6*6*6*6*6*701000000000".

The shortest path planning algorithm can also be applied to automated parallel parking as shown in Figure 6.34. We aim at placing the vehicle from the source (or starting) location to a designated (or destination) parking space. First, we use the shortest path planning algorithm to obtain a path from the source to an intermediate location, which is parallel to the parking space approximately and in the middle of upper and lower boundaries. From the intermediate location, the vehicle will move forward being still parallel to the parking space until it reaches its upper boundary. Next, the vehicle will turn outward 45° and move backward to enter the parking space as long as it does not hit the boundary as applied by the aforementioned erosion rule. Finally, the vehicle will adjust itself to the parking space by turning 45° inward and moving forward closer to its upper boundary. It is not necessary to assign the direction of driving and the direction of parking. The algorithm for automated parallel parking assumes that the vehicle can be parked in two directions depending on the vehicle direction in which it is oriented at the intermediate location. The chain codes in Figure 6.34(a) and (b) are obtained as "33444332111112333333333344*5*544" and "33444332111112334444444444445555556*7*766", respectively.

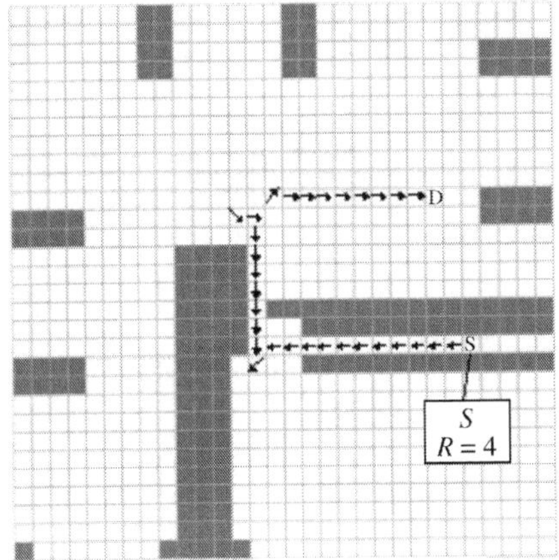

Figure 6.33 The shortest path planning with backward movement.

To extend the algorithm to three dimensions, the number of neighbors for a pixel p is expanded from 8 in 2D to 26 in 3D, as shown in Figure 6.35(a). The distances between p and its neighbors are $d(p,q_1)=d(p,q_3)=d(p,q_5)=d(p,q_7)=d(p,q_{19})=d(p,q_{29})=1$, $d(p,q_2)=d(p,q_4)=d(p,q_6)=d(p,q_8)=d(p,q_{11})=d(p,q_{13})=d(p,q_{15})=d(p,q_{17})=d(p,q_{21})=d(p,q_{23})=d(p,q_{25})=d(p,q_{27})=2$, and $d(p,q_{12})=d(p,q_{14})=d(p,q_{16})=d(p,q_{18})=d(p,q_{22})=d(p,q_{24})=d(p,q_{26})=d(p,q_{28})=3$. Meanwhile, the number of chain codes required for the representation is increased from 16 to 52, as shown in Figure 6.35 (b). To denote each 3D chain code in just one digit, the 26 lowercase alphabets are used, as shown in Table 6.1. Figure 6.36 illustrates our algorithm in the 3D space, where two obstacles, a cube and an octahedron, and an airplane model are used. We can obtain its chain codes as "rrrraaaaaaaaahhhabbbbbbtzzz".

The execution time for establishing the city-block distance map and collision-free codes is 2.56 s for the size of $30 \times 30 \times 30$. Once the required data are established, the time for obtaining the shortest path is 0.1 s.

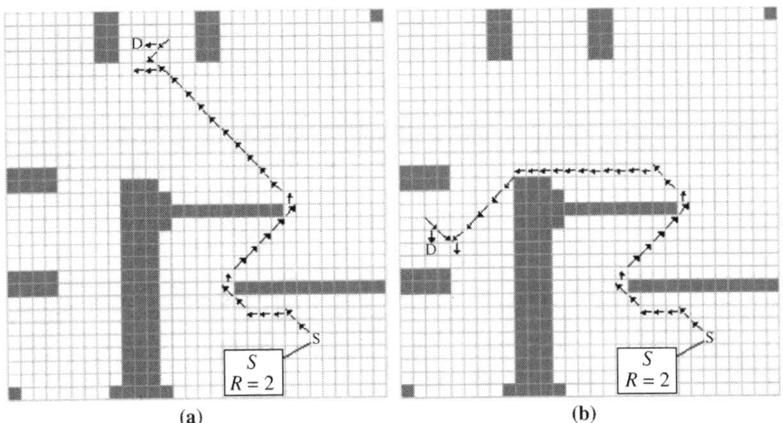

Figure 6.34 Shortest path planning for parallel parking.

TABLE 6.1 The 3D Chain-Code Representation by Using 26 Alphabets

Original	New	Original	New	Original	New	Original	New
0	a	13	n	*0	*a	*13	*n
1	b	14	o	*1	*b	*14	*o
2	c	15	p	*2	*c	*15	*p
3	d	16	q	*3	*d	*16	*q
4	e	17	r	*4	*e	*17	*r
5	f	18	s	*5	*f	*18	*s
6	g	19	t	*6	*g	*19	*t
7	h	20	u	*7	*h	*20	*u
8	I	21	v	*8	*i	*21	*v
9	j	22	w	*9	*j	*22	*w
10	k	23	x	*10	*k	*23	*x
11	l	24	y	*11	*l	*24	*y
12	m	25	z	*12	*m	*25	*z

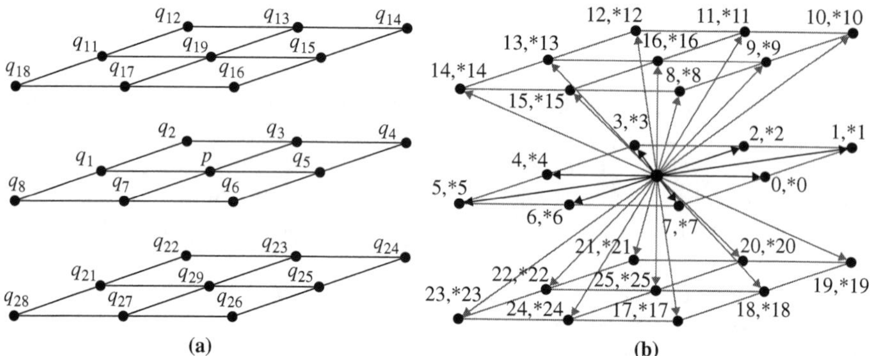

Figure 6.35 The 26 neighbors of a pixel p and the chain-code representation in the 3D space.

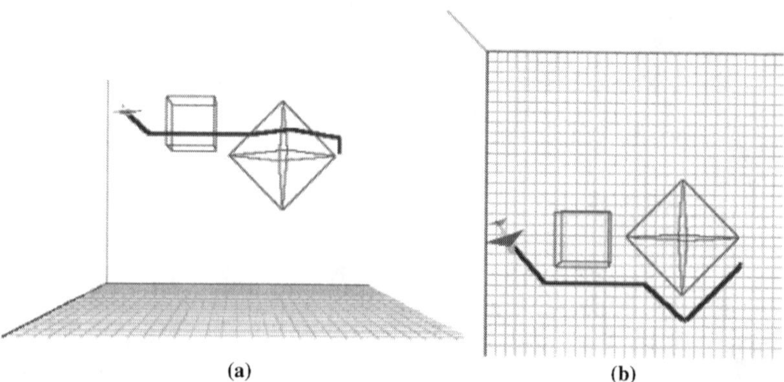

Figure 6.36 The example of the 3D shortest path planning.

REFERENCES

Arkin, E. M., Mitchell, J. S. B., and Piatko, C. D., "Bicriteria shortest path problems in the plane," Proceedings of the 3rd Canadian Conference on Computer Geometry, Vancouver, Canada, pp. 153–156, Aug. 1991.

Blum, H., "A transformation for extracting new descriptors of shape," *Proceedings of the Symposium on Models for the Perception of Speech and Visual Forms*, Boston, MA, MIT Press, pp. 362–380, Nov. 1967.

Borgefors, G., "Distance transformations in arbitrary dimensions," *Comput. Vision Graph. Image Process.*, vol. 27, no. 3, pp. 321–345, Sept. 1984.

Borgefors, G., "Distance transformations in digital images," *Comput. Vision Graph. Image Process.*, vol. 34, no. 3, pp. 344–371, June 1986.

Brooks, R. A., "Solving the find path problem by good representation for free apace," *IEEE Trans. Syst. Man Cybernet.*, vol. 13, no. 3, pp. 190–197, Mar. 1983.

Coeurjolly, D. and Montanvert, A., "Optimal separable algorithms to compute the reverse Euclidean distance transformation and discrete medial axis in arbitrary dimension," *IEEE Trans. Pattern Anal. Mach. Intell.*, vol. 29, no. 3, pp. 437–448, Mar. 2007.

Cuisenaire, O. and Macq, B., "Fast Euclidean distance transformation by propagation using multiple neighborhoods," *Comput. Vision Image Understand.*, vol. 76, no. 2, pp. 163–172, Nov. 1999.

Danielsson, P. E., "A new shape factor," *Comput. Graph. Image Process.*, vol. 7, no. 2, pp. 292–299, Apr. 1978.

Danielsson, P. E., "Euclidean distance mapping," *Comput. Graph. Image Process.*, vol. 14, no. 3, pp. 227–248, Nov. 1980.

Datta, A. and Soundaralakshmi, S., "Constant-time algorithm for the Euclidean distance transform on reconfigurable meshes," *J. Parallel Distrib. Comput.*, vol. 61, no. 10, pp. 1439–1455, Oct. 2001.

Diaz de Leon, S. and Sossa, A., "Automatic path planning for a mobile robot among obstacles of arbitrary shape," *IEEE Trans. Syst. Man Cybernet. B*, vol. 28, no. 3, pp. 467–472, June 1998.

Eggers, H., "Two fast Euclidean distance transformations in Z^2 based on sufficient propagation," *Comput. Vision Image Understand.*, vol. 69, no. 1, pp. 106–116, Jan. 1998.

Hilditch, J., "Linear skeletons from square cupboards," in *Machine Intelligence*, vol. 4 (B. Meltzer and D. Michie,eds.), American Elsevier Publishing Co., pp. 404–420, 1969.

Huang, C. T. and Mitchell, O. R., "A Euclidean distance transform using grayscale morphology decomposition," *IEEE Trans. Pattern Anal. Mach. Intell.*, vol. 16, no. 4, pp. 443–448, Apr. 1994.

Jan, G., Chang, K., and Parberry, I., "Optimal path planning for mobile robot navigation," *IEEE/ASME Trans. Mechatron.*, vol. 13, no. 4, pp. 451–460, Aug. 2008.

Kambhampati, S. K. and Davis, L. S., "Multiresolution path planning for mobile robots," *IEEE Trans. Robot. Automat.*, vol. 2, no. 3, pp. 135–145, Sept. 1994.

Lantuejoul, C., "Skeletonization in quantitative metallography," in *Issues in Digital Image Processing* (R. M. Haralick and J. C. Simon,eds.), Sijthoff & Noordhoff Publishers, pp. 107–135, 1980.

Latombe, J., *Robot Motion Planning*, Kluwer Academic Publishers, 1991.

Laumond, J. P., Jacobs, P. E., Taix, M., and Murray, R. M., "A motion planner for nonholonomic mobile robots," *IEEE Trans. Robot. Automat.*, vol. 10, no. 5, pp. 577–593, Oct. 1994.

Lozano-Perez, T., "Spatial planning: a configuration space approach," *IEEE Trans. Comput.*, vol. 32, no. 2, pp. 108–120, Feb. 1983.

Maurer, C. R., Qi, R., and Raghavan, V., "A linear time algorithm for computing exact Euclidean distance transforms of binary images in arbitrary dimensions," *IEEE Trans. Pattern Anal. Mach. Intell.*, vol. 25, no. 2, pp. 265–270, Feb. 2003.

Mezouar, Y. and Chaumette, F., "Path planning for robust image-based control," *IEEE Trans. Robot. Automat.*, vol. 18, no. 4, pp. 534–549, Aug. 2002.

Mitchell, J. S. B., Rote, G., and Woeginger, G., "Minimum-link paths among obstacles in the plane," Proceedings of the 6th ACM Symposium on Computer Geometry, Berkley, CA, pp. 63–72, 1990.

Pei, S.-C. and Horng, J.-H., "Finding the optimal driving path of a car using the modified constrained distance transformation," *IEEE Trans. Robot. Automat.*, vol. 14, no. 5, pp. 663–670, Oct. 1998.

Pei, S.-C., Lai, C.-L., and Shih, F. Y., "A morphological approach to shortest path planning for rotating objects," *Pattern Recogn.*, vol. 31, no. 8, pp. 1127–1138, Aug. 1998.

Reeds, J. A. and Shepp, R. A., "Optimal paths for a car that goes both forward and backward," *Pacific J. Math.*, vol. 145, no. 2, pp. 367–393, 1990.

Rosenfeld, A. and Kak, A. C., *Digital Picture Processing*, Academic Press, 1982.

Rosenfeld, A. and Pfaltz, J. L., "Sequential operations in digital picture processing," *J. ACM*, vol. 13, no. 4, pp. 471–494, Oct. 1966.

Saito, T. and Toriwaki, J.-I., "New algorithms for Euclidean distance transformation of an n-dimensional digitized picture with applications," *Pattern Recogn.*, vol. 27, no. 11, pp. 1551–1565, Nov. 1994.

Serra, J., *Image Analysis and Mathematical Morphology*, Academic Press, New York, 1982.

Shih, F. Y. and Mitchell, O. R., "Skeletonization and distance transformation by grayscale morphology," Proceedings of the SPIE Symposium on Automated Inspection and High Speed Vision Architectures, Cambridge, MA, pp. 80–86, Nov. 1987.

Shih, F. Y. and Mitchell, O. R., "Threshold decomposition of grayscale morphology into binary morphology," *IEEE Trans. Pattern Anal. Mach. Intell.*, vol. 11, no. 1, pp. 31–42, Jan. 1989.

Shih, F. Y. and Mitchell, O. R., "Decomposition of grayscale morphological structuring elements," *Pattern Recogn.*, vol. 24, no. 3, pp. 195–203, Mar. 1991.

Shih, F. Y. and Mitchell, O. R., "A mathematical morphology approach to Euclidean distance transformation," *IEEE Trans. Image Process.*, vol. 1, no. 2, pp. 197–204, Apr. 1992.

Shih, F. Y. and Wong, W., "An improved fast algorithm for the restoration of images based on chain codes description," *Comput. Vision Graph. Image Process.*, vol. 56, no. 4, pp. 348–351, July 1994.

Shih, F. Y. and Wong, W.-T., "An adaptive algorithm for conversion from quadtree to chain codes," *Pattern Recogn.*, vol. 34, no. 3, pp. 631–639, Mar. 2001.

Shih, F. Y. and Wu, H., "Optimization on Euclidean distance transformation using grayscale morphology," *J. Visual Commun. Image Represent.*, vol. 3, no. 2, pp. 104–114, June 1992.

Shih, F. Y. and Wu, Y., "The efficient algorithms for achieving Euclidean distance transformation," *IEEE Trans. Image Process.*, vol. 13, no. 8, pp. 1078–1091, Aug. 2004.

Shih, F. Y. and Wu, Y., "Decomposition of arbitrarily grayscale morphological structuring elements," *Pattern Recogn.*, vol. 38, no. 12, pp. 2323–2332, Dec. 2005.

Shih, F. Y., Wu, Y., and Chen, B., "Forward and backward chain-code representation for motion planning of cars," *Pattern Recogn. Artif. Intell.*, vol. 18, no. 8, pp. 1437–1451, Dec. 2004.

Sun, Z. and Reif, J. H., "On robotic optimal path planning in polygonal regions with pseudo-Euclidean metrics," *IEEE Trans. Syst. Man Cybernet., B*, vol. 37, no. 4, pp. 925–936, Aug. 2007.

Toriwaki, J. and Yokoi, S., "Distance transformations and skeletons of digitized pictures with applications," in *Progress in Pattern Recognition* (L. N. Kanal and A. Rosenfeld,eds.), North-Holland, pp. 187–264, 1981.

Vincent, L., "Exact Euclidean distance function by chain propagations," Proceedings of the IEEE Conference on Computer Vision and Pattern Recognition, Maui, Hawaii, pp. 520–525, June 1991.

Vossepoel, A. M., "A note on distance transformations in digital images," *Comput. Vision Graph. Image Process.*, vol. 43, no. 1, pp. 88–97, July 1988.

Willms, A. R. and Yang, S. X., "Real-time robot path planning via a distance-propagating dynamic system with obstacle clearance," *IEEE Trans. Syst. Man Cybernet. B*, vol. 38, no. 3, pp. 884–893, June 2008.

CHAPTER 7

IMAGE REPRESENTATION AND DESCRIPTION

Representation and manipulation of digital images are two important tasks in image processing, pattern recognition, pictorial database, computer graphics, geographic information systems (GIS), and other related applications. Image representation and description is critical for successful detection and recognition of objects in a scene. After an image has been segmented into object and background regions, one intends to represent and describe them in characteristic features for computer processing during pattern recognition or in quantitative codes for efficient storage during image compression. There are basically two types of characteristics: external and internal. An external representation focuses on object shape characteristics such as boundary. An internal representation focuses on object region characteristics such as texture, topology, and skeleton.

This chapter is organized as follows. Run-length coding is introduced in Section 7.1. Binary tree and quadtree representations are described in Section 7.2. In Section 7.3, we present different contour representation schemes, such as chain code, crack code, and mid-crack code. Skeletonization (or medial axis transformation) by thinning is presented in Section 7.4 and by maxima tracking on distance transform is presented in Section 7.5. Finally, in Section 7.6, we present object representation and tolerance by mathematical morphology.

7.1 RUN-LENGTH CODING

Pictures transmitted over communication channels and stored on disks or tapes in computer systems generally contain significant data redundancy. The technique or mechanism of image compression intends to represent pictures by as few bits as possible. The storage requirement for a digitized binary image, represented directly as a 2D $N \times N$ array of points, means that such a representation scheme requires N^2 bits. Different techniques have been used to reduce this storage requirement by eliminating some of the data correlation or redundancy that exists in an image at the expense of processing time, without the loss of any information and consequent introduction of misrepresentation in the recovered image. Many lossless image coding schemes have been proposed. In particular, run-length coding, chain coding,

Image Processing and Pattern Recognition by Frank Shih

crack coding, rectangular coding, quadtree representation, and medial axis transformation (MAT) are among those schemes that have been considered for use in image data compression.

Run-length coding is one of the most practical and simplest methods for binary images (Golomb, 1966). Each row of a picture consists of a sequence of maximal runs of points, such that the points in each run all have the same value. Thus, the row is completely determined, as far as a binary image is concerned, by specifying the value of the first run in the row and the lengths of all runs. The value of each run does not have to be specified because these values must alternate between 0 and 1. For an $N \times N$ picture, if the average number of runs in each row is R, the total number of bits required for the run-length representation is $N(1 + R \log N)$. High efficiency is obtained for images with a small number of runs.

Capon (1959) presented a probabilistic model for run-length coding of pictures. Tanaka and Leon-Garcia (1982) proposed an efficient method for the run-length coding of binary memoryless sources when the symbol probabilities are unknown. Montani and Scopigno (1990) developed the "STICKS" representation scheme data model based on a 3D extension of the run-length encoding methods. Bunke and Csirik (1995) proposed an improved algorithm for computing the edit distance of run-length coded strings. Berghorn et al. (2001) presented fast variable run-length coding for embedded progressive wavelet-based image compression.

Example 7.1 Represent the following figure using the run-length representation:

```
1 1 1 1 0 0 0 0
1 1 1 1 0 0 0 0
1 1 1 1 1 1 0 0
1 1 1 1 1 1 0 0
0 0 1 1 1 1 0 0
0 0 1 1 1 1 0 0
0 0 0 0 0 0 0 0
0 0 0 0 0 0 0 0
```

Answer: The run-length representation is

```
1 4 4
1 4 4
1 6 2
1 6 2
0 2 4 2
0 2 4 2
0 8
0 8
```

7.2 BINARY TREE AND QUADTREE

In computer science, a binary tree is a tree data structure where each node contains at most two children. We first introduce a binary tree to represent a row of an image. Let the row length be $n = 2^k$. If the row all has one value, it is labeled by the value; otherwise, we add two descendants to the root node representing two halves. If the half has one constant value, it is labeled with that value; otherwise, we give two descendants corresponding two halves of its half. The division is stopped when it reaches one pixel. Note that at level h in the tree, the nodes represent pieces of length 2^{k-h}.

Wang et al. (2003) proposed a region-based binary tree representation with adaptive processing of data structures for image classification. Robinson (1997) presented efficient general-purpose image compression with binary tree predictive coding, which uses a noncausal, shape-adaptive predictor to decompose an image into a binary tree of prediction errors and zero blocks. Robinson et al. (2000) presented video compression with binary tree recursive motion estimation and binary tree residue coding.

Example 7.2 Draw the binary tree for the following row:

10010110111111101001011111110110

Answer:

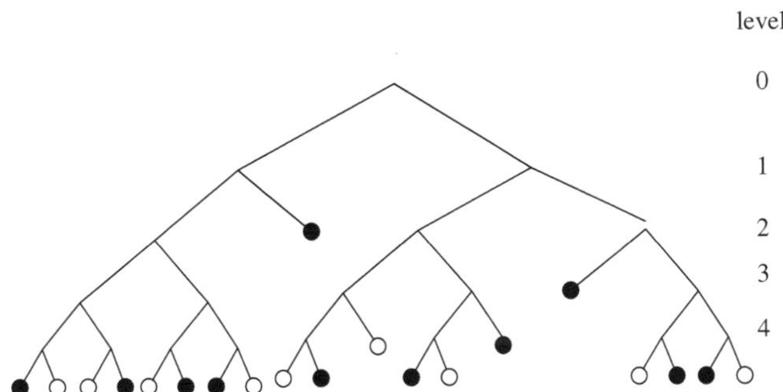

Quadtree is one of the compact hierarchical data structures in representing a binary image (Hunter and Steiglitz, 1979). Let the image size be $n = 2^k \times 2^k$. It is constructed by successively subdividing an image into four equal-size subimages in the NW (northwest), NE (northeast), SW (southwest), and SE (southeast) quadrants. A homogeneously colored quadrant is represented by a leaf node in the tree. Otherwise, the quadrant is represented by an internal node and further divided into four subquadrants until each subquadrant has the same color. The leaf node with a black (white) color is called the black (white) node, and the internal node is called the gray node. Note that at level h in the tree, the nodes represent blocks of size $2^{k-h} \times 2^{k-h}$. An example of quadtree representation is shown in Figure 7.1.

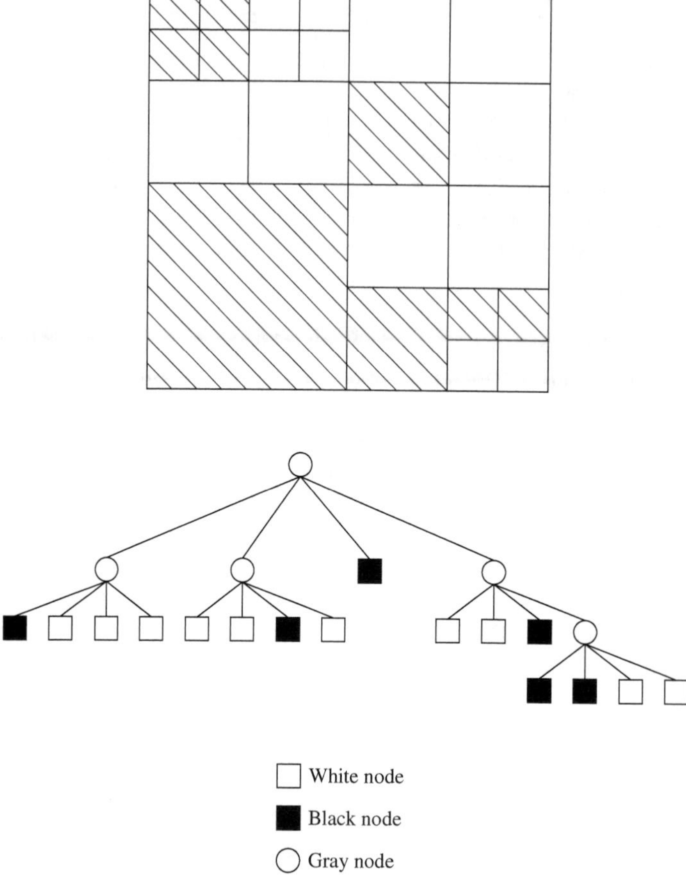

Figure 7.1 An image and its quadtree representation.

Quadtree representation has been used for a variety of image processing and image compression algorithms. Samet (1980) presented an algorithm for smoothing images represented by quadtrees. Shneier (1981) proposed an algorithm for computing geometric properties of binary images represented by quadtrees. Sullivan and Baker (1994) described the theory needed to construct quadtree data structures of images and videos that optimally allocate rate, given a set of quantizers. Shusterman and Feder (1994) presented image compression using improved quadtree decomposition algorithms. Knipe and Li (1998) proposed improvements over the reconstruction method described by Shusterman and Feder (1994). Laferte et al. (2000) presented discrete Markov image modeling and inference on the quadtree. Wong et al. (2006) presented thinning algorithms based on quadtree and octree representations.

Example 7.3 Draw the quadtree representation for the following image:

0	0	0	0	0	0	0	0
0	0	0	0	1	0	0	0
1	1	1	1	0	0	1	1
1	1	1	1	1	0	0	0
0	0	0	1	1	1	1	1
0	0	1	1	1	1	0	0
0	0	1	1	1	1	0	0
0	0	1	1	1	1	1	1

Answer: In the following figure, we only draw the white and black nodes. White nodes denote "0" and black nodes denote "1".

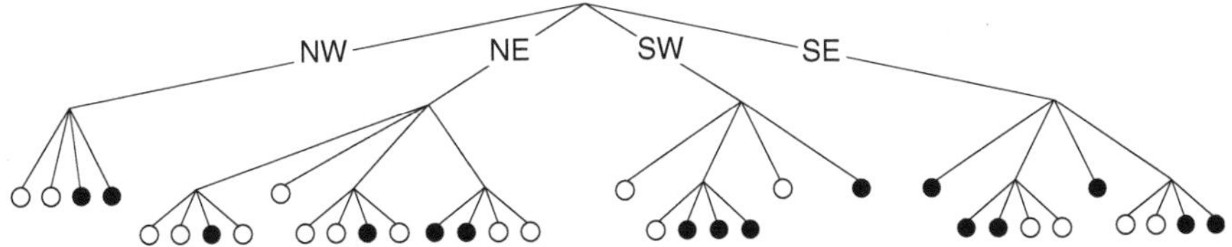

This tree is called *quadtree* since its nonleaf nodes all have degree 4. The storage requirement for the quadtree is proportional to the number of nodes. The quadtree is not data redundant. However, it is shift-variant; a simple position shift will cause a big variation in the quadtree representation. Therefore, it is difficult to match whether two images are congruent.

There are two widely used representations of quadtrees. A pointer-based quadtree uses the standard tree representation. A linear quadtree can be either a preorder traversal of the nodes of a quadtree or the sorted sequence of the quadtree's leaves. The quadtree's data structure has been applied to the representation of maps in the geographic information systems successfully. A fundamental operation intends to overlay two maps with union, intersection, and difference operators. Some research areas are parallel computational models of the quadtree algorithm.

In the *quadtree representation* case, maximal blocks can be of any size and at any position. They are analogous to the runs in the 1D case. If the picture has all one value, label the root node with that value and stop. Otherwise, four descendants are added to the root node, representing the four quadrants of the picture. The process is then repeated for each of these new nodes and so on. If the number of nodes in the tree is T, then $2T$ bits are required to represent the image. This scheme is conceptually clear and well structured, but it turns out that the compression efficiency of the scheme is not, in general, very high. Another disadvantage of the quadtree representation is that it is shift-variant. When a picture is shifted, it may generate a very different quadtree.

7.3 CONTOUR REPRESENTATION

In building the object model base, an object or multiple objects of a grayscale image are thresholded into a binary image, and object (or component) labeling is used to assign each object a unique nonzero label and reset the background pixels to zeros. Then a border operation is done by extracting all the boundary pixels in an object. Chain code, crack code, or mid-crack code can be used for contour representation.

There are other approaches for contour representation in the literature. Meer et al. (1990) proposed a hierarchical approach toward fast parallel processing of chain code contours. Saint-Marc et al. (1993) presented representation of edge contours by approximating B-splines and showed that such a representation facilitates the extraction of symmetries between contours. Hasan and Karam (2000) developed morphological reversible contour representation, in which a binary image is represented by a set of nonoverlapping multilevel contours and a residual image. Ghosh and Petkov (2005) proposed shape descriptors to incomplete contour representations.

7.3.1 Chain Code and Crack Code

The contour representation of a binary image is determined by specifying a starting point and a sequence of moves around the borders of each region. Methods often used for contour tracing are based on the chain code or crack code concept (Freeman, 1974; Rosenfeld and Kak, 1982; Wu, 1982). The chain code moves along a sequence of the centers of border points, while the crack code moves along a sequence of "cracks" between two adjacent border points. Typically, they are based on the 4- or 8-connectivity of the segments, where the direction of each segment is encoded by a numbering scheme, such as 3-bit numbers $\{i|i = 0, 1, \ldots, 7\}$ denoting an angle of $45i^\circ$ counterclockwise from the positive x-axis for a chain code as shown in Figure 7.2, or 2-bit numbers $\{i|i = 0, 1, 2, 3\}$ denoting an angle of $90i^\circ$ for a crack code. The elementary idea of the chain or crack coding algorithm is to trace the border pixels or cracks and sequentially generate codes by considering the neighborhood adjacency relationship.

Kaneko and Okudaira (1985) proposed an efficient encoding scheme for arbitrary curves based on the chain code representation. Zingaretti et al. (1998) presented a fast single-pass algorithm to convert a multivalued image from a raster-based representation into chain codes. Madhvanath et al. (1999) presented chain code contour processing for handwritten word recognition. Shih and Wong (1999) proposed a one-pass algorithm for local symmetry of contours

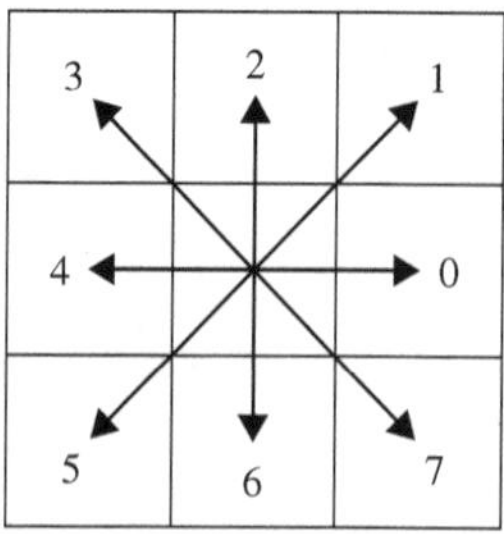

Figure 7.2 The chain code representation.

from chain codes. Bribiesca (2000) developed a chain code for representing 3D curves. Liu and Zalik (2005) proposed an efficient chain code with Huffman coding.

Example 7.4 What are the chain codes for the following binary region by considering 8-connectedness? Use the counterclockwise approach.

0	0	0	0	0	0	0
0	0	1	1	1	1	1
0	1	1	1	1	1	0
0	0	1	1	1	1	0
0	0	0	1	0	0	0

Answer: The chain codes are given as follows. Let the initial pixel be the upper-left pixel.

Starting pixel

0	0	0	0	0	0	0
0	0	1	1	1	1	1
0	1	1	1	1	1	0
0	0	1	1	1	1	0
0	0	0	1	0	0	0

Chain codes: 5, 7, 7, 1, 0, 2, 1, 4, 4, 4, 4. The crack codes are given as follows:

Example 7.5 What are the crack codes for the following binary region by considering 8-connectedness? Use the counterclockwise approach.

0	0	0	0	0	0	0
0	0	1	1	1	1	0
0	1	1	1	0	0	0
0	0	1	0	1	1	0
0	0	0	0	0	0	0

Answer: The crack code tracking is expressed as follows:

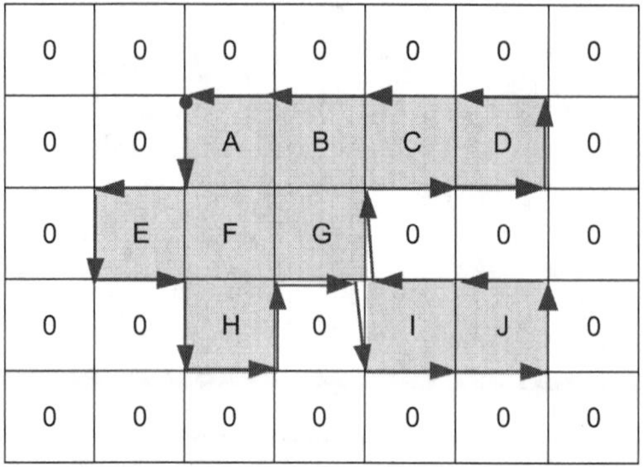

Therefore, the crack codes are $A(2, 3)$ 3 2 3 0 3 0 1 0 3 0 0 1 2 2 1 0 0 1 2 2 2 2.

7.3.2 Difference Chain Code

For any given chain code, a difference chain code can be constructed, for example, curvature, whose values represent the successive changes in direction. In the raster scan of an image, the first border point encountered is the starting point. The chain code is initialized as 0. According to the current chain code, it is checked whether the next point in the chain code direction is a border point. If it is, this point is appended to border linked list and the code is recorded. If it is not, the chain code is subtracted by 1 repeatedly until the next neighboring border point is found. Repeat the procedures until all the border points have been visited.

If the difference of successive unit slopes, that is, chain code, is used, a multiple of a 45° curvature will be always obtained. To obtain more continuous range of the curvature values, a smoothing procedure can be used. The *k-slope* of a chain code at a point P is defined as the average of the unit slopes at a sequence of k points centered at P. The k-curvature of a point P is defined as the difference between its left and right k-slopes. The first direction is determined by a point before the current point, and the second direction is the one after the current point. The curvature is computed from the first direction to the second direction counterclockwise. That is, if the direction changes counterclockwise, the curvature is positive, and negative otherwise.

Example 7.6 Give the chain code and the difference chain code of the following image:

$$\begin{matrix} 1 & 1 & 1 \\ 1 & 1 & 1 \\ 1 & 1 & 1 \end{matrix}$$

Answer: The chain code is 66002244. The difference chain code is 02020202.

Example 7.7 Compute the 1-curvature and 3-curvature of the following image:

$$\begin{matrix} 0 & 0 & 0 & 0 & 0 & 0 & 1 & 1 & 1 \\ 0 & 0 & 0 & 1 & 1 & 1 & 0 & 0 & 0 \\ 1 & 1 & 1 & 0 & 0 & 0 & 0 & 0 & 0 \end{matrix}$$

Answer: The 1-slope is 0° or 45°, and the 1-curvature is 0° or 45°. The 3-slope is $\tan^{-1}(1/3)$, and the 3-curvature is 0°.

7.3.3 Shape Signature

After the chain code has been extracted, the information pertaining to the slope and the curvature of boundary curves can be easily obtained. One natural characteristic that satisfies the conditions of a suitable shape signature is the curvature of a curve. It is well known that there is a one-to-one correspondence between a regular curve and its curvature function. The curvature is the rate of change of θ, the angle between the tangent vector to the curve and the horizontal axis, with respect to arc length s.

Let $\kappa(s)$ be the curvature function, then

$$\kappa(s) = \frac{d\theta}{ds}, \quad \text{where} \quad \theta = \tan^{-1}\left(\frac{dy/ds}{dx/ds}\right) \tag{7.1}$$

or

$$|\kappa(s)|^2 = \left[\frac{d^2x}{ds^2}\right]^2 + \left[\frac{d^2y}{ds^2}\right]^2 \tag{7.2}$$

One way to obtain the discrete approximation of the curvature is to draw the so called *arc-length-versus-turning-angle graph* of the curve (Kalvin et al., 1986) and to sample this graph at equally spaced (ΔS) points and at each point $S_i \quad (i = 1, 2, \ldots, n)$ to compute the difference such as

$$\Delta\theta(S_i) = \theta(S_i + \Delta S) - \theta(S_i) \tag{7.3}$$

7.3.4 The Mid-Crack Code

The chain code and crack code can be viewed as a connected sequence of straight line segments with specified lengths and directions. An obvious disadvantage of the chain code is observed when it is used to compute the area and perimeter of an object. As seen in Figure 7.3, the inside chain code appears to underestimate the area and perimeter, whereas the outside chain code overestimates them. The disadvantages in the crack code are that many more codes are generated and the perimeter is greatly overestimated. The mid-crack code (Dunkelberger and Mitchell, 1985), located in between, should make a more accurate computation of the geometric features.

A method of generating the chain code by using the run-length coding was presented in Kim et al. (1988). But, the COD-S and ABS-S tables must be produced in the first run-length step before the chain code generation and linking phases start. A useful concept for dealing with the inclusion relationship between boundaries

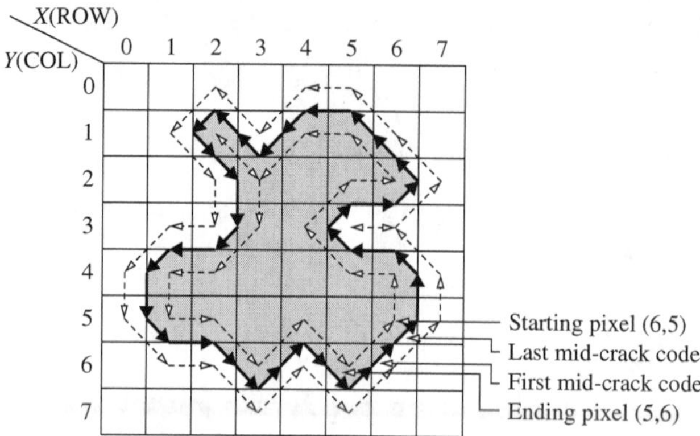

Figure 7.3 Silhouette with the inside and outside chain code contours (dashed lines) and the mid-crack code contour (solid line).

was also suggested. In addition, a raster-scan chain code introducing the max-point and min-point concept, especially a linking concept of relation links, was proposed in Cederberg (1979). Another single-pass algorithm (Chakravarty, 1981) for generating the chain code adopts the use of a step-by-step concept, such as chain link generation, a link segment data structure, and a junction of links.

The mid-crack code is a variation of and an improvement on the traditional tracing methods between the chain code and the crack code. In contrast to Freeman chain code, which moves along the center of pixels, the mid-crack code moves along the edge midpoint of a pixel, producing codes of links. For the horizontal and vertical moves, the length of a move is 1, and for diagonal moves, it is $\sqrt{2}/2$. If the crack is located in between two adjacent object pixels in the vertical direction, it is said to be on a vertical crack. Similarly, if the crack is located in between two adjacent object pixels in the horizontal direction, it is said to be on a horizontal crack.

Figure 7.4 shows the mid-crack code on the vertical and horizontal cracks. There are two restrictions on the moves in the mid-crack code. If the move is from the vertical crack, the codes 0 and 4 are not allowed. Similarly, the codes 2 and 6 are not allowed in the moves from the horizontal crack. The experimental verification of the mid-crack code in the area and perimeter computations is shown in Koplowitz (1981)

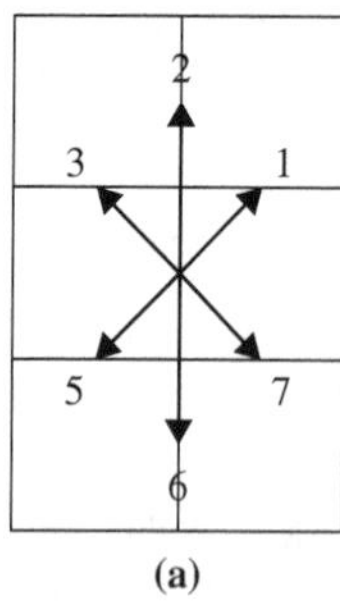

(a)

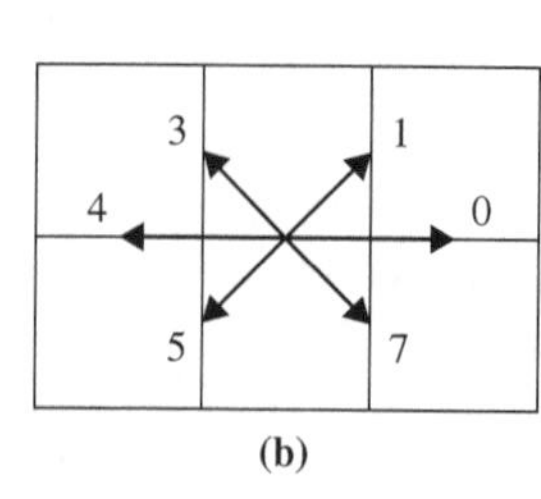

(b)

Figure 7.4 (a) The mid-crack code on the vertical crack and (b) the mid-crack code on the horizontal crack.

and Saghri and Freeman (1981), where the mean perimeter error value is −0.074% and the mean area error is −0.006%. Therefore, the mid-crack code is a desirable alternative method in contour tracing with its benefit in accuracy.

It is possible to invert a closed mid-crack code sequence into its corresponding silhouette. An internal boundary of the 8-connectedness object (or foreground) being traced counterclockwise can be reconstructed by treating it as the 4-connectedness background being traced clockwise. A disadvantage of the mid-crack code is that it is always longer than the Freeman chain code. An algorithm of conversion between a mid-crack code sequence and a chain code sequence is described in Dunkelberger and Mitchell (1985) to complement the defects as a compression process.

A fast algorithm for extracting the mid-crack code in a single-pass raster-scan fashion and its extension to parallel implementation are presented. The developed algorithm needs only one phase to generate all the code sequences even though a binary image is composed of several objects. Two types of boundaries, external and internal, are also considered, so that the inclusion relationship among region boundaries can be easily determined using the same algorithm.

An efficient algorithm to obtain the mid-crack codes can be referred in Shih and Wong (1992). In the contour description, the border points are represented in the form of a string of eight moving directional codes. The mid-crack code description can be seen as movement along the midpoints of border cracks that produce codes of links. The codes from 0 to 7 are assigned to indicate the eight moving directions from 0° to 315° at intervals of 45°. For the horizontal or vertical move, the length of a move is 1, and for the diagonal move, it is $\sqrt{2}/2$. If the crack is located in between two adjacent pixels in the vertical direction, it is said to be on a *vertical crack*. Similarly, if the crack is located in between two adjacent pixels in the horizontal direction, it is said to be on a *horizontal crack*. There are two restrictions on the moves in the mid-crack code as illustrated in Figure 7.4. If the move is from a vertical crack, the codes 0 and 4 are not allowed. Similarly, the codes 2 and 6 are not allowed in the moves from a horizontal crack.

The tracking rule of 8-connectedness counterclockwise is used. The description form of all m contours in a binary image can be expressed as follows:

$$
\begin{array}{l}
sx_{00}y_{00}d_{00}d_{01}\cdots\cdots\cdots\cdots\cdots\cdots\cdots\cdots d_{0n_0}e \\
sx_{10}y_{10}d_{10}d_{11}\cdots\cdots\cdots\cdots\cdots\cdots\cdots\cdots d_{1n_1}e \\
\cdots\cdots\cdots\cdots\cdots\cdots\cdots\cdots\cdots\cdots\cdots\cdots \\
\cdots\cdots\cdots\cdots\cdots\cdots d_{i(j-1)}d_{ij}\cdots\cdots\cdots\cdots \\
\cdots\cdots\cdots\cdots\cdots\cdots\cdots\cdots\cdots\cdots\cdots\cdots \\
sx_{m0}y_{m0}d_{m0}d_{m1}\cdots\cdots\cdots\cdots\cdots\cdots\cdots\cdots d_{mn_m}e\#
\end{array}
$$

where "#" is the end symbol of the description; "s" and "e" are the starting and ending symbols for each contour, respectively; "x" and "y" are the coordinates of the starting point; and "d" is the directional code. The subscript in "d_{ij}" indicates that d is the code in the ith contour and the jth move.

Each contour could be either an in-contour or an out-contour. In mid-crack code description, since the starting and ending pixels in each contour must be adjacent to each other, the opened curve occurring in the chain code description should not exist here. In addition, there are four cracks surrounding a pixel. Therefore, the problems in the chain code description of no code for an isolated point and redundantly tracing the same pixel twice for a multiway junction pixel will not happen again here.

A simple example of the mid-crack code description for a single-object image is illustrated in Figure 7.3. The resulting mid-crack codes are

$$s(6, 5)11771177076545677533554333101343 21e\#$$

Referring to Figure 7.3, the starting pixel (e.g., (6, 5)) is defined as the pixel that is pointed by the link of the first mid-crack code located at the ending pixel (e.g., (5, 6)) in a raster scan. As mentioned above, the counterclockwise tracing indicated by the arrow directions is adopted. Since the mid-crack codes are generated pixel by pixel according to a lookup table and then are concatenated using the code linking from the head to the tail of arrows, it is in nature that the generated sequences of the mid-crack code description are in a clockwise direction.

According to their algorithm, two results are shown below, which are considered using 8-connectedness for the object regions. In Figure 7.5, an image containing an object with a hole is shown. The result is shown as follows:

The total number of pixels = 29.

1. Starting pixel = (4, 6)
   Codes: 3571
   Perimeter = 5.66, area = 0.5, boundary type: internal
2. Starting pixel = (8, 8)
   Codes:
   33177117667121767111210765455555777111175555777753333553113 33-
   5557753355555311113333177771113333333
   Perimeter = 70.26, area = 29, boundary type: external

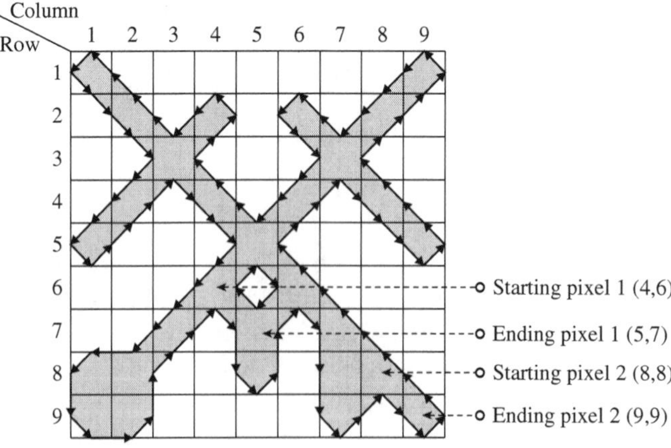

Figure 7.5 An image containing an object with a hole.

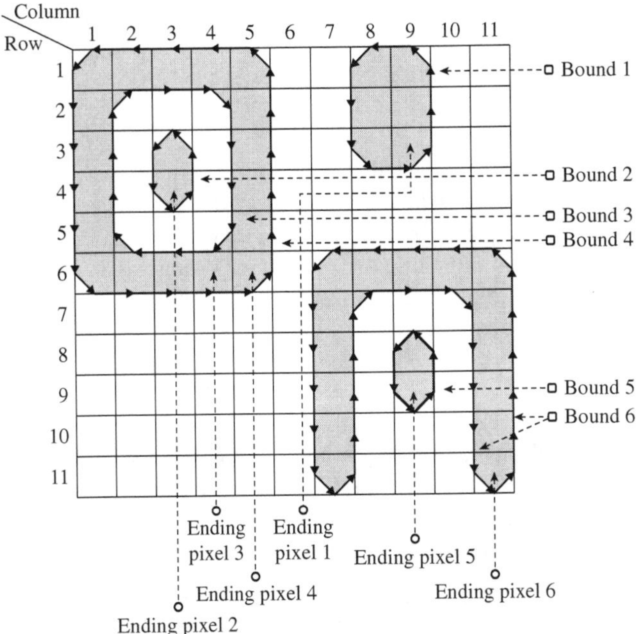

Figure 7.6 A five-object image with external and internal boundaries.

In Figure 7.6, an image with multiple objects is shown. The result is shown as follows:

The total number of pixels = 43.

1. Starting pixel = (9, 2)
   Codes: 2107665432
   Perimeter = 8.83, area = 5.5, boundary type: external
2. Starting pixel = (3, 3)
   Codes: 217653
   Perimeter = 4.83, area = 1.5, boundary type: external
3. Starting pixel = (3, 6)
   Codes: 45666700122234
   Perimeter = 12.83, area = 11.5, boundary type: internal
4. Starting pixel = (5, 5)
   Codes: 2100007666665444432222
   Perimeter = 20.83, area = 18, boundary type: external
5. Starting pixel = (9, 8)
   Codes: 217653
   Perimeter = 4.83, area = 1.5, boundary type: external
6. Starting pixel = (11, 10)
   Codes: 21766667001222217666665444432222
   Perimeter = 25.66, area = 14.5, boundary type: external

When the contours of multiple objects in an image are extracted, one of the most common problems is to fill the region inside each contour. A region consists of a group of adjacent, connected pixels. The task of filling primitives can be divided into two parts: the decision of which pixels to fill and the easier decision of with what value to fill them. If the image is binary, the same value is assigned to each pixel lying on a scan line running from the left edge to the right edge; that is, fill each span from $x_{\min}$ to $x_{\max}$. Spans exploit a primitive's *spatial coherence*, the fact that primitives often do not change from pixel to pixel within a span or from scan line to scan line. The coherence is obtained in general by looking for only those pixels at which changes occur. There are several algorithms performing region filling based on chain codes (Cai, 1988; Chakravarty, 1981) by the technique of parity checking (Ackland and Weste, 1981; Pavlidis, 1979) or a method of seed growing (Pavlidis, 1979; Shani, 1980). The technique of parity checking is often used because only a single scan is required instead of the iterations required in the seed-growing method. However, it is possible to produce an incorrect count of the number of intersections if points from two or more sides are mapped on the identical pixel. Also, a problem will arise if the test line is tangent to the contour.

A method of region filling by using discrete Green theorem was presented by Tang and Lien (1988). They reduced the region-filling problem to the inclusion problem and the intersection problem. The theorem is used to solve the inclusion problem if the spatial connectivity fails to make a decision. However, the decision causes an extra long computational time. A parity checking using the chain code contour (Cai, 1988; Chang and Leu, 1990) needs to deal with the characteristics of chain codes, such as the cases of opened contours, and isolated pixels, end-point pixels, internal line segment pixels, and multiway pixels (Ali and Burge, 1988), which makes the region-filling problem complicated.

In *chain coding*, each chain is composed of line segments connecting adjacent pixels in a horizontal, vertical, or diagonal direction. A closed boundary is chain coded from an arbitrary starting point. The initial border point coordinate is stored and the border around a region is then followed back to the first point, and it is repeated for each separate region. A region may be defined as a set of all connected points of one gray level, and a border of a region as a set of region's points that are adjacent to the points of different gray levels (in the binary image case: adjacent to the points of the region's complement). In moving from point to point around a border, a point is moved to one of its eight neighbors. The eight neighbors are labeled from 0 to 7 starting from the right-hand side and then tracking counterclockwise. Thus, each move is defined by an octal digit. It means that three bits are needed to indicate where the next border point is, starting from any particular border point. If the total number of regions in an image is K, and if the sum of the lengths of the borders for each region in the image is L, then the total storage requirement is $2K \log N + 3L$ bits. It is possible to improve chain coding technique by constructing *difference chain code*, whose values represent the successive changes in direction, for example, let 0 represent no turn, $+1/-1$ represent 45° right/left turns, $+2/-2$ and $+3/-3$ represent 90° and 135° right/left turns, respectively, and 4 represent a 180° turn. Thus, the difference chain code, like the chain code, has eight possible values and requires three bits per move.

However, the values are no longer equally likely, for example, 0 and $+1/-1$ should appear often, while 4 rarely appears. Hence, the variable-length coding technique can be used. High efficiency is obtained for images with short total border length and a small number of regions.

7.4 SKELETONIZATION BY THINNING

Thinning is a fundamental early processing step in pattern analysis such as industrial parts inspection (Shih and Mitchell, 1988), fingerprint recognition (Moayer and Fu, 1976), optical character recognition (Ogawa and Taniguchi, 1982), and biomedical diagnosis (Lu and Wang, 1986). The result of thinning a binary image is called the *skeleton*. One advantage of thinning is the reduction of memory space required for storage of the essential structural information presented in a pattern. Moreover, it simplifies the data structure required in pattern analysis.

Although there is no general agreement on the exact mathematical definition of thinness, a reasonable compromise has been reached that the generated skeleton must be essential to preserve the object's topology and to represent the pattern's shape informatively. The resulting skeleton must also be one pixel in width and lie approximately along the medial axes of the object regions. Many thinning algorithms are available in the literature. Different algorithms produce slightly different skeletons. Usually thinning algorithms are divided into two groups: sequential and parallel. A more general classification is as follows: sequential, parallel, hybrid, and others (Govindham and Shivaprasad, 1987). It is generally believed that a sequential algorithm will be faster than a parallel algorithm when they are both implemented on a serial computer (Rosenfeld and Pfaltz, 1966). However, the parallel thinning algorithms are best suited for parallel computers to gain efficiency.

In the parallel thinning approach, the new value of a given pixel at the nth pass depends on its neighbors over their neighborhood in the $(n-1)$th pass; therefore, all the pixels can be transformed simultaneously. To have a more efficient algorithm, the smaller neighborhood is considered. The 3×3 neighborhood is particularly desirable but has difficulty preserving the connectivity of the original shape. Thus, this problem is generally solved by partially serializing the parallel algorithms by the subiteration or subfield approaches. Subiteration approaches employ different thinning operators to the entire image in different subiterations. Subfield approaches partition the image space in some manner and a parallel operator is applied to different partitions of the image in different iterations.

To maintain the completely parallel processing, one branch of thinning is to use thinning operators on supports over the neighborhood larger than 3×3 and accomplish the fully parallel characteristics, which means that the same thinning operators are applied over the entire image at each iteration. An 11-pixel neighborhood thinning operator was developed as a fully parallel thinning algorithm by Guo and Hall (1989, 1992) that claims to be the optimally small support. This algorithm is very efficient because no subiteration is involved, but its resulting skeleton is highly affected by boundary noise. No matter how fast the algorithm is, the characteristics and consistency of the resulting skeletons are the first concern; moreover, the boundary noise fluctuation must not be neglected.

Ahmed and Ward (2002) presented a rule-based rotation-invariant thinning algorithm to produce a one-pixel-wide skeleton from a binary image. Rockett (2005) showed examples where Ahmed and Ward's algorithm fails on two-pixel-wide lines and proposed a modified method that corrects this shortcoming based on graph connectivity. Ji et al. (2007) presented a coarse-to-fine binary image thinning algorithm by proposing a template-based pulse-coupled neural network model.

So far in the existing literature, no precise standard evaluation of thinning algorithms has been established. However, it has been generally accepted that a good skeleton must possess the properties such as topology, shape, connectivity, and sensitivity to boundary noise. In other words, a good thinning algorithm should have the following characteristics:

1. The resulting skeleton must be topologically equivalent to the object.
2. It must run along the medial axes of the object regions.
3. It must be one pixel thick or as thin as possible.
4. It should preserve connectivity for both foreground and background.
5. It should be noise insensitive to small protrusions and indentations in the boundaries.
6. It should prevent extensive erosion and should not be completely deleted.
7. It should require as few iterations as possible.
8. It should delete pixels symmetrically to avoid bias in certain directions.

7.4.1 The Iterative Thinning Algorithm

In this section, we introduce the iterative thinning algorithm based on shrinking that deletes the object border pixels at each iteration, with consideration that such a deletion will not locally disconnect their neighborhoods (Rosenfeld and Kak, 1982). Let S denote the set of object pixels. If we delete all border points from S, then S will vanish completely. For example, in the following case:

$$\begin{matrix} 1 & 1 & 1 & 1 & 1 & 1 & 1 & 1 \\ 1 & 1 & 1 & 1 & 1 & 1 & 1 & 1 \end{matrix}$$

Instead, we delete only border points that lie on a given side of S at each iteration. To ensure the obtained skeleton close to "middle," we use opposite sides alternately, for example, north, south, east, and west. We provide the conditions under which a border pixel can be deleted without affecting the object connectivity. The border pixel P is called *simple* if the set of 8-neighbors of P that lie in S has exactly one component that is adjacent to P. In other words, deleting a simple pixel from S does not change the connectedness properties of either S or $\bar{S}$; that is, $S-\{P\}$ has the same component as S. For example, the following case is 8-simple, but not 4-simple because the 3×3 neighborhood contains one component originally and the deletion of P will cause this component to be disconnected into two components if considering 4-connectedness.

$$\begin{bmatrix} 0 & 1 & 1 \\ 1 & P & 0 \\ 1 & 1 & 0 \end{bmatrix}$$

According to the definition of simple point, the isolated point is not simple but the end point is simple. The iterative thinning algorithm is given below:

1. Delete all border points from a given side of S if they are simple and not end points.
2. Do this successively from the north, south, east, west, north, south, . . . sides of S until no further changes occur from a complete four-side of north, south, east, and west.

Example 7.8 Perform the iterative thinning algorithm on the following image using 8-connectedness:

$$\begin{array}{cccccccc} 1&1&1&1&1&1&&\\ 1&1&1&1&1&1&&\\ 1&1&1&1&1&1&&\\ 1&1&1&1&1&1&1&1\\ 1&1&1&&&1&&\\ 1&1&1&1&1&1&& \end{array} \xrightarrow{N} \begin{array}{cccccccc} 0&0&0&0&0&0&&\\ 1&1&1&1&1&1&&\\ 1&1&1&1&1&1&&\\ 1&1&1&1&1&1&1&1\\ 1&1&1&&&1&&\\ 1&1&1&1&1&1&& \end{array} \xrightarrow{S} \begin{array}{cccccccc} 0&0&0&0&0&0&&\\ 1&1&1&1&1&1&&\\ 1&1&1&1&1&1&&\\ 1&1&1&0&0&1&1&1\\ 1&1&1&&&1&&\\ 0&0&0&1&1&0&& \end{array}$$

$$\xrightarrow{E} \begin{array}{cccccccc} 0&0&0&0&0&0&&\\ 1&1&1&1&1&0&&\\ 1&1&1&1&1&0&&\\ 1&1&0&0&0&1&1&1\\ 1&1&1&&&1&&\\ 0&0&0&1&1&0&& \end{array} \xrightarrow{W} \begin{array}{cccccccc} 0&0&0&0&0&0&&\\ 0&1&1&1&1&0&&\\ 0&1&1&1&1&0&&\\ 0&1&0&0&0&1&1&1\\ 0&1&1&&&1&&\\ 0&0&0&1&1&0&& \end{array} \xrightarrow{N} \begin{array}{cccccccc} 0&0&0&0&0&0&&\\ 0&0&0&0&0&0&&\\ 0&1&1&1&1&0&&\\ 0&1&0&0&0&1&1&1\\ 0&1&1&&&1&&\\ 0&0&0&1&1&0&& \end{array}$$

$$\xrightarrow{S} \begin{array}{cccccccc} 0&0&0&0&0&0&&\\ 0&0&0&0&0&0&&\\ 0&1&1&1&1&0&&\\ 0&1&0&0&0&1&1&1\\ 0&0&1&&&1&&\\ 0&0&0&1&1&0&& \end{array} \xrightarrow{E} \text{Same} \xrightarrow{W} \begin{array}{cccccccc} 0&0&0&0&0&0&&\\ 0&0&0&0&0&0&&\\ 0&0&1&1&1&0&&\\ 0&1&0&0&0&1&1&1\\ 0&0&1&&&1&&\\ 0&0&0&1&1&0&& \end{array} \xrightarrow{N,S,E,W} \text{Same} \longrightarrow \text{Stop.}$$

7.4.2 The Fully Parallel Thinning Algorithm

Most of existing thinning algorithms do not have the ability to handle boundary noise. The lack of this capability will lead to the generation of spurious branches and the unpredictable accuracy of the resulting skeleton. This defect will not be avoided if only local 3×3, 3×4, or 11-pixel neighborhood is considered in the thinning algorithm. Such noise effects can be eliminated by a preprocessing step to smooth out the noisy boundary of the original shape before the application of the thinning or by a postprocessing step to delete the short branches that are shorter than a predefined length. Any of these additional steps is undesirable in some applications when real-time constraints are required.

Chin et al. (1987) developed an alternative that applies an additional set of noise cleaning templates when a set of thinning templates is used, and the templates at each iteration are restored to suppress the growth of spurious branch. However, the number of iterations must be predetermined. Yu and Tsai (1990) developed a thinning algorithm for grayscale images by applying the relaxation technique. A 9×9 neighborhood is considered in the computation to keep line straightness and to avoid noise affection. Nevertheless, the complicated formula/computation is time-consuming.

7.4.2.1 Definition of Safe Point A pixel in a binary image can be either black or white. Usually black pixels correspond to foreground and white pixels correspond to background. In a 3×3 neighborhood, if all eight neighbors are considered while a pixel is processed, it is called the *8-connectivity* for foreground. Similarly, if only four neighbors (north, south, west, and east) are used, it is called the *4-connectivity* for foreground. To avoid connectivity paradoxes, the 8-connectivity for foreground and 4-connectivity for background are used. Considering the eight neighbors, there exist 256 permutations in total. All the permutations can be analytically divided into 69 groups as shown in Figure 7.7. The number in parentheses denotes the group number from 1 to 69. The number of cases corresponds to the number of different permutations generated by rotating the pattern by 90°, 180°, and 270°.

The definitions of border point, end point, break point, and excessive erosion point are given as follows:

- A *border point* is an object point that has at least one of its 4-neighbors belonging to background.
- An *end point* is a border point that has only one 8-adjacent border point.
- A *break point* is a border point and its deletion will cause the loss of connectivity.
- An *excessive erosion point* is a border point and its deletion will cause the loss of the object's shape.

A *safe point* is defined as one of the end point, break point, or excessive erosion point but not the noise point. Safe points are preserved in the current iteration. Nonsafe points can be deleted in two ways: no condition or condition checking. The safe point status is effective in the present iteration and might be changed into nonsafe in the next iteration. The thinning algorithm is defined in terms of the deletion process that transforms certain ones (i.e., object pixels) to zeros (i.e., background pixels).

Referring to Figure 7.7, since the central points in cases (47), (56), (63), (65), (68), and (69) are not border points, in total 16 cases cannot be deleted; otherwise, a hole will be produced. Usually, the central points in cases (1) and (2) are defined as end points and in cases (3), (4), (13), and (66) are defined as excessive erosion points. In most thinning algorithms, the end points and excessive erosion points are not eliminated to avoid a totally deleted event and preserve the object's shape. However, it

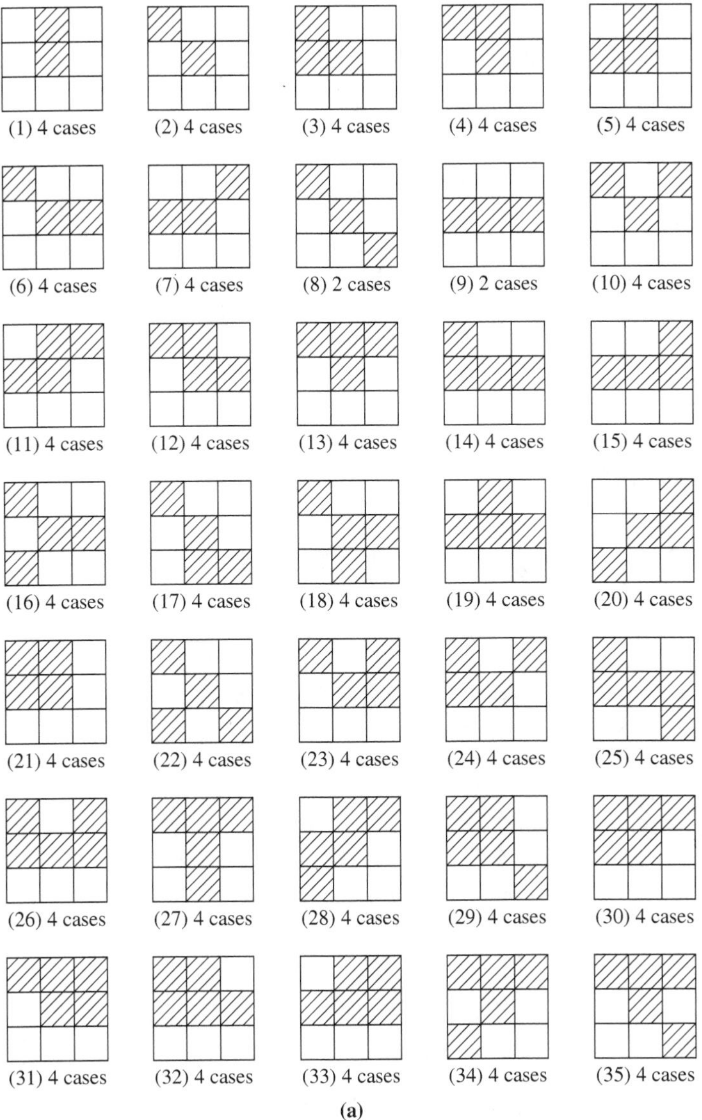

Figure 7.7 (a) Cases (1)–(35) of 69 groups from 256 permutations of 3×3 neighborhood. (b) Cases (36)–(69) of 69 groups from 256 permutations of 3×3 neighborhood.

is observed that if they are all preserved, the unnecessary branches would arise. In cases (2), (3), (4), and (13), the central points are not always safe points and will be deleted under certain deletability conditions. Because in cases (6)–(10), (14)–(18), (20), (22)–(27), (29), (34)–(39), (43), (45), (46), (48), (49), (51), (52), (57), (58), and (67) they are break points, in total 127 cases are safe points. The break points should be preserved in the resulting skeleton to ensure the connectivity.

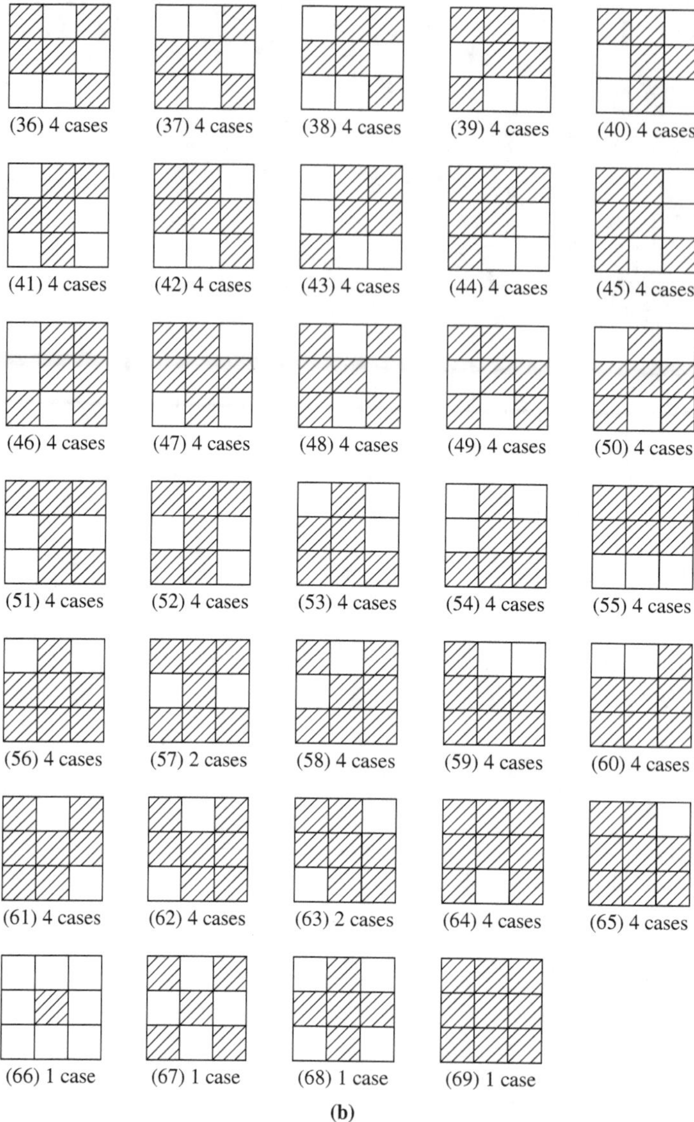

Figure 7.7 (*Continued*)

The remaining 92 cases may be divided into three types: preserved, deleted, and conditionally deleted. Because in cases (5), (19), (40)–(42), (50), (53), (54), (61), and (62) the deletion of the central point will cause the shape's distortion, in total 38 cases are preserved. Because the central points in cases (30), (31), and (44) are deletable without effects on distortion and connectivity, there are in total 12 cases to be deleted immediately. Since the parallel thinning algorithm is used, if all the central points are deleted in cases (11), (12), (21), (28), (32), (33), (55), (59), (60), and (64) at the same time, the loss of connectivity and shape will be encountered. Therefore, they are

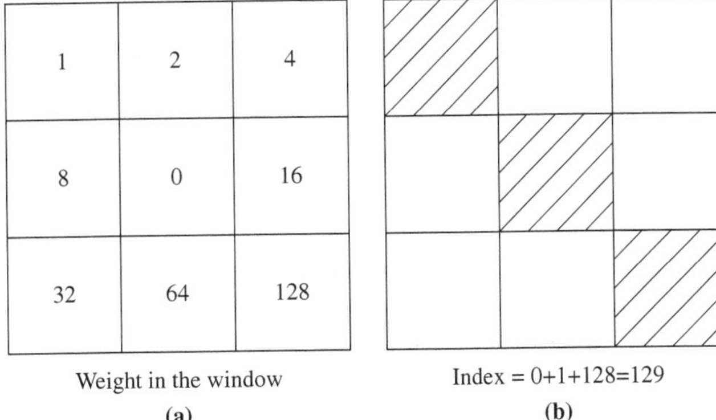

Figure 7.8 (a) A 3×3 template incorporating different weights at different neighbors and (b) an example of the index calculation.

conditionally deletable and some neighbors in the 5×5 neighborhood are checked. Overall 256 cases have been explored.

The definition of boundary noise that depends on the object size cannot be clearly given only over the 3×3 neighborhood of border points. The size of the neighborhood to be considered to determine the boundary noise increases with the size of an object. The noise can be classified into two types: *existent* and *postexistent*. Existent noise exists before the image is processed. For example, cases (2), (3), (4), and (13) of Figure 7.7 represent an existent noise on a specific condition in a large object. Postexistent noise is produced after the image has been processed. For example, cases (11), (12), (21), (32), and (33) represent a postexistent noise on a specific condition in a large object after the first iteration of thinning. In the experiments and observation, the boundary noise can be eliminated in a reasonable and consistent way if a 5×5 neighborhood is considered.

7.4.2.2 Safe Point Table A 3×3 weighted template shown in Figure 7.8(a) is used to indicate the variations of eight neighboring pixels. The correlation operation is performed by using the template over the 3×3 neighborhood to obtain the index value for table lookup. Note that the object pixel is represented as "1" and the background pixel as "0". An example of the index calculation is shown in Figure 7.8(b).

The safe point lookup table is shown in Table 7.1, which contains two items: *index* and *flag number*. The flag number is determined by looking at the 3×3 template whether the central object point is deletable or conditionally deletable. After the index is calculated and the specific conditions are checked, the object point can be determined whether to be deleted or preserved.

7.4.2.3 Deletability Conditions As mentioned earlier, the 3×3 neighborhood is not sufficient to describe the noise. There is an apparent trade-off to be considered here in order to use the 5×5 neighborhood. The strategy of safe point checking conditions is given as follows. The index of the 3×3 neighbors is computed first. If it equals any index number of the conditionally erasable points, some neighbors of the

TABLE 7.1 The Safe Point Lookup Table

Index	Flag number	Index	Flag number	Index	Flag number
1	2	75	2	210	2
3	2	84	1	212	1
4	2	86	2	214	1
6	2	96	2	215	1
7	2	104	2	216	2
9	2	105	1	224	2
11	1	106	2	232	1
14	2	107	2	233	1
15	1	111	2	235	2
19	2	112	1	239	2
20	2	116	1	240	1
22	1	120	2	244	1
23	1	128	2	246	1
27	2	144	2	247	1
30	2	146	2	248	2
31	1	148	2	249	2
32	2	150	1	252	2
40	2	151	1	253	2
41	2	159	1		
42	2	191	1		
43	1	192	2		
47	1	200	1		
63	1	201	1		
73	1	208	1		

Note: The flag number of erased points is 1 and of conditionally erased points is 2. Any index number that is not listed here belongs to the safe point.

5×5 neighborhood are necessarily checked. As illustrated in Figure 7.9, for example, the existent noise, three upper-left corner pixels of the 5×5 neighbors in case (2.1) are checked since the central pixel can be a noise if all three pixels are object pixels. Any case except case (28) that is not mentioned in Figure 7.9 is a nonsafe point.

In cases (3), (4), and (13), the conditions involve not only Boolean operation "AND" but also "OR". Referring to Figure 7.9, let P_{a1}, P_{a2}, and P_{a3} denote the neighbors with label "1", P_{b1} and P_{b2} denote the neighbors with label "2", and P_{c1}, P_{c2}, and P_{c3} denote the neighbors with label "0". The deletability Boolean operation can be represented as

$$(P_{a1} \wedge P_{a2} \wedge P_{a3}) \wedge (P_{b1} \vee P_{b2}) \wedge (\bar{P}_{c1} \wedge \bar{P}_{c2} \wedge \bar{P}_{c3})$$

For example, in case (3.1), if the neighbor denoted by "1" is an object pixel and the left-side neighbor denoted by "2" is an object pixel, then the above condition is true and the central pixel can be deleted. The conditions in cases (32) and (33) are similar, which are assumed to be a postexistent noise. In case (32.1), if three right-side

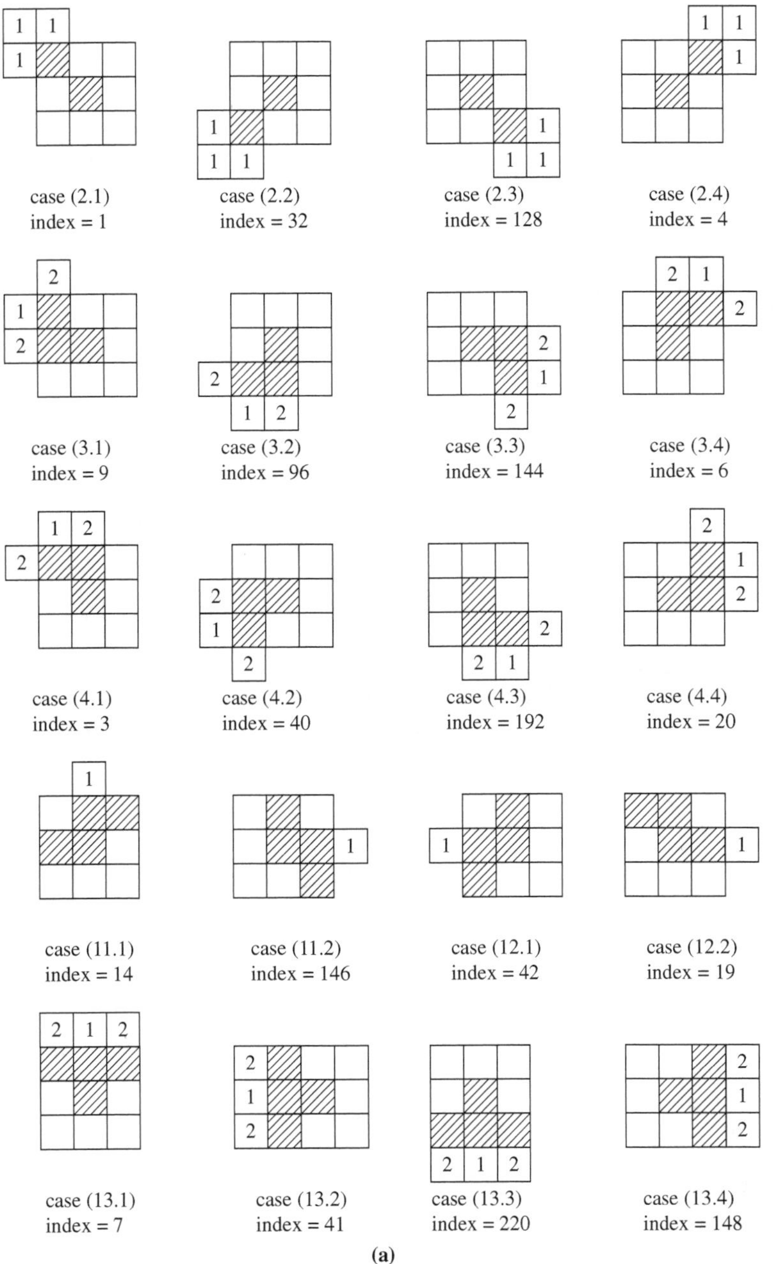

Figure 7.9 (a) Deletion checking conditions of cases (2), (3), (4), (11), (12), and (13). (b) Deletion checking conditions of cases (21), (32), (33), (55), (59), (60), and (64).

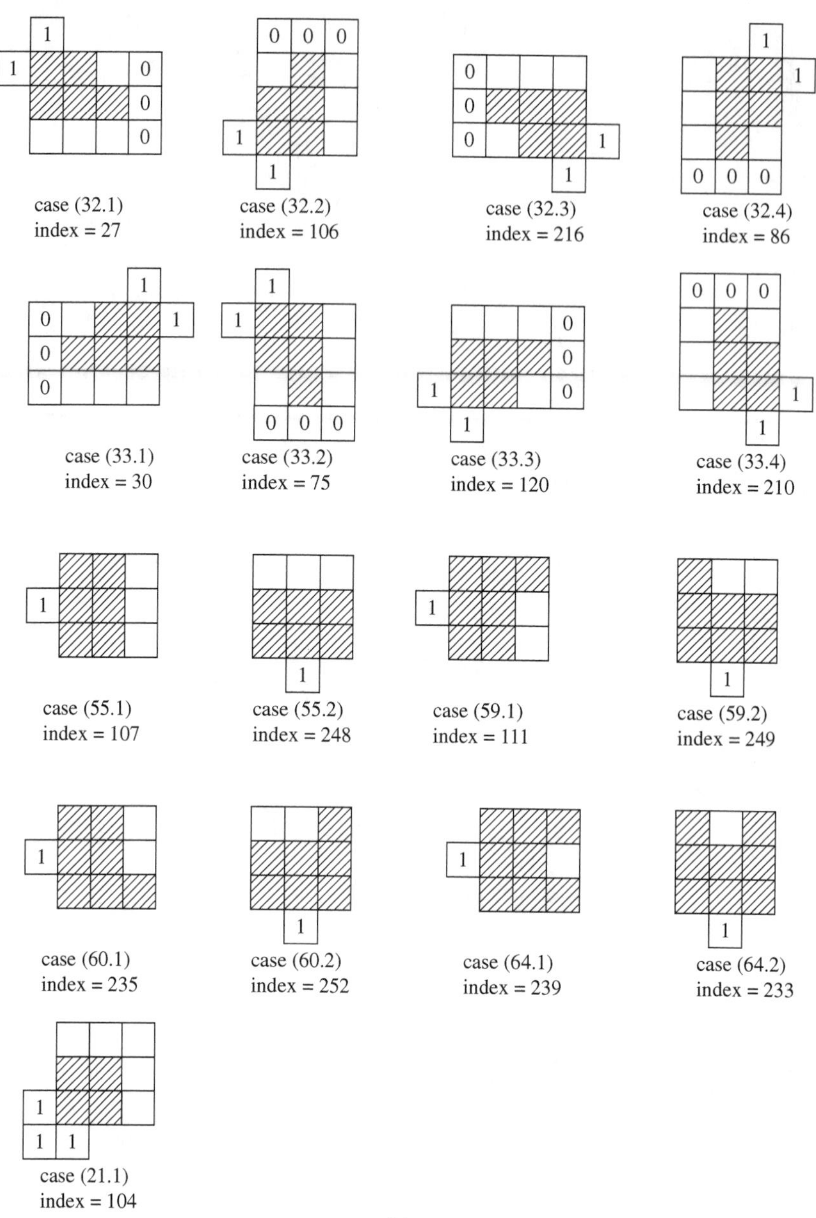

Figure 7.9 (*Continued*)

neighbors denoted by "0" belong to background and two neighbors denoted by "1" are object pixels, then the above condition is true and the central pixel can be deleted.

In two subcases of cases (11), (12), (55), (59), (60), and (64), only one more neighbor needs to be checked to preserve the connectivity, and the excessive erosion incident and the other two subcases are deletable immediately. In order to preserve the consistent connectivity of diagonal lines as in case (28), the two subcases with the index numbers 46 and 147 are safe points and the other two are nonsafe points, which can be deleted without condition checking. Three more neighbors in the subcase (21.1), as illustrated in the bottom of Figure 7.9(b), need to be checked, and the other three subcases of case (21) are erasable without condition checking. The reason is to preserve the consistent connectivity and avoid the excessive erosion. All the deletability conditions have been discussed and the critical concerns are not to conflict with each other during parallel processing and to fulfill the requirements of a good thinning algorithm.

7.4.2.4 The Fully Parallel Thinning Algorithm Simply applying the above safe point definition and deletability conditions in parallel on the entire image, the thinning result will not be as good as expected. The reason is that the noise definition does not contain all the possibilities that will produce noise after the first few iterations. In the experiments, two additional "expansions" are performed. These expansion steps will prevent the early generation of the safe points by adding an object pixel into the intermediate result. The expansion operation performs as follows: If the neighborhood satisfies the conditions shown in Figure 7.10 (*Note*: The index number is the same as defined before, and the pixel denoted by "0" belongs to background), then the central pixel will be turned on. The fully parallel thinning algorithm is described below (Shih and Wong, 1994, 1995):

1. The index number is calculated for each object pixel.
2. If the index number is one of the conditions for expansion, turn on the central pixel.
3. Repeat step 1.
4. Use the index number and lookup table to find the flag number.
5. Check the conditions if it is a conditionally deletable case; otherwise, preserve or delete the pixel accordingly.
6. If the deletability conditions are satisfied, erase the object pixel; otherwise preserve it.
7. Perform steps 1 and 2 once again.
8. Recursively perform steps 3–5 until no further deletions occur.

7.4.2.5 Experimental Results and Discussion In most of the existing thinning algorithms, a break point will be a safe point no matter what the circumstance is. It may cause trouble if the existent boundary noise is adjacent to any break point. Except when at least a whole 5×5 neighborhood is considered in the thinning algorithm, this problem cannot be solved. The results of the input images in Figure 7.11(a)–(d) are shown in Figure 7.11(e)–(h), respectively.

Three 37×31 input images shown in Figure 7.12(a), (b), and (c) describe a noisy rectangle and their corresponding results are shown in Figure 7.12(d), (e), and (f). The results using the proposed algorithm have the consistency that the similar input shapes will produce the similar output skeletons. The resulting skeleton of a

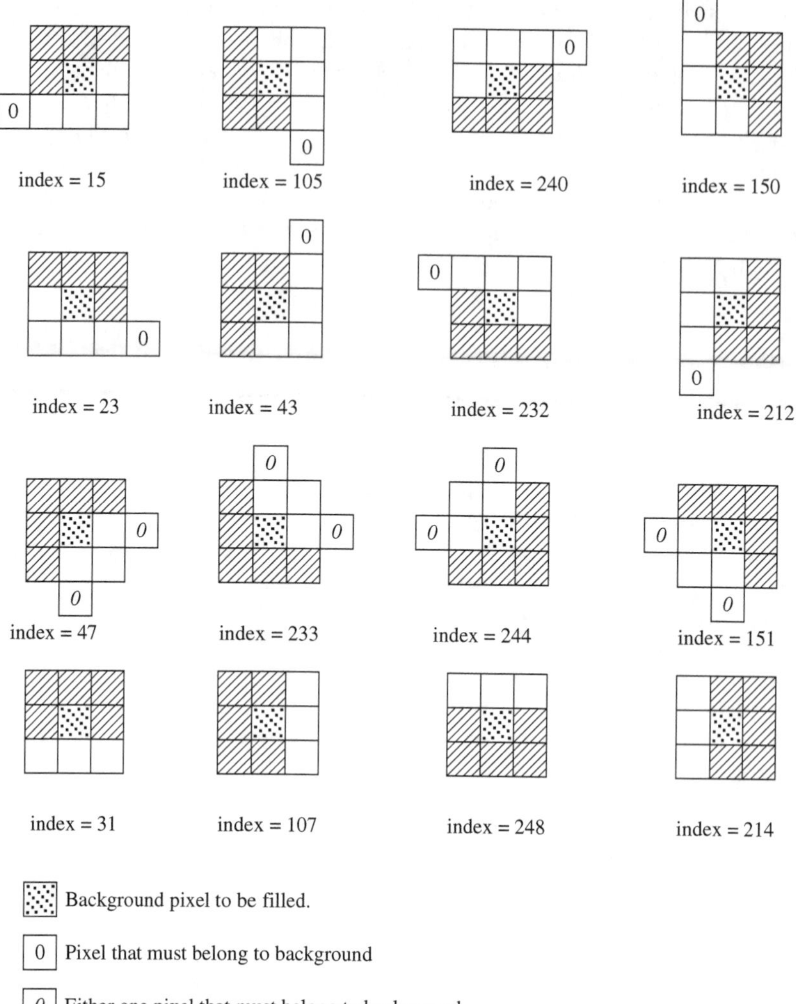

Figure 7.10 A set of conditions that are defined to have the expansion operation.

wrench using the proposed algorithm is shown in Figure 7.13(a) and using Guo and Hall's algorithm (1992) is shown in Figure 7.13(b), in which a lot of small branches are produced. The images of an English character "e" together with 30° and 45° rotations are shown in Figure 7.14. A set of Chinese characters is also tested as shown in Figure 7.15.

7.5 MEDIAL AXIS TRANSFORMATION

The set of centers and radii (values) of the maximal blocks is called the *medial axis* (or *symmetric axis*) *transformation*, abbreviated as MAT or SAT. Intuitively, if a

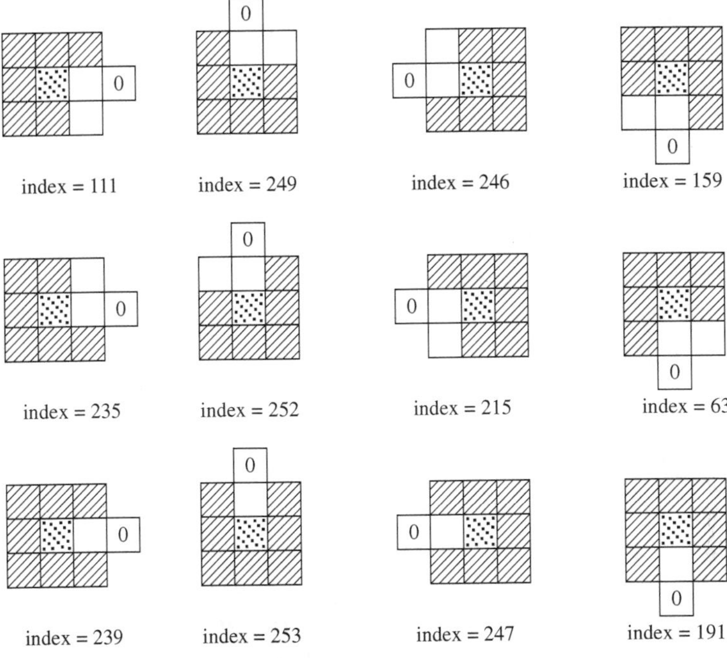

Figure 7.10 (*Continued*)

block S_P is maximal and is contained in the constant-value region S, it must touch the border of S in at least two places; otherwise, we could find a neighbor Q of P that was further away than P from the border of S, and then S_Q would contain S_P. Let the output of the distance transformation be d. A distance transformation converts a binary image, which consists of object (foreground) and nonobject (background) pixels, into an image where every object pixel has a value corresponding to the minimum distance to the background. Consider the slope variation in both X and Y axes for d. The union of the summit pixels of the slope in both X and Y directions will be the skeleton. From the distance transformation, one can see that distance change between two neighboring pixels is not more than 1. Hence, the four structuring elements are chosen as follows:

$$h_1 = \begin{bmatrix} 0 \\ 0 \\ \varepsilon \end{bmatrix}, \quad h_2 = \begin{bmatrix} \varepsilon \\ 0 \\ 0 \end{bmatrix}, \quad h_3 = [0 \quad 0 \quad \varepsilon], \quad h_4 = [\varepsilon \quad 0 \quad 0]$$

where ε denotes a small positive number less than 1.

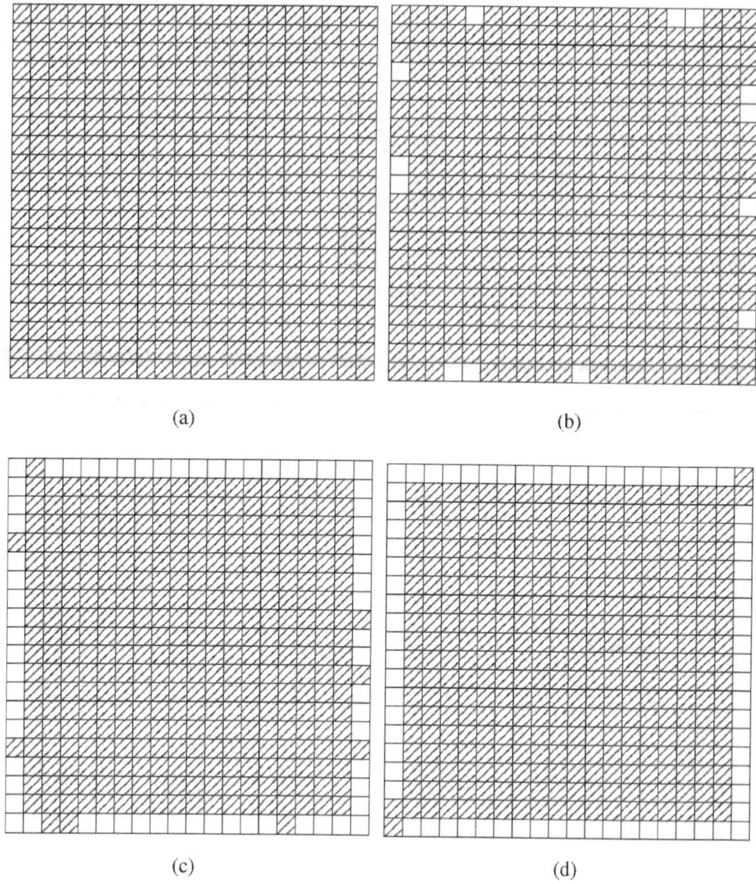

(a) (b)

(c) (d)

Figure 7.11 (e–h) The results of (a–d) by the thinning algorithm, respectively.

MAT Algorithm

1. Raster scan the distance image d. Do $d \oplus_g h_1, d \oplus_g h_2, d \oplus_g h_3$, and $d \oplus_g h_4$ at each pixel.
2. If any one of four resulting values is equal to the original gray value at this pixel, then the gray value is kept; otherwise, let the gray value of this pixel be zero. After all pixels are processed, the output is the skeleton with the gray value indicating the distance.

An example of the Euclidean distance transformation and the medial axis transformation is shown in Figure 7.16.

The distance transform from MAT: Let m denote the grayscale MAT. Using the distance structuring element k, the distance transform reconstruction from the MAT is achieved by the following algorithm:

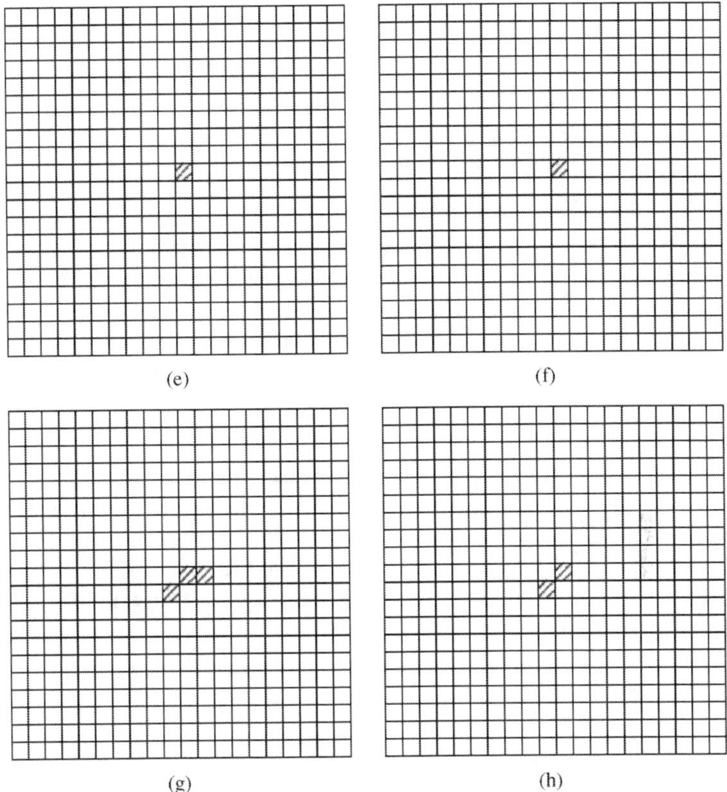

Figure 7.11 (*Continued*)

Reconstruction Algorithm

The grayscale MAT is dilated by the distance structuring element; that is, $m \oplus_g k$.

This reconstruction algorithm is quite simple. The result is exactly the distance transform. After thresholding at 1, the original binary image can be recovered.

In *MAT representation*, with each pixel P of an image, the set of upright squares of odd side length centered at P is associated. The largest such a square that is contained in the image and has a constant value is called *a maximal block*. It is easy to see that if the set of centers P and the radii $r(P)$ of the maximal blocks is specified, the image is completely determined. For a binary $N \times N$ image, $2 \log N$ bits are needed to specify the x- and y-coordinates of each block center and $\log(N/2)$ bits to specify the radius. Thus, if there are B blocks in MAT, the total number of bits required to specify is $B(3 \log N - 1)$. High efficiency is obtained for images consisting of only a few constant-valued regions that have simple shapes.

A *uniform-region representation* of binary images is usually constructed by introducing a set of basic shapes whose combinational operations construct the image. Although there are many possible basic shapes, the rectangle is one of the most appropriate elementary shapes for digitized binary images. In *rectangular coding*,

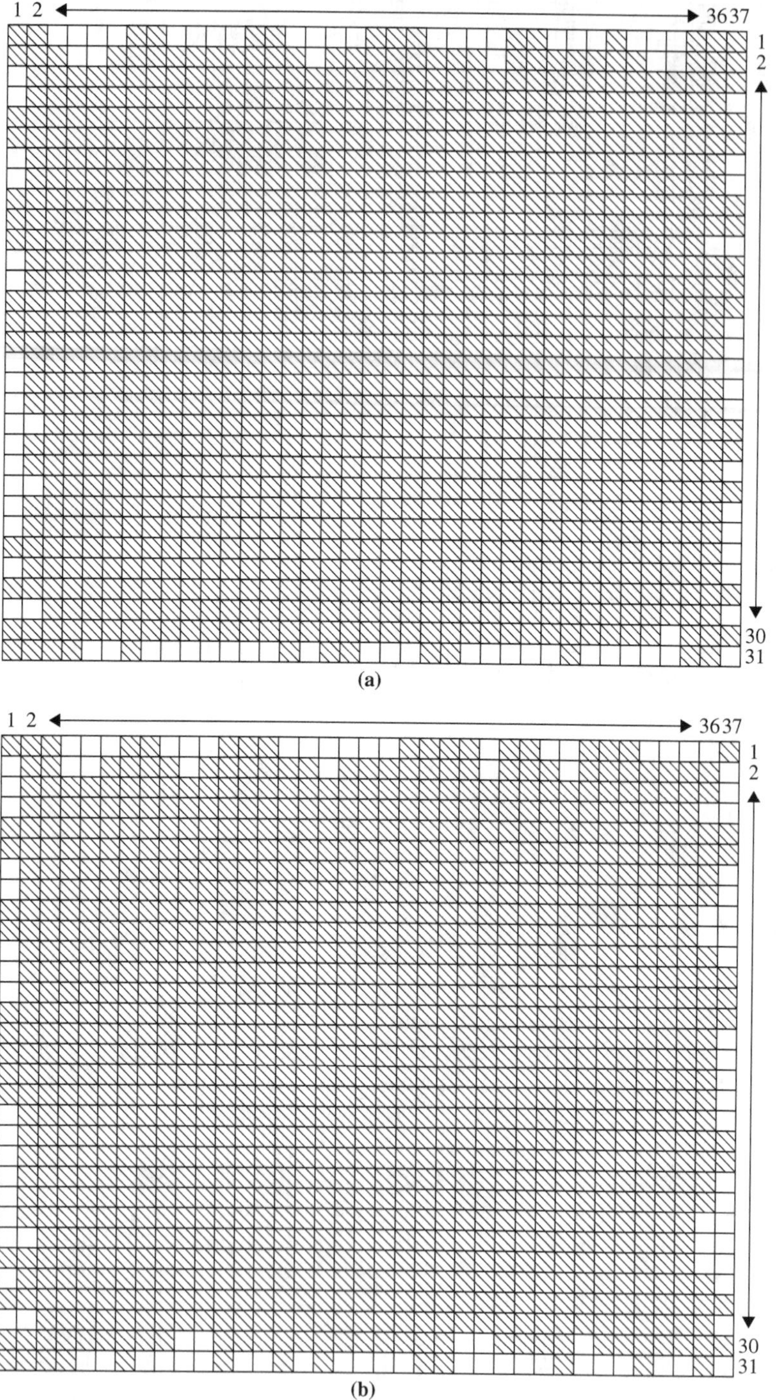

Figure 7.12 (a) A 37 × 31 input image describing a noisy rectangle. (b) A 37 × 31 input image describing a noisy rectangle. (c) A 37 × 31 input image describing a noisy rectangle. (d–f) The corresponding results of (a)–(c).

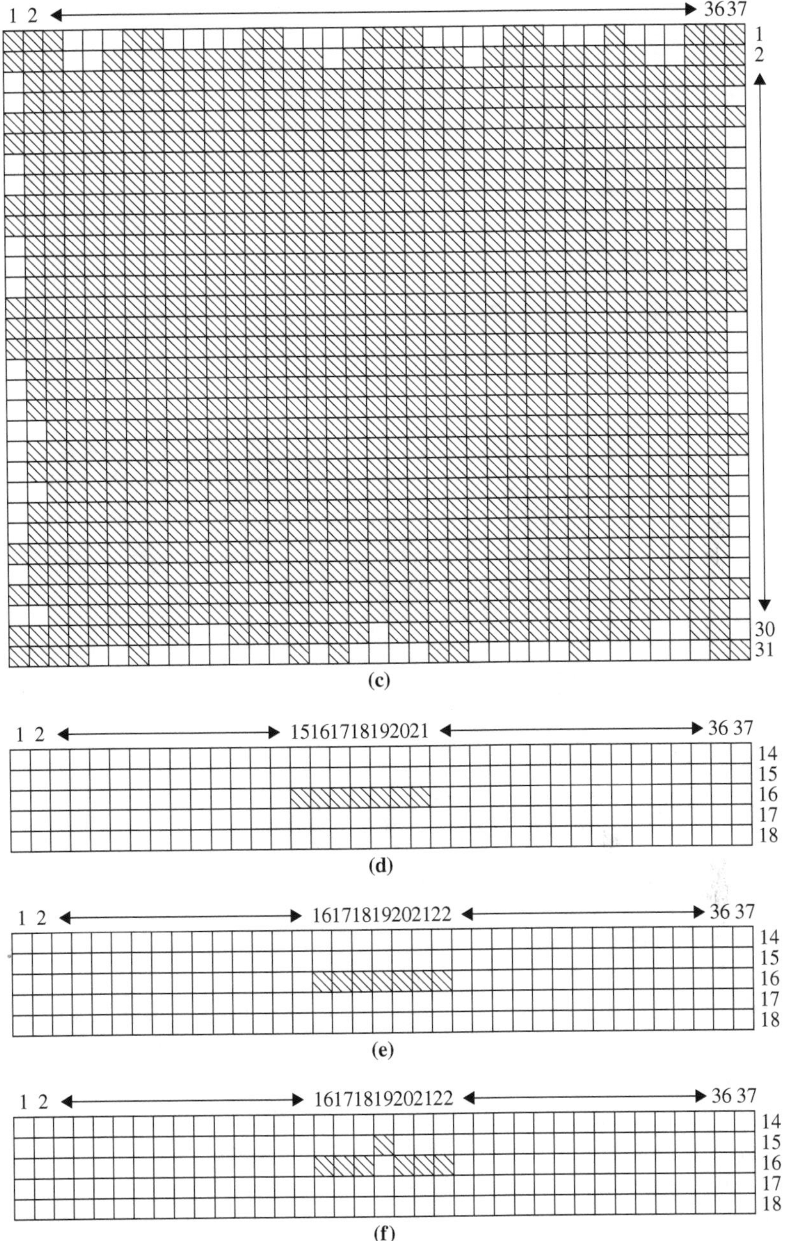

Figure 7.12 *(Continued)*

each region in an image is decomposed into a set of rectangles, and the rectangles are encoded to represent the image in an efficient way. Rectangular coding requires four parameters (the x- and y-coordinates of one vertex, and width and height of a rectangle). For an $N \times N$ picture, if the average number of rectangles is R, the total

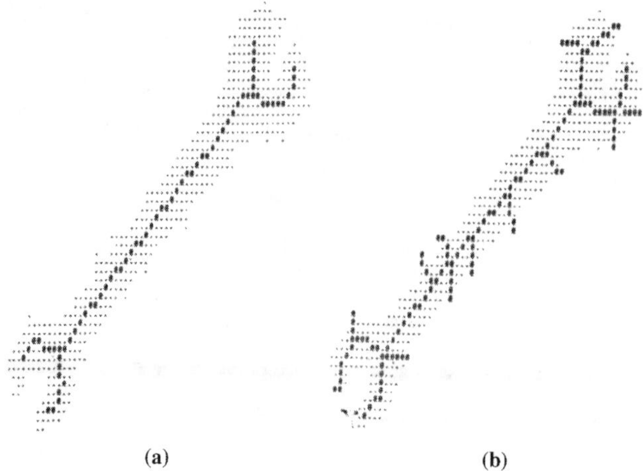

Figure 7.13 The resulting skeletons of a wrench (a) using the proposed algorithm and (b) using Guo and Hall's algorithm (1992).

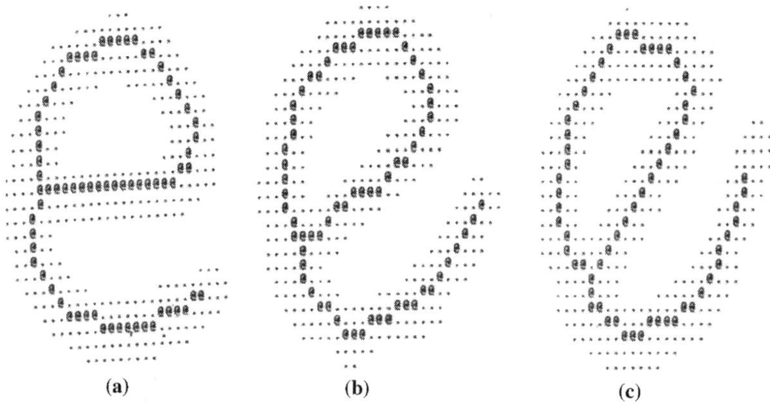

Figure 7.14 The resulting skeletons of (a) English character "e" together with (b) 30° and (c) 45° rotations.

number of bits required for representation is $4R \log N$, that is, $2 \log N$ for each vertex and $2 \log N$ for width and height of the maximal rectangle. An isolated point is considered to be a rectangle of zero width and height. High efficiency is obtained for images that can be decomposed into a small number of rectangles.

Since the above MAT algorithm produces a skeleton that may be disconnected into several short branches for an object, we present a simple and efficient algorithm using the maxima tracking approach on Euclidean distance transform to detect skeleton points. The advantages of the skeleton obtained are (1) connectivity preservation, (2) one pixel width, and (3) its locations as close as possible to the most symmetrical axes.

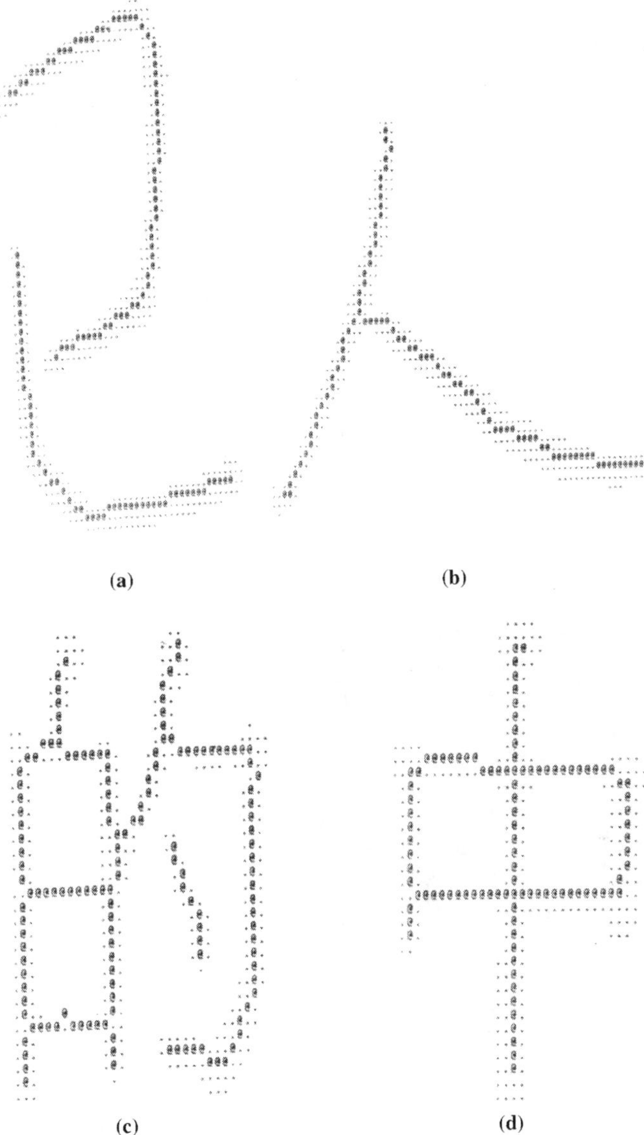

Figure 7.15 The resulting skeletons of a set of Chinese characters

7.5.1 Thick Skeleton Generation

A grayscale image is first segmented into a binary image, which is then applied by the Euclidean distance transformation algorithm. An example of the resulting Euclidean distance transformation is shown in Figure 7.17. With any pixel P, we associate a set of disks or circles with various radii centered at P. Let C_P be the largest disk that is completely contained in the object region, and let r_P be the radius of C_P. There may

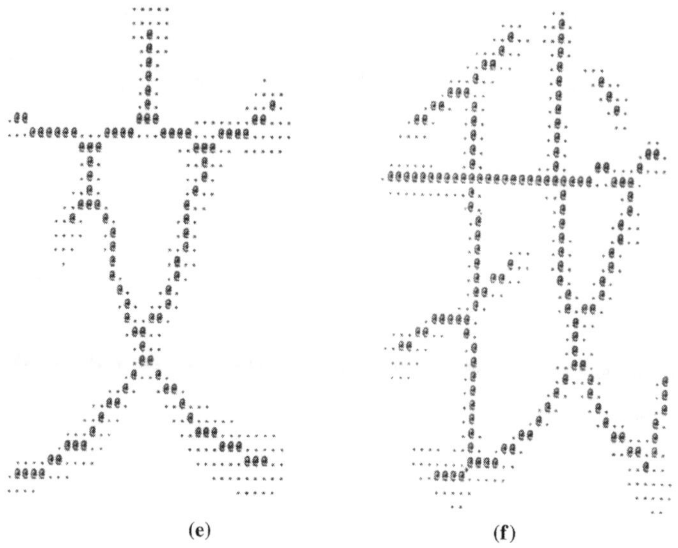

(e) (f)

Figure 7.15 (*Continued*)

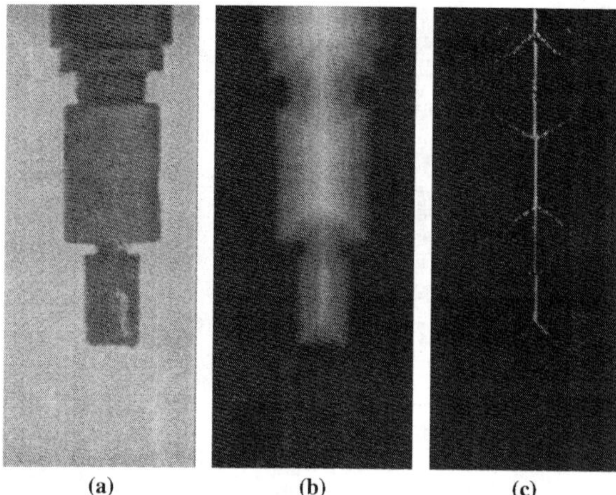

(a) (b) (c)

Figure 7.16 An example of Euclidean distance transformation and skeletonization: (a) tool picture, (b) distance transform, and (c) skeleton (medial axis).

-	-	-	-	-	-	-	-	-	-	-	-	-	-	-	-	-	-	-	-	-	-	-	-	-	-	-	-	-	-
-	**10**	**10**	10	10	10	10	10	10	10	10	10	10	10	10	10	10	10	10	10	10	10	10	10	10	10	**10**	**10**	**10**	-
-	**10**	**20**	**20**	20	20	20	20	20	20	20	20	20	20	20	20	20	20	20	20	20	20	20	20	20	**20**	**20**	**10**	-	-
-	10	**20**	**30**	**30**	**30**	**30**	**30**	**30**	**30**	**30**	**30**	**30**	30	30	30	30	30	30	30	30	30	30	**30**	**30**	**22**	**14**	**10**	-	-
-	10	20	**30**	**40**	**36**	28	22	20	20	22	28	**36**	**40**	**40**	**40**	40	40	40	40	40	**40**	**40**	**36**	**28**	20	10	-	-	-
-	10	20	30	**36**	28	22	14	10	10	14	22	28	36	**45**	**45**	**50**	50	50	50	**50**	**45**	36	28	**22**	**14**	**10**	-	-	-
-	10	20	**30**	**32**	22	14	10	-	-	10	14	22	32	41	41	**51**	**60**	**60**	**60**	**51**	41	32	22	14	10	-	-	-	-
-	10	20	**30**	**30**	20	10	-	-	-	-	10	20	30	40	40	50	**60**	**70**	**60**	50	40	30	20	10	-	-	-	-	-
-	10	20	**30**	**30**	20	10	-	-	-	-	10	20	30	40	40	50	**60**	**70**	**61**	51	41	32	22	14	10	10	**10**	**10**	-
-	10	20	**30**	**32**	22	14	10	-	-	10	14	22	32	41	41	**51**	**60**	**60**	**60**	**54**	45	36	28	**22**	**20**	**20**	**20**	**10**	-
-	10	20	30	**36**	28	22	14	10	10	14	22	28	36	**45**	**45**	**50**	50	50	50	**50**	**45**	36	28	**22**	**20**	**20**	**20**	**10**	-
-	10	20	**30**	**40**	**36**	28	22	20	20	22	28	**36**	**40**	**40**	**40**	40	40	40	40	40	**40**	**32**	22	14	10	10	**10**	**10**	-
-	10	**20**	**30**	**30**	**30**	**30**	**30**	**30**	**30**	**30**	**30**	**30**	30	30	30	30	30	30	30	30	30	**30**	**20**	10	-	-	-	-	-
-	**10**	**20**	**20**	20	20	20	20	20	20	20	20	20	20	20	20	20	20	20	20	20	20	**20**	**20**	**10**	-	-	-	-	-
-	**10**	**10**	10	10	10	10	10	10	10	10	10	10	10	10	10	10	10	10	10	10	10	10	**10**	**10**	-	-	-	-	-
-	-	-	-	-	-	-	-	-	-	-	-	-	-	-	-	-	-	-	-	-	-	-	-	-	-	-	-	-	-

Figure 7.17 The Euclidean distance function with the ridge points underlined. For expressional simplicity, we use 10 times the Euclidean distance in rounded numbers.

exist another pixel Q such that C_Q contains C_P. If no such Q exists, C_P is called a *maximal disk*. The set of the centers and radii (or Euclidean distance values) of these maximal inscribed disks is called the skeleton.

7.5.1.1 The Skeleton from Distance Function The distance function of pixel P in the object region S is defined as the smallest distance of P from all the background pixels $\bar{S}$. That is,

$$D(P) = \min_{Q \in \bar{S}} \{d(P, Q)\} \tag{7.4}$$

where $d(P, Q)$ is the Euclidean distance between P and Q. If we visualize the distance values as the altitude on a surface, then the "ridges" of the surface constitute the skeleton in which a tangent vector cannot be uniquely defined. It can be easily derived that if and only if a point P belongs to the skeleton of an object region S, the maximal disk with the radius $D(P)$ hits the contour of S at at least two places. Let us define the set

$$A(P) = \{Q | d(P, Q) = D(P), Q \in \bar{S}\} \tag{7.5}$$

If the set $A(P)$ contains more than one element, then P is a skeleton point.

7.5.1.2 Detection of Ridge Points Considering two neighbors of P in any of the horizontal, vertical, 45°-diagonal, and 135°-diagonal directions, if the altitude of P is higher than one neighbor and not lower than the other, then P is called the *ridge point*. Figure 7.17 shows the Euclidean distance transform with ridge points underlined. Note that the resulting skeleton on the right-hand side is disconnected from the major skeleton.

7.5.1.3 Trivial Uphill Generation The trivial uphill of a point P is the set of all the neighbors with a higher altitude. Figure 7.18 shows the resulting skeleton if we add the uphill of ridge points and continuously add the uphill of the new skeleton points until no further uphill is generated. Finally, a connected skeleton can be obtained, but a thick one. To make the skeleton one pixel wide, we need to get rid of the loose definition of ridge points and take into account the directional neighborhood.

-	-	-	-	-	-	-	-	-	-	-	-	-	-	-	-	-	-	-	-	-	-	-	-	-	-	-	-	-
-	10	10	10	10	10	10	10	10	10	10	10	10	10	10	10	10	10	10	10	10	10	10	10	10	10	10	10	-
-	10	20	20	20	20	20	20	20	20	20	20	20	20	20	20	20	20	20	20	20	20	20	20	20	20	10	-	-
-	10	20	30	30	30	30	30	30	30	30	30	30	30	30	30	30	30	30	30	30	30	30	30	22	14	10	-	-
-	10	20	30	40	36	28	22	20	20	22	28	36	40	40	40	40	40	40	40	40	40	36	28	20	10	-	-	-
-	10	20	30	36	28	22	14	10	10	14	22	28	36	45	50	50	50	50	50	45	36	28	22	14	10	-	-	-
-	10	20	30	32	22	14	10	-	-	10	14	22	32	41	51	60	60	60	51	41	32	22	14	10	-	-	-	-
-	10	20	30	30	20	10	-	-	-	-	10	20	30	40	50	60	70	60	50	40	30	20	10	-	-	-	-	-
-	10	20	30	30	20	10	-	-	-	-	10	20	30	40	50	60	70	61	51	41	32	22	14	10	10	10	10	-
-	10	20	30	32	22	14	10	-	-	10	14	22	32	41	51	60	60	60	54	45	36	28	22	20	20	20	10	-
-	10	20	30	36	28	22	14	10	10	14	22	28	36	45	50	50	50	50	50	45	36	28	22	20	20	20	10	-
-	10	20	30	40	36	28	22	20	20	22	28	36	40	40	40	40	40	40	40	40	32	22	14	10	10	10	10	-
-	10	20	30	30	30	30	30	30	30	30	30	30	30	30	30	30	30	30	30	30	30	20	10	-	-	-	-	-
-	10	20	20	20	20	20	20	20	20	20	20	20	20	20	20	20	20	20	20	20	20	20	10	-	-	-	-	-
-	10	10	10	10	10	10	10	10	10	10	10	10	10	10	10	10	10	10	10	10	10	10	10	-	-	-	-	-
-	-	-	-	-	-	-	-	-	-	-	-	-	-	-	-	-	-	-	-	-	-	-	-	-	-	-	-	-

Figure 7.18 The ridge points and their trivial uphill generation are underlined. It is indeed connected but too thick.

7.5.2 Basic Definitions

Prior to the discussion of our algorithm, the basic definitions of base point, apex point, directional uphill generation, and directional downhill generation are introduced below.

7.5.2.1 Base Point The base point is defined as a corner point that has the distance value 1 and is surrounded by a majority of background points. It belongs to one of the following three configurations as well as their variations up to eight 45° rotations:

$$\begin{array}{ccc} 1 & 0 & 0 \\ 1 & \underline{1} & 0 \\ 0 & 0 & 0 \end{array}, \quad \begin{array}{ccc} 1 & 1 & 0 \\ 1 & \underline{1} & 0 \\ 0 & 0 & 0 \end{array}, \quad \text{and} \quad \begin{array}{ccc} 1 & 1 & 1 \\ 1 & \underline{1} & 0 \\ 0 & 0 & 0 \end{array}$$

where the central underlined pixels represent the 45°, 90°, and 135° corner points, respectively. In a digital image, only the above three degrees are considered in a local 3×3 window.

If all the three configurations are considered as base points, then more nonsignificant short skeletal branches are produced. In other words, an approach based on the skeletal branches originating from all corner points in a 3×3 window would lead to unmanageable complexity in the skeletal structure. Therefore, an appropriate shape-informative skeleton should reach a compromise among the representativity of the connected object structure, the required reconstructivity, and the cost of deleting nonsignificant branches.

We consider sharper convexities as more significant. In detail, if the amplitude of the angle formed by two intersecting wavefronts of the fireline is viewed as the parameter characterizing the sharpness, the values smaller than or equal to 90° are identified as suitable convexities. More strictly, only 45° convexities are detected as base points. By using these points as the source to grow up the skeleton, the remaining procedures are acquired to preserve the skeletal connectedness.

7.5.2.2 Apex Point The apex point is the pixel representing the local maximum in its 3×3 neighborhood. Note that the local maximum pixels construct only a small portion of a disconnected skeleton. They occur as 45° corner points or interior elements that have the highest altitude locally. The 45° corner points also serve as base points. The base points and apex points are considered as sources or elementary cells of the skeleton that grows up emitting from them. Figure 7.19 shows the Euclidean distance function with the base points labeled "B" and the apex points underlined.

7.5.2.3 Directional Uphill Generation The set of points $\{P_i\}$ is called the *directional neighborhood* of P, denoted by D_P, if the points are in the 8-neighborhood of P and located within $\pm 45°$ slope changes from the current medial axis orientation of P. For example, using the 8-neighbors labeled $P_1, P_2, \ldots,$ and P_8 counterclockwise from the positive x-axis of P, if P_7 and P are the skeleton points, the points $P_2, P_3,$ and P_4 are the directional neighbors of P; that is, $D_P = \{P_2, P_3, P_4\}$. Several

-	-	-	-	-	-	-	-	-	-	-	-	-	-	-	-	-	-	-	-	-	-	-	-	-	-	-	-	-
-	B	10	10	10	10	10	10	10	10	10	10	10	10	10	10	10	10	10	10	10	10	10	10	10	10	10	B	-
-	10	20	20	20	20	20	20	20	20	20	20	20	20	20	20	20	20	20	20	20	20	20	20	20	20	10	-	-
-	10	20	30	30	30	30	$\underline{30}$	$\underline{30}$	$\underline{30}$	$\underline{30}$	30	30	30	30	30	30	30	30	30	30	30	30	30	22	14	10	-	-
-	10	20	30	$\underline{40}$	36	28	22	20	20	22	28	36	40	40	40	40	40	40	40	40	40	36	28	20	10	-	-	-
-	10	20	30	36	28	22	14	10	10	14	22	28	36	45	50	50	50	50	50	45	36	28	22	14	10	-	-	-
-	10	20	30	32	22	14	10	-	-	10	14	22	32	41	51	60	60	60	51	41	32	22	14	10	-	-	-	-
-	10	20	30	30	20	10	-	-	-	-	10	20	30	40	50	60	$\underline{70}$	60	50	40	30	20	10	-	-	-	-	-
-	10	20	30	30	20	10	-	-	-	-	10	20	30	40	50	60	$\underline{70}$	61	51	41	32	22	14	10	10	10	B	-
-	10	20	30	32	22	14	10	-	-	10	14	22	32	41	51	60	60	60	54	45	36	28	22	20	$\underline{20}$	$\underline{20}$	10	-
-	10	20	30	36	28	22	14	10	10	14	22	28	36	45	50	50	50	50	50	45	36	28	22	20	$\underline{20}$	$\underline{20}$	10	-
-	10	20	30	$\underline{40}$	36	28	22	20	20	22	28	36	40	40	40	40	40	40	40	40	32	22	14	10	10	10	B	-
-	10	20	30	30	30	30	$\underline{30}$	$\underline{30}$	$\underline{30}$	$\underline{30}$	30	30	30	30	30	30	30	30	30	30	30	20	10	-	-	-	-	-
-	10	20	20	20	20	20	20	20	20	20	20	20	20	20	20	20	20	20	20	20	20	20	10	-	-	-	-	-
-	B	10	10	10	10	10	10	10	10	10	10	10	10	10	10	10	10	10	10	10	10	10	B	-	-	-	-	-
-	-	-	-	-	-	-	-	-	-	-	-	-	-	-	-	-	-	-	-	-	-	-	-	-	-	-	-	-

Figure 7.19 The Euclidean distance function with the base points labeled "B" and the apex points underlined.

cases are illustrated below.

$$\begin{bmatrix} P_4 & P_3 & P_2 \\ \bullet & P & \bullet \\ \bullet & P_7 & \bullet \end{bmatrix}, \quad \begin{bmatrix} \bullet & \bullet & P_2 \\ P_5 & P & P_1 \\ \bullet & \bullet & P_8 \end{bmatrix}, \quad \begin{bmatrix} P_4 & P_3 & \bullet \\ P_5 & P & \bullet \\ \bullet & \bullet & P_8 \end{bmatrix}, \quad \text{and} \quad \begin{bmatrix} \bullet & P_3 & P_2 \\ \bullet & P & P_1 \\ P_6 & \bullet & \bullet \end{bmatrix}$$

Note that the set of directional neighbors always contains three elements. The directional uphill generation adds the point, which is the maximum of P, and its directional neighborhood. That is,

$$P^U_{\text{next}} = \max_{P_i \in D_P \cup \{P\}} \{P_i\}. \tag{7.6}$$

Figure 7.20 shows the result of directional uphill generation of Figure 7.19.

7.5.2.4 Directional Downhill Generation

From Figure 7.20, we observe that there should exist a vertical path connecting two apex points of value "40" underlined on the left-hand side, in which the altitude changes are not always increasing; instead, they are a mixture of decreasing and increasing values. The directional downhill generation, which is similar to the directional uphill generation except the maxima tracking of the set excluding the central point P, is used to produce this type of skeletal branch. That is,

$$P^D_{\text{next}} = \max_{P_i \in D_P} \{P_i\}. \tag{7.7}$$

-	-	-	-	-	-	-	-	-	-	-	-	-	-	-	-	-	-	-	-	-	-	-	-	-	-	-	-	-
-	$\underline{10}$	10	10	10	10	10	10	10	10	10	10	10	10	10	10	10	10	10	10	10	10	10	10	10	10	$\underline{10}$	$\underline{10}$	-
-	10	$\underline{20}$	20	20	20	20	20	20	20	20	20	20	20	20	20	20	20	20	20	20	20	20	20	20	$\underline{20}$	10	-	-
-	10	20	$\underline{30}$	30	30	$\underline{30}$	$\underline{30}$	$\underline{30}$	$\underline{30}$	$\underline{30}$	$\underline{30}$	30	30	30	30	30	30	30	30	30	30	30	$\underline{30}$	$\underline{22}$	14	10	-	-
-	10	20	30	$\underline{40}$	$\underline{36}$	28	22	20	20	22	28	$\underline{36}$	$\underline{40}$	40	40	40	40	40	40	40	$\underline{40}$	$\underline{36}$	28	20	10	-	-	-
-	10	20	30	36	28	22	14	10	10	14	22	28	36	$\underline{45}$	50	50	50	50	50	$\underline{45}$	36	28	22	14	10	-	-	-
-	10	20	30	32	22	14	10	-	-	10	14	22	32	41	$\underline{51}$	60	60	60	$\underline{51}$	41	32	22	14	10	-	-	-	-
-	10	20	30	30	20	10	-	-	-	-	10	20	30	40	50	$\underline{60}$	$\underline{70}$	$\underline{60}$	50	40	30	20	10	-	-	-	-	-
-	10	20	30	30	20	10	-	-	-	-	10	20	30	40	50	$\underline{60}$	$\underline{70}$	61	51	41	32	22	14	10	10	10	$\underline{10}$	-
-	10	20	30	32	22	14	10	-	-	10	14	22	32	41	$\underline{51}$	60	60	$\underline{60}$	$\underline{54}$	$\underline{45}$	$\underline{36}$	$\underline{28}$	$\underline{22}$	$\underline{20}$	$\underline{20}$	$\underline{20}$	10	-
-	10	20	30	36	28	22	14	10	10	14	22	28	36	$\underline{45}$	50	50	50	50	$\underline{50}$	45	36	28	22	20	$\underline{20}$	$\underline{20}$	10	-
-	10	20	30	$\underline{40}$	$\underline{36}$	28	22	20	20	22	28	$\underline{36}$	$\underline{40}$	40	40	40	40	40	40	$\underline{40}$	32	22	14	10	10	10	$\underline{10}$	-
-	10	20	$\underline{30}$	30	30	$\underline{30}$	$\underline{30}$	$\underline{30}$	$\underline{30}$	$\underline{30}$	$\underline{30}$	30	30	30	30	30	30	30	30	30	$\underline{30}$	20	10	-	-	-	-	-
-	10	$\underline{20}$	20	20	20	20	20	20	20	20	20	20	20	20	20	20	20	20	20	20	20	$\underline{20}$	10	-	-	-	-	-
-	$\underline{10}$	10	10	10	10	10	10	10	10	10	10	10	10	10	10	10	10	10	10	10	10	10	$\underline{10}$	-	-	-	-	-
-	-	-	-	-	-	-	-	-	-	-	-	-	-	-	-	-	-	-	-	-	-	-	-	-	-	-	-	-

Figure 7.20 The results of directional uphill generation of Figure 7.19.

-	-	-	-	-	-	-	-	-	-	-	-	-	-	-	-	-	-	-	-	-	-	-	-	-	-	-	-	-
-	10	10	10	10	10	10	10	10	10	10	10	10	10	10	10	10	10	10	10	10	10	10	10	10	10	10	10	-
-	10	20	20	20	20	20	20	20	20	20	20	20	20	20	20	20	20	20	20	20	20	20	20	20	20	10	-	-
-	10	20	30	30	30	30	30	30	30	30	30	30	30	30	30	30	30	30	30	30	30	30	30	22	14	10	-	-
-	10	20	30	40	36	28	22	20	20	22	28	36	40	40	40	40	40	40	40	40	40	36	28	20	10	-	-	-
-	10	20	30	36	28	22	14	10	10	14	22	28	36	45	50	50	50	50	50	45	36	28	22	14	10	-	-	-
-	10	20	30	32	22	14	10	-	-	10	14	22	32	41	51	60	60	60	51	41	32	22	14	10	-	-	-	-
-	10	20	30	30	20	10	-	-	-	-	10	20	30	40	50	60	70	60	50	40	30	20	10	-	-	-	-	-
-	10	20	30	30	20	10	-	-	-	-	10	20	30	40	50	60	70	61	51	41	32	22	14	10	10	10	10	-
-	10	20	30	32	22	14	10	-	-	10	14	22	32	41	51	60	60	60	54	45	36	28	22	20	20	20	10	-
-	10	20	30	36	28	22	14	10	10	14	22	28	36	45	50	50	50	50	50	45	36	28	22	20	20	20	10	-
-	10	20	30	40	36	28	22	20	20	22	28	36	40	40	40	40	40	40	40	40	32	22	14	10	10	10	10	-
-	10	20	30	30	30	30	30	30	30	30	30	30	30	30	30	30	30	30	30	30	30	20	10	-	-	-	-	-
-	10	20	20	20	20	20	20	20	20	20	20	20	20	20	20	20	20	20	20	20	20	20	10	-	-	-	-	-
-	10	10	10	10	10	10	10	10	10	10	10	10	10	10	10	10	10	10	10	10	10	10	10	-	-	-	-	-
-	-	-	-	-	-	-	-	-	-	-	-	-	-	-	-	-	-	-	-	-	-	-	-	-	-	-	-	-

Figure 7.21 The results of directional downhill generation of Figure 7.20.

The directional downhill generation is initialized from the apex points that cannot be further tracked by the directional uphill generation. Hence, the altitude of the next directional downhill should be lower in the beginning. However, the tracking procedure is continued without taking into account the comparison of neighbors with P. The next directional downhill altitude could be lower or even higher until a skeleton point appears in the directional neighborhood. Figure 7.21 shows the result of directional downhill generation of Figure 7.20. The skeleton is now connected and is one pixel wide except for two pixels having the same local maximum of Euclidean distance.

7.5.3 The Skeletonization Algorithm and Connectivity Properties

The skeletonization algorithm traces the skeleton points by choosing the local maxima on the Euclidean distance transform and takes into consideration the least slope changes in medial axes. The algorithm is described as follows:

1. The base points and apex points are detected as the initial skeleton points.
2. Starting with these skeleton points, the directional uphill generation in equation (8.6) is used to add more skeleton points and the directional uphill of the new skeleton points is continuously added until no further point is generated. The points that cannot be further tracked are marked.
3. Starting with the marked points, the directional downhill generation in equation (8.7) is used to complete the skeleton tracking.

Figure 7.22 illustrates the skeleton of a rectangle with holes and notches obtained by the skeletonization algorithm, where "M" indicates the skeleton pixel.

-	-	-	-	-	-	-	-	-	-	-	-	-	-	-	-	-	-	-	-	-	-	-	-	-	-	-	-	-
-	M	O	O	O	O	O	O	O	O	O	O	O	O	O	O	O	O	O	O	O	O	O	O	O	O	M	M	-
-	O	M	O	O	O	O	O	O	O	O	O	O	O	O	O	O	O	O	O	O	O	O	O	O	M	O	-	-
-	O	O	M	O	O	M	M	M	M	M	M	O	O	O	O	O	O	O	O	O	O	O	M	M	O	O	-	-
-	O	O	O	M	M	O	O	O	O	O	O	M	M	O	O	O	O	O	O	O	M	M	O	O	O	-	-	-
-	O	O	O	M	O	O	O	O	O	O	O	O	O	M	O	O	O	O	O	M	O	O	O	O	O	-	-	-
-	O	O	O	M	O	O	O	-	-	O	O	O	O	O	M	O	O	O	M	O	O	O	O	O	-	-	-	-
-	O	O	O	M	O	O	-	-	-	-	O	O	O	O	O	M	M	M	O	O	O	O	O	-	-	-	-	-
-	O	O	O	M	O	O	-	-	-	-	O	O	O	O	O	M	M	O	O	O	O	O	O	O	O	O	M	-
-	O	O	O	M	O	O	O	-	-	O	O	O	O	O	M	O	O	M	M	M	M	M	M	M	M	M	O	-
-	O	O	O	M	O	O	O	O	O	O	O	O	O	M	O	O	O	O	M	O	O	O	O	O	M	M	O	-
-	O	O	O	M	M	O	O	O	O	O	O	M	M	O	O	O	O	O	O	M	O	O	O	O	O	O	M	-
-	O	O	M	O	O	M	M	M	M	M	M	O	O	O	O	O	O	O	O	O	M	O	O	-	-	-	-	-
-	O	M	O	O	O	O	O	O	O	O	O	O	O	O	O	O	O	O	O	O	O	M	O	-	-	-	-	-
-	M	O	O	O	O	O	O	O	O	O	O	O	O	O	O	O	O	O	O	O	O	O	M	-	-	-	-	-
-	-	-	-	-	-	-	-	-	-	-	-	-	-	-	-	-	-	-	-	-	-	-	-	-	-	-	-	-

Figure 7.22 The resulting skeleton on a rectangle with holes and notches.

In the experiment, the 8-connectedness is applied for foreground and the 4-connectedness for background. The skeletonization algorithm possesses the following three connectivity properties:

C1. After skeletonization, an originally connected object will not be separated into two or more subobjects.

C2. After skeletonization, a connected object will not disappear at all.

C3. After skeletonization, the originally disconnected background components are not 4-connected.

The proofs of the above three properties are given below.

Proof of C1. *Induction hypothesis:* For any m ($m < n$) marked apex points of an object, the skeleton obtained will preserve the property C1, where n could be any number.

The base case is $m = 1$. The algorithm starts tracking from each of the base and the apex points, and it will stop when the current point is connected to another skeleton point or touches the marked point. This means every subskeleton branch starting from the base point will connect to some skeleton point or meet another branch at the marked apex point.

When $m = 2$, it is regarded as two subobjects, and each subobject contains a marked apex point. As discussed above, each subobject is 8-connected. The algorithm will trace from each marked apex and stop when it is connected to some skeleton points contained in the other skeleton subset. The reason why it would not connect to a skeleton point that belongs to the same skeleton subset is the directional neighbors that we use. Using directional neighbors will enforce the tracking direction to go toward the region that has not been tracked up to now. Finally, it will lead tracking toward another skeleton subset.

As the induction hypothesis claims, when $m = n$, the skeleton obtained is 8-connected. Now consider the case when $m = n + 1$. The $n + 1$ marked apex points can be regarded as two subobjects: one contains n marked apex points and the other contains only a marked apex (i.e., the $(n + 1)$th point). As discussed above, the one with n marked apex points is 8-connected, and the other with a marked apex is also 8-connected. Tracking from the $(n + 1)$th marked apex, it will lead the tracking to go toward the skeleton subset with n marked apex points. That means the whole object is 8-connected.

Proof of C2. Tracking starts from each base and apex points. The algorithm marks base points as the skeleton points. For any size of an object, there must be at least one base point or at least one apex point, and the skeleton obtained must have at least one point. For the ideal case of a circle, there is no base point but one apex point is present, which is the local maximum representing the center of the circle. That is, after skeletonization, an 8-connected object will not disappear at all, and the skeleton contains at least a pixel.

Proof of C3. According to the definition of the apex points, there exists at least one apex in the object between any two background components. As the algorithm is performed, it will trace from each apex point that is not 8-connected to any skeleton point. After tracking, there will be one skeleton branch that will make these two background components disconnected. Therefore, the skeleton obtained will not allow the originally disconnected background components to be 4-connected.

The algorithm including three procedures and two functions is described in pseudo-codes below.

```
Procedure SKEPIK
  /* trace starting from each base and apex points */
  for i, j in 1 ··· N, 1 ··· N loop
    if BASE_APEX (p(i, j)) then
        Uphill-Generation (p(i, j), p(i, j));
  end if
end loop
  /* trace from each marked apex point in the procedure
UpHill-Generation */
  while (marked_pixel)
    DownHill-Generation (marked_apex, marked_apex);
  end while
end SKEPIK
function APEX (p): Boolean
/* apex point is the local maximum point */
  if (p is local maxima) then
return TRUE
  else
return FALSE;
  end APEX
function BASE_APEX (p): Boolean
/* Base point is the point with distance 1 and
  has 4 or more zeros in its 8-neighborhood */
    if (distance of p = 1) then
find 8-neighbors of p;
if ((number of 8-neighbors with distance 0) > 4) then
  return TRUE;
else
  if APEX (p) then
return TRUE;
  else
return FALSE;
  end if
end if
  end if
end BASE_APEX
procedure Uphill-Generation (current-skeleton-pixel,
previous-skeleton-pixel)
  if (number of maximum in 8-neighbors > 1 and
```

```
distance of maximum ≥ distance of current-skeleton-pixel) then
  maximum-pixel = the maximum of the directional-neighbors;
    UpHill-Generation (maximum-pixel, current-skeleton-pixel);
  else
  if (number of maximum in 8-neighbors = 1 and
distance of maximum > distance of current-skeleton-pixel) then
maximum-pixel = the maximum;
UpHill-Generation (maximum-pixel, current-skeleton-pixel);
  else
mark-apex; /* mark current-skeleton-pixel for later processing */
  end if
  end if
end UpHill-Generation
procedure DownHill-Generation (current-skeleton-pixel,
previous-skeleton-pixel)
  maximum-pixel = the maximum of the directional-neighbors;
  if (maximum-pixel is not a skeleton point) then
DownHill-Generation (maximum-pixel, current-skeleton-pixel);
  end if
end DownHill-Generation
```

7.5.4 A Modified Algorithm

The modified maxima tracking algorithm will extract the skeleton by eliminating nonsignificant short skeletal branches that touch the object boundary at corners. It is different from the previous algorithm that detects the base and apex points as initial skeleton points. Instead, the maxima tracking starts from the apex points only. The algorithm will recursively repeat the same procedure that selects the maxima in the directional neighborhood as the next skeleton point until another apex point is reached.

The modified maxima tracking algorithm is given as follows (Shih and Pu, 1995):

1. The apex points are detected as the initial skeleton points.
2. Starting with each apex point, we use the directional uphill generation to generate the skeleton points. Recursively repeat this procedure until an apex point is reached, and then the apex point is marked.
3. Starting with these marked apex points, the directional downhill generation is used to track the new skeleton points.

Figure 7.23 illustrates the resulting skeleton by using the modified algorithm, where there is a closed curve indicating an inside hole.

Figure 7.24 illustrates that the character "e" and its rotations by 30°, 45°, and 90° using the modified algorithm will produce the identical rotated skeleton provided the digitization error is disregarded.

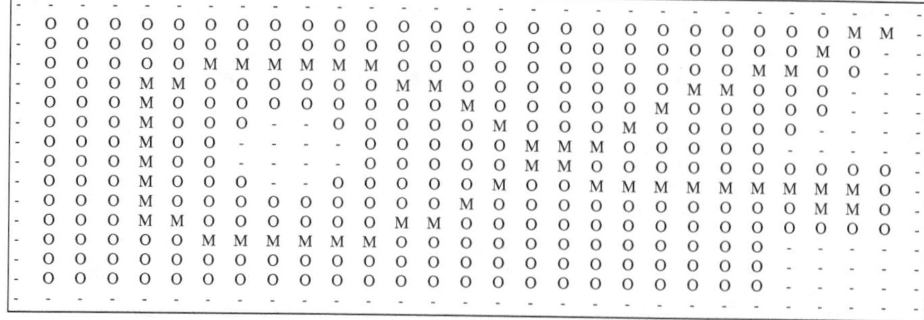

Figure 7.23 The resulting skeleton using the modified algorithm on a rectangle with holes and notches.

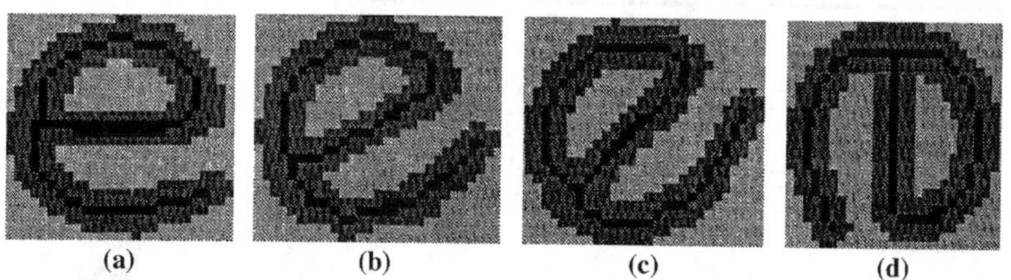

Figure 7.24 An example of the modified skeleton for a set of the character "e" in various rotations: (a) 0°, (b) 30°, (c) 45°, and (d) 90°.

7.6 OBJECT REPRESENTATION AND TOLERANCE

It has been well recognized that solid modeling is the foundation for CAD/CAM integration. The ultimate goal of planning for the automated manufacturing inspections and robotic assembly is to be able to generate a complete process plan automatically, starting from a CAD representation of the mechanical components. The representation must not only possess the nominal (or ideal) geometric shapes but also reason the geometric inaccuracies (or tolerances) into the locations and shapes of solid objects. Most of the existing solid modeling systems (in the market) have the very efficient graphical interactive input front end. Unfortunately, they are not satisfactory for such a fully integrated environment in the following two aspects:

1. *Information sufficiency problem:* The existing solid models do not support sufficient information for design and manufacturing, which require the specification of dimension, tolerance, surface finishing, data structure, and machining features.
2. *Input data redundancy problem:* The input data of the existing solid models are inadequate or more than the data required to specify a part.

Although many solid models have been proposed (Requicha, 1980), only boundary representation and CSG representation are popularly used as the internal database (Requicha and Voelcker, 1983). Boundary representation consists of two kinds of information—topological information and geometric information—which represent the vertex coordinates, surface equations, and the connectivity among faces, edges, and vertices. There are several advantages in boundary representation: large domain, unambiguity, unique, and explicit representation of faces, edges, and vertices. There are also several disadvantages: verbose data structure, difficult to create, difficult to check validity, and variational information unavailability.

The idea of CSG representation is to construct a complex part by hierarchically combining simple primitives using Boolean set operations. There are several advantages in CSG representation: large domain, unambiguity, easy to check validity, and easy to create. There are also several disadvantages: nonunique, difficult to edit graphically, input data redundancy, and variational information unavailability.

The mathematical framework used for modeling solids is mathematical morphology (Shih, 1991; Shih and Gaddipati, 2003, 2005). Adopting mathematical morphology as a tool, the theoretical research aims at studying the representation schemes for the dimension and tolerance of the geometric structure. This section is divided into three parts. The first part defines the representation framework for characterizing dimension and tolerance of solid objects. The second part then adopts the framework to represent several illustrated 2D and 3D objects. The third part describes the added tolerance information to control the quality of the parts and the interchangeability of the parts among assemblies. With the help of variational information, it is known how to manufacture, set up, and inspect to ensure the products within the required tolerance range.

7.6.1 Representation Framework: Formal Languages and Mathematical Morphology

There are rules that associate measured entities with features. Measured central planes, axes, and thinned components are associated with symmetric features. The representation framework is formalized as follows.

Let E^N denote the set of all points in the N-dimensional Euclidean space and $p = (x_1, x_2, \ldots, x_N)$ represent a point in E^N. In the following, any object is a subset of E^N. The formal model will be a context-free grammar, G, consisting of a 4-tuple (Fu, 1982; Ghosh, 1988):

$$G = (V_N, V_T, P, S)$$

where V_N is a set of nonterminal symbols, such as complicated shapes; V_T is a set of terminal symbols, which contains two sets: one is the decomposed primitive shapes, such as lines and circles, and the other is the shape operators; P is a finite set of rewrite rules or productions denoted by $A \rightarrow \beta$, where $A \in V_N$ and β is a string over $V_N \cup V_T$; and S is the start symbol, which is the solid object. The operators include morphological dilation and erosion, set union and intersection, and set subtraction.

Note that a production of such a form allows the nonterminal A to be replaced by the string β independent of the context in which the A appears. The grammar G is context-free, since for each production in P, the left part is a single nonterminal and the right part is a nonempty string of terminals and nonterminals. The languages generated by context-free grammars are called context-free languages. The object representation task may be reviewed as the task of converting a solid shape into a sentence in the language, while object recognition is the task of "parsing" a sentence.

There is no general solution for the primitive selection problem at this time. This determination will be largely influenced by the nature of the data, the specific application in question, and the technology available for implementing the system. The following requirements usually serve as a guide for selecting the shape primitives.

1. The primitives should serve as basic shape elements to provide a compact but adequate description of the object shape in terms of the specified structural relations (e.g., the concatenation relation).
2. The primitives should be easily extractable by the existing nonsyntactic (e.g., decision-theoretic) methods, since they are considered to be simple and compact shapes, and their structural information is not important.

7.6.2 Dimensional Attributes

Dimension and tolerance are the main components of the variational information. Dimensions are the control parameters that a designer uses to specify the shape of a part without redundancy, and tolerances are essentially associated with the dimensions to specify how accurate the part should be made. For a detailed description of dimension and tolerance refer to American National Standards Institute (1982).

7.6.2.1 The 2D Attributes The commonly used 2D attributes are rectangle, parallelogram, triangle, rhomb, circle, and trapezoid. These can be easily represented in the morphological way and some of them are illustrated in Figure 7.25. These expressions are not unique, but the evaluation depends on the simplest combination and least computational complexity. The common method is to decompose the attributes into smaller components and apply morphological dilation to "grow up" these components. Most geometric units can be decomposed into thinned small subunits (skeleton) with a certain size structuring element using recursive dilation. In most of the cases, the N-dimensional solid object with the morphological operation can be decomposed into one-dimensional lines.

The formal expressions of Figure 7.25 are written as follows:

For Figure 7.25(a): Rectangle $\rightarrow \vec{a} \oplus \vec{b}$

For Figure 7.25(b): Parallelogram $\rightarrow \vec{a} \oplus \vec{b}$

For Figure 7.25(c): Triangle $\rightarrow (\vec{a} \oplus \vec{b}) \cap (\vec{a} \oplus \vec{c})$

For Figure 7.25(d): Circle $\rightarrow (\cdots$ (unit circle $\oplus$ unit circle) $\oplus \cdots) \oplus$ unit circle $=$ (unit circle)$^{\oplus R}$, where the "unit circle" is a circle with the radius 1. Note that the sweep representation scheme becomes a special case for the model. This same analogy can be easily extended to three dimensions.

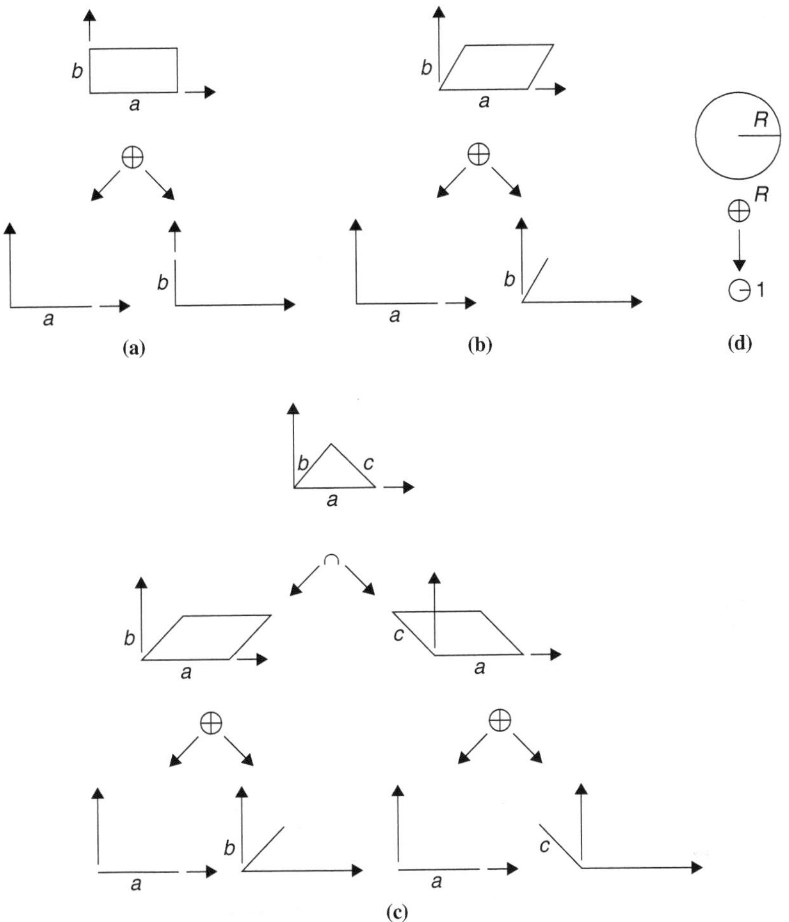

Figure 7.25 The decomposition of 2D attributes: (a) rectangle, (b) parallelogram, (c) triangle, and (d) circle.

7.6.2.2 The 3D Attributes The basic geometric entities in manufacturing are features. Simple surface features are subsets of an object's boundary that lie in a single surface, which typically is a plane, cylinder, cone, sphere, or torus. Composite surface features are aggregates of simple ones. The 3D attributes, which are similar to 2D attributes except addition of one more dimension, can also apply the similar decomposition method. A representation scheme for a 3D shape model is described in Figure 7.26, and its formal expression is

$$\text{shape} \rightarrow [(\vec{a} \oplus \vec{b}) - (\text{unit circle}^{\oplus R})] \oplus \vec{c}$$

7.6.2.3 Tolerancing Expression Tolerances constrain an object's features to lie within regions of space called *tolerance zones*. Tolerance zones in Rossignac and Requicha (1986) were constructed by expanding the nominal feature to obtain the

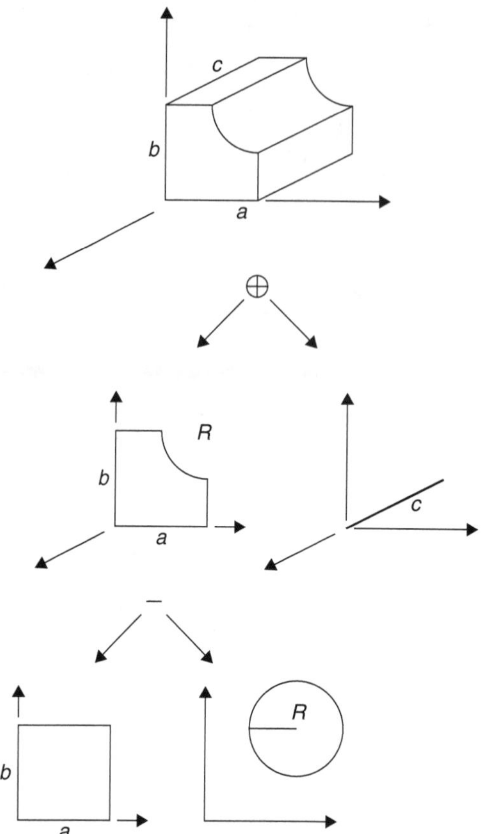

Figure 7.26 The decomposition of 3D attributes. The right-hand slide shows the decomposition subunits that construct the attributes through the morphological dilation.

region bounded by the outer closed curve, shrinking the nominal feature to obtain the region bounded by the inner curve, and then subtracting the two resulting regions. This procedure is equivalent to the morphological dilation of the offset inner contour with a tolerance-radius disked structuring element. Figure 7.27(a) shows an annular tolerance zone that corresponds to a circular hole, and Figure 7.27(b) shows a tolerance zone for an elongated slot. Both could be constructed by dilating the nominal contour with a tolerance-radius disked structuring element whose representation becomes simpler.

Mathematical rules for constructing tolerance zones depend on the types of tolerances, but are independent of the specific geometry of the features and of feature representation methods. The resulting tolerance zones, however, depend on features' geometrics. The tolerance zone for testing the size of a round hole is an annular region lying between two circles with the specified maximal and minimal diameters; the zone corresponding to a form constraint for the hole is also an annulus, defined by two concentric circles whose diameters must differ by a specified amount but are otherwise arbitrary.

The mathematical morphology does support the conventional limit ($\pm$) tolerances on "dimensions" that appear in many engineering drawings. The positive

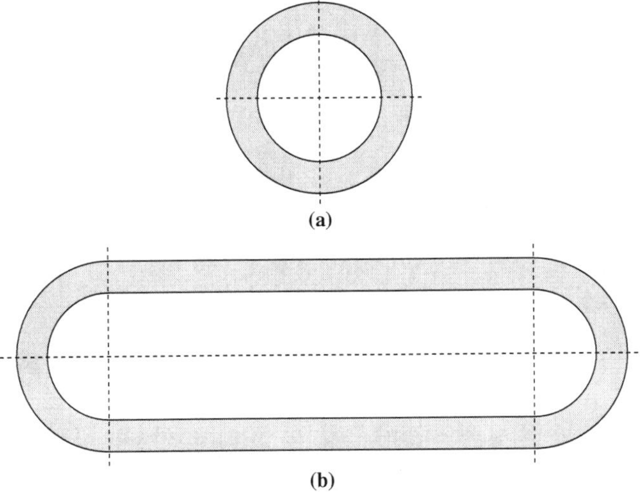

Figure 7.27 Tolerance zones. (a) An annular tolerance zone that corresponds to a circular hole and (b) a tolerance zone for an elongated slot.

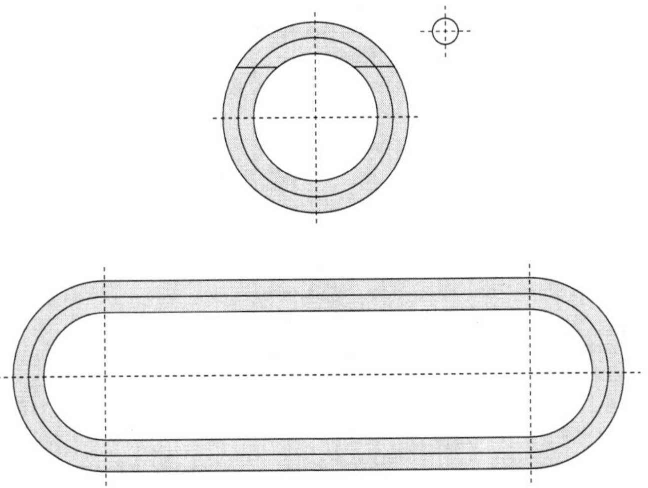

Figure 7.28 An example of adding tolerance by morphological dilation.

deviation is equivalent to the dilated result and the negative deviation is equivalent to the eroded result. The industrial parts adding tolerance information can be expressed using dilation with a circle (see Figure 7.28).

REFERENCES

Ackland, B. D. and Weste, N., "The edge flag algorithm: a fill method for raster scan display," *IEEE Trans. Comput.*, vol. 30, no. 1, pp. 41–48, Jan. 1981.

Ahmed, M. and Ward, R., "A rotation invariant rule-based thinning algorithm for character recognition," *IEEE Trans. Pattern Anal. Mach. Intell.*, vol. 24, no. 12, pp. 1672–1678, Dec. 2002.

Ali, S. M. and Burge, R. E., "A new algorithm for extracting the interior of bounded regions based on chain coding," *Comput. Vision Graph. Image Process.*, vol. 43, no. 2, pp. 256–264, Aug. 1988.

American National Standards Institute (ANSI), *Dimensioning and Tolerancing*, ANSI Standard Y14.5M, ASME, New York, 1982.

Berghorn, W., Boskamp, T., Lang, M., and Peitgen, H.-O., "Fast variable run-length coding for embedded progressive wavelet-based image compression," *IEEE Trans. Image Process.*, vol. 10, no. 12, pp. 1781–1790, Dec. 2001.

Bribiesca, E., "A chain code for representing 3D curves," *Pattern Recogn.*, vol. 33, no. 5, pp. 755–765, May 2000.

Bunke, H. and Csirik, J., "An improved algorithm for computing the edit distance of run-length coded strings," *Inform. Process. Lett.*, vol. 54, no. 2, pp. 93–96, Apr. 1995.

Cai, Z., "Restoration of binary images using contour direction chain codes description," *Comput. Vision Graph. Image Process.*, vol. 41, no. 1, pp. 101–106, Jan. 1988.

Capon, J., "A probabilistic model for run-length coding of pictures," *IRE Trans. Inform. Theory*, vol. 5, no. 4, pp. 157–163, Dec. 1959.

Cederberg, R. L. T., "Chain-link coding and segmentation for raster scan devices," *Comput. Graph. Image Process.*, vol. 10, no. 3, pp. 224–234, July 1979.

Chakravarty, I., "A single-pass, chain generating algorithm for region boundaries," *Comput. Graph. Image Process.*, vol. 15, no. 2, pp. 182–193, Feb. 1981.

Chang, L.-W. and Leu, K.-L., "A fast algorithm for the restoration of images based on chain codes description and its applications," *Comput. Graph. Image Process.*, vol. 50, no. 3, pp. 296–307, June 1990.

Chin, R. T., Wan, H.-K., Stover, D. L., and Iverson, R. D., "A one-pass thinning algorithm and its parallel implementation," *Comput. Vision Graph. Image Process.*, vol. 40, no. 1, pp. 30–40, Oct. 1987.

Dunkelberger, K. A. and Mitchell, O. R., "Contour tracing for precision measurement," Proceedings of the IEEE International Conference on Robotics and Automation, St. Louis, MO, vol. 2, pp. 22–27, Mar. 1985.

Freeman, H., "Computer processing of line drawing images," *Comput. Surv.*, vol. 6, no. 1, pp. 57–97, Mar. 1974.

Fu, K. S., *Syntactic Pattern Recognition and Applications*, Prentice-Hall, 1982.

Ghosh, P. K., "A mathematical model for shape description using Minkowski operators," *Comput. Vision Graph. Image Process.*, vol. 44, no. 3, pp. 239–269, Dec. 1988.

Ghosh, A. and Petkov, N., "Robustness of shape descriptors to incomplete contour representations," *IEEE Trans. Pattern Anal. Mach. Intell.*, vol. 27, no. 11, pp. 1793–1804, Nov. 2005.

Golomb, S. W., "Run length encodings," *IEEE Trans. Inform. Theory*, vol. 12, no. 3, pp. 399–401, July 1966.

Govindham, V. K. and Shivaprasad, A. P., "A pattern adaptive thinning algorithm," *Pattern Recogn.*, vol. 20, no. 6, pp. 623–637, Nov. 1987.

Guo, Z. and Hall, R. W., "Parallel thinning with two-subiteration algorithms," *Commun. ACM*, vol. 32, no. 3, pp. 359–373, Mar. 1989.

Guo, Z. and Hall, R. W., "Fast fully parallel thinning algorithms," *Comput. Vision Graph. Image Process.*, vol. 55, no. 3, pp. 317–328, May 1992.

Hasan, Y. M. Y. and Karam, L. J., "Morphological reversible contour representation," *IEEE Trans. Pattern Anal. Mach. Intell.*, vol. 22, no. 3, pp. 227–240, Mar. 2000.

Hunter, G. M. and Steiglitz, K., "Operations on images using quad trees," *IEEE Trans. Pattern Anal. Mach. Intell.*, vol. 1, no. 2, pp. 145–154, Apr. 1979.

Ji, L., Yi, Z., Shang, L., and Pu, X., "Binary fingerprint image thinning using template-based PCNNs," *IEEE Trans. Syst. Man Cybernet. B*, vol. 37, no. 5, pp. 1407–1413, Oct. 2007.

Kalvin, A., Schonberg, E., Schwartz, J. T., and Sharir, M., "Two dimensional model based boundary matching using footprints," *Int. J. Robot. Res.*, vol. 5, no. 4, pp. 38–55, 1986.

Kaneko, T. and Okudaira, M., "Encoding of arbitrary curves based on the chain code representation," *IEEE Trans. Commun.*, vol. 33, no. 7, pp. 697–707, July 1985.

Kim, S.-D., Lee, J.-H., and Kim, J.-K., "A new chain-coding algorithm for binary images using run-length codes," *Comput. Vision Graph. Image Process.*, vol. 41, no. 1, pp. 114–128, Jan. 1988.

Knipe, J. and Li, X., "On the reconstruction of quadtree data," *IEEE Trans. Image Process.*, vol. 7, no. 12, pp. 1653–1660, Dec. 1998.

Koplowitz, J., "On the performance of chain codes for quantization of line drawings," *IEEE Trans. Pattern Anal. Mach. Intell.*, vol. 3, no. 2, pp. 180–185, Mar. 1981.

Laferte, J.-M., Perez, P., and Heitz, F., "Discrete Markov image modeling and inference on the quadtree," *IEEE Trans. Image Process.*, vol. 9, no. 3, pp. 390–404, Mar. 2000.

Liu, Y. K. and Zalik, B., "An efficient chain code with Huffman coding," *Pattern Recogn.*, vol. 38, no. 4, pp. 553–557, Apr. 2005.

Lu, H. E. and Wang, P. S. P., "A comment on 'A fast parallel algorithm for thinning digital patterns'," *Commun. ACM*, vol. 29, no. 3, pp. 239–242, Mar. 1986.

Madhvanath, S., Kleinberg, E., and Govindaraju, V., "Holistic verification of handwritten phrases," *IEEE Trans. Pattern Anal. Mach. Intell.*, vol. 21, no. 12, pp. 1344–1356, Dec. 1999.

Meer, P., Sher, C. A., and Rosenfeld, A., "The chain pyramid: hierarchical contour processing," *IEEE Trans. Pattern Anal. Mach. Intell.*, vol. 12, no. 4, pp. 363–376, Apr. 1990.

Moayer, B. and Fu, K. S., "A tree system approach for fingerprint pattern recognition," *IEEE Trans. Comput.*, vol. 25, no. 3, pp. 262–274, Mar. 1976.

Montani, C. and Scopigno, R., "Rendering volumetric data using STICKS representation scheme," *ACM SIGGRAPH*, vol. 24, no. 5, pp. 87–93, Nov. 1990.

Ogawa, H. and Taniguchi, K., "Thinning and stroke segmentation for handwritten Chinese character recognition," *Pattern Recogn.*, vol. 15, no. 4, pp. 299–308, 1982.

Pavlidis, T., "Filling algorithms for raster graphics," *Comput. Graph. Image Process.*, vol. 10, no. 2, 126–141, 1979.

Requicha, A. A., "Representations for rigid solids: theory, method, and systems," *Comput. Surv.*, vol. 12, no. 4, pp. 437–464, Dec. 1980.

Requicha, A. A. and Voelcker, H. B., "Solid modelling: current status and research direction," *IEEE Comput. Graph. Appl.*, vol. 3, no. 7, pp. 25–37, Oct. 1983.

Robinson, J. A., "Efficient general-purpose image compression with binary tree predictive coding," *IEEE Trans. Image Process.*, vol. 6, no. 4, pp. 601–608, Apr. 1997.

Robinson, J. A., Druet, A., and Gosset, N., "Video compression with binary tree recursive motion estimation and binary tree residue coding," *IEEE Trans. Image Process.*, vol. 9, no. 7, pp. 1288–1292, July 2000.

Rockett, P. I., "An improved rotation-invariant thinning algorithm," *IEEE Trans. Image Process.*, vol. 27, no. 10, pp. 1671–1674, Oct. 2005.

Rosenfeld, A. Kak, A. C., *Digital Picture Processing*, vol. 2, Academic Press, 1982.

Rosenfeld, A. and Pfaltz, J. L., "Sequential operations in digital picture processing," *J. ACM*, vol. 13, no. 4, pp. 471–494, Oct. 1966.

Rossignac, J. R. and Requicha, A. A. G., "Offsetting operations in solid modelling," *Comput. Aided Geom. Des.*, vol. 3, no. 2, pp. 129–148, Aug. 1986.

Saghri, J. A. and Freeman, H., "Analysis of the precision of generalized chain codes for the representation of planar curves," *IEEE Trans. Pattern Anal. Mach. Intell.*, vol. 3, no. 5, pp. 533–539, Sept. 1981.

Saint-Marc, P., Rom, H., and Medioni, G., "B-spline contour representation and symmetry detection," *IEEE Trans. Pattern Anal. Mach. Intell.*, vol. 15, no. 11, pp. 1191–1197, Nov. 1993.

Samet, H., "Region representation: quadtrees from binary arrays," *Comput. Graph. Image Process.*, vol. 13, no. 1, pp. 88–93, May 1980.

Shani, U., "Filling regions in binary raster images," *SIGGRAPH Comput. Graph.*, vol. 14, no. 3, pp. 321–327, July 1980.

Shih, F. Y., "Object representation and recognition using mathematical morphology model," *Int. J. Syst. Integr.*, vol. 1, no. 2, pp. 235–256, Aug. 1991.

Shih, F. Y. and Gaddipati, V., "General sweep mathematical morphology," *Pattern Recogn.*, vol. 36, no. 7, pp. 1489–1500, July 2003.

Shih, F. Y. and Gaddipati, V., "Geometric modeling and representation based on sweep mathematical morphology," *Inform. Sci.*, vol. 171, no. 3, pp. 213–231, Mar. 2005.

Shih, F. Y. and Mitchell, O. R., "Industrial parts recognition and inspection by image morphology," Proceedings of the IEEE International Conference on Robotics and Automation, Philadelphia, PA, vol. 3, pp. 1764–1766, Apr. 1988.

Shih, F. Y. and Pu, C. C., "A skeletonization algorithm by maxima tracking on Euclidean distance transform," *Pattern Recogn.*, vol. 28, no. 3, pp. 331–341, Mar. 1995.

Shih, F. Y. and Wong, W., "A new single-pass algorithm for extracting the mid-crack codes of multiple regions," *J. Visual Commun. Image Represent.*, vol. 3, no. 3, pp. 217–224, Sept. 1992.

Shih, F. Y. and Wong, W., "Fully parallel thinning with tolerance to boundary noise," *Pattern Recogn.*, vol. 27, no. 12, pp. 1677–1695, Dec. 1994.

Shih, F. Y. and Wong, W., "A new safe-point thinning algorithm based on the mid-crack code tracing," *IEEE Trans. Syst. Man. Cybernet.*, vol. 25, no. 2, pp. 370–378, Feb. 1995.

Shih, F. Y. and Wong, W., "A one-pass algorithm for local symmetry of contours from chain codes," *Pattern Recogn.*, vol. 32, no. 7, pp. 1203–1210, July 1999.

Shneier, M., "Calculations of geometric properties using quadtrees," *Comput. Graph. Image Process.*, vol. 16, no. 3, pp. 296–302, July 1981.

Shusterman, E. and Feder, M., "Image compression via improved quadtree decomposition algorithms," *IEEE Trans. Image Process.*, vol. 3, no. 2, pp. 207–215, Mar. 1994.

Sullivan, G. J. and Baker, R. L., "Efficient quadtree coding of images and video," *IEEE Trans. Image Process.*, vol. 3, no. 3, pp. 327–331, May 1994.

Tanaka, H. and Leon-Garcia, A., "Efficient run-length encodings," *IEEE Trans. Inform. Theory*, vol. 28, no. 6, pp. 880–890, Nov. 1982.

Tang, G. Y. and Lien, B., "Region filling with the use of the discrete Green theorem," *Comput. Graph. Image Process.*, vol. 42, no. 3, pp. 297–305, June 1988.

Wang, Z., Feng, D., and Chi, Z., "Region-based binary tree representation for image classification," Proceedings of the International Conference on Neural Networks and Signal Processing, Nanjing, China, vol. 1, pp. 232–235, Dec. 2003.

Wong, W., Shih, F. Y., and Su, T., "Thinning algorithms based on quadtree and octree representations," *Inform. Sci.*, vol. 176, no. 10, pp. 1379–1394, May 2006.

Wu, L. D., "On the chain code of a line," *IEEE Trans. Pattern Anal. Mach. Intell.*, vol. 4, no. 3, pp. 347–353, May 1982.

Yu, S.-S. and Tsai, W.-H., "A new thinning algorithm for gray-scale images by the relaxation technique," *Pattern Recogn.*, vol. 23, no. 10, pp. 1067–1076, Oct. 1990.

Zingaretti, P., Gasparroni, M., and Vecci, L., "Fast chain coding of region boundaries," *IEEE Trans. Pattern Anal. Mach. Intell.*, vol. 20, no. 4, pp. 407–415, Apr. 1998.

CHAPTER 8

FEATURE EXTRACTION

To recognize or classify an object in an image, one must first extract some features out of the image, and then use these features inside a pattern classifier to obtain the final class. Feature extraction (or detection) aims to locate significant feature regions on images depending on their intrinsic characteristics and applications. These regions can be defined in global or local neighborhood and distinguished by shapes, textures, sizes, intensities, statistical properties, and so on. Local feature extraction methods are divided into intensity based and structure based. Intensity-based methods analyze local intensity patterns to find regions that satisfy desired uniqueness or stability criteria. Structure-based methods detect image structures such as edges, lines, corners, circles, ellipses, and so on. Feature extraction tends to identify the characteristic features that can form a good representation of the object, so as to discriminate across the object category with tolerance of variations.

In general, local feature-based region extractors show their advantages of robustness to object occlusion or image deformation. They can detect locally distributed image features and provide a distinctive, compact representation scheme. Using all the pixel values can be redundant and noisy, whereas feature descriptors extract, filter, and summarize the most distinctive features and build compact feature vectors. These significant descriptors can facilitate the pattern classifiers with their compactness and discrimination.

This chapter presents selected often-used feature extraction topics. It is organized as follows. Fourier descriptor (FD) and moment invariants are introduced in Section 8.1. Shape number and hierarchical features are described in Section 8.2. Corner detection is presented in Section 8.3. Hough transform is presented in Section 8.4. We describe principal component analysis (PCA) in Section 8.5 and linear discriminate analysis (LDA) in Section 8.6. In Section 8.7, we discuss feature reduction in input and feature spaces.

8.1 FOURIER DESCRIPTOR AND MOMENT INVARIANTS

Shape is one of the most critical image features because shape is discriminative to human perception. Human beings tend to view a scene as composed of individual objects that can be best described by their shapes. Pattern matching techniques measure shape similarity using pattern transformation and calculation of their degrees of resemblance. The similarity of two patterns is measured as the distance between their

Image Processing and Pattern Recognition by Frank Shih

feature vectors. Many shape representation methods (or shape descriptors) have been proposed in the literature that can be summarized into two categories: region based versus contour based. In region-based methods, all the pixels within a shape are taken into consideration to obtain the shape feature. An example is the moment descriptors. In contour-based methods, only shape boundary information is used. For examples, shape descriptors, shape signatures, and spectral descriptors belong to this class.

Spectral descriptors include FD and wavelet descriptor (WD). Wavelet descriptor has the advantage of describing shape features in both spatial and frequency domains (Chuang and Kuo, 1996; Hung, 2000). However, it requires intensive computation in shape matching because of not being rotation invariant. In FD, global shape features are attained by the first few low-frequency components, while high-frequency components describe finer features of the shape. There have been many FD methods applied in shape analysis, character recognition, shape coding, shape classification, and shape retrieval. We will focus on the Fourier descriptor.

Persoon and Fu (1977) proposed using Fourier descriptor as the shape representation. Rafiei and Mendelzon (2002) proposed an indexing technique for the fast retrieval of objects in 2D images based on similarity between their boundary shapes using Fourier descriptors. The shape signature is defined by a 1D continuous function $f(t)$. The Fourier transform of $f(t)$ is given in equation (8.1). Because discrete images are concerned, the discrete Fourier transform (DFT) is used. These were introduced in Section 2. For convenience, we briefly summarize them as follows. Let L be a closed polygonal line that consists of N vertices. Also, let $[x(t), y(t)]$, $t = 0, 1, \ldots, N-1$, be the Cartesian coordinates of each vertex. We consider the complex signal $f(t)$, where $f(t) = x(t) + iy(t)$, $t = 0, 1, \ldots, N-1$, by randomly selecting a vertex as a starting point and tracing the vertices in a clockwise or counterclockwise direction. The DFT of $f(t)$ is given by equation (8.2), where the value of $u \in \{0, 1, \ldots, N-1\}$. From Euler's equation, that is, $e^{-j2\pi ut/N} = \cos(2\pi ut/N) - j\sin(2\pi ut/N)$, equation (8.3) can be obtained. The coefficient $F(u)$ is called FD.

$$F(u) = \int_{-\infty}^{\infty} f(t)e^{-j2\pi ut}\,dt \tag{8.1}$$

$$F(u) = \frac{1}{N}\sum_{t=0}^{N-1} f(t)e^{-j2\pi ut/N} \tag{8.2}$$

$$F(u) = \frac{1}{N}\sum_{t=0}^{N-1} f(t)(\cos(2\pi ut/N) - j\sin(2\pi ut/N)) \tag{8.3}$$

Several shape signatures have been proposed to derive Fourier descriptor. Zhang and Lu (2002) proved empirically that the Fourier descriptor derived from the centroid distance (CeFD) can achieve significantly higher performance than other Fourier descriptors, such as the area FD, curvature FD, psi FD, position (complex) FD, affine FD, and chord-length FD. The centroid distance function $f(t)$ is defined as the distance of the boundary points to the centroid (x_c, y_c) as in equation (8.4). Only a half of Fourier descriptors $F(u)$ are used to index the corresponding shape because $f(t)$ is a set of real values. Therefore, equation (8.5) is obtained. The Fourier descriptors can be normalized to be independent of geometric translation, scaling, and rotation. A set of invariant descriptors CeFD can be obtained by equation (8.6). The first few terms of the invariant moments, like the first few terms of a Fourier series, capture the more general shape properties, while the later terms capture finer details. Since $F(0)$ is the largest coefficient, it can be used as the normalization factor.

$$f(t) = \sqrt{(x(t)-x_c)^2 + (y(t)-y_c)^2} \tag{8.4}$$

$$|F(u)| = \frac{1}{N}\sqrt{\left(\sum_{t=0}^{N-1} f(t)\cos(2\pi ut/N)\right)^2 + \left(\sum_{t=0}^{N-1} f(t)\sin(2\pi ut/N)\right)^2} \tag{8.5}$$

$$\mathrm{CeFD} = \frac{|F(u)|}{|F(0)|}, \quad \text{where} \quad u = 1, 2, \ldots, N/2 \tag{8.6}$$

Two-dimensional moment invariants for planar geometric figures have been investigated. Hu (1962) introduced a set of moment invariants using nonlinear combination based on normalized central moments. The computation of these invariants is required to engage all the pixels in the shape. Chen (1993) proposed an improved moment invariant technique based on boundary pixels to speed up the computation. In addition, Chen's normalized moment invariants were designed to be invariant to scaling, translation, and rotation (Sun et al., 2003). The definitions of Hu's and Chen's moment invariants are briefly introduced below.

Let $f(x, y)$ be a continuous shape-based image function. The (p, q)th central moment of $f(x, y)$ is defined as

$$\mu_{pq} = \int_{-\infty}^{\infty}\int_{-\infty}^{\infty} (x-x_c)^p (y-y_c)^q f(x,y)\,dx\,dy \tag{8.7}$$

where (x_c, y_c) are the coordinates of the centroid of the shape. The normalized (p, q)th central moment is defined as

$$\eta_{pq} = \frac{\mu_{pq}}{\mu_{00}^{\gamma}}, \quad \text{where} \quad \gamma = \frac{p+q+2}{2} \ \text{and} \ p, q = 2, 3, \ldots \tag{8.8}$$

Based on these moments, Hu (1962) derived the following seven moment invariants, Φ_i, where $i = 1, 2, \ldots, 7$.

$$\left.\begin{aligned}
\Phi_1 &= \eta_{20} + \eta_{02} \\
\Phi_2 &= (\eta_{20}-\eta_{02})^2 + 4\eta_{11}^2 \\
\Phi_3 &= (\eta_{30}-3\eta_{12})^2 + (\eta_{03}-3\eta_{21})^2 \\
\Phi_4 &= (\eta_{30}+\eta_{12})^2 + (\eta_{03}+\eta_{21})^2 \\
\Phi_5 &= (\eta_{30}-3\eta_{12})(\eta_{30}+\eta_{12})\{(\eta_{30}+\eta_{12})^2 - 3(\eta_{21}+\eta_{03})^2\} + \\
&\quad (3\eta_{21}-\eta_{03})(\eta_{21}+\eta_{03})\{3(\eta_{30}+\eta_{12})^2 - (\eta_{21}+\eta_{03})^2\} \\
\Phi_6 &= (\eta_{20}-\eta_{02})\{(\eta_{30}+\eta_{12})^2 - (\eta_{21}+\eta_{03})^2\} + 4\eta_{11}(\eta_{30}+\eta_{12})(\eta_{21}+\eta_{03}) \\
\Phi_7 &= (3\eta_{21}-\eta_{03})(\eta_{30}+\eta_{12})\{(\eta_{30}+\eta_{12})^2 - 3(\eta_{21}+\eta_{03})^2\} + \\
&\quad (3\eta_{21}-\eta_{30})(\eta_{21}+\eta_{03})\{3(\eta_{30}+\eta_{12})^2 - (\eta_{21}+\eta_{03})^2\}
\end{aligned}\right\} \tag{8.9}$$

Chen (1993) improved the central moments by applying the moment computation only on the boundary. Chen's (p, q)th central moment is defined as

$$\mu_{pq} = \int_S (x-x_c)^p (y-y_c)^q \, ds \tag{8.10}$$

For a digital shape-based image, the discrete form is described as

$$\mu_{pq} = \sum_{(x,y)\in S} (x-\bar{x})^p (y-\bar{y})^q \tag{8.11}$$

Chen's normalized (p, q)th central moment is defined as

$$\eta_{pq} = \frac{\mu_{pq}}{\mu_{00}^{\gamma}}, \quad \text{where } \gamma = p+q+1 \text{ and } p, q = 2, 3, \ldots \tag{8.12}$$

Another kind of shape descriptor, called *shell descriptor*, is introduced below. Assume that Γ is a counterclockwise, simple, closed contour with parametric representation $(x(t), y(t)) = z(t)$, where t is the observation sequence and $0 \leq t \leq T$ (T is the perimeter of the contour). Let $z(t_{\text{ref}})$ be a reference observation and $t_{\text{ref}} \neq 0$. The observations can be represented in a polar form by adopting $z(t_{\text{ref}})$ and $z(0)$ as the reference line. Denote $\theta(t)$ as the angular direction of $z(t)$ and $\iota(t)$ as the normalized spatial displacement of the observation. In the polar form, the observation contour Γ can be represented as $\{z(t) = \iota(t)e^{j\theta(t)}\}$, where $j = \sqrt{-1}$. The parameters $[\iota, \theta]$ are invariant to translation and rotation. To make the representation invariant to scaling, the length of the observations is normalized by the length of the reference line.

Let $\iota(0) = 0$ and $\theta(0) = 0$ be the initial values for the starting point $z(0)$. Let $\phi(r, s)$ be the angle between two lines, $\overline{z(0)z(r)}$ and $\overline{z(0)z(s)}$, as shown in Figure 8.1. The angle $\phi(r, s)$ is called a *sector* if $\iota(r) = \iota(s) = \iota$. Thus, a sector can be represented as $(\iota, \theta(r), \phi(r, s))$ in addition to $\phi(r, s)$. A shell is a collection of all sectors. To simplify and generalize the symbol, $(\iota_l, \theta_{lk}, \phi_{lk})$ is used to denote the kth sector of the lth shell. Thus, a contour Γ can be represented as

$$\{(\iota_l, \theta_{lk}, \phi_{lk}), 0 \leq \iota_l \leq 1\} \tag{8.13}$$

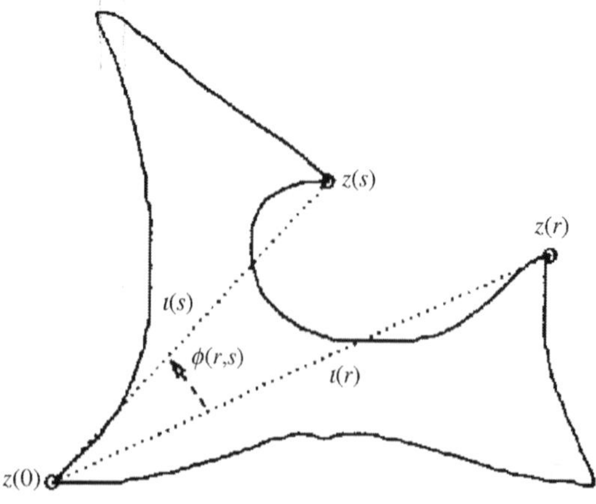

Figure 8.1 Sector notations.

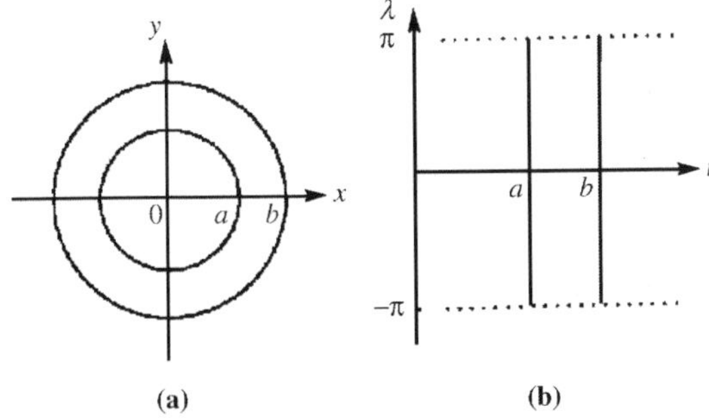

Figure 8.2 A mapping of a uniform polar angular grid onto a uniform (t, λ) grid.

The three-dimensional parametric space is then mapped onto a 2D plane (t, λ), where $t = \iota_l$ and $\lambda = \{x|\theta_{lk} \leq x \leq \theta_{lk} + \phi_{lk}\}$. A mapping of a uniform polar angular grid onto a uniform (t, λ) grid is illustrated in Figure 8.2. The values ι_l, θ_{lk}, and ϕ_{lk}, which are the shell descriptors, are named *sector coefficients*.

All simple closed curves with the starting point are mapped onto the rings of shell descriptors in such a way that all curves of the identical shape and the starting point go into the same descriptors. The shell space with descriptor parameters (t, λ) of Figure 8.1 is shown in Figure 8.3.

Many different descriptors have been proposed in the literature. It is unclear which descriptors are more appropriate and how their performance depends on the interest region detector. Mikolajczyk and Schmid (2005) compared the performance of descriptors computed for local interest regions. Kiranyaz et al. (2008) proposed a generic shape descriptor that can be extracted from the major object

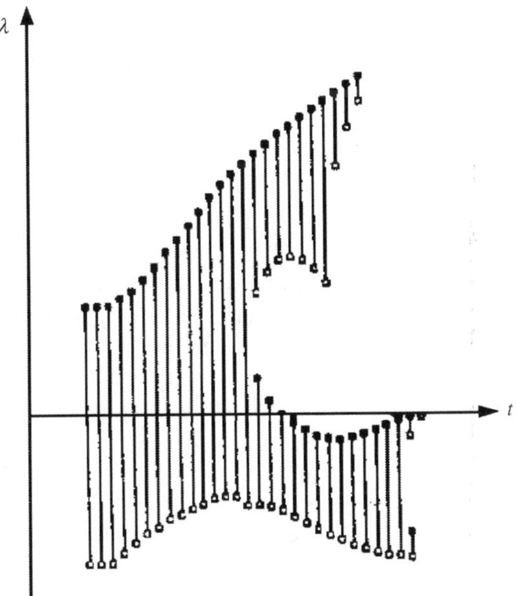

Figure 8.3 The shell descriptor of the shape shown in Figure 8.1.

edges automatically and used for the multimedia content-based retrieval in multimedia databases. Hung (2000) presented a generalized uniqueness property inherent in the 1D discrete periodized wavelet transformation. Bober (2001) proposed techniques and tools for shape representation and matching, developed in the context of MPEG-7 standardization. Manjunath et al. (2001) proposed a histogram of the main edge directions (vertical, horizontal, and two diagonals) within fixed size blocks. It is an efficient texture descriptor for the images with heavy textural presence. Shen and Ip (1998) presented discriminative wavelet shape descriptors for recognition of 2D patterns. Zhang and Lu (2002) presented a comparative study of Fourier descriptors for shape representation and retrieval. Wong et al. (2007) proposed shape-based image retrieval using Fourier descriptor and support vector machine.

Many different moment invariants have been proposed in the literature. Yap et al. (2003) proposed a new set of orthogonal moments based on the discrete classical Krawtchouk polynomials. The Krawtchouk polynomials (Krawtchouk, 1929) are scaled to ensure numerical stability, and thus creating a set of weighted Krawtchouk polynomials. Dudani et al. (1977) presented aircraft identification by moment invariants. Belkasim et al. (1991) presented a comparative study and new results of pattern recognition with moment invariants. Teh and Chin (1998) presented image analysis by the method of moments. Pawlak (1992) discussed the reconstruction aspect of moment descriptors. Liao and Pawlak (1996) presented image analysis by moments. Schlemmer et al. (2007) presented an approach for analyzing 2D flow field data based on the idea of invariant moments. Flusser and Suk (2006) proposed a set of moment invariants with respect to rotation, translation, and scaling suitable for recognition of objects having N-fold rotation symmetry. Mamistvalov (1998) presented n-dimensional moment invariants and conceptual mathematical theory of recognition n-dimensional solids. Rothe et al. (1996) developed a method of normalization to derive many sets of invariants. Flusser et al. (2003) presented the construction of combined blur and rotation moment invariants in arbitrary number of dimensions.

8.2 SHAPE NUMBER AND HIERARCHICAL FEATURES

The shape number of objects is designed to allow interpreting shape information into a number that is supposed to be independent of their location, size, and orientation. A set of hierarchical shape features includes a shape number, several layers of significant point radii (i.e., Euclidean distances to the boundary), and their x- and y-coordinates. Each local maximum point of the distance transform is used as a significant point that has the major contribution in representing features.

8.2.1 Shape Number

The shape number is computed from the total distance and the total number of all object points (Danielsson, 1978). It is a measure of shape compactness. Let N be the total number of the pixels belonging to the object, and let X_i be the distance of pixel i to

0	0	0	0	0	0	0	0	0	0	0	0	0
0	0	0	0	0	1	1	1	0	0	0	0	0
0	0	0	1	1	1	1	1	1	1	0	0	0
0	0	1	1	1	1	1	1	1	1	1	0	0
0	0	1	1	1	1	1	1	1	1	1	0	0
0	1	1	1	1	1	1	1	1	1	1	1	0
0	1	1	1	1	1	1	1	1	1	1	1	0
0	1	1	1	1	1	1	1	1	1	1	1	0
0	0	1	1	1	1	1	1	1	1	1	0	0
0	0	1	1	1	1	1	1	1	1	1	0	0
0	0	0	1	1	1	1	1	1	1	0	0	0
0	0	0	0	0	1	1	1	0	0	0	0	0
0	0	0	0	0	0	0	0	0	0	0	0	0

Figure 8.4 A 13 × 13 discrete circle.

the nearest pixel outside the object. The shape number is mathematically expressed as follows:

$$\text{Shape number} = \frac{N^3}{9\pi\left(\sum_{i=1}^{N} X_i\right)^2} \tag{8.14}$$

Note that the constant is chosen such that the shape number of an ideal circle is 1. Hence, the larger the shape number is, the less compact the object shape looks. A 13 × 13 discrete circle is shown in Figure 8.4, where the total number of object pixels is 89. Its city-block distance transform is shown in Figure 8.5. It is because of this fact that we have labeled all border pixels with the distance value $x = 1$ (actually, it is 0.5). For precise measurement, each distance value is subtracted by 0.5. The summation of total distance values is $218 - 0.5 \times 89 = 173.5$. Therefore, the shape number is 0.8283. Note that the shape number is not equal to 1 because of the digitization and the city-block distance used. If we select Euclidean distance, it will increase the computational precision. Figure 8.6 shows the Euclidean distance transform of Figure 8.4. In order to make the distance summation more accurate, it would be appropriate to decrease every x value by the average error $\sqrt{0.5}/2$. The summation of

0	0	0	0	0	0	0	0	0	0	0	0	0
0	0	0	0	0	1	1	1	0	0	0	0	0
0	0	0	1	1	2	2	2	1	1	0	0	0
0	0	1	2	2	3	3	3	2	2	1	0	0
0	0	1	2	3	4	4	4	3	2	1	0	0
0	1	2	3	4	5	5	5	4	3	2	1	0
0	1	2	3	4	5	6	5	4	3	2	1	0
0	1	2	3	4	5	5	5	4	3	2	1	0
0	0	1	2	3	4	4	4	3	2	1	0	0
0	0	1	2	2	3	3	3	2	2	1	0	0
0	0	0	1	1	2	2	2	1	1	0	0	0
0	0	0	0	0	1	1	1	0	0	0	0	0
0	0	0	0	0	0	0	0	0	0	0	0	0

Figure 8.5 The city-block distance transform of Figure 8.4.

0	0	0	0	0	0	0	0	0	0	0	0	0
0	0	0	0	0	1	1	1	0	0	0	0	0
0	0	0	1	1	$\sqrt{2}$	2	$\sqrt{2}$	1	1	0	0	0
0	0	1	$\sqrt{2}$	2	$\sqrt{5}$	$\sqrt{8}$	$\sqrt{5}$	2	$\sqrt{2}$	1	0	0
0	0	1	2	$\sqrt{8}$	$\sqrt{10}$	$\sqrt{13}$	$\sqrt{10}$	$\sqrt{8}$	2	1	0	0
0	1	$\sqrt{2}$	$\sqrt{5}$	$\sqrt{10}$	$\sqrt{17}$	$\sqrt{20}$	$\sqrt{17}$	$\sqrt{10}$	$\sqrt{5}$	$\sqrt{2}$	1	0
0	1	2	$\sqrt{8}$	$\sqrt{13}$	$\sqrt{20}$	$\sqrt{29}$	$\sqrt{20}$	$\sqrt{13}$	$\sqrt{8}$	2	1	0
0	1	$\sqrt{2}$	$\sqrt{5}$	$\sqrt{10}$	$\sqrt{17}$	$\sqrt{20}$	$\sqrt{17}$	$\sqrt{10}$	$\sqrt{5}$	$\sqrt{2}$	1	0
0	0	1	2	$\sqrt{8}$	$\sqrt{10}$	$\sqrt{13}$	$\sqrt{10}$	$\sqrt{8}$	2	1	0	0
0	0	1	$\sqrt{2}$	2	$\sqrt{5}$	$\sqrt{8}$	$\sqrt{5}$	2	$\sqrt{2}$	1	0	0
0	0	0	1	1	$\sqrt{2}$	2	$\sqrt{2}$	1	1	0	0	0
0	0	0	0	0	1	1	1	0	0	0	0	0
0	0	0	0	0	0	0	0	0	0	0	0	0

Figure 8.6 The Euclidean distance transform of Figure 8.4.

total distances is $188.973 - \sqrt{0.5}/2 \times 89 = 157.507$. Hence, the shape number is 1.005, which gives a more accurate measure.

One example of using the shape number to extract circle feature is described below. We first use the distance transformation algorithm to obtain the distance information. According to the calculated shape number, we conclude whether a region is a circle or not. If it is a circle, the maximum distance is the radius, and the location of that value is the center. If there is more than one point, we estimate the average of those points as the center.

Circle Detection Algorithm

1. A gray-level image is thresholded into a binary using a suitable thresholding algorithm.
2. Do a distance transformation algorithm.
3. Use the computed shape number to determine whether each region is a circle or not.
4. If the region is a circle, detect the maximum distance value in the region and its coordinates.
5. If there is more than one point with the maximum distance value, then select the average of the coordinates as the center and add 0.5 (approximated by a half of pixel unit length) to the maximum distance as the radius.

8.2.2 Significant Points Radius and Coordinates

Significant points are defined to be the local maxima of 8-neighborhood in the distance transform. If the Euclidean distance is used, then the circle drawn with the significant point as the center and the distance as the radius is the maximal inscribed circle in the local region.

Let d be the distance transform. The significant point extraction algorithm is

1. Execute $d \oplus_g k$. Here, k is a 3×3 structuring element with all zeroes.

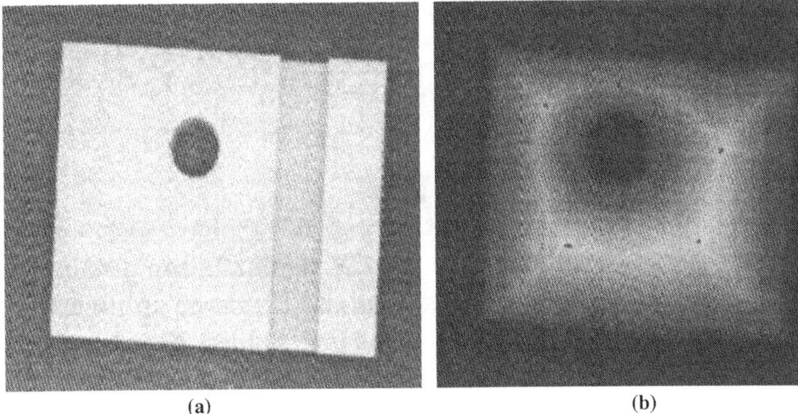

(a) (b)

Figure 8.7 (a) An industrial part with a hole and a slot. (b) The result of significant point extraction.

2. If the object pixel value in the output of step 1 is equal to that in d, then keep the value; otherwise, let this point be zero.
3. If the nonzero output points of step 2 are connected, then represent those points by a single average point.
4. List the values of radii and coordinates of all the nonzero points according to the decreasing order of the radius values.

Figure 8.7 shows a rectangular block with a hole and a slot and the extracted significant points.

Let s denote the shape number, and let r_i, x_i, y_i denote the radius and x, y coordinates in the decreasing order of radius, respectively. We can establish the shape database registered in the format $(s, r_1, x_1, y_1, r_2, x_2, y_2, \ldots, r_n, x_n, y_n)$. On comparing an unknown object with the registered database and computing the error value using equation (8.15), if the error is within the tolerance range, we classify the object to be identical as the registered object. Otherwise, we classify it as a different object. Let s', r_i', x_i', y_i' denote, respectively, shape number, radius, and x, y coordinates for the unknown object. The error is computed as

$$\text{Error} = \frac{1}{2}\left(\frac{|s'-s|}{s} + \frac{|r_1'-r_1| + \cdots + |r_n'-r_n|}{r_1 + \cdots + r_n}\right) \tag{8.15}$$

8.2.3 Localization by Hierarchical Morphological Band-Pass Filter

The problem considered in this section is how to quickly locate an object that was known and stored in the database. The morphological opening with an image will remove all of the pixels in regions that are too small to contain the probe. The opposite sequence (closing) will fill in holes and concavities that are smaller than the probe. Such filters can be used to suppress spatial features or discriminate against objects based on their size distribution. As an example, if a disk-shaped structuring element with radius h is used, then the morphological opening is equivalent to a low-pass filter.

The morphological opening residue is a high-pass filter. The difference of two morphological openings of an image with two nonequal radii is a band-pass filter. It can be expressed as follows:

$$\text{Low pass} = A \circ B^h$$
$$\text{High pass} = A - (A \circ B^h)$$
$$\text{Band-pass} = (A \circ B^{h_1}) - (A \circ B^{h_2}), \text{ where radii of } B,\ h_1 < h_2$$

If there are several unknown objects on the workbench and we want to pick up the one specified in the library database, the following algorithm is a simple and rapid approach:

The Localization Algorithm

In order to locate the position of objects, we change from the opening to the erosion operation. Let A denote a binary image that contains several objects and r_i denote the significant point radii in the library shape database.

1. Do $(A \ominus_b B^{r_1 - 1}) - (A \ominus_b B^{r_1 + 1})$.
2. If the output of step 1 is empty, it means that the library object is not on the workbench. If the output has at least two nonzero points spaced more than a certain interval, it means that at least two objects match the library object possibly. Hence, we need to do erosion with the second radius. Do $(A \ominus_b B^{r_2 - 1}) - (A \ominus_b B^{r_2 + 1})$. Apply this procedure recursively until only one point remains. This is the location of the object that matches the library object.

8.3 CORNER DETECTION

Corners are frequently employed in pattern recognition. They are the pixels whose slope changes abruptly; that is, the absolute curvature is high. Given an image, a typical approach to detect corners involves segmenting the object from background, representing the object boundary by chain codes, and searching for significant turns in the boundary. Local measurement such as nonmaxima suppression is used to obtain a binary corner map.

The algorithms of corner detection can be categorized into boundary-based and gray-level approaches. The boundary-based approach detects corners based on the boundary information of objects. Tsai et al. (1999) proposed a method using eigenvalues of the covariance matrix of data points on a curve segment. The gray-level approach directly works on gray-level images using the corner template matching or the gradients at edge pixels. Lee and Bien (1996) developed a real-time gray-level corner detector using fuzzy logic. Singh and Shneier (1990) proposed a fusion method by combining template-based and gradient-based techniques. Zheng et al. (1999) proposed a gradient-direction corner detector. Kitchen and Rosenfeld (1982) demonstrated that gray-level schemes perform better than boundary-based techniques.

Gao et al. (2007) presented two novel corner detection methods for gray-level images based on log-Gabor wavelet transform. Arnow and Bovik (2007) casted the problem of corner detection as a corner search process and developed principles of foveated visual search and automated fixation selection to accomplish the corner search. Ando (2000) presented image field categorization and edge/corner detection from gradient covariance. Basak and Mahata (2000) proposed a connectionist model for corner detection in binary and gray images. Rockett (2003) described a generic methodology for evaluating the labeling performance of feature detectors and proposed a method for generating a test set and then applied the methodology to the performance assessment of corner detectors. Paler et al. (1984) presented local ordered gray levels as an aid to corner detection. Harris and Stephens (1988) proposed a combined corner and edge detector.

Corners can be extracted using morphological operations. The basic idea is described below and further explored by regulated morphological operations. A suitable sized disk structuring element is selected. An image is opened with the disk structuring element and the result is subtracted from the original. The opening residue area varies with the angle of the corner. Then two structuring elements are selected: One is small enough to pass the residue area, and the other is too large. The opening residue is eroded with these two structuring elements and the two outputs are compared. This is equivalent to a shape band-pass filter. Different size structuring elements are used depending on the corner angles to be located. As the angle decreases, the opening residue area increases, and hence the structuring element area increases. The largest area structuring element is first tested. If nothing is obtained, then the area of the structuring element is decreased until something is obtained. Thus, the angle information can be obtained.

Corner Detection Algorithm

1. A binary image is dilated with a suitable size of the disk structuring element.
2. The output of step 1 is opened with a large size of the disk structuring element.
3. Subtract the output of step 2 from the output of step 1.
4. The output of step 3 is eroded with two disk structuring elements. The size of one structuring element is used in step 1 and the other size is a little bit larger.
5. Subtract the two eroded outputs of step 4. This is a band-pass filter. The remaining points are the corner points.

The size is determined by the largest area that can fit inside the opening residue corner (output of step 3). This is to estimate the location of the corner more accurately by correcting the bias introduced by the erosion in step 4. Figure 8.8 illustrates the corner detection algorithm, which was applied on a rectangular block with two holes.

In the corner detection based on morphological operations, Laganiere (1998) proposed a method using the *asymmetrical closing* by four different structuring elements including plus, lozenge, cross, and square. Lin et al. (1998) proposed a morphological corner detector to find convex and concave corner points using integer computation. However, their disadvantages are that (1) it is difficult to choose a suitable structuring element and (2) as the size of the structuring element increases, the computational cost increases. Zhang and Zhao (1995) developed a morphological corner detector, but it can only detect convex corners and the result is sensitive to the size of the structuring element.

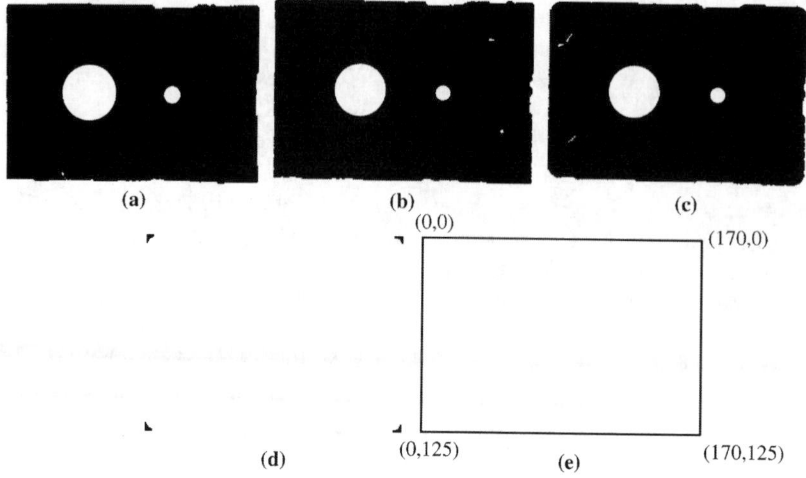

Figure 8.8 An example of a corner detector. Use morphology to detect the 90° corners. For clear display, the dark and white regions are reversed. (a) Original image, (b) output of step 1, (c) output of step 2, (d) output of step 3, (e) output of step 5.

8.3.1 Asymmetrical Closing for Corner Detection

Laganiere (1998) proposed corner detection using the *asymmetrical closing* that is defined as a dilation of an image by a structuring element followed by an erosion by another structuring element. The idea is to make dilation and erosion complementary and correspond to variant types of corners. Two structuring elements, cross " + " and lozenge "◇", are used. Let the asymmetrical closing of an image A by structuring elements $+$ and $\diamond$ be denoted by

$$A^{c}_{+\cdot\diamond} = (A \oplus +) \ominus \diamond \tag{8.16}$$

The corner strength is computed by

$$C_{+}(A) = |A - A^{c}_{+\cdot\diamond}| \tag{8.17}$$

For different types of corners, another corner strength (which is a 45°-rotated version) is computed as follows:

$$C_{\times}(A) = |A - A^{c}_{\times\cdot\square}| \tag{8.18}$$

By combining the four structuring elements in Figure 8.9, the corner detector is represented as

$$C_{+\cdot\times}(A) = |A^{c}_{+\cdot\diamond} - A^{c}_{\times\cdot\square}| \tag{8.19}$$

The disadvantages of the above corner detector are that it may miss obtuse-angle corners and remove sharp-angle corner pixels. Therefore, a modified corner

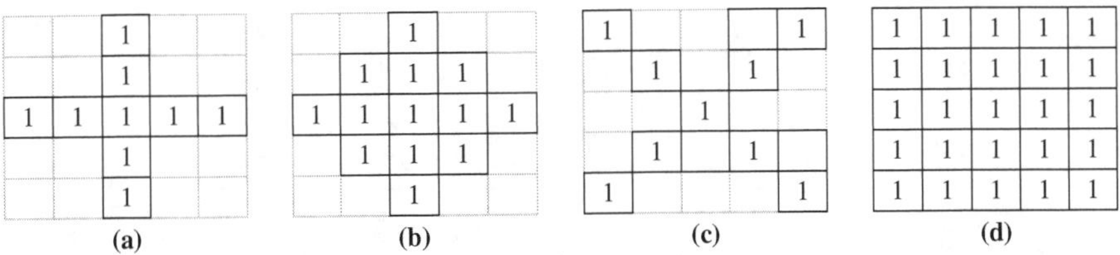

Figure 8.9 The four structuring elements: (a) + (b) ◇ (c) × (d) □.

detector uses $C_{+\cdot\times}(A) = |A - A^c_{+\cdot\diamond}| \cup |A - A^c_{\times\cdot\square}|$ (Shih et al., 2005). Two examples of comparing Laganiere's detector and the modified detector are shown in Figures 8.10 and 8.11. It is observed that the modified detector can locate the corner pixels more accurately.

8.3.2 Regulated Morphology for Corner Detection

Agam and Dinstein (1999) developed regulated morphological operators and showed how the fitting property can be adapted for analyzing the map and line-drawing images. Since regulated morphology inherits many properties of ordinary morphology, it is feasible to apply in image processing and optimize its strictness parameters using some criteria. Tian et al. (2002) extended the regulated morphological operators by adjusting the weights in the structuring element.

Note that the morphological hit-and-miss transform (Zhang and Zhao, 1995) can also be used to extract corners. For example, four different structuring elements can be designed to consist of four types of right-angle corners and perform the hit-and-miss transform with each structuring element. The four results are combined to obtain all the corners. However, the corners extracted are only the locations of all right-angle convex corners in four directions.

The *regulated dilation* of a set A by a structuring element set B with a strictness parameter s is defined by

$$A \oplus^s B = \{x \mid \#(A \cap (\hat{B})_x) \geq s, \quad s \in [1, \min(\#A, \#B)]\} \tag{8.20}$$

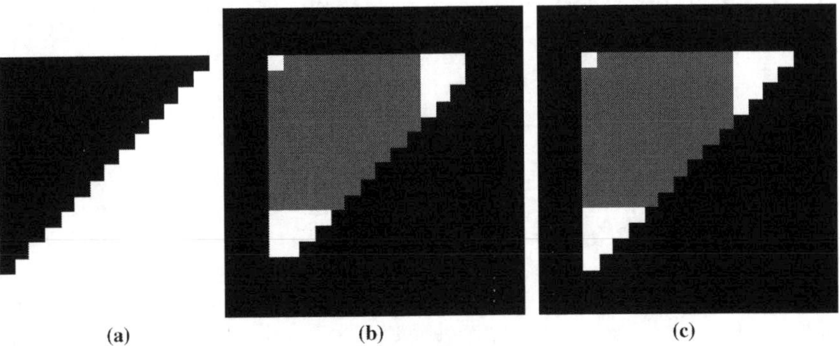

Figure 8.10 (a) A triangle image. (b) Laganiere's detector. (c) The modified detector.

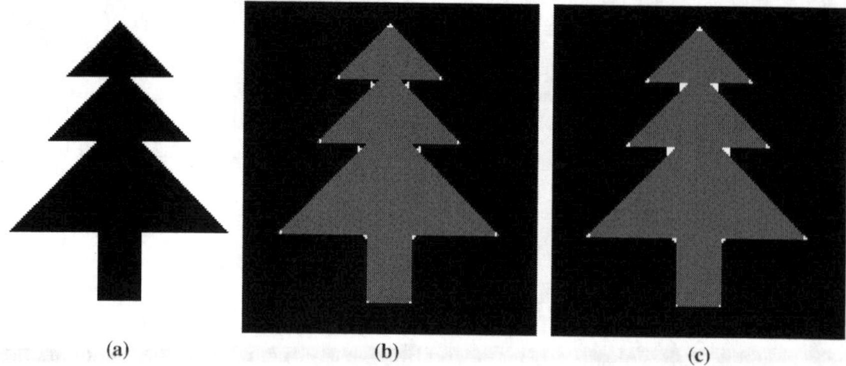

(a) (b) (c)

Figure 8.11 (a) A tree image. (b) Laganiere's detector. (c) The modified detector.

where the symbol "#" denotes the cardinality of a set. The *regulated erosion* of a set A by a structuring element set B with a strictness parameter s is defined by

$$A \ominus^s B = \{x | \#(A^c \cap (B)_x) < s, \quad s \in [1, \#B]\} \tag{8.21}$$

In order to maintain some properties the same as ordinary morphological operators, the *regulated closing* of a set A by a structuring element set B with a strictness parameter s is defined by

$$A \bullet^s B = ((A \underline{\oplus}^s B) \ominus B) \cup A \tag{8.22}$$

where $A \underline{\oplus}^s B = (A \oplus^s B) \cup A$ is defined as the *extensive regulated dilation*. The *regulated opening* of a set A by a structuring element set B with a strictness parameter s is defined by

$$A \circ^s B = ((A \underline{\ominus}^s B) \oplus B) \cap A \tag{8.23}$$

where $A \underline{\ominus}^s B = (A \ominus^s B) \cap A$ is defined as the *antiextensive regulated erosion*.

By adjusting the strictness parameter, the noise sensitivity problem and small intrusions or protrusions on the object boundary can be alleviated. For corner detection, regulated openings or closings are not used because the resulting image will stay the same as the original image after the strictness parameter exceeds a certain value. The modified regulated morphological corner detector is described as follows:

Step 1: $A_1 = (A \oplus^s B) \ominus^s B$. Corner strength: $C_1 = |A - A_1|$.

Step 2: $A_2 = (A \ominus^s B) \oplus^s B$. Corner strength: $C_2 = |A - A_2|$.

Step 3: Corner detector $= C_1 \cup C_2$.

The original image is performed by a regulated dilation using a 5×5 circular structuring element with a strictness s, and then followed by a regulated erosion by the same structuring element with the same strictness. The corner strength C_1 is computed by finding the absolute value of the difference between the original and resulting images. This step is to extract concave corners. By reversing the order such that the regulated erosion is applied first, and then followed by the regulated dilation, the convex corners can be extracted. Finally, both types of corners are combined.

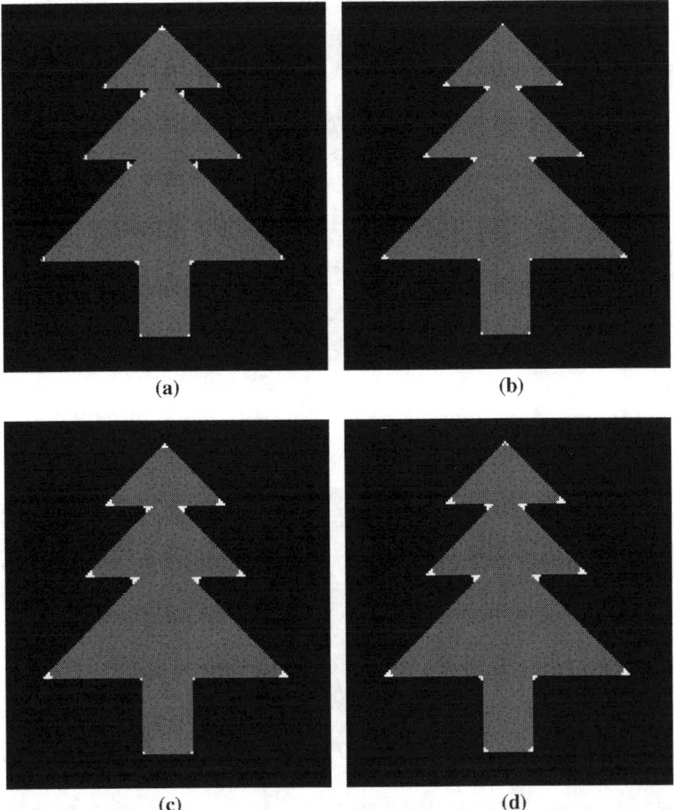

Figure 8.12 (a) Corner detection for a tree image using Laganiere's method. (b–d) Corner detection using the modified method at strictness = 2, 3, and 4, respectively.

8.3.3 Experimental Results

The comparisons of results by Laganiere's method and by the modified regulated morphology are shown in Figures 8.12–14. From Figures 8.12 and 8.13, the modified method can detect corners more accurately. From Figure 8.14, the modified method can detect corners more completely; for example, Laganiere's method can detect only

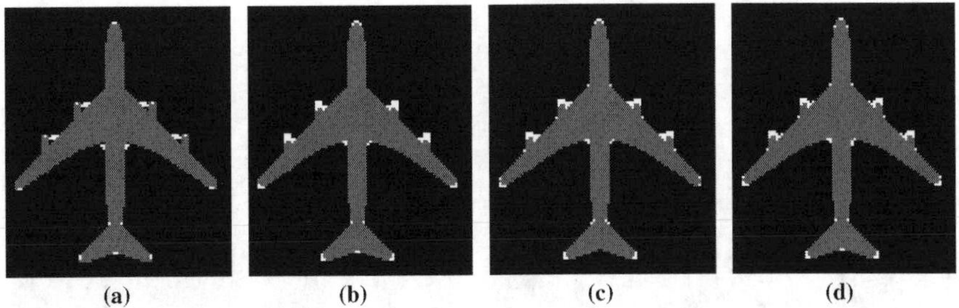

Figure 8.13 (a) Corner detection for an airplane image using Laganiere's method. (b–d) Corner detection using the modified method at strictness = 2, 3, and 4, respectively.

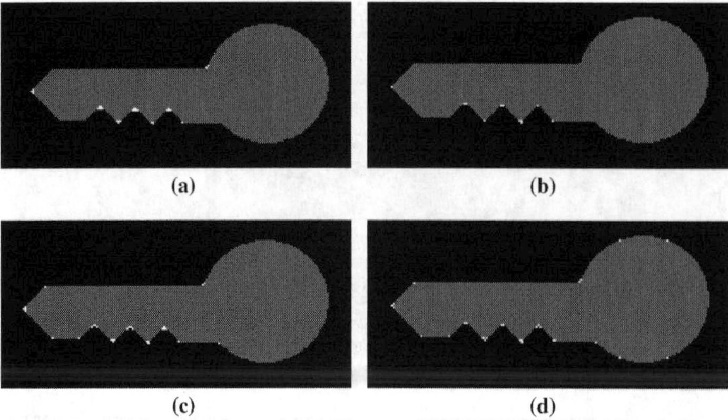

(a) (b)

(c) (d)

Figure 8.14 (a) Corner detection for a key image using Laganiere's method, (b), (c), and (d) Corner detection using the modified method at strictness = 2, 3, and 4, respectively.

Strictness = 2 Strictness = 3 Strictness = 4

Strictness = 2 Strictness = 3 Strictness = 4

Strictness = 2 Strictness = 3 Strictness = 4

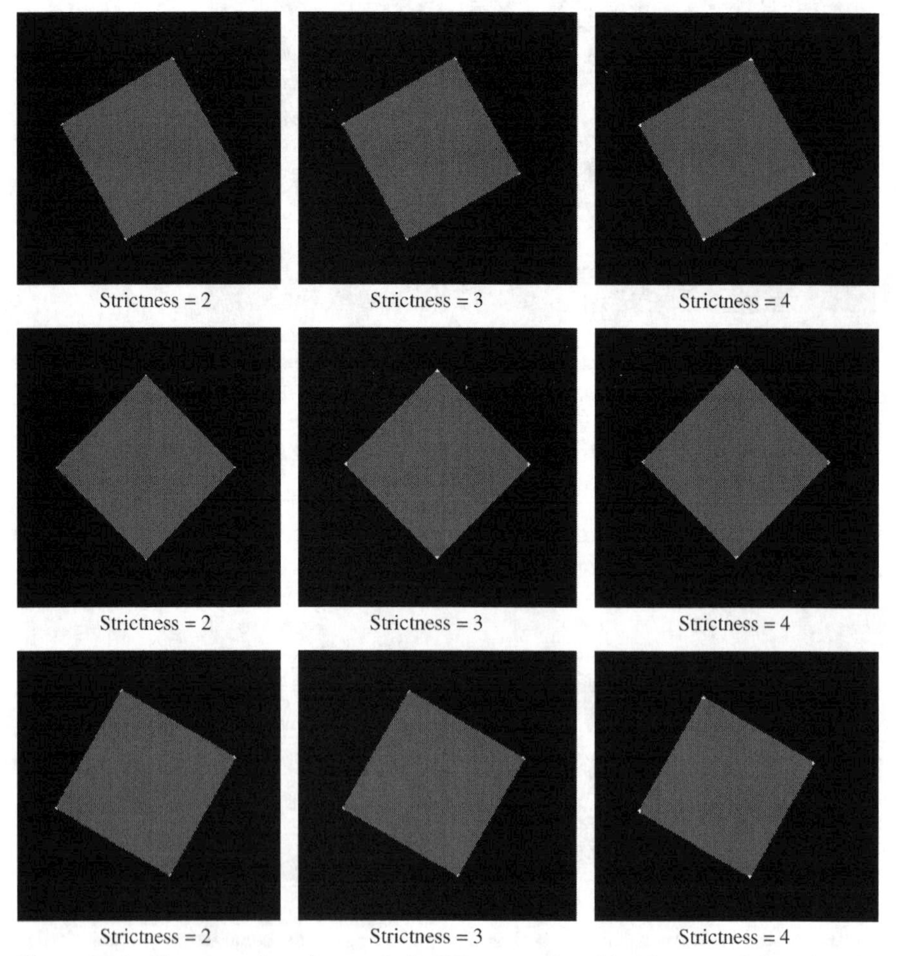

Figure 8.15 The corner detection results in different angles with strictness = 2, 3, and 4. The top-, middle-, and bottom-row images are rotated by 30°, 45°, and 60°, respectively.

TABLE 8.1 Number of Correctly Detected Corners, False Positives, and False Negatives

Test images	Figure 8.12	Figure 8.13	Figure 8.14
Total number of corners	15	24	12
Laganiere's Method			
Correctly detected corners	15	18	8
False positives	0	0	0
False negatives	0	6	4
Our Method at Strictness = 2			
Correctly detected corners	15	18	7
False positives	0	0	0
False negatives	0	6	5
Our Method at Strictness = 3			
Correctly detected corners	15	23	12
False positives	0	0	0
False negatives	0	1	0
Our Method at Strictness = 4			
Correctly detected corners	15	24	12
False positives	0	2	4
False negatives	0	0	0

8 corners, but the modified method can detect all 12 corners at strictness of 3. Implemented on the Platinum IV 2.0 Ghz PC, it takes 2.28 s to process Figure 8.14 by Laganiere's method; however, it spends 2.01 s by the proposed method.

A square is rotated by 30°, 45°, and 60° as illustrated in Figure 8.15. Experimental results show that the modified method can detect corners correctly. Table 8.1 summarizes the results of corner detection using Laganiere's and the modified methods with different values of strictness. From Table 8.1 and Figures 8.12–14, strictness of 3 is a better choice because it produces more accurate corners than strictness of 2 and generates less false positive corners than strictness of 4.

In the above experiments, a 5×5 circular structuring element is adopted. A 3×3 structuring element is applied on Figure 8.14, and the results are illustrated in Figure 8.16. It is observed that the 3×3 structuring element is worse than the 5×5 structuring element because some corners are missed.

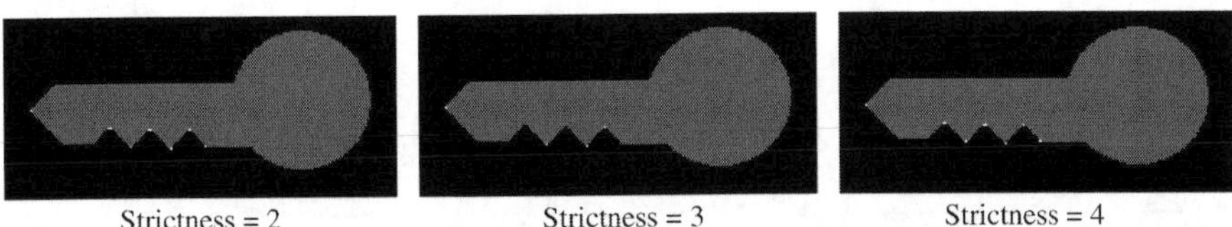

Figure 8.16 The detected corners using a 3×3 structuring element at strictness = 2, 3, and 4, respectively.

8.4 HOUGH TRANSFORM

Hough transform is a technique to identify specific shapes in an image. It converts all the points in a curve into a single location in another parametric space by coordinate transformation. This method intends to map global features into local features. This concept can also apply to detect a circle, an ellipse, or other geometric shapes.

Considering the detection straight lines, let all the lines passing through a point (x', y') be represented as $y' = mx' + c$, where m is the line slope and c is the intercept with y-axis. The line equation can be rearranged as $c = -x'm + y'$. A point in an image of (x, y) domain corresponds to a line in the parametric space of (m, c) domain. Therefore, the intersection of all lines in the (m, c) parametric space indicates the slope and intercept of those collinear points (i.e., the line) in the image. Since the intersections are determined approximately by partitioning the parameter space into accumulator cells and then using whichever cell collects the greatest number of intersections as a solution, they are not always exact. Besides, the slope of vertical lines goes to infinity. In practice, the (ρ, θ) parametric space is used instead of (m, c) by the equation $x \cos\theta + y \sin\theta = \rho$, where ρ is the normal distance to the line from the origin and θ is the angle of this normal. Figure 8.17 shows the transformation from the (x, y) domain to the (ρ, θ) parametric space.

Hough Transform to Identifying a Line

1. Let the parameters $(\rho_{\max}, \rho_{\min})$ and $(\theta_{\max}, \theta_{\min})$ represent the maximum and minimum distances from the origin and the maximum and minimum angles of the line, respectively.
2. Subdivide the parametric space into accumulator cells.
3. Initialize the accumulator cells to be all zeros.
4. For every point of interest (x, y), increment θ and solve for the corresponding ρ using the normal line equation, $x \cos\theta_j + y \sin\theta_j = \rho_i$, to find the line intersect at (ρ_i, θ_j). The cell determined by (ρ_i, θ_j) is associated with the $A(i,j)$ of the accumulator cells, which would be incremented by one when the equation is solved.

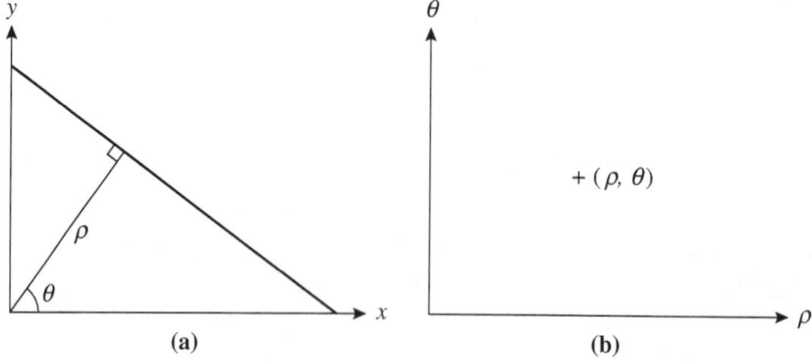

Figure 8.17 The transformation from the (x, y) domain to the (ρ, θ) parametric space.

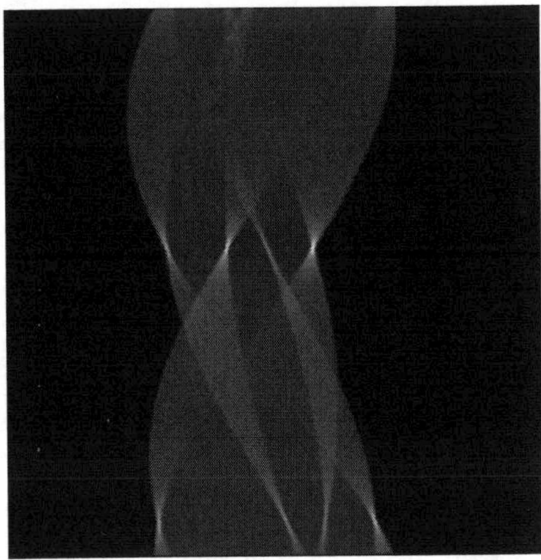

Figure 8.18 An example of Hough space with several lines.

To identify other geometric shapes, such as circles, arcs, squares, and triangles, using Hough transform, we extend the dimensions of the accumulator cells to a number of variables that need to be solved for the geometric shape equation. The Hough transform, when generalized, can be used to detect any geometric shape of the form $g(\mathbf{v}, \mathbf{c}) = 0$, where $\mathbf{v}$ and $\mathbf{c}$ are vectors of coordinates and coefficients. The procedure is similar to line detection except that all the curves through a single point will be represented by a surface in a parametric space that is higher dimensional. An example of Hough transform to detect an image of several lines is shown in Figure 8.18.

Example 8.1 Apply the Hough transform to the following image to detect lines:

0	0	0	0	0	0	0	1
0	0	0	0	0	0	1	0
0	0	0	0	0	1	0	0
0	0	0	0	1	0	0	0
0	0	0	1	0	0	0	0
0	0	1	0	0	0	0	0
0	1	0	0	0	0	0	0
1	0	0	0	0	0	0	0

Answer: Finding (m, c) for $c = y - mx$, we table the value of c as follows:

(x, y)	$m = -3$	$m = -1$	$m = 1$	$m = 3$
(0, 0)	0	0	0	0
(1, 1)	4	2	0	−2
(2, 2)	8	4	0	−4
(3, 3)	12	6	0	−6
(4, 4)	16	8	0	−8
(5, 5)	20	10	0	−10
(6, 6)	24	12	0	−12
(7, 7)	28	14	0	−14

The following graph shows the projection to (m, c) space, where A denotes the slope m and B denotes the intercept c.

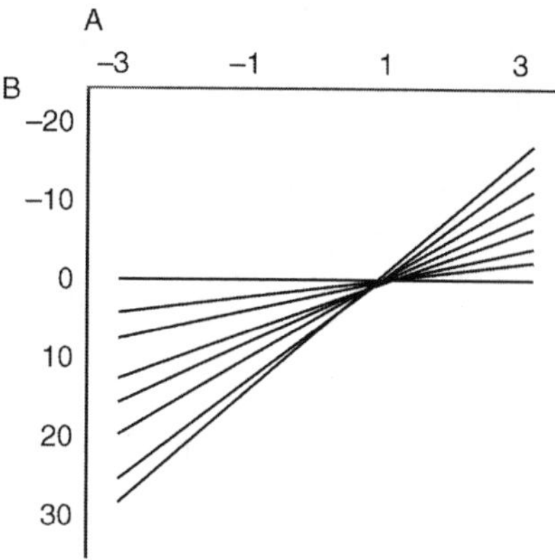

Since all lines cross $(m, c) = (1, 0)$ and the accumulator has the peak value of 8, it indicates that there is a line with slope 1 and intercept 0 passing through the eight points in the image.

Koshimizu et al. (1989) presented global feature extraction using efficient Hough transform. Bonci et al. (2005) explained how to associate a rigorous probability value with the main straight line features extracted from a digital image and proposed a Bayesian approach to the Hough Transform. Aggarwal and Karl (2006) posed the problem of detecting straight lines in grayscale images as an inverse problem and proposed a formulation based on the use of the inverse Radon operator, which relates the parameters determining the location and orientation of the lines in the image to the noisy input image. Satzoda et al. (2008) proposed a method called additive Hough transform (AHT) based on parallel processing to accelerate Hough transform computation.

8.5 PRINCIPAL COMPONENT ANALYSIS

PCA is known as the best data representation in the least-square sense for classical recognition (Stork et al., 2000). It is commonly used to reduce the dimensionality of images and retain most of information. The central idea behind PCA is to find an orthonormal set of axes pointing at the direction of maximum covariance in the data. It is often used in representing facial images. The idea is to find the orthonormal basis vectors, or the eigenvectors, of the covariance matrix of a set of images, with each image treated as a single point in a high-dimensional space. It is assumed that the facial images form a connected subregion in the image space. The eigenvectors map the most significant variations between faces and are preferred over other correlation techniques that assume that every pixel in an image is of equal importance (Kosugi, 1995).

Since each image contributes to each of the eigenvectors that resemble ghostlike faces when displayed, it is referred to as *holon* (Cottrell and Fleming, 1990) or *eigenfaces* (Turk and Pentland, 1991), and the new coordinate system is referred to as the *face space*. Individual images can be projected onto the face space and represented exactly as weighted combinations of the eigenface components. The resulting vector of weights that describes each face can be used in data compression and face classification. Data compression relies on the fact that the eigenfaces are ordered, with each one accounting for a different amount of variation among the faces. Compression is achieved by reconstructing images using only those few eigenfaces that account for the most variability (Sirovich and Kirby, 1987). It results in dramatic reduction of dimensionality. Classification is performed by projecting a new image onto the face space and comparing the resulting weight vector with the weight vectors of a given class.

According to statistics, a data set can be characterized by its mean value and its standard deviation, which is a measure of how spread out the data are. The principal components of a set of images can be derived directly as follows. Let $\mathbf{I}(x, y)$ be a two-dimensional array of intensity values of size $N \times N$. Let the set of facial images be $\Gamma_1, \Gamma_2, \Gamma_3, \ldots, \Gamma_M$. The average face of the set is

$$\Psi = \frac{1}{M}\sum_{k=1}^{M}\Gamma_k \tag{8.24}$$

The distance from a given facial image Γ_k to the mean image of the set is

$$\Phi_k = \Gamma_k - \Psi \tag{8.25}$$

PCA seeks the set of M orthonormal vectors, $\mathbf{u}_k$, and their associated eigenvalues, λ_k, which best describes the distribution of the image points. The vectors $\mathbf{u}_k$ and scalars λ_k are the eigenvectors and eigenvalues, respectively, of the covariance matrix

$$\mathbf{C} = \frac{1}{M}\sum_{k=1}^{M}\Phi_k\Phi_k^{\mathrm{T}} = \mathbf{A}\mathbf{A}^{\mathrm{T}} \tag{8.26}$$

where the matrix $\mathbf{A} = [\Phi_1, \Phi_2, \ldots, \Phi_M]$ (Turk and Pentland, 1991).The size of $\mathbf{C}$ is $N^2 \times N^2$ that for typical image sizes is an intractable task. However, since typically

$M < N^2$, that is, the number of images is less than the dimension, there will only be $N-1$ nonzero eigenvectors. Thus, the N^2 eigenvectors can be solved, in this case, by first solving for the eigenvectors of an $M \times M$ matrix, and then taking the appropriate linear combinations of the data points Φ.

PCA is closely associated with the singular value decomposition of a data matrix and can be decomposed as

$$\Phi = \mathbf{USV}^{\mathrm{T}} \tag{8.27}$$

where $\mathbf{S}$ is a diagonal matrix whose diagonal elements are the singular values, or eigenvalues, of Φ, and $\mathbf{U}$ and $\mathbf{V}$ are unary matrices. The columns of $\mathbf{U}$ are the eigenvectors of $\Phi\Phi^{\mathrm{T}}$ and are referred to as *eigenfaces*. The columns of $\mathbf{V}$ are the eigenvectors $\Phi^{\mathrm{T}}\Phi$ and are not used in this analysis.

Faces can be classified by projecting a new face Γ onto the face space as follows:

$$\omega_k = \mathbf{u}_k^{\mathrm{T}}(\Gamma_k - \Psi) \tag{8.28}$$

for $k = 1, \ldots, M'$ eigenvectors, with $M' \ll M$, if reduced dimensionality is desired. The weights form a vector $\Omega_k^{\mathrm{T}} = [\omega_1, \omega_2, \ldots, \omega_{M'}]$, which contains the projections onto each eigenvector. Classification is performed by calculating the distance of Ω_k from Ω, where Ω represents the average weight vector defining some class.

The applications of PCA include image compression, data mining, face detection (Sung and Poggio, 1998), face recognition, socioeconomic analysis, protein dynamics, and especially the areas requiring the classification of the high-dimensional data (Jolliffe, 1986).

Examples of eigenfaces are shown in Figure 8.19. Individual images can be projected onto the face space and represented exactly as weighted combinations of the eigenface components (see Fig. 8.20).

There are many PCA algorithms developed in the literature. Chatterjee et al. (2000) derived and discussed adaptive algorithms for PCA that are shown to converge faster than the traditional PCA algorithms. Weingessel and Hornik (2000) used a general framework to describe those PCA algorithms that are based on Hebbian learning. Martinez and Kak (2001) concluded that when the training data set is small,

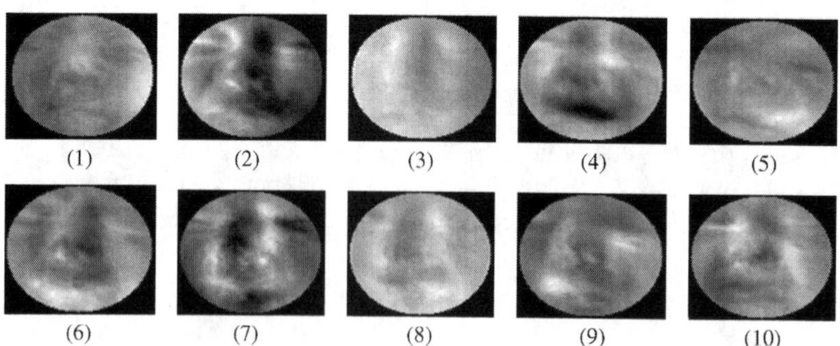

Figure 8.19 The first 10 eigenfaces of 204 neonate images. The eigenfaces are ordered by magnitude of the corresponding eigenvalue.

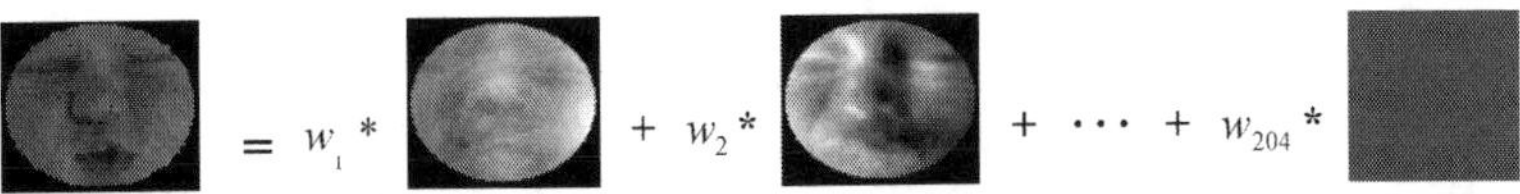

Figure 8.20 Illustration of the linear combination of eigenfaces. The face to the left can be represented as a weighted combination of eigenfaces.

PCA can outperform LDA and, also, that PCA is less sensitive to different training data sets by showing actual results on a face database. Kim et al. (2001) illustrated the potential of kernel PCA for texture classification. Kim et al. (2002) adopted the kernel PCA as a mechanism for extracting facial features. Liu (2004) presented a Gabor-based kernel PCA method by integrating the Gabor wavelet representation of face images and the kernel PCA method for face recognition. Xie and Lam (2006) proposed a Gabor-based kernel PCA with doubly nonlinear mapping for human face recognition.

Nishino et al. (2005) presented a framework for applying PCA to visual data that takes advantage of the spatiotemporal correlation and localized frequency variations that are typically found in such data. Zhao et al. (2006) proposed a novel incremental PCA and its application for face recognition. Vaswani and Chellappa (2006) presented a classification algorithm, called principal component null space analysis (PCNSA), which is designed for classification problems like object recognition where different classes have unequal and nonwhite noise covariance matrices. Pang et al. (2008) incorporated both 2DPCA-based image decomposition and fast numerical calculations based on Haar-like bases to develop binary 2DPCA (B-2DPCA) for fast training and testing procedures.

8.6 LINEAR DISCRIMINATE ANALYSIS

LDA is used to perform the feature extraction task and to classify samples of unknown classes based on training samples with known classes (Martinez and Kak, 2001). It provides a linear transformation of k-dimensional samples (or feature vectors in our experiments) into an m-dimensional space ($m < k$), so that samples belonging to the same class are close together, but samples from different classes are far apart from each other. This method maximizes the ratio of between-class variance to within-class variance in any data set; thereby, the theoretical maximum separation in the linear sense will be guaranteed. Since LDA seeks directions that are efficient for discrimination (Zhao et al., 1998), it is the optimal classifier for distinguishing classes that are Gaussian distribution and have equal covariance matrices.

After the PCA representation, each image is represented by k values as

$$y_i = (y_{i1}, y_{i2}, \ldots, y_{ik})^{\mathrm{T}} \tag{8.29}$$

LDA seeks a transformation matrix that in some sense maximizes the ratio of the between-scatter matrix to the within-scatter matrix. The within-scatter matrix is defined as

$$S_{\mathrm{w}} = \sum_{j=1}^{K} \sum_{i=1}^{N_j} (y_i^j - \mu_j)(y_i^j - \mu_j)^{\mathrm{T}} \tag{8.30}$$

where y_i^j is the ith sample of class j, μ_j is the mean of class j, K is the number of classes, and N_j is the number of samples in class j. The between-scatter matrix is defined as

$$S_{\mathrm{b}} = \frac{1}{K} \sum_{i=1}^{K} (\mu_i - \mu)(\mu_i - \mu)^{\mathrm{T}} \tag{8.31}$$

where μ is the mean of all classes.

Our goal is to maximize the between-class measure and to minimize the within-class measure (i.e., $\max\left(\frac{\det|S_{\mathrm{b}}|}{\det|S_{\mathrm{w}}|}\right)$). To achieve this, we can find a transformation matrix

$$W_{\mathrm{opt}} = \arg\max_{W} \left(\frac{W^{\mathrm{T}} S_{\mathrm{b}} W}{W^{\mathrm{T}} S_{\mathrm{w}} W}\right) = [w_1 w_2 \cdots w_m] \tag{8.32}$$

where $\{w_i\}$ is the set of generalized eigenvectors of S_{b} and S_{w}. We project y_i on W to find the LDA representation:

$$z_i = W^{-1}(y_i - \mu) \tag{8.33}$$

Once the transformation is found, the classification problem is simply a matter of finding the class whose transformed mean is closest to the transformed testing image.

PCA and LDA are two important feature extraction methods and have been widely applied in a variety of areas. A limitation of PCA and LDA is that when dealing with image data, the image matrices must be first transformed into vectors that are usually of very high dimensionality. This causes expensive computational cost and sometimes the singularity problem. Li and Yuan (2005) introduced a new approach for image feature extraction and representation called 2D-LDA. The difference between LDA and 2D-LDA is that in LDA we use the image vector to compute the between-class and within-class scatter matrices, but in 2D-LDA we use the original image matrix to compute the two matrices. They claimed that the 2D-LDA can achieve better results than other feature extraction methods, such as PCA, LDA, and 2D-PCA (Yang et al., 2004). The idea of 2D-LDA is described below.

Suppose that we have M training samples belonging to L classes $(L_1, L_2, \ldots, L_L)$. The training samples in each class are denoted as N_i $(i = 1, 2, \ldots, L)$. The size of each training image $\mathbf{A}_j$ $(j = 1, 2, \ldots, M)$ is $m \times n$. Our purpose is to find a good projection vector, $\mathbf{x}$, such that when $\mathbf{A}_j$ is projected onto $\mathbf{x}$, we can obtain the projected feature vector, $\mathbf{y}_j$, of the image $\mathbf{A}_j$.

$$\mathbf{y}_j = \mathbf{A}_j \mathbf{x} \quad j = 1, 2, \ldots, M \tag{8.34}$$

Similar to LDA, we can find the between-class scatter matrix, $\mathbf{TS}_{\mathrm{B}}$, and the within-class scatter matrix, $\mathbf{TS}_{\mathrm{W}}$, of the projected feature vectors by using training images. The criterion is to project the images onto a subspace that maximizes the between-class scatter and minimizes the within-class scatter of the projected data.

Since the total scatter of the projected samples can be represented by the trace of the covariance matrix of the projected feature vectors, the Fisher linear projection criterion can be described as

$$\mathbf{J}(\mathbf{x}) = \frac{\mathrm{tr}(\mathbf{TS_B})}{\mathrm{tr}(\mathbf{TS_W})} = \frac{\mathbf{x}^{\mathrm{T}}\mathbf{S_B}\mathbf{x}}{\mathbf{x}^{\mathrm{T}}\mathbf{S_W}\mathbf{x}} \tag{8.35}$$

where $\mathbf{S}_{\mathrm{B}} = \sum_{i=1}^{L} N_i(\bar{\mathbf{A}}_i - \bar{\mathbf{A}})^{\mathrm{T}}(\bar{\mathbf{A}}_i - \bar{\mathbf{A}})$ and $\mathbf{S}_{\mathrm{W}} = \sum_{i=1}^{L} \sum_{\mathbf{A}_k \in L_i} (\mathbf{A}_k - \bar{\mathbf{A}}_i)^{\mathrm{T}}(\mathbf{A}_k - \bar{\mathbf{A}}_i)$.

The optimal projection, $\mathbf{x}_{\mathrm{opt}}$, can be decided when the criterion is maximized. That is, $\mathbf{x}_{\mathrm{opt}} = \arg\ \max_x \mathbf{J}(\mathbf{x})$. We can find out the solution by solving the generalized eigenvalue problem. Therefore, if we choose d optimal projection axes $\mathbf{x}_1, \mathbf{x}_2, \ldots, \mathbf{x}_d$ corresponding to the first d largest eigenvalues, then we can extract feature for a given image $\mathbf{I}$. The following equation can be used for feature extraction:

$$\mathbf{y} = \mathbf{I}[\mathbf{x}_1, \mathbf{x}_2, \ldots, \mathbf{x}_d] \tag{8.36}$$

Since $\mathbf{I}$ is an $m \times n$ matrix and $[\mathbf{x}_1, \mathbf{x}_2, \ldots, \mathbf{x}_d]$ is an $n \times d$ matrix, we can form an $m \times d$ matrix, $\mathbf{y}$, to represent the original image, $\mathbf{I}$. We use the 2D-LDA representation of the original image as the input feeding into the classifier in the next stage.

Ye and Li (2005) proposed a two-stage LDA method that aims to overcome the singularity problems of classical LDA, while achieving efficiency and scalability simultaneously. Wang et al. (2006) showed that the matrix-based 2D algorithms are equivalent to special cases of image block-based feature extraction; that is, partition each image into several blocks and perform standard PCA or LDA on the aggregate of all image blocks.

8.7 FEATURE REDUCTION IN INPUT AND FEATURE SPACES

Feature extraction and reduction are two main issues in feature selection, which is important for many supervised learning problems. The goal of image feature extraction is to achieve a better classification rate by extracting new features to represent objects from raw pixel data. The goal of feature reduction is to select a subset of features while preserving or improving the classification rate of a classifier. Feature reduction can speed up the classification process by keeping the most important class-relevant features.

8.7.1 Feature Reduction in the Input Space

Principal component analysis is widely used in image representation for dimensionality reduction. To obtain m principal components, we need to multiply a transformation matrix of dimensions $m \times N$ with the input pattern $\mathbf{x}$ of dimensions $N \times 1$. In total, there are $m \times N$ multiplications and $m \times N$ summations. This is computation costly. In this section, we propose a method of selecting a subset of features to save time and maintain competitive performance as the PCA representation.

A straightforward method for features reduction is using Fisher's criterion to choose a subset of features that have large between-class and small within-class variances. For example, in face and nonface classification, the within-class variance in dimension i is calculated as

$$\sigma_i^2 = \frac{\sum_{j=1}^{l} (g_{j,i} - m_i)^2}{l-1}, \quad \text{for } i = 1, \ldots, N \tag{8.37}$$

where l is the number of samples, $g_{j,i}$ is the ith-dimensional gray value of sample j, and m_i is the mean value of ith dimension. For between-class measurement, we use the Fisher's score as

$$S_i = \left| \frac{m_{i,\text{face}} - m_{i,\text{nonface}}}{\sigma_{i,\text{face}}^2 + \sigma_{i,\text{nonface}}^2} \right| \tag{8.38}$$

By selecting the features with highest Fisher's scores, we can retain the most discriminative features between face and nonface classes.

Alternatively, we propose a feature reduction method for the second-degree support vector machine based on a decision function. For the second-degree polynomial SVM with kernel $K(\mathbf{x}, \mathbf{y}) = (1 + \mathbf{x} \cdot \mathbf{y})^2$, the decision function for a given pattern $\mathbf{x}$ is defined as

$$\begin{aligned} f(\mathbf{x}) &= \sum_{i=1}^{s} \alpha_i y_i (1 + \mathbf{x}_i \cdot \mathbf{x})^2 + b \\ &= \sum_{i=1}^{s} \alpha_i y_i (1 + x_{i,1}x_1 + x_{i,2}x_2 + \cdots + x_{i,k}x_k + \cdots + x_{i,N}x_N)^2 + b \end{aligned} \tag{8.39}$$

where s is the number of support vectors, $\mathbf{x}_i$ is the ith support vector, and $x_{i,k}$ and x_k are, respectively, the kth dimension for the support vector $\mathbf{x}_i$ and the pattern $\mathbf{x}$. The component related to the kth dimension (where $k = 1, 2, \ldots, N$) is

$$\begin{aligned} f(\mathbf{x}, k) &= \sum_{i=1}^{s} \alpha_i y_i [2x_k x_{i,k} (1 + x_{i,1}x_1 + \cdots + x_{i,k-1}x_{k-1} + x_{i,k+1}x_{k+1} \\ &\quad + \cdots + x_{i,N}x_N) + x_k^2 x_{i,k}^2] \\ &= \sum_{i=1}^{s} \alpha_i y_i [2x_k x_{i,k} (1 + x_{i,1}x_1 + \cdots + x_{i,N}x_N) - x_k^2 x_{i,k}^2] \end{aligned} \tag{8.40}$$

In order to select a subset of m features from the original N features, we use the largest m decision values. For the kth dimension, we need to estimate its contribution as

$$F(k) = \int_V f(\mathbf{x}, k) dP(\mathbf{x}) \tag{8.41}$$

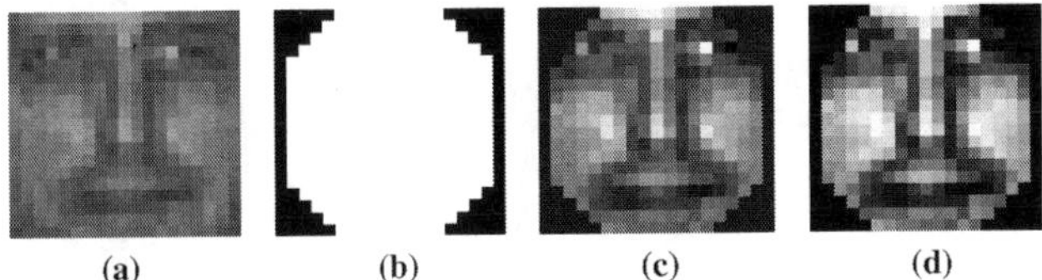

Figure 8.21 (a) Original face image, (b) the mask, (c) normalized image, and (d) histogram equalized image.

In fact, we cannot obtain $F(k)$ because the probability distribution function $P(\mathbf{x})$ is unknown. We approximate $F(k)$ with a sum over the support vectors

$$F(k) = \sum_{i=1}^{s}\sum_{j=1}^{s} |\alpha_j y_j [2x_{i,k}x_{j,k}(1 + x_{j,1}x_{i,1} + \cdots + x_{j,N}x_{i,N}) - x_{i,k}^2 x_{j,k}^2]| \qquad (8.42)$$

We adopt a face image database from the Center for Biological and Computational Learning at Massachusetts Institute of Technology (MIT), which contains 2429 face training samples, 472 face testing samples, and 23,573 nonface testing samples. We collected 15,228 nonface training samples randomly from the images that do not contain faces. The size of all these samples is 19×19. A second-degree polynomial SVM with kernel $K(\mathbf{x}, \mathbf{y}) = (1 + \mathbf{x} \cdot \mathbf{y})^2$ is used in our experiments.

In order to remove background pixels, a mask is applied to extract only the face. Prior to classification, we perform image normalization and histogram equalization. In normalization, we use zero mean and one variance. In Figure 8.21, (a) shows a face image, (b) shows the mask, and (c) and (d) are the images after normalization and histogram equalization, respectively. To calculate PCA values, we test the two methods: one is to combine face and nonface training samples, and the other is to use face training samples only.

Figure 8.22 shows the *receiver operating characteristic* (ROC) curves with using different numbers of principal components. The preprocessing is normalization. The ROC curve is defined as shifting the SVM hyperplane by changing the threshold value b. We perform face classification on the testing set and calculate the false positive and detection rates. The horizontal axis shows the false positive rate over

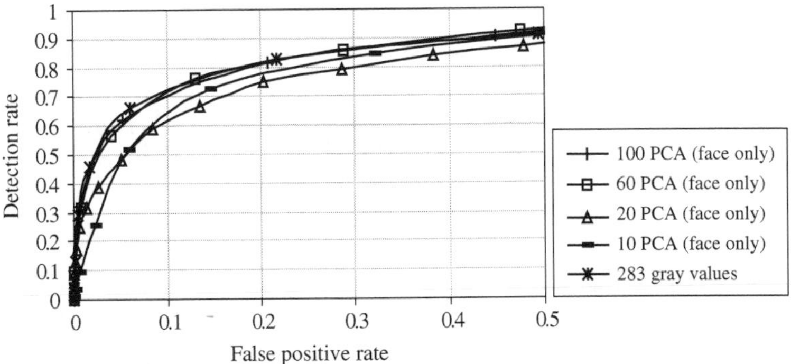

Figure 8.22 ROC curves for different numbers of PCA values. The preprocessing method is normalization and only face training samples are used to calculate PCA.

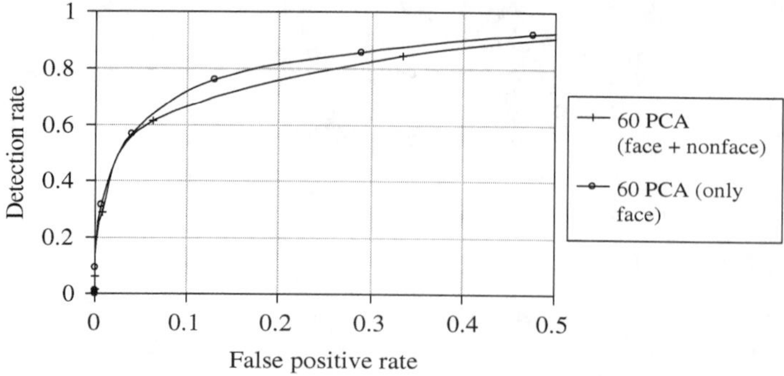

Figure 8.23 ROC curves for two methods to get PCA features. One method uses both face and nonface training samples, and the other uses only face training samples.

23,573 nonface testing samples. The vertical axis shows the detection rate over 472 face testing samples. The PCA values are calculated using only the face training samples. We observe that using 60 or 100 PCA values can achieve almost the same performance as using all 283 gray values. Figure 8.23 shows the performance of using only positive training samples that is better than using the combination of positive and negative training samples. Figure 8.24 shows the performance of using normalization or histogram equalization, and both obtain better results than that without preprocessing. In the following experiments, we will use normalization and positive training samples to calculate PCA values.

By using the normalized 2429 face and 15,228 nonface training samples and taking all 283 gray values as input to train the second-degree SVM, we obtain 252 and 514 support vectors for face and nonface classes, respectively. Using these support vectors and equation (8.42), we can obtain $F(k)$, where $k = 1, 2, \ldots, 283$. The performances of using 100 features selected by this ranking method, all 283 gray values, and 100 PCA features are shown in Figure 8.25. We also experiment on Fisher's scores using equations (8.37) and (8.38), and their performance is shown in Figure 8.25. We also compare the performance of using 100 features selected using Evgenious et al. (2003). Figure 8.25 shows that at low false positive rates, using 100 features selected by decision function ranking performs similarly as using all 283

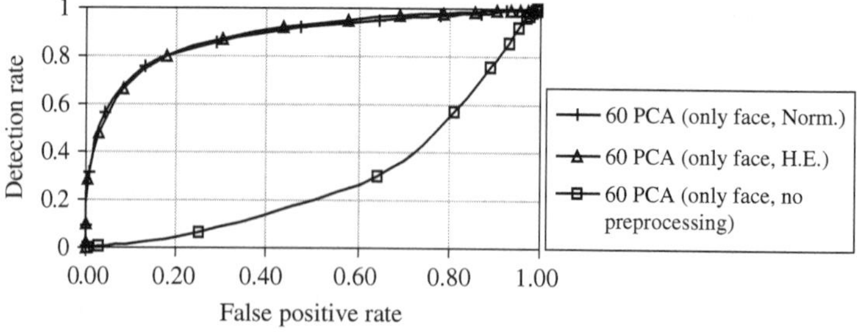

Figure 8.24 ROC curves for different preprocessing methods: normalization, histogram equalization, and without preprocessing. Only face training samples are used to calculate PCA values.

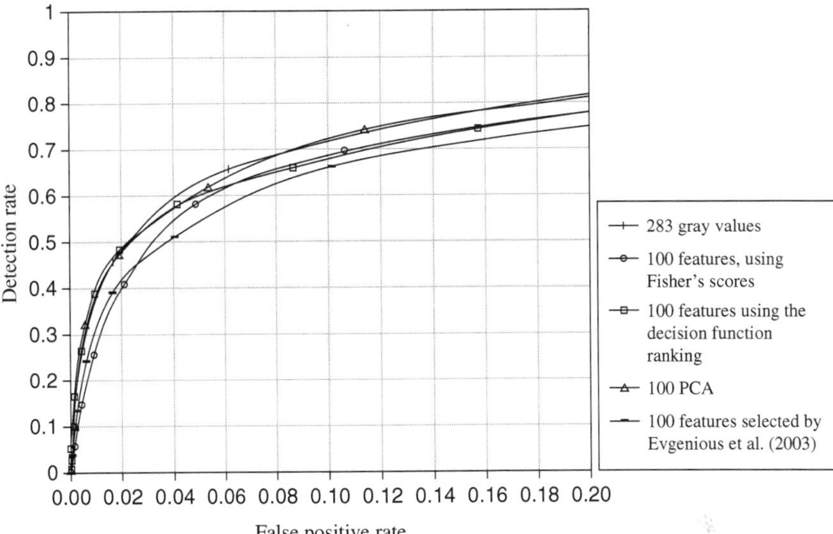

Figure 8.25 ROC curves for different features in input space. The preprocessing method is normalization and PCA values are calculated using only face training samples.

features and using 100 PCA values; however, it is significantly better than that of using 100 features by Fisher's scores and by Evgenious et al. (2003). This justifies the efficiency of our proposed feature reduction method.

8.7.2 Feature Reduction in the Feature Space

In feature space, the decision function $f(\mathbf{x})$ of SVM can be calculated as

$$f(\mathbf{x}) = \sum_{i=1}^{s} \alpha_i y_i (\Phi(\mathbf{x}) \cdot \Phi(\mathbf{x}_i)) + b = \mathbf{w} \cdot \Phi(\mathbf{x}) + b \tag{8.43}$$

where $\mathbf{w}$ is the support vector. For the second-degree polynomial SVM with input space of dimension N and kernel $K(\mathbf{x}, \mathbf{y}) = (1 + \mathbf{x} \cdot \mathbf{y})^2$, the feature space is given by

$$\Phi(\mathbf{x}) = \left(\sqrt{2}x_1, \ldots, \sqrt{2}x_N, x_1^2, \ldots, x_N^2, \sqrt{2}x_1x_2, \ldots, \sqrt{2}x_{N-1}x_N\right) \tag{8.44}$$

of dimension $P = N(N+3)/2$. Our goal is to select a subset of q features without significantly decreasing the performance.

Suppose that we train a second-degree SVM using face and nonface samples to obtain s support vectors. The support vector in the feature space can be represented as

$$\mathbf{w} = \sum_{i=1}^{s} \alpha_i y_i \, \Phi(\mathbf{x}_i) = (w_1, w_2, \ldots, w_P) \tag{8.45}$$

One simple method to select a subset of features is to rank according to $|w_k|$, for $k = 1, 2, \ldots, P$. However, we propose a better ranking method on

$\left|w_k \int_V |x_k^*| dp(x_k^*)\right|$, where x_k^* denotes the kth dimension of $\mathbf{x}$ in the feature space. Since the distribution function $dp(x_k^*)$ is unknown, we use

$$R(k) = \left| w_k \sum_{i=1}^{s} |x_{i,k}^*| \right| \tag{8.46}$$

where $x_{i,k}^*$ denotes the kth dimension of $\mathbf{x}_i$ in the feature space. The decision function of q features is calculated as

$$f(\mathbf{x}, q) = \mathbf{w}(q) \cdot \Phi(\mathbf{x}, q) + b \tag{8.47}$$

where $\mathbf{w}(q)$ is the selected q features of $\mathbf{w}$, and $\Phi(\mathbf{x}, q)$ is the corresponding q features of $\mathbf{x}$.

For a pattern $\mathbf{x}$, we calculate the difference of two decision values using all the features and using the subset of q features as

$$\Delta f_q(\mathbf{x}) = |f(\mathbf{x}) - f(\mathbf{x}, q)| \tag{8.48}$$

The quantity of the sum of difference values over all the support vectors is

$$\Delta F_q = \sum_{i=1}^{s} \Delta f_q(\mathbf{x}_i) \tag{8.49}$$

It can be used as a criterion to evaluate the efficiency of the methods.

In the experiments, we trained a second-degree polynomial SVM using 60 PCA values in the input space. The training samples are the same as before. The preprocessing method is normalization and the 60 PCA values are calculated from the transformation matrix obtained from the face training samples. We obtain 289 and 412 support vectors for face and nonface classes, respectively. The 1890 features in the feature space can be generated by equation (8.44). The support vector in feature space $\mathbf{w} = (w_1, w_2, \ldots, w_{1890})$ can be calculated by equation (8.45). The values of ΔF_q for different q can be calculated by equation (8.49). Figure 8.26 shows the variance of ΔF_q by varying the number of feature subsets of the two ranking methods

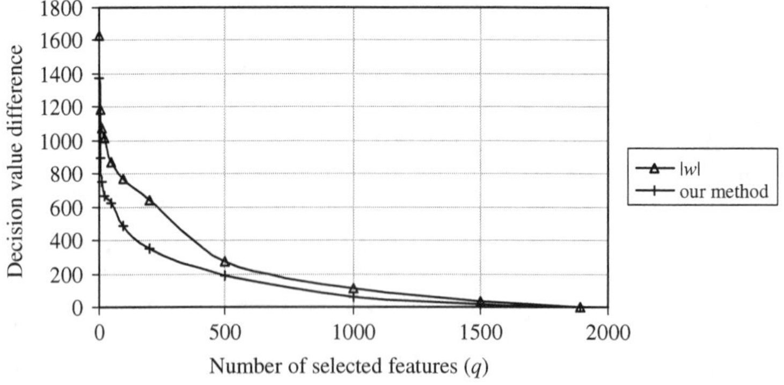

Figure 8.26 Decision value differences between using all the features and using a subset of q features in the feature space, with comparisons of two ranking methods.

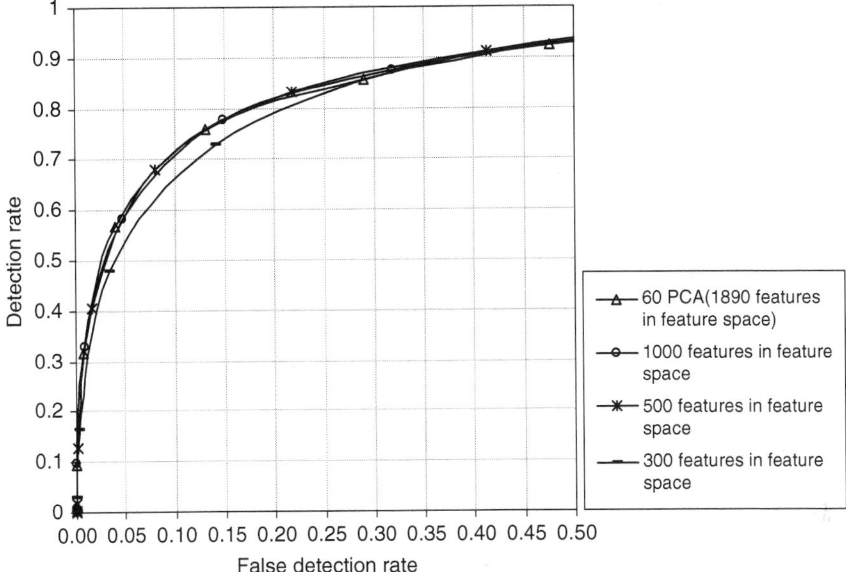

Figure 8.27 ROC curves for different numbers of features in the feature space. The preprocessing method is normalization and PCA values are calculated using only face training samples.

by $|w_k|$ and by our proposed method. We observe that our method leads to a faster decrease in difference; it means more efficient.

The ranking values of kth dimension in the feature space can be calculated by equation (8.46). Given a fixed number of q, for a pattern $\mathbf{x}$, the decision value of using the selected subset of q features can be calculated by equation (8.47). We take examples of $q = 300, 500, 1000$, and the results are shown in Figure 8.27. We observe that using 1000 features and 500 features selected by the ranking method can achieve almost the same performance as using all the 1890 features in the feature space. On the other hand, 300 features are insufficient to achieve good performance. Heisele et al. (2000) used the feature ranking according to $|w_k \sum_{i=1}^{s} y_i x_{i,k}^*|$. They used 1000 features and obtained similar performance as using all the features. When it is reduced to 500 features, the result is significantly worse.

8.7.3 Combination of Input and Feature Spaces

In this section, the performance of combining the feature reduction methods in input and feature space is evaluated. First, m features are chosen from the N input space and the second-degree polynomial SVM is trained. Next, q features are selected from $P = m(m+3)/2$ features in the feature space and the decision value is calculated.

There are two methods to compute the second-degree polynomial SVM, one by equation (8.39) in the input space and the other by equation (8.43) in the feature space. In equation (8.39), the number of multiplications required to calculate the decision

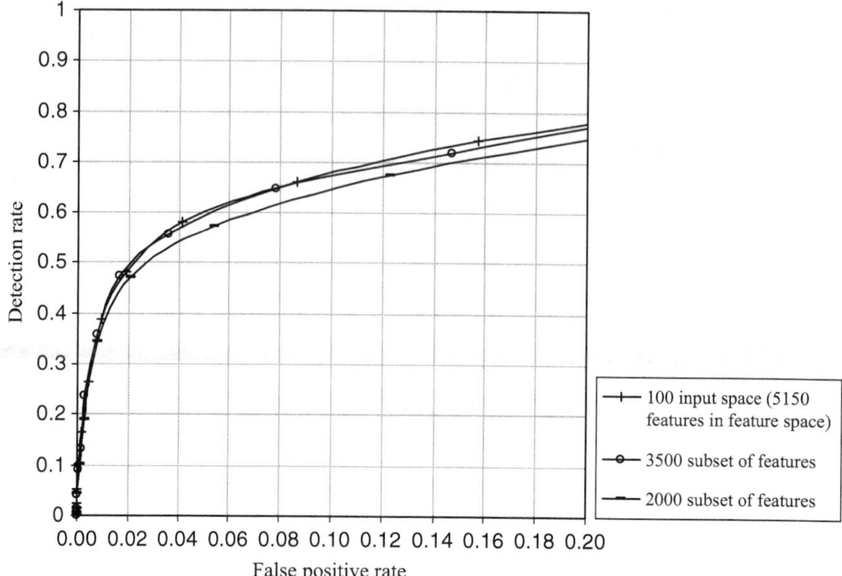

Figure 8.28 ROC curves of using different numbers of features in feature space. The 100 features in the input space are selected using the proposed ranking method.

function is $(N+1)s$. In equations (8.43) and (8.44), the number of multiplications required to calculate the decision function is $(N+3)N/2+(N+3)N/2=(N+3)N$. If $(N+1)s>(N+3)N$, it is more efficient to implement the second-degree polynomial SVM in the feature space. This is always true in our experiments because the number of support vectors is more than 700, that is, much larger than N. Note that $N=283, 60$, or 100 reflects all the gray value features, 60 PCA values, or 100 features, respectively.

We train SVM using the selected 100 features to obtain 222 and 532 support vectors, respectively, for face and nonface classes. The 5150 features in the feature space can be generated by equation (8.44). Figure 8.28 shows the results of using different subset of features in the feature space. Using 3500 features selected by our method can achieve similar performance as using all 5150 features. Figure 8.29 shows comparisons of using our combination method, 60 PCA values, and all 283 gray values. We observe that at low false positive rates, our method outperforms 60 PCA values.

We analyze computational complexity as follows. Using all 283 gray values in the input space, we need 80,938 multiplications to calculate the decision function for a pattern. Using 60 PCA values, we need 3780 multiplications to calculate the decision function and $m\times N=60\times 283=16,980$ multiplications to obtain 60 PCA values, which gives a total of 20,760 multiplications. If the feature selection method in feature space is applied and 1000 features are used, in total $1890+1000+16,980=19,870$ multiplications are required. If the proposed feature selection method in input space is applied and $m=100$ features are used, in total $(m+3)m=103\times 100=10,300$ multiplications are required. If the subset of $q=3500$ features in the feature space selected by the feature reduction method

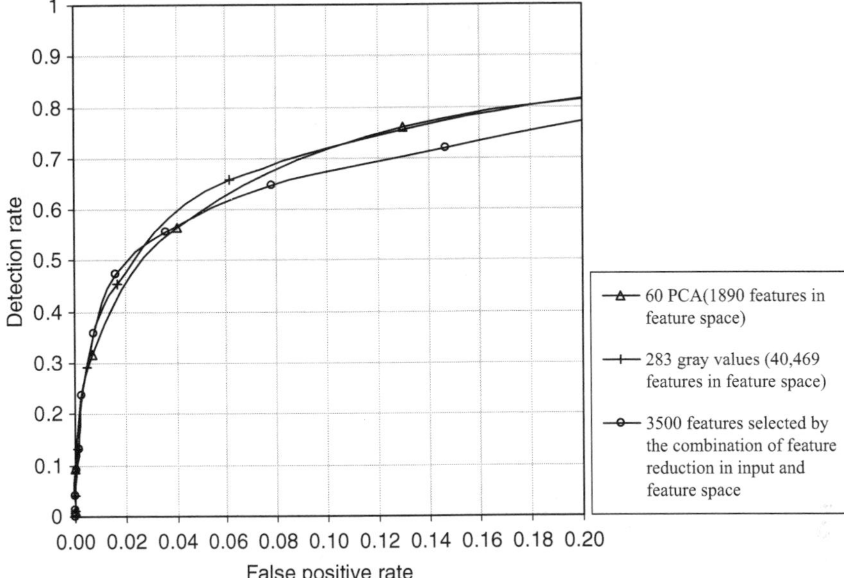

Figure 8.29 ROC curves of using the proposed feature reduction method, 60 PCA, and all the 283 gray values. The preprocessing method is normalization and PCA values are calculated using only face training samples.

are used, in total $(m+3)m/2+q = (100+3)100/2+3500 = 8650$ multiplications are needed. Table 8.2 summarizes the numbers of features used in input and feature space and the numbers of multiplication required to calculate the decision value for different methods.

Therefore, by using PCA for feature reduction in the input space and the feature reduction method in the feature space, a speedup factor of 4.08 can be achieved. By combining the feature reduction methods in the input and the feature space, a speedup factor of 9.36 can be achieved. In fact, once the features in the feature space are determined, we do not need to project the input space to the whole feature space, but to the selected q features. This can further speed up the computation, such that only $3500+3500 = 7000$ multiplications are required. We have shown that using the proposed feature reduction approaches, a fast nonlinear SVM classifier for face detection can be developed.

TABLE 8.2 Numbers of Features Used in Input and Feature Space and the Numbers of Multiplication Required to Calculate the Decision Value for Different Methods

Methods	Number of features in input space	Number of features in feature space	Number of multiplications
All gray values	283	40,469	80,938
PCA	60	1,890	20,760
Our method	100	3,500	8,650

REFERENCES

Agam, G. and Dinstein, I., "Regulated morphological operations," *Pattern Recognit.*, vol. 32, no. 6, pp. 947–971, June 1999.

Aggarwal, N. and Karl, W. C., "Line detection in images through regularized Hough transform," *IEEE Trans. Image Process.*, vol. 15, no. 3, pp. 582–591, Mar. 2006.

Ando, S., "Image field categorization and edge/corner detection from gradient covariance," *IEEE Trans. Pattern Anal. Mach. Intell.*, vol. 22, no. 2, pp. 179–190, Feb. 2000.

Arnow, T. L. and Bovik, A. C., "Foveated visual search for corners," *IEEE Trans. Image Process.*, vol. 16, no. 3, pp. 813–823, Mar. 2007.

Basak, J. and Mahata, D., "A connectionist model for corner detection in binary and gray images," *IEEE Trans. Neural Networks*, vol. 11, no. 5, pp. 1124–1132, Sep. 2000.

Belkasim, S. O., Shridhar, M., and Ahmadi, M., "Pattern recognition with moment invariants: a comparative study and new results," *Pattern Recognit.*, vol. 24, no. 12, pp. 1117–1138, Dec. 1991.

Bober, M., "MPEG-7 visual shape descriptors," *IEEE Trans. Circuits Syst. Video Technol.*, vol. 11, no. 6, pp. 716–719, June 2001.

Bonci, A., Leo, T., and Longhi, S., "A Bayesian approach to the Hough transform for line detection," *IEEE Trans. Syst. Man Cybern. A*, vol. 35, no. 6, pp. 945–955, Nov. 2005.

Chatterjee, C., Kang, Z., and Roychowdhury, V. P., "Algorithms for accelerated convergence of adaptive PCA," *IEEE Trans. Neural Networks*, vol. 11, no. 2, pp. 338–355, Mar. 2000.

Chen, C.-C., "Improved moment invariants for shape discrimination," *Pattern Recognit.*, vol. 26, no. 5, pp. 683–686, May 1993.

Chuang, G. C.-H. and Kuo, C.-C. J., "Wavelet descriptor of planar curves: theory and applications," *IEEE Trans. Image Process.*, vol. 5, no. 1, pp. 56–70, Jan. 1996.

Cottrell, G. W. and Fleming, M. K., "Face recognition using unsupervised feature extraction," *Proceedings of the International Conference on Neural Networks*, Paris, France, pp. 322–325, July 1990.

Danielsson, P. E., "A new shape factor," *Comput. Graph. Image Process.*, vol. 7, no. 2, pp. 292–299, Apr. 1978.

Dudani, S., Breeding, K., and McGhee, R., "Aircraft identification by moment invariants," *IEEE Trans. Comput.*, vol. 26, no. 1, pp. 39–45, Jan. 1977.

Evgenious, T., Pontil, M., Papageorgiou, C., and Poggio, T., "Image representations and feature selection for multimedia database search," *IEEE Trans. Knowl. Data Eng.*, vol. 15, no. 4, pp. 911–920, July 2003.

Flusser, J. and Suk, T., "Rotation moment invariants for recognition of symmetric objects," *IEEE Trans. Image Process.*, vol. 15, no. 12, pp. 3784–3790, Dec. 2006.

Flusser, J., Boldys, J., and Zitova, B., "Moment forms invariant to rotation and blur in arbitrary number of dimensions," *IEEE Trans. Pattern Anal. Mach. Intell.*, vol. 25, no. 2, pp. 234–246, Feb. 2003.

Gao, X., Sattar, F., and Venkateswarlu, R., "Multiscale corner detection of gray level images based on log-Gabor wavelet transform," *IEEE Trans. Circuits Syst. Video Technol.*, vol. 17, no. 7, pp. 868–875, July 2007.

Harris, C. and Stephens, M., "A combined corner and edge detector," *Proceedings of the Fourth Alvey Vision Conference*, Manchester, UK, pp. 147–151, 1988.

Heisele, B., Poggio, T., and Pontil, M., "Face detection in still gray images," A. I. Memo 1687, Center for biological and Computational Learning, MIT, Cambridge, MA, 2000.

Hu, M.-K., "Visual pattern recognition by moment invariants," *IRE Trans. Inform. Theory*, vol. 8, no. 2, pp. 179–187, Feb. 1962.

Hung, K.-C., "The generalized uniqueness wavelet descriptor for planar closed curves," *IEEE Trans. Image Process.*, vol. 9, no. 5, pp. 834–845, May 2000.

Jolliffe, I. T., *Principal Component Analysis*, Springer-Verlag, 1986.

Kim, K. I., Park, S. H., and Kim, H. J., "Kernel principal component analysis for texture classification," *IEEE Signal Process. Lett.*, vol. 8, no. 2, pp. 39–41, Feb. 2001.

Kim, K. I., Jung, K., and Kim, H. J., "Face recognition using kernel principal component analysis," *IEEE Signal Process. Lett.*, vol. 9, no. 2, pp. 40–42, Feb. 2002.

Kiranyaz, S., Ferreira, M., and Gabbouj, M., "A generic shape/texture descriptor over multiscale edge field: 2-D walking ant histogram," *IEEE Trans. Image Process.*, vol. 17, no. 3, pp. 377–391, Mar. 2008.

Kitchen, L. and Rosenfeld, A., "Gray-level corner detection," *Pattern Recognit. Lett.*, vol. 1, no. 2, pp. 95–102, Dec. 1982.

Koshimizu, H., Murakami, K., and Numada, M., "Global feature extraction using efficient Hough transform," *Proceedings of the International Workshop on Industrial Applications of Machine Intelligence and Vision*, Tokyo, Japan, pp. 298–303, 1989.

Kosugi, M., "Human-face search and location in a scene by multi-pyramid architecture for personal identification," *Syst. Comput. Jpn.*, vol. 26, no. 6, pp. 27–38, 1995.

Krawtchouk, M., "On interpolation by means of orthogonal polynomials," *Memoirs Agricultural Inst. Kyiv*, vol. 4, no. pp. 21–28, 1929.

Laganiere, R., "A morphological operator for corner detection," *Pattern Recognit.*, vol. 31, no. 11, pp. 1643–1652, Nov. 1998.

Lee, K.-J. and Bien Z., "A gray-level corner detector using fuzzy logic," *Pattern Recognit. Lett.*, vol. 17, no. 9, pp. 939–950, Aug. 1996.

Li, M. and Yuan, B., "2D-LDA: a statistical linear discriminant analysis for image matrix," *Pattern Recognit. Lett.*, vol. 26, no. 5, pp. 527–532, Apr. 2005.

Liao, S. X. and Pawlak, M., "On image analysis by moments," *IEEE Trans. Pattern Anal. Mach. Intell.*, vol. 18, no. 3, pp. 254–266, Mar. 1996.

Lin, R.-S., Chu, C.-H., and Hsueh, Y.-C., "A modified morphological corner detector," *Pattern Recognit. Lett.*, vol. 19, no. 3, pp. 279–286, Mar. 1998.

Liu, C., "Gabor-based kernel PCA with fractional power polynomial models for face recognition," *IEEE Trans. Pattern Anal. Mach. Intell.*, vol. 26, no. 5, pp. 572–581, May 2004.

Mamistvalov, A. G., "*N*-dimensional moment invariants and conceptual mathematical theory of recognition *n*-dimensional solids," *IEEE Trans. Pattern Anal. Mach. Intell.*, vol. 20, no. 8, pp. 819–831, Aug. 1998.

Manjunath, B. S., Ohm, J. R., Vasudevan, V. V., and Yamada, A., "Color and texture descriptors," *IEEE Trans. Circuits Syst. Video Technol.*, vol. 11, no. 6, pp. 716–719, Jun. 2001.

Martinez, A. M. and Kak, A. C., "PCA versus LDA," *IEEE Trans. Pattern Anal. Mach. Intell.*, vol. 23, no. 2, pp. 228–233, Feb. 2001.

Mikolajczyk, K. and Schmid, C., "A performance evaluation of local descriptors," *IEEE Trans. Pattern Anal. Mach. Intell.*, vol. 27, no. 10, pp. 1615–1630, Oct. 2005.

Nishino, K., Nayar, S. K., and Jebara, T., "Clustered blockwise PCA for representing visual data," *IEEE Trans. Pattern Anal. Mach. Intell.*, vol. 27, no. 10, pp. 1675–1679, Oct. 2005.

Paler, K., Föglein, J., Illingworth, J., and Kittler, J., "Local ordered grey levels as an aid to corner detection," *Pattern Recognit.*, vol. 17, no. 5, pp. 535–543, 1984.

Pang, Y., Tao, D., Yuan, Y., and Li, X., "Binary two-dimensional PCA," *IEEE Trans. Syst. Man Cybern. B*, vol. 38, no. 4, pp. 1176–1180, Aug. 2008.

Pawlak, M., "On the reconstruction aspect of moment descriptors," *IEEE Trans. Inf. Theory*, vol. 38, no. 6, pp. 1698–1708, Nov. 1992.

Persoon, E. and Fu, K. S., "Shape discrimination using Fourier descriptors," *IEEE Trans. Syst. Man Cybern.*, vol. 7, no. 2, pp. 170–179, Mar. 1977.

Rafiei, D. and Mendelzon, A. O., "Efficient retrieval of similar shapes," *Int. J Very Large Databases*, vol. 11, no. 1, pp. 17–27, Aug. 2002.

Rockett, P. I., "Performance assessment of feature detection algorithms: a methodology and case study on corner detectors," *IEEE Trans. Image Process.*, vol. 12, no. 12, pp. 1668–1676, Dec. 2003.

Rothe, I., Susse, H., and Voss, K., "The method of normalization to determine invariants," *IEEE Trans. Pattern Anal. Mach. Intell.*, vol. 18, no. 4, pp. 366–376, Apr. 1996.

Satzoda, R. K., Suchitra, S., and Srikanthan, T., "Parallelizing the Hough transform computation," *IEEE Trans. Signal Process. Lett.*, vol. 15, no. 1, pp. 297–300, Jan. 2008.

Schlemmer, M., Heringer, M., Morr, F., Hotz, I., Bertram, M.-H., Garth, C., Kollmann, W., Hamann, B., and Hagen, H., "Moment invariants for the analysis of 2D flow fields," *IEEE Trans. Visual. Comput. Graph.*, vol. 13, no. 6, pp. 1743–1750, Nov. 2007.

Shen, D. and Ip, H. S., "Discriminative wavelet shape descriptors for recognition of 2-D patterns," *Pattern Recognit.*, vol. 32, no. 2, pp. 151–165, Feb. 1998.

Shih, F. Y., Chuang, C., and Gaddipati, V., "A modified regulated morphological corner detector," *Pattern Recognit. Lett.*, vol. 26, no. 7, pp. 931–937, June 2005.

Singh, A. and Shneier, M., "Gray level corner detection a generalization and a robust real time implementation," *Comput. Vision Graph. Image Process.*, vol. 51, no. 1, pp. 54–69, July 1990.

Sirovich, L. and Kirby, M., "Low dimensional procedure for the characterization of human faces," *J. Opt. Soc. Am.*, vol. 4, no. 3, pp. 519–524, 1987.

Stork, D. G., Duda, R. O., and Hart, P. E., *Pattern Classification*, Wiley, 2000.

Sun, Y., Liu, W., and Wang, Y., "United moment invariants for shape discrimination," *Proceedings of the IEEE International Conference on Robotics, Intelligent Systems and Signal Processing*, Changsha, China, pp. 88–93, 2003.

Sung, K. K. and Poggio, T., "Example-based learning for view-based human face detection," *IEEE Trans. Pattern Anal. Mach. Intell.*, vol. 20, no. 1, pp. 39–51, Jan. 1998.

Teh, C.-H. and Chin, R. T., "On image analysis by the method of moments," *IEEE Trans. Pattern Anal. Mach. Intell.*, vol. 10, no. 4, pp. 496–513, Apr. 1998.

Tian, X.-H., Li, Q.-H., and Yan, S.-W., "Regulated morphological operations with weighted structuring element," *Proceedings of the First International Conference on Machine Learning and Cybernetics*, Beijing, China, vol. 2, pp. 768–771, Nov. 2002.

Tsai, D.-M., Hou, H.-T., and Su, H.-J., "Boundary-based corner detection using eigenvalues of covariance matrices," *Pattern Recognit. Lett.*, vol. 20, no. 1, pp. 31–40, Jan. 1999.

Turk, M. A. and Pentland, A. P., "Eigenfaces for recognition," *J. Cognit. Neurosci.*, vol. 3, no. 1, pp. 71–86, 1991.

Vaswani, N. and Chellappa, R., "Principal components null space analysis for image and video classification," *IEEE Trans. Image Process.*, vol. 15, no. 7, pp. 1816–1830, July 2006.

Wang, L., Wang, X., and Feng, J., "On image matrix based feature extraction algorithms," *IEEE Trans. Syst. Man Cybern. B*, vol. 36, no. 1, pp. 194–197, Feb. 2006.

Weingessel, A. and Hornik, K., "Local PCA algorithms," *IEEE Trans. Neural Networks*, vol. 11, no. 6, pp. 1242–1250, Nov. 2000.

Wong, W., Shih, F. Y., and Liu, J., "Shape-based image retrieval using Fourier descriptor and support vector machine," *Inform. Sci.*, vol. 177, no. 8, pp. 1878–1891, Apr. 2007.

Xie, X. and Lam, K.-M., "Gabor-based kernel PCA with doubly nonlinear mapping for face recognition with a single face image," *IEEE Trans. Image Process.*, vol. 15, no. 9, pp. 2481–2492, Sep. 2006.

Yang, J., Zhang, D., Frangi, A. F., and Yang, J.-Y., "Two-dimensional PCA: a new approach to appearance-based face representation and recognition," *IEEE Trans. Pattern Anal. Mach. Intell.*, vol. 26, no. 1, pp. 131–137, Jan. 2004.

Yap, P.-T., Paramesran, R., and Ong, S.-H., "Image analysis by Krawtchouk moments," *IEEE Trans. Image Process.*, vol. 12, no. 11, pp. 1367–1377, Nov. 2003.

Ye, J. and Li, Q., "Two-stage linear discriminant analysis via QR-decomposition," *IEEE Trans. Pattern Anal. Mach. Intell.*, vol. 27, no. 6, pp. 929–941, June 2005.

Zhang, D. S. and Lu, G., "A comparative study of Fourier descriptors for shape representation and retrieval," *Proceedings of the Fifth Asian Conference on Computer Vision*, Melbourne, Australia, pp. 646–651, 2002.

Zhang, X. and Zhao, D., "A morphological algorithm for detecting dominant points on digital curves," *SPIE Proceedings, Nonlinear Image Processing, 2424*, pp. 372–383, 1995.

Zhao, W., Chellappa, R., and Krishnaswamy, A., "Discriminant analysis of principal components for face recognition," *Proceedings of the Third IEEE International Conference on Automatic Face and Gesture Recognition*, Nara, Japan, pp. 336–341, Apr. 1998.

Zhao, H., Yuen, P. C., and Kwok, J. T., "A novel incremental principal component analysis and its application for face recognition," *IEEE Trans. Syst. Man Cybern. B*, vol. 36, no. 4, pp. 873–886, Aug. 2006.

Zheng, Z., Wang, H., and Teoh, E., "Analysis of gray level corner detection," *Pattern Recognit. Lett.*, vol. 20, no. 2, pp. 149–162, Feb. 1999.

CHAPTER 9

PATTERN RECOGNITION

What is pattern recognition? An English dictionary defines a "pattern" as an example or model—something that can be copied. A pattern is also an imitation of a model. A pattern, which describes various types of objects in the physical and abstract worlds, is any distinguishable interrelation of data (analog or digital), events, and/or concepts, for example, the shape of face, a table, the order of musical notes in a piece of music, the theme of a poem or a symphony, the tracks made on photographic plates by particles.

We can recognize a chair or a table from any direction, at any orientation of any color, and at a long distance. Furthermore, patterns can be perceived even when they are grossly distorted, ambiguous, incomplete, or severely affected by "noise." As for an example, if a letter "A" is distorted, it could be an "A" or an "H" or probably a new letter. But if the letter appears in the word "AOT," then it is most likely an "H," and if it appears in "CAT," then it is most likely an "A." Therefore, context is of crucial importance in pattern recognition.

The pattern recognition process can be viewed as a twofold task, namely, developing decision rules based on human knowledge (learning) and using them for decision making regarding an unknown pattern (classification). The problem of pattern recognition is divided into two parts. The first part is concerned with the study of recognition mechanism of patterns by human and other living organisms. This part is related to the disciplines such as physiology, psychology, biology, and so on. The second part deals with the development of theory and techniques for designing a device that can perform the recognition task automatically.

Applications of pattern recognition may be grouped as follows:

1. *Man-machine communication*: (a) automatic speech recognition, (b) speaker identification, (c) optical character recognition (OCR) systems, (d) cursive script recognition, (e) speech understanding, and (f) image understanding.
2. *Biomedical applications*: (a) electrocardiogram (ECG), electroencephalogram (EEG), electromyography (EMG) analysis, (b) cytological, histological, and other stereological applications, (c) X-ray analysis, and (d) diagnostics.
3. *Applications in physics*: (a) high-energy physics, (b) bubble chamber, and (c) other forms of track analysis.
4. *Crime and criminal detection*: (a) fingerprint, (b) handwriting, (c) speech sound, and (d) photographs.

Image Processing and Pattern Recognition by Frank Shih

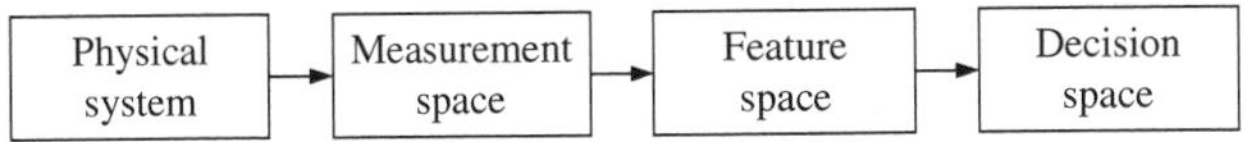

Figure 9.1 Operating stages in a recognition system.

5. *Natural resources study and estimation*: (a) agriculture, (b) hydrology, (c) forestry, (d) geology, (e) environment, (f) cloud pattern, and (g) urban quality.
6. *Stereological applications*: (a) metal processing, (b) mineral processing, and (c) biology.
7. *Military applications*: These include the above six areas of applications plus (a) detection of nuclear explosions, (b) missile guidance and detection, (c) radar and sonar signal detection, (d) target identification, (e) naval submarine detection, and (f) reconnaissance.
8. *Industrial applications*: (a) computer-assisted design and manufacture, (b) computer graphics simulation in product testing and assembly, (c) automatic inspection and quality control in factories, (d) nondestructive testing, and (e) information systems for the handicapped.
9. *Robotics and artificial intelligence*: (a) intelligent sensor technology and (b) natural language processing.

The operating stages necessary for developing and implementing the decision rule in a practical pattern recognition system are indicated in blocks of Figure 9.1. In this chapter, we will introduce the unsupervised clustering algorithm (UCA), support vector machine (SVM), neural networks, the adaptive resonance theory (ART) network, fuzzy sets, and image analysis.

This chapter aims to present an overview of selected often-used pattern recognition topics in image processing and analysis. The approaches of syntactic or statistical pattern recognition are not covered. This chapter is organized as follows. The unsupervised clustering algorithm is introduced in Section 9.1. The Bayes classifier is presented in Section 9.2. The support vector machine is described in Section 9.3. The neural network classifier is presented in Section 9.4. The adaptive resonance theory network is presented in Section 9.5. The fuzzy sets in image analysis are presented in Section 9.6.

9.1 THE UNSUPERVISED CLUSTERING ALGORITHM

The UCA is based on the *two-pass mode clustering algorithm* (Jenson, 1986). The UCA only requires processing the registered multispectral image twice. In the first pass, the cluster's mean vectors are generated. In the second pass, each pixel is assigned to a cluster that represents a single type in order to produce the hard partition map. The notations are introduced below.

B: the total number of bands used. The number defines the dimensionality in spectral space. For instance, if three bands are used, a 3D spectral space is constructed.

$C_{\max}$: the maximum number of clusters.

$r(P, k)$: the distance between the gray-value vector of a pixel P and the mean vector of the current cluster k. It can be expressed by

$$r(P,k) = \left(\sum_{i=1}^{B} (\text{MEAN}_i(k) - P_i)^2 \right)^{1/2} \tag{9.1}$$

where $1 \leq k \leq C_{\max}$, MEAN_i denotes the mean value of cluster k in band i, and P_i denotes the gray value in band i of the pixel P.

R: the constant radius in spectral space used to decide whether a new cluster's mean vector is needed. If r is greater than R, then a new cluster is created.

$d(k_1, k_2)$: the distance between the mean vectors of two distinct clusters k_1 and k_2. It can be expressed by

$$d(k_1,k_2) = \left(\sum_{i=1}^{B} (\text{MEAN}_i(k_1) - \text{MEAN}_i(k_2))^2 \right)^{1/2} \tag{9.2}$$

where $1 \leq k_1, k_2 \leq C_{\max}$ and $k_1 \neq k_2$.

D: the constant radius in spectral space used to determine whether two distinct clusters should be merged. If d is less than or equal to D, the two clusters are merged.

N: the constant representing the total number of pixels to be evaluated before the cluster merging.

$n(k)$: the total number of pixels accumulated in cluster k.

9.1.1 Pass 1: Cluster's Mean Vector Establishment

First, the gray-value vector of the first pixel is placed as the initial cluster's mean vector of cluster 1. The gray-value vector of the second pixel is then input to calculate its distance r from the mean vector of the first cluster. If r is greater than R, a new cluster is created with its mean vector equal to the gray-value vector. If r is less than or equal to R, the cluster's mean vector is adjusted using

$$\text{MEAN}_i(k)_{\text{new}} = \frac{\text{MEAN}_i(k)_{\text{old}} \times n(k) + P_i}{n(k)+1} \tag{9.3}$$

If equation (9.3) is used, then $n(k)$ and the total number of pixels n_{total} in evaluation are incremented by one. The procedure is repeated when a new input is placed. If n_{total} is greater than N or k is equal to $C_{\max}$, the process being repeated is terminated. Then the cluster merging process is activated to remove unnecessary clusters. Equation (9.2) will be used to calculate the distance between two mean vectors. The selected decision radius is dependent upon the condition that occurs to activate the cluster merging process. If the activation process is due to n_{total} being greater than N, then D is chosen as the decision radius; otherwise, R is the decision radius. If d is less than or equal to the decision radius, the two clusters are merged. The mean vector of the new cluster is adjusted by

$$\text{MEAN}_i(k) = \frac{\text{MEAN}_i(k_1) \times n(k_1) + \text{MEAN}_i(k_2) \times n(k_2)}{n(k_1) + n(k_2)} \tag{9.4}$$

where $1 \leq k, k_1, k_2 \leq C_{\max}$, and $k_1 \neq k_2$. The total number of pixels in the new cluster, $n(k)$, is the sum of $n(k_1)$ and $n(k_2)$. After the cluster merging process is completed, if the condition above results in n_{total} being greater than N, then n_{total} is set to zero; otherwise, n_{total} is left unchanged. The process of evaluating individual pixels and accumulating the clusters is repeated until n_{total} is greater than N or k is equal to $C_{\max}$.

Finally, a number of clusters are built after all the individual pixels are evaluated. In the second pass, these cluster's mean vectors will be used as the feature vectors to determine the attribute of each pixel.

9.1.2 Pass 2: Pixel Classification

Pass 2 is used to classify each pixel into one of these clusters built in pass 1. The minimum distance to means method is adopted. When each pixel is evaluated, equation (9.1) is used to compute the distance in spectral space between the gray-value vector of the pixel and each cluster's mean vector. After all the distances are determined, the cluster with the minimum distance is chosen; that is, the type of that cluster is assigned to the given pixel. By using the minimum distance to means method, the pixels with similar spectral features will be ensured to merge together.

On the basis of various theories, numerous clustering algorithms have been developed, and new clustering algorithms continue to appear in the literature. Yu (2005) proposed a unifying generative framework for partitional clustering algorithms, called a general *c*-means clustering model. Xu and Wunsch (2005) surveyed clustering algorithms for data sets appearing in statistics, computer science, and machine learning and illustrated their applications in some benchmark data sets, the traveling salesman problem, and bioinformatics, a new field attracting intensive efforts. Lian et al. (2007) proposed a framework for evaluating the performance of cluster algorithms for hierarchical networks. Fred and Jain (2005) explored the idea of evidence accumulation for combining the results of multiple clusterings. Das et al. (2008) proposed automatic clustering using an improved differential evolution algorithm.

For combining with fuzzy logic, numerous fuzzy clustering algorithms have been proposed. Corsini et al. (2004) proposed a fuzzy relational clustering algorithm based on a dissimilarity measure extracted from data. Liu et al. (2005) presented the fuzzy clustering analysis based on the axiomatic fuzzy sets theory. Yang et al. (2008) proposed the alpha-cut implemented fuzzy clustering algorithms, which allow the data points to be able to completely belong to one cluster. To deal with data patterns with linguistic ambiguity and with probabilistic uncertainty in a single framework, Lee et al. (2008) constructed an interpretable probabilistic fuzzy rule-based system that requires less human intervention and less prior knowledge than other state-of-the-art methods. Celikyilmaz and Turksen (2008) proposed enhanced fuzzy system models with improved fuzzy clustering algorithm.

Traditional clustering algorithms (e.g., the *K*-means algorithm and its variants) are used only for a fixed number of clusters. However, in many clustering applications, the actual number of clusters is unknown beforehand. To easily and effectively determine the optimal number of clusters and, at the same time, construct the clusters with good validity, Pan and Cheng (2007) proposed a framework of automatic clustering algorithms that do not require users to give each possible value of required parameters (including the number of clusters).

Nock and Nielsen (2006) presented weighting clustering that relies on the local variations of the expected complete log-likelihoods. Veenman et al. (2002) presented a partitional cluster algorithm that minimizes the sum-of-squared-error criterion while imposing a hard constraint on the cluster variance. By modifying a simple matching dissimilarity measure for categorical objects, a heuristic approach was developed in He et al. (2005) and San et al. (2004) that allows the use of the k-modes paradigm to obtain a cluster with strong intrasimilarity and to efficiently cluster large categorical data sets. Ng et al. (2007) derived the updating formula of the k-modes clustering algorithm with the new dissimilarity measure and the convergence of the algorithm under the optimization framework. Camastra and Verri (2005) presented a kernel method for clustering inspired by the classical k-means algorithm in which each cluster is iteratively refined using a one-class support vector machine.

9.2 BAYES CLASSIFIER

Consider a pattern classification problem where a sample x belongs to one of two classes, denoted as ω_1, ω_2. Assume the priori probabilities $P(\omega_1), P(\omega_2)$ are known. The density function $P(\omega_i|x)$ is obtained by

$$P(\omega_i|x) = \frac{p(x|\omega_i)P(\omega_i)}{p(x)} \tag{9.5}$$

According to Bayes (or Bayesian) theory, the probability of the classification error can be minimized by the following rule:

$$\begin{cases} \text{If } P(\omega_1|x) > P(\omega_2|x), & x \text{ is classified to } \omega_1 \\ \text{If } P(\omega_1|x) < P(\omega_2|x), & x \text{ is classified to } \omega_2 \end{cases} \tag{9.6}$$

If a pattern x in fact belongs to ω_1 but was classified into ω_2, it causes a loss, denoted as q_{12}; oppositely, it is denoted as q_{21}. The overall loss is

$$\varepsilon = P(\omega_1)q_{21}p(x|\omega_1) + P(\omega_2)q_{12}p(x|\omega_2) \tag{9.7}$$

The Bayes classifier intends to minimize the overall loss.

In the following, we will elaborate the binary class problem to multiple classes and extend Bayes theory to combine the results from multiple classifiers. Consider a pattern classification problem where a sample belongs to one of m classes, denoted as $\omega_1, \omega_2, \ldots, \omega_m$. According to Bayes theory, given all the decisions denoted as $j_1, j_2, \ldots, j_R$ from R classifiers for a sample, we have the combining decision rule as

$$i = \arg \max_i P(\omega_i|j_1 j_2 \cdots j_R) \tag{9.8}$$

where P denotes the posteriori probability and i denotes the joint decision.

In order to estimate posteriori probabilities$P(\omega_i|j_1 j_2 \cdots j_R)$, we build a confusion matrix by testing a total of N samples using R classifiers as

$$\underbrace{A[1 \cdots m][1 \cdots m] \cdots [1 \cdots m]}_{R+1} \tag{9.9}$$

Let $A[i,j_1,\ldots,j_R]$ denote the number of samples in class i to be given the decisions $(j_1,j_2,\ldots,j_R)$ by R classifiers. The posteriori probabilities can be calculated as

$$P\left(i \middle| j_1 \cdots j_R\right) = \frac{A[i,j_1,\ldots,j_R]}{\sum_{t=1}^{m} A[t,j_1,\ldots,j_R]} \tag{9.10}$$

Note that the sum of the posteriori probabilities of all m classes is equal to 1.

$$\sum_{i=1}^{m} P(i|j_1,\ldots,j_R) = 1 \tag{9.11}$$

By using the decision rule in equation (9.8), we can calculate the combined minimum error $P_{\text{comb\_min\_err}}$ as

$$\frac{P_{\text{comb\_min\_err}} = 1 - \sum_{j_1}\sum_{j_2} \cdots \sum_{j_R} \max_i A[i,j_1,\ldots,j_R]}{N} \tag{9.12}$$

For a single classifier s, its minimum error $P_{\text{min\_err}}(s)$ can be calculated as

$$P_{\text{min\_err}}(s) = 1 - \frac{\sum_{j_s}\left(\sum_{j_1} \cdots \sum_{j_R} A[j_s,j_1,\ldots,j_R]\right)}{N} \tag{9.13}$$

We can prove that the following inequality holds for any classifier s:

$$P_{\text{comb\_min\_err}} \le P_{\text{min\_err}}(s) \tag{9.14}$$

Proof:

$$1 - P_{\text{min\_err}}(s) = \frac{\sum_{j_s}\left(\sum_{j_1} \cdots \sum_{j_R} A[j_s,j_1,\ldots,j_R]\right)}{N}$$

$$\le \frac{\sum_{j_s}\max_i\left(\sum_{j_1} \cdots \sum_{j_R} A[i,j_1,\ldots,j_R]\right)}{N} \le \frac{\sum_{j_s}\sum_{j_1}\max_i\left(\sum_{j_2} \cdots \sum_{j_R} A[i,j_1,\ldots,j_R]\right)}{N}$$

$$\le \frac{\sum_{j_s}\sum_{j_1}\sum_{j_2}\max_i\left(\sum_{j_3} \cdots \sum_{j_R} A[i,j_1,\ldots,j_R]\right)}{N} \le \cdots$$

$$\le \frac{\sum_{j_s}\sum_{j_1} \cdots \sum_{j_R}\max_i A[i,j_1,\ldots,j_R]}{N}$$

$$= \frac{\sum_{j_1} \cdots \sum_{j_R} \max_i A[i, j_1, \ldots, j_R]}{N} = 1 - P_{\text{comb\_min\_err}} \Rightarrow P_{\text{comb\_min\_err}} \leq P_{\text{min\_err}}(s)$$

Therefore, the combined minimum error of multiple classifiers will not be greater than the minimum error of any single classifier. We illustrate the aforementioned property of Bayes combining method as follows. Given two classifiers and three classes, we assume that the confusion matrix of $A[1 \cdots 3], [1 \cdots 3], [1 \cdots 3]$ is given below. Note that the element $A[i, j, k]$ denotes the number of samples in class i to be assigned to class j by the first classifier and to class k by the second classifier. The total number of samples $N = 400$.

$$\underset{A[1,1\cdots3,1\cdots3],}{\begin{bmatrix} 90 & 2 & 5 \\ 1 & 5 & 5 \\ 5 & 1 & 5 \end{bmatrix}}, \underset{A[2,1\cdots3,1\cdots3],}{\begin{bmatrix} 2 & 15 & 5 \\ 5 & 90 & 5 \\ 5 & 5 & 3 \end{bmatrix}}, \underset{A[3,1\cdots3,1\cdots3]}{\begin{bmatrix} 1 & 5 & 10 \\ 5 & 10 & 5 \\ 10 & 5 & 95 \end{bmatrix}}$$

If the decisions from the two classifiers are $j_1 = j_2 = 1$, the posteriori probabilities can be calculated as

$$P(\omega_1 | j_1 j_2) = \frac{90}{90 + 2 + 1} = 0.9677$$

$$P(\omega_2 | j_1 j_2) = \frac{2}{90 + 2 + 1} = 0.0215$$

$$P(\omega_3 | j_1 j_2) = \frac{1}{90 + 2 + 1} = 0.0108$$

From these results, we make the final decision to be class 1 according to equation (9.8). The combined minimum error of two classifiers is

$$P_{\text{comb\_min\_err}} = 1 - \frac{\sum_{j_1} \sum_{j_2} \max_i A[i, j_1, j_2]}{N} = 1 - 0.8125 = 0.1875$$

The minimum errors of the first and second classifiers are, respectively,

$$P_{\text{min\_err}}(1) = 1 - \frac{\sum_{j_1} \sum_{j_2} A[j_1, j_1, j_2]}{N} = 1 - 0.7675 = 0.2325$$

$$P_{\text{min\_err}}(2) = 1 - \frac{\sum_{j_2} \sum_{j_1} A[j_2, j_1, j_2]}{N} = 1 - 0.79 = 0.21$$

Therefore, we obtain $P_{\text{comb\_min\_err}} \leq P_{\text{min\_err}}(1)$ and $P_{\text{comb\_min\_err}} \leq P_{\text{min\_err}}(2)$.

9.3 SUPPORT VECTOR MACHINE

The SVM, introduced by Vapnik (1995) and Cortes and Vapnik (1995), has attracted growing interest in pattern classification due to its competitive performance. It is a learning system that separates a set of input pattern vectors into two classes with an optimal separating hyperplane. The set of vectors is said to be optimally separated by the hyperplane if it is separated without error and the distance between the closest vectors to the hyperplane is maximal. SVM produces the pattern classifier by applying a variety of kernel functions (linear, polynomial, radial basis function (RBF), and so on) as the possible sets of approximating functions, by optimizing the dual quadratic programming problem, and by using structural risk minimization as the inductive principle, as opposed to classical statistical algorithms that maximize the absolute value of an error or of an error squared.

Originally, SVM was designed to handle dichotomic classes. Later, research has concentrated on expanding two-class classification to multiclass classification. Different types of SVM classifiers are used depending upon the type of input patterns: a linear maximal margin classifier is used for linearly separable data, a linear soft margin classifier is used for linearly nonseparable, or overlapping, classes, and a nonlinear classifier is used for classes that are overlapped as well as separated by nonlinear hyperplanes. All three classifiers are discussed in more detail below. It should be noted, however, that the linearly separable case is rare in real-world problems.

9.3.1 Linear Maximal Margin Classifier

The case where the training patterns can be linearly separated by a hyperplane, $\mathbf{w} \cdot \mathbf{x} + b = 0$, is the simplest case and provides a good foundation for the other two cases. The purpose of the SVM is to find the optimal values for $\mathbf{w}$ (e.g., $\mathbf{w}_0$) and b (e.g., b_0). After finding the optimal separating hyperplane, $\mathbf{w}_0 \cdot \mathbf{x} + b_0 = 0$, an unseen pattern, $\mathbf{x}_t$, can be classified by the decision rule $f(x) = \text{sign}(\mathbf{w}_0 \cdot \mathbf{x} + b_0)$, as shown below.

Suppose, there is a set of training data, $\mathbf{x}_1, \mathbf{x}_2, \ldots, \mathbf{x}_k$, where $\mathbf{x}_i \in \text{R}^n$ and $i = 1, 2, \ldots, k$. Each $\mathbf{x}_i$, belonging as it does to one of two classes, has a corresponding value y_i, where $y_i \in \{-1, 1\}$. The goal in this case is to build the hyperplane that maximizes the minimum distance between the two classes. Because the hyperplane is $\mathbf{w} \cdot \mathbf{x} + b = 0$, the training data can be divided into two classes such that

$$\begin{cases} \mathbf{w} \cdot \mathbf{x}_i + b \geq 1 & \text{if } y_i = 1 \\ \mathbf{w} \cdot \mathbf{x}_i + b \leq -1 & \text{if } y_i = -1 \end{cases} \tag{9.15}$$

where $\mathbf{w} \in \text{R}^n$ and $b \in \text{R}$. Then we obtain

$$y_i(\mathbf{w} \cdot \mathbf{x}_i + b) \geq 1 \quad \forall x_i, \quad i = 1, 2, \ldots, k \tag{9.16}$$

The distance between a point $\mathbf{x}$ and the hyperplane is $d(\mathbf{w}, b; \mathbf{x}) = |\mathbf{w} \cdot \mathbf{x} + b| \div ||\mathbf{w}||$.

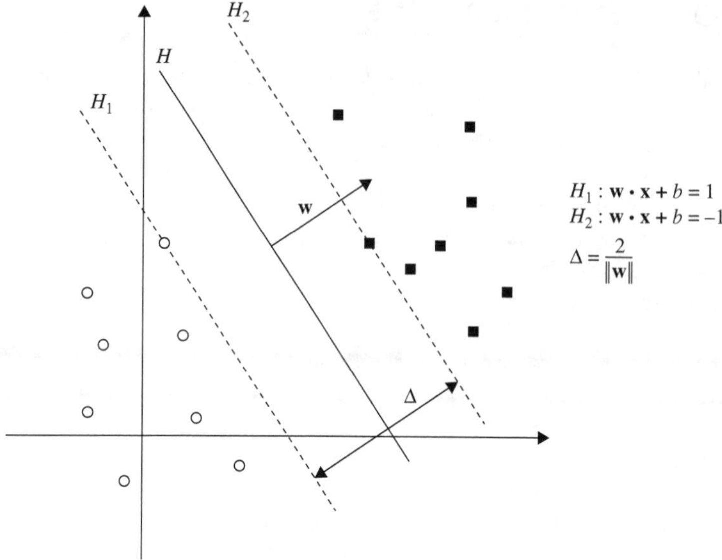

Figure 9.2 The model of SVM.

According to equation (9.15), the minimum distance between one of the two classes and the hyperplane is $1/||\mathbf{w}||$. The margin, *M*, which is the distance between the two classes, is $\Delta = 2/||\mathbf{w}||$. Figure 9.2 illustrates the model of SVM.

Finding the optimal separating hyperplane having a maximal margin requires that the following minimization problem be solved:

$$\text{Minimize} : \frac{1}{2}\mathbf{w} \cdot \mathbf{w}$$

$$\text{Subject to} : y_i(\mathbf{w} \cdot \mathbf{x}_i + b) \geq 1 \; \forall \mathbf{x}_i, \; i = 1, 2, \ldots, k$$

This nonlinear optimization problem with inequality constraints can be solved by the saddle point of the Lagrange function:

$$L(\mathbf{w}, b, \alpha) = \frac{1}{2}\mathbf{w} \cdot \mathbf{w} - \sum_{i=1}^{K} \alpha_i (y_i(\mathbf{w} \cdot \mathbf{x}_i + b) - 1) \tag{9.17}$$

where $\alpha_i \geq 0$ are the Lagrange multipliers.

By minimizing the Lagrange function with respect to $\mathbf{w}$ and b, as well as by maximizing with respect to α_i, the minimization problem above can be transformed to its dual problem, called the quadratic programming problem.

$$\left.\frac{\partial L(\mathbf{w}, b, \alpha)}{\partial \mathbf{w}}\right|_{\mathbf{w}=\mathbf{w}_0} = \mathbf{w}_0 - \sum_{i=1}^{K} \alpha_i y_i \mathbf{x}_i = 0 \tag{9.18}$$

$$\left.\frac{\partial L(\mathbf{w}, b, \alpha)}{\partial b}\right|_{b=b_0} = \sum_{i=1}^{K} \alpha_i y_i = 0 \tag{9.19}$$

By plugging equation (9.18) and equation (9.19) into equation (9.17), we obtain

$$L(\alpha) = \sum_{i=1}^{K} \alpha_i - \frac{1}{2} \sum_{i=1}^{K} \sum_{j=1}^{K} \alpha_i \alpha_j y_i y_j \mathbf{x}_i \mathbf{x}_j \tag{9.20}$$

The dual problem is then described as

$$\text{Maximize} : \sum_{i=1}^{K} \alpha_i - \frac{1}{2} \sum_{i=1}^{K} \sum_{j=1}^{K} \alpha_i \alpha_j y_i y_j \mathbf{x}_i \mathbf{x}_j$$

$$\text{Subject to} : \sum_{i=1}^{K} y_i \alpha_i = 0, \ \alpha_i \geq 0$$

By solving the dual problem, we can determine the optimal separating hyperplane by

$$\mathbf{w}_0 = \sum_{i=1}^{K} \alpha_i y_i \mathbf{x}_i \tag{9.21}$$

$$b_0 = y_i - \mathbf{w}_0 \cdot \mathbf{x}_i \tag{9.22}$$

where $\mathbf{x}_i$ belongs to support vectors and $y_i \in \{-1, 1\}$.

The unseen test data $\mathbf{x}_t$ can be classified by simply computing

$$f(x) = \text{sign}(\mathbf{w}_0 \cdot \mathbf{x}_t + b_0) \tag{9.23}$$

It can be seen that the hyperplane is determined by all the training data, $\mathbf{x}_i$, that have the corresponding attributes of $\alpha_i > 0$. We call this kind of training data as *support vectors*. Thus, the optimal separating hyperplane is not determined by the training data per se but rather by the support vectors.

9.3.2 Linear Soft Margin Classifier

As mentioned above, patterns are rare to be linearly separable in real-world problems. In this section, we expand SVM to handle input patterns that are overlapping or linearly nonseparable. In this case, our objective is to separate the two classes of training data with a minimal number of errors.

To accomplish this, we introduce some nonnegative slack variables ξ_i, $i = 1, 2, \ldots, k$ to the system. Thus, equations (9.15) and (9.16), in the linearly separable case above, can be rewritten as equations (9.24) and (9.25):

$$\begin{cases} \mathbf{w} \cdot \mathbf{x}_i + b \geq 1 - \xi_i & \text{if } y_i = 1 \\ \mathbf{w} \cdot \mathbf{x}_i + b \leq -1 + \xi & \text{if } y_i = -1 \end{cases} \tag{9.24}$$

$$y_i(\mathbf{w} \cdot \mathbf{x}_i + b) \geq 1 - \xi_i \quad \forall \mathbf{x}_i, \quad i = 1, 2, \ldots, k \tag{9.25}$$

Just as we obtained the optimal separating hyperplane in the linearly separable case, obtaining the soft margin hyperplane in the linearly nonseparable case requires that the following minimization problem be solved:

$$\text{Minimize}: \frac{1}{2}\mathbf{w}\cdot\mathbf{w}+C\left(\sum_{i=1}^{K}\xi_i\right)$$

$$\text{Subject to}: y_i(\mathbf{w}\cdot\mathbf{x}_i+b)\geq 1-\xi_i \text{ and } \xi_i\geq 0,\ i=1,2,\ldots,k$$

where C is a penalty or regularization parameter.

By minimizing the Lagrange function with respect to $\mathbf{w}$, b, and ξ_i, as well as by maximizing with respect to α_i, the minimization problem above can be transformed to its dual problem as

$$\text{Maximize}: \sum_{i=1}^{K}\alpha_i-\frac{1}{2}\sum_{i=1}^{K}\sum_{j=1}^{K}\alpha_i\alpha_j y_i y_j \mathbf{x}_i\mathbf{x}_j$$

$$\text{Subject to}: \sum_{i=1}^{K}y_i\alpha_i=0,\ 0\leq\alpha_i\leq C$$

By solving the dual problem, we determine the soft margin hyperplane by

$$\mathbf{w}_0=\sum_{i=1}^{K}\alpha_i y_i\mathbf{x}_i \tag{9.26}$$

$$b_0=y_i-\mathbf{w}_0\cdot\mathbf{x}_i \tag{9.27}$$

where $\mathbf{x}_i$ belongs to margin vectors, $y_i\in\{-1,1\}$. It can be seen that the hyperplane is determined by all the training data, $\mathbf{x}_i$, that have the corresponding attributes of $\alpha_i>0$. These support vectors can be divided into two categories. The first category has the attribute of $\alpha_i<C$. In this category, $\xi_i=0$, and these support vectors lie at the distance $1/||\mathbf{w}||$ from the optimal separating hyperplane. We call these support vectors as *margin vectors*. The second category has the attributes of $\alpha_i=C$. In this category, either the support vectors are correctly classified with a distance smaller than $1/||\mathbf{w}||$ from the optimal separating hyperplane (if $0<\xi_i\leq 1$) or they are misclassified (if $\xi_i>1$). The support vectors in the second category are regarded as errors.

9.3.3 Nonlinear Classifier

Sometimes, the input vectors cannot be linearly separated in the input space. In this case, kernel functions, such as the polynomial or RBF, are used to transform the input space to a feature space of higher dimensionality. In the feature space, a linear separating hyperplane is obtained to separate the input vectors into two classes.

If $\mathbf{x}\in R^n$ is in the input space, we can map the input vector $\mathbf{x}$ from the n-dimensional input space to a corresponding N-dimensional feature space through a function, ϕ. After the transformation, we know $\phi(\mathbf{x})\in R^n$. Following the steps described in the case of linearly separable training patterns and the case of linearly nonseparable training patterns, the hyperplane and decision rule for nonlinear training patterns can be established.

In a manner similar to obtaining the hyperplane for the linearly separable training patterns in equations (9.21) and (9.22) and the hyperplane for the linearly nonseparable training patterns in equations (9.26) and (9.27), we can obtain the hyperplane for the nonlinear training pattern as

$$\mathbf{w}_0 \cdot \phi(\mathbf{x}) + b_0 = \left(\sum_{i=1}^{K} \alpha_i y_i \phi(\mathbf{x}_i) \right) \cdot \phi(\mathbf{x}) + b_0 \tag{9.28}$$

In equation (9.28), we see that the original dot products of input variables can be replaced by afunction ϕ. That is, kernel function $K(\mathbf{x}_i, \mathbf{x}) = \phi(\mathbf{x}_i) \cdot \phi(\mathbf{x})$. The decision rule for nonlinear training patterns can be established as

$$f(\mathbf{x}) = \text{sign}\left(\sum_{i=1}^{K} \alpha_i y_i K(\mathbf{x}_t, \mathbf{x}) + b_0 \right) \tag{9.29}$$

9.3.4 SVM Networks

The SVM is originally designed as a binary (or two-class) classifier. Researchers have proposed extensions to multiclass (or multigroup) classification by combining multiple SVMs, such as one-against-all, one-against-one, and DAG-SVM (Platt et al., 2000). One-against-all method is to learn k binary SVMs during the training stage. For the ith SVM, we label all training samples by $y(j)_{j=i} = +1$ and $y(j)_{j\neq i} = -1$. In the testing, the decision function (Hsu and Lin, 2002) is obtained by

$$\text{Class of } \vec{\mathbf{x}} = \arg\max_{i=1,2,\ldots,k}((w^i)^{\mathrm{T}} \phi(\vec{\mathbf{x}}) + b^i) \tag{9.30}$$

where $(w^i)^{\mathrm{T}} \phi(\vec{\mathbf{x}}) + b^i = 0$ represents the hyperplane for the ith SVM. It was shown that we can solve the k-group problem simultaneously (Herbrich, 2001).

One-against-one method is to construct $k(k-1)/2$ SVMs in a tree structure. Each SVM represents a distinguishable pairwise classifier from different classes in the training stage. In the testing, each leaf SVM pops up one desirable class label that propagates to the upper level in the tree until it processes the root SVM of the tree.

The directed acyclic graph (DAG) SVM (Platt et al., 2000) is to combine multiple binary one-against-one SVM classifiers into one multiclass classifier. It has $k(k-1)/2$ internal SVM nodes, but in the testing phase, it starts from the root node and moves to the left or right subtree depending on the output value of the binary decision function until a leaf node is reached, which indicates the recognized class. There are two main differences between one-against-one and DAG methods. One is that one-against-one needs to evaluate all $k(k-1)/2$ nodes, while DAG only needs to evaluate k nodes due to the different testing phase schema. The other is that one-against-one uses the bottom-up approach, but DAG uses the top-down approach.

The performance of above three methods depends on the quality of each individual binary SVM. There is no error tolerance that can adjust the system. If one SVM fails, an incorrect result could be produced. Each SVM is a stand-alone black box. Other than the conveyed class label, no further information is propagated to other SVMs. Shih and Zhang (2005) proposed a combined SVM and neural network method to overcome this problem.

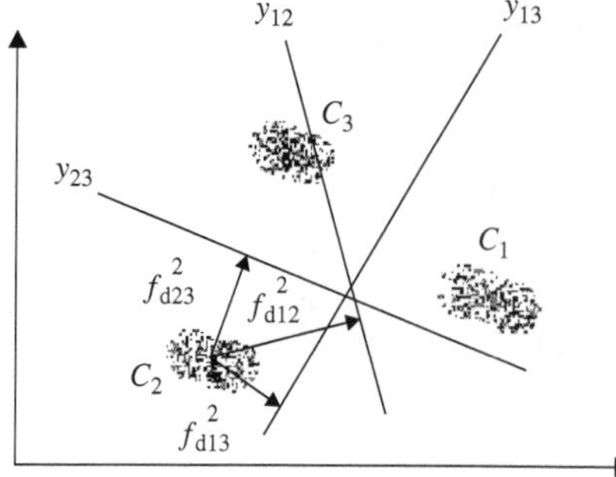

Figure 9.3 Decision function of a three-class SVM.

The decision function of SVM provides not only the class label but also useful information, such as class density and distribution knowledge. We can change equation (9.16) to

$$f_{\text{d}}(\vec{\mathbf{x}}) = \vec{\mathbf{w}}_0 \cdot \vec{\mathbf{x}} + b_0 \tag{9.31}$$

where $f_{\text{d}}(\vec{\mathbf{x}})$ represents the distance between the sample $\vec{\mathbf{x}}$ and the optimal hyperplane $\vec{\mathbf{w}}_0 \cdot \vec{\mathbf{x}} + b_0 = 0$. As shown in Figure 9.3, the distances $f^2_{\text{d}12}$, $f^2_{\text{d}13}$, and $f^2_{\text{d}23}$ compose a vector $\vec{\mathbf{f}}^2_{\text{d}} = (f^2_{\text{d}12}, f^2_{\text{d}13}, f^2_{\text{d}23})^{\text{T}}$ to identify the position of class C_2, and y_{ij} denotes the hyperplane between classes i and j. Therefore, we can use $\vec{\mathbf{f}}^i_{\text{d}}$ to illustrate the distribution of class i.

We extend this idea to a layered neural network to construct a new framework as shown in Figure 9.4. For the k-class problem, we create k hidden neurons, and each of which is a one-against-all SVM having the decision function $f_{\text{d}}(\vec{\mathbf{x}}) = \vec{\mathbf{w}}_0 \cdot \vec{\mathbf{x}} + b_0$ and k output nodes, which compose an output vector $\vec{\mathbf{Y}} = (y_1, y_2, \ldots, y_k)^{\text{T}}$.

For a training sample $\vec{\mathbf{x}}$ from class i, it serves all the SVMs. We obtain a new distance vector $\vec{\mathbf{f}}_{\text{d}}(\vec{\mathbf{x}}) = (f^1_{\text{d}}, f^2_{\text{d}}, \ldots, f^k_{\text{d}})^{\text{T}} \in R^k$ to represent the sample $\vec{\mathbf{x}}$ in R^k space. Finally, the desired output vector $\vec{\mathbf{Y}}(\vec{\mathbf{x}}) = (y_1, y_2, \ldots, y_k)^{\text{T}} = \vec{\mathbf{f}}_{\text{d}}(\vec{\mathbf{x}}) \cdot W$, where W is a $k \times k$ weight matrix, and $y_i = 1$ and $y_{j \neq i} = 0$. Because $\vec{\mathbf{Y}}(\vec{\mathbf{x}})$ and $\vec{\mathbf{f}}_{\text{d}}(\vec{\mathbf{x}})$ are known, the weight matrix can be obtained by

$$W = F^+_{\text{d}} Y = (F^{\text{T}}_{\text{d}} F_{\text{d}})^{-1} F^{\text{T}}_{\text{d}} Y \tag{9.32}$$

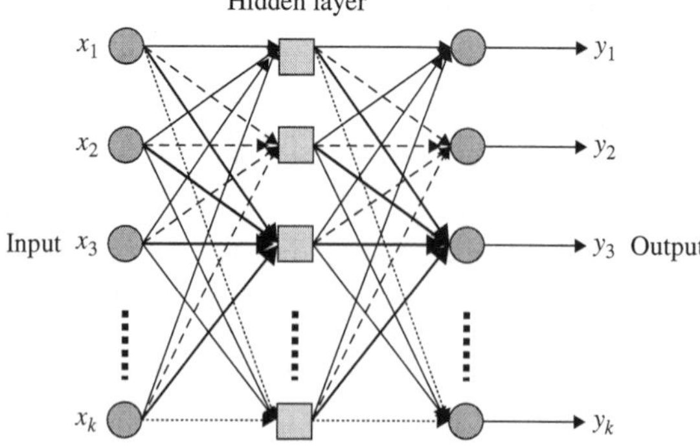

Figure 9.4 The architecture of our framework.

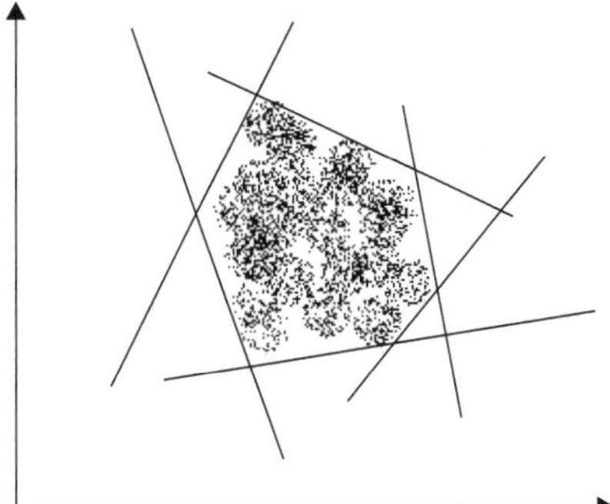

Figure 9.5 Hyperplanes divide the space.

where F_{d} and Y are derived by all training samples, and F_{d}^{+} denotes the pseudo-inverse of F_{d}. Besides, we can imagine that for a k-class problem, the whole space is divided by k hyperplanes through such a framework as shown in Figure 9.5.

Traditional transformation maps the original image vectors to another high-dimensional space associated with a fixed origin point. But our proposed framework maps them to a kinetic coordinate system whose degree is the number of classes. There are three phases in our framework. The first is the constructing phase, the second is the training phase, and the third is the testing phase. The constructing phase is to build all SVMs. The dimensionality of all samples is reduced from $R^m \rightarrow R^n$ through the PCA procedure. The $\vec{\mathbf{x}} \in R^n$ will serve as the input vector into our framework. Supposedly, there are l samples for each class in the training set. For the ith SVM M_i^1, we select all training samples in class i, and any one sample from all other classes $j \neq i$ to obtain the training set T_i^1. After M_i^1 is trained, we test the complete training set. There exist some misclassified samples. We add them to T_i^1 with negative labels as a new training set T_i^2, so that we can obtain a new trained SVM called M_i. If no misclassified samples are found, then $M_i^1 = M_i$. Through such a procedure, M_i can enhance its generalization. The SVM is trained by all training samples just once. After all of M_i's are established, the target outputs will be created. For a training sample $\vec{\mathbf{x}}$ from the class i, the desired output vector is $\vec{\mathbf{Y}}(\vec{\mathbf{x}}) = (y_1, y_2, \ldots, y_k)^{\mathrm{T}}$, where $y_i = 1$ and $y_{j \neq i} = 0$. The overall procedures are given in Figure 9.6.

The objective of the training phase is to compute the weight matrix. For a complete training set, the system generates $F_{\mathrm{d}} W = Y$, where $F_{\mathrm{d}} = (\vec{\mathbf{f}}_{\mathrm{d}}^{1}, \vec{\mathbf{f}}_{\mathrm{d}}^{2}, \ldots, \vec{\mathbf{f}}_{\mathrm{d}}^{k \times l})$ and $Y = [\vec{\mathbf{Y}}(\vec{\mathbf{x}}_{1,1}), \vec{\mathbf{Y}}(\vec{\mathbf{x}}_{1,2}), \ldots, \vec{\mathbf{Y}}(\vec{\mathbf{x}}_{k,l})]$. The output vector only contains a single "1" in the corresponding output node to identify the desired class number. We have $\vec{\mathbf{Y}}(\vec{\mathbf{x}}_{i,m}) = (y_1, y_2, \ldots, y_k)$, where $y_i = 1$ and $y_{j \neq i} = 0$. We expect that the output vector only contains a single "1" in the corresponding output node to identify the desired class. Once F_{d} and Y are known, the weight matrix can be obtained by equation (9.32). Finally, the whole system is established.

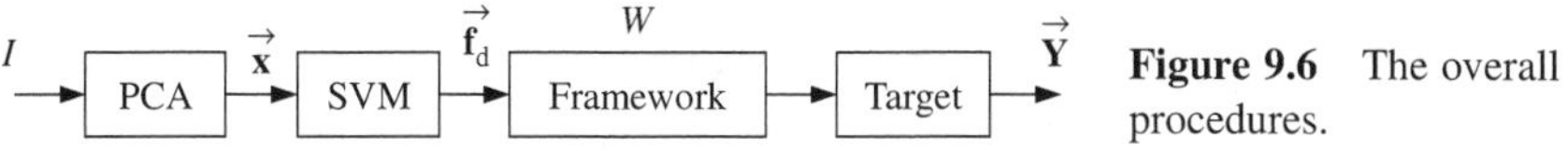

Figure 9.6 The overall procedures.

In the testing phase, for an unknown sample $\vec{\mathbf{x}}$, the system outputs a vector $\vec{\mathbf{Y}}(\vec{\mathbf{x}})$, where the location of the maximum of output nodes, as given in equation (9.33), indicates the recognized class. In other words, it is the most likely class.

$$\text{Class of } \vec{\mathbf{x}} = \arg\max_{i=1,2,\ldots,k} y_i(\vec{\mathbf{x}}) \tag{9.33}$$

Numerous improvements on SVMs and their applications have been proposed. Haasdonk (2005) presented feature space interpretation of SVMs with indefinite kernels. Lin and Lin (2003) presented a study on reduced support vector machine, which is an alternative of the standard SVM, by resolving the difficulty on handling large data sets using SVM with nonlinear kernels. Xia and Wang (2004) presented a one-layer recurrent neural network for SVM learning in pattern classification and regression. Jayadeva et al. (2007) proposed twin SVM, a binary SVM classifier that determines two nonparallel planes by solving two related SVM-type problems, each of which is smaller than in a conventional SVM.

Liu and Zheng (2007) presented soft SVM and its application in video-object extraction. Li et al. (2006) developed a machine learning technique, multitraining SVM, which combines the merits of the cotraining technique and a random sampling method in the feature space for image retrieval. Navia-Vazquez et al. (2006) proposed distributed support vector machines. Tao et al. (2008) proposed a recursive SVM, in which several orthogonal directions that best separate the data with the maximum margin are obtained. During the usual SVM biclassification learning process, the bias is chosen *a posteriori* as the value halfway between separating hyperplanes. Gonzalez-Abril et al. (2008) presents different approaches on the calculation of the bias when SVM is used for multiclassification.

Fei and Liu (2006) presented a new architecture, named binary tree of SVM, in order to achieve high classification efficiency for multiclass problems. Anguita et al. (2003) proposed a digital architecture for SVM learning and discussed its implementation on a field programmable gate array. Anguita et al. (2006) proposed a coordinate rotation digital computer (CORDIC)-like algorithm for computing the feedforward phase of a SVM in fixed point arithmetic, using only shift and add operations and avoiding resource-consuming multiplications.

9.4 NEURAL NETWORKS

The neural network approach exhibits two major characteristics in the field of problem solving. First, neural network architectures are parallel and connectionist. The connections between pairs of neurons play a primary role in the function performed by the whole system. The second characteristic is that the prescriptive computer programming routine is replaced by heuristic learning processes in establishing a system to solve a particular problem. It means that a neural network is provided with representative data, also called *training data*, and some learning process that induces the data regularities. The modification of the transfer function at each neuron is derived empirically according to the training data set in order for maximizing the goal.

Two operating phases, learning and execution, are always encountered in neural networks. During the *learning phase*, the networks are programmed to take input from training set and adjust the connection weights to achieve the desired association or classification. During the *execution phase*, the neural networks are to process input data from outside world to retrieve corresponding outputs. These two phases are different in both the processing types and the time constraints involved.

Neural networks can be considered as signal processors that are "value passing" architectures, conceptually similar to traditional analog computers. Each neuron may receive a number of inputs from the input layer or some other layers, and then produces a single output that can be an input to some other neurons or a global network output. Many neural network models, such as Rosenblatt's and Minsky's perceptron (Minsky and Papert, 1969; Rosenblatt, 1962) and Fukushima's cognitron (Fukushima, 1975), which use binary inputs and produce only binary outputs, can implement Boolean functions.

In feedforward layered neural networks, which form a simple hierarchical architecture, execution of each layer is performed in a single step. The input neurons in the first layer can be viewed as entry points for sensor information from outside world, or called sensoring neurons, and the output neurons in the last layer are considered as decision points control, or called decision neurons. The neurons in the intermediate layers, often called hidden neurons, play the role of interneurons. The network implicitly memorizes an evaluated mapping from the input vector and the output vector at the execution time, each time the network traversed. The mapping, represented as $\mathbf{X} \rightarrow \mathbf{Y}$, is just a transfer function that takes the n-dimensional input vectors $\mathbf{X} = (x_1, x_2, \ldots, x_n)$ and then outputs m-dimensional input vectors $\mathbf{Y} = (y_1, y_2, \ldots, y_m)$.

9.4.1 Programmable Logic Neural Networks

The concept of building neural network structure with dynamically programmable logic modules (DPLMs) is based on early efforts aimed at providing general design of logic circuits by using universal logic modules (Armstrong and Gecsei, 1979; Vidal, 1988). Neural networks are constructed with node modules that have only a limited number of inputs. Figure 9.7 shows a module of 2-bit input DPLM that consists of a commonly used 4-to-1 multiplexer and a 4-bit control register. In this circuit, the role of 4-bit input and 2-bit control is exchanged. The 2-bit control ($\mathbf{X}$) is used to select any bit of the 4-bit value in the register (C) as the output ($\mathbf{Y}$).

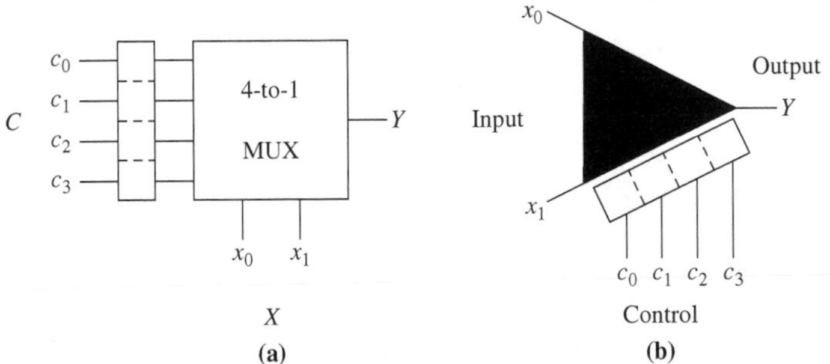

Figure 9.7 (a) A neuron cell consists of a 4-to-1 multiplexer and a 4-bit control register. (b) Symbol for a 2-input DPLM with a 4-bit control register.

TABLE 9.1 Truth Tables of 16 Possible Boolean Functions

		Input (x_0x_1)			
Function	Control code	00	01	10	11
FALSE	0000	0	0	0	0
AND	0001	0	0	0	1
LFT	0010	0	0	1	0
LFTX	0011	0	0	1	1
RIT	0100	0	1	0	0
RITX	0101	0	1	0	1
XOR	0110	0	1	1	0
OR	0111	0	1	1	1
NOR	1000	1	0	0	0
XNOR	1001	1	0	0	1
NRITX	1010	1	0	1	0
NRIT	1011	1	0	1	1
NLFTX	1100	1	1	0	0
NLFT	1101	1	1	0	1
NAND	1110	1	1	1	0
TRUE	1111	1	1	I	1

In Figure 9.7, $\mathbf{X} = (x_0, x_1)$ denotes the input vector having four combinations: $\{00, 01, 10, 11\}$. The control register C determines the value of the single-bit output $\mathbf{Y}$ from the input vector. Therefore, by loading a 4-bit control word, one of the 16 (i.e., 2^4) possible Boolean functions for two input variables can be chosen. These Boolean functions and associated control codes (i.e., the truth table) are given in Table 9.1. A two-input DPLM with all 16 possible Boolean functions is called a *universal module*. If the 16 possible functions are not fully utilized, the size of the control register could be reduced (Vidal, 1988).

The 2-input modules are the simplest primitive units. Figure 9.8 illustrates a general-purpose 3-input DPLM composed of six 2-input modules. The desired

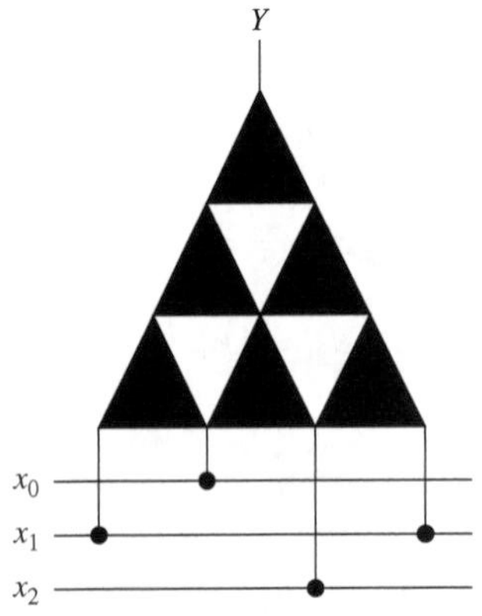

Figure 9.8 A 3-input primitive unit composed of six 2-input DPLMs.

Figure 9.9 A 3-input primitive unit composed of three 2-input DPLMs.

function is selected by the corresponding control codes loaded to the six 2-input modules. For example, the AND codes loaded to all six modules acts as the AND function of the three inputs; this is called CONVERGE. Similarly, by loading the OR codes, it performs the OR function of the inputs, called DIVERGE. The accumulative combinations of primitive blocks can be used to achieve more complex architectures for various applications.

The structure of a 3-input DPLM in Figure 9.8 could be reduced to the structure in Figure 9.9. In the neighborhood operations, only three 2-input modules are needed in the pyramid structure. That is, the module in the upper level will receive inputs from the two in the lower level. Instead of loading the same control code to all modules, the new architecture can be loaded with different control codes in each level so that more functions could be performed. A number of these functions are discussed in the next section.

9.4.2 Pyramid Neural Network Structure

To illustrate the process of pyramid model, a simple example constructed with 3-input DPLMs is used. Figure 9.10 shows the structure of the network. In the network, every node (C_{k+1}) in the $(k + 1)$th layer receives inputs from the three neighbors (nodes L_k, C_k, and R_k) in the kth layer and all nodes in the same layer are assigned the same

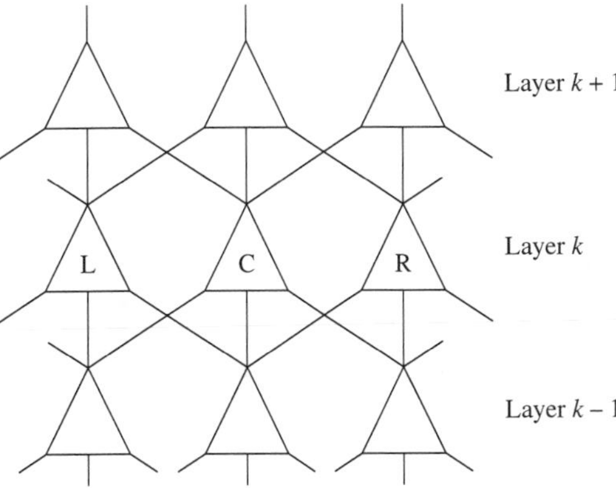

Figure 9.10 The structure of 2D pyramid neural network composed of 3-input DPLMs.

TABLE 9.2 Boolean Functions Performed by 3-Input DPLMs

Function names	Equations	Control	
		Upper	Lower
CONVERGE	$C_{k+1} \leftarrow L_k\, C_k\, R_k$	AND	AND
DIVERGE	$C_{k+1} \leftarrow L_k + C_k + R_k$	OR	OR
NONE	$C_{k+1} \leftarrow \overline{L}_k\, \overline{C}_k\, \overline{R}_k$	AND	NOR
CENTER-ON	$C_{k+1} \leftarrow C_k$	RITX	LFTX
		LFTX	RITX
RIGHT-ON	$C_{k+1} \leftarrow R_k$	RITX	RITX
LEFT-ON	$C_{k+1} \leftarrow L_k$	LFTX	LFTX

function. Hence, only one control register is required for each layer. Some functions are illustrated in Table 9.2.

It has to be noted that the CONVERGE and DIVERGE operations are actually the AND and the OR Boolean functions, respectively. They are equivalent to the morphological erosion and dilation. The RIGHT-ON operation is used to detect the right neighbor and the LEFT-ON operation the left neighbor. Similarly, the CENTER-ON operation detects the center pixel.

9.4.3 Binary Morphological Operations by Logic Modules

To implement morphological operations, a new neural architecture by the use of the pyramid model is constructed. The connections among the four 3-input DPLMs shown in Figure 9.11 form a 9-input 3D pyramid module. The layers k and $k+1$ are called *regular layer* and the layer $k + 1/2$ is called *intermediate layer*.

In the structure of 9-input module, each neuron $U_{k+1/2}(i,j)$ in the intermediate layer $k + 1/2$ receives three inputs of its column-major neighbors $(U_k(i,$

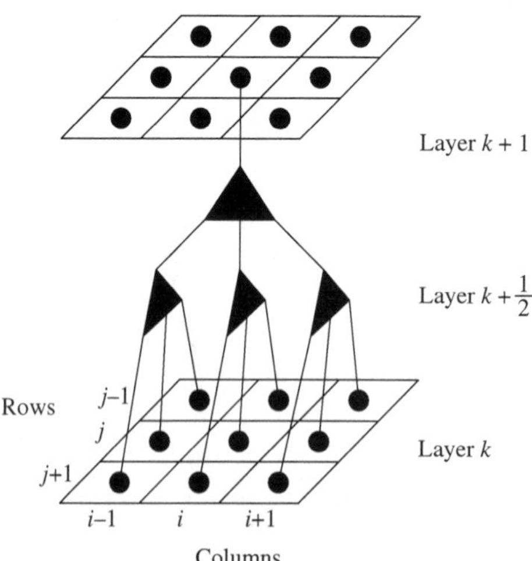

Figure 9.11 The 9-input pyramid module composed of four 3-input DPLMs.

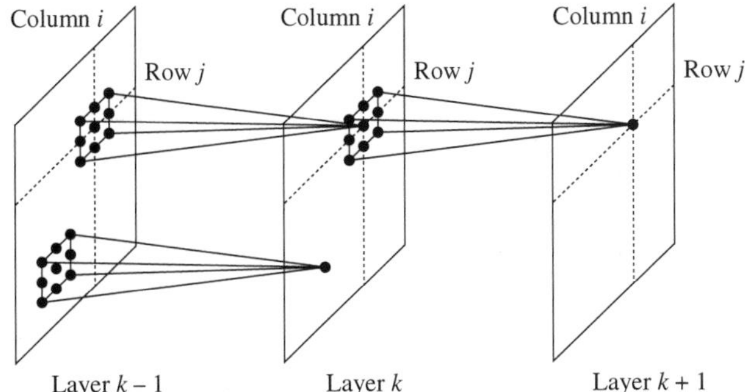

Figure 9.12 The overall structure of neural network formed with 3×3 input DPLMs for image morphological processing.

$j-1), U_k(i,j), U_k(i,j+1))$ in the preceding layer k. The neuron $U_{k+1}(i,j)$ then collects the three outputs of its three row-major neighbors $(U_{k+1/2}(i-1, j), U_{k+1/2}(i,j), U_{k+1/2}(i+1,j))$ in the preceding intermediate layer $k + 1/2$. In other words, the module will extract 1D local features from each column, and then combine the three adjacent 3×1 features into 3×3 2D local feature. The overall structure of the network is shown in Figure 9.12. There is no limitation for the number of layers in a neural network; that is, one could concatenate as many layers as desired and then assign different Boolean functions to the nodes for specific applications.

Since only one operation is applied on the whole original image at a certain time, loading different control codes to DPLMs on the same major dimension is not required. Only two control registers (one for the upper level and the other for the lower level) are connected for each column in any intermediate layer (layer $k + 1/2$) in the same way as for each row in any regular layer (layer k), that is, to apply a certain function with respect to a column or a row. Some examples of morphological operations performed by this neural network for extracting the local features are shown in Table 9.3.

As seen in Table 9.3, for all dilations, the upper level of the modules should be loaded with the DIVERGE operation, and for all erosions, the CONVERGE operation is employed. The lower level is loaded with different operations for different structuring elements. Suppose that an erosion is applied on a binary image with the following 3×3 structuring element:

1	1	1
1	1	1
1	1	1

in the layer k and pass the result to the layer $k + 1$. One may simply load CONVERGE to all 3-input DPLMs in both intermediate layer $k + 1/2$ and regular layer $k + 1$. The neurons in layer $k + 1/2$ will output a "1" if the three neighbors in column are all 1's. The neurons in layer $k + 1$ will output a "1" if the three neighboring intermediate neurons in row have all 1's; that is, the nine (3×3) neighboring pixels are all 1's. Similarly, for dilation with the same structuring element, DIVERGE may be loaded to all neurons.

TABLE 9.3 Morphological Operations Using Pyramid Networks

Structuring elements	Morphological operations	Intermediate layers	Regular layers
111	DILATION	DIVERGE	DIVERGE
111			
111	EROSION	CONVERGE	CONVERGE
000	DILATION	CENTER-ON	DIVERGE
111			
000	EROSION	CENTER-ON	CONVERGE
010	DILATION	NONE, CONVERGE, NONE	DIVERGE
010			
010	EROSION	NONE, CONVERGE, NONE	CONVERGE
010	DILATION	CENTER-ON. CONVERGE, CENTER-ON	DIVERGE
111			
010	EROSION	CENTER-ON, CONVERGE. CENTER-ON	CONVERGE
100	DILATION	LEFT-ON, CENTER-ON, RIGHT-ON	DIVERGE
010			
001	EROSION	LEFT-ON, CENTER-ON. RIGHT-ON	CONVERGE
001	DILATION	RIGHT-ON, CENTER-ON. LEFT-ON	DIVERGE
010			
100	EROSION	RIGHT-ON, CENTER-ON. LEFT-ON	CONVERGE

If the structuring element is changed to

0	0	0
1	1	1
0	0	0

the erosion requires applying CENTER-ON to neurons in the layer $k + 1/2$ and CONVERGE to neurons in layer $k + 1$, and the dilation requires applying CENTER-ON to neurons in the layer $k + 1/2$ and DIVERGE to neurons in the layer $k + 1$. Another example using a different structuring element

1	0	0
0	1	0
0	0	1

is illustrated as follows. The LEFT-ON may be loaded to the first columns, CENTER-ON to the second columns, and RIGHT-ON to the third columns of each pyramid module in the layer $k + 1/2$. At the same time, CONVERGE is loaded in the layer $k + 1$. Therefore, any occurrence of line segments with 135° is extracted in the layer $k + 1$. Similarly, DIVERGE is also applied to the layer $k + 1$ to dilate any occurrence of "1" in the layer k to a diagonal line segment with the length 3.

A more complicated neural network could be constructed by extending DPLMs and their connections. By assigning various combinations of Boolean functions, a neural network may act in quite different behaviors. Thus, DPLM networks can be viewed as a class of *digital perceptron* that can support discrete forms of distributed parallel problem solving. They offer a new conceptual framework as well as a promising technology for developing artificial neural networks.

9.4.4 Multilayer Perceptron as Processing Modules

Multilayer perceptrons are feedforward nets with one or more layers, called *hidden layers*, of neural cells, called *hidden neurons*, between the input and output layers. The learning process of MLPs is conducted with the error back-propagation learning algorithm derived from the *generalized delta rule* (Rumelhart et al., 1986). According to Lippmann (1987), no more than three layers of neurons are required to form arbitrarily complex decision regions in the hyperspace spanned by the input patterns. By using the sigmoidal nonlinearities and the decision rule in selecting the largest output, the decision regions are typically bounded by smooth curves instead of line segments. If training data are sufficiently provided, an MLP can be trained to discriminate input patterns successfully. A discussion on limits of the number of hidden neurons in multilayer perceptrons can be found in Huang and Huang (1991).

Most of image processing operations, such as smoothing, enhancement, edge detection, noise removal, and morphological operations, require checking the values of neighboring pixels. Two basic morphological operations, dilation (similar to "expansion") and erosion (similar to "shrink"), are often combined in sequences for image filtering and feature extraction. The size of neighborhood may vary with applications. Eight-neighbor along with the center pixel is often used. The neighborhood processing can be implemented by lookup tables to associate the relationship between input and output values. The input–output association can be realized by neural network models with 9-input neurons (assuming that the 8-neighbor area is used) and one output neuron with one or more layers of hidden neurons in between. By iteratively providing input vectors to compute output from the lookup table and comparing with the desired output, the error will be used to adjust the weights of connections. Therefore, MLPs can gradually and adaptively learn the input–output association. If the training MLP converges with a set of weights, it can be used to perform the same transformation on any image.

Figure 9.13 shows the MLP modules designed as the general-purpose image processing modules. The more hidden layers are used and the more complicated discriminant regions it forms in the domain space spanned with input vectors. There is a trade-off between the complexity of MLP connections and the converge time of the MLP. Since the general-purpose MLP module is concerned, reasonably short training time to adapting the MLP and big capacity of connection weights are expected.

The error back-propagation training algorithm was given in Rumelhart et al. (1986). A sigmoidal logistic nonlinearity

$$f(\alpha) = \frac{1}{1 + e^{-(\alpha-\theta)}} \tag{9.34}$$

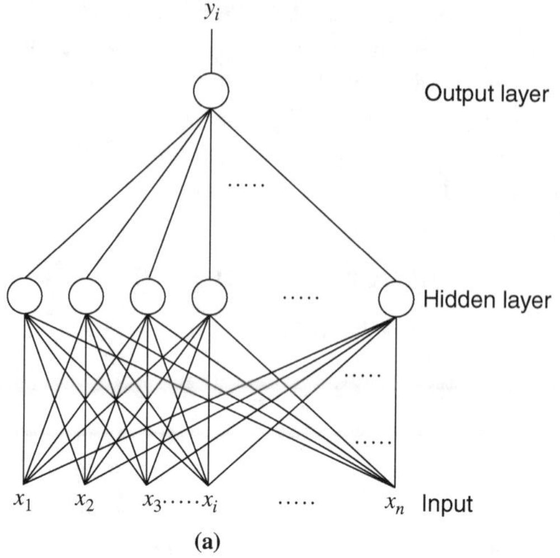

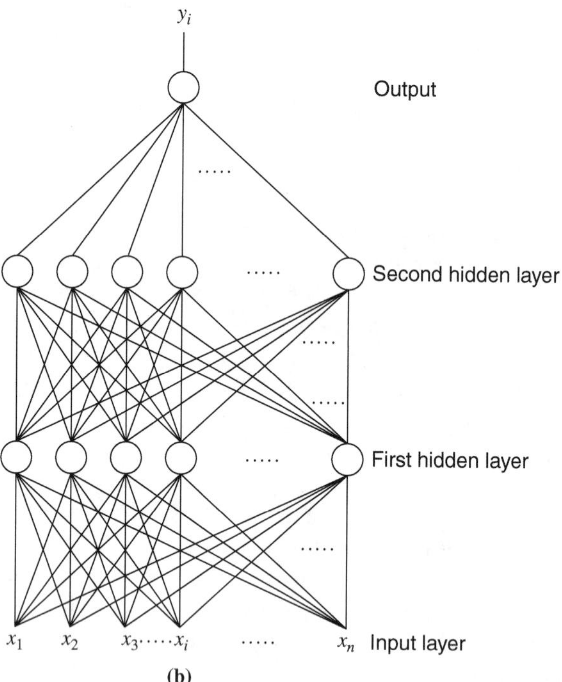

Figure 9.13 Multilayer perceptron modules for the general-purpose image processing. (a) A multilayer perceptron includes only one hidden layer. (b) A three-layer perceptron includes two hidden layers.

is used as the output function of neurons, where θ is a bias (or a threshold). Let $x_i^{(n)}$ denote the output of neuron i in layer n. Also, let w_{ij} denote the connection weight from neuron i in a lower layer to neuron j in the immediately higher layer. The activation values are calculated by

$$x_j^{(n+1)} = f\left(\sum_i w_{ij}x_i^{(n)} - \theta_j\right) \tag{9.35}$$

The major part of the algorithm is to adjust the connection weights according to the difference between the actual output of equation (9.35) and the desired output provided in the training patterns.

The equation for weight adjustment is

$$w_{ij}(t+1) = w_{ij}(t) + \eta\delta_j x_i^{(n)} \tag{9.36}$$

where η is the gain term (or called learning rate), δ_j is the error term for neuron j, and x_i is either the output of neuron i or an input when $n=1$. The error term δ_j is

$$\delta_j = \begin{cases} x_j^{(n+1)}\left(1-x_j^{(n+1)}\right)\left(d_j-x_j^{(n+1)}\right) & \text{if neuron } j \text{ is in the output layer} \\ x_j^{(n+1)}\left(1-x_j^{(n+1)}\right)\sum_k \delta_k w_{jk} & \text{if neuron } j \text{ is in a hidden layer} \end{cases} \tag{9.37}$$

where k is with respect to all the neurons in the layer where neuron j is located.

The training patterns can be generated by all the variations in a local window of the specified low-level image operations. For example, in a 3×3 neighboring area, there are up to $2^9 = 512$ different training patterns, with a vector of nine binary values and one desired output. For instance, a morphological erosion with the structuring element shown in Figure 9.14 will respond output 1 at the central pixel for the 16 input patterns in Figure 9.15 and output 0 for other input patterns.

0	1	0
1	1	1
0	1	0

Figure 9.14 A morphological structuring element.

0	1	0	1	1	0	0	1	0	0	1	0	0	1	1	1	1	0	0	1	0	0	1	1
1	1	1	1	1	1	1	1	1	1	1	1	1	1	1	1	1	1	1	1	1	1	1	1
0	1	0	0	1	0	1	1	0	0	1	1	0	1	0	1	1	0	1	1	1	0	1	1

1	1	1	1	1	0	0	1	1	1	1	0	0	1	1	1	1	1	1	1	1	1	1	1
1	1	1	1	1	1	1	1	1	1	1	1	1	1	1	1	1	1	1	1	1	1	1	1
0	1	0	0	1	1	1	1	0	1	1	1	1	1	1	0	1	1	1	1	0	1	1	1

Figure 9.15 Patterns having desired output 1 with erosion by the structuring element in Figure 9.14.

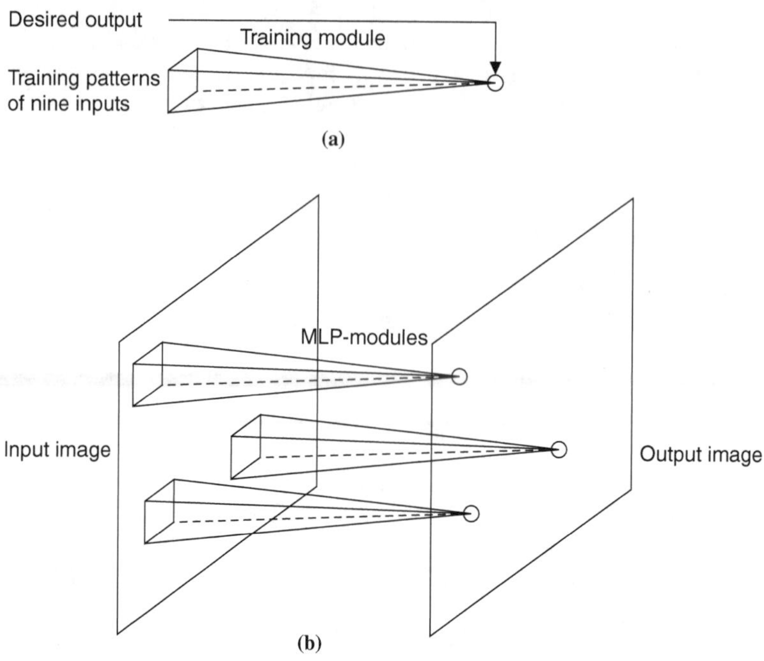

Figure 9.16 Overall system architecture with MLP modules for low-level image operations. (a) The pyramid is used for training tasks and storage of the trained weights. (b) Each pyramid represents an MLP module and receives m^2 inputs from its $m \times m$ neighboring pixels.

Figure 9.16 shows the overall architecture consisting of MLP modules for low-level image operations. There is one MLP module corresponding to each pixel in the input image. Every MLP module applies the operation, which has been trained to the pixel and its $m \times m$ neighboring pixels, where m is the window size of neighborhood. The window size can be changed depending upon the desired operations for training.

Since the same operation is simultaneously applied to all the modules for all the pixels in the input image, the set of connection weights are considered the same for all MLP modules. That is, all MLP modules in the same stage can share the same set of trained connection weights. This will significantly reduce the required number of local memory for connections weights in MLP module. In Figure 9.16a, the pyramid, called the training module, is used for training task and provides shared memory for connection weights. Thus, no local memory is required for individual MLP modules in between the input and output images to hold the connection weights. The training phase is first turned up at the training module. After the weight vector in the training module converges, the connection weights are frozen, shared, and retrieved by all the MLP modules to determine the activation value of each pixel after the operation is applied.

With the introduction of the error back-propagation learning algorithm, the MLP modules are capable of adapting themselves to the expected operations. It is also more flexible than those designed for dilation and erosion operations only (Krishnapuram and Chen, 1993).

1	1	1
1	1	1
1	1	1

Figure 9.17 The structuring element with all 1's.

During the experiments with different image operations, the training sets for some operations, such as dilation, erosion, and noise removal, have a small number of patterns with the desired output 1 or 0. For example, an erosion with the structuring element in Figure 9.14 has only 16 patterns with output 1, and an erosion with the structuring element shown in Figure 9.17 expects an output 1 only if all nine inputs are 1's.

By checking the training patterns, the values of some m^2 inputs do not even affect the output. Without considering a certain value of the central pixel, the training pattern set can be reduced. The same effect can be achieved by connecting the output neuron directly with the central input neuron and the MLP module operation is bypassed when the central pixel equals 0 or 1. For instance, applying an erosion to the central pixel with value 0 will never output 1. Also, applying a dilation to the central pixel with value 1 will always output 1. Such bypassing connections can be specified with the desired output as weights shared in all the MLP modules. The bypassing connections are intended to perform an exclusive OR between the central pixel and the unchanged input that is defined with respect to operations. The operation of an MLP is disable if the exclusive OR gets output 1. It means that the central pixel has the same value as specified in the shared module. Figure 9.18 illustrates the bypassing connection and expresses how it affects the operation of an MLP module. The bypassing control is to enable/disable the activation of neurons in the modules and is determined by

$$E = x_c \quad \mathbf{XOR} \quad U \tag{9.38}$$

where E means the enable/disable control to neurons, x_c is the value of the central pixel, and U is the expected input. For instance, $U = 1$ is set for dilation and $U = 0$ for erosion. The dashed lines with arrows in Figure 9.18 show the enable/disable controls to neurons, while the dotted lines indicate the bypassing connection for passing the value of the central pixel directly to the output. The output is defined by

$$\text{Output} = y \cdot E + x_c \cdot \bar{E} \tag{9.39}$$

where y is the output of the MLP module.

The proposed architecture can be used as a basis to organize multistage or recurrent networks. One may stack up with more than one stage as shown in Figure 9.19. By training the stages with a different set of patterns, one can provide and apply a sequence of image operations to the input image. The outputs of the last stage may be connected to the inputs of the first stage to form a recurrent architecture so that the same operations in sequence can be applied to the image for more than one iteration.

A layered neural network is a network of neurons organized in the form of layers. The first layer is the input layer. In the middle, there are one or more hidden

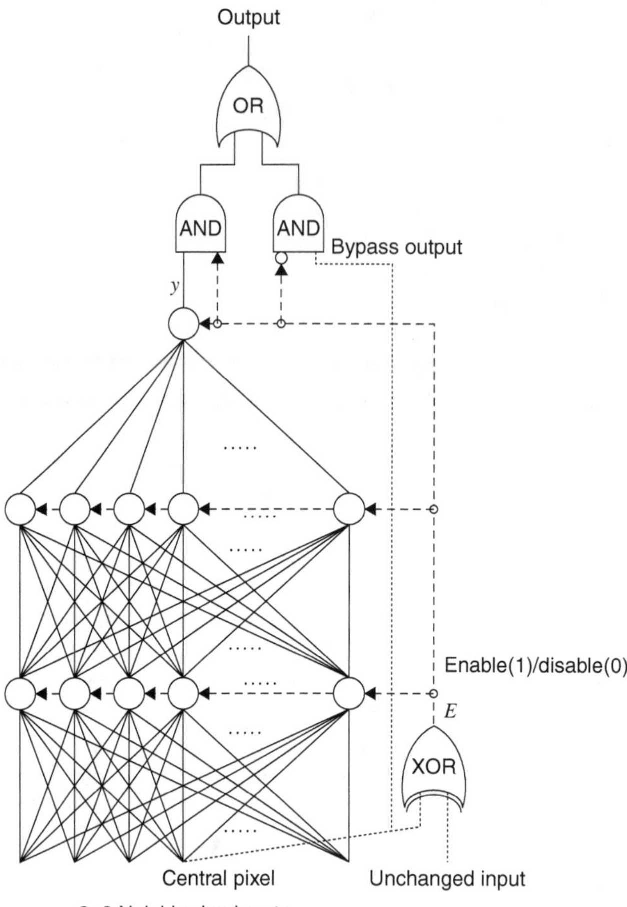

Figure 9.18 MLP module with a bypass control to enable/disable the activation of neurons in the modules. The dashed lines with arrows show the enable/disable controls to neurons, while the dotted lines indicate the bypassed connection for passing the value of the center pixel to the output.

layers. The last one is the output layer. The architecture of a layered neural network is shown in Figure 9.20. The function of the hidden neurons is to intervene between the external input and the network output. The output vector $\vec{\mathbf{Y}}(\vec{\mathbf{x}})$ is

$$\vec{\mathbf{Y}}(\vec{\mathbf{x}}) = \sum_{i=1}^{l} f_i(\vec{\mathbf{x}}) w_{ik} = \vec{\mathbf{F}} \cdot W \tag{9.40}$$

where $\vec{\mathbf{F}}$ is the hidden layer's output vector corresponding to a sample $\vec{\mathbf{x}}$, W is the weight matrix with dimensions of $l \times k$, l is the number of hidden neurons, and k is the number of output nodes. In such a structure, all neurons cooperate together to generate a single output vector. In other words, the output depends on not only each individual neuron, but also all of them.

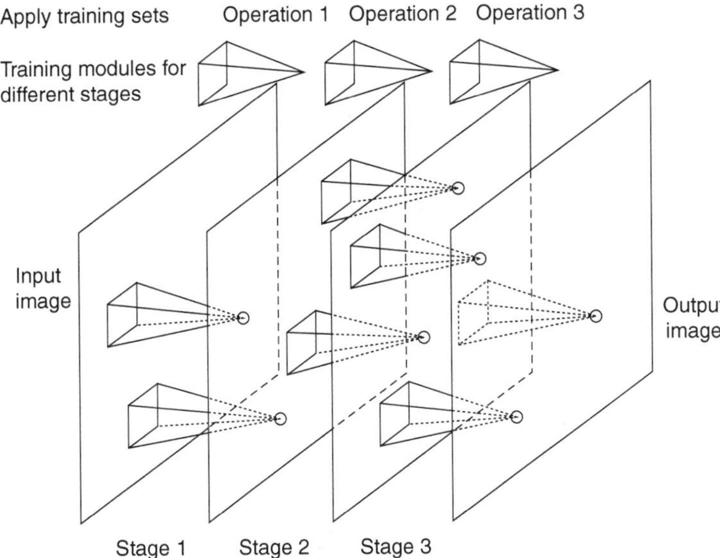

Figure 9.19 Stacked MLP-based architecture for multiple image operations. Only some MLP modules are shown to simplify the figure. The separated pyramids at the top of the figure are the training modules for different stages. There may be more stages stacked on the architecture.

Numerous neural network algorithms and architectures have been proposed. Hou et al. (2007) presented a recurrent neural network for hierarchical control of interconnected dynamic systems. Guo and Huang (2006) employed Lyapunov functions to establish some sufficient conditions ensuring existence, uniqueness, global asymptotic stability, and even global exponential stability of equilibria for the Cohen and Grossberg (1983) neural networks with and without delays. Yang (2006) presented a radial basis function neural network for discriminant analysis. Jin and Liu (2008) developed the wavelet basis function neural networks. Wang and

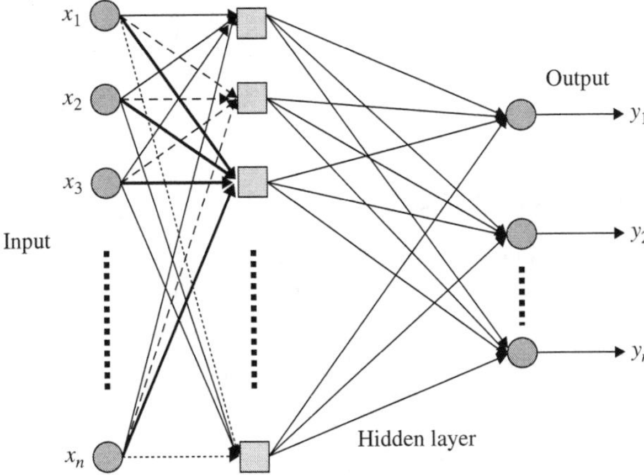

Figure 9.20 Architecture of a layered neural network.

Wang (2008) used neural networks for password authentication to overcome the shortcomings of traditional approaches.

Aizenberg et al. (2008) presented blur identification by multilayer neural network based on multivalued neurons. Wang and Peterson (2008) proposed constrained least absolute deviation neural networks. Nandedkar and Biswas (2007) proposed a fuzzy min–max neural network classifier with compensatory neurons that uses hyperbox fuzzy sets to represent the pattern classes. Goh et al. (2008) presented a geometrical measure based on the singular value decomposition to estimate the necessary number of neurons to be used in training a single-hidden-layer feedforward neural network. Phung and Bouzerdoum (2007) proposed a neural architecture for classification of visual patterns that is motivated by the two concepts of image pyramids and local receptive fields. Savran (2007) proposed multifeedback-layer neural network. Palmes et al. (2005) proposed mutation-based genetic neural network.

9.5 THE ADAPTIVE RESONANCE THEORY NETWORK

In this section, we are mainly concerned with the model of adaptive resonance theory (ART), which was developed by Carpenter and Grossberg (1987a, 1987b, 1990). This model, which forms clusters and is trained without a supervisor, can achieve the self-organization of input codes for pattern classification in response to random sequences of input patterns. With the concept of such an architecture, the process of adaptive pattern recognition is a special case of the more general cognitive process of hypothesis discovery, testing, searching, classification, and learning. This property makes it feasible to apply in more general problems of adaptively processing a big set of information sources and databases.

The ART model grew out of the analysis of a simple type of adaptive pattern recognition network, often called the *competitive learning paradigm* (Carpenter and Grossberg, 1988). Such a combination of adaptive filtering and competition is shared by many other models of pattern recognition and associative learning. There are three classes of ART architectures: ART1 (Carpenter and Grossberg, 1987a) for classifying binary patterns, ART2 (Carpenter and Grossberg, 1987b) for analog patterns, and ART3 (Carpenter and Grossberg, 1990) for parallel search of learned pattern recognition codes. Hwang and Kim (2006) developed a fast and efficient method for computing ART.

9.5.1 The ART1 Model and Learning Process

Figure 9.21 shows the typical architecture of an ART1-based neural network (Stork, 1989). The network consists of two subsystems: the *attentional subsystem* and the *orienting subsystem*, which is also called *novelty detector*. In the attentional subsystem, there are two layers of neurons, F and E, shown in Figure 9.21, which are called *feature representation field* and *exemplar* (*category*) *representation field*, respectively. Neurons in these two fields are fully connected by adaptive weights analogous to synapses in animals' nervous system. The stimuli in F pass through the

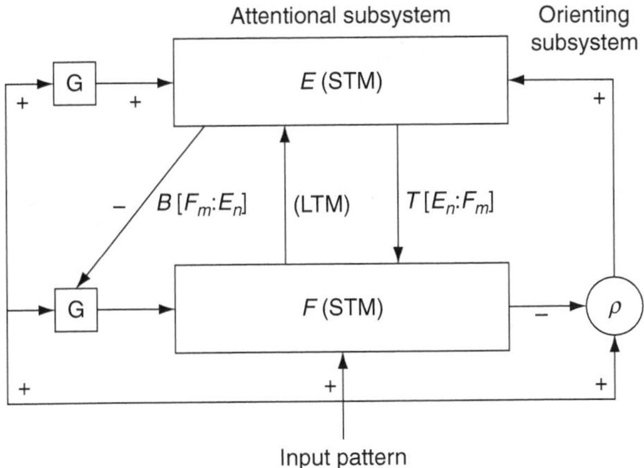

Figure 9.21 Typical ART1 architecture. Rectangles represent the attentional subsystem containing two fields where STM patterns are stored. The weights of connections between neurons in two layers determine the LTM traces. Squares with the label "G" indicate the gain control devices. The orienting subsystem with vigilance parameter ρ is represented by the circle.

connections $B[F_i : E_j]$ to the neurons in E and are multiplied by their corresponding weights. Each neuron in E sums up all of these weighted activities and responses accordingly. That is, the output μ_j of the neuron j is equal to

$$\mu_t[j] = \sum_i B_t[F_i : E_j]x_i \tag{9.41}$$

where $B_t[F_i : E_j]$ denotes weights of bottom-up connections from neuron F_i in layer F to the neuron E_j in layer E at time t, and x_i is the ith component of input pattern. The layer E is capable of choosing the neuron that receives the largest total input (i.e., "winner-take-all"). The MAXNET (Lippmann, 1987), for example, with heavy lateral inhibitory may choose the maximum from the input signals.

A set of interconnection weights acts as a filter, or *mask*, which is a term commonly used in the field of image processing. It transforms the activity of all neurons in layer F into a set of input values to E. Each neuron in E represents a different category. The winning neuron in E then triggers its own associative pattern learning within the vector of LTM traces that sends its inputs through the adaptive filter to E. This is very useful for applications with categorical perception, that is, classifying each input pattern as being in one and only one category.

During learning process, the first pattern is classified into the first category represented by the first neuron in the output layer, and the succeeding patterns are classified into either a newly created category or a previously created category. In addition to bottom-up connections, for computing response (i.e., sum of weighted inputs), the network is also designed to have a set of feedback (also called top-down) connections for verifying the distance between the input pattern and the exemplar related to the winning neuron. The set of weights of connections from any neuron in layer E to all neurons in layer F represents a template (or a set of critical features) for its related category.

A matching threshold (ρ) called *vigilance parameter* (ranged between 0.0 and 1.0) was introduced to determine how close a new input pattern must be to a stored exemplar to be considered similar. (Conversely, the value $1-\rho$ determines how large a mismatch can be tolerated.) An orienting system, represented by a circle with ρ in Figure 9.21, acts as a novelty detector and handles the vigilance testing tasks. It receives input signals from both the input pattern with excitatory connections and the overall activity in layer F with inhibitory connections.

The vigilance test performed by this subnetwork can be expressed by

$$\frac{\sum_i T[E_j : F_i]x_i}{||\mathbf{X}||} > \rho \tag{9.42}$$

where $T[E_j : F_i]$ is the weight of the connection from neuron E_j in E to neuron F_i in F, x_i the activation value of neuron i in layer F, and $||\mathbf{X}||$ the norm (length) of the input pattern vector. For ART1, $||\mathbf{X}|| = \sum_i |x_i|$. If the input pattern satisfies the above inequality, then it is classified into the category corresponding to the winning neuron. Otherwise, the subsystem sends a reset signal through the excitatory connection to the winning neuron in order to temporarily disable its output so that the system may choose another winning neuron. Such a task will be repeated until either an exemplar similar to the input pattern is chosen or no more neuron is there to choose, and then a new neuron is created in E and assigned to represent the new category.

A top-down connection linking two active neurons approaches a strength of 1.0 because of the following dynamic equations:

$$T_0[E_j : F_i] = 1 \quad \text{and} \quad T_{t+1}[E_j : F_i] = T_t[E_j : F_i]x_i \tag{9.43}$$

where $T_0[E_j : F_i]$ is the initial weight and $T_t[E_j : F_i]$ is the weight of top-down connection from neuron E_j in E to neuron F_i in F at time t. It will be unchanged if both neurons are inactive. Suppose that a new pattern $\mathbf{X}$, whose ith component x_i is 0, is classified into a category represented by neuron E_j. If the original exemplar of the category kept in the top-down connections makes the strength of $T[E_j : F_i]$ equal to 1.0, then $T[E_j : F_i]$ decays toward 0.0 after adaptation. Therefore, the exemplar of each category will keep only the common features of all the input patterns classified into the same category.

The learning rule for the bottom-up connections is similar to top-down except that the increasing value for a connection linking two active neurons is not 1.0. Instead, the value depends on the number of active neurons. The following equations give the initial bottom-up weight and the adaptation value during learning process for the bottom-up connection from neuron F_i in F to neuron E_j in E (suppose $q \times q$ images are used as inputs):

$$\begin{cases} B_0\left[E_j : F_i\right] = \dfrac{1}{1+q^2} \\ B_{t+1}\left[E_j : F_i\right] = \dfrac{T_t[E_j : F_i]x_i}{0.5 + \sum_i T_t[E_j : F_i]x_i} \end{cases} \tag{9.44}$$

Therefore, we may see that for each training pattern, the final strength of bottom-up connections between two active neurons will equally divide the maximum output (i.e. 1.0) of a neuron approximately. That is, at any time, the total strength of bottom-up connections to any neuron in E will be close to 1.0. This ensures that the response value of the neurons in E is between 0.0 and 1.0.

9.5.2 The ART2 Model

The adaptive resonance theory 2 (ART2) family of neural network architectures proposed by Carpenter and Grossberg (1987b) has been proven to be proficient in recognizing both binary- and analog-valued patterns. A typical ART2 neural architecture is composed of two layers of fully interconnected neurons. Each neuron has many connections emanating from and terminating on it. The adaptive connections between neurons store the long-term memory (LTM) traces of the network. The LTM represents the information that the network has learned, that is, the structure of the categories, into which it has segregated the input patterns. The connections between neurons may be either excitatory that increase a neuron's activation, or inhibitory that decrease a neuron's activation.

The two layers (or fields) of neurons in the ART2 architecture form the attentional subsystem. The first field is named the feature representation field or F_1. Each neuron in F_1 contains six processing elements (PEs) that form three intra-PE sublayers. The main function of the feature representation field is to enhance the current input pattern's salient features while suppressing noise (Atal, 1976). This is achieved through pattern normalization and thresholding. Normalization allows an equitable comparison to be made between the input pattern and the patterns stored in the network's LTM traces (Shih et al., 1992). Thresholding maps the infinite domain of the input patterns to a prescribed range. Both normalization and thresholding are required for the processing of analog patterns. They are absent from the ART1 model since it is only capable of processing binary-valued patterns.

The second layer in the attentional subsystem is called the category representation field or F_2. Each neuron in this field represents a category (or class) that has been learned by the network. The connections emanating from a particular F_2 neuron store the exemplar of the category it represents. The exemplar represents the essential features of the category, that is, those characteristics exhibited by all the patterns belonging to this class.

9.5.2.1 Learning in the ART2 Model ART2 utilizes an unsupervised competitive learning technique, in which patterns are represented by points in an

N-dimensional feature space, where N denotes the number of attributes in the patterns. Let $(a_1, a_2, \ldots, a_N)$ and $(b_1, b_2, \ldots, b_N)$ represent two patterns. Pattern similarity is assessed on the basis of a distance measurement in the feature space. The most common distance measurement is the Euclidean distance (d_e) that is computed as

$$d_w[(a_1, a_2, \ldots, a_N), (b_1, b_2, \ldots, b_N)] = \sqrt{(a_1-b_1)^2 + (a_2-b_2)^2 + \cdots + (a_N-b_N)^2} \tag{9.45}$$

Patterns that are deemed to be sufficiently close to one another are placed in the same category. The N-dimensional location of the centroid of a class represents the class's exemplar. An unsupervised learning procedure attempts to automatically discover the distributions and centroids of the categories for the patterns presented. Optimally, ART2 will discover pattern classes with minimum radii and maximum intercentroid distances; that is, intraclass distance is minimized and interclass distance is maximized. However, this goal is not always attainable when two classes overlap with each other as illustrated in Figure 9.22.

The learning technique is considered to be competitive since each F_2 neuron competes with the others for including the current input pattern in its category code. ART2 utilizes a "winner-take-all" classification strategy, such as MAXNET (Haykin, 1992) that operates in the following manner:

1. An input pattern is presented to the feature representation field where it is normalized and thresholded.
2. The resultant signal, which is called the short-term memory (STM), is passed through the bottom-up connections to the category representation field.
3. Each established class in F_2 responds to the signal with an activation level that it sends to itself through excitatory connections and to all its neighbors through inhibitory connections.

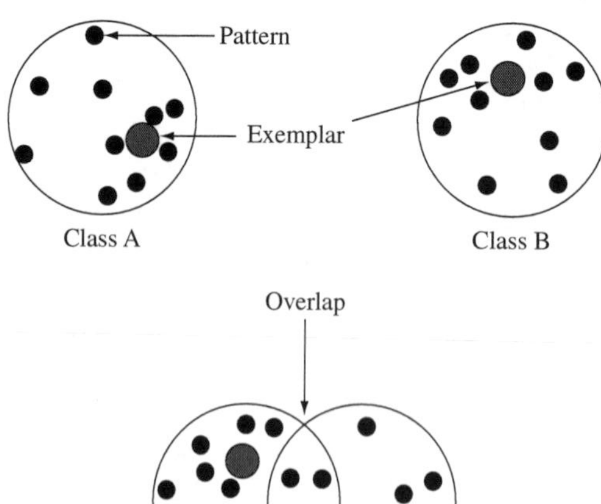

Figure 9.22 Illustration of the class separation problem.

4. Eventually, the F_2 neuron with the highest activation will manage to inhibit the others sufficiently, so that they will become disabled. The sole remaining active F_2 neuron is considered to be the winner; that is, it is the category whose exemplar most resembles the current input pattern.

Having selected the winner in this manner, the second subsystem of ART2, named the *orienting subsystem*, is activated. The orienting subsystem determines whether the exemplar that is stored in the winning neuron's LTM traces sufficiently resembles the circulating STM pattern to be considered a match. The degree of match between two patterns is related to the cosine of the angle between them in feature space. Patterns that are very similar are nearly parallel to each other, while greatly dissimilar patterns are orthogonal to each other.

A matching threshold called the vigilance parameter (which ranges from 0.0 to 1.0) determines how similar the input pattern must be to the exemplar to be considered a match. If the degree of match computed by the orienting subsystem exceeds the vigilance parameter, a state of resonance is attained and the STM pattern at F_1 is merged onto the winning neuron's LTM traces. Otherwise, the orienting subsystem sends a reset signal through the excitatory connections to the winning neuron to inhibit it from competing again for the current input pattern. This searching process is repeated until either an F_2 neuron passes the vigilance test or all established F_2 neurons have failed the test. In the latter case, a new category is established in the next available F_2 neuron.

The stability of pattern code learning in ART2 is achieved via readout of the winning neuron's learned category exemplar. The F_2 neuron that wins the competition for the current input pattern sends an STM signal representing the category exemplar through the top-down connections to F_1. The structure of the F_1 neurons allows the bottom-up (input) and top-down (exemplar) STM signals to be combined. This mixing process is essential for achieving stability since without it confusion between useful pattern differences and spurious baseline fluctuations could easily upset the matching process and cause spurious reset events to occur.

The actual learning process, whereby the current input pattern is encoded into the network's memory, involves modification of the bottom-up and top-down LTM traces that join the winning F_2 neuron to the feature representation field. Learning either refines the code of a previously established class based on any new information that is contained in the input pattern, or initiates code learning in a previously uncommitted F_2 neuron. In either case, learning only occurs when the system is in a resonant state. This property ensures that an input pattern does not obliterate information that has been previously stored in an established class.

9.5.2.2 *Functional-Link Net Preprocessor* Performance of the ART2 model is measured by its ability to separate pattern clusters in the feature space. Unless a considerable degree of pattern separation is attained, there is likely to be much overlap of these clusters. Pao (1989) describes an approach toward increasing the separability of patterns by enhancing them in a linearly independent manner. In essence, the input patterns are represented in a space of increased dimensions; this facilitates ART2's ability to learn hyperplanes for separation.

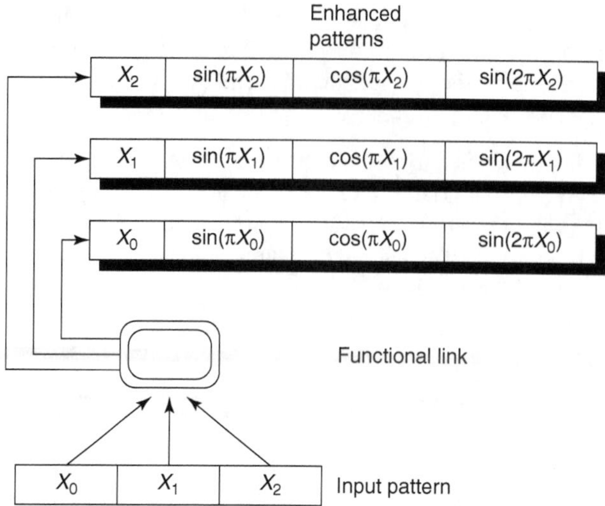

Figure 9.23 Functional-link net: functional expansion model.

The functional-link net enhances the input patterns without actually introducing any new information. Rather, it expands each attribute in the input pattern by evaluating a set of linearly independent functions with the attribute as the argument. There are two variations of the functional-link net and each produces a different effect. In the functional expansion model shown in Figure 9.23, the functional link evaluates a set of three functions that span a portion of feature space for each attribute in the input pattern. The functional expansion model was demonstrated by Pao (1989) to be effective in training a network to learn a function from a limited set of sample points. In this implementation, each attribute X_i in the input pattern is mapped into the following set of 10 functions:

$$\{X_i, X_i^2, X_i^3, \sin(\pi X_i), \cos(\pi X_i), \sin(2\pi X_i), \cos(2\pi X_i), \sin(3\pi X_i), \cos(3\pi X_i), \sin(4\pi X_i)\}$$

In the tensor or outer-product model shown in Figure 9.24, each attribute of the input pattern multiplies the remainder of that pattern. To limit the size of these derived patterns, attribute X_i only multiplies attribute X_j if $j \geq i$. The outer-product model introduces higher order terms in the enhanced representation in the sense that many of the terms represent joint activations (Pao, 1989).

We incorporate the functional-link net into the ART2 model as an input pattern preprocessor. The functional-link net operates directly on the original input pattern and passes the enhanced pattern representation to the input layer of the network. An additional benefit of the functional-link net is that it allows both supervised and unsupervised learning to be performed by the same network. Furthermore, the degree of pattern enhancement that can be attained through the net eliminates the need for network architectures that contain hidden layers.

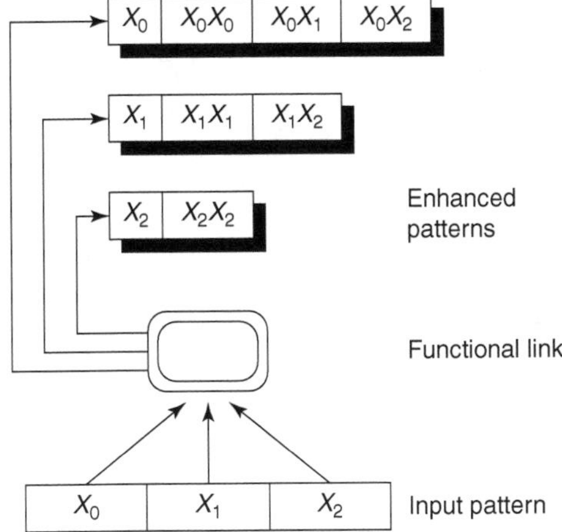

Figure 9.24 Functional-link net: outer-product expansion model.

9.5.3 Improvement of ART Model

In implementing the algorithm of Carpenter and Grossberg by software simulation dealing with optical character recognition, we obtained the same result of category classification as described in Carpenter and Grossberg (1987a), Carpenter (1989), and Stork (1989). However, we have observed that the input patterns are classified with variance to their training sequences. For instance, if the sequence is {**E**, **F**, **L**}, "**F**" is classified as the same category with the exemplar "**E**" and replaces "**E**" as a new exemplar, and then "**L**" is classified as another new category. If the sequence is {**E**, **L**, **F**}, "**L**" belongs to the category with the exemplar "**E**" and replaces it as a new exemplar, and then "**F**" is classified as another new category. If the sequence is {**F**, **L**, **E**}, then all three characters are in different categories. Even though the vigilance parameter is adjusted, the variation of categories still exists.

In this section, we will first give the problem analysis and then propose an improved model and illustrate the experimental results.

9.5.3.1 Problem Analysis In the ART model, each neuron in the higher layer E sums up all the weighted activities from neurons in the lower layer F. The activation is calculated by

$$\mu_{E_j}^{(t)} = \sum_i B_{F_i,E_j}^{(t)} \cdot x_i \tag{9.46}$$

where $B_{F_i,E_j}^{(t)}$ denotes weights of bottom-up connections from neuron F_i to neuron E_j at time t and x_i is the ith component of the input pattern. When a winning neuron E_{j*} is chosen based on the maximum activation, the orienting subsystem activates for the vigilance test, which can be expressed as

$$\frac{\sum_i T_{E_{j*},F_i}^{(t)} \cdot x_i}{||\mathbf{X}||} \geq \rho \tag{9.47}$$

where $T_{E_{j*},F_i}^{(t)}$ denotes weights of top-down connections from the winning neuron E_{j*} to neuron F_i at time t, and $||\mathbf{X}||$ is the norm (or magnitude) of the input pattern vector, that is, $||\mathbf{X}|| = \sum_i |x_i|$. In equation (9.47) for vigilance testing, the numerator picks up all common pixels between the input pattern and the exemplar it is comparing with, and the denominator is the norm of the input vector. This ratio represents the percentage of the pixels of the input pattern existing in the exemplar. Since "**F**" is a subset of "**E**", the ratio is 1.0; that is, all the pixels of "**F**" are included in "**E**". But, how can we take into account the pixels of "**E**" that are not in "**F**" (the line at the bottom)? The orienting subsystem is supposed to be able to count not only the pixels in "**F**" and not in "**E**" but also the pixels in "**E**" and not in "**F**" in order to give a fair judgment of the similarity.

Another issue is that for each new training pattern the network repeats choosing the winning neuron and performing vigilance testing until a category is chosen. Whereas if none of the existing neurons in E satisfies the vigilance test, a new neuron is created to represent the new category using the pattern as an exemplar. This procedure is time-consuming and the performance is the worst when a new input pattern is quite different from all the existing exemplars. To reduce the searching time of the model, a minimum activation response may be set up so that the model need not go through the neurons with the lower response.

9.5.3.2 *An Improved ART Model for Pattern Classification*

The task of vigilance testing is now modified. We use a bidirectional testing to determine the similarity between an input pattern and the exemplar of a chosen category. In addition to equation (9.47), another inequality is introduced as follows to determine how similar an exemplar is to the input pattern. That is,

$$\frac{\sum_i T^{(t)}_{E_{j*},F_i} \cdot x_i}{||\mathbf{T}||} \geq \rho \tag{9.48}$$

where $||\mathbf{T}||$ is norm (or magnitude) of the exemplar vector, that is, sum of top-down weights from the winning neuron E_{j*} to all the neurons in F. The difference between equation (9.47) and equation (9.48) is the denominator; one is the norm of the input vector and the other is the norm of the exemplar vector.

The ratio represents the percentage of the pixels of the exemplar existing in the input pattern. Only if these two equations are satisfied, the new pattern is associated and combined with the exemplar, otherwise, the next exemplar is chosen to compare with. It has to be noted that the vigilance parameters in equations (9.47) and (9.48) could be selected differently. Therefore, an improved ART model shown in Figure 9.25 includes a set of new devices that accumulates the total weights of top-down connections of an active neuron E_j and sends them to the orienting subsystem.

The algorithm of the improved ART model is described as follows:

Step 1: Initialize weights.

$$T^{(0)}_{E_j,F_i} = 1.0 \tag{9.49}$$

$$B^{(0)}_{F_i,E_j} = \frac{1}{1+N} \tag{9.50}$$

where $1 \leq i \leq N$ and $1 \leq j \leq M$ (N is number of elements in the input vector and M is number of existing exemplars). Also, the vigilance parameter, ρ, where $0.0 \leq \rho \leq 1.0$, is selected.

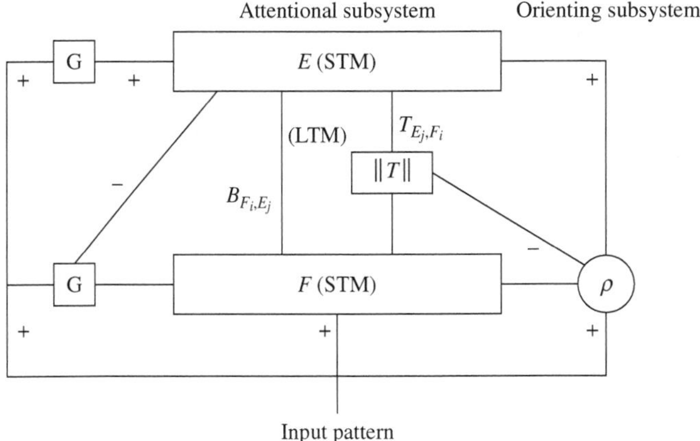

Figure 9.25 An improved ART model. The activation in layers F and E forms a STM and the connections including bottom-up and top-down construct a LTM. Squares with the label "**G**" indicate the gain control devices and the orienting subsystem with vigilance parameter ρ is represented by a circle.

Step 2: Read input pattern **X**.

Step 3: Compute neurons' responses.

The output of the neuron E_j is computed by

$$\mu_{E_j}^{(t)} = \sum_{i=1}^{N} B_{F_i,E_j}^{(t)} \cdot x_i \tag{9.51}$$

where $B_{F_i,E_j}^{(t)}$ denotes weights of bottom-up connections from F_i to E_j at time t, and x_i is the ith component of the input pattern vector **X** that can be either 0 or 1 (binary).

Step 4: Select a winning neuron with the highest response.

Let C be the total number of currently used exemplar neurons.

$$\mu_{E_{j*}}^{(t)} = \max\{\mu_{E_j}^{(t)} | 1 \leq j \leq C \quad \text{and} \quad j = C+1\} \tag{9.52}$$

where the plus 1 is the first available unused neuron and E_{j*} is the winning neuron.

Step 5: Bidirectional vigilance.

Let $||\mathbf{X}|| = \sum_i |x_i|$, $||\mathbf{T}|| = \sum_i T_{E_{j*},F_i}$, and $||\mathbf{T}\cdot\mathbf{X}|| = \sum_i T_{E_{j*},F_i} \cdot x_i$. If $||\mathbf{T}\cdot\mathbf{X}||/||\mathbf{X}|| \geq \rho$ and $||\mathbf{T}\cdot\mathbf{X}||/||\mathbf{T}|| \geq \rho$, go to step 7, otherwise, go to step 6.

Step 6: Disable unmatched category.

Disable temporarily the output of the winning neuron chosen and go to step 4 for choosing the next winning neuron.

Step 7: Adjust weights of the winning neuron.

$$T_{E_{j*},F_i}^{(t+1)} = T_{E_{j*},F_i}^{(t)} \cdot x_i \tag{9.53}$$

$$B_{E_{j^*},F_i}^{(t+1)} = \frac{T_{E_{j^*},F_i}^{(t)} \cdot x_i}{0.5 + \sum_{i=1}^{N-1} T_{E_{j^*},F_i}^{(t)} \cdot x_i} \tag{9.54}$$

Step 8: Enable all neurons and repeat the whole procedure by going to step 2.

The intention of adjusting the weights in the top-down connections is to store the information with respect to the exemplar of each category. Equation (9.53) shows that the new weight is derived from the intersection of the existing exemplar and the input pattern. Since the top-down weights are initialized to be all 1's, the first classified exemplar is always the same as the input pattern; the updated exemplar will be the structure of the common pixels of all the patterns in each category. This step will ensure that the classified categories are independent of the input sequences.

Equation (9.50) initializes the weights of all bottom-up connections to be $1/(1+N)$, where N is the number of elements in $\mathbf{X}$. Thus, when an $\mathbf{X}$ is presented, the magnitude of any unused neuron's response will be $||\mathbf{X}||/(1+N)$. Referring to equations (9.47) and (9.48), a perfectly matched pattern will cause the neuron to respond 1.0, and a pattern with totally unmatched pixels will respond with the value 0.0. The higher the response value is, the more matched pixels the pattern has. Therefore, the searching procedure may skip checking the neurons with responses lower than those of unused neurons. In step 4 of the algorithm, the searching always chooses the maximum of all active E neurons plus the first available unused neuron. If all active neurons with higher response than the unused neuron could not pass the vigilance testing, the unused neuron is assigned to represent the new created category. This can save significant time in searching most of the unmatched categories.

9.5.3.3 *Experimental Results of the Improved Model*

The input patterns used in this experiment are 16×16 optical characters. Therefore, a 256-neuron layer is designed for F and a 26-neuron layer for E. The vigilance parameter is set to be 0.8. Figure 9.26 shows the result of character classification and the exemplar in each category.

While the character "**F**" is trained, the existing exemplars already have "**A**", "**B**", "**C**", "**D**", and "**E**", and the exemplar "**E**" responds with the highest value to be the winning neuron. Thus, the vigilance testing is

$$\frac{||\mathbf{T} \cdot \mathbf{X}||}{||\mathbf{X}||} = 1.00 > \rho = 0.8$$

and

$$\frac{||\mathbf{T} \cdot \mathbf{X}||}{||\mathbf{T}||} = 0.72 < \rho = 0.8$$

Since the pattern "**F**" is a subset of the pattern "**E**", the first ratio having value 1.00 signifies that the testing passes. But, on the other hand, "**E**" has more pixels than "**F**" has. This means that the second testing fails. The second highest responding neuron, that is, the category "**B**", is next chosen and the vigilance testing still fails. Thus, the system creates a new category to represent "**F**".

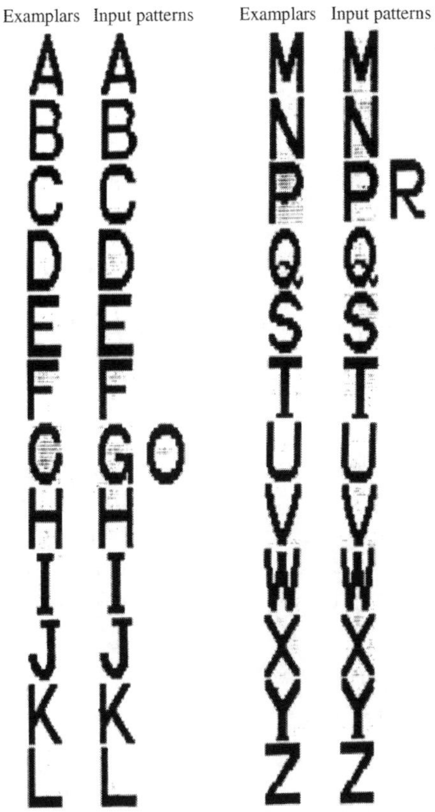

Figure 9.26 Result of character classification and the exemplar in each category.

With $\rho = 0.8$, the two pairs of characters {"**G**", "**O**"} and {"**P**", "**R**"} are classified into the same category, respectively. To achieve 100% classification rate, the higher vigilance parameter, for example, 0.9, is assigned. In addition, certain noisy patterns are experimented to test the system performance. The weights of bottom-up and top-down connections are fixed. As shown in Figure 9.27, these noisy patterns can be recognized and classified into their proper categories.

Numerous ART algorithms and architectures have been proposed. Baraldi and Alpaydin (2002) proposed constructive feedforward ART clustering networks. Vasilic and Kezunovic (2005) presented an advanced pattern recognition algorithm for classifying the transmission line faults, based on supervised ART neural network with fuzzy decision rule. Araujo (2006) presented prune-able fuzzy ART neural architecture for robot map learning and navigation in dynamic environments. Chen et al. (2005) presented incremental communication for ART networks. Projective adaptive resonance theory (PART) neural network developed by Cao and Wu (2002) recently has been shown to be very effective in clustering data sets in high-dimensional spaces. Cao and Wu (2004) further provided a rigorous proof of these regular dynamics of the PART model when the signal functions are special step functions.

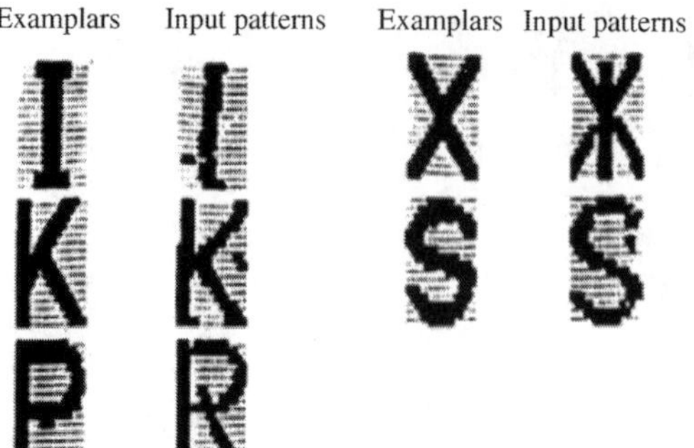

Figure 9.27 Testing patterns with noise and variances are also properly recognized and classified.

9.6 FUZZY SETS IN IMAGE ANALYSIS

Ever since Zadeh introduced the concept of fuzzy set theory in 1965, it has found many applications in a variety of fields. The fuzzy logic has been developed in order to capture the uncertainties associated with human cognitive processes such as in thinking, reasoning, perception, and so on. A conventional set is defined as a collection of elements that have some common properties. The sets are known as *crisp sets* and are defined by a characteristic function as

$$f(x) = \begin{cases} 1 & \text{if } x \in A \\ 0 & \text{if } x \notin A \end{cases} \tag{9.55}$$

where A is any finite set. However, the object classes generally encountered in the real world are not so "precisely" or "crisply" defined. In most cases, several ambiguities arise in the determination of whether a particular element belongs to a set or not. A good example mentioned by Zadeh is a class of animals. This class clearly includes dogs, cats, tigers, and so on and excludes rocks, plants, houses, and so on. However, an ambiguity arises in the context of objects such as bacteria and starfish with respect to the class of animals.

Unlike a crisp set in which each object is assigned a value of 0 or 1, each object in a fuzzy set is given a certain "degree of membership" that denotes the degree of belongingness of the object to the set. Therefore, a fuzzy set can be considered as a class of objects with a continuum of membership grades. The membership function assigns to each object a degree of membership value ranging from 0 to 1. The closer the membership value is to 1, the more the object belongs to the set, and vice versa.

9.6.1 Role of Fuzzy Geometry in Image Analysis

Image processing inherently bears some fuzziness in nature. This is due to the fact that the regions in the image are not always crisply defined. The approach to the analysis or interpretation of an image requires traditionally to segment the image into meaningful regions, extract the features of each region, and finally construct the relationships among the regions. However, due to the fuzzy behavior of the images, it is convenient to regard the regions of the image as fuzzy subsets of the image. Rosenfeld (1984) generalized the standard geometrical properties to the properties of fuzzy image geometry, such as topological connectedness, adjacency, surroundedness, area, perimeter, and so on based on which many significant algorithms utilizing the concepts of fuzzy geometry in the field of image analysis have been developed.

9.6.2 Definitions of Fuzzy Sets

A fuzzy set of a grayscale image S is a mapping m from S into the values ranging [0, 1]. For any pixel $p \in S$, $\mu(p)$ is called *the degree of membership* of p. The mapping function μ is any function that follows the properties of symmetry and ambiguity around the crossover point. From the definition above, one may assume that the fuzzy set is identical to the probability function. However, there are significant differences between these two. For example, the summation of all the values of the probability function should be equal to 1, which is not the case in fuzzy sets. The summation of all the values of the degree of membership in a fuzzy set need not be equal to 1.

Several definitions in fuzzy sets are given below.

(a) *Empty set:* A fuzzy set μ is said to be empty, denoted by $A = \phi$, iff $\mu(x) = 0$, $\forall x \in S$.

(b) *Equality:* Two fuzzy sets μ and ν are said to be equal, denoted by $\mu = \nu$, iff $\mu(x) = \nu(x)$, $\forall x \in S$.

(c) *Support:* The support of a fuzzy set μ is an ordinary subset of S; that is, $\text{Supp}\,(\mu) = \{x \in S | \mu(x) > 0\}$.

(d) *Crossover points:* The elements $x \in S$ such that $\mu(x) = 1/2$ are called crossover points of μ.

(e) *Normalized fuzzy set:* μ is said to be normalized iff $\exists\, x \in S$, $\mu(x) = 1$.

(f) *Height of* $\boldsymbol{\mu}$*:* The height of a fuzzy set μ is defined as $\text{hgt}(\mu) = \sup\{\mu(x)\}$; that is, the height of $\mu$is defined to be the least upper bound of $\mu(x)$. It is therefore obvious that for a normalized fuzzy set m, $\text{hgt}(\mu) = 1$.

(g) *Cardinality of a fuzzy set:* When X is a finite set, the cardinality $|A|$ of a fuzzy set A is defined as $|A| = \sum \mu(x)$, which is called the *power* of A.

(h) *Convexity of a fuzzy set:* There are several alternative definitions to the convexity property of fuzzy sets. Listed below are three of major ones.

Definition 9.1. A fuzzy set m is said to be convex if for every p, q belonging to S, and all r on the line segment pq, it is true that $\mu(r) \geq \min[\mu(p), \mu(q)]$. This implies that μ is fuzzily convex if it is convex in the ordinary sense.

Definition 9.2. A fuzzy set m of X is convex iff the sets T defined by $T_\alpha = \{x \in S | \mu(x) \geq \alpha\}$ are convex for all in the interval [0,1]. The sets T_α are called the "level sets" of μ.

Definition 9.3. A direct definition of convexity is the following: μ is convex iff

$$\mu[\lambda x_1 + (1-\lambda)x_2] \geq \min\,[\mu(x_1), \mu(x_2)]$$

for every x_1, x_2 belonging to S and in [0,1].

(i) *Cross-Section:* The cross-section of m by a line l is defined as the restriction of μ to l.

(j) *Star-shapedness:* μ is star-shaped from p if its cross sections by lines through p are all convex.

9.6.3 Set Theoretic Operations

(a) *Union:* The union of two fuzzy sets A and B with their respective membership functions $f_A(x)$ and $f_B(x)$ is a fuzzy set C, denoted as $C = A \cup B$, whose membership function is

$$f_C(x) = \max[f_A(x), f_B(x)], \forall x \in X$$

The union of two fuzzy sets A and B can be interpreted as the smallest fuzzy set containing both A and B. Let D be any fuzzy set containing both A and B. Since $f_D \geq f_A$ and $f_D \geq f_B$, $f_D \geq \max[f_A, f_B] = f_C$.

(b) *Intersection:* The intersection of two fuzzy sets A and B with their respective membership functions f_A and f_B is a fuzzy set C, denoted as $C = A \cap B$, whose membership function is $f_C(x) = \min[f_A(x), f_B(x)], x \in X$. The intersection of two fuzzy sets can also be interpreted as the largest fuzzy set that is contained in both.

Corollary: A and B are disjoint if their intersection C is empty, that is, $f_C(x) = 0$.

(c) *Complement:* The complement of a fuzzy set μ is denoted by $\bar{\mu}$ and is defined as $\bar{\mu}(x) = 1 - \mu(x), \forall x \in S$.

(d) *Containment:* A fuzzy set μ is said to be contained in another fuzzy set ν iff $\mu(x) \leq \nu(x), \forall x \in S$.

Properties:

(a) *Commutativity:* $A \cup B = B \cup A$; $A \cap B = B \cap A$.

(b) *Associativity:* $(A \cup B) \cup C = A \cup (B \cup C)$; $(A \cap B) \cap C = A \cap (B \cap C)$.
Note: The above two properties follow immediately for union and intersection because their corresponding operators of max and min are associative and commutative.

(c) *Idempotency:* $A \cup A = A$; $A \cap A = A$.

(d) *Distributivity:* $A \cup (B \cap C) = (A \cup B) \cap (A \cup C)$;
$A \cap (B \cup C) = (A \cap B) \cup (A \cap C)$.

(e) $A \cap \phi = \phi$; $A \cup X = X$.

(f) *Identity:* The identity element with respect to union is ϕ and with respect to intersection is X.

(g) *Absorption:* $A \cup (A \cap B) = A$; $A \cap (A \cup B) = A$.

(h) *De Morgan's laws:* $(\overline{A \cup B}) = \bar{A} \cap \bar{B}$; $(\overline{A \cap B}) = \bar{A} \cup \bar{B}$.

(**i**) *Involution:* $A = A$.

(**j**) *Equivalence formula:* $(A \cup B) \cap (A \cup C) = (A \cap B) \cup (A \cap C)$.

(**k**) *Symmetrical difference formula:* $(A \cap B) \cup (A \cap C) = (A \cup B) \cap (A \cup C)$.

Alternative operators (probabilistic operators):

1. *Intersection:* For every $x \in X$, $\mu_{A \bullet B}(x) = \mu_A(x) \bullet \mu_B(x)$.
2. *Union:* For every $x \in X$, $\mu_{A+B}(x) = \mu_A(x) + \mu_B(x) - \mu_A(x) \bullet \mu_B(x)$.

Numerous fuzzy classification methods have been proposed. Sousa et al. (2007) proposed a fuzzy classification system to perform word indexing in ancient printed documents. Hamilton-Wright et al. (2007) proposed fuzzy classification using pattern discovery. Mansoori et al. (2008) considered the automatic design of fuzzy rule-based classification systems from labeled data and developed a steady-state genetic algorithm for extracting fuzzy classification rules from data. Nowicki (2008) presented an approach to fuzzy classification in the case of missing features. Cloete and van Zyl (2006) introduced a new class of induction algorithms based on fuzzy set covering principles.

REFERENCES

Aizenberg, I., Paliy, D. V., Zurada, J. M., and Astola, J. T., "Blur identification by multilayer neural network based on multivalued neurons," *IEEE Trans. Neural Networks*, vol. 19, no. 5, pp. 883–898, May 2008.

Anguita, D., Boni, A., and Ridella, S., "A digital architecture for support vector machines: theory, algorithm, and FPGA implementation," *IEEE Trans. Neural Networks*, vol. 14, no. 5, pp. 993–1009, Sep. 2003.

Anguita, D., Pischiutta, S., Ridella, S., and Sterpi, D., "Feed-forward support vector machine without multipliers," *IEEE Trans. Neural Networks*, vol. 17, no. 5, pp. 1328–1331, Sep. 2006.

Araujo, "Prune-able fuzzy ART neural architecture for robot map learning and navigation in dynamic environments," *IEEE Trans. Neural Networks*, vol. 17, no. 5, pp. 1235–1249, Sep. 2006.

Armstrong, W. W. and Gecsei, J., "Adaptation algorithms for binary tree networks," *IEEE Trans. Syst. Man Cybern.*, vol. 9, no. 5, pp. 276–285, May 1979.

Atal, B. S., "Automatic recognition of speakers from their voices," *Proc. IEEE*, vol. 64, no. 4, pp. 460–475, Apr. 1976.

Baraldi, A. and Alpaydin, E., "Constructive feedforward ART clustering networks," *IEEE Trans. Neural Networks*, vol. 13, no. 3, pp. 645–661, May 2002.

Camastra, F. and Verri, A., "A novel kernel method for clustering," *IEEE Trans. Pattern Anal. Mach. Intell.*, vol. 27, no. 5, pp. 801–805, May 2005.

Cao, Y. and Wu, J., "Projective ART for clustering data sets in high dimensional spaces," *Neural Networks*, vol. 15, no. 1, pp. 105–120, Jan. 2002.

Cao, Y. and Wu, J., "Dynamics of projective adaptive resonance theory model: the foundation of PART algorithm," *IEEE Trans. Neural Networks*, vol. 15, no. 2, pp. 245–260, Mar. 2004.

Carpenter, G. A., "Neural network models for pattern recognition and associative memory," *Neural Networks*, vol. 2, no. 4, pp. 243–257, 1989.

Carpenter, G. A. and Grossberg, S., "A massively parallel architecture for a self-organizing neural pattern recognition machine," *Comput. Vis. Graph. Image Process.*, vol. 37, no. 1, pp. 54–115, Jan. 1987a.

Carpenter, G. A. and Grossberg, S., "ART 2: self-organization of stable category recognition codes for analog input patterns," *Appl. Opt.*, vol. 26, no. 23, pp. 4919–4930, Dec. 1987b.

Carpenter, G. A. and Grossberg, S., "The ART of adaptive pattern recognition by a self-organization neural network," *IEEE Comput.*, vol. 21, no. 3, pp. 77–88, Mar. 1988.

Carpenter, G. A. and Grossberg, S., "ART3: hierarchical search using chemical transmitters in self-organizing pattern recognition architectures," *Neural Networks*, vol. 3, no. 2, pp. 129–152, 1990.

Celikyilmaz, A. and Turksen, B., "Enhanced fuzzy system models with improved fuzzy clustering algorithm," *IEEE Trans. Fuzzy Syst.*, vol. 16, no. 3, pp. 779–794, June 2008.

Chen, M., Ghorbani, A. A., and Bhavsar, V. C., "Incremental communication for adaptive resonance theory networks," *IEEE Trans. Neural Networks*, vol. 16, no. 1, pp. 132–144, Jan. 2005.

Cloete, I. and van Zyl, J., "Fuzzy rule induction in a set covering framework," *IEEE Trans. Fuzzy Syst.*, vol. 14, no. 1, pp. 93–110, Feb. 2006.

Cohen, M. A. and Grossberg, S., "Absolute stability and global pattern formation and parallel memory storage by competitive neural networks," *IEEE Trans. Syst. Man Cybern.*, vol. 13, no. 7, pp. 815–821, July 1983.

Corsini, P., Lazzerini, B., and Marcelloni, F., "A fuzzy relational clustering algorithm based on a dissimilarity measure extracted from data," *IEEE Trans. Syst. Man Cybern. B*, vol. 34, no. 1, pp. 775–781, Feb. 2004.

Cortes, C. and Vapnik, V., "Support-vector network," *Mach. Learn.*, vol. 20, no. 3, pp. 273–297, Sep. 1995.

Das, S., Abraham, A., and Konar, A., "Automatic clustering using an improved differential evolution algorithm," *IEEE Trans. Syst. Man Cybern. A*, vol. 38, no. 1, pp. 218–237, Jan. 2008.

Fei, B. and Liu, J., "Binary tree of SVM: a new fast multiclass training and classification algorithm," *IEEE Trans. Neural Networks*, vol. 17, no. 3, pp. 696–704, May 2006.

Fred, A. L. N. and Jain, A. K., "Combining multiple clusterings using evidence accumulation," *IEEE Trans. Pattern Anal. Mach. Intell.*, vol. 27, no. 6, pp. 835–850, June 2005.

Fukushima, K., "Cognitron: a self-organizing multilayered neural network," *Biol. Cybern.*, vol. 20, no. 3, pp. 121–136, Sep. 1975.

Goh, C. K., Teoh, E. J., and Tan, K. C., "Hybrid multiobjective evolutionary design for artificial neural networks," *IEEE Trans. Neural Networks*, vol. 19, no. 9, pp. 1531–1548, Sep. 2008.

Gonzalez-Abril, L., Angulo, C., Velasco, F., and Ortega, J. A., "A note on the bias in SVMs for multiclassification," *IEEE Trans. Neural Networks*, vol. 19, no. 4, pp. 723–725, Apr. 2008.

Guo, S. and Huang, L., "Stability analysis of Cohen-Grossberg neural networks," *IEEE Trans. Neural Networks*, vol. 17, no. 1, pp. 106–117, Jan. 2006.

Haasdonk, B., "Feature space interpretation of SVMs with indefinite kernels," *IEEE Trans. Pattern Anal. Mach. Intell.*, vol. 27, no. 4, pp. 482–492, Apr. 2005.

Hamilton-Wright, A., Stashuk, D. W., and Tizhoosh, H. R., "Fuzzy classification using pattern discovery," *IEEE Trans. Fuzzy Syst.*, vol. 15, no. 5, pp. 772–783, Oct. 2007.

Haykin, S., *Neural Networks*, The IEEE Press, 1992.

He, Z., Deng, S., and Xu, X., "Improving k-modes algorithm considering frequencies of attribute values in mode," *Proceedings of the International Conference on Computational Intelligence and Security*, Xi'an, China, pp. 157–162, 2005.

Herbrich, R., *Learning Kernel Classifiers Theory and Algorithms*, The MIT Press, 2001.

Hou, Z.-H., Gupta, M. M., Nikiforuk, P. N., Tan, M., and Cheng, L., "A recurrent neural network for hierarchical control of interconnected dynamic systems," *IEEE Trans. Neural Networks*, vol. 18, no. 2, pp. 466–481, Mar. 2007.

Hsu, C. W. and Lin, C. J., "A comparison of methods for multiclass support vector machines," *IEEE Trans. Neural Networks*, vol. 13, no. 2, pp. 415–425, Mar. 2002.

Huang, S. C. and Huang, Y. F., "Bounds on the number of hidden neurons in multilayer perceptrons," *IEEE Trans. Neural Networks*, vol. 2, no. 1, pp. 47–55, Jan. 1991.

Hwang, S.-K. and Kim, W.-Y., "Fast and efficient method for computing ART," *IEEE Trans. Image Process.*, vol. 15, no. 1, pp. 112–117, Jan. 2006.

Jayadeva, Khemchandani, R, and Chandra, S., "Twin support vector machines for pattern classification," *IEEE Trans. Pattern Anal. Mach. Intell.*, vol. 29, no. 5, pp. 905–910, May 2007.

Jenson, J. R., *Introductory Digital Image Processing—A Remote Sensing Perspective*, Prentice-Hall, New Jersey, 1986.

Jin, N. and Liu, D., "Wavelet basis function neural networks for sequential learning," *IEEE Trans. Neural Networks*, vol. 19, no. 3, pp. 523–528, Mar. 2008.

Krishnapuram, R. and Chen, L. F., "Implementation of parallel thinning algorithms using recurrent neural networks," *IEEE Trans. Neural Networks*, vol. 4, no. 1, pp. 142–147, Jan. 1993.

Lee, H.-E., Park, K.-H., and Bien, Z. Z., "Iterative fuzzy clustering algorithm with supervision to construct probabilistic fuzzy rule base from numerical data," *IEEE Trans. Fuzzy Syst.*, vol. 16, no. 1, pp. 263–277, Feb. 2008.

Li, J., Allinson, N., Tao, D., and Li, X., "Multitraining support vector machine for image retrieval," *IEEE Trans. Image Process.*, vol. 15, no. 11, pp. 3597–3601, Nov. 2006.

Lian, J., Naik, K., and Agnew, G. B., "A framework for evaluating the performance of cluster algorithms for hierarchical networks," *IEEE/ACM Trans. Networking*, vol. 15, no. 6, 1478–1489, Dec. 2007.

Lin, K.-M. and Lin, C.-J., "A study on reduced support vector machines," *IEEE Trans. Neural Networks*, vol. 14, no. 6, pp. 1449–1459, Nov. 2003.

Lippmann, R. P., "An introduction to computing with neural nets," *IEEE ASSP Mag.*, vol. 4, no. 2, pp. 4–22, Apr. 1987.

Liu, Y. and Zheng, Y. F., "Soft SVM and its application in video-object extraction," *IEEE Trans. Signal Process.*, vol. 55, no. 7, pp. 3272–3282, July 2007.

Liu, X., Wang, W., and Chai, T., "The fuzzy clustering analysis based on AFS theory," *IEEE Trans. Syst. Man Cybern. B*, vol. 35, no. 5, pp. 1013–1027, Oct. 2005.

Mansoori, E. G., Zolghadri, M. J., and Katebi, S. D., "SGERD: a steady-state genetic algorithm for extracting fuzzy classification rules from data," *IEEE Trans. Fuzzy Syst.*, vol. 16, no. 4, pp. 1061–1071, Aug. 2008.

Minsky, M. L. and Papert, S. A., *Perceptrons*, MIT Press, 1969.

Nandedkar, A. V. and Biswas, P. K., "A fuzzy min-max neural network classifier with compensatory neuron architecture," *IEEE Trans. Neural Networks*, vol. 18, no. 1, pp. 42–54, Jan. 2007.

Navia-Vazquez, A., Gutierrez-Gonzalez, D., Parrado-Hernandez, E., and Navarro-Abellan, J. J., "Distributed support vector machines," *IEEE Trans. Neural Networks*, vol. 17, no. 4, pp. 1091–1097, July 2006.

Ng, M. K., Li, M. J., Huang, J. Z., and He, Z., "On the impact of dissimilarity measure in k-modes clustering algorithm," *IEEE Trans. Pattern Anal. Mach. Intell.*, vol. 29, no. 3, pp. 503–507, Mar. 2007.

Nock, R. and Nielsen, F., "On weighting clustering," *IEEE Trans. Pattern Anal. Mach. Intell.*, vol. 28, no. 8, pp. 1223–1235, Aug. 2006.

Nowicki, R., "On combining neuro-fuzzy architectures with the rough set theory to solve classification problems with incomplete data," *IEEE Trans. Knowl. Data Eng.*, vol. 20, no. 9, pp. 1239–1253, Sep. 2008.

Palmes, P. P., Hayasaka, T., and Usui, S., "Mutation-based genetic neural network," *IEEE Trans. Neural Networks*, vol. 16, no. 3, pp. 587–600, May 2005.

Pan, S.-M. and Cheng, K.-S., "Evolution-based Tabu search approach to automatic clustering," *IEEE Trans. Syst. Man Cybern. C*, vol. 37, no. 5, pp. 827–838, Sep. 2007.

Pao, Y.-H., *Adaptive Pattern Recognition and Neural Networks*, Addison-Wesley, New York, 1989.

Phung, S. L. and Bouzerdoum, A., "A pyramidal neural network for visual pattern recognition," *IEEE Trans. Neural Networks*, vol. 18, no. 2, pp. 329–343, Mar. 2007.

Platt, J. C., Cristianini, N., and Shawe-Taylor, J., "Large margin DAGs for multiclass classification," in *Advances in Neural Information Processing Systems*, MIT Press, vol. 12, pp. 547–553, 2000.

Rosenblatt, F., *Principles of Neurodynamics*, Spartan, New York, 1962.

Rosenfeld, A., "The fuzzy geometry of image subsets," *Pattern Recognit. Lett.*, vol. 2, no. 5, pp. 311–317, Sep. 1984.

Rumelhart, D. E., Hinton, G. E., and Williams, R. J., "Learning representations by back-propagating errors," *Nature*, vol. 323, pp. 533–536, Oct. 1986.

San, O., Huynh, V., and Nakamori, Y., "An alternative extension of the k-means algorithm for clustering categorical data," *Int. J. Appl. Math. Comput. Sci.*, vol. 14, no. 2, pp. 241–247, 2004.

Savran, A., "Multifeedback-layer neural network," *IEEE Trans. Neural Networks*, vol. 18, no. 2, pp. 373–384, Mar. 2007.

Shih, F. Y. and Zhang, K., "Support vector machine networks for multi-class classification," *Pattern Recognit. Artif. Intell.*, vol. 19, no. 6, pp. 775–786, Sep. 2005.

Shih, F. Y., Moh, J., and Chang, F.-C., "A new ART-based neural architecture for pattern classification and image enhancement without prior knowledge," *Pattern Recognit.*, vol. 25, no. 5, pp. 533–542, May 1992.

Sousa, J. M. C., Gil, J. M., and Pinto, J. R. C., "Word indexing of ancient documents using fuzzy classification," *IEEE Trans. Fuzzy Syst.*, vol. 15, no. 5, pp. 852–862, Oct. 2007.

Stork, D. G., "Self-organization, pattern recognition, and adaptive resonance networks," *J. Neural Network Comput.*, vol. 1, pp. 26–42, Summer 1989.

Tao, Q., Chu, D., and Wang, J., "Recursive support vector machines for dimensionality reduction," *IEEE Trans. Neural Networks*, vol. 19, no. 1, pp. 189–193, Jan. 2008.

Vapnik, V, *The Nature of Statistical Learning Theory*, Springer-Verlag, 1995.

Vasilic, S. and Kezunovic, M., "Fuzzy ART neural network algorithm for classifying the power system faults," *IEEE Trans. Power Deliv.*, vol. 20, no. 2, pp. 1306–1314, Apr. 2005.

Veenman, C. J., Reinders, M. J. T., and Backer, E., "A maximum variance cluster algorithm," *IEEE Trans. Pattern Anal. Mach. Intell.*, vol. 24, no. 9, pp. 1273–1280, Sep. 2002.

Vidal, J. J., "Implementing neural nets with programmable logic," *IEEE Trans. ASSP*, vol. 36, no. 7, pp. 1180–1190, July 1988.

Wang, Z. and Peterson, B. S., "Constrained least absolute deviation neural networks," *IEEE Trans. Neural Networks*, vol. 19, no. 2, pp. 273–283, Feb. 2008.

Wang, S. and Wang, H., "Password authentication using Hopfield neural networks," *IEEE Trans. Syst. Man Cybern. C*, vol. 38, no. 2, pp. 265–268, Mar. 2008.

Xia, Y. and Wang, J., "A one-layer recurrent neural network for support vector machine learning," *IEEE Trans. Syst. Man Cybern. B*, vol. 34, no. 2, pp. 1261–1269, Apr. 2004.

Xu, R. and Wunsch, D., "Survey of clustering algorithms," *IEEE Trans. Neural Networks*, vol. 16, no. 3, pp. 645–678, May 2005.

Yang, Z. R., "A novel radial basis function neural network for discriminant analysis," *IEEE Trans. Neural Networks*, vol. 17, no. 3, pp. 604–612, May 2006.

Yang, M.-S., Wu, K.-L., Hsieh, J.-N., and Yu, J., "Alpha-cut implemented fuzzy clustering algorithms and switching regressions," *IEEE Trans. Syst. Man Cybern. B*, vol. 38, no. 3, pp. 588–603, June 2008.

Yu, J., "General C-means clustering model," *IEEE Trans. Pattern Anal. Mach. Intell.*, vol. 27, no. 8, pp. 1197–1211, Aug. 2005.

PART II

APPLICATIONS

CHAPTER 10

FACE IMAGE PROCESSING AND ANALYSIS

Face recognition is an interdisciplinary field that integrates techniques from image processing, pattern recognition, computer vision, computer graphics, psychology, and evaluation approaches. In general, the computerized face recognition includes four steps. First, a face image is enhanced and segmented. Second, the face boundary and facial features are detected. Third, the extracted facial features are matched against the features stored in the database. Fourth, the classification of the face image into one or more persons is achieved.

Automatic extraction of human head, face boundaries, and facial features plays an important role in the areas of access control, criminal identification, security and surveillance systems, human computer interfacing, and model-based video coding. To extract facial features, one needs to first detect human faces in images. Yang et al. (2002) classified face detection methods into four categories: knowledge-based, feature invariant, template matching, and appearance-based. Song et al. (2007) proposed a multiple maximum scatter difference discriminant criterion for facial feature extraction. Sung and Poggio (1998) used Gaussian clusters to model the distribution of face and nonface patterns. Rowley et al. (1998) designed a neural network-based algorithm to detect frontal-view faces in grayscale images. Schneiderman and Kanade (2000) applied Bayesian classifier to estimate the joint probability of local appearance and the position of face patterns. Li and Zhang (2004) proposed multiview face detection algorithms using the FloatBoost learning method. Wu et al. (1999) built two fuzzy models to extract skin and hair colors using Farnsworth's perceptually uniform color system. Hsieh et al. (2002) applied clustering-based splitting to extract face regions and performed statistics-based face verification.

Researchers have also developed techniques for locating facial features, such as eyes and mouths. Lam and Yan (1996) used the snake model, corner detection, and cost functions to detect eyes. However, they need to initialize the snake points manually and can only handle the images that have a single person or a plain background. Feng and Yuen (2001) proposed eye detection using three cues: relatively low intensity of eye regions, orientation of the line connecting two eye centers, and response of convolving the eye variance filter with the face image. Huang and Wechsler (1999) developed optimal wavelet packets for eye representation and radial basis function neural network (RBFNN) for eye classification. Cai et al. (1998) used template matching to locate eyes and mouths. Wang and Yuan (2001) applied

Image Processing and Pattern Recognition by Frank Shih

wavelet decomposition to detect facial features and three-layer backpropagation neural networks to classify eyes. You et al. (2006) presented a method of face representation using nontensor product wavelets. Shih et al. (2008b) presented extraction of faces and facial features from color images.

Li et al. (2007) presented an active near-infrared imaging system that is able to produce face images of good condition regardless of visible light in the environment. Liu and Liu (2008) presented a hybrid color and frequency features method for face recognition. Park and Savvides (2007) proposed a tensor approach based on an individual modeling method and nonlinear mappings for face recognition. Zhang et al. (2007b) proposed Kullback–Leibler divergence-based local Gabor binary patterns for partially occluded face recognition. Buddharaju et al. (2007) presented a framework for face recognition based on physiological information. Mpiperis et al. (2007) proposed a geodesic polar parameterization of the face surface.

Zhao and Yuen (2008) presented incremental linear discriminant analysis (LDA) for face recognition. Adler and Schuckers (2007) compared human and automatic face recognition performance. O'Toole et al. (2007) compared seven state-of-the-art face recognition algorithms with humans on a face-matching task. Lu et al. (2007) presented a method for face recognition based on parallel neural networks. Zhang et al. (2007a) proposed a novel object descriptor, histogram of Gabor phase pattern (HGPP), for robust face recognition. Cook et al. (2007) proposed multiscale representation for 3D face recognition. Zou et al. (2007) presented a comparative study of local matching approach for face recognition. Dai and Yuen (2007) presented face recognition by regularized discriminant analysis. Chai et al. (2007) proposed a simple, but efficient, locally linear regression method, which generates the virtual frontal view from a given nonfrontal face image. Shih et al. (2004) proposed a hybrid two-phase algorithm for face recognition. Shih et al. (2005) presented multiview face identification and pose estimation using B-spline interpolation.

In this chapter, we introduce the application of image processing and pattern recognition on face images. In Section 10.1, face and facial feature extraction using support vector machine (SVM) is presented. In Section 10.2, the extraction of head and face boundaries and facial features is described. Section 10.3 presents the recognition of facial action units. Section 10.4 presents facial expression recognition.

10.1 FACE AND FACIAL FEATURE EXTRACTION

Figure 10.1 shows the outline of the overall design scheme. First, a color image is segmented into skin and nonskin regions by a 2D Gaussian skin color model. Second, the mathematical morphology technique is applied to remove noises, and the region-filling technique is used to fill holes. Third, the information on shape, size, and principal component analysis (PCA) is used to verify face candidates. Fourth, an ellipse model is used to locate the area of interest (AOI) of eyes and mouths. Fifth, automatic thresholding and morphological opening are applied to discard nonfacial feature pixels. Sixth, the SVM classification is performed on the AOI of eyes and mouths. Finally, the eye and mouth candidates are verified based on knowledge rules.

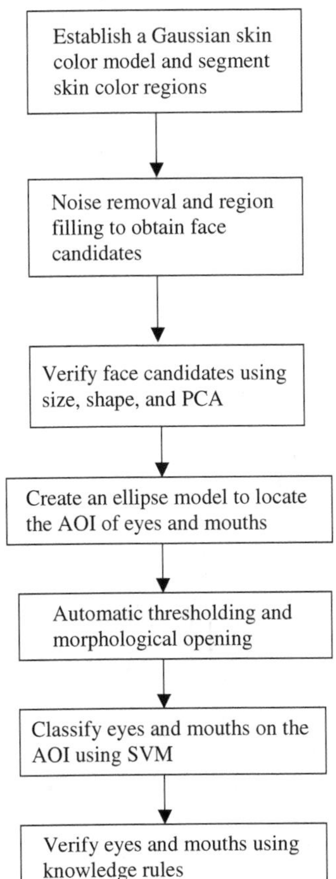

Figure 10.1 Outline of the overall design scheme.

10.1.1 Face Extraction

The YC_bC_r color space is adopted for quick skin color segmentation because it is perceptually uniform. The input color image is converted into a luminance part Y and two chrominance parts C_b and C_r, where C_b represents the difference between the blue component and a reference value, and C_r represents the difference between the red component and a reference value. Y, C_b, and C_r can be obtained from RGB values in the range [0, 1] as

$$\begin{cases} Y = 16 + 65.481R + 128.553G + 24.966B \\ C_b = 128 - 39.797R - 74.203G + 112B \\ C_r = 128 + 112R - 93.786G - 18.214B \end{cases} \tag{10.1}$$

Central limit theorem (Chai and Bouzerdoum, 2000) is very important in probability theory. It states that under various conditions, the distribution for the sum of d independent random variables approaches a particular limiting form, known as the normal or Gaussian distribution. There is a close relationship between the normal distribution and entropy. The *entropy* of a distribution $p(x)$ is given by

$$H(p(x)) = -\int p(x) \ln p(x) dx \tag{10.2}$$

The entropy measures the fundamental uncertainty in the values of points selected randomly from a distribution. It can be shown that the normal distribution has the maximum entropy of all distributions having a given mean and variance. The general multivariate normal density in d dimensions is expressed as

$$p(\mathbf{x}) = \frac{1}{(2\pi)^{d/2}|\Sigma|^{1/2}} \exp\left[-\frac{1}{2}(\mathbf{x}-\boldsymbol{\mu})'\Sigma^{-1}(\mathbf{x}-\boldsymbol{\mu})\right] \tag{10.3}$$

where $\mathbf{x}$ is a d-component column vector, $\boldsymbol{\mu}$ is the d-component mean vector, $\boldsymbol{\Sigma}$ is the $d \times d$ covariance matrix, and $|\boldsymbol{\Sigma}|$ and $\boldsymbol{\Sigma}^{-1}$ are its determinant and inverse matrix, respectively. For example, for a 2D normal density the form is

$$\begin{aligned} p(x_1, x_2) = \frac{1}{2\pi\sigma_1\sigma_2\sqrt{1-\rho^2}} \exp\Bigg\{ -\frac{1}{2(1-\rho^2)} \Bigg[\left(\frac{x_1-\mu_1}{\sigma_1}\right)^2 \\ -2\rho\left(\frac{x_1-\mu_1}{\sigma_1}\right)\left(\frac{x_2-\mu_2}{\sigma_2}\right) + \left(\frac{x_2-\mu_2}{\sigma_2}\right)^2 \Bigg] \Bigg\} \end{aligned} \tag{10.4}$$

where x_1 and x_2 are the random variables in the 2D space, $\boldsymbol{\mu}_1$ and μ_2 are the mean values, σ_1 and σ_2 are the standard deviations, and ρ is the correlation coefficient.

To establish the skin model, we downloaded 80 color face images randomly from the Internet. A face skin patch of size 20 × 20 is selected to establish the 2D Gaussian skin color model in C_b and C_r color spaces. Figure 10.2 shows the histogram of overall C_b and C_r components. The parameters are calculated using the maximum likelihood method as

$$\boldsymbol{\mu} = \begin{bmatrix} \bar{C}_b \\ \bar{C}_r \end{bmatrix} = \begin{bmatrix} 116.88 \\ 158.71 \end{bmatrix}, \quad \boldsymbol{\Sigma} = \begin{bmatrix} 74.19 & -43.73 \\ -43.73 & 82.76 \end{bmatrix}, \quad \rho = -0.5581$$

where $\boldsymbol{\mu}$ is the mean vector, $\boldsymbol{\Sigma}$ is the covariance matrix, and ρ is the correlation coefficient. By placing these values in equation (10.4), we obtain the 2D normal density function.

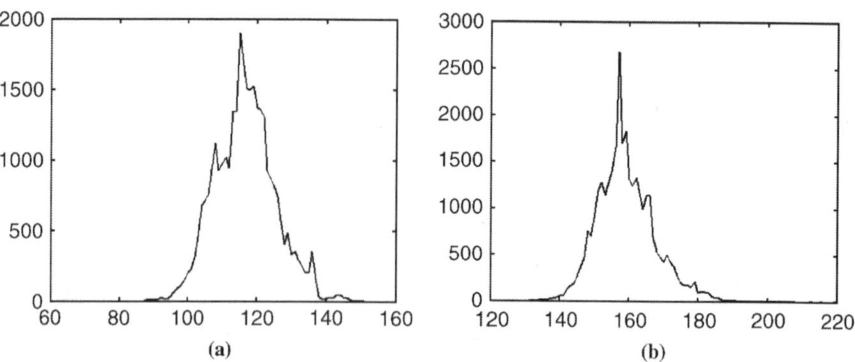

Figure 10.2 The histograms of (a) C_b and (b) C_r.

Figure 10.3 (a) An input color image, (b) the skin likelihood image, (c) the segmented skin image, (d) the image after morphological opening and region filling, and (e) the result after PCA classification.

For each input pixel, a probability value is calculated to indicate its likelihood belonging to the skin color. The skin likelihood probabilities for the whole image are normalized in the range [1, 100]. Figure 10.3(a) shows an input color image, and Figure 10.3(b) shows the skin likelihood image. The next step is to segment the likelihood image into skin and nonskin regions. Otsu's algorithm (1979) is used for automatic threshold selection. The threshold is chosen to maximize the discrimination ratio of σ_B^2/σ_W^2, where σ_B^2 is the between-class variance and σ_W^2 is the within-class variance. Figure 10.3(c) shows the segmented skin image.

Since the segmented skin image contains noises and holes, we need to conduct image cleaning. Mathematical morphology provides an effective tool in image processing and analysis (Serra, 1982; Shih and Mitchell, 1989). Mathematical morphology, based on set theory, can explore object shapes by designing variant structuring elements. The basic morphological operators include dilation, erosion, opening, and closing. We apply a morphological opening by a disk structuring element of 5×5 to remove small objects from the image, while preserving the shape and size of larger objects. The initial erosion not only removes small details, but also darkens the image. The subsequent dilation increases the brightness of the image without reintroducing the details removed by the erosion. Then a region-filling algorithm is used to fill holes. The result is shown in Figure 10.3(d).

Since a face region consists of a large group of pixels, the connected components of small sizes are removed. Based on experimental results, the threshold value of 100 pixels is used. Furthermore, the width-to-length ratio of the smallest rectangle that encloses the component is calculated. If the ratio is in the range [0.6, 2.2], the component is considered as a face candidate. Sometimes, the face candidates are connected to other skin regions; for example, in Figure 10.4, the woman's face is connected to hands.

To extract face regions correctly, multiple squared windows of different sizes are used. The dimension of the first squared window is chosen as the smaller number of the maximum width and the maximum height. The maximum width is defined as the maximum number of connected pixels in the horizontal direction. The maximum height is defined as the maximum number of connected pixels in the vertical direction. The interval of the window size is set to 10%; that is, the size of the next window is decreased to 90% of the previous window. The search processes continue until the size reaches the smaller number of the minimum width and the minimum height.

Furthermore, the ratio of the area of the skin component to the area of the rectangle for each face candidate is calculated. If the ratio is smaller than a threshold, the face candidate is removed. Note that the ratio of 55% is used according to experiments.

The color components C_b and C_r are used only once in the 2D Gaussian skin model for skin classification. Then, only grayscale images are processed. Eigenfaces are used to represent face features because face images do not change radically in projection onto a face space, whereas the projection of nonface images appears quite different.

For training, a total of 450 faces of size 32×32 representing 50 people with nine poses ($+60°$, $+40°$, $+25°$, $+15°$, $0°$, $-15°$, $-25°$, $-40°$, and $-60°$) per person are used in our experiment. Figure 10.5 shows an example of a face image with nine poses. Let each of 450 face vectors be $\mathbf{x}_i$ ($i = 1, 2, \ldots, 450$) of dimension 1024×1. The mean and the standard deviation are calculated, and each vector is normalized to be mean 0 and variance 1 as

$$\frac{\mathbf{x}_i - \text{mean}(\mathbf{x}_i)}{\text{std}(\mathbf{x}_i)} \tag{10.5}$$

where $\text{mean}(\mathbf{x}_i)$ indicates the mean value of all components of vector $\mathbf{x}_i$, and $\text{std}(\mathbf{x}_i)$ indicates the standard deviation of all components of vector $\mathbf{x}_i$.

The mean of all normalized vectors is calculated as

$$\bar{\mathbf{x}} = \frac{1}{450}\sum_{i=1}^{450} \mathbf{x}_i \tag{10.6}$$

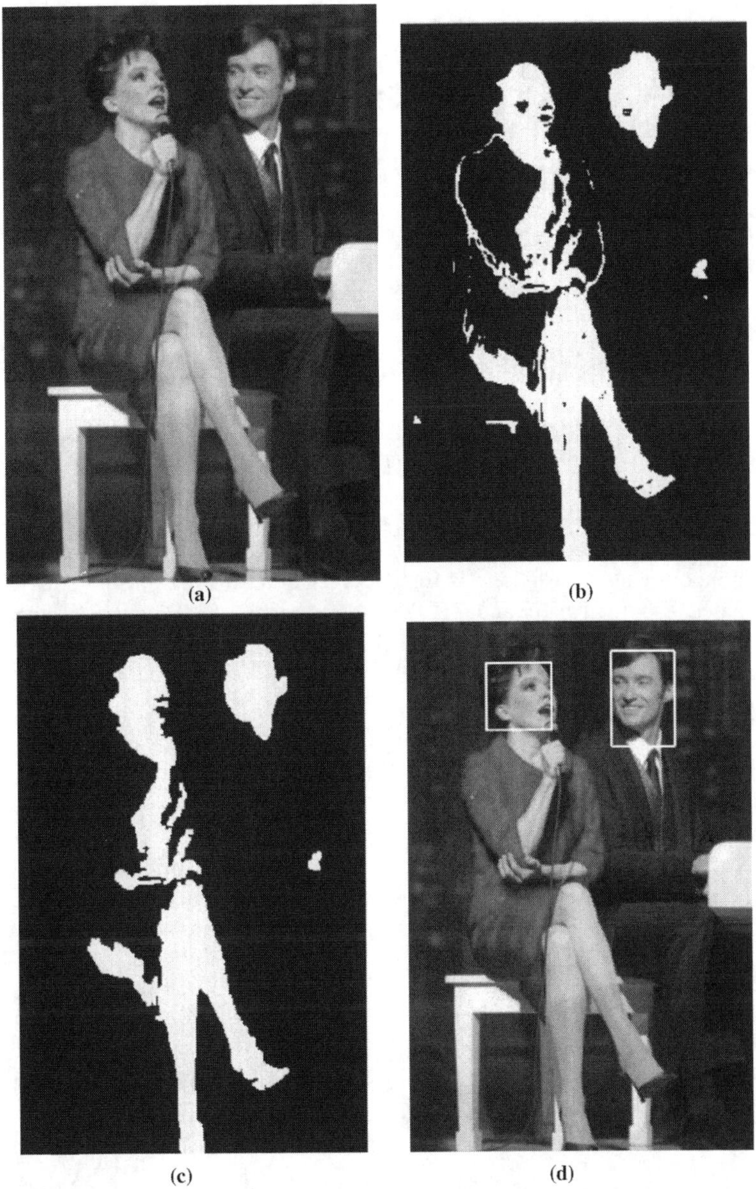

(a) (b)

(c) (d)

Figure 10.4 (a) An input color image, (b) the skin likelihood image, (c) the image after morphological opening and region filling, and (d) the result after PCA classification.

Figure 10.5 An example of a face image with nine poses.

The covariance matrix of all normalized vectors is calculated as

$$\mathbf{Y} = \sum_{i=1}^{450} (\mathbf{x}_i - \bar{\mathbf{x}})(\mathbf{x}_i - \bar{\mathbf{x}})^{\mathrm{T}} \tag{10.7}$$

where "T" denotes the transpose of a matrix. Then, the eigenvectors and eigenvalues of $\mathbf{Y}$ are calculated.

A matrix $\mathbf{U}$ of size 1024×20 is constructed, which contains the first 20 most important eigenvectors. These eigenvectors represent the most face-like images. Some eigenfaces are shown in Figure 10.6. Each face, as a vector of dimension 20×1, is calculated by

$$\text{face}\,\{i\} = \mathbf{U}^{\mathrm{T}}(\mathbf{x}_i - \bar{\mathbf{x}}) \tag{10.8}$$

To test a face candidate image, its PCA representation is obtained by equation (10.8). For face candidates of different sizes, they are normalized into 32×32. The distances between the face candidate and each of 450 training faces are calculated. If the minimum distance is larger than a threshold (i.e., 90 is used according to experiments), the image is removed. Therefore, some false positives can be eliminated. Figures 10.3(e) and (10.4)(d) show the face detection results after the PCA classification.

10.1.2 Facial Feature Extraction

After face images are extracted, the detection of facial features, such as eyes and mouths, is performed. To locate facial features and eliminate false positives quickly, an ellipse model is developed to fit in the face. Let w denote the width of the face image extracted from the face detection algorithm. An ellipse is inserted in the face image by using w as the minor axis and $h = 1.55w$ as the major axis. The eye rectangle whose horizontal lines are located in between $\frac{2}{10}h$ and $\frac{9}{20}h$ away from the top of the ellipse is

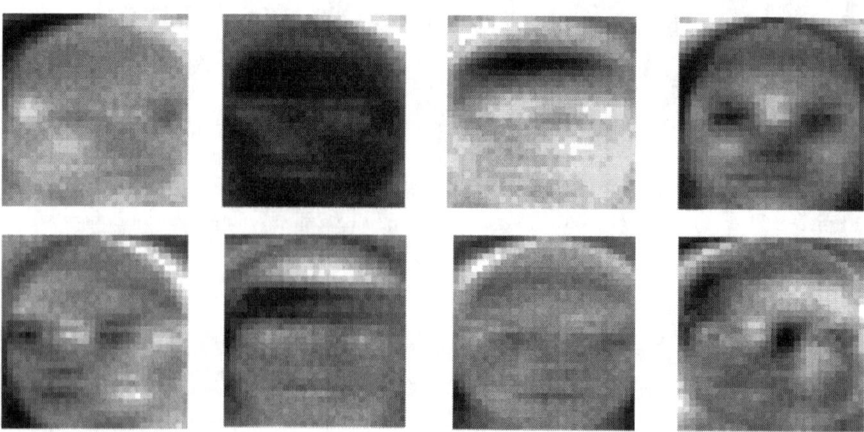

Figure 10.6 Examples of eigenfaces.

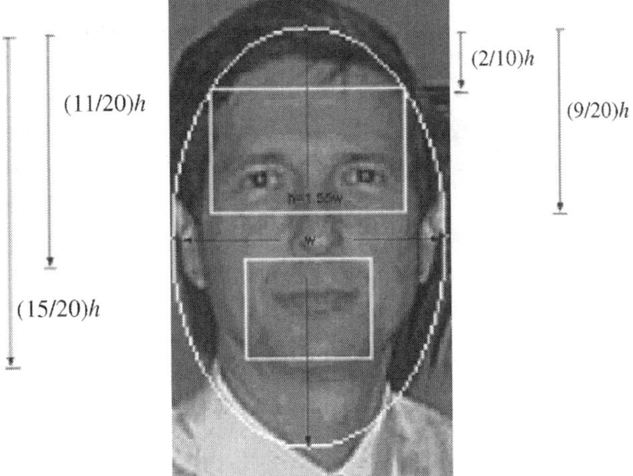

Figure 10.7 The ellipse and the rectangles of facial features.

used, as illustrated in Figure 10.7. Similarly, the mouth rectangle whose horizontal lines are located in between $\frac{11}{20}h$ and $\frac{15}{20}h$ is used. By connecting two eyes and the mouth, a triangle is formed to enclose the nose. Note that compared to eyes and mouth, the nose plays a less important role in face identification.

After extracting the rectangles of facial features, we apply automatic thresholding and morphological opening to discard nonfacial feature pixels. It is observed that facial features are darker than nonfacial features. Therefore, nonfacial feature pixels are removed to speed up the succeeding SVM verification and decrease the existence of false positives. Figure 10.8 shows the resulting facial feature pixels.

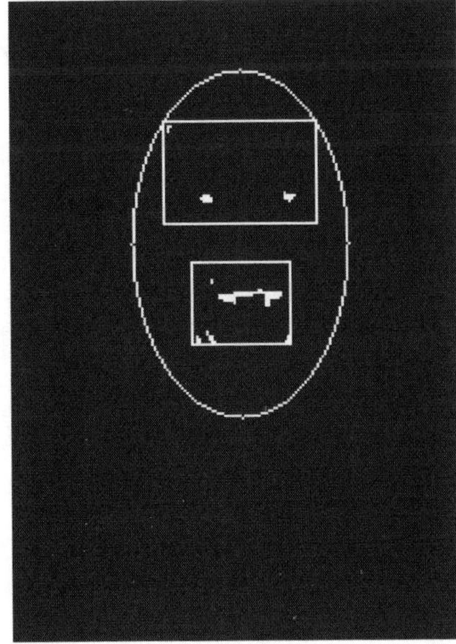

Figure 10.8 The resulting facial feature pixels.

Figure 10.9 The training set of different pose variations.

The FERET face database (Phillips et al., 2000) is used to train eyes and mouths in the SVM. It contains different pose variations from $-90°$ (profile left) to $90°$ (profile right). In experiments, face images from $-25°$ to $25°$ are selected as the training set. Figure 10.9 shows some images of different pose variations in the training set.

The SVM performs pattern classification by applying a variety of kernel functions as the possible sets of approximating functions, optimizing the dual quadratic programming problem, and applying structural risk minimization as the inductive principle. Different types of SVM classifiers are used depending upon input patterns. A linear maximal margin classifier is used for linearly separable data, a linear soft margin classifier is used for overlapping classes, and a nonlinear classifier is used for classes that are overlapped as well as separable by nonlinear hyperplanes. The linearly nonseparable SVM is adopted as the classification tool for facial feature classification since it can produce the best result.

Suppose that there is a set of training data, $\mathbf{x}_1, \mathbf{x}_2, \ldots, \mathbf{x}_k$. Each $\mathbf{x}_i$ belongs to one of the two classes $y_i \in \{-1, 1\}$. The objective is to separate the training data into two classes with a minimal error. To accomplish this, nonnegative slack variables ξ_i, $i = 1, 2, \ldots, 4000$, are used. Thus, the two-class input data can be constructed using equations (10.9) and (10.10), where $\mathbf{w}$ denotes a vector and b denotes a scalar.

$$\begin{cases} \mathbf{w}\cdot\mathbf{x}_i + b \geq 1 - \xi_i & \text{if } y_i = 1 \\ \mathbf{w}\cdot\mathbf{x}_i + b \leq -1 + \xi_i & \text{if } y_i = -1 \end{cases} \tag{10.9}$$

$$y_i(\mathbf{w}\cdot\mathbf{x}_i + b) \geq 1 - \xi_i, \quad i = 1, 2, \ldots, 4000 \tag{10.10}$$

Obtaining the soft margin hyperplane in the linear nonseparable case requires that the following minimization problem be solved:

$$\text{Minimize}: \quad \frac{1}{2}\mathbf{w}\cdot\mathbf{w} + C\left(\sum_{i=1}^{4000}\xi_i\right)$$

$$\text{Subject to}: \quad y_i(\mathbf{w}\cdot\mathbf{x}_i + b) \geq 1 - \xi_i, \quad i = 1, 2, \ldots, 4000$$

$$\xi_i \geq 0, \quad i = 1, 2, \ldots, 4000$$

where C denotes a penalty or regularization parameter.

Minimizing the Lagrangian function with respect to $\mathbf{w}$, b, and ξ_i, its dual problem is obtained:

$$\text{Maximize}: \quad \sum_{i=1}^{4000} \alpha_i - \frac{1}{2} \sum_{i=1}^{4000} \sum_{j=1}^{4000} \alpha_i \alpha_j y_i y_j x_i x_j$$

$$\text{Subject to}: \quad \sum_{i=1}^{4000} y_i \alpha_i = 0, \quad 0 \leq \alpha_i \leq C$$

By solving the dual problem, the soft margin hyperplane is obtained:

$$\mathbf{w}_0 = \sum_{i=1}^{4000} \alpha_i y_i \mathbf{x}_i \tag{10.11}$$

$$b_0 = y_i - \mathbf{w}_0 \cdot \mathbf{x}_i \tag{10.12}$$

From equation (10.11), the hyperplane is obtained by all the training data $\mathbf{x}_i$ that have the corresponding attributes of $\alpha_i > 0$. These support vectors can be divided into two categories. The first category has the attribute of $\alpha_i <$ C and $\xi_i = 0$. These support vectors lie at the distance $1/||\mathbf{w}||$ from the optimal separating hyperplane. The second category has the attribute of $\alpha_i =$ C. The support vectors are either correctly classified with a distance smaller than $1/||\mathbf{w}||$ from the optimal separating hyperplane if $0 < \xi_i \leq 1$, or misclassified if $\xi_i > 1$.

In the SVM training of eye classification, 800 eye and 10,000 noneye images from the FERET database are selected. These images are of size 19×39, which is approximately the ratio 1:2 in terms of height and width. To determine the number of training samples that is most suitable for eye classification, five cases of 3000, 4000, 6000, 8000, and 10000 samples are tested. From experiments, the best eye classification rate of 99.85% is obtained by using 4000 samples.

The training and test images are of size 19×39, which is equivalent to 741 dimensions in classification. PCA is used to reduce the dimensionality. To determine the number of eigenvalues, seven cases of 20, 30, 40, 50, 60, 70, and 80 eigenvalues are tested. From experimental results, it is concluded that 60 eigenvalues are the best trade-off between computational time and accuracy rate.

Furthermore, polynomial kernel functions of different degrees, such as 2, 3, and 4, are tested using 60 eigenvalues on 4000 samples, and their respective accuracy rates are obtained as 98.42%, 98.48%, and 98.51%. From results, it is observed that the linear SVM (of degree 2) performs almost the same as nonlinear SVMs. Therefore, the linear SVM is chosen to save computational time.

For the SVM training in mouth classification, 400 mouth and 10,000 nonmouth images from the FERET database are selected. These images are of size 31×62, which is approximately the ratio 1:2 in terms of height and width. Five cases of 3000, 4000, 6000, 8000, and 10,000 samples are tested. From experiments, the best mouth classification rate of 96.50% is obtained by using 4000 samples.

PCA is also used to reduce the dimensionality in mouth classification. Seven cases of 20, 30, 40, 50, 60, 70, and 80 eigenvalues are tested. From results, using 70 eigenvalues can reduce computational time and maintain the classification rate of

95.50%. Furthermore, polynomial kernel functions of different degrees are tested. It is concluded that the linear SVM classifier has the best accuracy rate.

For mouth classification, four different sizes are chosen due to the size variation of facial features in a face image. Let the width of the face rectangle be w. The four widths of the facial feature rectangle are used: $\frac{1}{3}w$, $\frac{1}{4}w$, $\frac{1}{5}w$, and $\frac{1}{6}w$. The height is approximately a half of its width. To allow tolerance, vertical shifts of 2-pixel up and 2-pixel down the rectangle are included.

Furthermore, the following six rules are applied to eliminate false positives:

1. From the four sizes used in each eye and mouth, the one having the largest occurrence is chosen. For example, if two of $\frac{1}{3}w$, one of $\frac{1}{5}w$, and one of $\frac{1}{6}w$ are detected as eyes, $\frac{1}{3}w$ is chosen as the eye size.
2. Both eyes have similar gray levels. The average intensity is calculated from a set of eye samples. If the intensity of an eye candidate is within $\pm 30\%$ of the average intensity, it is considered to be a true eye.
3. The distance between both eyes is within a certain range of head width. It should be within the range of 30–80% of the width of the face rectangle.
4. The line connecting both eyes is nearly horizontal (i.e., the line slope is within $\pm 15°$).
5. The mouth is vertically located in between two eyes.
6. The three lines connecting two eyes and a mouth form an approximate isosceles triangle.

Figure 10.10 shows some results of facial feature extraction.

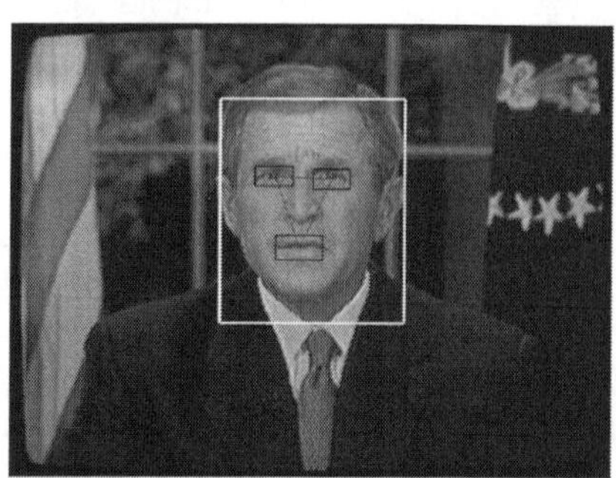
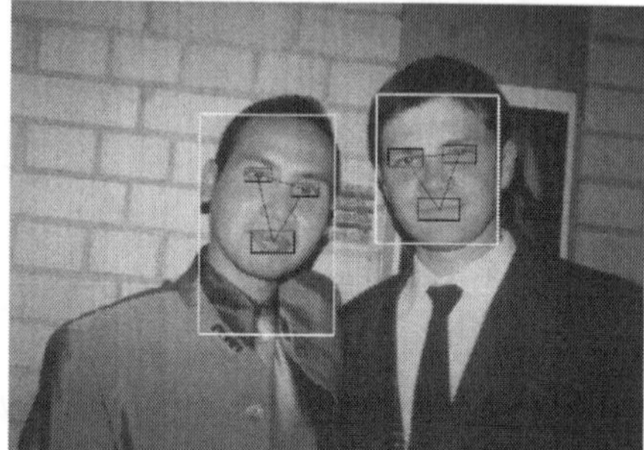
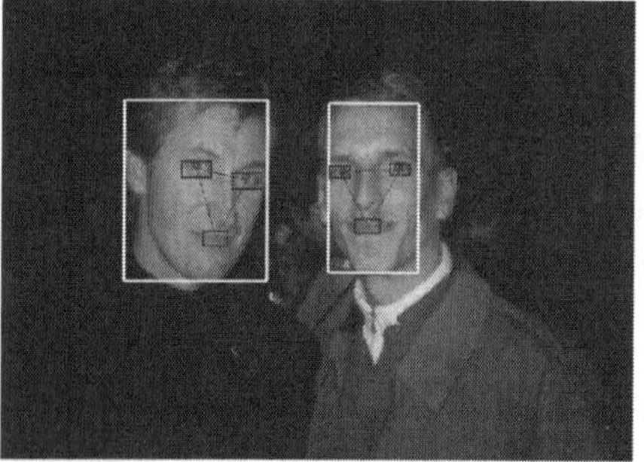

Figure 10.10 Some results of facial feature extraction.

TABLE 10.1 Face Extraction Results

	No. of images	No. of faces	No. of CD	No. of FP	No. of FN	DR (%)	FPR (%)	FNR (%)
One-person	110	110	107	7	3	97.3	6.4	2.7
Two-person	46	92	90	10	2	97.8	10.9	2.2
Three-person	26	78	73	9	5	93.6	11.5	6.4
Others	31	206	200	10	6	97.1	4.8	2.9
Total	213	486	470	36	16	96.7	7.4	3.3

CD: correct detection; FP: false positive; FN: false negative; DR: detection rate; FPR: false positive rate; FNR: false negative rate.

10.1.3 Experimental Results

There are 213 color images randomly selected from the Internet. They consist of 110 one-person, 46 two-person, 26 three-person, and 31 four-or-more-person images. These images include outdoor and indoor scenery, shadow, and lighting variations. The background varies from uniform wall to complex natural scene. The faces appear at different positions in images and with different sizes. Some faces have poses up to 45° and orientations up to 15°. The facial expressions include smile, angry, sad, and surprise. Some images showing these variations are given in Figure 10.9.

Table 10.1 summarizes the face extraction results. The number of *correct detections* (CDs) is based on 213 color images, which contain 486 faces with variations in size, pose, expression, and illumination. The *detection rate* (DR) is defined as the ratio of the number of correctly detected faces to the total number of faces in all the images. The *false positive rate* (FPR) is defined as the ratio of the number of detected false positives to the total number of faces. The *false negative rate* (FNR) is defined as the ratio of the number of false negatives to the total number of faces. The face detection rate of 96.7% is obtained over a total of 486 faces with an FPR of 7.4% and an FNR of 3.3%. Figure 10.11 shows the detected results in different cases: (a) many faces in an image, (b) faces of different sizes, (c and d) faces with pose and orientation variations, and (e) a shadowed face. Note that these faces also present different expressions.

Table 10.2 lists the comparison of the proposed method and five other methods. The CMU and MIT methods use grayscale images. The method in Hsieh et al. (2002) uses color images, but its testing data are not available. Therefore, we create the testing database by randomly downloading color images from the Internet. Note that the proposed method need not search for face candidates at all locations with different sizes as the methods in Sung and Poggio (1998), Rowley et al. (1998), and Schneiderman and Kanade (2000). Compared to Hsieh et al. (2002), the proposed method can detect faces connected with other skin-like regions as demonstrated in Figure 10.4. The CMU face detection system is

Figure 10.11 Some face extraction examples showing variations in face images.

available at http://www.vasc.ri.cmu.edu/cgi-bin/demos/findface.cgi. The grayscale images were sent to the CMU system for processing and the obtained results are shown in Figure 10.12, where some false positives in nonskin regions are present.

TABLE 10.2 Performance Comparisons of the Proposed Method and Five Other Methods

System	Detection rate (%)	False positives (%)	Test set
MIT distribution-based (Sung and Poggio, 1998)	81.9	8.4	23 grayscale images containing 155 frontal faces
CMU neural network (Rowley et al., 1998)	84.5	5.2	23 grayscale images containing 155 frontal faces
CMU Bayes (Schneiderman and Kanade, 1998)	91.2	8.8	20 grayscale images containing 136 faces
CMU Bayes (Schneiderman and Kanade, 2000)	85.5	20.63	208 grayscale images containing 441 faces
Color clustering (Hsieh et al., 2002)	70.3	6.6	60 color images containing 118 faces
Our system	96.7	7.4	213 color images containing 486 faces

Table 10.3 summarizes the results of eye and mouth detection. The accuracy rate is obtained by using the ratio of the total number of correctly detected eyes and mouths to the total number of eyes and mouths, which is equal to 1269:1410.

There are some situations that the proposed face detection algorithm may fail. First, some face region might not be detected as a skin region due to its color distortion caused by lighting or shadowing. Second, some part of background might be detected as a skin region. Third, if two or more faces are too close to be merged into a connected component, the error may occur. Figure 10.13(a) shows that a face is missed due to its face region being detected as a nonskin region, and Figure 10.13(b) shows that the hand region is detected as a false positive.

For facial feature extraction, the proposed eye and mouth extraction algorithm works well on one-person images (i.e., the detection rate of 95.3%). The proposed algorithm can detect eyes and mouths on tilted heads. The detection rate decreases as the number of persons increases in an image. Other factors causing misclassification may be that eyes are closed, too small, or too much tilted.

TABLE 10.3 Eye and Mouth Detection Results

Image	Detection rate (%)
One-person	95.3
Two-person	92.8
Three-person	90.4
Others	85.8
Total	90.0

Figure 10.12 Two results obtained using the CMU face detection system show that false positives occur in nonskin color regions.

10.2 EXTRACTION OF HEAD AND FACE BOUNDARIES AND FACIAL FEATURES

To detect faces and locate facial features correctly, researchers have proposed a variety of methods that can be divided into two categories. One is based on gray-level

(a)

(b)

Figure 10.13 (a) A face is missed because the face region is not detected as a skin region, and (b) the hand is detected as a false positive.

template matching, and the other is based on computation of geometric relationship among facial features.

In the geometric relationship aspect, Jeng et al. (1998) proposed an efficient face detection approach based on the configuration of facial features. Using this method, one can detect the images with frontal-view faces as well as with tilted faces. However, it may fail on the images with face sizes smaller than 80×80 pixels or with multiple faces. Wong et al. (2001) developed an algorithm for face detection and facial feature extraction based on genetic algorithms and eigenface. Lin and Fan (2001) presented a face detection algorithm to detect multiple faces in complex background. They assume that in the frontal-view face images, the centers of two eyes and the center of mouth form an isosceles triangle, and in the side-view face images, the center of one eye, the center of one ear hole, and the center of mouth form a right triangle. The algorithm may fail when an image is too dark or eyes are occluded by hair.

In the template matching aspect, Ryu and Oh (2001) proposed an algorithm based on eigenfeatures and neural networks for the extraction of eyes and mouth using rectangular fitting from gray-level face images. They use eigenfeatures and sliding windows, so there is no need for a large training set. However, their algorithm will fail on the face images with glasses or beard.

Besides the aforementioned two categories, some researchers used motion active contour or snakes to detect the face contours. Kass et al. (1988) proposed the snake algorithm, which has been widely used to detect contours. Yow and Cipolla (1998) used the active contour model to enhance the feature-based approach to detect the face boundary. Sobottka and Pitas (1998) used snakes to trace face contour on a number of image sequences. Because each method has its own advantages, Nikolaidis and Pitas (2000) developed a combined approach of using adaptive Hough transform, template matching, active contour model, and projective geometry properties. They used adaptive Hough transform for curve detection, template matching for inner facial features' location, active contour model for inner face contour detection, and projective geometry properties for accurate pose determination.

Goldmann et al. (2007) presented an approach for automatic and robust object detection using a component-based approach that combines techniques from both statistical and structural pattern recognition domains. Huang et al. (2007) proposed a series of innovative methods to construct a high-performance rotation invariant multiview face detector. Jang and Kim (2008) proposed evolutionary pruning to reduce the number of weak classifiers in AdaBoost-based cascade detector, while maintaining the detection accuracy.

10.2.1 The Methodology

In this section, the proposed method is introduced to process the frontal-view face images for the extraction of head boundary, face boundary, and facial features including eyes with eyebrows, nostrils, and mouth. Head boundary is the outer profile of head including shoulders. Face boundary is the face contour that excludes hair, shoulders, and neck. Rectangular boxes are used to locate facial features.

10.2.1.1 Smoothing and Thresholding The scheme diagram of the double-threshold method is shown in Figure 10.14. The first step is to reduce noise using a

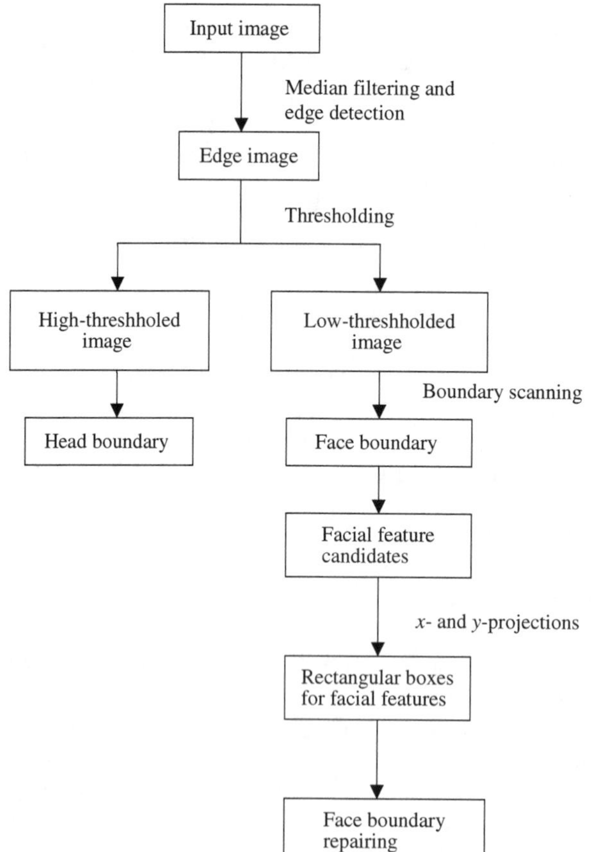

Figure 10.14 The scheme diagram of double-threshold method.

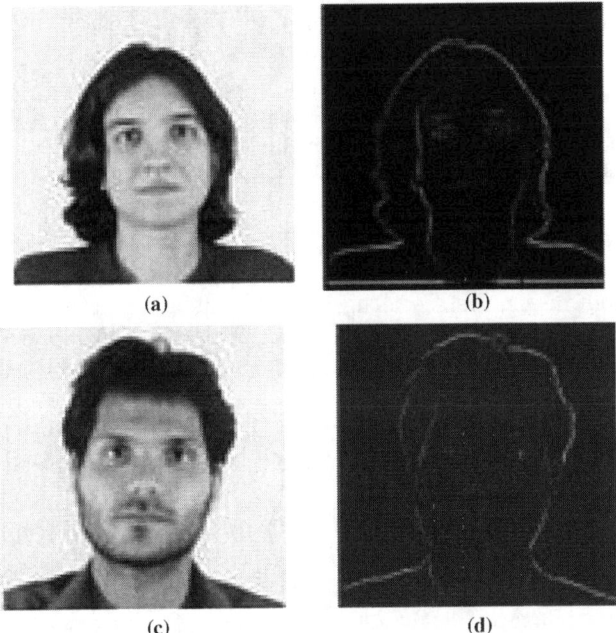

Figure 10.15 The outputs of Wechsler and Kidode's method: (a) and (c) are original images and (c) and (d) are edge detection images.

3×3 median filter. Then, an edge operator is applied. The edge detection technique by Wechsler and Kidode (1977) is tested and the result is shown in Figure 10.15. The edge output appears too thin, and the top-face boundary is too weak to be detected in the later thresholding procedure.

To obtain fat boundary, Hu et al. (1999) proposed four masks, horizontal (size 3×7), vertical (7×3), 45°-diagonal (9×3), and 135°-diagonal (9×3), to detect image edges and select the maximum as the edge strength. Instead of using large sizes, smaller sizes of masks are used as shown in Figure 10.16 for edge detection. Experimental results show that the proposed method is as good as their method. Thresholding is performed to transform the gray-level edge images into binary

–1	–1	0	1	1
–2	–2	0	2	2
–1	–1	0	1	1

(a)

–1	–2	–1
–1	–2	–1
0	0	0
1	2	1
1	2	1

(b)

1	0	1	2	2
–2	–1	0	1	2
–2	–2	–1	0	1

(c)

–2	–2	–1	0	1
–2	–1	0	1	2
–1	0	1	2	2

(d)

Figure 10.16 Edge detection masks: (a) vertical mask, (b) horizontal mask, (c) 135°-diagonal mask, and (d) 45°-diagonal mask.

images. The high threshold value is determined by choosing all the high intensity pixels that occupy 5% of the entire pixels. The low threshold value is decided by choosing all the high intensity pixels that occupy 25% of the entire pixels. The high-thresholded image is used to obtain the head boundary and the low-thresholded image is used to produce the face boundary. These thresholding percentages are determined based on the empirical data to achieve the best results.

10.2.1.2 Tracing Head and Face Boundaries To trace head boundary, the high-thresholded image is divided into left and right halves. When the edge image is scanned from top to bottom, the first and second layers of the contour are the head and face boundaries, respectively. For tracing the head boundary, a starting point is located at the first white pixel on the first layer of the left half. From the starting point, the left and right profiles of head are traced. Because the outer border of the first layer is shifted outward from the actual boundary for a few pixels (say, p), the edge of the left half to the right and the edge of the right half to the left are adjusted by p pixels, respectively.

Because some face profiles disappear in the high-thresholded image, the low-thresholded image is used to trace face boundary. The head borders are removed and a morphological opening is used to eliminate unnecessary noises. Then, the image is scanned in four directions (right to left, left to right, top to bottom, and bottom to top) to produce the face boundary.

10.2.1.3 Locate Facial Features To identify facial features, their candidates are first extracted. The candidates are extracted by overlaying the face boundary obtained from the previous section on the binary edge image and converting all the white pixels in the binary edge image that are on or outside the face boundary to black. Then, x- and y-projections are applied to locate facial features. In the candidate image, x-projection is used to obtain the facial features' horizontal locations and y-projection to obtain their vertical locations. By combining the horizontal and vertical locations, four rectangular boxes are obtained: two for eyes, one for nostrils, and one for mouth.

10.2.1.4 Face Boundary Repairing Sometimes, the chin edge is too low-contrasted to be completely detected by edge detection. Two approaches are used to repair the chin line of the face boundary. In the first approach, the center point of chin (i.e., the average point of the available chin edge pixels) is used as the initial point in the grayscale image to trace the chin line. The algorithm for tracing the chin line is described below.

1. *From the center point of the chin to its right*: Let the gray level of the center point of the chin be $f(x, y)$. The maximum of $\{f(x + 1, y + 1), f(x + 1, y), f(x + 1, y - 1)\}$ is chosen to be the next connected point, and the procedure is repeated until it reaches the right part of the face boundary. As an example, Figure 10.17(a) shows an image of facial feature candidates, Figure 10.17(b) shows the unrepaired face boundary, and Figure 10.17(c) shows the repaired face boundary.

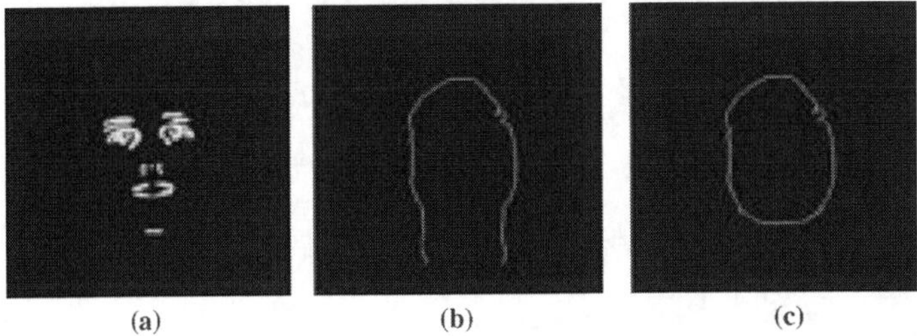

Figure 10.17 (a) An image of facial feature candidates, (b) the unrepaired face boundary, and (c) the repaired face boundary.

2. *From the center point of the chin to its left*: The maximum of $\{f(x-1, y+1), f(x-1, y), f(x-1, y+1)\}$ is chosen to be the next connected point, and the procedure is repeated until it reaches the left part of the face boundary.

In the second approach, an elliptic model is adopted using the major axis to the minor axis ratio of 1.5. Figure 10.18(a) and (c) illustrates the missing chin lines due to the face shadows. By using the lower end points of the left and right face boundaries, the horizontal distance between the two end points is chosen as the minor axis in the elliptic model to repair the chin line. Figure 10.18(b) and (d) shows the results after repairing.

10.2.2 Finding Facial Features Based on Geometric Face Model

10.2.2.1 Geometric Face Model Sometimes, the facial feature candidates are too clustered, and the x- and y-projections cannot work well. Under these

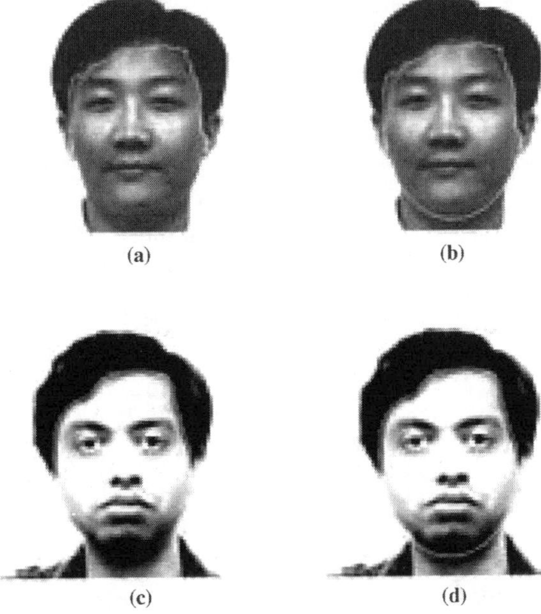

Figure 10.18 The examples of repairing: (a) and (c) are images before repairing, and (b) and (d) are images after repairing.

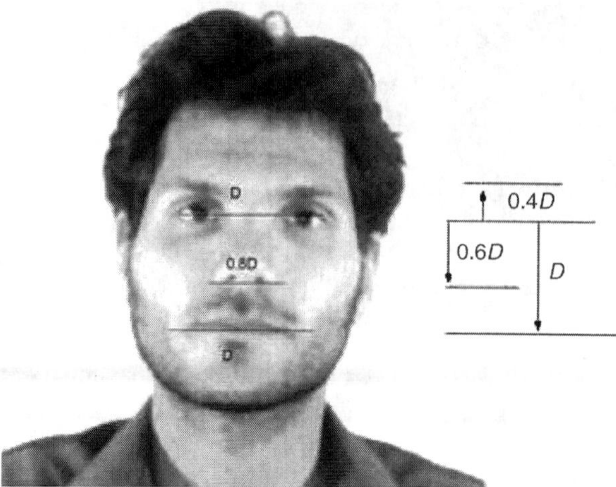

Figure 10.19 Geometric face model.

circumstances, the geometric face model is applied to locate facial features. The model uses the configuration among eyes, nostrils, and mouth. It is assumed that in most of the faces, the vertical distances between eyes and nose and between eyes and mouth are proportional to the horizontal distance between the two centers of eyes.

Referring to Figure 10.19, let the distance between the two centers of eyes be D. The geometric face model and related distances are described below.

1. The vertical distance between two eyes and the center of mouth is D.
2. The vertical distance between two eyes and the center of the nostrils is $0.6D$.
3. The width of the mouth is D.
4. The width of nose is $0.8D$.
5. The vertical distance between eyes and eyebrows is $0.4D$.

The procedure of locating facial features by the geometric face model is illustrated in Figure 10.20.

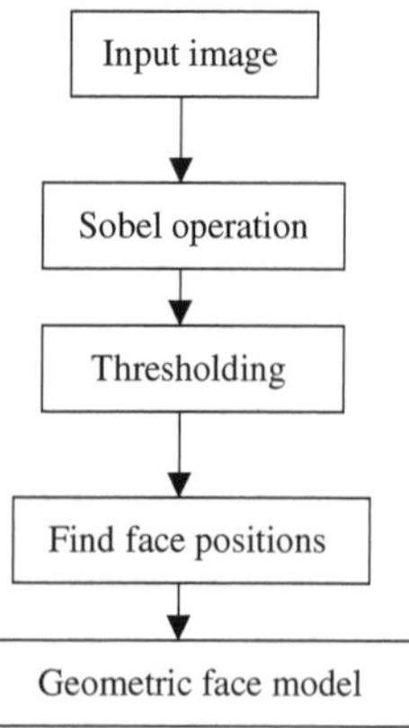

Figure 10.20 The procedure of finding facial features.

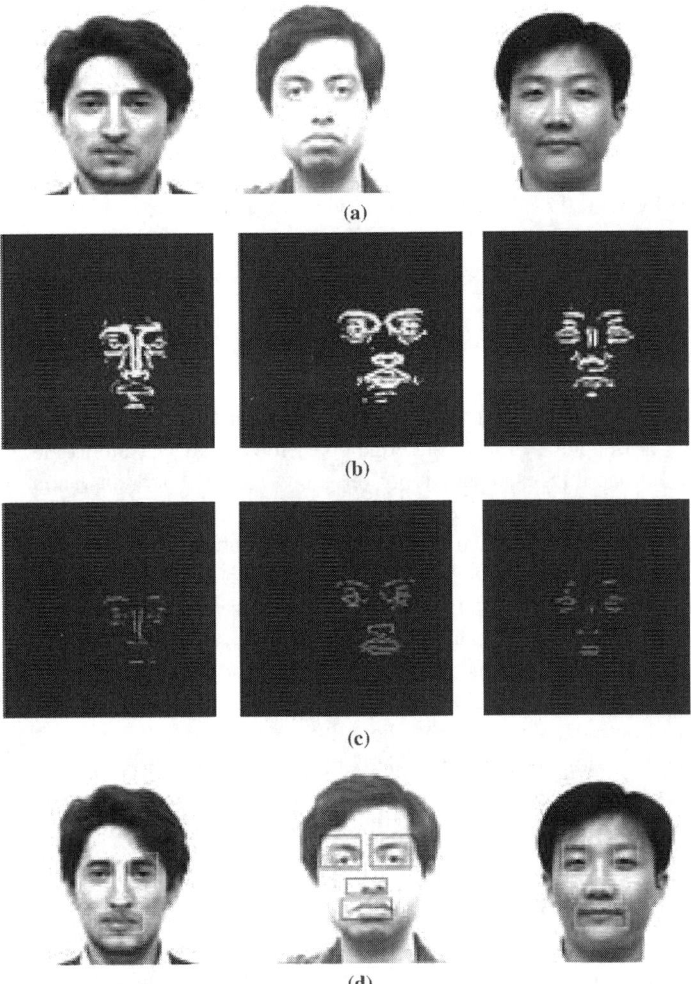

Figure 10.21 Examples of geometric face model: (a) original images, (b) facial feature candidates after thresholding, (c) facial feature candidates after Sobel operation, and (d) extractions of facial features.

Figure 10.21 shows three examples of using the geometric face model. Figure 10.21(a) shows the original images and Figure 10.21(b) shows the resulting facial feature candidates. Because the facial feature candidates are clustered, x- and y-projections cannot be applied. Figure 10.21(c) shows the facial feature candidates after Sobel operation and thresholding, and Figure 10.21(d) shows the results after applying the geometric face model.

10.2.2.2 Geometrical Face Model Based on Gabor Filter For particularly difficult cases such as poor lighting and shadows, two eyes can be located using Gabor filter and then the elliptic model can be applied to extract face boundary. The three steps for locating two eyes are described as follows:

1. Apply Gabor filter to the original image.
2. Apply Gaussian weighting to the filtered image.
3. Locate peaks in the image and find the locations of eyes.

After two eyes are located, the geometric face model is used to extract other facial features. Let the distance between two eyes be D. The center of the ellipse is placed at $0.4D$ below the midpoint of the two eyes. The length of the major axis is $3D$ and that of the minor axis is $2D$.

10.2.3 Experimental Results

Figure 10.22 shows the results of using the double-threshold method. In Figure 10.22, (a) shows the original image, (b) shows the edge detection image, (c) and (d) show high- and low-thresholded images, respectively, (e) shows the head boundary, and (f) shows the face boundary. Figure 10.22(g) shows the facial feature candidates. Figure 10.22(h) shows the rectangular boxes for facial features. Figure 10.22(i) shows the overlay of the head and face boundaries with the original image. Figure 10.22(j) shows the overlay of facial features with the original image. The experiments are conducted on more than 100 FERET face images. A partial set of the experimental results using the geometric face model is shown in Figure 10.23. Compared to Nikolaidis and Pitas (2000), the proposed approaches demonstrate better results in head boundary tracing, chin boundary repairing, and facial feature detection. The programs using Matlab are implemented on Sun Ultra 60 workstation. The execution time is about 170 s, including 15 s for the geometric face model to process an image of size 384×256 on the FERET face database.

10.3 RECOGNIZING FACIAL ACTION UNITS

Facial expression plays a principal role in human interaction and communication since it contains critical information regarding emotion analysis. Its applications include human–computer interface, human emotion analysis, and medical care and cure. The task of automatically recognizing different facial expressions in human–computer environment is significant and challenging. To facilitate this research, Kanade et al. (2000) established a comprehensive, heterogeneous database, called *Cohn–Kanade expression database*, for classifying the upper or lower face action units (AUs).

Tong et al. (2007) conducted facial action unit recognition by exploiting their dynamic and semantic relationships. Tian et al. (2001) developed an automatic face analysis (AFA) system to analyze individual action units based on both permanent and transient facial features in frontal face image sequences. Their recognition rate is 95.6% on Cohn–Kanade expression database. Donato et al. (1999) used different techniques for classifying six upper and lower facial action units on Ekman–Hager facial action exemplars. They found that the best performance is achieved by adopting Gabor wavelet decomposition. Their recognition rate is 96.9%. Bazzo and

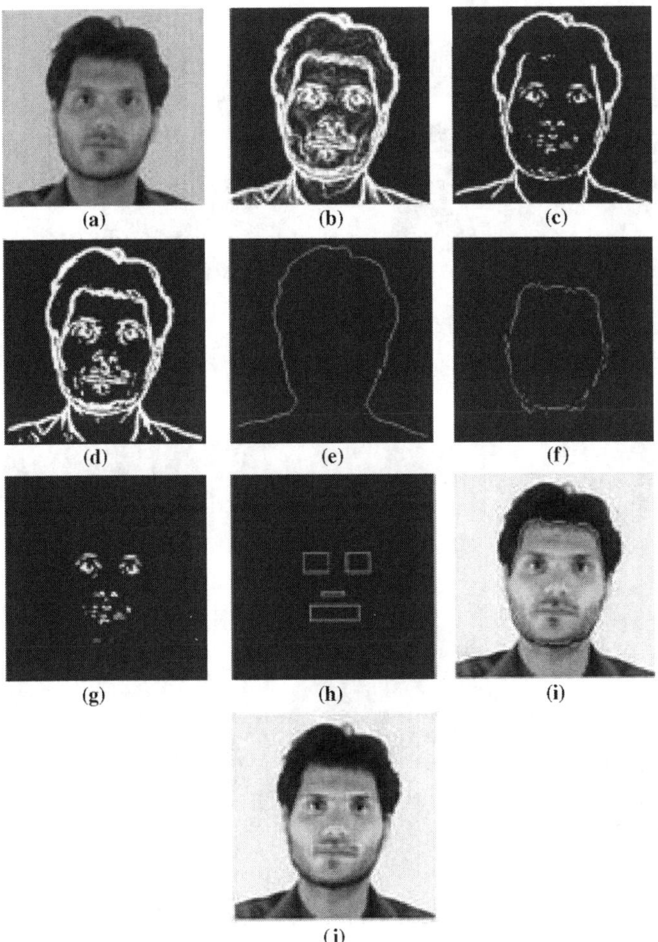

Figure 10.22 The experimental results of double-threshold method: (a) original image, (b) edge detection image, (c) high-thresholded image, (d) low-thresholded image, (e) head boundary, (f) face boundary, (g) facial feature candidates, (h) rectangular boxes for facial features, (i) detection of contours, and (j) extraction of facial features.

Lamar (2004) invented a preprocessing step based on the neutral face average difference. Their system used a neural network-based classifier combined with Gabor wavelet and obtained the recognition rates of 86.55% and 81.63% for the upper and the lower faces, respectively. Chuang and Shih (2006) proposed recognizing facial action units using independent component analysis and support vector machines.

10.3.1 Facial Action Coding System and Expression Database

Ekman and Friesen (1978) designed the facial action coding system (FACS) for characterizing facial expressions by action units. This system is a human observed

Figure 10.23 Partial experimental results.

Figure 10.23 (*Continued*)

system developed to explain the subtle changes of facial expressions. There are 44 AUs in total. Among them, 30 are related to facial muscle contraction including 12 for upper faces and 18 for lower faces. For example, action unit 1 is related to frontalis and pars medialis describing the inner corner of eyebrow raised, and action unit 27 is related to pterygoids and digastric depicting mouth stretched open. The remaining AUs are attributed to miscellaneous actions. For example, action unit 21 portrays the status of neck tightening.

The action units can exist individually or in combinations, which have additive or nonadditive effects. Additive combination means that the combination does not alter the appearance of comprised AUs. An example is AU12 + AU25 indicating smile with mouth opened. Nonadditive combination means that the appearance of comprised AUs is modified. It represents difficulty and complication for the recognition task. An example is AU12 + AU15 indicating that the lip corner of AU12 is changed by the downward motion of AU15.

The facial expression image database used in the experiment is the Cohn–Kanade AU-Coded Face Expression Image Database (Kanade et al., 2000. This database is a representative, comprehensive, and robust test bed for comparative studies of facial expression. It contains image sequences of 210 adult subjects ranging from ages of 18 to 50. For gender classification, there are 69% females and 31% males. For racial classification, there are 81% Euro-American, 13% Afro-American, and 6% other groups. Lighting conditions and context are relatively uniform. The image sequences also include in-plane and out-of-plane head motion from small to mild. The image resolution is 640×480 pixels for 8-bit grayscale and 640×490 pixels for 24-bit color images.

10.3.2 The Proposed System

The proposed system includes histogram equalization for lighting normalization, independent component analysis (ICA) for feature extraction and representation, and SVM for classification measure.

ICA is a statistical and computational technique for finding the hidden factors that are representative and favorable for separating different sets of images, sounds, telecommunication channels, or signals. ICA was originally designed to process the cocktail-party problem and belongs to a class of *blind source separation* (BSS) methods for separating data into underlying representative components. ICA is a general-purpose statistical and unsupervised technique, where the observed random vectors are linearly transformed into components that are minimally dependent upon each other. The concept of ICA is an extension of the PCA, which can only impose independence up to the second order and consequently define the directions that are orthogonal.

SVMs, introduced by Vapnik, are learning systems that separate sets of input pattern vectors into two classes using an optimal hyperplane. The set of vectors is said to be optimally separated by the hyperplane if it is separated without an error and the distance between the closest vector and the hyperplane is maximal. SVMs produce a pattern classifier in three steps:

1. applying a variety of kernel functions (e.g., linear, polynomial, and radial basis function (RBF)) as the possible sets of approximating functions,
2. optimizing the dual quadratic programming problem, and
3. using structural risk minimization as the inductive principle, as opposed to classical statistical algorithms that maximize the absolute value of an error or its square.

Different types of SVM classifiers are used according to the type of input patterns. A linear maximal margin classifier is used for linearly separable classes, a linear soft margin classifier is used for linearly nonseparable classes, and a nonlinear classifier is used for overlapping classes.

The automatic facial expression processing and analysis system includes face detection, facial component extraction and representation, and facial expression recognition. The head and face boundary extraction algorithm in the previous section is used to automatically detect face regions in still images. Facial component extraction and representation are targeted to extract the most representative information derived from facial expression changes to represent the original detected faces. The advantages are lower dimensionality of the detected faces from the previous stage and higher computation speed in the next stage.

Facial expression recognition is intended to identify different facial expressions accurately and promptly. The facial expressions to be recognized can be categorized into two types. The first type is emotion-specified expressions, such as happy, angry, and surprise; the second type is facial action. In this section, the recognition of facial action units is concerned.

The proposed system is outlined in Figure 10.24. The first step is to divide the detected face into upper and lower parts. Histogram equalization is then applied to normalize lighting effect. The ICA is used to extract and represent the subtle changes of facial expressions, and the linear SVM is adopted to recognize the individual action units and their combinations.

10.3.3 Experimental Results

Four experiments were conducted. The first experiment was intended to recognize six individuals and their combinations of the upper face action units including AU4, AU6, AU1 + AU2, AU1 + AU4, AU4 + AU7, and AU1 + AU2 + AU5. The second experiment was constructed to classify six individuals and their combinations of the lower face action units containing AU17, AU9 + AU17, AU12 + AU25, AU15 + AU17, AU20 + AU25, and AU25 + AU27. The third experiment was designed to categorize four combinations of action units on the whole face, which include neutral, AU1 + AU2 + AU5 + AU25 + AU27, AU6 + AU12 + AU25, and AU4 + AU17.

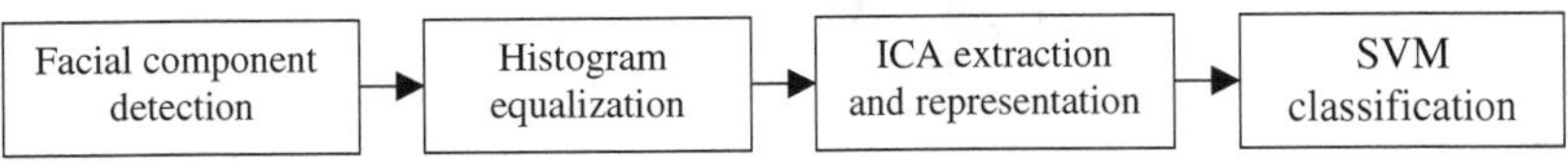

Figure 10.24 The overall system.

The fourth experiment was used to measure the effect of gender factor on the aforementioned three experiments.

There are 27 subjects randomly selected from the Cohn–Kanade AU-Coded Face Expression Image Database in terms of action units. Among them, 20 are female and 7 are male. In each image sequence of 27 subjects, the first three and the last three images are selected as neutral and their corresponding action units, respectively. In total, there are 141 images for the upper part of action units, 126 for the lower part, and 135 for the whole face. Since the number of images in the dataset is limited, 90% are used as the training set and the remaining as the test set. The same procedure is repeated 10 times and the recognition rates of 10 independent tests for each experiment are averaged. Tables 10.4–10.6 list the recognition rates for different action units. Table 10.7 compares the recognition rates between genders on action units.

From Table 10.4, it is observed that most misclassified patterns are derived from the combination of AU4 + AU7. This category at times is misclassified as AU6 because the combination of AU4 + AU7 makes the eyes narrow similar to AU6. From Table 10.5, it can be seen that most misclassification stems from the combination of

TABLE 10.4 Recognition Rates for the Upper Part of Faces

Pattern	Image	No. of samples	Recognition rate
AU4	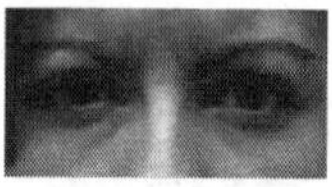	12	19/20 = 95%
AU6	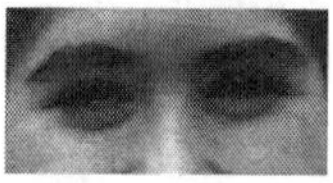	48	50/50 = 100%
AU1 + AU2	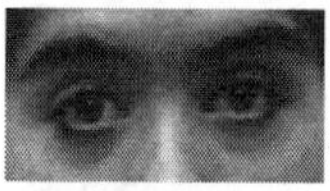	12	19/20 = 95%
AU1 + AU4	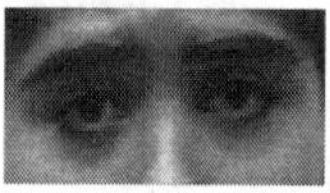	9	10/10 = 100%
AU4 + AU7	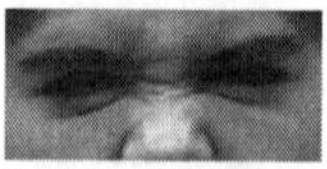	12	17/20 = 85%
AU1 + AU2 + AU5	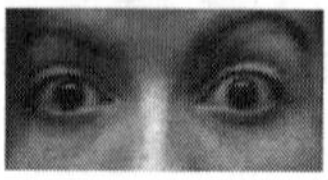	48	50/50 = 100%
Total		141	165/170 = 97.06%

TABLE 10.5 Recognition Rates for the Lower Part of Faces

Pattern	Image	No. of samples	Recognition rate
AU17		12	20/20 = 100%
AU9 + AU17		9	10/10 = 100%
AU12 + AU25		39	40/40 = 100%
AU15 + AU17		9	10/10 = 100%
AU20 + AU25		12	16/20 = 80%
AU25 + AU27		45	40/40 = 100%
Total		126	136/140 = 97.13%

AU20 + AU25. This group is sometimes recognized as the group of AU12 + AU25 since the difference between these two groups is simply the motion of lip corners. From Tables 10.4 and 10.5, it is found that there is no significant difference for the proposed system to recognize the upper and the lower parts of faces. Their recognition rates are 97.06% and 97.13%, respectively.

From Table 10.6, we observe that the recognition rate of 100% is achieved in classifying the action units of the whole face. The combination of facial feature components can produce better results than individual components in facial expression recognition. The gender effect on the classification of action units is explored. From Table 10.7, it is clear that males are capable of expressing the action units more accurately than females.

The proposed method is compared with those of Tian et al. (2001), Donato et al. (1999), and Bazzo and Lamar (2004), and the results are summarized in Table 10.8. The proposed system receives the highest recognition rates: 97.06% on the upper part of faces, 97.13% on the lower part of faces, and 100% on the whole faces. The proposed system is implemented in Matlab on a Pentium IV 2.80 GHz PC with Windows XP. It takes only 1.8 ms for classifying a test image.

TABLE 10.6 Recognition Rates of the Whole Face

Pattern	Image	No. of samples	Recognition rate (%)
Neutral	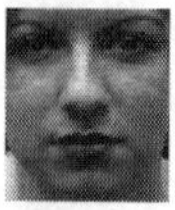	48	100
AU4 + AU17	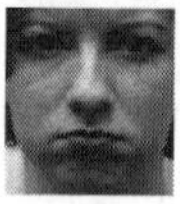	15	100
AU6 + AU12 + AU25		39	100
AU1 + AU2 + AU5 + AU25 + AU27	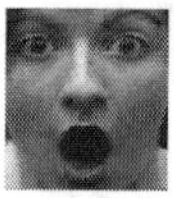	33	100
Total		135	100

TABLE 10.7 Recognition Rates Between Genders on Action Units

	Male	Female
Action units on the upper face	100%	95.38%
Action units on the lower face	100%	93.33%
Action units on the whole face	100%	99.29%

10.4 FACIAL EXPRESSION RECOGNITION IN JAFFE DATABASE

Facial expression plays a principal role in human interaction and communication since it contains critical and necessary information regarding emotion. The task of automatically recognizing different facial expressions in human–computer environment is significant and challenging. A variety of systems have been developed to perform facial expression recognition. These systems possess some common characteristics. First, they classify facial expressions using adult facial expression databases. For instance, Lyons et al. (1999) used the JAFFE (Japanese female facial expression) database to recognize seven main facial expressions: happy, neutral, angry, disgust, fear, sad, and surprise. Chen and Huang (2003) used AR database to classify three facial expressions: neutral, smile, and angry. Second, most systems have two stages: feature extraction and expression classification. For feature

TABLE 10.8 Performance Comparison of Different Systems

Systems	Database	AUs to be recognized	Correct rate (%)
Tian et al. (2001)	Cohn–Kanade	AU9, AU10, AU12, AU15, AU17, AU20, AU25, AU26, AU27, AU23 + AU24	95.6
Donato et al. (1999)	Ekman–Hager	AU1, AU2, AU4, AU5, AU6, AU7	96.9
Bazzo and Lamar (2004)	Cohn–Kanade	Upper AU0, AU6, AU1 + AU2, AU4 + AU7, AU1 + AU2 + AU5, AU4 + AU6 + AU7 + AU9, AU4 + AU7 + AU9	86.55
		Lower AU0, AU25, AU26, AU27, AU12 + AU25, AU15 + AU17, AU20 + AU25	81.63
Proposed system	Cohn–Kanade	Upper AU4, AU6, AU1 + AU2, AU1 + AU4, AU4 + AU7, AU1 + AU2 + AU5	97.06
		Lower AU17, AU9 + AU17, AU12 + AU25, AU15 + AU17, AU20 + AU25, AU25 + AU27	97.13
		Whole face AU neutral, AU4 + AU17, AU6 AU12 + AU25, AU1 + AU2 + AU5 + AU25 + AU27	100

extraction, Gabor filter (Zhang et al., 1998), PCA, and ICA are used. For expression classification, LDA, SVM, two-layer perceptron, and HMM (hidden Markov model) (Zhu et al., 2002) are used.

Sung and Kim (2008) proposed a pose-robust face tracking and facial expression recognition method using a view-based 2D plus 3D active appearance model. Aleksic and Katsaggelos (2006) presented automatic facial expression recognition using facial animation parameters and multistream hidden Markov models. Mpiperis et al. (2008) explored bilinear models for jointly addressing 3D face and facial expression recognition. Kotsia and Pitas (2007) presented facial expression recognition in image sequences using geometric deformation features and support vector machines. Yeasin et al. (2006) presented a spatiotemporal approach in recognizing six universal facial expressions from visual data and using them to compute levels of interest. Anderson and McOwan (2006) presented a fully automated, multistage system for real-time recognition of facial expression.

Facial expression recognition can also be applied to medical treatment of patients. Dai et al. (2001) proposed to monitor patients on bed by using the facial expression recognition to detect the status of patients. Gagliardi et al. (2003) investigated the facial expression ability for individuals with Williams syndrome. Sprengelmeyer et al. (2003) explored facial expression recognition of emotion for people with medicated and unmedicated Parkinson's disease. Brahnam et al. (2006a) proposed machine recognition and representation of neonatal facial displays of acute pain. Brahnam et al. (2006b) presented SVM classification of neonatal facial images of pain. Brahnam et al. (2007) presented machine assessment of neonatal facial expressions of acute pain.

Since JAFFE database is commonly used in measuring the performance of facial expression recognition systems, the proposed system is also applied on this database. Lyons et al. (1999) made use of Gabor filter at different scales and orientations and applied 34 fiducial points for each convolved image to construct the feature vector for representing each facial image. Then, PCA is applied to reduce the dimensionality of feature vectors, and LDA is used to identify seven different facial expressions. Their recognition rate is 92% for JAFFE database. Zhang et al. (1998) adopted Gabor wavelet coefficients and geometric positions to construct the feature vector for each image and applied two-layer perceptron to distinguish seven different facial expressions. Their recognition rate is 90.1%. Xiang and Huang (2006) proposed a recursive procedure for extracting discriminant features, termed recursive cluster-based linear discriminant (RCLD). They showed significant improvement of the proposed algorithm over other feature extraction methods on various types of Yale, Olivetti Research Laboratory, and JAFFE databases. Shih et al. (2008a) presented performance comparisons of facial expression recognition in JAFFE database.

Buciu et al. (2003) tested different feature extraction methods, such as Gabor filter and ICA combined with SVM using three different kernels, linear, polynomial, and radial basis function, to check which combination can produce the best result. From their experiments, the best recognition rate is 90.34% by using Gabor wavelet at high frequencies combined with the polynomial kernel SVM of degree 2. Dubuisson et al. (2002) combined the sorted PCA as feature extractor and the LDA as the classifier to recognize facial expressions. Their recognition rate is 87.6%. Shinohara and Otsu (2004) used higher order local autocorrelation (HLAC) features and LDA to test the performance. Unfortunately, their system is unreliable since the correct rate is only 69.4%.

10.4.1 The JAFFE Database

The image database used in the experiments is the JAFFE database (Lyons et al., 1999). The database contains 10 Japanese females. There are seven different facial expressions, such as neutral, happy, angry, disgust, fear, sad, and surprise. Each female has two to four examples for each expression. In total, there are 213 grayscale facial expression images in this database. Each image is of size 256 × 256. Figure 10.25 shows two expressers containing seven different facial expressions from the JAFFE database.

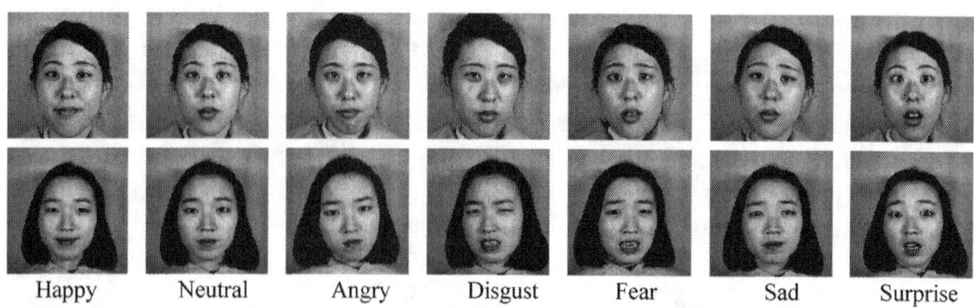

Figure 10.25 Samples of two expressers containing seven different facial expressions.

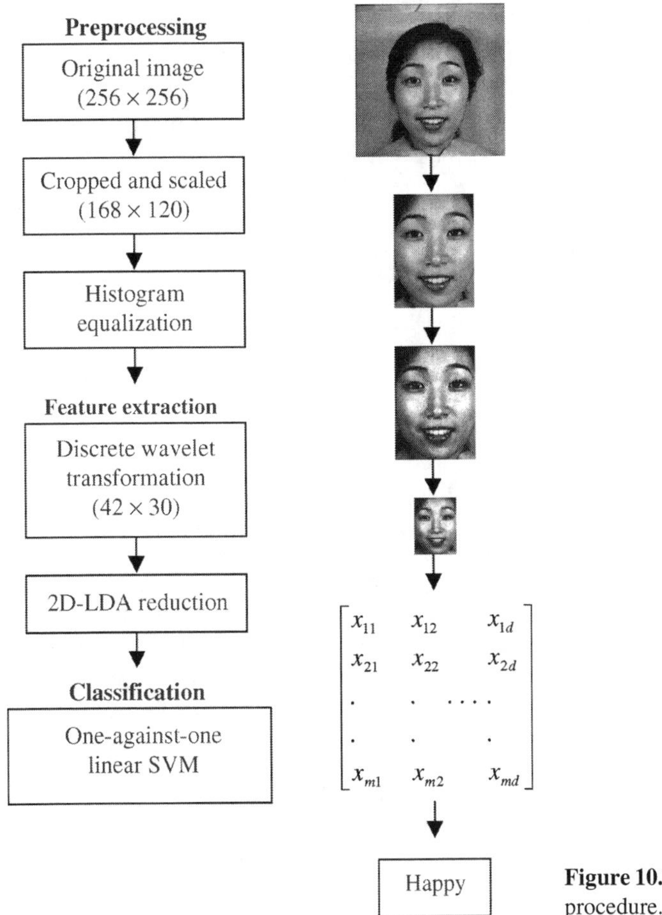

Figure 10.26 The experimental procedure.

10.4.2 The Proposed Method

The proposed method is divided into three stages, preprocessing, feature extraction, and expression classification, as illustrated in Figure 10.26.

10.4.2.1 Preprocessing For comparison, the original image of size 256 × 256 is cropped into 168 × 120 by removing the background influences. Since the illumination condition of the images in the JAFFE database is varied, we apply histogram equalization to eliminate lighting effects.

10.4.2.2 Feature Extraction DWT is applied on the cropped images two times and the LL component is used. Then, 2D-LDA is used to extract important features from each image. Other feature representations, such as PCA, LDA, ICA, and 2D-PCA, are also used for comparisons.

10.4.2.3 Expression Classification The linear SVM is used to identify seven facial expressions in the JAFFE database. To handle the multiclass problem, the tree-based one-against-one SVM as in (Hsu and Lin, 2002) is constructed. Different kernels of the SVM, such as linear, polynomial, and radar basis function, are also tested to compare the performance with the linear SVM.

10.4.3 Experimental Results and Performance Comparisons

The cross-validation strategy is applied as in (Lyons et al., 1999) and the leave-one-out strategy as in (Buciu et al., 2003) to perform comparisons with other existing systems. For the cross-validation strategy, the database is randomly divided into 10 segments in terms of different facial expressions. At each time, 9 out of the 10 segments are trained and the remaining segment is tested. The same procedure of training and testing is repeated 30 times. Finally, all the 30 recognition rates are averaged as the final performance of the proposed system. For the leave-one-out strategy, only one image in each class is tested at each time and the remaining images are used for training.

The experimental results show that the proposed system can successfully meet the criteria of accuracy and efficiency for identifying different facial expressions. For accuracy, the proposed method can outperform other existing systems based on the same database. The recognition rate of the proposed system is 95.71% by using the leave-one-out strategy and 94.13% by using the cross-validation strategy. For efficiency, it takes only 0.0357 s to process an input image of size 256×256.

Table 10.9 shows the performance comparisons among the proposed system and the existing systems using the same JAFFE database. From Table 10.9, it is observed that no matter which strategy is used, the proposed system outperforms the others. The effects of different kernels of SVM, such as polynomial and radial basis functions, are also tested. In the experiments, the linear SVM is best suitable for the JAFFE database. It is because the feature vectors extracted by 2D-LDA are clustered in each class and can be separated by the linear SVM. The results are shown in Table 10.10, from which we observe that 2D-LDA is superior to PCA, LDA, and 2D-PCA as the feature extraction method. The performances of different feature extraction methods and different classifiers are also compared.

Tables 10.11 and 10.12 list the comparisons of using PCA, LDA, 2D-PCA, ICA, and 2D-LDA with SVM and RBFN. It is observed that SVM is the best classifier; however, RBFN is unreliable since the recognition rate is unsatisfactory. Tables 10.13

TABLE 10.9 Performance Comparisons in the JAFFE Database

The existing system	Strategy	Generalization rate (%)
Lyons et al. (1999)	Cross-validation	92.00
Zhang et al. (1998)	Cross-validation	90.10
Buciu et al. (2003)	Leave-one-out	90.34
Dubuisson et al. (2002)		87.60
Shinohara and Otsu (2004)	Cross-validation	69.40
Proposed system	Cross-validation	95.71
	Leave-one-out	94.13

TABLE 10.10 Performance Comparisons of Using Different Kernels of SVM

	Recognition rate (%)	
Kernel function	Cross-validation	Leave-one-out
Linear	94.13	95.71
Polynomial of degree 2	91.43	92.38
Polynomial of degree 3	92.86	94.29
Polynomial of degree 4	92.22	93.33
Radial basis function	85.71	87.14

TABLE 10.11 Comparisons of PCA, LDA, 2D-PCA, ICA, and 2D-LDA

	Recognition rate (%)		
Feature extraction method	Cross-validation	Leave-one-out	Testing speed per image (s)
LDA + SVM	91.27	91.90	0.0367
2D-PCA + SVM	92.06	93.33	0.0357
ICA + SVM	93.35	93.81	0.0359
PCA + SVM	93.43	94.84	0.0353
2D-LDA + SVM	94.13	95.71	0.0357

TABLE 10.12 Comparisons of PCA, LDA, 2D-PCA, ICA, and 2D-LDA

	Recognition rate (%)		
Feature extraction method	Cross-validation	Leave-one-out	Testing speed per image (s)
LDA + RBFN	26.67	27.71	0.0351
2D-PCA + RBFN	25.24	26.67	0.0349
ICA + RBFN	27.14	27.43	0.0347
PCA + RBFN	36.67	37.21	0.0346
2D-LDA + RBFN	37.14	37.71	0.0347

TABLE 10.13 Confusion Matrix Using Cross-Validation Strategy

	Angry	Disgust	Fear	Happy	Neural	Sad	Surprise	Total
Angry	30							30
Disgust	1	28						29
Fear		2	27		1		2	32
Happy				30	1			31
Neural					30			30
Sad			3	1		27		31
Surprise				2			28	30
Total								213

TABLE 10.14 Confusion Matrix Using Leave-One-Out Strategy

	Angry	Disgust	Fear	Happy	Neural	Sad	Surprise	Total
Angry	30							30
Disgust	1	28						29
Fear		1	29			1	1	32
Happy				29		1	1	31
Neural					30			30
Sad			2	1		28		31
Surprise							30	30
Total								213

and 10.14 show the confusion matrices of the correct and false numbers under cross-validation and leave-one-out strategies for facial expressions. All the experiments are implemented in the Matlab 7 environment on a Pentium IV 2.80 GHz PC with Windows XP. The speed of the proposed method is fast. It only takes about 0.0357 s to process a test image.

REFERENCES

Adler, A. and Schuckers, M. E., "Comparing human and automatic face recognition performance," *IEEE Trans. Syst. Man Cybernet. B*, vol. 37, no. 5, pp. 1248–1255, Oct. 2007.

Aleksic, P. S. and Katsaggelos, A. K., "Automatic facial expression recognition using facial animation parameters and multistream HMMs," *IEEE Trans. Inform. Forensics Security*, vol. 1, no. 1, pp. 3–11, Mar. 2006.

Anderson, K. and McOwan, P. W., "A real-time automated system for the recognition of human facial expressions," *IEEE Trans. Syst. Man Cybernet. B*, vol. 36, no. 1, pp. 96–105, Feb. 2006.

Bazzo, J. J. and Lamar, M. V., "Recognizing facial actions using Gabor wavelets with neutral face average difference," *Proceedings of the IEEE International Conference on Automatic Face and Gesture Recognition*, Seoul, Korea, pp. 505–510, May 2004.

Brahnam, S., Chuang, C., Shih, F. Y., and Slack, M., "Machine recognition and representation of neonatal facial displays of acute pain," *Int. J. Artif. Intell. Med.*, vol. 36, no. 3, pp. 211–222, Mar. 2006a.

Brahnam, S., Chuang, C., Shih, F. Y., and Slack, M., "SVM classification of neonatal facial images of pain," in *Fuzzy Logic and Applications* (I. Bloch, A. Petrosino, and A. G. B. Tettamanzi, eds.), *Lecture Notes in Computer Science*, vol. 3849, pp. 121–128, 2006b.

Brahnam, S., Chuang, C., Sexton, R., and Shih, F. Y., "Machine assessment of neonatal facial expressions of acute pain," *J. Decision Support Syst.*, Special Issue on Decision Support in Medicine, vol. 43, no. 4, pp. 1242–1254, Aug. 2007.

Buciu, I., Kotropoulos, C., and Pitas, I., "ICA and Gabor representation for facial expression recognition," *Proceedings of the IEEE International Conference on Image Processing*, Barcelona, Spain, pp. 855–858, Sept. 2003.

Buddharaju, P., Pavlidis, I. T., Tsiamyrtzis, P., and Bazakos, M., "Physiology-based face recognition in the thermal infrared spectrum," *IEEE Trans. Pattern Anal. Mach. Intell.*, vol. 29, no. 4, pp. 613–626, Apr. 2007.

Cai, J., Goshtasby, A., and Yu, C., "Detecting human faces in color images," *Proceedings of the International Workshop on Multimedia Database Management Systems*, Dayton, OH, pp. 124–131, Aug. 1998.

Chai, D. and Bouzerdoum, A., "A Bayesian approach to skin color classification in YCbCr color space," *Proceedings of TENCON 2000*, Kuala Lumpur, Malaysia, vol. 2, pp. 421–424, Sept. 2000.

Chai, X., Shan, S., Chen, X., and Gao, W., "Locally linear regression for pose-invariant face recognition," *IEEE Trans. Image Process.*, vol. 16, no. 7, pp. 1716–1725, July 2007.

Chen, X.-W. and Huang, T., "Facial expression recognition: a clustering-based approach," *Pattern Recogn. Lett.*, vol. 24, no. 9, pp. 1295–1302, June 2003.

Chuang, C. and Shih, F. Y., "Recognizing facial action units using independent component analysis and support vector machine," *Pattern Recogn.*, vol. 39, no. 9, pp. 1795–1798, Sept. 2006.

Cook, J., Chandran, V., and Sridharan, S., "Multiscale representation for 3-D face recognition," *IEEE Trans. Inform. Forensics Security*, vol. 2, no. 3, pp. 529–536, Sept. 2007.

Dai, D.-Q. and Yuen, P. C., "Face recognition by regularized discriminant analysis," *IEEE Trans. Syst. Man Cybernet. B*, vol. 37, no. 4, pp. 1080–1085, Aug. 2007.

Dai, Y., Shibata, Y., Ishii, T., katamachi, K., Noguchi, K., Kakizaki, N., and Cai, D., "An associate memory model of facial expression and its application in facial expression recognition of patients on bed," *Proceedings of the IEEE International Conference on Multimedia and Expo*, Tokyo, Japan, pp. 772–775, Aug. 2001.

Donato, G., Bartlett, M. S., Hager, J. C., Ekman, P., and Sejnowski, T. J., "Classifying facial actions," *IEEE Trans. Pattern Anal. Mach. Intell.*, vol. 21, no. 10, pp. 974–985, Oct. 1999.

Dubuisson, S., Davoine, F., and Masson, M., "A solution for facial expression representation and recognition," *Signal Process.: Image Commun.*, vol. 17, no. 9, pp. 657–673, Oct. 2002.

Ekman, P. and Friesen, W. V., *Facial Action Coding System: A Technique for the Measurement of Facial Movement*, Consulting Psychologists Press, San Francisco, CA, 1978.

Feng, G. C. and Yuen, P. C., "Multi cues eye detection on gray intensity image," *Pattern Recogn.*, vol. 34, no. 5, pp. 1033–1046, May 2001.

Gagliardi, C., Frigerio, E., Buro, D., Cazzaniga, I., Pret, D., and Borgatti, R., "Facial expression recognition in Williams syndrome," *Neuropsychologia*, vol. 41, no. 6, pp. 733–738, 2003.

Goldmann, L., Monich, U. J., and Sikora, T., "Components and their topology for robust face detection in the presence of partial occlusions," *IEEE Trans. Inform. Forensics Security*, vol. 2, no. 3, pp. 559–569, Sept. 2007.

Hsieh, I.-S., Fan, K.-C., and Lin, C., "A statistic approach to the detection of human faces in color nature scene," *Pattern Recogn.*, vol. 35, no. 7, pp. 1583–1596, July 2002.

Hsu, C.-W. and Lin, C.-J., "A comparison of methods for multiclass support vector machines," *IEEE Trans. Neural Networks*, vol. 13, no. 2, pp. 415–425, Mar. 2002.

Hu, J., Yan, H., and Sakalli, M., "Locating head and face boundaries for head–shoulder images," *Pattern Recogn.*, vol. 32, no. 8, pp. 1317–1333, Aug. 1999.

Huang, J. and Wechsler, H., "Eye detection using optimal wavelet packets and radial basis functions (RBFs)," *Pattern Recogn. Artif. Intell.*, vol. 13, no. 7, pp. 1009–1026, Nov. 1999.

Huang, C., Ai, H., Li, Y., and Lao, S., "High-performance rotation invariant multiview face detection," *IEEE Trans. Pattern Anal. Mach. Intell.*, vol. 29, no. 4, pp. 671–686, Apr. 2007.

Jang, J.-S. and Kim, J.-H., "Fast and robust face detection using evolutionary pruning," *IEEE Trans. Evol. Comput.*, vol. 12, no. 5, pp. 562–571, Oct. 2008.

Jeng, S.-H., Yuan, H., Liao, M., Han, C. C., Chern, M. Y., and Liu, Y., "Facial feature detection using geometrical face model: an efficient approach," *Pattern Recogn.*, vol. 31, no. 3, pp. 273–282, Mar. 1998.

Kanade, T., Cohn, J., and Tian, Y., "Comprehensive database for facial expression analysis," *Proceedings of the IEEE International Conference on Automatic Face and Gesture Recognition*, Grenoble, France, pp. 46–53, Mar. 2000.

Kass, M., Witkin, A., and Terzopoulos, D., "Snakes: active contour models," *Int. J. Comput. Vision*, vol. 1, no. 4, pp. 321–331, Jan. 1988.

Kotsia, I. and Pitas, I., "Facial expression recognition in image sequences using geometric deformation features and support vector machines," *IEEE Trans. Image Process.*, vol. 16, no. 1, pp. 172–187, Jan. 2007.

Lam, K.-M. and Yan, H., "Locating and extracting the eye in human face images," *Pattern Recogn.*, vol. 29, no. 5, pp. 771–779, May 1996.

Li, S. Z. and Zhang, Z. Q., "FloatBoost learning and statistical face detection," *IEEE Trans. Pattern Anal. Mach. Intell.*, vol. 26, no. 9, pp. 1112–1123, Sept. 2004.

Li, S. Z., Chu, R., Liao, S., and Zhang, L., "Illumination invariant face recognition using near-infrared images," *IEEE Trans. Pattern Anal. Mach. Intell.*, vol. 29, no. 4, pp. 627–639, Apr. 2007.

Lin, C. and Fan, K.-C., "Triangle-based approach to the detection of human face," *Pattern Recogn.*, vol. 34, no. 6, pp. 1271–1284, June 2001.

Liu, Z. and Liu, C., "A hybrid color and frequency features method for face recognition," *IEEE Trans. Image Process.*, vol. 17, no. 10, pp. 1975–1980, Oct. 2008.

Lu, J., Yuan, X., and Yahagi, T., "A method of face recognition based on fuzzy *c*-means clustering and associated sub-NNs," *IEEE Trans. Neural Networks*, vol. 18, no. 1, pp. 150–160, Jan. 2007.

Lyons, M., Budynek, J., and Akamatsu, S., "Automatic classification of single facial images," *IEEE Trans. Pattern Anal. Mach. Intell.*, vol. 21, no. 12, pp. 1357–1362, Dec. 1999.

Mpiperis, I., Malassiotis, S., and Strintzis, M. G., "3-D face recognition with the geodesic polar representation," *IEEE Trans. Inform. Forensics Security*, vol. 2, no. 3, pp. 537–547, Sept. 2007.

Mpiperis, I., Malassiotis, S., and Strintzis, M. G., "Bilinear models for 3-D face and facial expression recognition," *IEEE Trans. Inform. Forensics Security*, vol. 3, no. 3, pp. 498–511, Sept. 2008.

Nikolaidis, A. and Pitas, I., "Facial feature extraction and pose determination," *Pattern Recogn.*, vol. 33, no. 11, pp. 1783–1791, Nov. 2000.

O'Toole, A. J., Phillips, P. J., Jiang, F., Ayyad, J., Penard, N., and Abdi, H., "Face recognition algorithms surpass humans matching faces over changes in illumination," *IEEE Trans. Pattern Anal. Mach. Intell.*, vol. 29, no. 9, pp. 1642–1646, Sept. 2007.

Otsu, N., "A threshold selection method from gray-level histogram," *IEEE Trans. Syst. Man Cybernet.*, vol. 9, no. 1, pp. 62–66, Jan. 1979.

Park, S. W. and Savvides, M., "Individual kernel tensor-subspaces for robust face recognition: a computationally efficient tensor framework without requiring mode factorization," *IEEE Trans. Syst. Man Cybernet. B*, vol. 37, no. 5, pp. 1156–1166, Oct. 2007.

Phillips, P. J., Moon, H., Rauss, P. J., and Rizvi, S., "The FERET evaluation methodology for face-recognition algorithms," *IEEE Trans. Pattern Anal. Mach. Intell.*, vol. 22, no. 10, pp. 1090–1100, Oct. 2000.

Rowley, H., Baluja, S., and Kanade, T., "Neural network-based face detection," *IEEE Trans. Pattern Anal. Mach. Intell.*, vol. 20, no. 1, pp. 23–38, Jan. 1998.

Ryu, Y.-S. and Oh, S.-Y., "Automatic extraction of eye and mouth field from a face image using eigenfeatures and multilayer perceptrons," *Pattern Recogn.*, vol. 34, no. 12, pp. 2459–2466, Dec. 2001.

Schneiderman, H. and Kanade, T., "Probabilistic modeling of local appearance and spatial relationships for object recognition," *Proceedings of the IEEE Conference on Computer Vision and Pattern Recognition*, Santa Barbara, CA, pp. 45–51, June 1998.

Schneiderman, H. and Kanade, T., "A statistical method for 3D object detection applied to faces and cars," *Proceedings of the IEEE Conference on Computer Vision and Pattern Recognition*, Hilton Head Island, SC, pp. 746–751, June 2000.

Serra, J., *Image Analysis and Mathematical Morphology*, Academic Press, New York, 1982.

Shih, F. Y. and Mitchell, O. R., "Threshold decomposition of grayscale morphology into binary morphology," *IEEE Trans. Pattern Anal. Mach. Intell.*, vol. 11, no. 1, pp. 31–42, Jan. 1989.

Shih, F. Y., Zhang, K., and Fu, Y., "A hybrid two-phase algorithm for face recognition," *Pattern Recogn. Artif. Intell.*, vol. 18, no. 8, pp. 1423–1435, Dec. 2004.

Shih, F. Y., Fu, Y., and Zhang, K., "Multi-view face identification and pose estimation using B-spline interpolation," *Inform. Sci.*, vol. 169, no. 3, pp. 189–204, Feb. 2005.

Shih, F. Y., Chuang, C., and Wang, P., "Performance comparisons of facial expression recognition in JAFFE database," *Pattern Recogn. Artif. Intell.*, vol. 22, no. 3, pp. 445–459, May 2008a.

Shih, F. Y., Cheng, S., Chuang, C., and Wang, P., "Extracting faces and facial features from color images," *Pattern Recogn. Artif. Intell.*, vol. 22, no. 3, pp. 515–534, May 2008b.

Shinohara, Y. and Otsu, N., "Facial expression recognition using Fisher weight maps," *Proceedings of the IEEE International Conference on Automatic Face and Gesture Recognition* Seoul, Korea, pp. 499–504, May 2004.

Sobottka, K. and Pitas, I., "A novel method for automatic face segmentation, facial feature extraction and tracking," *Signal Process.: Image Commun.*, vol. 12, no. 3, pp. 263–281, June 1998.

Song, F., Zhang, D., Mei, D., and Guo, Z., "A multiple maximum scatter difference discriminant criterion for facial feature extraction," *IEEE Trans. Syst. Man Cybernet. B*, vol. 37, no. 6, pp. 1599–1606, Dec. 2007.

Sprengelmeyer, R., Young, A., Mahn, K., Schroeder, U., Woitalla, D., Butter, T., Kuhn, W., and Przuntek, H., "Facial expression recognition in people with medicated and unmedicated Parkinson's disease," *Neuropsychologia*, vol. 41, no. 8, pp. 1047–1057, 2003.

Sung, J. and Kim, D., "Pose-robust facial expression recognition using view-based 2D + 3D AAM," *IEEE Trans. Syst. Man Cybernet.*, vol. 38, no. 4, pp. 852–866, July 2008.

Sung, K.-K. and Poggio, T., "Example-based learning for view-based human face detection," *IEEE Trans. Pattern Anal. Mach. Intell.*, vol. 20, no. 1, pp. 39–51, Jan. 1998.

Tian, Y.-L., Kanade, T., and Cohn, J. F., "Recognizing action units for facial expression analysis," *IEEE Trans. Pattern Anal. Mach. Intell.*, vol. 23, no. 2, pp. 97–115, Feb. 2001.

Tong, Y., Liao, W., and Ji, Q., "Facial action unit recognition by exploiting their dynamic and semantic relationships," *IEEE Trans. Pattern Anal. Mach. Intell.*, vol. 29, no. 10, pp. 1683–1699, Oct. 2007.

Wang, Y. and Yuan, B., "A novel approach for human face detection from color images under complex background," *Pattern Recogn.*, vol. 34, no. 10, pp. 1983–1992, Oct. 2001.

Wechsler, H. and Kidode, M., "A new detection technique and its implementation," *IEEE Trans. Syst. Man Cybernet.*, vol. 7, no. 12, pp. 827–835, Dec. 1977.

Wong, K.-W., Lam, K.-M., and Siu, W.-C., "An efficient algorithm for human face detection and facial feature extraction under different conditions," *Pattern Recogn.*, vol. 34, no. 10, pp. 1993–2004, Oct. 2001.

Wu, H., Chen, Q., and Yachida, M., "Face detection from color images using a fuzzy pattern matching method," *IEEE Trans. Pattern Anal. Mach. Intell.*, vol. 21, no. 6, pp. 557–563, June 1999.

Xiang, C. and Huang, D., "Feature extraction using recursive cluster-based linear discriminant with application to face recognition," *IEEE Trans. Image Process.*, vol. 15, no. 12, pp. 3824–3832, Dec. 2006.

Yang, M.-H., Kriegman, D. J., and Ahuja, N., "Detecting faces in images: a survey," *IEEE Trans. Pattern Anal. Mach. Intell.*, vol. 24, no. 1, pp. 34–58, Jan. 2002.

Yeasin, M., Bullot, B., and Sharma, R., "Recognition of facial expressions and measurement of levels of interest from video," *IEEE Trans. Multimedia*, vol. 8, no. 3, pp. 500–508, June 2006.

You, X., Zhang, D., Chen, Q., Wang, P., and Tang, Y. Y., "Face representation by using nontensor product wavelets," *Proceedings of the International Conference on Pattern Recognition*, Hong Kong, pp. 503–506, Aug. 2006.

Yow, K. C. and Cipolla, R., "Enhancing human face detection using motion and active contours," *Proceedings of the 3rd Asian Conference on Computer Vision*, Hong Kong, vol. 1, pp. 515–522, 1998.

Zhang, B., Shan, S., Chen, X., and Gao, W., "Histogram of Gabor phase patterns (HGPP): a novel object representation approach for face recognition," *IEEE Trans. Image Process.*, vol. 16, no. 1, pp. 57–68, Jan. 2007a.

Zhang, W., Shan, S., Chen, X., and Gao, W., "Local Gabor binary patterns based on Kullback–Leibler divergence for partially occluded face recognition," *IEEE Trans. Signal Process.*, vol. 14, no. 11, pp. 875–878, Nov. 2007b.

Zhang, Z., Lyons, M., Schuster, M., and Akamatsu, S., "Comparison between geometry-based and Gabor-wavelets-based facial expression recognition using multi-layer perceptron," *Proceedings of the IEEE International Conference on Automatic Face and Gesture Recognition*, Nara, Japan, pp. 454–459, Apr. 1998.

Zhao, H. and Yuen, P. C., "Incremental linear discriminant analysis for face recognition," *IEEE Trans. Syst. Man Cybernet. B*, vol. 38, no. 1, pp. 210–221, Feb. 2008.

Zou, J., Ji, Q., and Nagy, G., "Comparative study of local matching approach for face recognition," *IEEE Trans. Image Process.*, vol. 16, no. 10, pp. 2617–2628, Oct. 2007.

Zhu, Y., De Silva, L. C., and Ko, C. C., "Using moment invariants and HMM in facial expression recognition," *Pattern Recogn. Lett.*, vol. 23, no. 1, pp. 83–91, Jan. 2002.

CHAPTER 11

DOCUMENT IMAGE PROCESSING AND CLASSIFICATION

In every office, there is an urgent need to encode huge amounts of daily received documents automatically for further computer processing. In a document processing system, the data in any text content take on a variety of sizes and fonts of characters, graphics, and pictures. The interpretation of graphics and pictures is achieved by image processing and computer vision accompanied with rule-based technologies alluded to in the descriptive paragraphs. Therefore, the document processing system is the state-of-the-art enterprise of automating and integrating a wide range of processes and representations for document perception. It integrates many techniques involved in computer graphics, image processing, computer vision, and pattern recognition.

Numerous algorithms have been proposed for document processing. Lee (2002) presented substitution deciphering based on hidden Markov models with applications to compressed document processing. Marinai et al. (2005) surveyed the most significant problems in the area of offline document image processing, where connectionist-based approaches have been applied. Diligenti et al. (2003) proposed hidden tree Markov models for document image classification. Dawoud and Kamel (2004) presented iterative multimodel subimage binarization for handwritten character segmentation. Xue and Govindaraju (2002) presented a performance model that views word recognition as a function of character recognition and statistically discovers the relation between a word recognizer and the lexicon.

An overall organization of the proposed document processing system is shown in Figure 11.1. The office documents are digitized and thresholded into binary images by a scanner. In order to encode information from a mixed-mode document that contains text, graphics, and pictures, the first step is to segment the document image into individual blocks, namely, the text blocks, graphics blocks, and picture blocks. The text blocks are further separated into isolated characters that are passed through a character recognition subsystem. The graphics blocks are further divided into text description and graphical primitives, such as lines and curves. This text description can be dealt with in the same way as the text blocks. The graphical primitives are further encoded into parametric representations or symbolic words, or passed through a picture compressor. For instance, the logical AND gate circuit diagram is converted into the name "AND." The logo picture discriminated from the halftone picture is

Image Processing and Pattern Recognition by Frank Shih

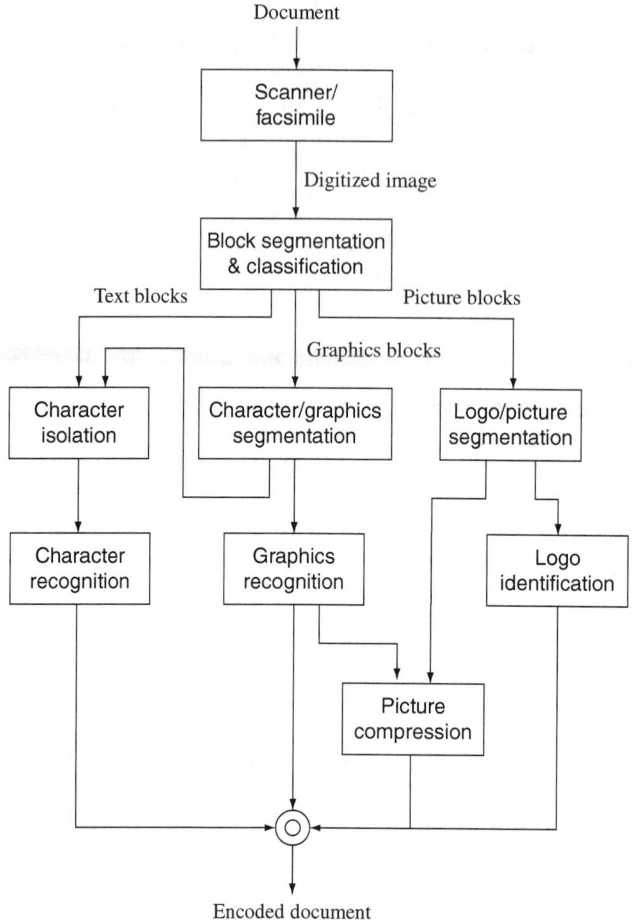

Figure 11.1 An overall organization of the proposed document processing system.

identified and transformed into its symbolic words. The picture blocks, consisting of high transitional density of white-to-black pixels, can be compressed for efficient storage using existing compression techniques.

This chapter is organized as follows. Block segmentation and classification are presented in Section 11.1. A rule-based character recognition system (CRS) is proposed in Section 11.2. Logo identification is described in Section 11.3. Fuzzy typographical analysis (FTA) for character preclassification is presented in Section 11.4. Fuzzy model for character classification is presented in Section 11.5.

11.1 BLOCK SEGMENTATION AND CLASSIFICATION

Document processing, which aims at the transformation of any information into an equivalent symbolic representation, can be viewed as an inverse processing from an image form to a processible form. The documents are digitized and/or thresholded into binary images by a scanner or a facsimile. The document images need to be first segmented into individual blocks of text, graphics, or picture.

Several techniques for block segmentation and classification have been developed. Wahl et al. (1982) used a run-length smoothing algorithm (RLSA) to segment a document into text, horizontal/vertical lines, or pictures. They classified graphics and pictures into the same category. Fisher et al. (1990) developed a rule-based algorithm consisting of smearing a document image and computing statistical properties of connected components. Bley (1984) decomposed a connected component into subcomponents that makes the further recognition process more complicated. The resulting algorithm is sensitive to font and size variations. Fletcher and Kasturi (1988) applied Hough transform to link connected components into a logical character string in order to discriminate from graphics. Their algorithm is relatively independent of variations in font, size, and the string orientation of text. Shih and Chen (1996b) presented adaptive document block segmentation and classification.

11.1.1 An Improved Two-Step Algorithm for Block Segmentation

The office documents are digitized and thresholded into black-and-white (or binary) images through a scanner or a facsimile, where white pixels are represented as 0's and black pixels as 1's. A RLSA (Fisher et al., 1990; Wong et al., 1982) can be applied to the binary sequences in the row-by-row or column-by-column scan. A set of adjacent 0's or 1's is called a *run*. The algorithm converts a binary input sequence f into an output sequence g by using the rule: the 0's in f are changed to 1's in g if the run length is of 0's; that is, the number of adjacent 0's is less than or equal to a predefined threshold value C. For example, if the threshold $C = 5$ is chosen and the input is

$$f: 00110000000100111000110000010 0$$

then the output will be

$$g: 11110000000111111111100000111$$

The smoothing rule will merge two runs of 1's together if the interval between them is not sufficiently spaced. Since the spacings of document elements are different horizontally and vertically, different values of C are used in row-by-row and column-by-column processing. With appropriate selection of C's, the merged runs will construct various blocks of a common data class. The RLSA consists of the following four steps:

1. A horizontal smoothing is applied to the document image by a threshold C_{h}.
2. A vertical smoothing is applied to the document image by a threshold C_{v}.
3. The results of steps 1 and 2 are combined by a logical AND operation.
4. An additional horizontal smoothing is applied to the output of step 3 using a relatively small threshold C_{a}.

The original RLSA requires scanning an image four times. However, the reduction of scanning the image from four to two can be achieved in the following way. The order of steps 1 and 2 can be exchanged. That is, if the vertical smoothing is performed first, then the horizontal smoothing can be merged together with step 3 by using algebra theory of $A \cap B = A - \bar{B}$, where A and B are sets. Therefore, the new three-step algorithm is

1. A vertical smoothing is applied on the original document image by a threshold C_{v}.
2. If the run length of 0's in the horizontal direction of the original image is greater than C_{h}, then the corresponding pixels in the output of step 1 are set 0's; otherwise, they remain unchanged.
3. An additional horizontal smoothing is applied to the output of step 2 by a relatively small threshold C_{a}.

Because $C_{\mathrm{a}} < C_{\mathrm{h}}$, it is observed that if the number of horizontally consecutive 0's of the original image is between C_{h} and C_{a}, then the corresponding pixels in the output of step 1 are checked whether they are subdivided into smaller segments and the spacings between segments are compared with C_{a} to determine whether 0's or 1's are to be assigned. Thus, the improved two-step algorithm is

1. A vertical smoothing is applied to the original document image by a threshold C_{v}.
2. If the run length of 0's in the horizontal direction of the original image (denoted by RL) is greater than C_{h}, then set the corresponding pixels in the output of step 1 to 0's. If $\mathrm{RL} \leq C_{\mathrm{a}}$, then set the corresponding pixels in the output of step 1 to 1's. If $C_{\mathrm{a}} < \mathrm{RL} \leq C_{\mathrm{h}}$ and the run length of horizontally consecutive 0's in the output of step 1 is less than or equal to C_{a}, then set the corresponding pixels in the output of step 1 to 1's.

11.1.2 Rule-Based Block Classification

This section presents a block classification algorithm for classifying blocks into one of text, horizontal/vertical line, graphics, or picture classes. A 2D plane consisting of the mean value of block height versus the run length of block mean black pixels was used to classify document blocks into text, nontext, horizontal line, or vertical line by Wong et al. (1982). Fisher et al. (1990) applied a rule-based classification based on the features of height, aspect ratio, density, perimeter, and perimeter/width ratio. Wang and Srihari (1989) used the black–white pair run-length matrix and the black–white–black combination run-length matrix to derive three features (1) short run emphasis, (2) long run emphasis, and (3) extra long run emphasis for newspaper classification. The above techniques present a problem of misclassification when the font and size of characters or the scanning resolution is varied. They adopted a set of chosen threshold values that cannot be adjusted to adapt the situations for coping with different types of documents. In the sequel, we aim at overcoming these difficulties.

Let the upper-left corner of an image be the origin of coordinates. The following measures are applied on each block:

- Minimum x- and y-coordinates and the width and height of a block $(x_{\min}, y_{\min}, \Delta x, \Delta y)$.
- Number of black pixels corresponding to the block of the original image (N).
- Number of horizontal transitions of white-to-black pixels corresponding to the block of the original image (TH).
- Number of vertical transitions of white-to-black pixels corresponding to the block of the original image (TV).
- Number of columns in which any black pixel exists corresponding to the block of the original image (δx).

The following features are adopted for block classification:

- Height of each block, $H = \Delta y$.
- Ratio of width to height (or aspect ratio), $R = \Delta x / \Delta y$.
- Density of black pixels in a block, $D = N/\Delta x \Delta y$.
- Horizontal transitions of white-to-black pixels per unit width, $\mathrm{TH}_x = \dfrac{\mathrm{TH}}{\delta x}$.
- Vertical transitions of white-to-black pixels per unit width, $\mathrm{TV}_x = \dfrac{\mathrm{TV}}{\delta x}$.
- Horizontal transitions of white-to-black pixels per unit height, $\mathrm{TH}_y = \dfrac{\mathrm{TH}}{\Delta y}$.

Because a document usually contains the characters of a most popular size and font, the mean value of all the block heights approximates the most popular block height. The width-to-height ratio can be used to detect horizontal or vertical lines. TH_x and TV_x are used for text and nontext discrimination. It is observed that both are independent of fonts and sizes of characters. Let $\mathrm{TH}_x^{\max}$ and $\mathrm{TH}_x^{\min}$ denote the maximum and minimum values for text discrimination, and let $\mathrm{TV}_x^{\max}$ and $\mathrm{TV}_x^{\min}$ denote for nontext discrimination. Let H_m be the average height of the most popular blocks. The rule-based block segmentation is described as follows:

- *Rule 1:* If $c_1 H_{\mathrm{m}} < H < c_2 H_{\mathrm{m}}$, the block belongs to text.
- *Rule 2:* If $H < c_1 H_{\mathrm{m}}$ and $c_{\mathrm{h}1} < \mathrm{TH}_x < c_{\mathrm{h}2}$, the block belongs to text.
- *Rule 3:* If $\mathrm{TH}_x < c_{\mathrm{h}3}$, $R > c_R$, and $c_3 < \mathrm{TV}_x < c_4$, the block is a horizontal line.
- *Rule 4:* If $\mathrm{TH}_x > 1/c_{\mathrm{h}3}$, $R < 1/c_R$, and $c_3 < \mathrm{TH}_y < c_4$, the block is a vertical line.
- *Rule 5:* If $H > c_2 H_{\mathrm{m}}$, $c_{\mathrm{h}1} < \mathrm{TH}_x < c_{\mathrm{h}2}$, and $c_{\mathrm{v}1} < \mathrm{TH}_x < c_{\mathrm{v}2}$, the block belongs to text.
- *Rule 6:* If $D < c_5$, the block belongs to graphics.
- *Rule 7:* Otherwise, the block belongs to a picture.

Rule 1 is used to extract most of the text lines. Rule 2 is used to capture the text lines with smaller sizes, such as footnotes or remarks. Rules 3 and 4 determine the horizontal and vertical lines, respectively. Rule 5 extracts the text with larger sizes, such as titles and headings. The remaining blocks are classified as nontext. Rule 6 is used to separate graphics from pictures by using the block density, such that graphics has lower density than pictures.

11.1.3 Parameters Adaptation

In our experiment, 100 documents with character size varying from 6 to 15 points were used. For each document, the mean values of height H_m and the standard deviation sd are derived from blocks of the most popular height. The ratios of sd_H/H_m are distributed within the range of 0.027 and 0.044 with an average 0.034. For reliability, the tolerance of text height is selected to be six times of the average ratio, that is, 0.2. Therefore, $c_1 = 1-0.2 = 0.8$ and $c_2 = 1+0.2 = 1.2$. This allows most of the text blocks to be extracted except those with significantly different sizes (e.g., title of text).

A set of text blocks with character sizes varying from 6 to 36 points, mixed with four commonly used fonts, Roman, italic, boldface, and Courier, were collected. Results of the mean values of TH_x and TV_x for different fonts and sizes are shown in Figure 11.2. For all fonts, the mean value is slightly decreased as the character size decreases because the gaps between objects are relatively small and could be filled up easily in printing or scanning. The mean values of Roman, italic, and boldface fonts are almost the same. The mean value of $TH_x$of Courier font is slightly lower than the others because other fonts are printed in a proportional style, while Courier font is printed in a fixed pitch so that it consists of less vertical-like strokes within a unit length of text. On the contrary, the mean value of $TV_x$of Courier font is slightly higher than the other fonts because horizontal serifs of Courier font are longer. The standard deviation of TH_x and TV_x for different fonts and sizes is less than 0.1. Therefore, the mean white-to-black transitions are reliable for text extraction with variant fonts and sizes. As a result of considering tolerances, $c_{h1} = 1.2, c_{h2} = 3.0, c_{v1} = 1.2$, and $c_{v2} = 2.6$ are selected.

Different resolutions (dots per inch) of a scanner were also experimented. As shown in Figure 11.3, the mean values of TH_x and TV_x are slightly changed. The standard deviation is lower than 0.1, so that the mean transitions are reliable with respect to the scanning resolution.

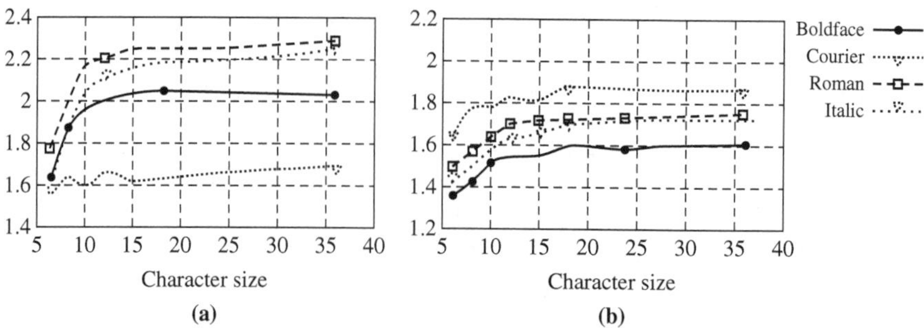

Figure 11.2 Mean values of (a) TH_x and (b) TV_x for different fonts and sizes.

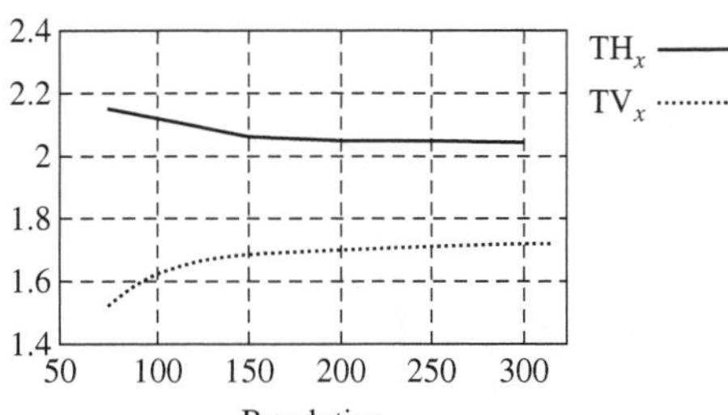

Figure 11.3 Mean values of TH_x and TV_x for different resolutions.

Results of TH_x and TV_x for blocks of text, vertical lines, graphics, and pictures are shown in Figure 11.4, where a great deal of graphics and pictures are out of the display range using the scale. The big cluster centered at (1.82, 1.60) represents text blocks, and the cluster centered at (0.12, 1.00) corresponds to vertical lines. Note that for vertical lines, TV_x is always very close to 1 and the standard deviation is 0.0004, which means TV_x is very stable for vertical line extraction. Similarly, for horizontal lines, TH_y is very close to 1 and is stable for horizontal line extraction. Therefore, $c_3 = 0.95$ and $c_4 = 1.05$ are selected. Disregarding the linewidth, the aspect ratio of a horizontal line R approximates $\cot\theta$, where θ denotes the tilt angle of the scanned document. Therefore, c_R is set to 5, allowing the document to be tilted up to 11.3. For a horizontal line, TH_x is within the range of $\pm\tan\theta$. Therefore, c_{h3} is selected as 0.2.

Because graphics and pictures are sparsely distributed on the TH_x–TV_x plane, it is not suited to use the mean transitions for discrimination. In principle, graphics composed of lines has lower density than pictures. From experiments, the mean value of density for graphics is 0.061 and the standard deviation is 0.033, while for pictures the mean value is 0.853 and the standard deviation is 0.108. In practice, the value in the range of (0.15, 0.25) is sufficient to separate graphics and pictures. Therefore, $c_5 = 0.2$ is selected.

11.1.4 Experimental Results

The proposed document block segmentation and classification algorithms were implemented on a SUN SPARC workstation under UNIX operating system. Document images are scanned at the resolution of 300 dots per inch. There are 100 documents randomly selected that consist of a mixture of text, graphics, and pictures. The algorithms perform the block segmentation and the classification successfully.

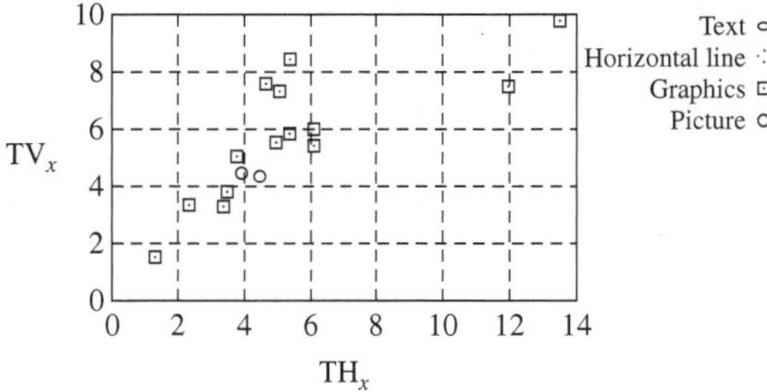

Figure 11.4 The projected TH_x-TV_x plane.

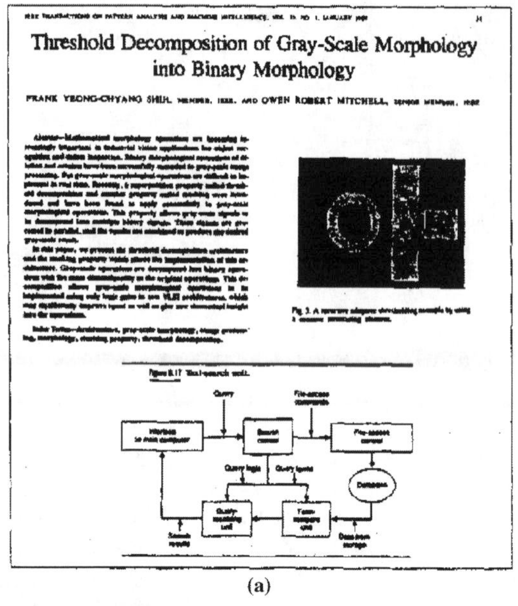

Threshold Decomposition of Gray-Scale Morphology into Binary Morphology

FRANK YEONG-CHYANG SHIH, ... AND OWEN ROBERT MITCHELL, ...

(a)

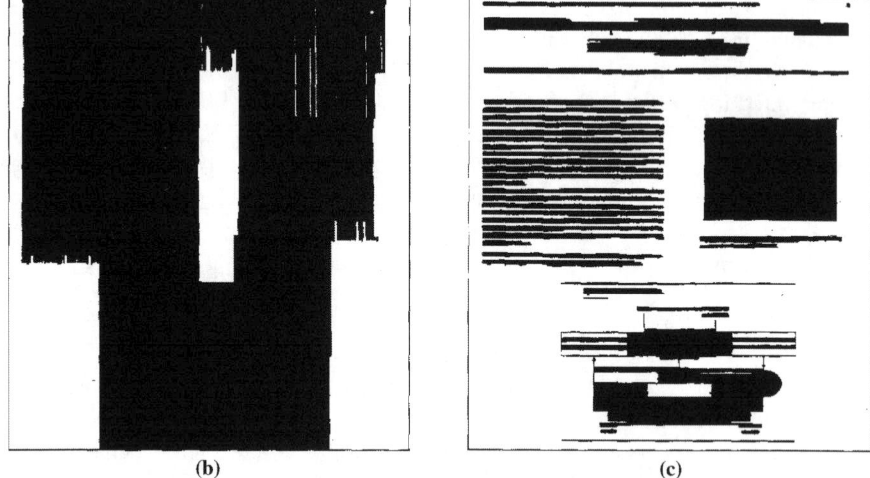

(b) (c)

Figure 11.5 An example of our improved two-step smoothing algorithm: (a) the document image, (b) the resulting image after step 1, and (c) the resulting image after step 2.

Figure 11.5a shows a document image that contains text with different fonts and sizes, a flowchart, a picture, and two horizontal lines. Figure 11.5b and c illustrate the resulting images after steps 1 and 2 of our improved two-step smoothing algorithm. The result of the block classification method in Figure 11.5a is shown in Table 11.1, where classes t, p, g, and h denote text, picture, graphics, and horizontal line, respectively. The result of applying Wong et al. (1982) Wong et al.'s algorithm (1982) is shown in Table 11.2 for comparison, where classes t, h, n denote text,

TABLE 11.1 The Result of Block Classification of the Proposed Algorithm

No.	x_{min}	Δx	y_{min}	Δy	δx	$\delta x/\Delta x$	TH	TV	TH	TV	N	D	Class
1	81	3591	38	23	1118	0.7027	2191	1676	1.96	1.50	9557	0.26117	t
2	2173	26	40	24	19	0.7308	47	33	2.47	1.74	197	0.31571	t
3	88	2101	127	101	2848	0.8796	3954	3210	2.14	1.74	43263	0.20388	t
4	676	936	247	94	836	0.6932	1888	1446	2.26	1.73	20493	0.23292	t
5	88	3101	408	45	1599	0.7611	3079	2707	1.93	1.69	15257	0.19310	t
6	88	1027	588	34	803	0.7819	1478	3166	1.84	1.45	8917	0.25537	t
7	77	1038	630	33	797	0.7678	1580	1210	1.98	152	9447	027579	t
8	77	1037	672	33	827	0.7575	1667	1254	2.02	152	10114	0.29555	t
9	1336	795	687	552	792	0.9962	19117	20627	24.14	26.04	338680	0.73198	P
10	77	1037	713	34	833	0.8033	1589	1286	1.91	1.54	9634	0.27296	t
11	77	1036	755	32	843	0.8137	3635	1292	1.94	1.53	10130	0030556	t
12	77	1035	796	33	824	0.7961	1594	1206	1.93	1.46	9971	0.291193	t
13	76	1036	537	34	802	0.7741	1526	1177	1.90	1.47	9447	0.26820	t
14	76	1036	879	33	722	0.6969	1394	1073	1.93	1.49	8671	0.25363	t
15	76	1035	920	35	820	0.7923	1584	1263	1.93	1.54	9709	0.26802	t
16	75	1035	962	35	774	0.7478	1526	1208	1.97	1.56	9374	0.25877	t
17	74	1036	1004	34	811	0.7828	1577	1219	1.94	1.50	3765	0.27723	t
18	74	253	1046	31	208	0.6221	375	341	1.80	1.64	2365	0.30154	t
19	74	1035	1087	35	794	0.7671	1507	1163	1.90	1.46	9391	0.35924	t
20	72	1037	1329	34	820	0.7907	1606	1206	1.96	1.47	9764	0.27693	1
21	71	1038	1170	36	822	0.7919	1528	1217	1.86	1.48	9360	0.25048	t
22	71	1038	1212	34	309	0.7794	1608	1226	1.99	1.52	9706	0.27502	t
23	70	1039	1254	34	744	0.7161	1439	1125	1.93	1.51	8782	0.24860	t
24	70	1036	1294	34	802	0.7741	1581	1212	1.97	1.51	9536	0.27072	t
25	70	1039	1337	34	800	0.7700	1618	1261	2.02	1.58	9795	0.27727	t
26	1313	833	1351	38	634	0.7611	1315	1074	2.0?	1.69	6102	0.19277	t
27	70	286	1378	32	232	0.8112	447	338	1.93	1.46	2794	0.30529	t
28	1314	462	1392	33	345	0.7468	677	566	1.96	1.64	3130	0.20530	t
29	70	1034	1444	35	823	0.7959	1443	1239	1.75	1.51	8820	0.24371	t
30	69	921	1485	35	747	0.8111	1450	1096	1.94	1.47	8698	0.26983	t.
31	521	1350	1609	11	1350	1.0000	93	1350	0.07	1.00	11991	0.80747	h
32	649	461	1640	39	343	0.7440	673	554	1.96	1.62	3857	0.21453	t
33	647	145	1693	3	145	1.0000	17	145	0.12	1.00	371	0.85287	h
34	951	535	1737	32	213	0.3981	410	385	1.92	1.81	2241	0.13090	t
35	517	1359	1774	630	1359	1.0000	7236	7972	5.36	5.87	51075	0.05966	g
36	740	94	2413	25	83	0.8830	172	139	2.07	1.67	948	0.40340	t
37	1548	107	3417	29	94	0.8785	183	181	1.95	1.93	1054	0.33967	t
38	514	1350	2478	5	1349	0.9593	143	1349	0.11	1.00	2436	0.36089	h

TABLE 11.2 The Result of Block Classification of Wong et al.'s Algorithm

No.	x_{min}	Δx	y_{min}	Δy	TH	N	N/TH	$\Delta x/\Delta y$	Class
1	31	1591	38	23	2191	9557	4.36	69.17	t
2	2173	26	40	24	47	197	4.19	1.08	t
3	88	2101	127	101	3954	43263	10.94	20.80	t
4	676	936	247	94	1888	20493	10.85	9.96	t
5	88	2101	408	45	3079	18257	5.93	46.69	I
6	88	1027	585	34	1478	8917	6.03	30.21	t
7	77	1038	630	33	1580	9447	5.98	31.45	t
8	77	1037	672	33	1667	10114	6.07	31.42	t
9	1336	795	687	582	19117	338680	17.72	1.37	n
10	77	1037	713	34	1589	9624	6.06	30.50	t
11	77	1036	755	32	1635	10130	6.20	32.38	t
12	77	1035	796	33	1594	9971	6.26	31.36	I
13	76	103G	837	34	1526	9447	6.19	30.47	t
14	76	1036	879	33	1394	8671	6.22	31.39	t
15	76	1035	920	35	1584	9709	6.13	29.57	t
16	75	1035	962	35	1526	9374	6.14	29.57	t
17	74	1036	1004	34	1577	9765	6.19	30.47	t
13	74	253	1046	31	375	2365	6.31	8.16	t
19	74	1035	1087	35	1507	9391	6.23	29.57	t
20	72	1037	1129	34	1606	9764	6.08	30.50	t
21	71	1038	1170	36	1528	9360	6.13	28.83	t
22	71	1038	1212	34	1608	9706	6.04	30.53	t
23	70	1039	1254	34	3439	8782	6.10	30.56	t
24	70	1036	1294	34	1581	9536	6.03	30.47	t
25	70	1039	1337	34	1618	9795	6.05	30.56	t
26	1313	833	1351	38	1315	6102	4.64	21.92	t
27	70	286	1378	32	447	2794	6.25	8.94	t
28	1314	462	1392	33	677	3130	4.62	14.00	t
29	70	1034	1444	35	1443	8820	6.11	29.54	t
30	69	921	1485	35	1450	8698	6.00	26.31	t
31	523	1350	1609	11	93	11991	128.94	122.73	h
32	649	461	1640	39	672	3857	5.74	11.82	t
33	647	145	1693	3	17	371	21.82	48.33	h
34	951	535	1737	32	410	2241	5.47	16.72	t
35	517	1359	1774	630	7286	51075	7.01	2.16	n
36	740	94	2413	25	172	948	5.51	3.76	t
37	1548	107	2417	29	183	1054	5.76	3.69	t
38	514	1350	2478	5	143	2436	17.03	270.00	t

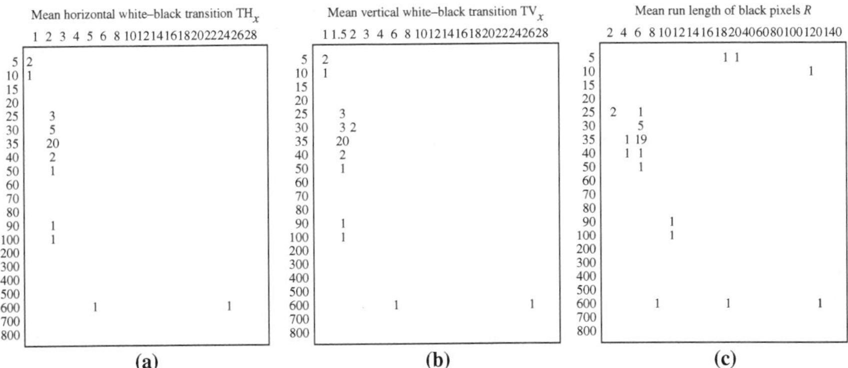

Figure 11.6 (a) The projected TH_x–H plane of our algorithm, (b) the projected TV_x–H plane of our algorithm, and (c) the projected R–H plane of Wong et al. (1982) Wong et al.'s algorithm (1982).

horizontal line, and nontext, respectively. Note that in Table 11.2, block 38, which should be a horizontal line, is misclassified as text.

Figure 11.6a and b shows the projected TH_x–H plane and TV_x–Hplane of our algorithm, and Figure 11.6c shows the projected R–H plane of Wong et al.'s algorithm. The proposed algorithm has three advantages over their algorithm. First, the features of TH_x and TV_x are more suitable in classifying text and nontext blocks than the feature of R. It is observed that TH_x and TV_x are more convergent than R for text blocks. Furthermore, the range of R of graphics blocks is mixed with that of text blocks. However, this effect does not occur in TH_x and TV_x.

Second, the described method is better in classifying solid lines than Wong et al. (1982) Wong et al.'s algorithm (1982). A thin line could be misclassified as text by using their algorithm. From observation, TV_x of a horizontal line is convergent to 1.0 and is also for TH_y of a vertical line. Third, TH_x and TV_x are independent of character sizes and fonts, while R is not. As the character size increases, R increases. Wong et al. (1982) restricted the character size up to three times of the most popular text blocks, which limits the generality of documents.

In the described block segmentation, step 1 is identical to step 2 of the original algorithm. However, step 2 is more complicated and takes about 10–15% overhead. Generally, our algorithm takes about 60–65% of the computational time of the original algorithm. For example, the document shown in Figure 11.5a takes 3.7432 s to perform the block segmentation using the original algorithm and takes 2.3682 s using the proposed algorithm. Therefore, it saves 37% of the computational time in this example.

11.2 RULE-BASED CHARACTER RECOGNITION SYSTEM

Research in optical character recognition (OCR) has a long history and has achieved considerable progress. Many techniques have been reported in existing literature. The fundamental task is to extract useful features from the input image and then to perform classification by comparing them with preestablished rules. Traditionally, the binary image of characters is thinned and vectorized before the character recognition begins (Pavlidis, 1986; Smith, 1987). Some approaches, such as analysis of strokes, can avoid the vectorization step (Bozinovic and Srihari, 1989) by scanning the pattern repeatedly in different directions. It is also possible to avoid thinning and vectorization, and can achieve a high recognition rate by using the topological description (Shridber and Badreldin, 1984), which is for contour or stroke detection.

Park et al. (2000) described hierarchical OCR, a character recognition methodology that achieves high speed and accuracy by using a multiresolution and hierarchical feature space. Dawoud (2007) proposed iterative cross section sequence graph for handwritten character segmentation. Cannon et al. (2003) introduced a learning problem related to the task of converting printed documents to ASCII text files.

The proposed approach presents a special way of recognizing segmented printed or handwritten characters based on a rule-based module. Rules defined by a module can detect character features and are grouped into character recognizers that accomplish character recognition. In addition to knowledge-based rules, a set of control rules are also employed. These control rules represented by a hashing table embody inferences into the order in which the character recognizers are executed.

In the proposed CRS, the image is regarded as an object that has the defined "methods" to implement the image processing routines and to produce the normalized, outbounded, and smoothed images (Shih et al., 1992). The rule-based class, which consists of many well-defined rules produced by tracking the processed image (raster tracking segmentation), is used to compose the relational rules into character recognition functions that can detect the processed image row by row (or column by column) and recognize what character it is.

The CRS is an object-oriented software system that is built on a set of classes characterizing the behaviors of all underlying data in the system. Objects from each class are manipulated by invoking the methods of the class, that is, sending messages to these objects. These messages represent the actions that are taken on the set of objects. Object-oriented programming focuses on the data to be manipulated rather than on the procedures that do the manipulating. The data form the basis for the software decomposition. The main challenge of object-oriented software design is the decomposition of a software system into underlying data types, classes or subclasses, and the definition of properties of the basic classes or subclasses.

In order to recognize an unknown character, the input image is examined by each recognizer until the character is classified. For example, if the input character is "Z," we need to go through 26 character recognizers. Thus, this process requires a costly computation. We can group certain characters into subsets based on their topological structures. The controller is implemented by a hashing table. During the recognition step, the controller determines the order in which these recognizers are executed. The recognition phase consists of two steps. The first step is to check whether the input image satisfies the common properties of the group. If not, all characters in this group are skipped and the checking continues in another group. The second step is to match the input image against each recognizer in the group. In each group, characters are arranged in the order that those with more strokes are in front of those with less strokes. This is a model-driven and data-driven composition. Using this concept, the recognition process can speed up.

The recognition procedure described includes three steps: contour extraction, stroke detection, and rule-based classification (or recognition). First, each character image passes through the contour extraction stage. This serves to extract outbounded boundary of characters. The stroke detection and rule-based classification are driven by two sets of rules: one is the stroke definition that specifies features of strokes, and the other is the character definition that specifies spatial interrelationships of strokes. The second step is to detect strokes of characters. On the basis of the detection, the third step performs the character recognition by following the classification rules. After the recognition, the character recognized is stored in the format of ASCII code. The flow chart of the character recognition subsystem is shown in Figure 11.7.

A text block from the block segmentation consists of a text line. Character separation is a process to isolate single characters from the text block. In general, the algorithms for character separation can be categorized into two types. One is the top-down method (Fletcher and Kasturi, 1988; Kahan et al., 1987), which is based on

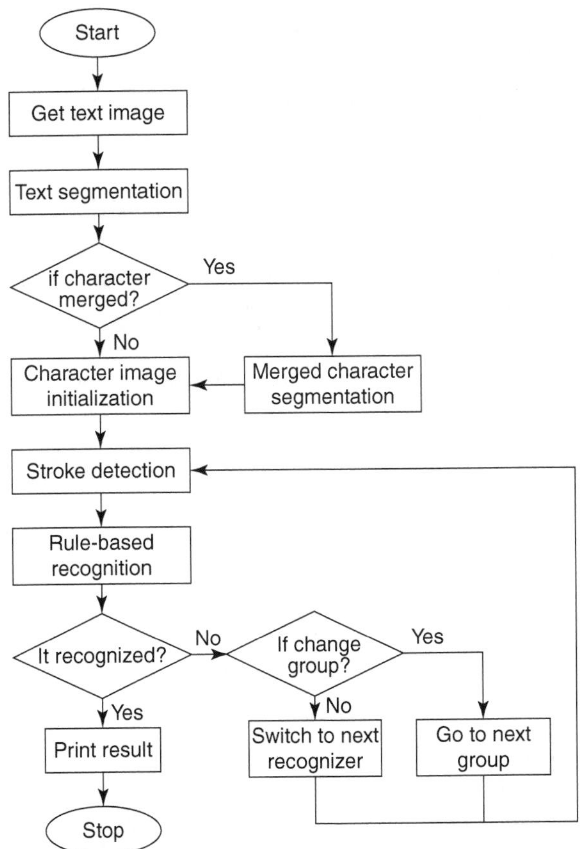

Figure 11.7 System flow for the character recognition.

projection profiles obtained by projecting an image onto specific axes. The profiles show the spatial structure of the document image. The other is the bottom-up method (Srihari and Zack, 1986), in which the image is grouped into connected components, such as characters, and then merged into words, lines, paragraph, and so forth. In the CRS, the top-down method is adopted.

By detecting those valleys on the projection profile onto x-axis, each text block can be separated into words and further into characters. If the ratio of character width to block height is greater than a threshold, the situation of merged characters occurs. The merged characters are cut in the place with a maximum value in the first derivative of the projection profile (Kahan et al., 1987).

The character image enclosed by a smallest rectangle is divided into four quadrants. In each quadrant, all the outbounded borders are extracted by checking two neighbors with respect to the quadrant's location. For instance, in the upper-right (or first) quadrant, if a "0" exists in anyone of the upper and right neighbors of a black pixel, it is an outbounded border. After a single raster scan is completed for all four quadrants, the outbounded contour is extracted. An example of the outbounded contour of the character "D" is shown in Figure 11.8b.

The existence of noise or broken lines arises in the outbounded contour. Therefore, the contour enhancement is performed to delete the noise and to fill in gaps of broken lines. An isolated point is considered as noise. A broken vertical line is detected if at least three of these points are darker than their left and right neighbors by examining five vertically consecutive points (x, y), $(x, y \pm 1)$, and $(x, y \pm 2)$. The similar procedure applies to a broken horizontal line. A resulting image after the contour enhancement of Figure 11.8b is shown in (c).

The stroke detection is to detect the strokes such as horizontal/vertical lines, left/right slash lines, and left/right opened curves, based on a set of rules. Characters consist of structural strokes and have certain relationship between strokes. Corresponding to a character recognizer, each rule consists of a set of detectors. Each detector is a function call of its parent class, which is the stroke detection. Once a rule is matched, the input character is recognized; otherwise, the system will go to the next character recognizer.

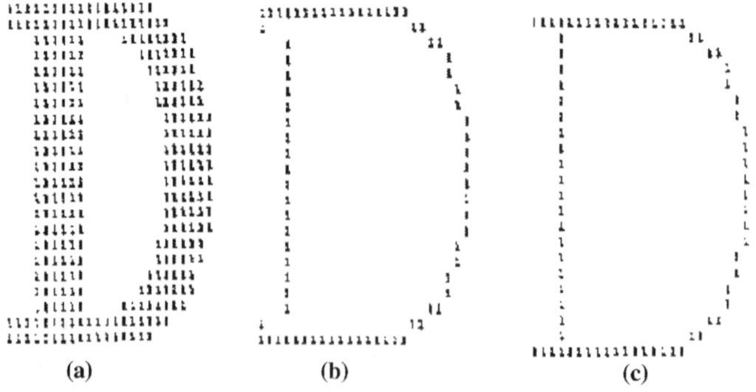

Figure 11.8 (a) A character image "D." (b) The extracted outbounded contour. (c) The image after the contour enhancement.

11.3 LOGO IDENTIFICATION

Logo pictures can be identified and converted into representative symbols. The logo identification presented in this section is based on the combination of local and global features (Shih et al., 1992). Many recognition techniques have been presented in literature. In general, they can be summarized into two categories: global and local features, and can use the object contour, silhouette, intensity profile, or range information in order to calculate the features such as Fourier description (Zahn and Roskies, 1972), moments of the silhouette (Reddi, 1981), and autoregressive model (Dubois and Glanz, 1986).

The primary difference between global and local features is that the former relies on the entire shape for the discrimination among objects and the later depends on only the local shape. The distortion or occlusion of a subregion of the shape will result in changes of global features. The local features describing a limited subregion are not affected by the other subregions. A new approach for constructing global features combined with local properties is described below.

Assume that Γ is a counterclockwise, simple, closed contour with parametric representation $(x(t), y(t)) = z(t)$, where t is the observation sequence and $0 \leq t \leq T$ (T is the perimeter of the contour). Let $z(t_{\text{ref}})$ be a reference observation and $t_{\text{ref}} \neq 0$. The observations can be represented in a polar form by adopting $z(t_{\text{ref}})$ and $z(0)$ as the reference line. Denote $\theta(t)$as the angular direction of $z(t)$ and $\iota(t)$ as the normalized spatial displacement of the observation. In the polar form, the observation contour Γ can be represented as $\{z(t) = \iota(t)e^{j\theta(t)}\}$, where $j = \sqrt{-1}$. The parameters $[\iota, \theta]$ are invariant to translation and rotation. To make the representation invariant to scaling, the length of the observations is normalized by the length of the reference line.

Let $\iota(0) = 0$ and $\theta(0) = 0$ be the initial values for the starting point $z(0)$. Let $\phi(r, s)$ be the angle between two lines, $\overline{z(0)z(r)}$ and $\overline{z(0)z(s)}$, as shown in Figure 11.9. The angle $\phi(r, s)$ is called a *sector* if $\iota(r) = \iota(s) = \iota$. Thus, a sector can be represented as $(\iota, \theta(r), \phi(r, s))$ in addition to $\phi(r, s)$. A shell is a collection of all sectors. To simplify and generalize the symbol, $(\iota_l, \theta_{lk}, \phi_{lk})$ is used to denote the kth sector of the

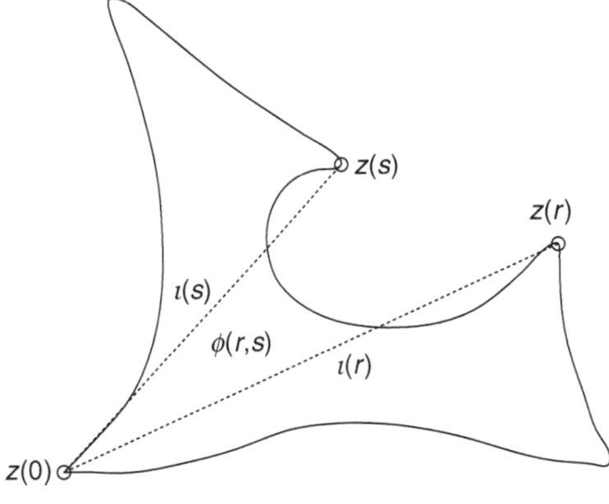

Figure 11.9 Definitions of information extraction.

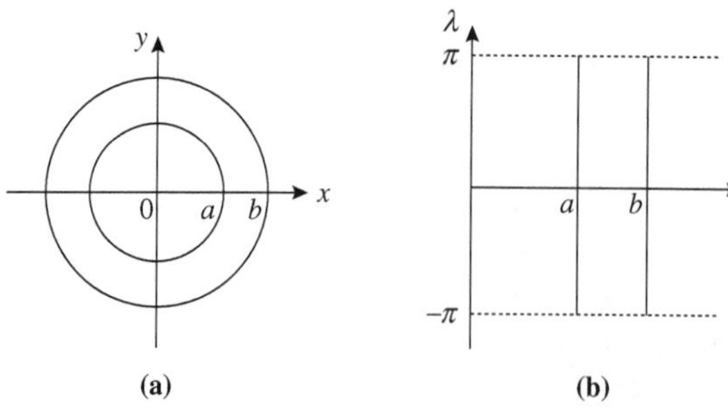

Figure 11.10 A mapping of a uniform polar angular grid onto a uniform (t, λ) grid.

lth shell. Thus, a contour Γ can be represented as

$$(\iota_l, \theta_{lk}, \phi_{lk}), \quad 0 \leq \iota_l \leq 1 \tag{11.1}$$

The three-dimensional parametric space is then mapped onto a 2D plane (t, λ), where $t = \iota_l$ and $\lambda = \{x | \theta_{lk} \leq x \leq \theta_{lk} + \phi_{lk}\}$. A mapping of a uniform polar angular grid onto a uniform (t, λ) grid is illustrated in Figure 11.10. The values ι_l, θ_{lk}, and ϕ_{lk} are the shell descriptors and are called *sector coefficients*.

In summary, all simple closed curves with the starting point are mapped onto the rings of shell descriptors in such a way that all curves of the identical shape and the starting point go into the same descriptors. The shell space with descriptor parameters (t, λ) of Figure 11.9 is shown in Figure 11.11. Figure 11.12 shows the shell descriptors of an "NJIT" logo from three different viewing directions.

Given two logos Γ and Γ' with the shell descriptors $S_n = \{t_n, [\lambda_p, \Lambda_p]_n\}$ and $S_m = \{t_m, [\lambda_q, \Lambda_q]_m\}$, respectively, where λ_p, λ_q are the lowest points in the sector and $\lambda_p + \Lambda_p, \lambda_q + \Lambda_q$ are the highest points. For measuring the similarities of all sector pairs (p, q), the followings are computed:

$$B_{mn} = \begin{cases} 0 & \text{if } t_n \neq t_m \\ \Psi & \text{elsewhere} \end{cases} \tag{11.2}$$

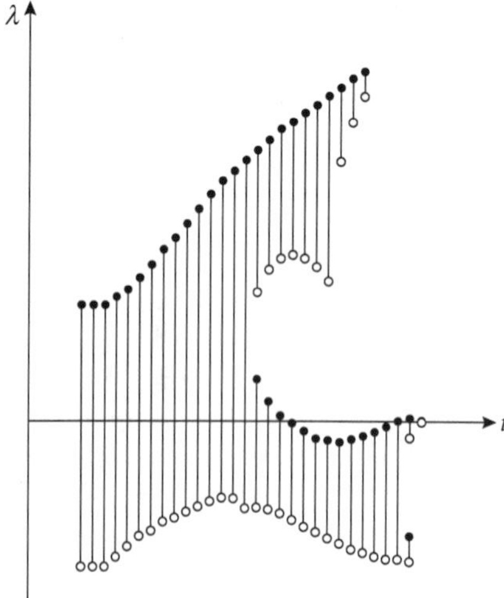

Figure 11.11 The shell descriptor of the shape shown in Figure 11.9.

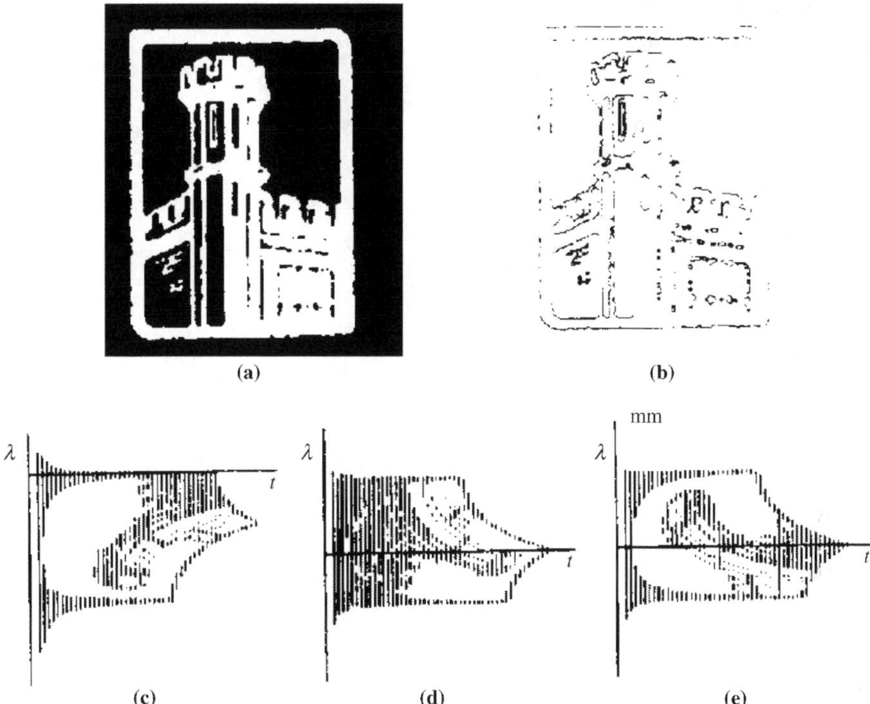

Figure 11.12 (a) Binary image of "NJIT" logo. (b) Logo contour. The shell descriptors of the "NJIT" logo in which the reference point is located at (c) upper left corner, (b) lower right corner, and (e) upper right corner.

where $\Psi = \min(\lambda_p + \Lambda_p, \lambda_q + \Lambda_q) - \max(\lambda_p, \lambda_q)$.

$$C_{mn} = \Lambda_p - B_{mn} \tag{11.3}$$

and

$$R_{mn} = \Lambda_q - B_{mn} \tag{11.4}$$

To measure the similarity between S_n and S_m, the above measures indicate the common phase bandwidth B_{mn}, the cost C_{mn}, and the risk R_{mn}. Two sectors are almost identical in phase if the sum of the cost and the risk are nearly zero.

Let S_n and S_m be the sector descriptors from the unknown image and the model object, respectively. For a predetermined tolerance τ, if the maximum cost C_{mn} and the maximum risk R_{mn} for similarity check are less than the threshold, then the unknown sector S_n is assigned to the model sector S_m. If the cost is high but the risk is less than the tolerance, then S_m is partially matched to S_m.

In the character recognition subsystem, 260 characters including 26 uppercase Roman letters are tested. A 95% recognition rate is achieved. In the remaining 5% misclassification, most of them are the influence of noise and bad quality of input images. Figure 11.13 shows a template of the "NJIT" logo used to demonstrate the logo identification. To match against the logo shown in Figure 11.12a, a cost of 6.22

Figure 11.13 A template of the "NJIT" logo used for logo identification.

and a risk of 3.78 are obtained. It implies that the match between Figure 11.12a and Figure 11.13 is more than 93%.

11.4 FUZZY TYPOGRAPHICAL ANALYSIS FOR CHARACTER PRECLASSIFICATION

Computerized document processing has been growing up rapidly since 1980s because of the exponentially increasing amount of daily received documents and the more powerful and affordable computer systems. Intuitively, the conversion of textual blocks into ASCII codes represents one of the most important tasks in document processing. Our strategy of preclassifying character is to incorporate the typographical structure analysis that categorizes characters in the first step to reduce the scope of character recognition (Chen et al., 1995). As illustrated in Figure 11.14, a text line can be decomposed into three stripes: the upper, middle, and lower zones. They are delimited by a top line, an upper baseline, a baseline, and an underline. The middle zone, being the primary part of a text line, is about twice as high as the other two zones and can be further split by a midline.

Explicitly, the baseline appears in a text line. The upper baseline may not be present in the case of a short text composed of ascenders only, and the top line and the underline may not exist if only centered characters appear. Therefore, it is essential to

Figure 11.14 Typographical structure of a text line.

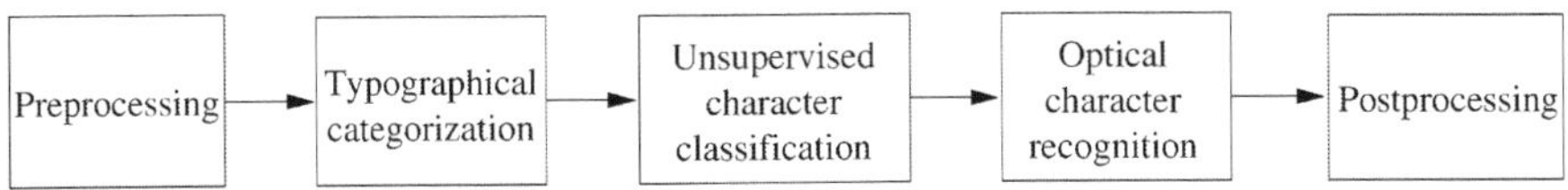

Figure 11.15 Overall procedure of our document processing system.

locate the baseline. The baseline can also be used for document skew normalization and for determining interline spacing to be more computationally efficient than the traditional Hough and Fourier transform approaches.

Our baseline detection algorithm based on a line of text is more reliable and efficient than the one based on a single word. The remaining virtual reference lines are extracted by a clustering technique. To allow the unpredictable noise and deformation, the tolerance analysis is included. To ensure the robustness and flexibility, a fuzzy logic approach is used to assign a membership to each typographical category for ambiguous classes. A linear mapping function is adopted and its boundary conditions are derived to preserve the continuity.

The typographical categorization is a part of our document processing system (Shih et al., 1992) as shown in Figure 11.15. The office documents are first digitized and thresholded into binary images by a scanner. The textual blocks are distinguished from graphics and pictures by the preprocessing that includes a block segmentation and classification. The unsupervised character classification classifies characters into a set of fuzzy prototypes based on a nonlinear weighted similarity function. The optical character recognition is intended to recognize the set of fuzzy prototypes. Finally, the postprocessing intends to correct the errors by means of dictionary checking or semantics understanding.

11.4.1 Character Typographical Structure Analysis

According to the stripes the characters intercept with, they are categorized into primary and secondary classes, which are in turn divided into further categories as follows:

1. *Primary class:* This class consists of large symbols including most of Latin alphabets.
   - **(a)** *Ascender:* Characters fully intercept the upper and middle zones, such as "A", "B", and "d".
   - **(b)** *Descender:* Characters fully intercept the lower and middle zones, such as "g", "q", "y", and "p".
   - **(c)** *Centered:* Characters fully lie in the middle zone only, such as "a", "c", "e", "m", and "x".
   - **(d)** *Full ranged:* Characters span all the three zones, such as "j", "(", ")", "{", and "f".
2. *Secondary class:*
   - **(e)** *Subscript:* Symbols lie around the baseline, such as ".", ",", "_", and the shifted characters as subscripts.

(f) *Superscript:* Symbols lie around the upper baseline, such as """, "∼", "∧", and the shifted characters as superscripts.

(g) *Internal:* Characters are located around the midline and partially intercept the middle zone, such as "-".

Typographical structure analysis aims to categorize each character unit into one of the seven categories. As described earlier, the baseline detection is crucial to the typographical analysis. However, the typographical structure analysis is sensitive to skew even in a very little angle. Despite the image was skew normalized, the text line may not be exactly horizontal due to some error in printing or scanning. In the next section, an efficient algorithm is presented to locate the baseline.

11.4.2 Baseline Detection

In order to correctly analyze the typographical structure, the virtual baseline must be detected. An efficient algorithm to locate the baseline is given as follows:

1. Let a text line (denoted as $\mathbf{T}$) of n character units (ch_i) appear in the text line from left to right sequentially. That is, $\mathbf{T} = (\text{ch}_1, \text{ch}_2, \ldots, \text{ch}_n)$. Let $p_i = (x_i, y_i)$ denote x- and y-coordinates of the bottom center point p_i at the bounding box of the ith character. The set of those bottom center points is $\mathbf{P} = (p_1, p_2, \ldots, p_n)$. Because most of the characters are of ascenders or centered, they are aligned on the baseline. Therefore, $\mathbf{P}$ provides the basis to find out the virtual baseline.
2. Let $\mathbf{Y}' = (y'_1, y'_2, \ldots, y'_{n-1})$ be the set of all the slopes, where the slope of the line segment $\overline{p_i p_{i+1}}$ is $y'_i = (y_{i+1} - y_i)/(x_{i+1} - x_i)$. Mostly, the slopes are near zero. That is, the contiguous characters ch_i and ch_{i+1} are frequently aligned on the baseline or the underline.
3. Let $\mathbf{Y}'_{\text{mp}}$ denote the set of the most popular slope obtained by clustering analysis from the set $\mathbf{Y}'$. That is,

$$\mathbf{Y}'_{\text{mp}} = \left\{ y'_i \middle| \left| y'_i - y'_{\text{mp}} \right| \leq \varepsilon, \quad i = 1, 2, \ldots, n-1 \right\} \tag{11.5}$$

   where y'_{mp} denotes the most popular slope and ε denotes the clustering factor. The initial slope approximation of the whole baseline is derived as

$$m_{\text{appr}} = \frac{\sum_{y'_i \in \mathbf{Y}'_{\text{mp}}} y'_i \Delta x_i}{\sum_{y'_i \in \mathbf{Y}'_{\text{mp}}} \Delta x_i}, \quad \text{where } \Delta x_i = x_{i+1} - x_i \tag{11.6}$$

4. A line can be expressed in the form $y = mx + b$, where m is slope and b is the intercept of the line with y-axis. Let $\mathbf{B}$ denote the set of intercepts, where the lines pass through the points in $\mathbf{P}$ with the slope m_{appr}. That is,

$$B = \left\{ b_i \middle| b_i = y_i - m_{\text{appr}} x_i, \quad i = 1, 2, \ldots, n \right\} \tag{11.7}$$

   where b_i denotes the intercept of a line with the slope m_{appr} passing through p_i. Since the baseline slope is approximately equal to m_{appr}, the bottom center points of the characters that are aligned on the baseline will be collinear and tend to cluster with respect to b. Some other small clusters, which may be sparsely distributed, represent some characters that are aligned in the same orientation as baseline. For example, "g" and "y" are clustered on the underline. Similar to $\mathbf{Y}'_{\text{mp}}$, the set of the most popular intercepts can be derived as

$$\mathbf{B}_{\text{mp}} = \{b_i \| b_i - b_{\text{mp}}| \le \delta\} \quad (11.8)$$

5. $\mathbf{B}_{\text{mp}}$ represent all the points located at the baseline $\mathbf{P}_{\text{bl}}$, where

$$\mathbf{P}_{\text{bl}} = \{p_i | b_i \in \mathbf{B}_{\text{mp}}\} \quad (11.9)$$

In order to obtain the precise baseline expressed by $y = m_{\text{bl}}x + b_{\text{bl}}$, a linear regression is performed on the points in $\mathbf{P}_{\text{bl}}$ by using the least-square-error approach. The equations are

$$m_{\text{bl}} = \frac{\begin{vmatrix} \sum_{p_i \in \mathbf{P}_{\text{bl}}} x_i y_i & \sum_{p_i \in \mathbf{P}_{\text{bl}}} x_i \\ \sum_{p_i \in \mathbf{P}_{\text{bl}}} y_i & N_{\mathbf{P}_{\text{bl}}} \end{vmatrix}}{\begin{vmatrix} \sum_{p_i \in \mathbf{P}_{\text{bl}}} x_i^2 & \sum_{p_i \in \mathbf{P}_{\text{bl}}} x_i \\ \sum_{p_i \in \mathbf{P}_{\text{bl}}} x_i & N_{\mathbf{P}_{\text{bl}}} \end{vmatrix}} \quad (11.10)$$

and

$$b_{\text{bl}} = \frac{\begin{vmatrix} \sum_{p_i \in \mathbf{P}_{\text{bl}}} x_i^2 & \sum_{p_i \in \mathbf{P}_{\text{bl}}} x_i y_i \\ \sum_{p_i \in \mathbf{P}_{\text{bl}}} x_i & \sum_{p_i \in \mathbf{P}_{\text{bl}}} y_i \end{vmatrix}}{\begin{vmatrix} \sum_{p_i \in \mathbf{P}_{\text{bl}}} x_i^2 & \sum_{p_i \in \mathbf{P}_{\text{bl}}} x_i \\ \sum_{p_i \in \mathbf{P}_{\text{bl}}} x_i & N_{\mathbf{P}_{\text{bl}}} \end{vmatrix}} \quad (11.11)$$

where $N_{\mathbf{P}_{\text{bl}}} = \sum_{p_i \in \mathbf{P}_{\text{bl}}} x_i^0$, denoting the number of the elements in $\mathbf{P}_{\text{bl}}$.

11.4.3 Tolerance Analysis

Typographical analysis is a statistical approach for locating the virtual reference lines in a text line. Due to the unavoidable noise in an image, a variable tolerance must be allowed in the baseline detection algorithm and in the typographical categorization described in the next section. As illustrated in Figure 11.16, the tolerance is defined as $\tau = 2\delta$, and the shaded areas denote the tolerance of the reference lines. Suppose that the tolerance of each reference line is equal. Let H be the height of the text line including the tolerance, and let H_l be the height between baseline and underline. Assume $H_l = 4\delta$. Since the heights of the upper zone and the lower zone are the same and approximately half of the middle zone, we have $H = 18\delta$.

Figure 11.16 Tolerances of the reference lines.

Let $\mathbf{Q} = \{q_1, q_1, \cdots, q_n\}$ denote the set of points with $q_i = (x_i, y_i)$ representing the upper center x- and y-coordinates of the character ch_i. The set of intercepts of the lines passing through the points in $\mathbf{Q}$ with y-axis can be defined as

$$\mathbf{B_Q} = \{b_i | b_i = y_i - m_{\mathrm{appr}} x_i, \quad \text{where } (x_i, y_i) \in \mathbf{Q}\} \tag{11.12}$$

The height of the text line is derived as

$$H = \frac{1}{\sqrt{1 + m_{\mathrm{appr}}^2}} (\max\{\omega | \omega \in \mathbf{B_P}\} - \min\{\omega | \omega \in \mathbf{B_Q}\}) \tag{11.13}$$

When m_{appr} is very small,

$$H \approx \max\{\omega | \omega \in \mathbf{B_P}\} - \min\{\omega | \omega \in \mathbf{B_Q}\} \tag{11.14}$$

The tolerance $\delta = H/18$ is then computed and used in step 4.

For example of an image with the resolution 300 dpi (where 1 dot = 1 pixel), the height of a text line is equal to $s \times 300/72$ pixels, where s denotes a font size, because the point size of 72 is 1 in. high. If the point size is 12 and $H = 50$ pixels, the upper and lower zones are approximately equal to 12 pixels and $\tau = 6$.

Assume that the distance between the centers of the adjacent characters is no more than twice of the text height. Let ϕ be the angle of the virtual baseline with respect to x-axis. If two adjacent characters, say ch_i and ch_{i+1}, are baseline aligned, the line $\overline{p_i p_{i+1}}$ may exist in the range of the orientations from $\phi + \theta$ to $\phi - \theta$ as shown in Figure 11.17, where

$$\theta = \tan^{-1} \frac{\tau}{2H} = \tan^{-1} \frac{\delta}{H} \tag{11.15}$$

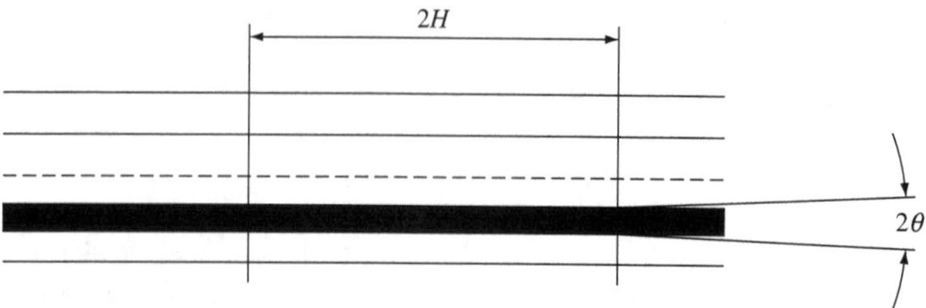

Figure 11.17 Tolerance of the slope of the baseline.

In other words, the slope varies from $m_{\phi+\theta}$ to $m_{\phi-\theta}$, where

$$m_{\phi+\theta} = \tan(\phi+\theta) = \frac{\tan\phi + \tan\theta}{1 - \tan\phi\tan\theta} \tag{11.16}$$

$$m_{\phi-\theta} = \tan(\phi-\theta) = \frac{\tan\phi - \tan\theta}{1 + \tan\phi\tan\theta} \tag{11.17}$$

When ϕ and θ are small, the term "$\tan\phi\tan\theta$" is negligible. Equations (11.16) and (11.17) can be simplified as

$$m_{\phi+\theta} \approx \tan\phi + \tan\theta = m + \varepsilon \tag{11.18}$$

$$m_{\phi-\theta} \approx \tan\phi - \tan\theta = m - \varepsilon \tag{11.19}$$

where

$$\varepsilon = \tan\theta = \frac{\delta}{H} = \frac{1}{18} \tag{11.20}$$

11.4.4 Fuzzy Typographical Categorization

The detection of other reference lines can be easily achieved by deriving from the baseline. First, the upper baseline is detected by projecting in parallel to the baseline the set of upper center points **Q** of the characters whose sizes are larger than a threshold. The threshold is used to exclude the characters in the secondary class. Since the centered characters have the smallest height in the primary class, the threshold is selected as $\frac{H}{2} - \frac{H}{18} \times 2 = \frac{7}{18}H$ by considering the tolerance. The projected points onto y-axis tend to have two clusters. The most popular cluster corresponds to the upper baseline, while the other cluster corresponds to the top line. These results can be verified if the height of upper zone approximates to a half of the height of middle zone. In other words, $b_{ub} - b_{tl} \approx (b_{bl} - b_{ub})/2$.

Similarly, the underline can be detected by projecting to the y-axis the set of bottom center points of the characters whose sizes are larger than the above threshold and whose bottom center points are not located at the baseline.

The projected locations of upper center points would appear only one cluster if a textual block contains only ascenders (e.g., a title with capital letters) or centered characters (e.g., the last line of a paragraph with a few characters). The block is hypothesized as the ascender if it does not exist in a paragraph or the number of characters in the block is reasonably large.

A character is assigned to one of the seven typographical categories based on its location listed in Table 11.3, where y_q and y_p denote y-coordinates of the upper center and lower center points of a character, respectively. The tolerance zones of the top line, upper baseline, midline, baseline, and underline are denoted as r_1, r_3, r_5, r_7, and r_9, respectively, and the ranges in between are denoted as r_2, r_4, r_6, and r_8 as illustrated in Figure 11.18. The seven categories are denoted as $\uparrow$ (superscript), $\downarrow$ (subscript), A (ascender), D (descender), C (center), F (full range), and I (internal). Most of the characters belong to the deterministic classes, denoted as (y_q, y_p)'s pair such as $(1, 7)$, $(1, 9)$, $(2, 7)$, and $(2, 9)$, which are classified as the ascender, full range, center, and

TABLE 11.3 The Decision Table for Typographical Categorization

$y_p \backslash y_q$	r_1	r_2	r_3	r_4	r_5	r_6	r_7	r_8	r_9
r_1	↑	–	–	–	–	–	–	–	–
r_2	↑	↑	–	–	–	–	–	–	–
r_3	↑	↑	↑	–	–	–	–	–	–
r_4	↑	↑	↑, (*I*)	↑, (*I*)	–	–	–	–	–
r_5	↑	↑	↑, *I*	*I*, (↑)	*I*	–	–	–	–
r_6	↑, *A*	↑, *A*, *C*	*I*, *C*, (↑)	*C*, *I*	*I*, (↓)	*I*, ↓	–	–	–
r_7	*A*	*A*, *C*	*C*	*C*, *I*, (↓)	*I*, ↓	↓, (*I*)	↓	–	–
r_8	*A*, *F*	*A*, *C*, *F*, *D*	*C*, *D*	*C*, *D*, ↓	↓	↓	↓	↓	–
r_9	*F*	*F*, *D*	*D*	*D*, ↓	↓	↓	↓	↓	↓

descender, respectively. For other classes containing more than one category, an uncertainty for each class is detected. In Table 11.3, the category enclosed in parentheses represents a weak class.

Due to unpredictable noise or deformation, a character may be detected in an ambiguous position, for example, y_q in r_1 and y_p in r_8, which could be an ascender or a full-ranged character. In this case, the memberships are assigned with the degrees of the character belonging to typographical categories.

Definition 11.1. The fuzzy typographical categorization of a character alpha is a list of ordered pairs such that $\alpha = \{(\Omega, \chi(\Omega))\}$, where $\Omega \in \{\uparrow, \downarrow, A, D, C, F, I\}$, and $\chi(\Omega)$, ranging in [0, 1], represents the grade of membership the character belongs to the category Ω.

Essentially, the memberships are characterized by size and position of characters, which in turn are determined by y_p and y_q. The size is $y_p - y_q$ and the position can be indicated by $y_p + y_q$. Let $y_0, y_1, \ldots, y_9$ be the y-coordinates that delimit $r_0, r_1, \ldots, r_9$, respectively, as shown in Figure 11.18. For simplicity, the tolerance ranges $r_i, i = 1, \cdots, 9$ are normalized and denoted as r_i'. The normalized y-coordinate of $y_j, j = p, q$, in r_i' will be $y_j' = (y_j - y_{j-1})/(y_i - y_{i-1})$ by linear interpolation. For example, the memberships in the class (1, 8) are given as

$$\begin{cases} \chi_{(1,8)}(A) = 1 - y_p' = \dfrac{y_8 - y_p}{y_8 - y_7} \\ \chi_{(1,8)}(F) = y_p' = \dfrac{y_p - y_7}{y_8 - y_7} \end{cases} \tag{11.21}$$

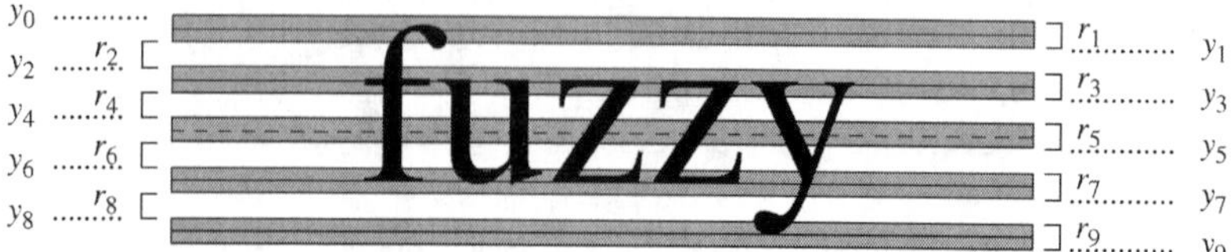

Figure 11.18 The illustration for the tolerance ranges used in Table 11.3.

where (1, 8) denotes the decision class with $y_q \in r_1$ and $y_p \in r_8$. Note that the membership functions in equation (11.21) are continuous with $\chi_{(1,7)}(\Omega)$ and $\chi_{(1,9)}(\Omega)$.

Similarly, the membership functions $\chi_{(2,7)}(\Omega), \chi_{(2,9)}(\Omega)$, and $\chi_{(3,8)}(\Omega)$ can be defined as

$$\begin{cases} \chi_{(2,7)}(A) = 1-y'_q = \dfrac{y_2-y_q}{y_2-y_1} \\ \chi_{(2,7)}(C) = y'_q = \dfrac{y_q-y_1}{y_2-y_1} \end{cases} \tag{11.22}$$

$$\begin{cases} \chi_{(2,9)}(D) = y'_q = \dfrac{y_q-y_1}{y_2-y_1} \\ \chi_{(2,9)}(F) = 1-y'_q = \dfrac{y_2-y_q}{y_2-y_1} \end{cases} \tag{11.23}$$

and

$$\begin{cases} \chi_{(3,8)}(D) = y'_p = \dfrac{y_p-y_7}{y_8-y_7} \\ \chi_{(3,8)}(C) = 1-y'_p = \dfrac{y_8-y_p}{y_8-y_7} \end{cases} \tag{11.24}$$

Now, the membership in the most ambiguous decision class (2, 8) will be determined by the characteristics of the typographical categories and the boundary conditions which are ensured the continuity with the memberships in classes (1, 8), (2, 7), (2, 9), and (3, 8). In class (2, 8), the character could be one of the ascender, descender, centered, and full-ranged characters. First, let us discuss the membership of the full-ranged character, which has the following boundary conditions:

$$\chi_{(2,8)}(F)_{y_q=y_1} = y'_p \tag{11.25}$$

and

$$\chi_{(2,8)}(F)_{y_p=y_8} = 1-y'_q \tag{11.26}$$

The membership of the full-ranged character is characterized by the size only. Let $s = y'_p - y'_q$ and $t = y'_p + y'_q$. The membership function, which is linear with respect to the size, has the following characteristics:

$$\frac{\partial \chi_{(2,8)}(F)}{\partial s} = c_1 \tag{11.27}$$

where $\chi_{(2,8)}(F) > 0$, c_1 is a positive constant, and

$$\frac{\partial \chi_{(2,8)}(F)}{\partial t} = 0 \tag{11.28}$$

From equations (11.25)–(11.28), the membership function of the full-ranged character can be derived as

$$\chi_{(2,8)}(F) = \begin{cases} y'_p - y'_q, & \text{if } y'_p > y'_q \\ 0, & \text{otherwise} \end{cases} \tag{11.29}$$

Similarly, the membership function of the centered character, which also simply depends on the character size, can be formulated as

$$\chi_{(2,8)}(C) = \begin{cases} y'_q - y'_p, & \text{if } y'_p < y'_q \\ 0, & \text{otherwise} \end{cases} \tag{11.30}$$

The memberships of the ascender and descender are more complicated since they depend on size and position. For the ascender, the membership function has the following boundary conditions:

$$\chi_{(2,8)}(A)_{y_q=y_1} = 1 - y'_p \tag{11.31}$$

$$\chi_{(2,8)}(A)_{y_p=y_7} = 1 - y'_q \tag{11.32}$$

On the other hand, the membership function should have the following characteristics:

$$\frac{\partial \chi_{(2,8)}(A)}{\partial t} = c_2 \tag{11.33}$$

and

$$\frac{\partial \chi_{(2,8)}(A)}{\partial s} = \begin{cases} c_3, & \text{if } y'_p < y'_q \\ -c_3, & \text{otherwise} \end{cases} \tag{11.34}$$

where c_2 and c_3 are positive constants.

From equations (11.31)–(11.34), the membership function can be obtained as

$$\chi_{(2,8)}(A) = \begin{cases} 1 - y'_q, & \text{if } y'_p \le y'_q \\ 1 - y'_p, & \text{otherwise} \end{cases} \tag{11.35}$$

Similarly, the membership function of the descender can be derived as

$$\chi_{(2,8)}(D) = \begin{cases} y'_q, & \text{if } y'_p \ge y'_q \\ y'_p, & \text{otherwise} \end{cases} \tag{11.36}$$

The membership functions in class (2, 8) are continuous together with neighboring categories since the boundary conditions are considered. The membership functions in class (2, 8) are shown in Figure 11.19. Note that the sum of the memberships is equal to one.

The membership functions of other ambiguous decision classes can be similarly derived. The entire membership functions for each typographical category are shown in Figure 11.20. Note that the membership functions of the internal, centered, and

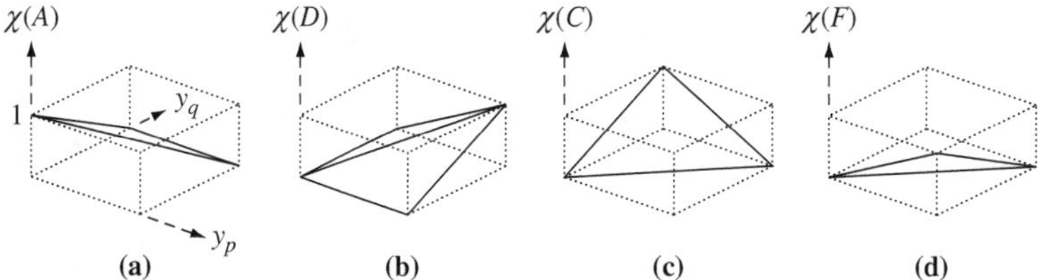

Figure 11.19 The membership functions in class (2, 8) of (a) ascender, (b) descender, (c) center, and (d) full range.

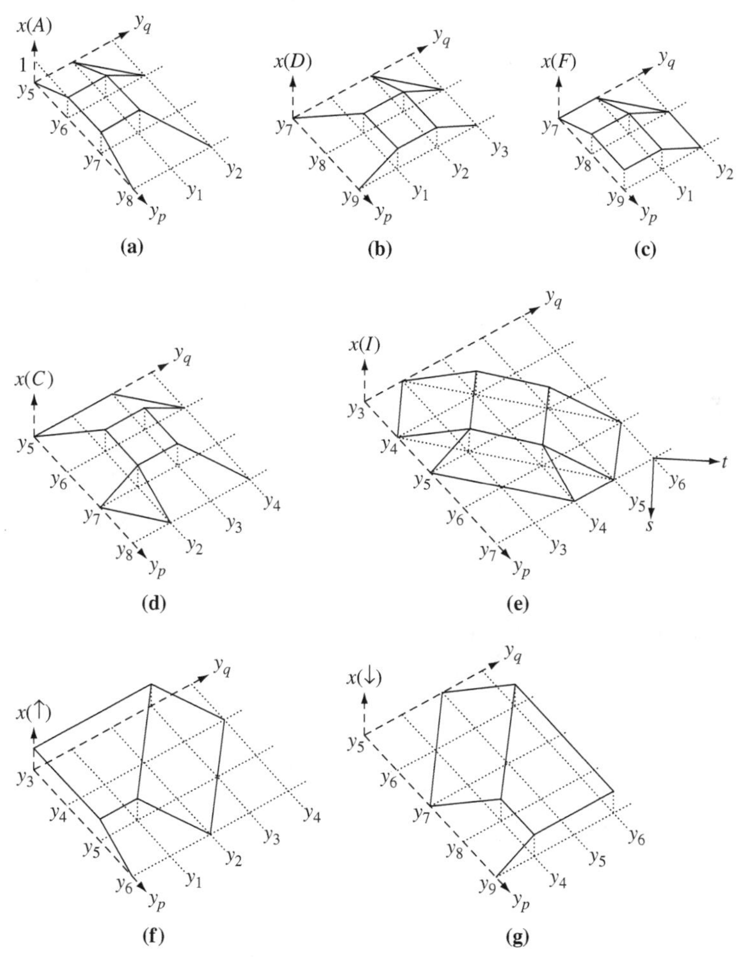

Figure 11.20 The membership functions of (a) ascender, (b) descender, (c) full range, (d) center, (e) internal, (f) superscript, and (g) subscript.

Threshold Decomposition of Gray-Scale Morphology

(a)

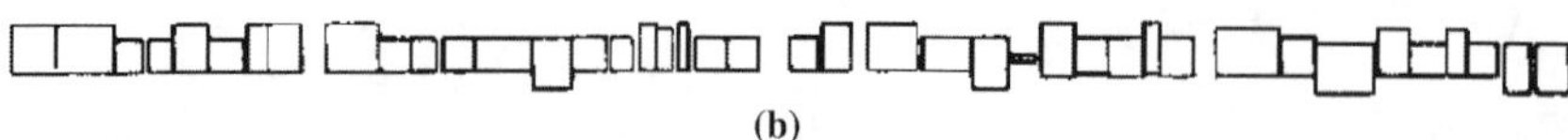

(b)

Threshold Decomposition of Gray-Scale Morphology

(c)

Figure 11.21 The typographical analysis of a textual block. (a) A sample text image, (b) the bounding boxes for each corresponding character, and (c) the virtual reference lines of the text line.

full-ranged characters are diagonally symmetric in the normalized (y_p, y_q) plane. In addition, the membership functions of the ascender and superscript are diagonally symmetric with those of the descender and the subscript.

11.4.5 Experimental Results

The raw image of a document is scanned and decomposed into blocks of text, graphics, pictures, horizontal lines, and vertical lines. The textual blocks are extracted for typographical analysis.

The result of applying the virtual reference lines detection is shown in Figure 11.21. Various character sizes from 5 to 12 points, each consisting of 500 text lines, have been tested. The baseline is 100% correctly detected for the character size larger than five points. For 5-point characters, the error rate of 2.5% for baseline detection is found because the merged characters are highly increased. As shown in Table 11.4, 48.65% of characters are merged characters, which may contain more characters as their sizes decrease. In our experiment, 26.28% of merged characters contain more than three characters, which makes the slope approximation inaccurate.

Beyond the seven basic typographical categories, one more category, that is, the semiascender that corresponds to the class with $y_q \in r_2$ and $y_p \in r_7$ as listed in

TABLE 11.4 Experimental Results from Our Algorithm

Character size	Error rate for baseline detection	Ambiguous character	Merged character	Average run time (μs/ch)
5	2.5%	0.50%	48.65%	85
6	0%	0.27%	28.23%	82
7	0%	0.10%	17.17%	84
8	0%	0.21%	17.59%	84
9	0%	0.19%,	13.39%	81
10	0%	0.62%	5.56%	81
11	0%	0.68%	1.90%	79
12	0%	0.41%,	1.45%	80

IEEE TRANSACTIONS ON PATTERN ANALYSIS AND MACHINE. VOL. II. NO. I. JANUARY 1989

Figure 11.22 A text line with a small character size.

Table 11.3, has been included. This category includes character "t". In the experiment of a sample document with the 7-point character size, 78 input patterns that are character "t" in Roman or Italic fonts are classified as semiascenders, and 10 input patterns are classified as ascenders due to noise.

When the baseline is correctly located, the overall ability for typographical categorization is 100%, in which more than 99% of the characters are classified from the deterministic classes. According to experimental results, there are a few special symbols that may fall into the ambiguous classes. For example, the symbols "/", "(", ")", "[", and "]" are sometimes detected in class (1, 8) and ";" is detected in class (3, 8). When the fuzzy logic is applied, "/" is classified as ascender, ";" is classified as descender, and the remainders are classified as full range if the category with highest membership is selected. Other ambiguous characters happen when merged characters contain these symbols or some characters belonging to the secondary class. For example, merged character "t," and "t;" have been detected as belonging to class (2, 8). This typographical knowledge may help to separate them in the recognition phase.

The FTA has been tested on 4000 text lines containing 307,221 characters. The average running time (denoted as t) is independent of the number and the size of characters. When FTA is executed on each text line individually, $t = 160.5$ μs/ch (microsecond per character). However, when a whole document is processed, the initial slope of the baseline need not be estimated every time. Instead, the slope of the previous text line is used. Therefore, the running time can be reduced to $t = 82$ μs/ch. In other words, for 1 s, FTA can process 12,195.12 characters. Comparing to the speed of character recognition, it is worthy of preclassifying the characters by FTA, so that OCR can be simplified and improved.

One problem that has to be considered is the tolerance estimation. When the character size is small, the tolerance will be relatively small. Therefore, a small noise may induce misclassification. For example, the text line shown in Figure 11.22 is first estimated with 21.6 pixel high and the tolerance $\delta = 1.2$. Therefore, $\delta = 2$ is enforced when the derived tolerance is smaller than 2. However, the tolerance of top line and upper baseline will be overlapped in the case of the text height that is smaller than 16, that is, smaller than a 4-point size character. Fortunately, it is rarely used for normal documents. The results show that the text line can be correctly categorized for the character size as small as six points.

Another problem is occurred if a text line contains different sizes of characters. This special case is illustrated in Figure 11.23, where the characters belonging to the string "IEEE MEMBER" are categorized as centers. In this example, every word contains only a single type of characters, that is, the ascender or the center.

FRANK YEONG-CHYANG SHIH, IEEE, MEMBER, AND OWEN ROBERT MITCHELL, SENIOR MEMBER, IEEE

Figure 11.23 A special case of text contains different sizes of characters.

Figure 11.24 Sample sets of handwritten characters for fuzzy typographical analysis.

The FTA can also be applied on handwritten characters. Particularly, it is significant to determine that a character should be an uppercase or lowercase. Several examples are illustrated in Figure 11.24. For the first example of the left column, the skew angle is also detected. For the second example of the left column, the word could be recognized as "RosemARiE" and corrected as "Rosemarie" by the FTA.

11.5 FUZZY MODEL FOR CHARACTER CLASSIFICATION

This section presents a fuzzy logic approach to efficiently perform unsupervised character classification for improvement in robustness, correctness, and speed of a character recognition system (Chen et al., 1994). The characters are first split into seven typographical categories. The classification scheme uses pattern matching to classify the characters in each category into a set of fuzzy prototypes based on a nonlinear weighted similarity function. The fuzzy unsupervised character classification, which is natural in the representation of prototypes for character matching, is developed, and a weighted fuzzy similarity measure is explored. The characteristics of the fuzzy model are discussed and used in speeding up the classification process. After classification, the character recognition, which is simply applied on a smaller set of the fuzzy prototypes, becomes much easier and less time-consuming. Shih and Chen (1996a) presented skeletonization for fuzzy degraded character images. Kim and Bang (2000) proposed a handwritten numeral character classification method based on the tolerant rough set that extends the existing equivalent rough set. Liu et al. (2004) proposed a discriminative learning algorithm to optimize the parameters of the modified quadratic discriminant function with an aim to improve the handwriting classification accuracy while preserving the superior noncharacter resistance.

11.5.1 Similarity Measurement

The fuzzy unsupervised character classification divides the characters in each typographical category into a set of fuzzy prototypes. Initially, the set of fuzzy prototypes in each category is empty. The first input character to the category is set to the first element of the corresponding fuzzy prototype set. The next input character is matched against the first element. If it is matched, the input character is grouped together with the first element; otherwise, it is placed as the second element of the set of fuzzy prototypes. The procedure is repeated until all the inputs are classified.

Similarity is an abstract concept of fuzziness that provides a quantitative measure to the relationship between two variables. Given two patterns A and B, let $|A|$ denote the cardinality of set A. The similarity measurement or correlation coefficient, denoted as $\xi(A, B)$, can be expressed in the following ways:

$$\xi(A,B) = \frac{|A \cap B|}{|A \cup B|} \tag{11.37}$$

$$\xi(A,B) = \frac{\sum a_{ij} b_{ij}}{\sqrt{\sum a_{ij}^2 \sum b_{ij}^2}} \tag{11.38}$$

$$\xi(A,B) = \frac{\sum (a_{ij} - a'_{ij}) b_{ij}}{\sqrt{\sum a_{ij}^2 \sum b_{ij}^2}} \tag{11.39}$$

$$\xi(A,B) = \frac{\sum (a_{ij} - \bar{a})(b_{ij} - \bar{b})}{\sqrt{\sum (a_{ij} - \bar{a})^2 \sum (b_{ij} - \bar{b})^2}} \tag{11.40}$$

where a'_{ij} denotes the complement a_{ij}, $\bar{a} = \sum a_{ij}/|A|$, and $\bar{b} = \sum b_{ij}/|B|$. Assume A and B are images of size $m = \text{cols} \times \text{rows}$. That is, $A = \{a_{ij} | 1 \le i \le \text{cols}, 1 \le j \le \text{rows}\}$, and $B = \{b_{ij} | 1 \le i \le \text{cols}, 1 \le j \le \text{rows}\}$. If they are binary images, let 0 or 1 represent background and foreground, respectively. Equations (11.37)–(11.39) are specifically for binary images. In order to preserve the symmetricity; that is, $\xi(A,B) = \xi(B,A)$, equation (11.39) is modified as

$$\xi(A,B) = \frac{\sum(a_{ij}b_{ij} - \frac{1}{2}a'_{ij}b_{ij} - \frac{1}{2}a_{ij}b'_{ij})}{\sqrt{\sum a_{ij}^2 \sum b_{ij}^2}} \tag{11.41}$$

The weight 1/2 is added to normalize the similarity ranging between −1 and 1. Equation (11.40) is a general form for grayscale images, which is known as the Pearson product moment correlation.

Let us discuss the characteristics of the aforementioned similarity functions. Let n_A and n_B denote the cardinality of sets A and B, respectively. That is, $n_A = |A|$ and $n_B = |B|$. Also, let $x = |A \cap B|$, which denotes the number of common elements in A and B, and let m denote the total number of elements in the domain of A and B; that is, $m = |A| + |A'| = |B| + |B'| = \text{cols} \times \text{rows}$. Equations (11.37), (11.38), (11.41), and (11.40) can be rewritten respectively as

$$\xi = \frac{|A \cap B|}{|A \cup B|} = \frac{x}{n_A + n_B - x} \tag{11.42}$$

$$\xi(A,B) = \frac{\sum a_{ij}b_{ij}}{\sqrt{\sum a_{ij}^2 \sum b_{ij}^2}} = \frac{x}{\sqrt{n_A n_B}} \tag{11.43}$$

$$\xi(A,B) = \frac{\sum(a_{ij}b_{ij} - \frac{1}{2}a'_{ij}b_{ij} - \frac{1}{2}a_{ij}b'_{ij})}{\sqrt{\sum a_{ij}^2 \sum b_{ij}^2}} = \frac{2x - n_A/2 - n_B/2}{\sqrt{n_A n_B}} \tag{11.44}$$

and

$$\xi(A,B) = \frac{\sum(a_{ij} - \bar{a})(b_{ij} - \bar{b})}{\sqrt{\sum (a_{ij} - \bar{a})^2 \sum (b_{ij} - \bar{b})^2}} = \frac{mx - n_A n_B}{\sqrt{(mn_A - n_A^2)(mn_B - n_B^2)}} \tag{11.45}$$

Equation (11.42) is a parabola function and equations (11.43)–(11.45) are linear functions. Figure 11.25 shows the relationship of correlation coefficient and the percentage of the intersection when $n_A = n_B = n$.

Equation (11.45) is not suitable for matching since the total number of elements in the domain image m affects the value of ξ. Equation (11.44) seems more appropriate because it measures the weighted differences of the matched elements (or equality measure) and the unmatched elements, that is,

$$\frac{x}{\sqrt{n_A n_B}} - \left(\frac{1}{2}\right)\frac{n_A + n_B - 2x}{\sqrt{n_A n_B}}$$

However, if the weights are the same, the calculation of the unmatched is redundant since it is implied from the matched.

A major problem occurs in equations (11.42)–(11.45). Since their matching is based on the percentage of intersection of two patterns, the locations of unmatched

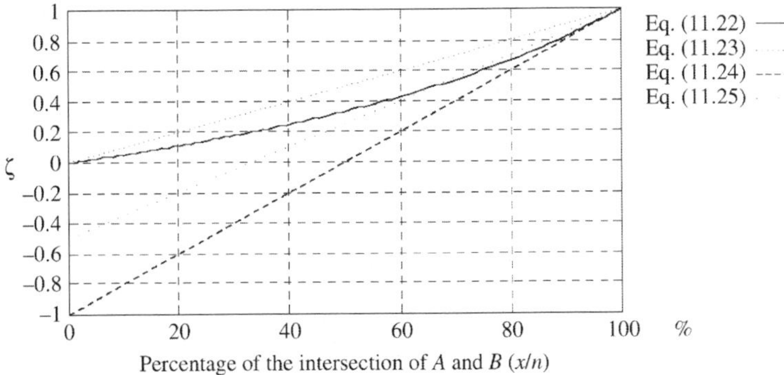

Figure 11.25 The relationship of ξ and x/n when $n_A = n_B = n$.

elements are not taken into account. By observing three images A_1, A_2, and A_3 in Figure 11.26, it is found that A_2 and A_3 have the same number of foreground pixels, and the total number of common pixels of A_1 and A_2 is the same as that of A_1 and A_3. From human's perception, A_1 and A_2 look more similar than A_1 and A_3 since the differences of A_1 and A_2 are the boundary pixels, but the differences of A_1 and A_3 are the significant noise. However, the results of using equations (11.42)–(11.45) show that the correlation of A_1 and A_2 is identical to that of A_1 and A_3.

It is concluded that the similarity measure must include the location differences. The differences lying along the pattern's boundaries should be considered less significant than lying far away from the object. We propose a nonlinear weighted similarity function that can be expressed as

$$\xi(A, B) = \frac{\sum \left(a_{ij}b_{ij} - \frac{1}{2}\omega_{ij}b_{ij} - \frac{1}{2}\nu_{ij}a_{ij}\right)}{\sqrt{\sum a_{ij}^2 \sum b_{ij}^2}} \tag{11.46}$$

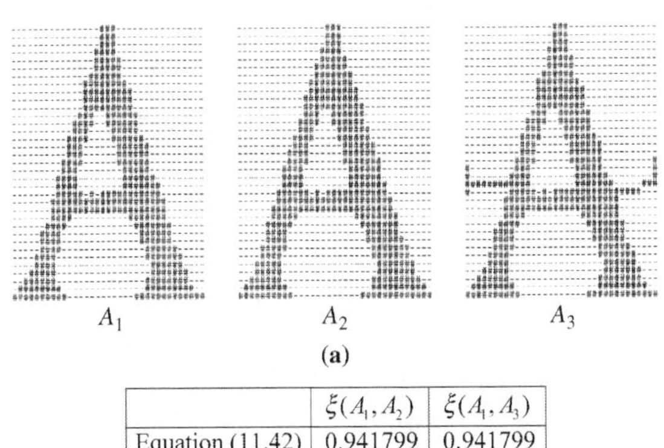

(a)

	$\xi(A_1, A_2)$	$\xi(A_1, A_3)$
Equation (11.42)	0.941799	0.941799
Equation (11.43)	0.970463	0.970463
Equation (11.44)	0.940477	0.940477
Equation (11.45)	0.958100	0.958100

(b)

Figure 11.26 (a) Sample images A_1, A_2, and A_3. (b) Correlation coefficients by equations (11.42)–(11.45).

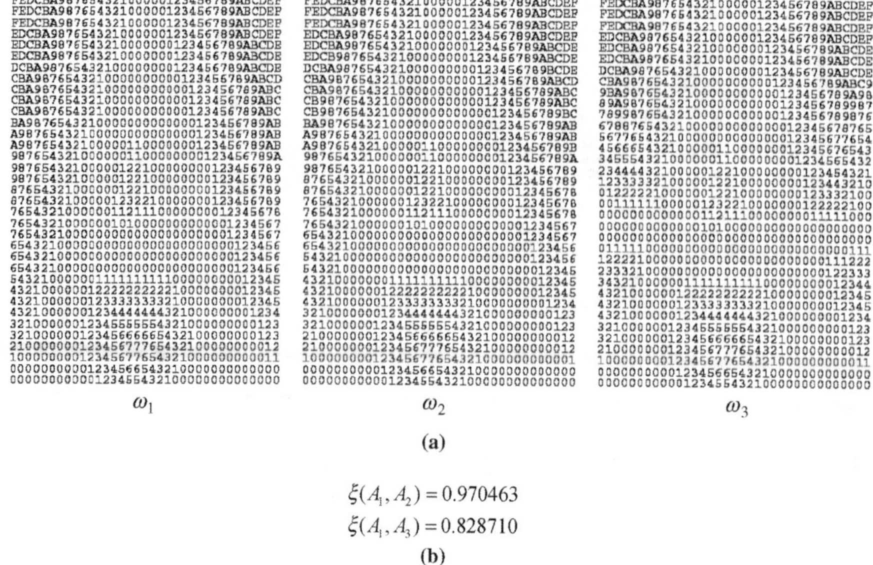

(a)

$$\xi(A_1, A_2) = 0.970463$$
$$\xi(A_1, A_3) = 0.828710$$

(b)

Figure 11.27 (a) The weights of A_1, A_2, and A_3. (b) Their correlation coefficients by equation (11.46).

where ω_{ij} and ν_{ij} are the weights representing a distance measure of pixel (i, j) to objects A and B, respectively. Since the 8-connectivity for objects and 4-connectivity for background are used, we adopt a simple city-block distance measure. It can be formulated as

$$\omega = \{\omega_{ij} | \omega_{ij} = \max(d_4(a_{ij}) - 1, 0)\} \tag{11.47}$$

$$\nu = \{\nu_{ij} | \nu_{ij} = \max(d_4(b_{ij}) - 1, 0)\} \tag{11.48}$$

where $d_4(a_{ij})$ denotes the city-block distance of pixel a_{ij} from the object A. Note that 1 is subtracted because a pixel along the contour of A is considered as a reasonable tolerance of noise. Figure 11.27 shows the weights of the patterns in Figure 11.26a and their correlations by equation (11.46). Note that $\xi(A_1, A_2)$ has the same value as equation (11.43) is applied, but $\xi(A_1, A_3)$ decreases because the noisy pixels away from the boundary receive higher weights in the subtraction of equation (11.46).

11.5.2 Statistical Fuzzy Model for Classification

The set of fuzzy prototypes is constructed based on statistical analysis by grouping similar patterns into a single class. An image of a fuzzy prototype is a matrix of pixels where each element is associated with a membership representing the degree of the pixel belonging to the object. The definitions, propositions, and theorems pertaining to fuzzy matching are described below.

Definition 11.2. Let E^2 denote two-dimensional Euclidean space. A fuzzy model λ in E^2 is a matrix of ordered pairs such that $\lambda = \{(\rho_{ij}, \chi_{ij})\}$, where ρ_{ij} represents a pixel and χ_{ij}, in the range [0, 1], represents the grade of membership of the pixel belonging to the object.

Note that the memberships of the pixels outside the matrix are considered as 0. In addition, a binary image can be regarded as a special case in fuzzy model whose membership has values 1 or 0. A fuzzy model is natural to represent the fuzziness of the image boundary. Readers should be aware of the concept of the fuzzy model being different from the grayscale image whose element represents the gray levels in brightness. In the following, a set of "merged" binary images indicates those images to be classified as the same class.

Proposition 11.1. Assume a fuzzy prototype λ is composed of a set of merged binary images $\{A_1, A_2, \ldots, A_m\}$, where their elements are denoted by a_{ij}. The membership of each pixel in λ is computed as

$$\chi_{ij} = \frac{\sum_{a_{ij}\in A_1}^{A_m} a_{ij}}{m} \tag{11.49}$$

The properties of the fuzzy model extended from the crisp model should be also fuzzy (Kandel, 1986). Some properties pertaining to character classification are discussed in this section.

Definition 11.3. Let $\lambda = \{(p_1, \chi_1), (p_2, \chi_2), \ldots, (p_n, \chi_n)\}$ be a fuzzy prototype. The *cardinality* of λ, denoted as σ_λ, is a fuzzy number and can be formulated as $\sigma_\lambda = \{(i, \psi_i) | i = 0, 1, \ldots, n\}$, where ψ_i denotes the membership of the cardinality that is equal to i.

From probability theory, it is plausible to formulate ψ_i as

$$\begin{aligned} \psi_0 &= \prod_{i=1}^{n} \chi_i', \psi_1 = \sum_{i=1}^{n} \left[\chi_i \left(\prod_{j=1, j\neq i}^{n} \chi_j' \right) \right], \ldots \\ \psi_{n-1} &= \sum_{i=1}^{n} \left[\left(\prod_{j=1, j\neq i}^{n} \chi_j \right) \chi_i' \right], \psi_n = \prod_{i=1}^{n} \chi_i \end{aligned} \tag{11.50}$$

It can be expressed in a general form as

$$\psi_m = \sum_{i_1 \neq i_2 \neq \cdots \neq i_m} \left[\chi_{i_1} \chi_{i_2} \cdots \chi_{i_m} \left(\prod_{j\neq i_1, j\neq i_2, \ldots, j\neq i_m} \chi_j' \right) \right] \tag{11.51}$$

Note that the fuzzy cardinality has the property that $\sum_{i=0}^{n} \psi_i = 1$. The expected value of the cardinality can be derived as

$$E(\sigma_\lambda) = \sum_{i=1}^{n} i\psi_i = \sum_{i=1}^{n} \chi_i \tag{11.52}$$

Details of the derivation are omitted here. It is significant to simplify the cardinality of a fuzzy set to a unique value, that is, the expected value of the fuzzy cardinality, which is the sum of the membership of the fuzzy set.

Example 11.1 Let $\lambda = \{(p_1, 1/2), (p_2, 1), (p_3, 1/4)\}$. The fuzzy cardinality is $\sigma_\lambda = \{(0, 0), (1, 3/8), (2, 1/2), (3, 1/8)\}$. The expected value of cardinality is $0 \times 0 + 1 \times 3/8 + 2 \times 1/2 + 3 \times 1/8 = 7/4$.

Proposition 11.2. The cardinality of a fuzzy prototype λ in E^2 is equal to the sum of the membership values. That is, $\sigma_\lambda = \sum \chi_{ij}$.

Similar to the derivation in cardinality, the centroid of a fuzzy prototype can be simplified to be a unique number.

Proposition 11.3. The centroid of the first moment of a fuzzy prototype λ in E^2 is formulated as

$$x_{c_\lambda} = \frac{\sum j\chi_{ij}}{\sum \chi_{ij}}, \quad y_{c_\lambda} = \frac{\sum i\chi_{ij}}{\sum \chi_{ij}} \tag{11.53}$$

where all summations are with respect to ij.

Proposition 11.4. The width of a fuzzy prototype λ in E^2 is the summation of the maximal memberships in columns, and the height of a fuzzy prototype λ in E^2 is the summation of the maximal memberships in rows; that is,

$$w_\lambda = \sum_j \left(\max_i(\chi_{ij}) \right), \qquad h_\lambda = \sum_i \left(\max_j(\chi_{ij}) \right) \tag{11.54}$$

Theorem 11.1. If a fuzzy prototype λ is composed of a set of merged binary images $\{A_1, A_2, \ldots, A_m\}$ with widths $\{w_1, w_2, \ldots, w_m\}$ and heights $\{h_1, h_2, \ldots, h_m\}$, then the width of λ is less than or equal to the average width of the set of the images, and the height of λ is less than or equal to the average height of the set of the images; that is,

$$w_\lambda \leq \sum_{i=1}^{m} \frac{w_i}{m}, \qquad h_\lambda \leq \sum_{i=1}^{m} \frac{h_i}{m} \tag{11.55}$$

Proof. The proof is given by the induction hypothesis. For the case when $m = 1$, we have $w_\lambda = w_i$, which satisfies equation (11.55).

Assume that $m = x$ also satisfies equation (11.55), such that $w_\lambda \leq (1/x)\sum_{i=1}^{x} w_i$. Let $\alpha_j = \max_i(\chi_{ij}^\lambda)$ denote the maximal membership value of jth column in the fuzzy prototype λ, $\beta_j = \max_i(\chi_{ij}^{A_{x+1}})$ denote the maximal membership value of jth column in A_{x+1}, which is either 1 or 0, and ρ_j denote the maximal membership value of jth column in the new fuzzy prototype λ'. When a new image A_{x+1} is merged, the following inequality is derived:

$$\rho_j \leq \frac{x\alpha_j + \beta_j}{x+1}$$

The width of the new fuzzy prototype will be

$$w_{\lambda'} = \sum \rho_j \leq \frac{\sum x\alpha_j + \beta_j}{x+1} = \frac{x}{x+1} w_\lambda + \frac{1}{x+1} w_{x+1}$$

$$\leq \frac{x}{x+1}\left(\frac{1}{x}\sum_{i=1}^{x} w_i\right) + \frac{1}{x+1} w_{x+1} = \frac{1}{x+1}\sum_{i=1}^{x+1} w_i$$

Therefore, equation (11.55) for the width is proved. The proof for the height can be similarly derived. Because the fuzzy prototype is a group of similar patterns, the difference between w_λ and $(1/m)\sum_{i=1}^{m} w_i$ is small. Thus, the latter can be applied to approximate the width of the fuzzy prototype for simplicity.

11.5.3 Similarity Measure in Fuzzy Model

The similarity measure between two binary patterns can be extended to the fuzzy model. Similar to equation (11.46), the similarity measure of two fuzzy prototypes is proposed below.

Proposition 11.5. Let $\lambda_1 = \{(p_{ij}, \chi_{ij}^{(1)})\}$ and $\lambda_2 = \{(p_{ij}, \chi_{ij}^{(2)})\}$ denote two fuzzy prototypes in E^2, and $\gamma_{\lambda_1} = \{\gamma_{ij}^{(1)}\}$ and $\gamma_{\lambda_2} = \{\gamma_{ij}^{(2)}\}$ represent the weight functions associated with λ_1 and λ_2, respectively. The similarity measure of λ_1 and λ_2 is defined as

$$\xi(\lambda_1, \lambda_2) = \frac{\sum(\chi_{ij}^{(1)} \wedge \chi_{ij}^{(2)} - \frac{1}{2}\gamma_{ij}^{(1)}\chi_{ij}^{(2)} - \frac{1}{2}\gamma_{ij}^{(2)}\chi_{ij}^{(1)})}{\sqrt{\sum \chi_{ij}^{(1)^2} \chi_{ij}^{(2)^2}}} \tag{11.56}$$

where $\wedge$ is the symbol for minimum representing the intersection on fuzzy sets; that is,

$$\lambda_1 \cap \lambda_2 = \left\{\left(p_{ij}, \chi_{ij}^{(1)} \wedge \chi_{ij}^{(2)}\right)\right\} \tag{11.57}$$

where $\chi_{ij}^{(n)^2}$ $(n = 1, 2)$denotes the self-intersection $\chi_{ij}^{(n)} \wedge \chi_{ij}^{(n)}$.

The reason why the denominator $\chi_{ij}^{(n)} \wedge \chi_{ij}^{(n)}$ is used instead of $\chi_{ij}^{(n)} \times \chi_{ij}^{(n)}$ comes from the viewpoint of fuzzy properties. Consider a membership, $\chi_{ij} = 0.8$, which represents a concept that 80% of the area in a pixel p_{ij} belongs to the object (i.e., has value "1"), and 20% of the area belongs to the background (i.e., has value "0"). Therefore, χ_{ij}^2 should be carried out as $0.8 \times 1^2 + 0.2 \times 0^2 = 0.8$ instead of $0.8^2 = 0.64$. Moreover, the first term of the numerator is $\chi_{ij}^{(1)} \wedge \chi_{ij}^{(2)}$ instead of $\chi_{ij}^{(1)} \times \chi_{ij}^{(2)}$ because two fuzzy subsets are equal if their membership functions are equal; that is,

$$\lambda_1 = \lambda_2 \quad \text{iff } \chi_{ij}^{(1)} = \chi_{ij}^{(2)} \tag{11.58}$$

Therefore,

$$\xi(\lambda_1, \lambda_2) = \frac{\sum\left(\chi_{ij}^{(1)} \wedge \chi_{ij}^{(1)} - \frac{1}{2}\gamma_{ij}^{(1)}\chi_{ij}^{(1)} - \frac{1}{2}\gamma_{ij}^{(1)}\chi_{ij}^{(1)}\right)}{\sqrt{\sum \chi_{ij}^{(1)^2} \chi_{ij}^{(1)^2}}} = 1 \tag{11.59}$$

but

$$(\lambda_1, \lambda_2) \neq \frac{\sum \left(\chi_{ij}^{(1)} \times \chi_{ij}^{(1)} - \frac{1}{2}\gamma_{ij}^{(1)}\chi_{ij}^{(1)} - \frac{1}{2}\gamma_{ij}^{(1)}\chi_{ij}^{(1)} \right)}{\sqrt{\sum \chi_{ij}^{(1)^2} \chi_{ij}^{(1)^2}}} \leq 1 \tag{11.60}$$

Equation (11.56) is the extended form of equation (11.46) except that $\chi_{ij}^{(1)}$ and $\chi_{ij}^{(2)}$ are in the interval [0, 1], and $\gamma_{ij}^{(1)}$ and $\gamma_{ij}^{(2)}$ are real numbers instead of integers. The weight function γ_λ, which represents a fuzzy distance measure from the object, is also fuzzy. Similarly, it should be derived from the weight function ω of all the patterns comprising the prototype.

Proposition 11.6. Assume that λ, a fuzzy prototype, is composed of a set of merged binary images $\{A_1, A_2, \ldots, A_m\}$ associated with the weight functions $\{\omega_{A_1}, \omega_{A_2}, \ldots, \omega_{A_m}\}$, which are obtained from equation (11.47). The weight function, γ, of λ is defined as

$$\gamma_{ij} = \frac{\sum_{\omega_{ij} \in \omega_{A_1}}^{\omega_{A_m}} \omega_{ij}}{m} \tag{11.61}$$

where ω_{ij}'s are the elements in $\omega_{A_1}, \omega_{A_2}, \ldots, \omega_{A_m}$.

11.5.4 Matching Algorithm

Given two fuzzy patterns A and B, the way to find the best matching is to shift A around B, calculate the correlation coefficient for every position, and select the highest value. However, it is inefficient. If two patterns are similar, they should have similar geometric properties. If two patterns are dissimilar, finding the best matching does not make sense. A simple way is to calculate the centroids of both patterns, and the similarity measure is calculated by matching the centroids. Some allowance must be considered due to noise. The algorithm is described below.

1. Calculate c_A and c_B, which represent the centroids of A and B. That is,

$$c_A = \left(\frac{\sum j\chi_{ij}^{(A)}}{\sum \chi_{ij}^{(A)}}, \frac{\sum i\chi_{ij}^{(A)}}{\sum \chi_{ij}^{(A)}} \right) \quad \text{and} \quad c_B = \left(\frac{\sum j\chi_{ij}^{(B)}}{\sum \chi_{ij}^{(B)}}, \frac{\sum i\chi_{ij}^{(B)}}{\sum \chi_{ij}^{(B)}} \right) \tag{11.62}$$

2. Compute $\xi(A, B)$ with minimum distance $\overline{c_A c_B}$. In other words, $\xi_{\alpha\beta}(A, B)$ is derived from shifting pattern A with (α, β):

$$\xi_{\alpha\beta}(A, B) = \frac{\sum \left(\chi_{ij}^{(A)} \wedge \chi_{i+\alpha,j+\beta}^{(B)} - \frac{1}{2}\gamma_{ij}^{(A)}\chi_{i+\alpha,j+\beta}^{(B)} - \frac{1}{2}\gamma_{i+\alpha,j+\beta}^{(B)}\chi_{ij}^{(A)} \right)}{\sqrt{\sum \chi_{ij}^{(A)^2} \chi_{ij}^{(B)^2}}} \tag{11.63}$$

where $\alpha = \text{round}(x_{c_B} - x_{c_A})$ and $\beta = \text{round}(y_{c_B} - y_{c_A})$, which are approximately x and y components of $\overline{c_A c_B}$. If the correlation coefficient is higher than the threshold, say 0.9, A and B are considered to be the same.

3. If ξ is in a critical range, say from 0.8 to 0.9, the values of $\xi_{\alpha\beta}(A, B)$ are also calculated with α and β in the range of

$$|\alpha - (x_{c_B} - x_{c_A})| \leq 1 \quad \text{and} \quad |\beta - (y_{c_B} - y_{c_A})| \leq 1 \tag{11.64}$$

Normally, if the values of $x_{c_B} - x_{c_A}$ and $y_{c_B} - y_{c_A}$ are not integers, there are three more positions for both α and β to be matched in this step. If there is a match, that is, higher than 0.9, A and B are set to the same class. Otherwise, the two patterns are considered as distinct objects.

Step 3 represents the fuzzy reasoning since the similarity is ambiguous in a critical range. The similarity between two similar patterns could be measured in this range because of the distortion of the centroid of the image due to noise, and a better match could be obtained by shifting one pixel for pattern A. Figure 11.28 illustrates an example of two primitive images. The ambiguous similarity calculated in step 2 is shown in Figure 11.28b, and a better match found by step 3 is shown in Figure 11.28c. Note that Figure 11.28b and c shows the superposition of images α_1 and α_2 in their

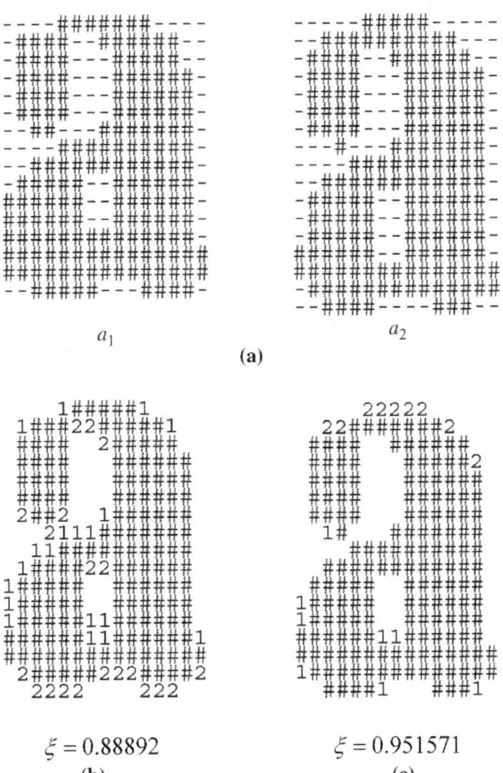

Figure 11.28 (a) The sample images α_1 and α_2, (b) a critical similarity is measured, and (c) a good matching is found by shifting α_1 1 pixel down.

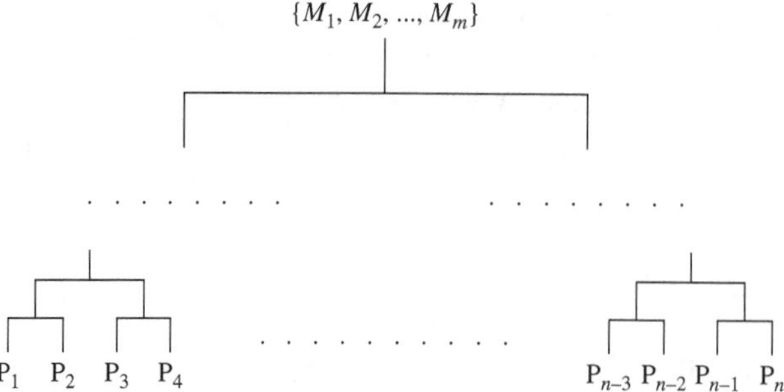

Figure 11.29 The general view of hierarchical classification.

matching positions. The symbols "#", "1", and "2" are used to represent, respectively, the common pixel of two images, the pixel in image α_1 only, and the pixel in image α_2 only.

11.5.5 Classification Hierarchy

It is inefficient if the matching is performed against all the character prototypes in an arbitrary sequence. In this section, a hierarchical-tree approach is proposed as illustrated in Figure 11.29. The leaves in the classification tree represent the primitive patterns, that is, the raw character images. The intermediate nodes represent the fuzzy model of prototypes that comprise their descendant subtrees. The root is finally the set of prototypes that is composed of the characters in the whole document.

There are two advantages in the hierarchical approach. First, the searching time for matching is less. Consider two subsets of fuzzy prototypes, X and Y, that consist of the same set of n prototypes. Let the two fuzzy subsets X and Y be constructed from two primitive subsets A and B, respectively, each of which containing m primitive images. Assume that the number of searching for matching an input through a set of x distinct prototypes is x. Therefore, to merge fuzzy sets X and Y requires $(n(n+1))/2$ times because the matched prototypes are removed from the set sequentially. Let $T(A)$ and $T(B)$ denote the total searching times in classifying A into X and B into Y, respectively. The worst case for $T(A)$ or $T(B)$ is when the first n patterns are distinct and the sequential classification is applied, the searching times for the total m patterns are $0, 1, \ldots, n-1, n, n, \ldots, n$, which sum up to be $(n(n-1))/2 + n(m-n)$. The comparison of searching times using hierarchical and sequential methods is expressed as

$$T_{\text{hier}} = T(A) + T(B) + n + (n-1) + \cdots + 1 \tag{11.65}$$

$$\leq T(A) + \frac{n(n-1)}{2} + n(m-n) + \frac{n(n+1)}{2} \tag{11.66}$$

$$= T(A) + nm = T_{\text{seq}} \tag{11.67}$$

where T_{seq} and T_{hier} denote the searching times required for sequential and hierarchical approaches, respectively.

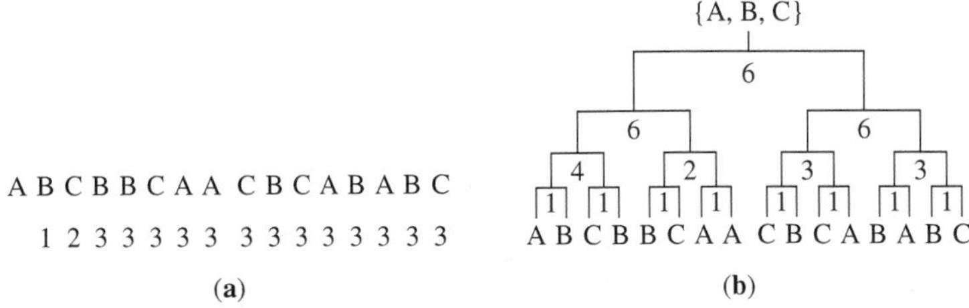

Figure 11.30 Illustration of searching times (a) $T_{seq} = 42$ and (b) $T_{hier} = 38$.

Figure 11.30 shows an example of 16 inputs with three categories by (a) sequential classification and (b) hierarchical classification. The number under each input in Figure 11.30a represents the searching time for the classification, and the number under each node in Figure 11.30b represents the times for merging two sets of fuzzy prototypes.

Another advantage of the hierarchical classification is the capability of being processed in parallel. Two kinds of parallelism can be carried out. First, each node in classification tree can be performed in parallel. Second, the merging of two fuzzy subsets can be implemented in parallel since all elements in a fuzzy subset are distinct and can be processed at the same time.

A text block after segmentation typically represents a line of characters. Usually, the sizes of characters in a line are uniform. Therefore, a text line is considered as a unit in classification. The set of fuzzy prototypes associated with each line is grouped hierarchically. This facilitates the comparison of the character sizes. If the sizes of two text lines are different, the merged prototype set is split into two disjoint subsets correspondingly and the matching is not performed. The hierarchy of our classification is shown in Figure 11.31, where $\mathbf{TL}_i$ denotes the set of fuzzy prototypes corresponding to a text line.

11.5.6 Preclassifier for Grouping the Fuzzy Prototypes

For the lower levels in the hierarchical classification tree, the representative subsets of fuzzy prototypes contain only a few elements. Therefore, it is simple to search for

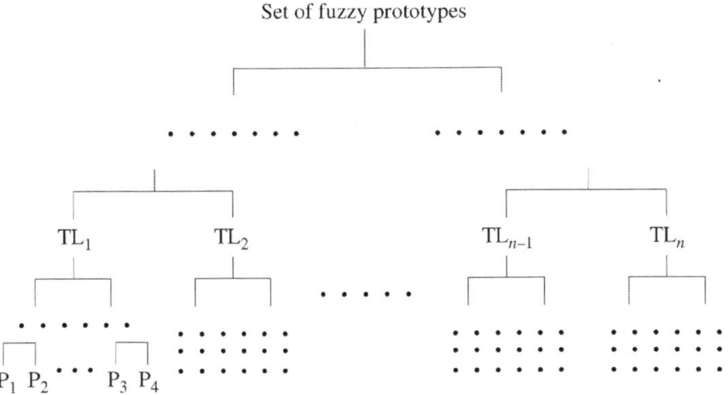

Figure 11.31 The hierarchy of our classification.

similar objects in two subsets. However, in the higher levels, it becomes impractical to perform matching one by one since the size of the libraries increases. Therefore, a preclassifier for finding the possible matching prototypes is used to reduce computation.

Casey et al. (1982) applied a binary decision network with each interior node representing a comparison of a specific pixel location to the input character and having two branches, which are known as "black" and "white," leading to a terminal node that represents a possibly matched prototype. However, the reliability is questionable since the reliable pixels of each prototype are different. In addition, noise may cause errors in decision making.

To solve the aforementioned problems, a rule-based preclassifier is used. Considering two fuzzy patterns λ_1 and λ_2, the similarity measure is computed using equation (11.56). The correlation coefficient, ξ, can be divided into two terms of equality measure E and inequality measure I.

$$E = \frac{\sum \chi_{ij}^{(1)} \wedge \chi_{ij}^{(2)}}{\sqrt{\sum \chi_{ij}^{(1)^2} \chi_{ij}^{(2)^2}}} = \frac{\sigma_{\lambda_1 \cap \lambda_2}}{\sqrt{\sigma_{\lambda_1} \sigma_{\lambda_2}}} \tag{11.68}$$

$$I = \frac{\Upsilon_{ij}^{(1)} \chi_{ij}^{(2)} + \Upsilon_{ij}^{(2)} \chi_{ij}^{(1)}}{2\sqrt{\sigma_{\lambda_1} \sigma_{\lambda_2}}} \tag{11.69}$$

Let ξ_t be the threshold of the similarity measure, λ_1 be the input model, and λ_2 be the fuzzy variable in the library. For $\sigma_{\lambda_1} > \sigma_{\lambda_2}$, the best equality measure happens if $\lambda_1 \supset \lambda_2$; that is, $\chi_{ij}^{(1)} \geq \chi_{ij}^{(2)}$. Therefore, λ_2 is a possible match prototype only if

$$E = \frac{\sigma_{\lambda_1 \cap \lambda_2}}{\sqrt{\sigma_{\lambda_1} \sigma_{\lambda_2}}} = \frac{\sigma_{\lambda_2}}{\sqrt{\sigma_{\lambda_1} \sigma_{\lambda_2}}} = \sqrt{\frac{\sigma_{\lambda_2}}{\sigma_{\lambda_1}}} \geq \xi_t \tag{11.70}$$

Similarly, for $\sigma_{\lambda_1} < \sigma_{\lambda_2}$, λ_2 is a possible match prototype only if

$$\sqrt{\frac{\sigma_{\lambda_1}}{\sigma_{\lambda_2}}} \geq \xi_t \tag{11.71}$$

Therefore, the first rule is concluded using equations (11.70) and (11.71).

Rule 1: λ_2 is a possible match prototype of λ_1 iff $\xi_t^2 \sigma_{\lambda_1} \leq \sigma_{\lambda_2} \leq \sigma_{\lambda_1}/\xi_t^2$.Since similar prototypes possess similar features, additional heuristic rules based on the features of the prototypes are needed.

Rule 2: Two fuzzy prototypes are impossible to match if the difference between their widths exceeds a threshold w_t.

Note that the height is not taken into consideration since the prototypes in the same typographical category have the similar heights.

Rule 3: Two fuzzy prototypes are impossible to match if the difference of two prototypes in the total number of columns of the left or right region to the centroid is greater than a threshold c_1.

Rule 4: Two fuzzy prototypes are impossible to match if the difference of two prototypes in the total number of rows of the upper or lower region to the centroid is greater than a threshold c_2.

In the proposed system, each library of prototypes is sorted by its cardinalities. The set of prototypes to be possibly matched is extracted by rule 1. Rules 2, 3, and 4 filter out the prototypes that are impossibly matched. Finally, a rough estimation of the similarity measure based on the projection profile is applied to extract the prototypes to be possibly matched prior to the two-dimensional pattern matching.

Rule 5: Let γ_i^{λ} denote the summation of the membership values on *i*th column of fuzzy prototype λ. Two fuzzy prototypes λ_1 and λ_2 are possibly matched iff the following condition holds:

$$\xi_t\sqrt{\sigma_{\lambda_1}\sigma_{\lambda_2}} \leq \sum(\gamma_i^{\lambda_1} \wedge \gamma_{i+\alpha}^{\lambda_2}) \quad \text{for} \quad |\alpha-(x_{c_B}-x_{c_A})| \leq 1 \qquad (11.72)$$

where $\wedge$ denotes the symbol of minimum.

11.5.7 Experimental Results

The raw image of a document is scanned and thresholded into binary. It is decomposed into blocks of text, graphics, pictures, horizontal lines, and vertical lines. The characters in each textual block are extracted and typographically analyzed and categorized.

The unsupervised character classification is performed on each text line and a subset of fuzzy prototypes is generated correspondingly. Note that a fuzzy set contains eight subsets of the typographical categories including the semiascender. Semiascender is an additional typographical category for English alphabets, such as "t", which occupies the middle zone and a half of the upper zone. The subsets of the similar height are grouped hierarchically. Finally, several sets of fuzzy prototypes corresponding to different sizes of characters are produced. A sample document image is shown in Figure 11.32. The set of fuzzy prototypes for the characters in Figure 11.32 is illustrated in Figure 11.33, including 25 ascenders, 6 descenders, 23 centereds, 1 full ranged, 2 subscripts, 2 internals, and 2 semiascenders. Figure 11.34 shows a few prototype examples, where "#" and "-" denote the membership greater than 0.95 and less than 0.05 respectively, and an integer *i* represents the membership between $i/10 - 0.05$ and $i/10 + 0.05$. It is obvious that the fuzzy prototypes are more reliable to be recognized since each pixel is analyzed based on a set of merged images and the pixels with low membership values can be considered as noise to be removed. Some of the merged characters are illustrated in Figure 11.35. It is observed that the membership values of the linked pixels are lower when more patterns are included, so they can be separated. Those that do not have lower values on the linked pixels because the prototype contains too few patterns can be split easily using the splitting algorithm, partial matching, and dictionary lookup.

We choose a suitable threshold for the similarity measure. For the bigger character size, the boundary between distinct character categories is clear. However,

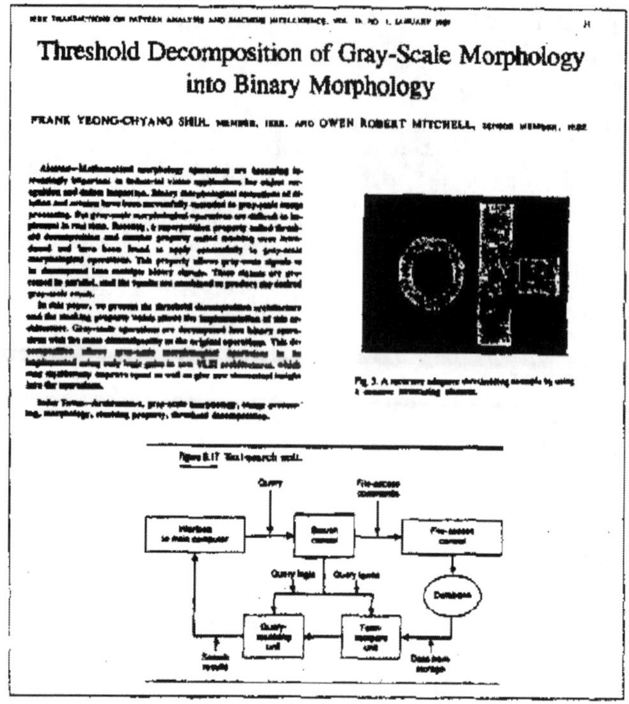

Threshold Decomposition of Gray-Scale Morphology into Binary Morphology

FRANK YEONG-CHYANG SHIH, AND OWEN ROBERT MITCHELL

Figure 11.32 A sample document image.

itl/fITVbLATSkAbddhGfiBRMffi

ypgpegypo

rscvrscexexoaanuwnscecocmrm

Figure 11.33 The set of fuzzy prototypes for the main text in Figure 11.32.

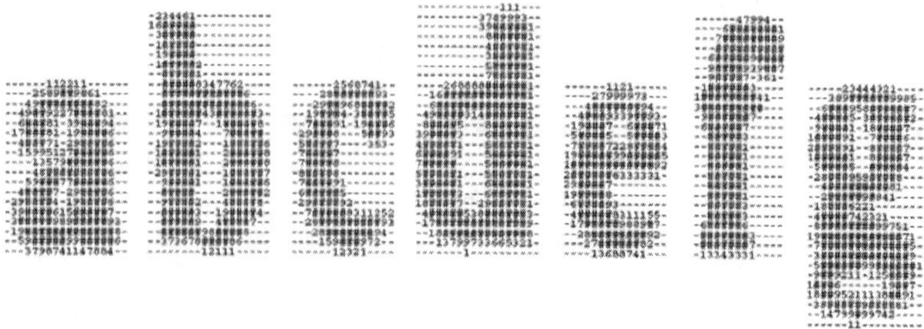

Figure 11.34 Examples of some fuzzy prototypes.

Figure 11.35 Examples of fuzzy prototypes with merged characters.

when the characters are smaller, the scope of different categories may be overlapped. There are two reasons. First, for smaller size the cardinality of the patterns is relatively small. However, the tolerance of noise, which may be caused by scanning or printing, is not affected by character sizes. Therefore, the possible minimal equality measure for the same category will be lower. Second, the inequality measure may be smaller for different categories since the different part of two different patterns in smaller size is smaller, and the weight is linear to the distance from the object. For example, the inequality measure between character "i" and "l" in smaller size is usually less than that in bigger size for the same font because the latter contains more pixels of difference and has a higher weight. Therefore, the threshold must be adaptive to the character size. One approach is to emphasize the inequality measure such as

$$E - I \geq \xi_t + cI \quad \text{or} \quad E - (c+1)I \geq \xi_t \tag{11.73}$$

where c is an coefficient. For the sample image, $c = 4$ and $\xi_t = 0.86$ are adopted in classifying the main text with about 30-pixel high (point size 7). By emphasizing the inequality measure, different categories are ensured to be separated, which is the principle of the classification. While the patterns belonging to the same category may be split due to noise, the fuzzy set may contain more prototypes than it really has. We perform the similarity comparison for the final set. Since noise effect is reduced for the fuzzy set, the similar prototypes separated before will be probably merged. In the main text of the sample document, there are 12 prototypes that are finally merged.

REFERENCES

Bley, H., "Segmentation and preprocessing of electrical schematics using picture graphs," *Comput. Vis. Graph. Image Process.*, vol. 28, no. 3, pp. 271–288, Dec. 1984.

Bozinovic, R. M. and Srihari, S. N., "Off-line cursive script word recognition," *IEEE Trans. Pattern Anal. Mach. Intell.*, vol. 11, no. 1, pp. 68–83, Jan. 1989.

Cannon, M., Fugate, M., Hush, D. R., and Scovel, C., "Selecting a restoration technique to minimize OCR error," *IEEE Trans. Neural Networks*, vol. 14, no. 3, pp. 478–490, May 2003.

Casey, R. G., Chai, S. K., and Wong, K. Y., "Unsupervised construction of decision networks for pattern classification," Research Report RJ 4264, IBM Research Lab., San Jose, CA, 1982.

Chen, S. S., Shih, F. Y., and Ng, P. A., "A fuzzy model for unsupervised character classification," *Int. J. Inf. Sci. Appl.*, vol. 2, no. 3, pp. 143–165, Nov. 1994.

Chen, S.-S., Shih, F. Y., and Ng, P. A., "Fuzzy typographical analysis for character preclassification," *IEEE Trans. Syst. Man Cybern.*, vol. 25, no. 10, pp. 1408–1413, Oct. 1995.

Dawoud, A., "Iterative cross section sequence graph for handwritten character segmentation," *IEEE Trans. Image Process.*, vol. 16, no. 8, pp. 2150–2154, Aug. 2007.

Dawoud, A. and Kamel, M. S., "Iterative multimodel subimage binarization for handwritten character segmentation," *IEEE Trans. Image Process.*, vol. 13, no. 9, pp. 1223–1230, Sep. 2004.

Diligenti, M., Frasconi, P., and Gori, M., "Hidden tree Markov models for document image classification," *IEEE Trans. Pattern Anal. Mach. Intell.*, vol. 25, no. 4, pp. 519–523, Apr. 2003.

Dubois, S. R. and Glanz, F. H., "An autoregressive model approach to two-dimensional shape classification," *IEEE Trans. Pattern Anal. Mach. Intell.*, vol. 8, no. 1, pp. 55–66, Jan. 1986.

Fisher, J. L., Hinds, S. C., and D'Amato, D. P., "A rule-based system for document image segmentation," IEEE Proceedings of the Tenth International Conference on Pattern Recognition, Atlantic City, NJ, pp. 567–572, June 1990.

Fletcher, L. A. and Kasturi, R., "A robust algorithm for text string separation from mixed text/graphics images," *IEEE Trans. Pattern Anal. Mach. Intell.*, vol. 10, no. 6, pp. 910–918, Nov. 1988.

Kahan, S., Pavlidis T., and Baird, H. S., "On the recognition of printed characters of any font and size," *IEEE Trans. Pattern Anal. Mach. Intell.*, vol. 9, no. 2, pp. 274–288, Mar. 1987.

Kandel, A., *Fuzzy Mathematical Techniques with Applications*, Addison-Wesley, Reading, MA, 1986.

Kim, D. and Bang, S.-Y., "A handwritten numeral character classification using tolerant rough set," *IEEE Trans. Pattern Anal. Mach. Intell.*, vol. 22, no. 9, pp. 923–937, Sep. 2000.

Lee, D.-S., "Substitution deciphering based on HMMs with applications to compressed document processing," *IEEE Trans. Pattern Anal. Mach. Intell.*, vol. 24, no. 12, pp. 1661–1666, Dec. 2002.

Liu, C.-L., Sako, H., and Fujisawa, H., "Discriminative learning quadratic discriminant function for handwriting recognition," *IEEE Trans. Neural Networks*, vol. 15, no. 2, pp. 430–444, Mar. 2004.

Marinai, S., Gori, M., and Soda, G., "Artificial neural networks for document analysis and recognition," *IEEE Trans. Pattern Anal. Mach. Intell.*, vol. 27, no. 1, pp. 23–35, Jan. 2005.

Park, J., Govindaraju, V., and Srihari, S. N., "OCR in a hierarchical feature space," *IEEE Trans. Pattern Anal. Mach. Intell.*, vol. 22, no. 4, pp. 400–407, Apr. 2000.

Pavlidis, T., "A vectorizer and feature extractor for document recognition," *Comput. Vis. Graph. Image Process.*, vol. 35, no. 1, pp. 111–127, July 1986.

Reddi, J. J., "Radial and angular moment invariants for image identification," *IEEE Trans. Pattern Anal. Mach. Intell.*, vol. 3, no. 2, pp. 240–242, Mar. 1981.

Shih, F. Y. and Chen, S., "Skeletonization for fuzzy degraded character images," *IEEE Trans. Image Process.*, vol. 5, no. 10, pp. 1481–1485, Oct. 1996a.

Shih, F. Y. and Chen, S., "Adaptive document block segmentation and classification," *IEEE Trans. Syst. Man Cybern.*, vol. 26, no. 5, pp. 797–802, Oct. 1996b.

Shih, F. Y., Chen, S.-S., Hung, D. D., and Ng, P. A., "A document segmentation, classification and recognition system," Proceedings of the International Conference on System Integration, Morristown, NJ, pp. 258–267, June 1992.

Shridber, M. and Badreldin, A., "High accuracy character recognition algorithm using Fourier and topological descriptors," *Pattern Recognit.*, vol. 17, no. 5, pp. 515–524, 1984.

Smith, R. W., "Computer processing of line images: a survey," *Pattern Recognit.*, vol. 20, no. 1, pp. 7–15, 1987.

Srihari, S. N. and Zack, G. W., "Document image analysis," *Proceedings of the IEEE Eighth International Conference on Pattern Recognition*, Paris, France, pp. 434–436, Oct. 1986.

Wahl, F. M., Wong, K. Y., and Casey, R. G., "Block segmentation and text extraction in mixed text/image documents," *Comput. Vis. Graph. Image Process.*, vol. 20, no. 4, pp. 375–390, Dec. 1982.

Wang, D. and Srihari, S. N., "Classification of newspaper image blocks using texture analysis," *Comput. Vis. Graph. Image Process.*, vol. 47, no. 3, pp. 327–352, Sep. 1989.

Wong, K. Y., Casey, R. G., and Wahl, F. M., "Document analysis system," *IBM J. Res. Dev.*, vol. 6, pp. 642–656, Nov. 1982.

Xue, H. and Govindaraju, V., "On the dependence of handwritten word recognizers on lexicons," *IEEE Trans. Pattern Anal. Mach. Intell.*, vol. 24, no. 12, pp. 1553–1564, Dec. 2002.

Zahn, C. T. and Roskies, R. Z., "Fourier descriptors for plane closed curves," *IEEE Trans. Comput.*, vol. 21, no. 3, pp. 269–281, Mar. 1972.

CHAPTER 12

IMAGE WATERMARKING

With the fast increasing number of electronic commerce web sites and applications, intellectual property protection is an extremely important concern for content owners who exhibit digital representations of photographs, books, manuscripts, and original artwork on the Internet. Moreover, as available computing power continues to rise, there is an increasing interest in protecting video files from attack. The applications are widespread including electronic publishing, advertisement, merchandise ordering and delivery, picture galleries, digital libraries, online newspapers and magazines, digital video and audio, personal communication, and so on.

Watermarking is not a brand new phenomenon. For nearly 1000 years, watermarks on paper have often been used to visibly indicate a particular publisher and to discourage counterfeiting in currency. A watermark is a design impressed on a piece of paper during production and used for copyright identification. The design may be a pattern, a logo, or an image. In the modern era, as most of the data and information are stored and communicated in a digital form, proving authenticity plays an increasingly important role. Digital watermarking is a process whereby arbitrary information is encoded into an image in such a way that the additional payload is imperceptible to image observers.

Image watermarking has been proposed as a suitable tool to identify the source, creator, owner, distributor, or authorized consumer of a document or an image (Podilchuk and Delp, 2001). It can also be used to detect a document or an image that has been illegally distributed or modified. Encryption, used in cryptography, is a process of obscuring information to make it unreadable to observers without specific keys or knowledge. This technology is sometimes referred to as data scrambling. Watermarking complemented by encryption can serve a large number of purposes including copyright protection, broadcast monitoring, and data authentication.

Xiang et al. (2008) presented an image watermarking scheme by using two statistical features (the histogram shape and the mean) in the Gaussian filtered low-frequency component of images. Wang et al. (2007) proposed a feature-based image watermarking scheme against desynchronization attacks. Lee et al. (2007) proposed a high-capacity reversible image watermarking scheme based on integer-to-integer wavelet transformation. Lu et al. (2005) presented a multipurpose digital image watermarking method based on the multistage vector quantizer structure, which can be applied to image authentication and copyright protection. Macq et al. (2004) discussed the issues related to image watermarking benchmarking and scenarios

Image Processing and Pattern Recognition by Frank Shih

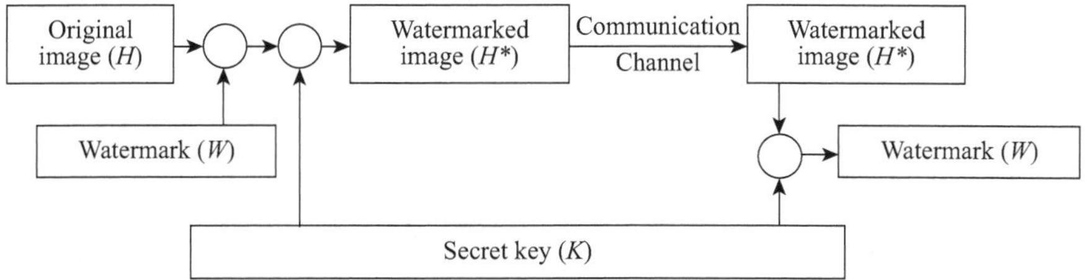

Figure 12.1 A general digital watermarking system.

based on digital rights management requirements. Seo and Yoo (2006) proposed a content-based image watermarking method based on invariant regions of an image. Pi et al. (2006) proposed a watermarking method to hide a binary watermark into image files compressed by fractal block coding.

In digital world, a watermark is a pattern of bits inserted into a digital medium that can identify the creator or authorized users. The digital watermark, unlike the printed visible stamp watermark, is designed to be invisible to viewers. The bits embedded into an image are scattered all around to avoid identification or modification. Therefore, a digital watermark must be robust enough to survive the detection, compression, and operations that are applied on.

Figure 12.1 depicts a general digital watermarking system. A *watermark message* W is embedded into a media message, which is defined as the *host image* H. The resulting image is the *watermarked image* H^*. In the embedding process, a *secret key* K, for example, a random number generator, is sometimes involved to generate a more secure watermark. The watermarked image H^* is then transmitted along a communication channel. The watermark can be detected or extracted later by the receiver.

This chapter is organized as follows. We introduce watermarking classification in Section 12.1. Spatial-domain watermarking and frequency-domain watermarking are described in Sections 12.2 and 12.3, respectively. Fragile watermark and robust watermark are presented in Sections 12.4 and 12.5, respectively. The method of combinational domain digital watermarking is proposed in Section 12.6.

12.1 WATERMARKING CLASSIFICATION

This section categorizes digital watermarking technologies into five classes according to the characteristics of embedded watermarks. These five classes are blind versus nonblind, perceptible versus imperceptible, private versus public, robust versus fragile, and spatial domain versus frequency domain.

12.1.1 Blind Versus Nonblind

A watermarking technique is called *blind* if it does not require access to the original unwatermarked data (e.g., image, video, audio, etc.) to recover the watermark. Conversely, a watermarking technique is called *nonblind* if the original data are needed in the extraction of watermark. For example, the algorithms of Barni et al. (1998) and Nikolaidis and Pitas (1998) belong to blind, and those of Cox et al. (1997) and Swanson et al. (1996) belong to nonblind. Bi et al. (2007) proposed a blind image watermarking algorithm based on the multiband wavelet transformation and the empirical mode decomposition. In general, the nonblind scheme is more robust than the blind one because it is obvious that the watermark can be easily extracted by knowing the unwatermarked data. However, in most applications, the unmodified host signal is not available to the watermark detector. Since the blind scheme does not need the original data, it is more useful than the nonblind one in most applications.

12.1.2 Perceptible Versus Imperceptible

A watermark is called *perceptible* if the embedded watermark is intended to be visible, for example, a logo inserted into a corner of the image. A good perceptible watermark is difficult to remove by an unauthorized person and can resist falsification. Since it is relatively easy to embed a pattern or a logo into a host image, we must make sure the perceptible watermark was indeed the one inserted by the author. In contrast, an *imperceptible* watermark is embedded into a host image by sophisticated algorithms and is perceptually invisible to any person. However, it can be extracted by a computer. Huang and Wu (2004) proposed an attacking scheme against current visible image watermarking techniques. Hu et al. (2006) presented an algorithm for removable visible watermarking.

12.1.3 Private Versus Public

A watermark is called *private* if only authorized users can detect it. In other words, the private watermarking techniques invest all efforts to make it impossible for unauthorized users to extract the watermark, for instance, using a private pseudorandom key. The private key indicates a watermark's location in the host image, allowing insertion and removal of the watermark if the secret location is known. In contrast, the watermarking techniques allowing anyone to read the watermark are called *public*. Public watermarks are embedded in a location known to everyone, so the watermark detection software can easily extract the watermark by scanning the whole image. In general, private watermarking techniques are more robust than public ones, in which an attacker can easily remove or destroy the message once the embedded code is known.

There is also *asymmetric* watermarking (or public watermarking), which has the property that any user can read the watermark without being able to remove it. We call this an asymmetric cryptosystem. In this case, the detection process (and in particular the detection key) is known to everyone, so only a public key is needed for verification and a private key is used for embedding.

12.1.4 Robust Versus Fragile

Watermark robustness accounts for the capability of the hidden watermark to survive legitimate daily usage or image processing manipulation, such as intentional or unintentional attacks. The intentional attacks aim at destroying the watermark, while the unintentional attacks do not explicitly intend to alter the watermark. For embedding purposes, watermarks can be categorized into three types: robust, fragile, and semifragile. *Robust* watermarks (Lin and Chen, 2000; Deguillaume et al., 2003) are designed to survive intentional (malicious) and unintentional (nonmalicious) modifications of the watermarked image. The unfriendly intentional modifications include unauthorized removal or alternation of the embedded watermark and unauthorized embedding of any other information. The unintentional modifications are image processing operations, such as scaling, cropping, filtering, and compression. They are usually used for copyright protection to declare the rightful ownership.

On the contrary, for the purpose of authentication, *fragile* watermarks (Wolfgang and Delp, 1997; Wong, 1998a, 1998b; Celik et al., 2002) are adopted to detect any unauthorized modification. The fragile watermarking techniques are concerned with complete integrity verification. The slightest modification of the watermarked image will alter or destroy the fragile watermark. Different from robust and fragile watermarks, the *semifragile* watermarks (Sun and Chang, 2002) are designed for detecting any unauthorized modification, at the same time allowing some image processing operations. They are used for selective authentication that detects illegitimate distortion, while ignoring applications of legitimate distortion. In other words, the semifragile watermarking techniques can discriminate common image processing and small content-preserving noise, such as lossy compression, bit error, or salt-and-pepper noise, from malicious content modification. Ahmed and Moskowitz (2004) proposed a binary-phase-only-filter (BPOF)-based watermarking technique for image authentication, where a watermark is embedded upon the BPOF of the Fourier spectrum of an image into the corresponding magnitude. Sang and Alam (2008) proposed to use the identical ratio between the extracted watermark and the computed BPOF to detect watermark and measure the degree of authenticity.

12.1.5 Spatial Domain Versus Frequency Domain

There are two image domains for embedding watermarks: one is spatial domain and the other is frequency domain. In the spatial domain (Berghel and O'Gorman, 1996; Karybali and Berberidis, 2006), we can simply insert watermark into a host image by changing the gray levels of some pixels in the host image. It has the advantages of low complexity and easy implementation, but the inserted information may be easily detected using computer analysis or could be easily attacked. We can embed the watermark into the coefficients of a transformed image in the frequency domain (Cox et al., 1996, 1997). The transformations include discrete cosine transform (DCT) (Briassouli and Strintzis, 2004), discrete Fourier transform (DFT) (Tsui et al., 2008), and discrete wavelet transform (DWT) (Kumsawat et al., 2005; Ghouti et al., 2006). However, if we embed too much data in the frequency domain, the image quality will be degraded significantly.

The spatial-domain watermarking techniques are usually less robust to attacks such as compression and noise addition. However, they have much lower computational complexity and usually can survive the cropping attack, which usually the frequency-domain watermarking techniques fail. Another technique combines both spatial-domain watermarking and frequency-domain watermarking to become more robust and less complex.

12.2 SPATIAL-DOMAIN WATERMARKING

Spatial-domain watermarking modifies pixel values directly on the spatial domain of an image. In general, spatial-domain watermarking schemes are simple and do not need the original image to extract the watermark. They also provide a better compromise among robustness, capacity, and imperceptibility. However, they have a disadvantage of not being robust against image processing operations because the embedded watermark does not spread around the entire image and the operations can easily destroy the watermark. The purpose of watermarking is to embed a secret message at the content level under the constraints of imperceptibility, security, and robustness to attacks. We can categorize most watermark embedding algorithms into substitution by a codebook element or additive embedding.

12.2.1 Substitution Watermarking in the Spatial Domain

Substitution watermarking in the spatial domain (Wolfgang and Delp, 1997) is the simplest watermarking algorithm. Basically, the embedding locations, such as the specific bits of all pixels, are predefined before watermark embedding. Once the receiver obtains the watermarked image, he/she knows the exact locations to extract the watermark. During the watermark embedding procedure, the watermark is first converted into a bit stream. Then, each bit of the bit stream is embedded into the specific bit of selected locations of the host image. Figure 12.2 shows an example of substitution watermarking in the spatial domain, in which three LSB (least significant bit) substitutions are performed, that is, embedding 1, 0, and 1 into 50, 50, and 48, respectively.

During the watermark extraction procedure, the specific pixel locations of the watermarked image are already known. Then, each pixel value is converted into its binary format. Finally, the watermark is collected from the bit, where the watermark is embedded. Figure 12.3 shows an example of the watermark extracting procedure of substitution watermarking in the spatial domain.

In general, the watermark capacity of substitution watermarking in the spatial domain is larger than those of other watermarking approaches. Its maximum watermark capacity is eight times the host image size. However, for the purpose of imperceptibility, a reasonable watermark capacity is three times the host image size. If the watermark is embedded into the least three significant bits, human beings cannot distinguish the original image from the watermarked image. Figure 12.4 shows an example of embedding a watermark into different bits. It is clear that the watermarked image can be distinguished by human eyes if the watermark is embedded into the seventh and fifth bits. However, if the watermark is embedded

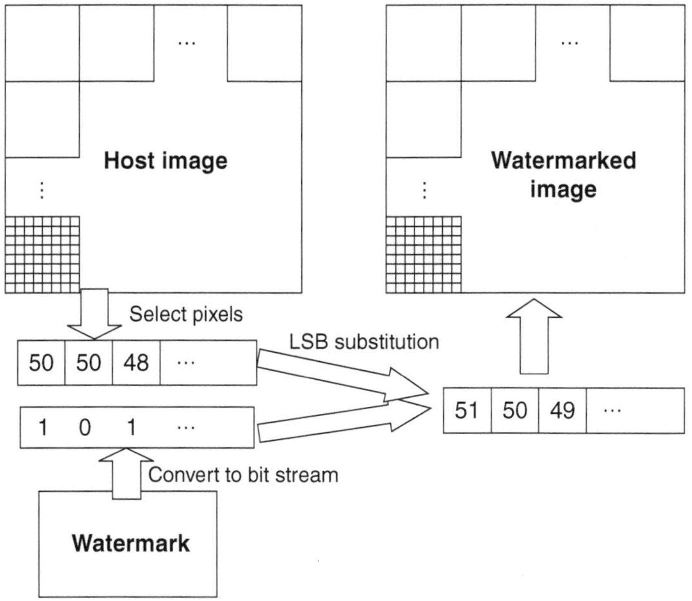

Figure 12.2 The watermark embedding procedure of substitution watermarking in the spatial domain.

into the third bit, it would be difficult to tell whether there is a watermark hidden in the image.

Substitution watermarking is simple to implement. However, the embedded watermark is not robust against collage or lossy compression attack. Therefore, some improved spatial-domain watermarking algorithms have been proposed to target robustness, for example, the hash function method (Wong, 1998b) and the bipolar *M*-sequence method (Wolfgang and Delp, 1996). In contrast to robust watermarking,

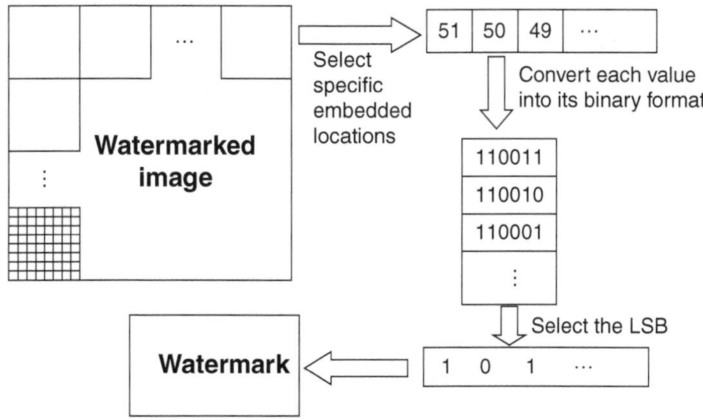

Figure 12.3 The watermark extracting procedure of substitution watermarking in the spatial domain.

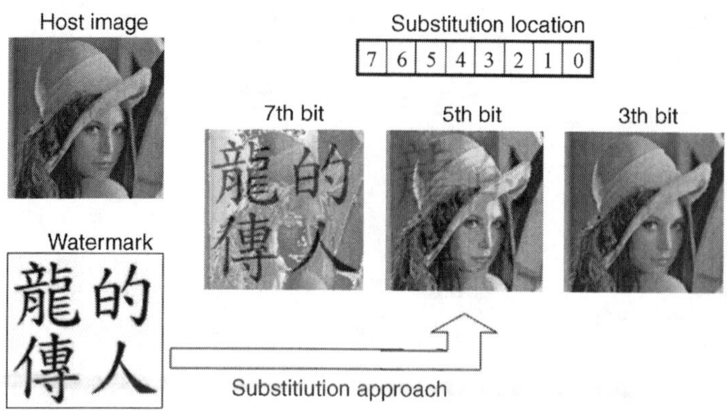

Figure 12.4 An example of embedding watermark into different bits.

the fragile watermark can embed breakable watermark into an image to detect whether it was altered.

12.2.2 Additive Watermarking in the Spatial Domain

Different from the substitution approach, the additive watermarking approach does not consider specific bits of a pixel. Instead, it adds an amount into a pixel to perform the embedding approach. Let H be the original gray-level host image and W be the binary watermark image. Let $\{h(i,j)\}$ and $\{w(i,j)\}$ denote their respective pixels. We can embed W into H to obtain the watermarked image H^* as

$$h^*(i,j) = h(i,j) + a(i,j) \cdot w(i,j) \tag{12.1}$$

where $\{a(i,j)\}$ denotes the scaling factor. Basically, the larger $a(i, j)$ is, the more robust the watermarking algorithm is. Figure 12.5 shows an example of additive watermarking in the spatial domain, in which if the embedded watermark is 1, then $a(i, j) = 100$; otherwise, $a(i, j) = 0$. Therefore, after watermark embedding, the original values are changed from 50, 50, and 48 to 150, 50, and 148 using watermarks 1, 0, and 1, respectively.

It is obvious that if we have a large $a(i, j)$, the watermarked image would be distorted as shown in Figure 12.4. It is then difficult to achieve high imperceptibility. Two critical issues to consider in additive watermarking are imperceptibility and the need of the original image to identify the embedded message.

To enhance imperceptibility, a big value is not embedded into a single pixel; instead, it is embedded into a block of pixels (Lin and Delp, 2004; Mukherjee et al., 2004). Figure 12.6 shows an example of block-based additive watermarking. First, a 3×3 block is selected for embedding a watermark by adding 99. Then, the amount is divided by 9.

The second issue is the need of the original image in extracting the embedded watermark. When a receiver obtains the watermarked image, it is difficult to determine the embedded message since he/she does not know the locations of the block for embedding. Therefore, the reference (or original) image is usually required to extract the watermark as shown in Figure 12.7.

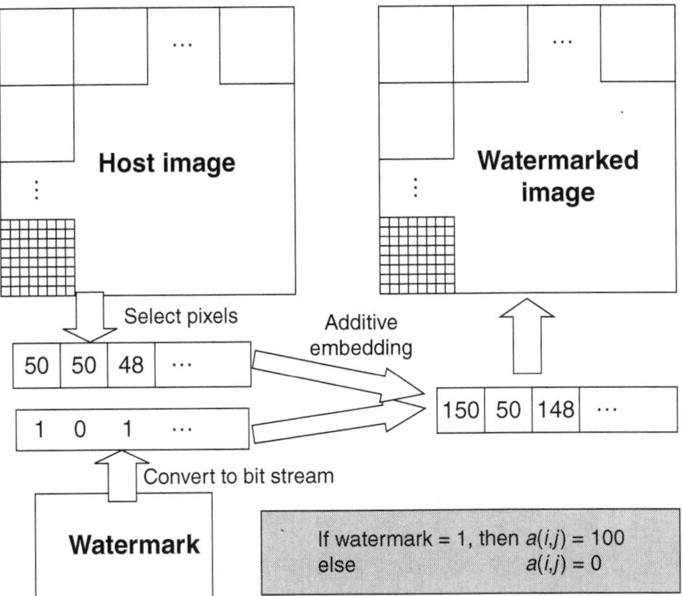

Figure 12.5 The watermark embedding procedure of additive watermarking in the spatial domain.

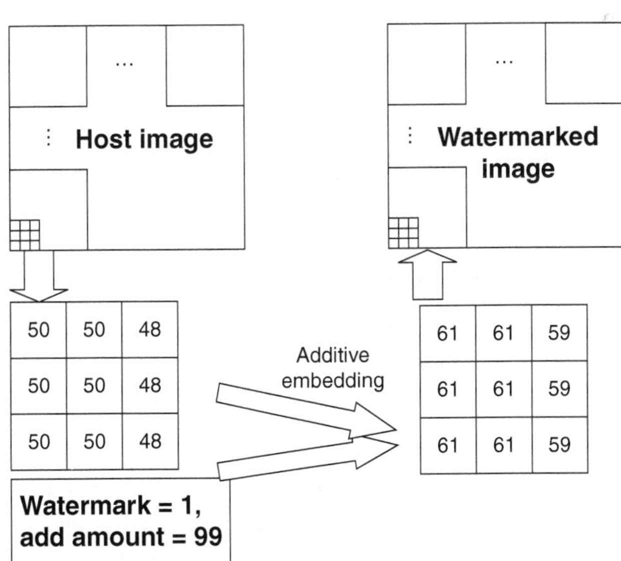

Figure 12.6 An example of block-based additive watermarking in the spatial domain.

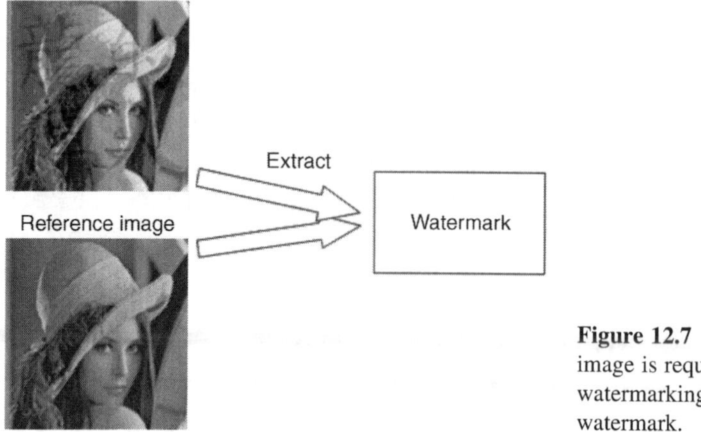

Figure 12.7 The reference image is required in additive watermarking to extract the watermark.

12.3 FREQUENCY-DOMAIN WATERMARKING

In frequency (or spectrum)-domain watermarking, we can insert a watermark into frequency coefficients of the transformed image by DFT, DCT, or DWT. Because the frequency transforms usually decorrelate spatial relationship of pixels, majority of the energy concentrates on the low-frequency components. When we embed the watermark into the low or middle frequencies, these changes will be distributed around the entire image. When image processing operations are applied on the watermarked image, they are less affected. Therefore, compared to the spatial-domain watermarking method, the frequency-domain watermarking technique is relatively more robust.

In this section, we will describe the substitution and multiplicative watermarking methods in the frequency domain. Then, we will introduce the watermarking scheme based on vector quantization. Finally, we will present a rounding error problem in the frequency-domain methods.

12.3.1 Substitution Watermarking in the Frequency Domain

The substitution watermarking scheme in the frequency domain is basically similar to that in the spatial domain except that the watermark is embedded into frequency coefficients of the transformed image. Figure 12.8 shows an example of embedding a 4-bit watermark into the frequency domain of an image. Figure 12.8(a) shows an 8×8 gray-level host image and Figure 12.8(b) shows the transformed image by DCT. Figure 12.8(c) shows a binary watermark, in which "0" and "1" are the embedded values and the dash sign "–" indicates no change in its position. We obtain Figure 12.8 (d) by embedding (c) into (b) using the LSB substitution. Figure 12.9 shows the details of the LSB substitution approach.

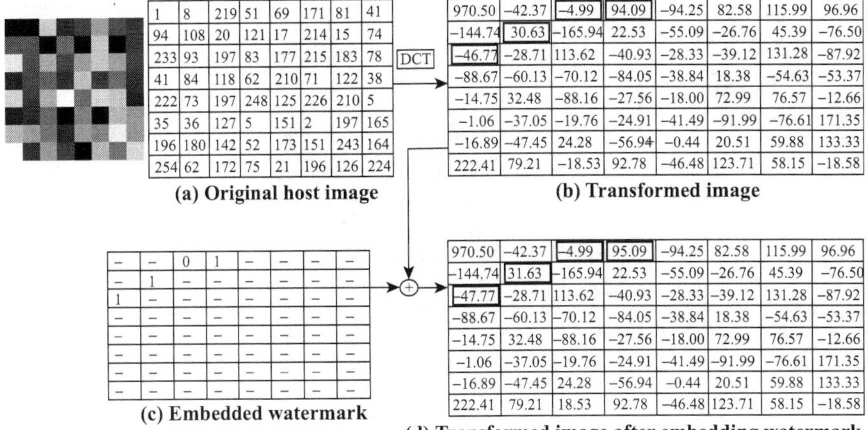

1	8	219	51	69	171	81	41
94	108	20	121	17	214	15	74
233	93	197	83	177	215	183	78
41	84	118	62	210	71	122	38
222	73	197	248	125	226	210	5
35	36	127	5	151	2	197	165
196	180	142	52	173	151	243	164
254	62	172	75	21	196	126	224

(a) Original host image

970.50	−42.37	−4.99	94.09	−94.25	82.58	115.99	96.96
−144.74	30.63	−165.94	22.53	−55.09	−26.76	45.39	−76.50
−46.77	−28.71	113.62	−40.93	−28.33	−39.12	131.28	−87.92
−88.67	−60.13	−70.12	−84.05	−38.84	18.38	−54.63	−53.37
−14.75	32.48	−88.16	−27.56	−18.00	72.99	76.57	−12.66
−1.06	−37.05	−19.76	−24.91	−41.49	−91.99	−76.61	171.35
−16.89	−47.45	24.28	−56.94	−0.44	20.51	59.88	133.33
222.41	79.21	−18.53	92.78	−46.48	123.71	58.15	−18.58

(b) Transformed image

–	–	0	1	–	–	–	–
–	1	–	–	–	–	–	–
1	–	–	–	–	–	–	–
–	–	–	–	–	–	–	–
–	–	–	–	–	–	–	–
–	–	–	–	–	–	–	–
–	–	–	–	–	–	–	–
–	–	–	–	–	–	–	–

(c) Embedded watermark

970.50	−42.37	−4.99	95.09	−94.25	82.58	115.99	96.96
−144.74	31.63	−165.94	22.53	−55.09	−26.76	45.39	−76.50
−47.77	−28.71	113.62	−40.93	−28.33	−39.12	131.28	−87.92
−88.67	−60.13	−70.12	−84.05	−38.84	18.38	−54.63	−53.37
−14.75	32.48	−88.16	−27.56	−18.00	72.99	76.57	−12.66
−1.06	−37.05	−19.76	−24.91	−41.49	−91.99	−76.61	171.35
−16.89	−47.45	24.28	−56.94	−0.44	20.51	59.88	133.33
222.41	79.21	18.53	92.78	−46.48	123.71	58.15	−18.58

(d) Transformed image after embedding watermark

Figure 12.8 An example of embedding a 4-bit watermark in the frequency domain of an image.

12.3.2 Multiplicative Watermarking in the Frequency Domain

The frequency-domain embedding is used to insert the watermark into a prespecified range of frequencies in the transformed image. The watermark is usually embedded in the perceptually significant portion (which is the significant frequency component) of the image to make it robust to resist attack. The watermark is scaled according to the magnitude of the particular frequency component. The watermark consists of a random, Gaussian distributed sequence. This kind of embedding is called *multiplicative watermarking* (Cox et al., 1997). Let H be the DCT coefficients of the host image and W be the random vector. Let $\{h(m,n)\}$ and $\{w(i)\}$ denote their respective pixels. We can embed W into H to obtain the watermarked image H^* as

$$h^*(m,n) = h(m,n)(1 + \alpha(i) \cdot w(i)) \tag{12.2}$$

Note that a large $\{\alpha(i)\}$ would produce a higher distortion in the watermarked image. It is usually set to 0.1 to provide a good trade-off between imperceptibility

LSB (Least Significant Bit) Substitution

W	Original coefficient	Integer part	Binary format	Watermarked	
				Binary	Coefficient
1	−46.77	46	00101110	00101111	−47.77
1	30.63	30	00011110	00011111	31.63
0	−4.99	4	00000100	00000100	−4.99
1	94.09	94	01011110	01011111	94.09

Figure 12.9 Embedding a watermark into coefficients of the transformed image in Figure 12.8.

Figure 12.10 The watermarked image.

and robustness. An alternative embedding formula of using logarithm of the original coefficients is

$$h^*(m,n) = h(m,n) \cdot e^{\alpha(i) \cdot w(i)} \tag{12.3}$$

For example, let the watermark be a Gaussian sequence of 1000 pseudorandom real numbers. We select the 1000 largest coefficients in the DCT domain. We embed the watermark into a Lena image to obtain the watermarked image as shown in Figure 12.10. The watermark can be extracted using the inverse embedding formula as

$$w'(i) = \frac{h^*(m,n) - h(m,n)}{\alpha(i) \cdot h(m,n)} \tag{12.4}$$

For comparison of the extracted watermark sequence $\mathbf{w}'$ and the original watermark $\mathbf{w}$, we can use the similarity measure as

$$\text{sim}(\mathbf{w}', \mathbf{w}) = \frac{\mathbf{w}' \cdot \mathbf{w}}{|\mathbf{w}'|} \tag{12.5}$$

The dot product of $\mathbf{w}' \cdot \mathbf{w}$ will be distributed according to the distribution of a linear combination of variables that are independent and normally distributed as

$$N(0, \mathbf{w}' \cdot \mathbf{w}) \tag{12.6}$$

Thus, $\text{sim}(\mathbf{w}', \mathbf{w})$ is distributed according to $N(0, 1)$. We can then apply the standard significance tests for the normal distribution. Figure 12.11 shows the response of the watermark detector to 500 randomly generated watermarks of which only one

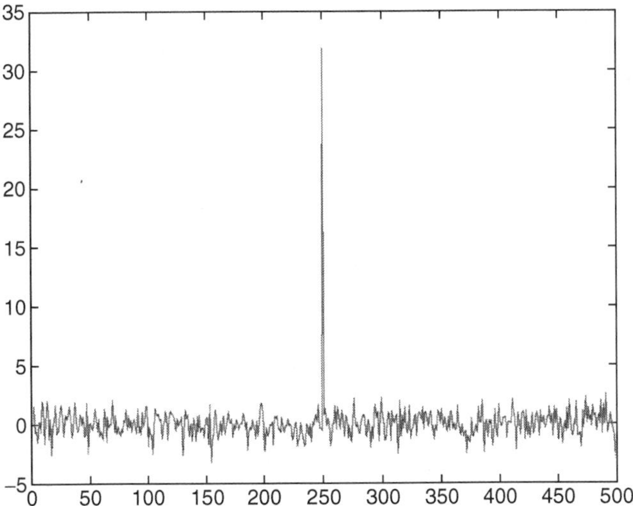

Figure 12.11 Responses of 500 randomly generated watermarks of which only one exactly matches the original watermark.

matches the watermark. The positive response due to the correct watermark is very much stronger than the response to incorrect watermarks. If $\mathbf{w}'$ is created independently from $\mathbf{w}$, then it is extremely unlikely that $\mathrm{sim}(\mathbf{w}', \mathbf{w}) > 6$. This suggests that the embedding technique has very low false positive and false negative response rates. By using equation (12.5) on Figure 12.10, we obtain the similarity measure to be 29.8520.

12.3.3 Watermarking Based on Vector Quantization

A comprehensive review on vector quantization (VQ) can be referred in Gray (1984). It is a lossy block-based compression technique in which the vectors instead of scalars are quantized. An image is partitioned into 2D blocks. Each input block is mapped to a finite set of vectors forming the codebook, shared between the encoder and the decoder. The index of the vector that best matches the input block based on some cost functions is sent to the decoder. While traditional vector quantization is a fixed dimension scheme, a more general variable dimension vector quantization (VDVQ) has been developed.

A codebook of codewords is used to store all the vectors. Given an input vector $\mathbf{v}$, Euclidean distance is used in the search process to measure the distance between two vectors as

$$k = \arg\min_j \sqrt{\sum_i (\mathbf{v}(i) - \mathbf{s}_j(i))^2}, \quad \text{where } j = 0, 1, \ldots, N-1 \qquad (12.7)$$

The closest Euclidean distance codeword $\mathbf{s}_k$ is selected and transmitted to the receiver. With the same codebook, the decomposition procedure can easily extract the vector $\mathbf{v}$ by table lookup.

Lu and Sun (2000) presented an image watermarking technique based on VQ. They divided the codebook into a few clusters of close codevectors with the Euclidean distance less than a given threshold. The scheme requires the original image for watermark extraction and uses the codebook partition as the secret key. When the received index is not in the same cluster as the best match index, a tampering attack is detected.

12.3.4 Rounding Error Problem

The rounding error problem exists in substitution watermarking in frequency domain. A receiver may confront a problem of correctly extracting the embedded watermark because the embedded message is changed due to modification of coefficient values. Ideally, the data after performing the transformation such as DCT or DWT, followed by an inverse transformation such as IDCT or IDWT, should be exactly the same as the original data. However, if some data in the transformed image are changed, we will not obtain an image with all integers. We illustrate this situation in Figure 12.12, where (a) shows the transformed image of Figure 12.8(a). If we apply IDCT to Figure 12.12(a), the original image can be reconstructed as (b). If the data enclosed by the bold rectangular boxes in Figure 12.12(c) are changed, the pixel values of the reconstructed image will not be integers as in (d). Note that the changes between Figure 12.12(a) and (c) are small.

Some researchers suggested that the rounding technique be used to convert the real numbers to integers. Therefore, after adopting the rounding approach to Figure 12.13(a), we show the rounded image in (b). If the receiver wants to extract the watermark from the rounded image, he/she will perform a DCT to the rounded image and obtain its transformed image as shown in Figure 12.13(c). Unfortunately, the receiver cannot correctly extract the watermark from the locations, where the watermark is embedded even though there is only one difference, enclosed by the bold rectangle, between Figure 12.13(b) and its original image in Figure 12.8(a).

970.50	−42.37	−4.99	94.09	−94.25	82.58	115.99	96.96
−144.74	−30.63	−165.94	22.53	−55.09	−26.76	45.39	−76.50
−46.77	−28.71	113.62	−40.93	−28.33	−39.12	131.28	−87.92
−88.67	−60.13	−70.12	−84.05	−38.84	18.38	−54.63	53.37
−14.75	−32.48	−38.16	−27.56	−18.00	72.99	−76.57	−12.66
−1.06	−37.05	−19.76	−24.91	−41.49	−91.99	−76.61	172.35
−16.89	−47.45	24.28	−56.94	−0.44	20.51	59.88	133.33
222.41	79.21	−18.53	92.78	−46.48	123.71	58.15	−18.58

(a) Transformed image

1	8	219	51	69	171	81	41
94	108	20	121	17	214	15	74
233	93	197	83	177	215	183	78
41	84	118	62	210	71	122	38
222	73	197	248	125	226	210	5
35	36	127	5	151	2	197	165
196	180	142	52	173	151	243	164
254	62	172	75	21	196	126	224

(b) An image after IDCT from (a)

970.50	−42.37	−4.99	95.09	−94.25	82.58	115.99	96.96
−144.74	−31.63	−165.94	22.53	−55.09	−26.76	45.39	−76.50
−47.77	−28.71	113.62	−40.93	−28.33	−39.12	131.28	−87.92
−88.67	−60.13	−70.12	−84.05	−38.84	18.38	−54.63	53.37
−14.75	−32.48	−38.16	−27.56	−18.00	72.99	−76.57	−12.66
−1.06	−37.05	−19.76	−24.91	−41.49	−91.99	−76.61	172.35
−16.89	−47.45	24.28	−56.94	−0.44	20.51	59.88	133.33
222.41	79.21	−18.53	92.78	−46.48	123.71	58.15	−18.58

(c) Transformed image after embedding watermark

1.22	8.01	218.80	50.79	68.89	170.87	80.67	40.45
94.28	108.07	19.87	120.87	16.99	213.99	14.79	73.58
233.35	93.15	196.97	83.00	177.14	215.16	182.99	77.78
41.36	84.17	118.02	62.07	210.25	71.31	122.16	37.97
222.26	73.09	196.96	248.06	125.27	226.36	210.24	5.06
35.08	35.92	126.82	4.94	151.19	2.32	197.22	165.06
195.88	179.73	141.64	51.79	173.07	151.22	243.14	163.99
253.74	61.60	171.53	74.69	20.98	196.15	126.08	223.93

(d) An image after IDCT from (c)

Figure 12.12 An illustration that the pixel values in the reconstructed image will not be integers if some coefficient values are changed.

1.22	8.01	218.80	50.79	68.89	170.87	80.67	40.45
94.28	108.07	19.87	120.87	16.99	213.99	14.79	73.58
233.35	93.15	196.97	83.00	177.14	215.16	182.99	77.78
41.36	84.17	118.02	62.07	210.25	71.31	122.16	37.97
222.26	73.09	196.96	248.06	125.27	226.36	210.24	5.06
35.08	35.92	126.82	4.94	151.19	2.32	197.22	165.06
195.88	179.73	141.64	51.79	173.07	151.22	243.14	163.99
253.74	61.60	171.53	74.69	20.98	196.15	126.08	223.93

(a) Image after IDCRT transform

1	8	219	51	69	171	81	40
94	108	20	121	17	214	15	74
233	93	197	83	177	215	183	78
41	84	118	62	210	71	122	38
222	73	197	248	125	226	210	5
35	36	127	5	151	2	197	165
196	180	142	52	173	151	243	164
254	62	172	75	21	196	126	224

(b) Translate real numbers into integers by ROUND

970.38	−42.20	−5.16	94.24	−94.38	82.68	115.92	96.99
−144.91	−30.87	−166.17	22.73	−55.26	−26.62	45.30	−76.46
−46.94	−28.49	113.41	−40.74	−28.49	−38.99	131.19	−87.88
−88.82	−59.93	−70.31	−83.88	−38.99	18.49	−54.71	53.41
−14.88	32.65	−88.32	−27.41	−18.13	73.09	−76.50	−12.62
−1.16	−36.91	−19.89	−24.79	−41.58	−91.91	−76.66	171.37
−16.95	−47.36	24.19	−56.86	−0.51	20.56	59.84	133.35
222.37	79.26	−18.58	92.82	−46.51	123.73	58.13	−18.57

(c) Transform rounded image by DCT

–	–	1	0	–	–	–	–
–	0	–	–	–	–	–	–
0	–	–	–				
–	–	–	–				
–	–	–	–				
–	–	–	–				
–	–	–	–				
–	–	–	–	–	–	–	–

Original watermark 1101

(d) Extract the embedded watermark

Figure 12.13 An example to show that the watermark cannot be correctly extracted.

Table 12.1 shows the results of embedding different kinds of watermarks into the same transformed image in Figure 12.8(d). The embedding rule is shown in Figure 12.14. For a 4-bit watermark "1234", we insert the most significant bit "1" into position A, "2" into position B, "3" into position C, and "4" into position D. There are 2^4 possible embedding combinations. Among the 16 possibilities, only two cases of watermarks can be extracted correctly.

TABLE 12.1 The Total Possible Embedded Watermarks

Embedded	Extracted	Error bits	Embedded	Extracted	Error bits
0000	0000	0	1000	0000	1
0001	0000	1	1001	0000	2
0010	0000	1	1010	0000	2
0011	0000	2	1011	0000	3
0100	0000	1	1100	0000	2
0101	0000	2	1101	0010	4
0110	0000	2	1110	1110	0
0111	1110	2	1111	1110	1

–	–	C	D	–	–	–	–
–	B	–	–	–	–	–	–
A	–	–	–	–	–	–	–
–	–	–	–	–	–	–	–
–	–	–	–	–	–	–	–
–	–	–	–	–	–	–	–
–	–	–	–	–	–	–	–
–	–	–	–	–	–	–	–

Figure 12.14 The positions where the watermark is embedded.

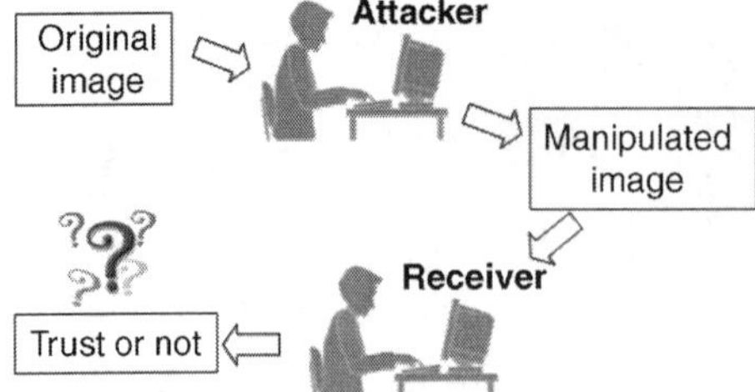

Figure 12.15 A receiver has a problem in judging whether a received image is altered.

12.4 FRAGILE WATERMARK

Due to the popularity of computerized image processing, a large number of software tools are available to modify images. Therefore, it becomes a critical problem for a receiver to judge whether the received image is altered by an attacker. This situation is illustrated in Figure 12.15.

The fragile watermark provides a solution to ensure if a received image can be trusted. A receiver will evaluate the hidden fragile watermark to see whether the image is altered. Therefore, the fragile watermarking scheme is used to detect any unauthorized modification.

12.4.1 The Block-Based Fragile Watermark

Wong (1998b) presented a block-based fragile watermarking algorithm that is capable of detecting changes such as pixel values and image size by adopting the *Rivest–Shamir–Adleman* (RSA) public key encryption algorithm (Rivest et al., 1978) and *Message Digest 5* (MD5) (Rivest, 1992) for the hash function. Wong's watermarking insertion and extraction algorithms are introduced below, and their flowcharts are shown in Figures 12.16 and 12.17, respectively.

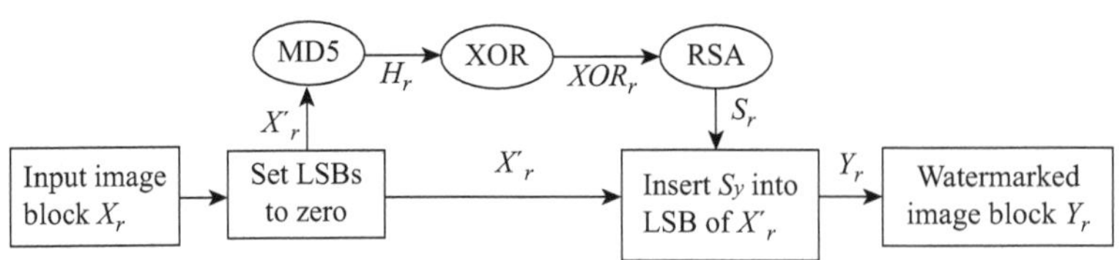

Figure 12.16 Wong's watermarking insertion algorithm.

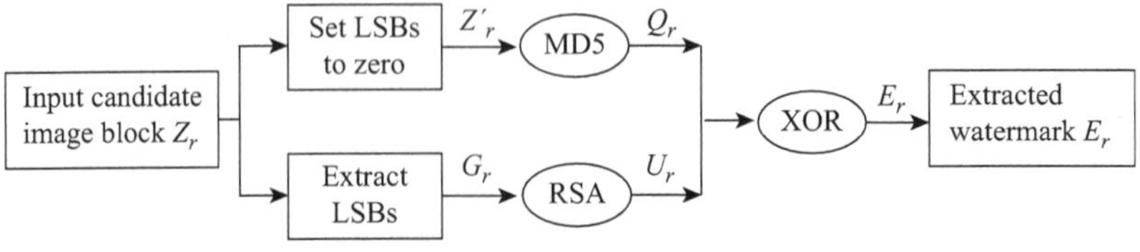

Figure 12.17 Wong's watermarking extraction algorithm.

Wong's Watermarking Insertion Algorithm

1. Divide the original image X into subimages X_r.
2. Divide the watermark W into subwatermarks W_r.
3. For each subimage X_r, we obtain X'_r by setting LSB to 0.
4. For each X'_r, we obtain the corresponding codes H_r by a cryptographic hash function, for example, MD5.
5. XOR_r is obtained by adopting the XOR operation of H_r and W_r.
6. S_r is obtained by encrypting XOR_r using the RSA encryption with a private key K'.
7. We obtain the watermarked subimage Y_r by embedding S_r into LSB of X'_r.

Wong's Watermarking Extraction Algorithm

1. Divide the candidate image Z into subimages Z_r.
2. Z'_r is obtained by setting LSB of Z_r to 0.
3. G_r is obtained by extracting LSB of Z_r.
4. U_r is obtained by decrypting G_r using RSA with a public key K.
5. For each Z'_r, we obtain the corresponding codes Q_r by a cryptographic hash function, for example, MD5 or SHA.
6. E_r is the extracted watermark by adopting the XOR operation of Q_r and U_r.

12.4.2 Weakness of the Block-Based Fragile Watermark

Holliman and Memon (2000) developed a VQ counterfeiting attack to forge the fragile watermark by exploiting the blockwise independence of Wong's algorithm. By exploitation, we can rearrange blocks of an image to form a new collage image in which the embedded fragile watermarks are not altered. Therefore, given a large database of watermarked images, the VQ attack can approximate a counterfeit collage image with the same visual appearance as the original unwatermarked image from a codebook. Note that the VQ attack does not need to have the knowledge of the embedded watermarks.

Let D be the large database of watermarked images and C be the codebook with n items generated from D. The algorithm of the VQ attack is described below, and its flowchart is shown in Figure 12.18.

VQ Counterfeiting Attack Algorithm

1. Generate the codebook C from D.
2. Divide the original image X into subimages X_r.
3. For each subimage X_r, we obtain X_r^c by selecting the data from the codebook with the minimum distance (difference) to X_r. Finally, we obtain a counterfeit collage image X^c with the same visual appearance as the original unwatermarked image.

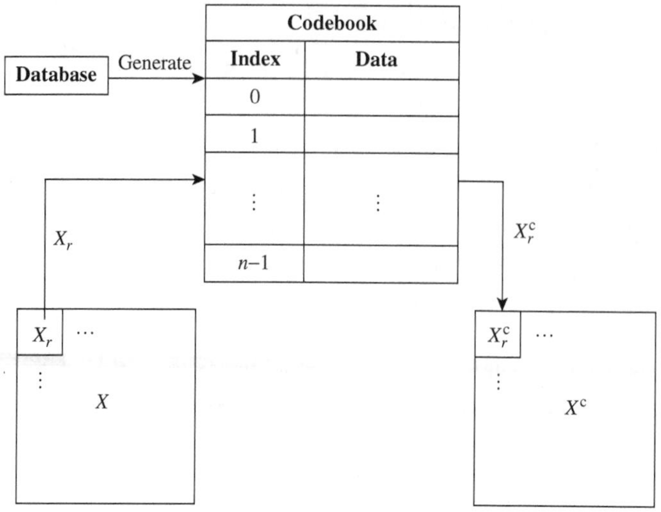

Figure 12.18 The VQ counterfeit attack.

12.4.3 The Hierarchical Block-Based Fragile Watermark

In order to defeat the VQ counterfeiting attack, a hierarchical block-based watermarking technique was developed by Celik et al. (2002). The idea of the hierarchical watermarking approach is simply to break the blockwise independence of Wong's algorithm. That is, the embedded data not only include the information of its corresponding block, but also possess the relative information of its higher level blocks. Therefore, the VQ attack cannot approximate an image based on the codebook that only records the information of each watermarked block.

Let $X_{i,j}^l$ denote a block in the hierarchical approach, where (i, j) represents the spatial position of the block and l is the level to which the block belongs. The total number of levels in the hierarchical approach is denoted by L. At each successive level, the higher level block is divided into 2×2 lower level blocks. That is, for $l = L - 1$ to 2

$$\begin{bmatrix} X_{2i,2j}^{l+1} & X_{2i,2j+1}^{l+1} \\ X_{2i+1,2j}^{l+1} & X_{2i+1,2j+1}^{l+1} \end{bmatrix} = X_{i,j}^l$$

For each block $X_{i,j}^l$, we obtain its corresponding *ready-to-insert data* (RID), $S_{i,j}^l$, after the processes of MD5, XOR, and RSA. Then, we construct a payload block $P_{i,j}^L$ based on the lowest level block $X_{i,j}^L$. For each payload block, it contains both RID and the data belonging to the higher level block of $X_{i,j}^L$.

Figure 12.19 illustrates an example of the hierarchical approach. Figure 12.19 (a) denotes an original image X and its three-level hierarchical results. Note that $X_{0,0}^1$ is the top level of the hierarchy consisting of only one block X. Figure 12.19(b) shows the corresponding RID of each block $X_{i,j}^l$. In Figure 12.19(c), when dealing with the block $X_{3,3}^3$, we consider the following three RIDs: $S_{3,3}^3$, $S_{1,1}^2$, and $S_{0,0}^1$. After generating the payload block $P_{3,3}^3$, we embed it into LSB of the original image X on $X_{3,3}^3$.

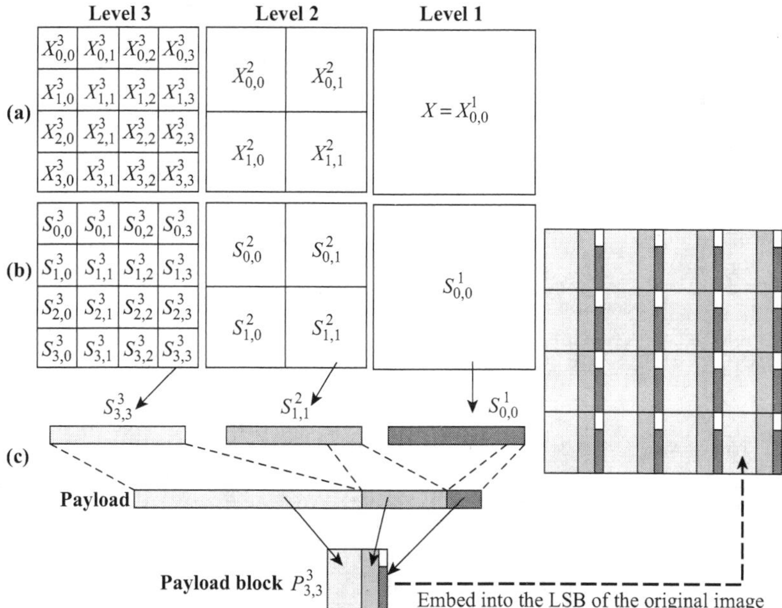

Figure 12.19 An example of the hierarchical approach with three levels.

12.5 ROBUST WATERMARK

The computer technology allows people to easily duplicate and distribute the digital multimedia through the Internet. However, these benefits come with concomitant risks of data piracy. One solution to provide the security in copyright protection is robust watermarking, which ensures the survival of embedded messages under some attacks, such as JPEG compression, Gaussian noise, and low-pass filter. Figure 12.20 illustrates the purpose of robust watermarking.

12.5.1 The Redundant Embedding Approach

It is obvious that the embedded message will be distorted due to some image processing procedures from attackers. The extracted watermark from watermarked image

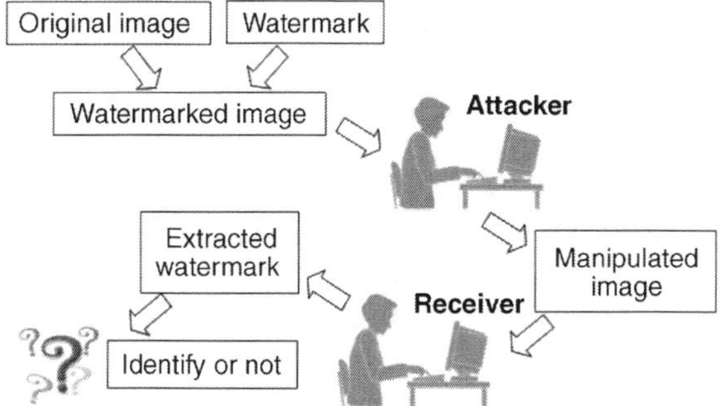

Figure 12.20 The purpose of robust watermarking.

215	201	177	145	111	79	55	41
201	177	145	111	79	55	41	215
177	145	111	79	55	41	215	201
145	111	79	55	41	215	201	177
111	79	55	41	215	201	177	145
79	55	41	215	201	177	145	111
55	41	215	201	177	145	111	79
41	215	201	177	145	111	79	55

(a)

1024.0	0.0	0.0	0.0	0.0	0.0	0.0	0.0
0.0	110.8	244.0	–10.0	35.3	–2.6	10.5	–0.5
0.0	244.0	–116.7	–138.2	0.0	–27.5	0.0	–6.4
0.0	–10.0	–138.2	99.0	72.2	0.2	13.3	0.3
0.0	35.3	0.0	72.2	–94.0	–48.2	0.0	–7.0
0.0	–2.6	–27.5	0.2	–48.2	92.3	29.9	0.4
0.0	10.5	0.0	13.3	0.0	29.9	–91.3	–15.7
0.0	–0.5	–6.4	0.3	–7.0	0.4	–15.7	91.9

(b)

Figure 12.21 An example of significant coefficients.

followed by image processing procedures may not be identified. In order to achieve robustness, we embed redundant watermarks in a host image. Epstein and McDermott (2006) hold a patent for copy protection using redundant watermark encoding.

12.5.2 The Spread Spectrum Approach

Cox et al. (1997) presented a spread spectrum-based robust watermarking algorithm, in which the embedded messages are spread out all over the image. Their algorithm first translates an image into its frequency domain. Then, the watermarks are embedded into significant coefficients of the transformed image. Note that the significant coefficients are the locations of large absolute values of the transformed image as shown in Figure 12.21. Figure 12.21(a) shows an original image, and Figure 12.21(b) shows its transformed image by DCT. Let the threshold be 70. The bold rectangles in Figure 12.21(b) are called *significant coefficients* since the absolute values are larger than 70.

12.6 COMBINATIONAL DOMAIN DIGITAL WATERMARKING

Digital watermarking plays a critical role in current state-of-the-art information hiding and security. It allows the embedding of an imperceptible watermark in the multimedia data to identify the ownership, trace the authorized users, and detect the malicious attacks. Previous research states that the embedding of watermark into the least significant bits or the low-frequency components is reliable and robust. Therefore, we can categorize digital watermarking techniques into two embedding domains: one is the spatial domain and the other is the frequency domain.

In spatial-domain watermarking, one can simply insert watermark into a host image by changing the gray levels of some pixels in the host image. The scheme is simple and easy to implement, but tends to be not robust against attacks. In the frequency domain, we can insert watermark into the coefficients of a transformed image, for example, using DFT, DCT, and DWT. This scheme is generally considered to be robust against attacks. However, one cannot embed a high-capacity watermark in the frequency domain since the image quality after watermarking would be degraded significantly.

To provide high-capacity watermarks and minimize image distortion, we present the technique of applying combinational spatial and frequency domains. The idea is to split the watermark image into two parts, which are used for spatial and frequency insertions, respectively, relying on the user's preference and data importance. The splitting strategy can be designed even more complicated for a user to be unable to compose the watermark. Furthermore, to enhance robustness, a random permutation of the watermark is used to defeat the attacks of image processing operations, such as image cropping.

12.6.1 Overview of Combinational Watermarking

In order to insert more data into a host image, the simple way is to embed them in the spatial domain of the host image. However, the disadvantage is that the inserted data could be detected by some simple extraction skills. How can we insert more signals, keeping visual effect unperceivable? We present a new strategy of embedding a high-capacity watermark into a host image by splitting the watermark image into two parts: one is embedded in the spatial domain of a host image, and the other is embedded in the frequency domain.

Let H be the original gray-level host image of size $N \times N$ and W be the binary watermark image of size $M \times M$. Let W^1 and W^2 denote the two separated watermarks from W, and let H^{S} denote the image combined from H and W^1 in the spatial domain. H^{DCT} is the image where H^{S} is transformed into the frequency domain by DCT. H^{F} is the image where H^{DCT} and W^2 are combined in the frequency domain. Let $\oplus$ denote the operation that substitutes bits of watermark for least significant bits of the host image.

The algorithm of combinational image watermarking is presented below, and its flowchart is shown in Figure 12.22.

1. Split the watermark into two parts:

$$\begin{aligned} W &= \{w(i,j), 0 \le i,j < M\}, && \text{where} \quad w(i,j) \in \{0,1\} \\ W^1 &= \{w^1(i,j), 0 \le i,j < M_1\}, && \text{where} \quad w^1(i,j) \in \{0,1\} \\ W^2 &= \{w^2(i,j), 0 \le i,j < M_2\}, && \text{where} \quad w^2(i,j) \in \{0,1\} \\ M &= M_1 + M_2 \end{aligned}$$

2. Insert W^1 into the spatial domain of H to obtain H^{S} as

$$H^{\mathrm{S}} = \{h^S(i,j) = h(i,j) \oplus w^1(i,j), 0 \le i,j < N\},$$
$$\text{where } h(i,j) \text{ and } h^S(i,j) \in \{0,1,2,\ldots,2^L-1\}$$

and L is the number of bits used as in the gray level of pixels.

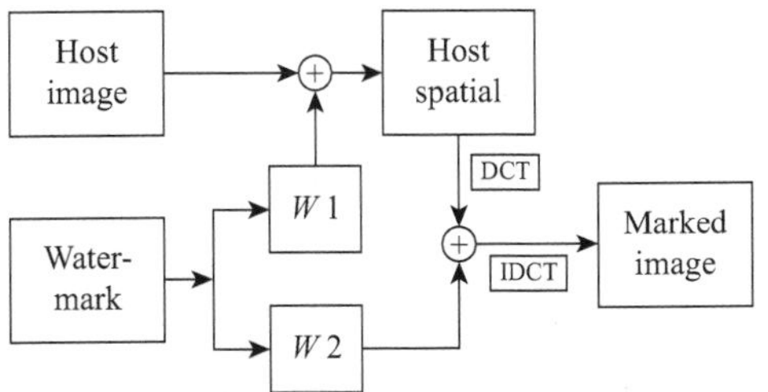

Figure 12.22 The flowchart in combinational spatial and frequency domains.

3. Transform H^{S} by DCT to obtain H^{DCT}.
4. Insert W^2 into the coefficients of H^{DCT} to obtain H^{F} as

$$H^F = \{h^{\mathrm{F}}(i,j) = h^{\mathrm{DCT}}(i,j) \oplus w^2(i,j), 0 \le i,j < N\},$$
$$\text{where} \quad h^F(i,j) \in \{0, 1, 2, \ldots, 2^L - 1\}$$

5. Transform the embedded host image by inverse DCT.

The criteria of splitting the watermark image into two parts, which are individually inserted into the input image in the spatial and frequency domains, depend on user's requirements and applications. In principle, the most important information appears at the center of an image. Therefore, a simple way of splitting is to select the central window in the watermark image to be inserted into the frequency domain. With the user's preference, we can crop the most private data to be inserted into the frequency domain.

12.6.2 Watermarking in the Spatial Domain

There are many ways of embedding a watermark into the spatial domain of a host image, for example, substituting the least significant bits of some pixels (Wolfgang and Delp, 1997), changing the paired pixels (Pitas and Kaskalis, 1995), and coding by textured blocks (Caronni, 1995). The most straightforward method is the LSB. Given a sufficiently high channel capacity in data transmission, a smaller object may be embedded multiple times. So even if most of the watermarks are lost or damaged due to malicious attacks, a single surviving watermark would be considered as a success.

Despite its simplicity, the LSB substitution suffers a major drawback. Any noise addition or lossy compression is likely to defeat the watermark. A simple attack is to simply set the LSB bits of each pixel to 1. Moreover, the LSB insertion can be altered by an attacker without noticeable change. An improved method would be to apply a pseudorandom number generator to determine the pixels to be used for embedding based on a designed key.

As shown in Figure 12.23, watermarking can be implemented by modifying the bits of some pixels in the host image. Let H^* be the watermarked image. The algorithm is presented below.

1. Obtain pixels from the host image.

$$H = \{h(i,j), 0 \le i,j < N\}, h(i,j) \in \{0, 1, 2, \ldots, 2^L - 1\}$$

2. Obtain pixels from the watermark.

$$W = \{w(i,j), 0 \le i,j < M\}$$

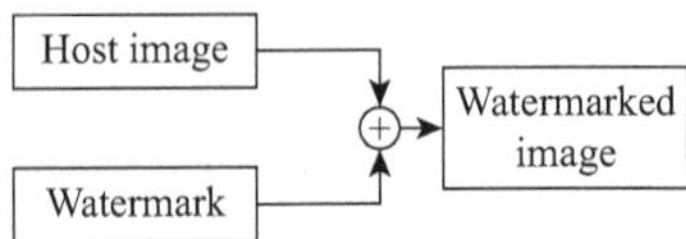

Figure 12.23 The flowchart in spatial domains.

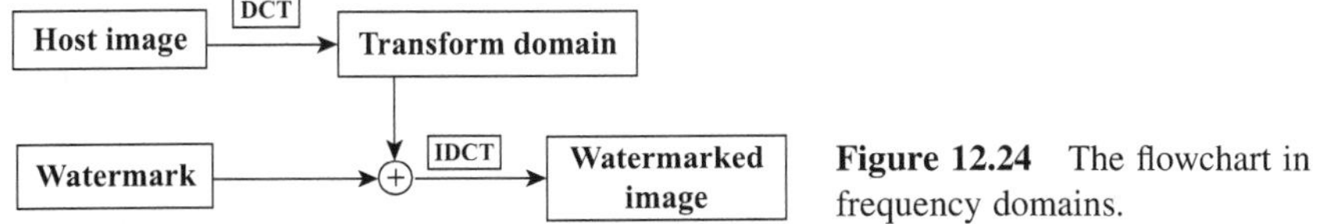

Figure 12.24 The flowchart in frequency domains.

3. Substitute the pixels of the watermark into the LSB pixels of the host image.

$$H^* = \{h^*(i,j) = h(i,j) \oplus w(i,j), 0 \leq i,j < N\}, h^*(i,j) \in \{0, 1, 2, \ldots, 2^L - 1\}$$

12.6.3 Watermarking in the Frequency Domain

Watermarking can be applied in the frequency domain by first applying a transform. Similar to spatial-domain watermarking, the values of specific frequencies can be modified from the original. Since high-frequency components are often lost by compression or scaling, the watermark is embedded into lower frequency components, or alternatively embedded adaptively into the frequency components that include critical information of the image. Since the watermarks applied to the frequency domain will be dispersed entirely over everywhere in the spatial image after the inverse transform, this method is not as susceptible to defeat by cropping as the spatial-domain approaches.

Several approaches can be used in frequency-domain watermarking, for example, JPEG-based (Zhao, 1994), spread spectrum (Cox et al., 1997), and content-based approaches (Bas et al., 2002). The often-used transformation functions are DCT, DWT, and DFT. They allow an image to be separated into different frequency bands, making it much easier to embed watermark by selecting frequency bands of an image. One can avoid the most important visual information in the image (i.e., low frequencies) without overexposing them for removal through compression and noise attacks (i.e., high frequencies).

Generally, we can insert the watermark into the coefficients of a transformed image as shown in Figures 12.24 and 12.25. The important consideration is which locations are the best places for embedding watermark in the frequency domain to avoid distortion.

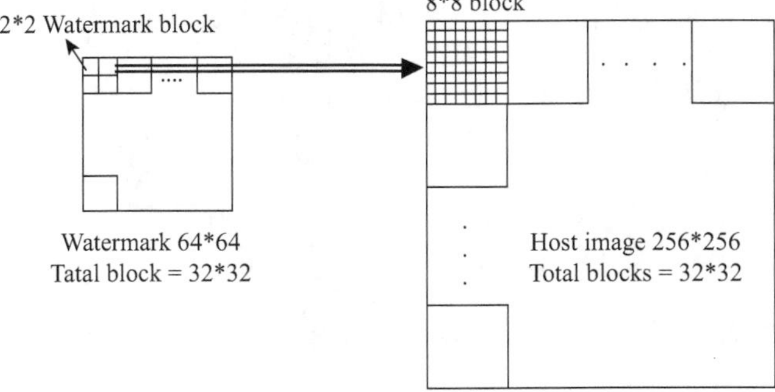

Figure 12.25 The embedding skill in frequency domain.

Let H^m and W^n be the subdivided images from H and W, respectively, H^{m_DCT} be the image transformed fromH^m by DCT, and H^{m_F} be the image combined by H^{m_DCT} and W^n in the frequency domain. The algorithm is presented below.

1. Divide the host image into a set of 8×8 blocks.

$$H = \{h(i,j),\ 0 \le i,j < N\}$$
$$H^m = \{h^m(i,j), 0 \le i,j < 8\}$$

where $h^m(i,j) \in \{0, 1, 2, \ldots, 2^L - 1\}$ and m is the total number of the 8×8 blocks.

2. Divide the watermark image into a set of 2×2 blocks.

$$W = \{w(i,j), 0 \le i,j < M\}$$
$$W^n = \{w^n(i,j), 0 \le i,j < 2\}$$

where $w^n(i,j) \in \{0, 1\}$ and n is the total number of the 2×2 blocks.

3. Transform H^m to H^{m_DCT} by DCT.
4. Insert W^m into the coefficients of H^{m_DCT}.

$$H^{m\_\mathrm{F}} = \{h^{m\_\mathrm{F}}(i,j) = h^{m\_\mathrm{DCT}}(i,j) \oplus w^m(i,j), 0 \le i,j < 8\},$$

where $h^{m\_\mathrm{DCT}}(i,j) \in \{0, 1, 2, \ldots, 2^L - 1\}$

5. Transform the embedded host image, H^{m_F}, by inverse DCT.

The criterion for embedding the watermark image into the frequency domain of a host image is that the total number of 8×8 blocks in the host image must be larger than the total number of 2×2 blocks in the watermark image.

12.6.4 Experimental Results

Figure 12.26 illustrates that some parts of a watermark are important to be split for security purpose. For example, people cannot view who the writer is in the images.

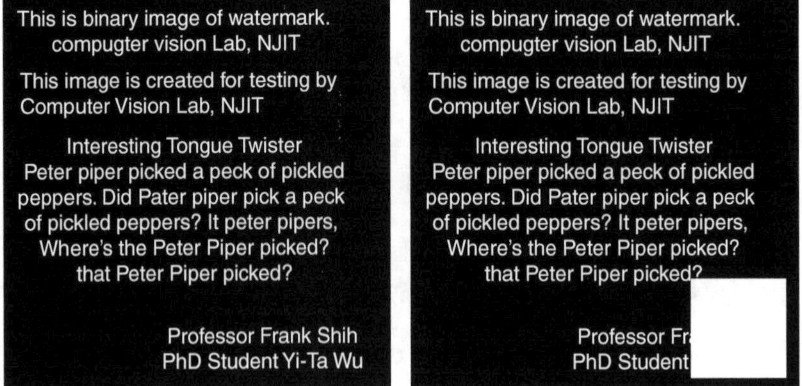

Figure 12.26 A 64×64 square area on the lower-right corner is cut from the original 256×256 image.

Figure 12.27 A Lena image.

Therefore, it is embedded into the frequency domain and the rest is embedded into the spatial domain. In this way, we not only enlarge the capacity but also secure the information that is concerned.

Figure 12.27 shows the original Lena image of size 256 × 256. Figure 12.28 is the traditional watermarking technique of embedding a 64 × 64 watermark into the frequency domain of a host image. Figure 12.28(a) shows the original 64 × 64 watermark image and Figure 12.28(b) shows the watermarked Lena image by embedding (a) into the frequency domain of Figure 12.27. Figure 12.28(c) shows the extracted watermark image from (b).

Figure 12.29 demonstrates the embedding of a large watermark, a 128 × 128 image, into a host image. Figure 12.29(a) shows the original 128 × 128 watermark image, and Figure 12.29(b) and (c) shows the two divided images from the original watermark. We obtain Figure 12.29(e) by embedding (b) into the spatial domain of Lena image and obtain (f) by embedding (c) into the frequency domain of the

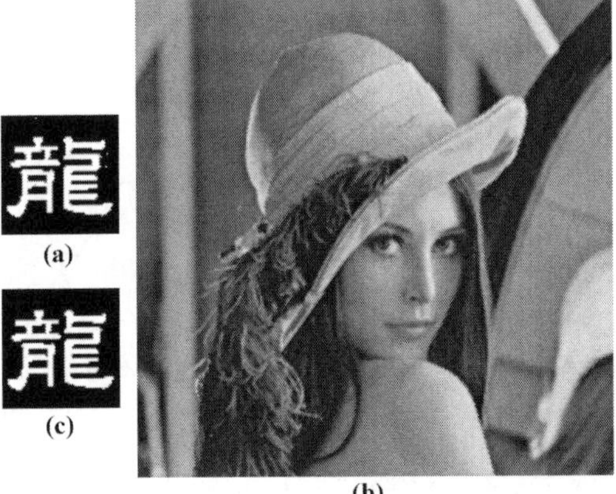

Figure 12.28 The traditional technique embedding a 64 × 64 watermark into Lena image.

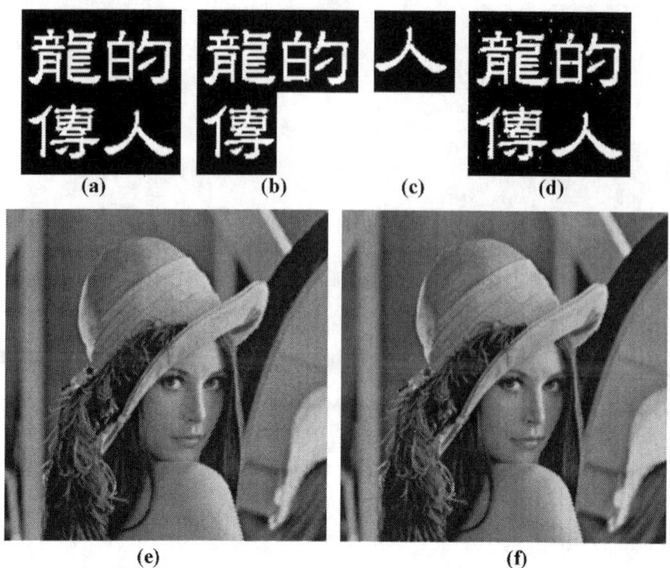

Figure 12.29 The results of embedding a 128 × 128 watermark into Lena image.

watermarked Lena image in (e). Figure 12.29(d) shows the extracted watermark from (f).

Figure 12.30(a) shows a larger watermark, a 256 × 256 image, which is split into two parts as in Figure 12.25. Figure 12.30(c) and (d) shows the watermarked images after embedding watermark in the spatial and frequency domains of a host image, respectively. Figure 12.30(b) shows the extracted watermark from (d).

We can apply the error measures, such as *normalized correlation* (NC) and *peak signal-to-noise ratio* (PSNR), to compute the image distortion after watermarking. The correlation between two images is often used in feature detection. Normalized correlation can be used to locate a pattern on a target image that best matches the specified reference pattern from the registered image base. Let $h(i,j)$ denote the original image and $h^*(i,j)$ denote the modified image. The normalized correlation is defined as

$$\mathrm{NC} = \frac{\sum_{i=1}^{N}\sum_{j=1}^{N} h(i,j)h^*(i,j)}{\sum_{i=1}^{N}\sum_{j=1}^{N} [h(i,j)]^2} \tag{12.8}$$

PSNR is often used in engineering to measure the signal ratio between the maximum power and the power of corrupting noise. Because signals possess a wide dynamic range, we apply the logarithmic decibel scale to limit its variation. It can measure the quality of reconstruction in image compression. However, it is a rough quality measure. In comparing two video files, we can calculate the mean PSNR.

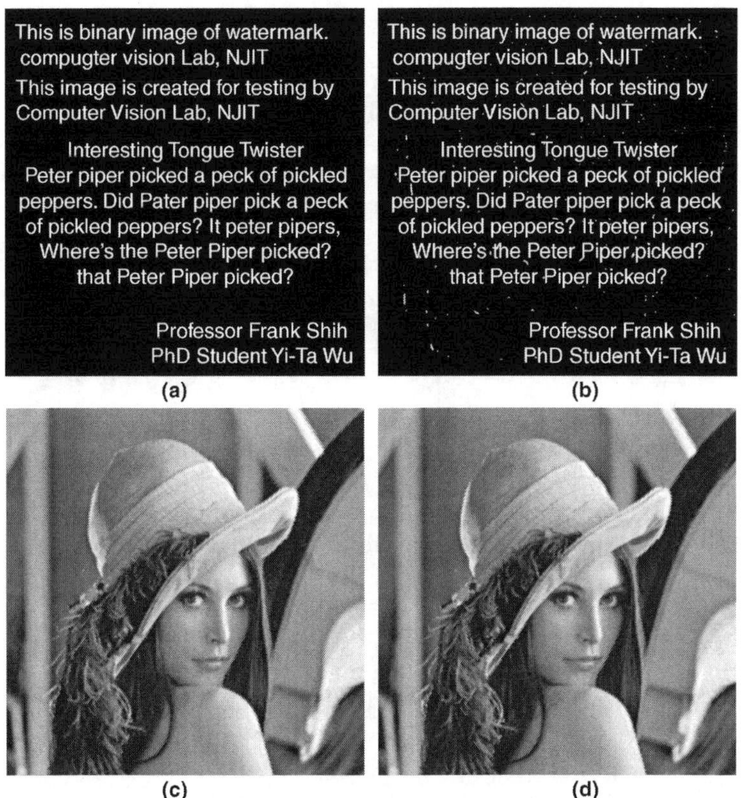

Figure 12.30 The results of embedding a 256 × 256 watermark into Lena image.

The peak signal-to-noise ratio is defined as

$$\text{PSNR} = 10\log_{10}\left(\frac{\sum_{i=1}^{N}\sum_{j=1}^{N}[h^*(i,j)]^2}{\sum_{i=1}^{N}\sum_{j=1}^{N}[h(i,j)-h^*(i,j)]^2}\right) \tag{12.9}$$

Table 12.2 presents the results of embedding watermarks of different sizes into a host image. The PSNR in the first step compares both the original and the embedded Lena images in the spatial domain. The PSNR in the second step compares both

TABLE 12.2 The Comparisons When Embedding Different Sized Watermarks into a Lena Image of Size 256 × 256

	64 × 64	128 × 128	256 × 256
PSNR in the first step	None	56.58	51.14
PSNR in the second step	64.57	55.93	50.98
NC	1	0.9813	0.9644

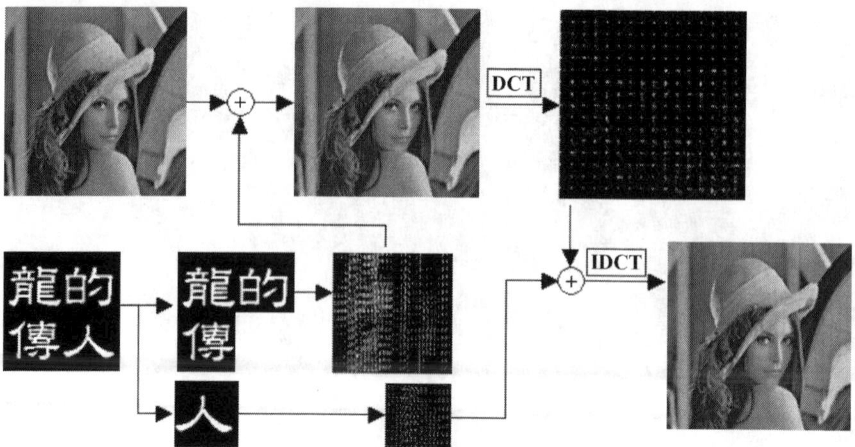

Figure 12.31 The procedures of randomly permuting the watermark.

images in the frequency domain. The NC compares both the original and the extracted watermarks.

12.6.5 Further Encryption of Combinational Watermarking

For the purpose of enhancing robustness, a random permutation of the watermark is used to defeat the attacks of image processing operations, such as image cropping. The procedure is illustrated in Figure 12.31.

A random sequence generator is used to relocate the order of sequential numbers (Lin and Chen, 2000). For example, the 12-bit random sequence generator is used to relocate the order of a watermark of size 64×64, as shown in Figure 12.32. We rearrange the bits 9 and 6 to the rear of the whole sequence, and the result is shown in Figure 12.32(b).

Figure 12.33 shows the result when a half of the Lena image is cropped, where (a) shows the original 128×128 watermark, (b) shows the cropped Lena image, and (c) shows the extracted watermark.

Table 12.3 shows the results when parts of an embedded host image with a watermark of size 128×128 are cropped. For example, if the size of 16×256 is cropped from a 256×256 embedded host image, the NC is 0.92. If a half of the embedded host image is cropped (i.e., eight of 16×256) as shown in Figure 12.32(b), the NC is 0.52.

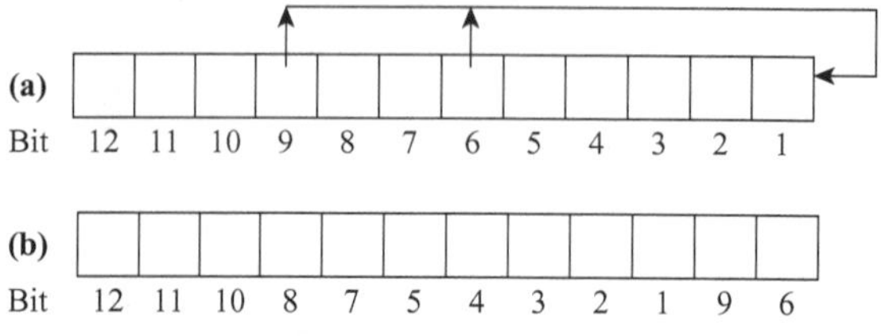

Figure 12.32 A 12-bit random sequence generator.

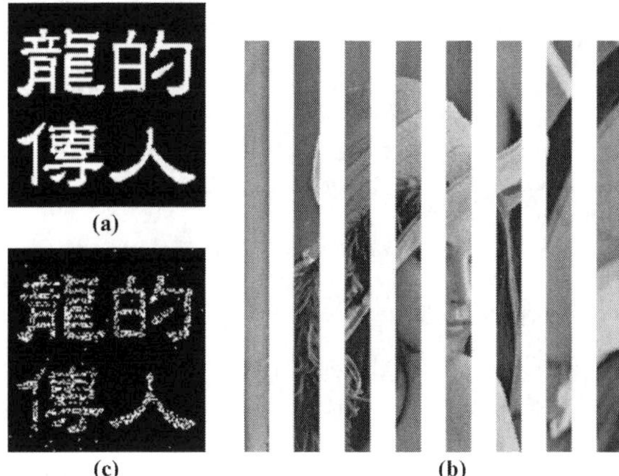

Figure 12.33 The results of cropping half of a Lena image.

TABLE 12.3 The Comparisons When Cropping Different Sizes of a Lena Image

Times of 64 × 64	1	2	3	4	5	6	7	8
NC	0.92	0.88	0.81	0.73	0.68	0.61	0.58	0.52

Generally, the important part of an image is not very large in size. We can cut the important part and embed it into the frequency domain and the rest into the spatial domain of a host image. Therefore, we can not only enlarge the size of watermark but also retain the property of security and imperceptibility. Combinational image watermarking possesses the following advantages (Shih and Wu, 2003). More watermark data can be inserted into the host image, so the capacity is increased. The splitting of the watermark into two parts makes the degree of protection double. The splitting strategy can be designed even more complicated to be unable to compose.

Another scheme of combining spatial-domain watermarking with wavelet-domain watermarking can be referred in Tsai et al. (2000). Employing the error control codes can increase the robustness of spatial-domain watermarking (Lancini et al., 2002), but the watermark capacity is reduced since the error control codes add some redundancy to the watermark.

REFERENCES

Ahmed, F. and Moskowitz, I. S., "Correlation-based watermarking method for image authentication applications," *Opt. Eng.*, vol. 43, no. 8, pp. 1833–1838, Aug. 2004.

Barni, M., Bartolini, F., Cappellini, V., and Piva, A., "A DCT-domain system for robust image watermarking," *Signal Process.*, vol. 66, no. 3, pp. 357–372, May 1998.

Bas, P., Chassery, J.-M., and Macq, B., "Image watermarking: an evolution to content based approaches," *Pattern Recogn.*, vol. 35, no. 3, pp. 545–561, Mar. 2002.

Berghel, H. and O'Gorman, L., "Protecting ownership rights through digital watermarking," *IEEE Comput. Mag.*, vol. 29, no. 7, pp. 101–103, July 1996.

Bi, N., Sun, Q., Huang, D., Yang, Z., and Huang, J., "Robust image watermarking based on multiband wavelets and empirical mode decomposition," *IEEE Image Process.*, vol. 16, no. 8, pp. 1956–1966, Aug. 2007.

Briassouli, A. and Strintzis, M. G., "Locally optimum nonlinearities for DCT watermark detection," *IEEE Image Process.*, vol. 13, no. 12, pp. 1604–1617, Dec. 2004.

Caronni, G., "Assuring ownership rights for digital images," *Proceedings of Reliable IT Systems*, Vieweg Publishing Company, Germany, pp. 251–263, Jan. 1995.

Celik, M. U., Sharma, G., Saber, E., and Tekalp, A. M., "Hierarchical watermarking for secure image authentication with localization," *IEEE Trans. Image Process.*, vol. 11, no. 6, pp. 585–595, June 2002.

Cox, I. J., Kilian, J., Leighton, T., and Shamoon, T., "Secure spread spectrum watermarking for images audio and video," Proceedings of the IEEE International Conference on Image Processing, Lausanne, Switzerland, vol. 3, pp. 243–246, Sept. 1996.

Cox, I. J., Kilian, J., Leighton, F. T., and Shamoon, T., "Secure spread spectrum watermarking for multimedia," *IEEE Trans. Image Process.*, vol. 6, no. 12, pp. 1673–1687, Dec. 1997.

Deguillaume, F., Voloshynovskiy, S., and Pun, T., "Secure hybrid robust watermarking resistant against tampering and copy attack," *Signal Process.*, vol. 83, no. 10, pp. 2133–2170, Oct. 2003.

Epstein, M. and McDermott, R., "Copy protection via redundant watermark encoding," U.S. Patent 7,133,534, 2006 (http://www.patentgenius.com/patent/71335312.html).

Ghouti, L., Bouridane, A., Ibrahim, M. K., and Boussakta, S., "Digital image watermarking using balanced multiwavelets," *IEEE Trans. Signal Process.*, vol. 51, no. 4, pp. 1519–1536, Apr. 2006.

Gray, R. M., "Vector quantization," *IEEE ASSP Mag.*, vol. 1, no. 2, pp. 4–29, Apr. 1984.

Holliman, M. and Memon, N., "Counterfeiting attacks on oblivious block-wise independent invisible watermarking schemes," *IEEE Trans. Image Process.*, vol. 9, no. 3, pp. 432–441, Mar. 2000.

Hu, Y., Kwong, S., and Huang, J., "An algorithm for removable visible watermarking," *IEEE Trans. Circuits Syst. Video Technol.*, vol. 16, no. 1, pp. 129–133, Jan. 2006.

Huang, C.-H. and Wu, J.-L., "Attacking visible watermarking schemes," *IEEE Trans. Multimedia*, vol. 6, no. 1, pp. 16–30, Feb. 2004.

Karybali, I. G. and Berberidis, K., "Efficient spatial image watermarking via new perceptual masking and blind detection schemes," *IEEE Trans. Inform. Forensics Security*, vol. 1, no. 2, pp. 256–274, June 2006.

Kumsawat, P., Attakitmongcol, K., and Srikaew, A., "A new approach for optimization in image watermarking by using genetic algorithms," *IEEE Trans. Signal Process.*, vol. 53, no. 12, pp. 4707–4719, Dec. 2005.

Lancini, R., Mapelli, F., and Tubaro, S., "A robust video watermarking technique for compression and transcoding processing," *Proceedings of the IEEE International Conference on Multimedia and Expo*, Lausanne, Switzerland, vol. 1, pp. 549–552, Aug. 2002.

Lee, S., Yoo, C. D., and Kalker, T., "Reversible image watermarking based on integer-to-integer wavelet transform," *IEEE Trans. Inform. Forensics Security*, vol. 2, no. 3, pp. 321–330, Sept. 2007.

Lin, D. and Chen, C.-F., "A robust DCT-based watermarking for copyright protection," *IEEE Trans. Consum. Electron.*, vol. 46, no. 3, pp. 415–421, Aug. 2000.

Lin, E. and Delp, E., "Spatial synchronization using watermark key structure," *Proceedings of the SPIE Conference on Security, Steganography, and Watermarking of Multimedia Contents*, San Jose, CA, pp. 536–547, Jan. 2004.

Lu, Z. and Sun, S., "Digital image watermarking technique based on vector quantization," *IEEE Electron. Lett.*, vol. 36, no. 4, pp. 303–305, Feb. 2000.

Lu, Z., Xu, D., and Sun, S., "Multipurpose image watermarking algorithm based on multistage vector quantization," *IEEE Trans. Image Process.*, vol. 14, no. 6, pp. 822–831, June 2005.

Macq, B., Dittmann, J., and Delp, E. J., "Benchmarking of image watermarking algorithms for digital rights management," *Proc. IEEE*, vol. 92, no. 6, pp. 971–984, June 2004.

Mukherjee, D., Maitra, S., and Acton, S., "Spatial domain digital watermarking of multimedia objects for buyer authentication," *IEEE Trans. Multimedia*, vol. 6, no. 1, pp. 1–15, Feb. 2004.

Nikolaidis, N. and Pitas, I., "Robust image watermarking in the spatial domain," *Signal Process.*, vol. 66, no. 3, pp. 385–403, May 1998.

Pi, H., Li, H., and Li, H., "A novel fractal image watermarking," *IEEE Trans. Multimedia*, vol. 8, no. 3, pp. 488–499, June 2006.

Pitas, I. and Kaskalis, T., "Applying signatures on digital images," *Proceedings of the IEEE Workshop on Nonlinear Signal and Image Processing*, Halkidiki, Greece, pp. 460–463, June 1995.

Podilchuk, C. I. and Delp, E. J., "Digital watermarking: algorithms and applications," *IEEE Signal Process. Mag.*, vol. 18, no. 4, pp. 33–46, July 2001.

Rivest, R. L., "The MD5 message digest algorithm," RFC 1321, MIT Laboratory for Computer Science and RSA Data Security, Inc., 1992.

Rivest, R. L., Shamir, A., and Adleman, L., "A method for obtaining digital signatures and public-key cryptosystems," *Commun. ACM*, vol. 21, no. 2, pp. 120–126, Feb. 1978.

Sang, J. and Alam, M. S., "Fragility and robustness of binary-phase-only-filter-based fragile/semifragile digital image watermarking," *IEEE Trans. Instrum. Meas.*, vol. 57, no. 3, pp. 595–606, Mar. 2008.

Seo, J. S. and Yoo, C. D., "Image watermarking based on invariant regions of scale-space representation," *IEEE Trans. Signal Process.*, vol. 54, no. 4, pp. 1537–1549, Apr. 2006.

Shih, F. Y. and Wu, S. Y. T., "Combinational image watermarking in the spatial and frequency domains," *Pattern Recogn.*, vol. 36, no. 4, pp. 969–975, Apr. 2003.

Sun, Q. and Chang, S.-F., "Semi-fragile image authentication using generic wavelet domain features and ECC," *Proceedings of the IEEE International Conference on Image Processing*, Rochester, NY, vol. 2, pp. 901–904, Sept. 2002.

Swanson, M. D., Zhu, B., and Tewfik, A. H., "Transparent robust image watermarking," *Proceedings of the IEEE International Conference on Image Processing*, Lausanne, Switzerland, vol. 3, pp. 211–214, Sept. 1996.

Tsai, M. J., Yu, K. Y., and Chen, Y. Z., "Joint wavelet and spatial transformation for digital watermarking," *IEEE Trans. Consum. Electron.*, vol. 46, no. 1, pp. 237–241, Feb. 2000.

Tsui, T. K., Zhang, X.-P., and Androutsos, D., "Color image watermarking using multidimensional Fourier transforms," *IEEE Trans. Inform. Forensics Security*, vol. 3, no. 1, pp. 16–28, Mar. 2008.

Wang, X., Wu, J., and Niu, P., "A new digital image watermarking algorithm resilient to desynchronization attacks," *IEEE Trans. Inform. Forensics Security*, vol. 2, no. 4, pp. 655–663, Dec. 2007.

Wolfgang, R. and Delp, E., "A watermark for digital images," *Proceedings of the IEEE International Conference on Image Processing*, Lausanne, Switzerland, pp. 219–222, Sept. 1996.

Wolfgang, R. and Delp, E., "A watermarking technique for digital imagery: further studies," *Proceedings of the International Conference on Imaging Science, Systems and Technology*, Las Vegas, NV, pp. 279–287, June 1997.

Wong, P. W., "A watermark for image integrity and ownership verification," *Proceedings of the IS&T PICS Conference*, Portland, OR, pp. 374–379, 1998a.

Wong, P. W., "A public key watermark for image verification and authentication," *Proceedings of the IEEE International Conference on Image Processing*, Chicago, IL, vol. 1, pp. 455–459, Oct. 1998b.

Xiang, S., Kim, H. J., and Huang, J., "Invariant image watermarking based on statistical features in the low-frequency domain," *IEEE Trans. Circuits Syst. Video Technol.*, vol. 18, no. 6, pp. 777–790, June 2008.

Zhao, K. E., "Embedding robust labels into images for copyright protection," Technical Report, Fraunhofer Institute for Computer Graphics, Darmstadt, Germany, 1994.

CHAPTER 13

IMAGE STEGANOGRAPHY

Digital steganography aims at hiding digital information into covert channels, so that one can conceal the information and prevent the detection of the hidden message. Steganalytic systems are used to detect whether an image contains a hidden message. By analyzing various image features of stego images (i.e., the images containing hidden messages) and cover images (i.e., the images containing no hidden messages), a steganalytic system is able to detect stego images. Steganography and cryptography are used for hiding data. Cryptography is the practice of scrambling a message to an obscured form to prevent others from understanding it. Steganography is the study of obscuring the message, so that it cannot be detected.

As shown in Figure 13.1, a famous classic steganographic model presented by Simmons (1984) is the prisoners' problem that Alice and Bob in a jail plan to escape together. All communication between them is monitored by Wendy, a warden. Therefore, they must hide the messages in other innocuous-looking medium (cover object) to obtain the stego object. Then, the stego object is sent through the public channel. Wendy is free to inspect all the messages between Alice and Bob with two options, passive or active. The passive way is to inspect the message to determine whether it contains a hidden message and then take a proper action. On the other hand, the active way is always to alter messages though Wendy may not perceive any trace of a hidden message. An example of the active method uses image processing operations, such as lossy compression, quality factor alteration, format conversion, palette modification, and low-pass filtering.

For steganographic systems, the fundamental requirement is that the stego object should be perceptually indistinguishable to the degree that it does not raise suspicion. In other words, the hidden information introduces only slight modification to the cover object. Most passive warden distinguishes the stego images by analyzing their statistic features. Steganalysis is the art of detecting the hidden information. In general, steganalytic systems can be categorized into two classes: spatial-domain steganalytic system (SDSS) and frequency-domain steganalytic system (FDSS). The SDSS (Avcibas et al., 2003; Westfeld and Pfitzmann, 1999) is adopted for checking the lossless compressed images by analyzing the spatial-domain statistic features. For the lossy compressed images such as JPEG, the FDSS (Farid, 2001; Fridrich et al., 2003) is used to analyze the frequency-domain statistic features. Westfeld and Pfitzmann (1999) presented two SDSSs based on visual and chi-square attacks. The visual attack uses human eyes to inspect stego images by checking their

Image Processing and Pattern Recognition by Frank Shih

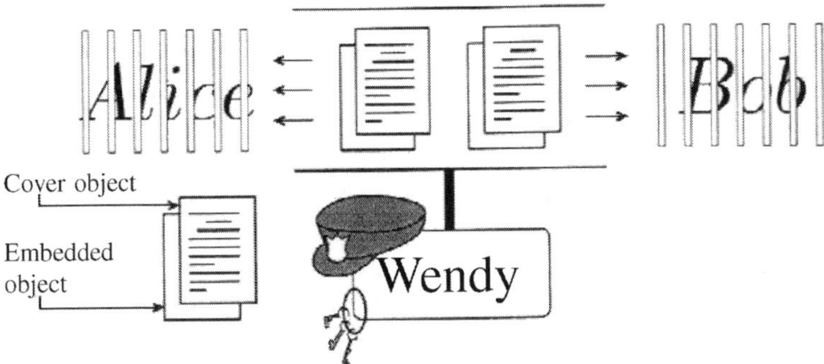

Figure 13.1 A classic steganographic model.

lower bit planes. The chi-square attack can automatically detect the specific characteristic generated by the least significant bit (LSB) steganographic technique.

Wang and Chen (2006) presented an image steganography method that utilizes a two-way block-matching procedure to search for the highest similarity block for each block of the important image. Martin et al. (2005) experimentally investigated if stego images, bearing a secret message, are statistically "natural." Dumitrescu and Wu (2005) proposed a general framework for the detection of the least significant bit steganography using digital media files as cover objects. Satish et al. (2004) proposed a scheme of chaos-based spread spectrum image steganography involving the use of chaotic encryption and chaotic modulation in spread spectrum image steganography. Marvel et al. (1999) presented a method of digital steganography, called spread spectrum image steganography.

The term "steganography" is derived from two Greek words "steganos," meaning "covered," and "graphein," meaning "to write." It is intended to hide the information in a medium in such a manner that no one else except the anticipated recipient knows about the existence of the information. Being a child, we might experience writing an invisible message in lemon juice and asking people to find out the secret message by holding it against a light bulb. Hence, we might have already used the technology of steganography in the past.

The history of steganography could be traced back to around 440 B.C., when the Greek historian Herodotus described in his writings about two events: one used wax to cover secret messages and the other used shaved heads. In ancient China, military generals and diplomats hid secret messages on thin sheets of silk or paper. One famous story on the successful revolt of Han Chinese against the Mongolian during the Yuan dynasty is the use of the steganographic technique. During the mid-Autumn festival, the Han people inserted a message (as hidden information) inside moon cakes (as cover objects) and gave these cakes to their members to deliver the information of the planned revolt. Nowadays, with the advent of modern computer technology, a large number of steganographic algorithms took the old-fashioned steganographic techniques to reform the state-of-the-art information hiding, watermarking, and steganography literature.

Modern techniques in steganography have far more powerful tools. Many software tools allow a paranoid sender to embed messages in digitized information, typically audio, video, or still image files that are sent to a recipient. Although steganography has attracted great interests from the military and governmental organizations, there is even a big interest shown by commercial companies to safeguard their information from piracy. Today, steganography is often used to transmit information safely and embed trademarks securely in images and music files to assure copyright. In this chapter, we will introduce types of steganography, applications of steganography, embedding security and imperceptibility, examples of steganography software, and genetic algorithm (GA)-based steganography.

This chapter is organized as follows. Section 13.1 presents types of steganography, and Section 13.2 describes the applications of steganography. Section 13.3 states embedding security and imperceptibility. Section 13.4 introduces some examples of steganography software. Section 13.5 proposes a GA-based steganography method.

13.1 TYPES OF STEGANOGRAPHY

Steganography can be divided into three types: technical steganography, linguistic steganography, and digital steganography. Technical steganography applies scientific methods to conceal the secret messages while linguistic steganography uses written natural language. Digital steganography, developed with the advent of computers, employs computer files or digital multimedia data. In this section, we describe each type separately.

13.1.1 Technical Steganography

Technical steganography applies scientific methods to conceal a secret message, for example, the use of invisible ink, microdots, and shaved heads. The invisible ink is the simplest method to write secret information on a piece of paper, because when it dries, the written information disappears. The paper looks the same as the original blank piece of paper. We could use organic compounds, such as milk, urine, vinegar, or fruit juice, for writing. When we apply heat or ultraviolet light, the dry message will become dark and visible to human eyes. In 1641, Bishop John Wilkins invented the invisible ink using onion juice, alum, and ammonia salts. Modern invisible inks fluoresce under ultraviolet light and are used as counterfeit detection devices.

With the advent of photography, microfilm was created as a medium of recording a great amount of information in a very small piece. In World War II, the Germans used "microdots" to convey secret information. The technique is famous as "the enemy's masterpiece of espionage" called by J. Edgar Hoover. The secret message was first photographed and then reduced to the size of a printed period to be pasted onto innocuous host documents, such as newspapers or magazines. Microdots are printed as many small dots. Because their sizes are very small, people would not pay attention to them. However, with the transmission of enormous amounts of microdots, one can hide secret messages, drawings, or even images. Another method is to use printers to print a large amount of small yellow dots that are hardly visible under normal white light illumination. The pattern of small yellow dots can construct secret messages. When we place it under light of different color, for example, blue light, the pattern of yellow dots becomes visible.

Steganography using shaved heads started when Histaeus, the ruler of Miletus, planned to send a message to his friend Aristagorus, urging his revolt against the Persians. He shaved the head of his most trusted slave as the messenger, and then tattooed a message or a symbol on the messenger's head. After the hair grew back to normal, the messenger ran to Aristagorus to deliver the hidden message. When he arrived, his head was shaved again to reveal the secret message.

Later in Herodotus' history, the Spartans received the secret message from Demaratus that Xerxes of a Persian army was preparing to invade Greece. Demaratus was a Greek in exile in Persia. Fearing discovery, he securely concealed the secret message by scraping the wax off one of the writing tablets, inscribing the message on the wood, and then recovering the tablet with wax to make it look normal. These tablets were delivered to the Spartans who upon receiving the message hastened to rescue Greece.

13.1.2 Linguistic Steganography

Linguistic steganography utilizes written natural language to hide information. It can be categorized into semagrams and open codes.

Semagrams hide secret messages using visual symbols or signs. Modern updates to this method use computers to hide the secret message making them less noticeable. For example, we can alter the font size of specific letters or change their locations a little bit higher or lower to conceal messages. We can also add extra spaces in specific locations of the text or adjust spacing of lines. Semagrams can also be encoded in pictures. The most well known form of this is the secret message in *The Adventure of the Dancing Men* by Sir Arthur Conan Doyle.

Semagrams as described above are sometimes not secure enough. Another secret method is used when spies want to set up meetings or pass information to their networks. This includes hiding information in everyday matters, such as newspapers, dates, clothing, or conversation. Sometimes meeting times can be hidden in reading materials. In one case, an advertisement for a car was placed in a city newspaper during a specific week, which read "Toyota Camry, 1996, needs engine work, sale for $1500." If the spy saw this advertisement, he/she knew that the Japanese contact person wanted to have personal communication immediately.

Open codes hide the secret messages in a specifically designed pattern on the document that is not obvious to the average reader. An example of an open code used by a German spy in World War II is

> "Apparently neutral's protest is thoroughly discounted and ignored. Isman hard hit. Blockade issue affects pretext for embargo on by-products, ejecting suets and vegetable oils."

If we collect the second letter in each word, we can extract the secret message as

> "Pershing sails from NY June 1."

13.1.3 Digital Steganography

Computer technology has made steganography a lot easier to hide messages and more difficult to discover the messages. Digital steganography is the science of hiding secret messages within digital media, such as digital images, audio files, or video files. There are many different methods for digital steganography including least significant bit substitution, message scattering, masking and filtering, and image processing functions.

Secret information or images can be hidden in image files because image files are often very large. When a picture is scanned, a sampling process is conducted to quantize the picture into a discrete set of real numbers. These samples are the gray levels at an equally spaced array of pixels. The pixels are quantized to a set of discrete gray-level values, which are also taken to be equally spaced. The result of sampling and quantizing produces a digital image. For example, if we use 8-bit quantization (i.e., 256 gray levels) and 500-line squared sampling, it would generate an array of 250,000 8-bit numbers. It means that a digital image will need 2 million bits.

We can also hide the secret messages on a hard drive in a secret partition. The hidden partition is invisible although it can be accessed by disk configuration and other tools. Another method is to use the network protocols. By using the secret protocol, which includes the sequence number field in transmission control segments and the identification field in Internet protocol packets, we can construct the covert communication channels.

Modern digital steganographic software employs sophisticated algorithms to conceal secret messages. It reflects a particularly significant threat nowadays due to a large number of digital steganography tools freely available on the Internet that can be used to hide any digital file inside another digital file. With the easy access and simple usage, criminals are inclined to conceal their activities in cyberspace. It has been reported that even the Al Qaeda terrorists used digital steganography tools to deliver messages. With the devastating attack on the World Trade Center in New York City on September 11, 2001, there were indications that the Al Qaeda terrorists had used steganography to conceal their correspondence during the planning of the attack. Therefore, digital steganography presents a grand challenge to law enforcement as well as industry because detecting and extracting hidden information is very difficult.

13.2 APPLICATIONS OF STEGANOGRAPHY

13.2.1 Covert Communication

A famous classic steganographic model presented by Simmons (1984) is the prisoners' problem although it was originally introduced to describe a cryptography scenario. Two inmates, Alice and Bob, are locked in separate jail cells and intend to communicate a secret plan to escape together. All communication between them is monitored by Wendy, a warden. Alice and Bob know that Wendy will definitely terminate their communication if she discovers something strange in communication channels.

Fitting in the steganographic model, Alice and Bob must hide the escape messages in other innocuous-looking medium, called cover object, to construct the stego object for Wendy's inspection. Then, the stego object is sent through the public channel for communication. Wendy is free to inspect all the messages between Alice and Bob with two options, *passive* or *active*. The passive way is to inspect the stego object carefully to determine whether it contains a hidden message and then take a proper action. On the other hand, the active way is always to alter the stego object for destroying hidden messages, although Wendy may not perceive any trace of a hidden message.

The basic applications of steganography are related to secret communication. Modern computer and networking technology allows individual, group, and company to host a web page that may contain secret information meant for another. Anyone can download the web page; however, the hidden information is invisible and does not draw any attention. The extraction of secret information requires specific software with the correct keys. Adding encryption to the secret information would further enhance its security. The situation is similar to that of hiding some important documents or valuable merchandize in a very secure safe. Furthermore, we hide the safe in a secret place that is difficult to find.

All digital data files can be used for steganography, but the files containing a superior degree of redundancy are more appropriate. Redundancy is defined as the number of bits for an object to provide necessarily accurate representation. If we remove the redundant bits, the object would look the same. Digital images and audios mostly contain a great amount of redundant bits. Therefore, they are often taken as the cover objects.

13.2.2 One-Time Pad Communication

The *one-time pad* was developed in cryptography to provide a random private key that can be used only once for encrypting a message, and then we can decrypt it using a one-time matching pad and key. It is generated using a string of numbers or characters having the length of the longest message sent. The random number generator is used to generate the string of values randomly. These values are stored on a pad or a device for someone to use. Then, the pads are delivered to the sender and the receiver. Usually, a collection of secret keys is delivered and can be used only once in such a manner that there is one key for each day in a month, and the key will expire at the end of each day.

Messages encrypted by a randomly generated key possess the advantage that there is no solution theoretically to uncover the code by analyzing a series of the messages. All encryptions are different in nature and bear no relationship. This kind of encryption is referred to as a 100% noise source for hiding the message. Only the sender and receiver are able to delete the noise. We should note that the one-time pad can be used only once for security reasons. If it is reused, someone may perform comparisons of multiple messages to extract the key for deciphering the messages. One-time pad technology was adopted notably in secret communication during World War II and the Cold War. In today's Internet communication, it is also used in the public key cryptography.

The one-time pad is the only encryption method that can be mathematically proved to be unbreakable. The technique can be used to hide an image by splitting it into two random layers of dots. When they are superimposed, the image appears. We can also generate two innocuous-looking images in such a way that putting both images together reveals the secret message. The encrypted data or images are perceived as a set of random numbers. It is very obvious that no warden will permit the prisoners to exchange encrypted random malicious-looking messages. Therefore, the one-time pad encryption, though statistically secure, will be practically impossible to use in this scenario.

13.3 EMBEDDING SECURITY AND IMPERCEPTIBILITY

A robust steganographic system must be extremely secure from all kinds of attacks (i.e., steganalysis) and must not degrade the visual quality of cover images. We can combine several strategies to ensure the security and imperceptibility. For security, we can use the chaotic mechanism (CM), frequency hopping (FH) structure, pseudo-random number generator (PNG), and patchwork locating algorithm (PLA). For imperceptibility, we can use the set partitioning in hierarchical trees (SPIHT) coding, discrete wavelet transform (DWT), and parity check embedding (PCE).

Benchmarking of steganographic techniques is a complicated task that requires examination of a set of mutually dependent performance indices. A benchmarking tool should be able to interact with different performance aspects, such as visual quality and robustness. The input images to the benchmarking system should contain images that vary in size and frequency contents since these factors affect the system performance.

We should also evaluate the execution time of the embedding and detection modules. Mean, maximum, and minimum execution times for the two modules might be evaluated over the set of all keys and messages for each host image. We should measure the perceptual quality of stego images by subjective quality evaluation and by the quantitative way that correlates well with the way human observers perceive image quality. We should also evaluate the maximum number of information bits that can be embedded per host image pixel.

Since steganography is expected to be robust to host image manipulations, tests for judging the performance of a steganographic technique when applied on distorted images constitute an important part of a benchmarking system. The set of attacks available in a benchmarking system should include all operations that the average user or an intelligent pirate can use to make the embedded messages undetectable. It should also include signal processing operations and distortions that occur during normal image usage, transmission, storage, and so on.

13.4 EXAMPLES OF STEGANOGRAPHY SOFTWARE

Steganography embeds secret messages into innocuous files. Its goal is to not only hide the message but also let the stego object pass without causing any suspicion.

There are many steganographic tools available for hiding messages in images, audio files, and video files. In this section, we briefly introduce S-Tools, StegoDos, EzStego, and JSteg-Jpeg.

13.4.1 S-Tools

The S-Tools (Brown, 1994), standing for steganography tools, was written by Andy Brown to hide secret messages in BMP, GIF, and WAV files. It is a combined steganographic and cryptographic product since the messages to be hidden are encrypted using symmetric keys. It uses the LSB substitution in files that employ lossless compression, such as 8- or 24-bit color and pulse code modulation. It also applies a pseudorandom number generator to make the extraction of secret messages more difficult and provides the encryption and decryption of hidden files with several different encryption algorithms.

We can open up a copy of S-Tools and drag images and audios across it. To hide files we can drag them over open audio or image windows. We can hide multiple files in one audio or image and compress the data before encryption. The multithreaded operation provides the flexibility of many hide/reveal operations running simultaneously without interfering with other work. We can also close the original image or audio without affecting the ongoing threads.

13.4.2 StegoDos

StegoDos (Wolf, 2006), also known as Black Wolf's Picture Encoder, consists of a set of programs to capture an image, encode a secret message, and display the stego image. This stego image may also be captured again into another format using a third-party program. Then, we can recapture it and decode the hidden message. It works only with 320×200 images with 256 colors. It also uses the LSB substitution to hide messages.

13.4.3 EzStego

EzStego, developed by Romana Machado, simulates the invisible ink for Internet communication. It hides an encrypted message in a GIF format image file by modulating the least significant bits. It begins by sorting the colors in the palette, so closest colors fall next to each other. Then similar colors are paired up, and for each pair one color will represent 1 while the other will represent 0. It encodes the encrypted message by replacing the LSB. EzStego compares the bits it wants to hide with the color of each pixel. If the pixel already represents the correct bit, it remains untouched. If the pixel represents the incorrect bit, the color would be changed to its pair color. The changes of least significant bits are so small that they are undetectable to the human eyes. EzStego treats the colors as cities in the three-dimensional RGB space and intends to search for the shortest path through all of the stops.

13.4.4 JSteg-Jpeg

JSteg-Jpeg (Korejwa, 2006), developed by Derek Upham, can read multiple format images and embed the secret messages to be saved as the JPEG format images.

It utilizes the splitting of the JPEG encoding into lossy and lossless stages. The lossy stages use DCT and a quantization step to compress the image data, and the lossless stage uses Huffmann coding to further compress the image data. It therefore inserts the secret message into the image data between these two steps. It also modulates the rounding processes in the quantized DCT coefficients. The image degradation is affected by the embedded amount as well as by the quality factor used in the JPEG compression. The trade-off between the two can be adjusted to allow the imperceptible requirement.

13.5 GENETIC ALGORITHM-BASED STEGANOGRAPHY

For steganographic systems, the fundamental requirement is that the stego object should be perceptually indistinguishable to the degree that it does not raise suspicion. In other words, the hidden information introduces only slight modification to the cover object. The most passive warden distinguishes the stego images by analyzing their statistic features. Since the steganalytic system analyzes certain statistic features of an image, the idea of developing a robust steganographic system is to generate the stego image by avoiding changes in the statistic features of the cover image. In the literature, several papers have presented the algorithms for steganographic and steganalytic systems. Very few papers have discussed the algorithms for breaking the steganalytic systems. Recently, Chu et al. (2004) presented a DCT-based steganographic system by utilizing the similarities of DCT coefficients between the adjacent image blocks where the embedding distortion is spread. Their algorithm can allow random selection of DCT coefficients to maintain key statistic features. However, the drawback of their approach is that the capacity of the embedded message is limited, that is, only two bits for an 8×8 DCT block.

In this chapter, we present a GA-based method for breaking steganalytic systems. The emphasis is shifted from traditionally avoiding the change of statistic features to artificially counterfeiting the statistic features. Our idea is based on the following: to manipulate the statistic features for breaking the inspection of steganalytic systems, the GA-based approach is adopted to counterfeit several stego images (candidates) until one of them can break the inspection of steganalytic systems.

13.5.1 Overview of the GA-Based Breaking Methodology

The genetic algorithm, introduced by Holland (1975) in his seminal work, is commonly used as an adaptive approach that provides a randomized, parallel, and global search based on the mechanics of natural selection and genetics to find solutions to a problem.

In general, the genetic algorithm starts with some randomly selected genes as the first generation, called *population*. Each individual in the population corresponding to a solution in the problem domain is called *chromosome*. An objective, called *fitness function*, is used to evaluate the quality of each chromosome. The chromosomes of high quality will survive and form a new population of the next generation. By using the three operators, reproduction, crossover, and mutation, we recombine

g_0	g_1	g_2	g_3	g_4	g_5	g_6	g_7
g_8	g_9	g_{10}	g_{11}	g_{12}	g_{13}	g_{14}	g_{15}
g_{16}	g_{17}	g_{18}	g_{19}	g_{20}	g_{21}	g_{22}	g_{23}
g_{24}	g_{25}	g_{26}	g_{27}	g_{28}	g_{29}	g_{30}	g_{31}
g_{32}	g_{33}	g_{34}	g_{35}	g_{36}	g_{37}	g_{38}	g_{39}
g_{40}	g_{41}	g_{42}	g_{43}	g_{44}	g_{45}	g_{46}	g_{47}
g_{48}	g_{49}	g_{50}	g_{51}	g_{52}	g_{53}	g_{54}	g_{55}
g_{56}	g_{57}	g_{58}	g_{59}	g_{60}	g_{61}	g_{62}	g_{63}

(a)

0	0	1	0	0	2	0	1
1	2	0	0	1	1	0	1
2	1	1	0	0	−2	0	1
3	1	2	5	0	0	0	0
0	2	0	0	0	1	0	0
0	−1	0	2	1	1	4	2
1	1	1	0	0	0	0	0
0	0	0	0	1	−2	1	3

(b)

Figure 13.2 The numbering positions corresponding to 64 genes.

a new generation to find the best solution. The process is repeated until a predefined condition is satisfied or a constant number of iterations are reached. The predefined condition in this paper is the situation when we can correctly extract the desired hidden message.

To apply the genetic algorithm for embedding messages into the frequency domain of a cover image to obtain the stego image, we use the chromosome ξ consisting of n genes as $\xi = g_0, g_1, g_2, \ldots, g_n$. Figure 13.2 shows an example of a chromosome ($\xi \in Z^{64}$) containing 64 genes ($g_i \in Z$ (integers)). Figure 13.2(a) shows the distribution order of a chromosome in an 8×8 block, and Figure 13.2(b) shows an example of the corresponding chromosome.

The chromosome is used to adjust the pixel values of a cover image to generate a stego image, so the embedded message can be correctly extracted and at the same time the statistic features can be retained to break steganalytic systems. A fitness function is used to evaluate the embedded message and statistic features.

Let C and S, respectively, denote the cover and the stego images of size 8×8. We generate the stego images by adding the cover image and the chromosome as

$$S = \{s_i | s_i = c_i + g_i, \quad \text{where } 0 \leq i \leq 63\} \tag{13.1}$$

Fitness function: To embed messages into DCT-based coefficients and avoid the detection of steganalytic systems, we develop a fitness function to evaluate the following two terms:

1. Analysis(ξ, C): The analysis function evaluates the difference between the cover image and the stego image to maintain the statistic features. It is related to the type of the steganalytic systems used and will be explained in Sections 13.2 and 13.3.
2. BER(ξ, C): The *bit error rate* (BER) sums up the bit differences between the embedded and the extracted messages. It is defined as

$$\text{BER}(\xi, C) = \frac{1}{|\text{Message}^{\text{H}}|} \sum_{i=0}^{\text{all pixels}} |\text{Message}_i^{\text{H}} - \text{Message}_i^{\text{E}}| \tag{13.2}$$

where $\text{Message}^{\text{H}}$ and $\text{Message}^{\text{E}}$ denote the embedded and the extracted binary messages, respectively, and $|\text{Message}^{\text{H}}|$ denotes the length of the message. For example, if $\text{Message}^{\text{H}} = 11111$ and $\text{Message}^{\text{E}} = 10101$, then $\text{BER}(\xi, C) = 0.4$.

We use a linear combination of the analysis and the bit error rate to be the fitness function as

$$\text{Evaluation}(\xi, C) = \alpha_1 \times \text{Analysis}(\xi, C) + \alpha_2 \times \text{BER}(\xi, C) \tag{13.3}$$

where α_1 and α_2 denote weights. The weights can be adjusted according to the user's demand on the degree of distortion to the stego image or the extracted message.

Reproduction:

$$\text{Reproduction}(\Psi, k) = \{\xi_i | \text{Evaluation}(\xi_i, C) \leq \Omega \quad \text{for } \xi_i \in \Psi\} \tag{13.4}$$

where Ω is a threshold for sieving chromosomes, and $\Psi = \{\xi_1, \xi_2, \ldots, \xi_n\}$. It is used to reproduce k better chromosomes from the original population for higher qualities.

Crossover:

$$\text{Crossover}(\Psi, l) = \{\xi_i \Theta \xi_j | \xi_i, \xi_j \in \Psi\} \tag{13.5}$$

where Θ denotes the operation of generating chromosomes by exchanging genes from their parents: ξ_i and ξ_j. It is used to gestate l better offspring by inheriting healthy genes (i.e., higher qualities in the fitness evaluation) from their parents. The often-used crossovers are one-point, two-point, and multipoint crossovers. The criteria of selecting a suitable crossover depend on the length and structure of chromosomes. We adopt the one- and two-point crossovers as shown in Shih and Wu (2005).

Mutation:

$$\text{Mutation}(\Psi, m) = \{\xi_i \circ j | \, 0 \leq j \leq |\xi_i| \text{ and } \xi_i \in \Psi\} \tag{13.6}$$

where $\circ$ denotes the operation of randomly selecting a chromosome ξ_i from Ψ and changing the jth bit from ξ_i. It is used to generate m new chromosomes. The mutation is usually performed with a probability p $(0 < p \leq 1)$, meaning only p portion of the genes in a chromosome will be selected to be mutated. Since the length of a chromosome is 64 in this chapter and there are only one or two genes to be mutated when the GA mutation operator is performed, we select p to be 1/64 or 2/64.

Note that in each generation the new population is generated by the above three operations. The new population is actually of the same size $(k + l + m)$ as the original population. Note that, to break the inspection of the steganalytic systems, we use a straightforward GA selection method that the new generation is generated based on the chromosomes having superior evaluation values in the current generation. For example, only top 10% of the current chromosomes will be considered for the GA operations, such as reproduction, crossover, and mutation.

Mutation tends to be more efficient than crossover if a candidate solution is close to the real optimum solution. To enhance the performance of our GA-based methodology, we generate a new chromosome with desired minor adjustment when the previous generation is close to the goal. Therefore, the strategy of dynamically determining the ratio of three GA operations is utilized. Let R_P, R_M, and R_C denote the ratios of production, mutation, and crossover, respectively. In the beginning, we select small R_P and R_M and a large R_C to enable the global search. After certain iterations, we will decrease R_C and increase R_M and R_P to shift focus on local search if the current generation is better than the old one; otherwise, we will increase R_C and decrease R_M and R_P to enlarge the range of global search. Note that the following property must be satisfied: $R_P + R_M + R_C = 100\%$.

The recombination strategy of our GA-based algorithm is presented below. We apply the same strategy in recombining the chromosome through this chapter except that the fitness function is differently defined with respect to the properties of individual problem.

Algorithm for Recombining the Chromosomes

1. Initialize the base population of chromosomes, R_P, R_M, and R_C.
2. Generate candidates by adjusting pixel values of the original image.
3. Determine the fitness value of each chromosome.
4. If a predefined condition is satisfied or a constant number of iterations are reached, the algorithm will stop and output the best chromosome to be the solution; otherwise, go to the following steps to recombine the chromosomes.
5. If a certain number of iterations are reached, go to step 6 to adjust R_P, R_M, and R_C; otherwise, go to step 7 to recombine the new chromosomes.
6. If 20% of the new generation is better than the best chromosome in the preceding generation, then $R_C = R_C - 10\%$, $R_M = R_M + 5\%$, and $R_P = R_P + 5\%$; otherwise, $R_C = R_C + 10\%$, $R_M = R_M - 5\%$, and $R_P = R_P - 5\%$.
7. Obtain the new generation by recombining the preceding chromosomes using production, mutation, and crossover.
8. Go to step 2.

13.5.2 The GA-Based Breaking Algorithm on SDSS

As mentioned earlier, to generate a stego image to pass though the inspection of SDSS, the messages should be embedded into the specific positions for maintaining the statistic features of a cover image. In the spatial-domain embedding approach, it is difficult to select such positions since the messages are distributed regularly. On the other hand, if the messages are embedded into specific positions of coefficients of a transformed image, the changes in the spatial domain are difficult to predict. In this section, we intend to find the desired positions on the frequency domain that produce minimum statistic feature disturbance on the spatial domain. We apply the genetic algorithm to generate the stego image by adjusting the pixel values on the spatial domain using the following two criteria:

1. Evaluate the extracted messages obtained from the specific coefficients of a stego image to be as close as possible to the embedded messages.
2. Evaluate the statistic features of the stego image and compare them with those of the cover image so that the differences are as small as possible.

13.5.2.1 Generating the Stego Image on the Visual Steganalytic System

Westfeld and Pfitzmann (1999) presented a *visual steganalytic system* (VSS) that uses an assignment function of color replacement, called the *visual filter*, to efficiently detect a stego image by translating a grayscale image into binary. Therefore, to break the VSS, the two results VF^{C} and VF^{S} of applying the visual filter on the cover and the stego images, respectively, should be as identical as possible. The VSS was originally designed to detect the GIF format images by reassigning the color in the color palette of an image. We extend their method to detect the BMP format images as well by setting the odd- and even-numbered grayscales to black and white, respectively. The $\mathrm{Analysis}(\xi, C)$ to VSS indicating the sum of difference between $\mathrm{LSB}^{\mathrm{C}}$ and $\mathrm{LSB}^{\mathrm{S}}$ is defined as

$$\mathrm{Analysis}(\xi, C) = \frac{1}{|C|} \sum_{i=0}^{\text{all pixels}} (\mathrm{VF}_i^{\mathrm{C}} \oplus \mathrm{VF}_i^{\mathrm{S}}) \tag{13.7}$$

where $\oplus$ denotes the exclusive OR (XOR) operator. Our algorithm is described below.

Algorithm for Generating a Stego Image on VSS

1. Divide a cover image into a set of cover images of size 8×8.
2. For each 8×8 cover image, we generate a stego image based on the GA to perform the embedding procedure as well as to ensure that $\mathrm{LSB}^{\mathrm{C}}$ and $\mathrm{LSB}^{\mathrm{S}}$ are as identical as possible.
3. Combine all the 8×8 stego images together to form a complete stego image.

13.5.2.2 Generating the Stego Image on the IQM-Based Steganalytic System (IQM-SDSS)

Avcibas et al. (2003) proposed a steganalytic system by analyzing the *image quality measures* (IQMs) (Avcibas and Sankur, 2002) of the cover and the stego images. The IQM-SDSS consists of two phases: training and testing. In the training phase, the IQM is calculated between an image and its filtered image using a low-pass filter based on the Gaussian kernel. Suppose there are N images and q IQMs in the training set. Let x_{ij} denote the score in the ith image and the jth IQM, where $1 \leq i \leq N$ and $1 \leq j \leq q$. Let y_i be -1 or 1 indicating the cover or stego image, respectively. We can represent all the images in the training set as

$$\begin{aligned} y_1 &= \beta_1 x_{11} + \beta_2 x_{12} + \cdots + \beta_q x_{1q} + \varepsilon_1 \\ y_2 &= \beta_1 x_{21} + \beta_2 x_{22} + \cdots + \beta_q x_{2q} + \varepsilon_2 \\ &\vdots \\ y_N &= \beta_1 x_{N1} + \beta_2 x_{N2} + \cdots + \beta_q x_{Nq} + \varepsilon_N \end{aligned} \tag{13.8}$$

where $\varepsilon_1, \varepsilon_2, \ldots, \varepsilon_N$ denote random errors in the linear regression model (Rencher, 1995). Then the linear predictor $\beta = [\beta_1, \beta_2, \ldots, \beta_q]$ can be obtained from all the training images.

In the testing phase, we use the q IQMs to compute y_i to determine whether it is a stego image. If y_i is positive, the test image is a stego image; otherwise, it is a cover image.

We first train the IQM-SDSS to obtain the linear predictor, that is, $\beta = [\beta_1, \beta_2, \ldots, \beta_q]$, from our database. Then, we use β to generate the stego image by our GA-based algorithm, so that it can pass through the inspection of IQM-SDSS. Note that the GA procedure is not used in the training phase. The Analysis(ξ, C) to the IQM-SDSS is defined as

$$\text{Analysis}(\xi, C) = \beta_1 x_1 + \beta_2 x_2 + \cdots + \beta_q x_q \tag{13.9}$$

Algorithm for Generating a Stego Image on IQM-SDSS

1. Divide a cover image into a set of cover images of size 8×8.
2. Adjust the pixel values in each 8×8 cover image based on GA to embed messages into the frequency domain and ensure that the stego image can pass through the inspection of the IQM-SDSS. The procedures for generating the 8×8 stego image are presented next after this algorithm.
3. Combine all the 8×8 embedded images together to form a completely embedded image.
4. Test the embedded image on IQM-SDSS. If it passes, it is the desired stego image; otherwise, repeat steps 2–4.

Procedure for Generating an 8 × 8 Embedded Image on IQM-SDSS

1. Define the fitness function, the number of genes, the size of population, the crossover rate, the critical value, and the mutation rate.
2. Generate the first generation by a random selection.
3. Generate an 8×8 embedded image based on each chromosome.
4. Evaluate the fitness value for each chromosome by analyzing the 8×8 embedded image.
5. Obtain the better chromosome based on the fitness value.
6. Recombine new chromosomes by crossover.
7. Recombine new chromosomes by mutation.
8. Repeat steps 3–8 until a predefined condition is satisfied or a constant number of iterations are reached.

13.5.3 The GA-Based Breaking Algorithm on FDSS

Fridrich et al. (2003) presented a steganalytic system for detecting JPEG stego images based on the assumption that the histogram distributions of some specific AC DCT coefficients of a cover image and its cropped image should be similar. Note that, in the DCT coefficients, only the zero frequency (0, 0) is the DC component, and the remaining frequencies are the AC components. Let $h_{kl}(d)$ and $\overline{h}_{kl}(d)$, respectively, denote the total number of AC DCT coefficients in the 8×8 cover and its corresponding 8×8 cropped images with the absolute value equal to d at location (k, l), where $0 \leq k, l \leq 7$. Note that the 8×8 cropped images, defined in

Fridrich et al. (2003), are obtained in the same way as the 8×8 cover images with a horizontal shift by four pixels.

The probability, ρ_{kl}, of the modification of a nonzero AC coefficient at (k, l) can be obtained by

$$\rho_{kl} = \frac{\bar{h}_{kl}(1)\left[h_{kl}(0)-\bar{h}_{kl}(0)\right] + \left[h_{kl}(1)-\bar{h}_{kl}(1)\right]\left[\bar{h}_{kl}(2)-\bar{h}_{kl}(1)\right]}{\left[\bar{h}_{kl}(1)\right]^2 + \left[\bar{h}_{kl}(2)-\bar{h}_{kl}(1)\right]^2} \tag{13.10}$$

Note that the final value of the parameter ρ is calculated as an average over the selected low-frequency DCT coefficients $(k, l) \in \{(1,2),(2,1),(2,2)\}$, and only 0, 1, and 2 are considered when checking the coefficient values of the specific frequencies between a cover and its corresponding cropped image.

Our GA-based breaking algorithm on the JFDSS is intended to minimize the differences between the two histograms of a stego image and its cropped image. It is presented below.

Algorithm for Generating a Stego Image on JFDSS

1. Compress a cover image by JPEG and divide it into a set of small cover images of size 8×8. Each is performed by DCT.
2. Embed the messages into the specific DCT coefficients and decompress the embedded image by IDCT.
3. We select a 12×8 working window and generate an 8×8 cropped image for each 8×8 embedded image.
4. Determine the overlapping area between each 8×8 embedded image and its cropped image.
5. Adjust the overlapping pixel values by making the coefficients of some specific frequencies (k, l) of the stego image and its cropped image as identical as possible and the embedded messages are not altered.
6. Repeat steps 3–6 until all the 8×8 embedded images are generated.

Let $\text{Coef}^{\text{Stego}}$ and $\text{Coef}^{\text{Crop}}$ denote the coefficients of each 8×8 stego image and its cropped image, respectively. The Analysis(ξ, C) to the JFDSS is defined as

$$\text{Analysis}(\xi, C) = \frac{1}{|C|} \sum_{i=0}^{\text{all pixels}} \left(\text{Coef}_i^{\text{Stego}} \otimes \text{Coef}_i^{\text{Crop}}\right) \tag{13.11}$$

where $\otimes$ denotes the operator defined by

$$\begin{cases} \text{Coef}_i^{\text{Stego}} \otimes \text{Coef}_i^{\text{Crop}} = 1 & \text{if } (\text{Coef}_i^{\text{Stego}} = 0 \quad \text{and} \quad \text{Coef}_i^{\text{Crop}} \neq 0) \text{ or} \\ & (\text{Coef}_i^{\text{Stego}} = 1 \quad \text{and} \quad \text{Coef}_i^{\text{Crop}} \neq 1) \text{ or} \\ & (\text{Coef}_i^{\text{Stego}} = 2 \quad \text{and} \quad \text{Coef}_i^{\text{Crop}} \neq 2) \text{ or} \\ & (\text{Coef}_i^{\text{Stego}} \neq 0,1,2 \quad \text{and} \quad \text{Coef}_i^{\text{Crop}} = 0,1,2) \\ \text{Coef}_i^{\text{Stego}} \otimes \text{Coef}_i^{\text{Crop}} = 0 & \text{otherwise} \end{cases} \tag{13.12}$$

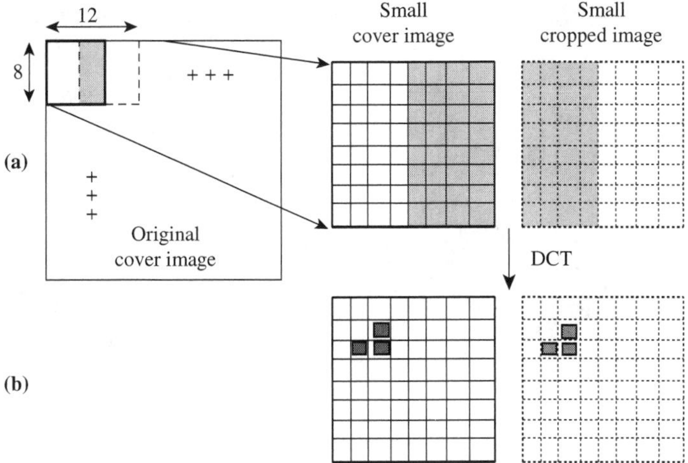

Figure 13.3 An example of our GA-based algorithm on the JFDSS.

Note that 0, 1, and 2 denote the values in the specific frequencies obtained by dividing the quantization table. We consider only the values of the desired frequencies to be 0, 1, 2, or some values in equation (13.12) because of the strategy of JFDSS in equation (13.10).

Figure 13.3 shows an example of our GA-based algorithm on the JFDSS. In Figure 13.3(a), we select a 12×8 working window for each 8×8 stego image and generate its 8×8 cropped image. Note that the shaded pixels indicate their overlapping area, and the three black boxes in Figure 13.3(b) are the desired locations.

13.5.4 Experimental Results

In this section, we provide experimental results to show that our GA-based steganographic system can successfully break the inspection of steganalytic systems. For testing our algorithm, we use a database of 200 grayscale images of size 256×256. All the images were originally stored in the BMP format.

13.5.4.1 The GA-Based Breaking Algorithm on VSS We test our algorithm on VSS. Figure 13.4(a) and (f) shows a stego image and a message image of sizes 256×256 and 64×64, respectively. We embed four bits into the 8×8 DCT coefficients on frequencies (0, 2), (1, 1), (2, 0), and (3, 0) to avoid distortion. Note that the stego image in Figure 13.4(a) is generated by embedding Figure 13.4(f) into the DCT coefficients of the cover image using our GA-based algorithm. Figure 13.4(b)–(j), respectively, displays the bit planes from 7 to 0. Figure 13.5 shows the stego image and its visual filtered result. It is difficult to determine that Figure 13.5(a) is a stego image.

Figure 13.6 shows the relationship of the average iteration for adjusting an 8×8 cover image versus the correct rate of the visual filter. The correct rate is the

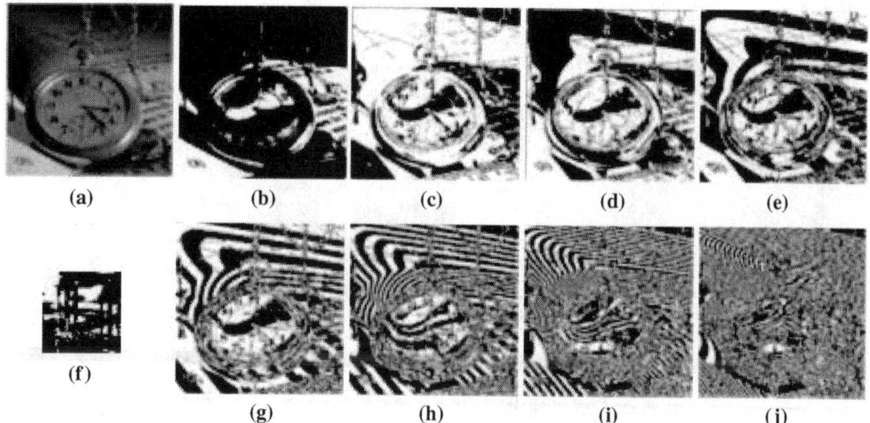

Figure 13.4 A stego image generated by our GA-based algorithm and its eight bit planes.

percentage of similarity between the transformed results of the cover and the stego images using the visual filter. Note that the BERs in Figure 13.5 are all 0%.

13.5.4.2 The GA-Based Breaking Algorithm on IQM-SDSS

We generate three stego images as the training samples for each cover image by Photoshop plug-in Digimarc (PictureMarc, 2006), Cox et al. (1997) Cox's technique (1997), and S-Tools (Brown, 1994). Therefore, there are in total 800 images of size 256 × 256, including 200 cover images and 600 stego images. The embedded message sizes are 1/10, 1/24, and 1/40 of the cover image size for Digimarc, Cox's technique, and S-Tools, respectively. Note that the IQM-SDSS (Avcibas et al., 2003) can detect the stego images containing the message size of 1/100 of the cover image. We develop the following four training strategies to obtain the linear predictors:

1. Train all the images in our database to obtain the linear predictor β^{A}.
2. Train 100 cover images and 100 stego images to obtain the linear predictor β^{B}, in which the stego images are obtained by Cox's technique.

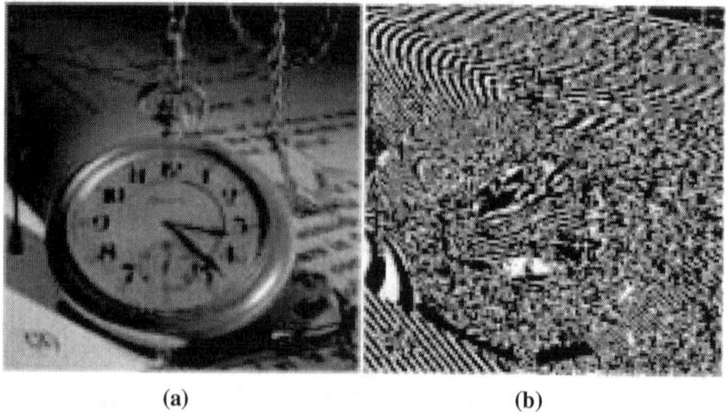

Figure 13.5 A stego image and its visual filtered result.

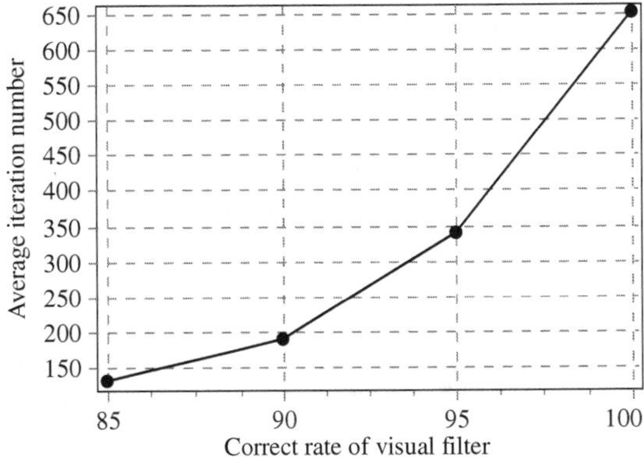

Figure 13.6 The relationship of the average iteration versus the correct rate of the visual filter.

3. Train 100 cover images and 100 stego images to obtain the linear predictor β^C, in which the stego images are obtained by Photoshop plug-in Digimarc.
4. Train 100 cover images and 100 stego images to obtain the linear predictor β^D, in which the stego images are obtained by S-Tools.

In the testing phase, we generate 50 stego images for each linear predictor. Therefore, we will obtain four types of stego images SI_A, SI_B, SI_C, and SI_D corresponding to β^A, β^B, β^C, and β^D, respectively. It is obvious that all the SI_A images will pass through the inspection of IQM-SDSS with the linear predictor β^A, but may fail with other linear predictors; similarly for the SI_B, SI_C, and SI_D as well.

Table 13.1 shows the false negative (type I error) and false positive (type II error) rates of testing each type of stego and cover images under different linear predictors. For example, the false negative (FN) rate of testing 50 SI_A stego images under β^B is 94% indicating that the β^B-IQMSS falsely decides that 94% of our stego images belong to the cover image. On the other hand, the false positive (FP) rate of testing 50 cover images under β^B is 10% indicating that the β^B-IQMSS decides that 10% of cover images belong to the stego images. Note that the cover images used in the testing phase are different from those used in the training phase.

13.5.4.3 The GA-Based Breaking Algorithm on JFDSS Figure 13.7 shows an example of adjusting an 8×8 embedded image to obtain an 8×8 stego image for breaking the JEDSS. Figure 13.7(a) shows a 12×8 working window, where the

TABLE 13.1 The Experimental Results of GA-Based Algorithm on IQM-SDSS

		β^A	β^B	β^C	β^D	Average
50 SI_A images	FN		94%	94%	100%	96%
50 SI_B images	FN	84%		84%	80%	83%
50 SI_C images	FN	82%	86%		88%	85%
50 SI_D images	FN	82%	86%	82%		83%
50 cover images	FP	6%	10%	16%	16%	12%

169	166	162	158	157	158	161	163	157	158	161	163
166	164	162	159	159	159	161	162	159	159	161	162
162	162	161	161	161	161	161	161	161	161	161	161
158	159	161	162	162	162	160	160	162	162	160	160
157	159	161	163	163	161	160	158	163	161	160	158
158	159	161	162	162	160	158	157	162	160	158	157
160	161	161	160	160	158	157	156	160	158	157	156
162	162	161	159	158	157	156	156	158	157	156	156

(a)

169	166	162	158	164	163	164	159	157	158	161	163
166	164	162	159	162	162	162	161	159	159	161	162
162	162	161	161	162	162	161	161	161	161	161	161
158	159	161	162	160	162	161	161	162	162	160	160
157	159	161	163	160	158	159	160	163	161	160	158
158	159	161	162	160	160	157	159	162	160	158	157
160	161	161	160	161	159	161	158	160	158	157	156
162	162	161	159	161	161	158	159	158	157	156	156

(f)

169	166	162	158	157	158	161	163
166	164	162	159	159	159	161	162
162	162	161	161	161	161	161	161
158	159	161	162	162	162	160	160
157	159	161	163	163	161	160	158
158	159	161	162	162	160	158	157
160	161	161	160	160	158	157	156
162	162	161	159	158	157	156	156

(b)

127	1	0	0	0	0	0	0
1	0	1	0	0	0	0	0
0	1	1	0	0	0	0	0
0	0	0	0	0	0	0	0
0	0	0	0	0	0	0	0
0	0	0	0	0	0	0	0
0	0	0	0	0	0	0	0
0	0	0	0	0	0	0	0

(d)

169	166	162	158	164	163	164	159
166	164	162	159	162	162	162	161
162	162	161	161	162	162	161	161
158	159	161	162	160	162	161	161
157	159	161	163	160	158	159	160
158	159	161	162	160	160	157	159
160	161	161	160	161	159	161	158
162	162	161	159	161	161	158	159

(g)

127	1	0	0	0	0	0	0
1	0	1	0	0	0	0	0
1	1	1	0	0	0	0	0
0	0	0	0	0	0	0	0
0	0	0	0	0	0	0	0
0	0	0	0	0	0	0	0
0	0	0	0	0	0	0	0
0	0	0	0	0	0	0	0

(i)

157	158	161	163	157	158	161	163
159	159	161	162	159	159	161	162
161	161	161	161	161	161	161	161
162	162	160	160	162	162	160	160
163	161	160	158	163	161	160	158
162	160	158	157	162	160	158	157
160	158	157	156	160	158	157	156
158	157	156	156	158	157	156	156

(c)

126	0	0	0	0	0	0	0
1	−1	0	−1	0	0	0	0
−1	0	0	0	0	0	0	0
0	0	0	0	0	0	0	0
0	0	0	0	0	0	0	0
0	0	0	0	0	0	0	0
0	0	0	0	0	0	0	0
0	0	0	0	0	0	0	0

(e)

164	163	164	159	157	158	161	163
162	162	162	161	159	159	161	162
162	162	161	161	161	161	161	161
160	162	161	161	162	162	160	160
160	158	159	160	163	161	160	158
160	160	157	159	162	160	158	157
161	159	161	158	160	158	157	156
161	161	158	159	158	157	156	156

(h)

127	1	0	0	0	0	0	0
1	0	1	0	0	0	0	0
0	1	1	0	0	0	0	0
0	0	0	0	0	0	0	0
0	0	0	0	0	0	0	0
0	0	0	0	0	0	0	0
0	0	0	0	0	0	0	0
0	0	0	0	0	0	0	0

(j)

Figure 13.7 An example of adjusting an 8×8 embedded image.

overlapping area is enclosed. Figure 13.7(b) and (c) shows the original 8×8 embedded and cropped images, respectively. We embed "1" on (1, 2), (2, 1), and (2, 2) by compressing Figure 13.7(b) using JPEG under 70% compression quality to obtain Figure 13.7(d). Note that the top-left pixel is (0, 0) and the messages can be embedded into any frequency of the transformed domain, so that the embedding capacity could be sufficiently high. Since the JFDSS checks only frequencies (1, 2), (2, 1), and (2, 2), we show an example of embedding three bits into these three frequencies. Similarly, Figure 13.7(e) is obtained by compressing Figure 13.7(c) using JPEG under the same compression quality. By evaluating the frequencies (1, 2), (2, 1), and (2, 2), the JFDSS can determine whether the embedded image is a stego image. Therefore, to break the JFDSS, we obtain the stego image as in Figure 13.7(f). Figure 13.7(g) and (h) shows the new embedded and cropped images, respectively. Similarly, we obtain Figure 13.7(i) and (j) by compressing Figure 13.7(g) and (h), respectively, using JPEG under 70% compression quality. Therefore, the JFDSS cannot distinguish the frequencies (1, 2), (2, 1), and (2, 2).

Let QTable(i,j) denote the standard quantization table, where $0 \leq i,j \leq 7$. The new quantization table, NewTable(i,j), with $x\%$ compression quality can be obtained by

$$\text{NewTable}(i,j) = \frac{\text{QTable}(i,j) \times \text{factor} + 50}{100} \tag{13.13}$$

where the factor is determined by

16	11	10	16	24	40	51	61
12	12	14	19	26	58	60	55
14	13	16	24	40	57	69	56
14	17	22	29	51	87	80	62
18	22	37	56	68	109	103	77
24	35	55	64	81	104	113	92
49	64	78	87	103	121	120	101
72	92	95	98	112	100	103	99

(a)

10.1	7.1	6.5	10.1	14.9	24.5	31.1	37.1
7.7	7.7	8.9	11.9	16.1	35.3	36.5	33.5
8.9	8.3	10.1	14.9	24.5	34.7	41.9	34.1
8.9	10.7	13.7	17.9	31.1	52.7	48.5	37.7
11.3	13.7	22.7	34.1	41.3	65.9	62.3	46.7
14.9	21.5	33.5	38.9	49.1	62.9	68.3	55.7
29.9	38.9	47.3	52.7	62.3	73.1	72.5	61.1
43.7	55.7	57.5	59.3	67.7	60.5	62.3	59.9

(b)

Figure 13.8 The quantization table of JPEG.

$$\begin{cases} \text{factor} = \dfrac{5000}{x} & \text{if } x \leq 50 \\ \\ \text{factor} = 200{-}2x & \text{otherwise} \end{cases} \tag{13.14}$$

Figure 13.8(a) and (b) shows the quantization tables of the standard and 70% compression quality, respectively.

13.5.5 Complexity Analysis

In general, the complexity of our GA-based algorithm is related to the size of the embedded message and the position of embedding (Shih and Wu, 2003). Figure 13.9 shows the relationship between the embedded message and the required iterations in which the cover image is of size 8×8 and the message is embedded into the LSB of the cover image by the zigzag order starting from the DC (zero-frequency) component. We observe that the more the messages embedded into a stego image, the more the iterations required in our GA-based algorithm. Figure 13.10 shows an example

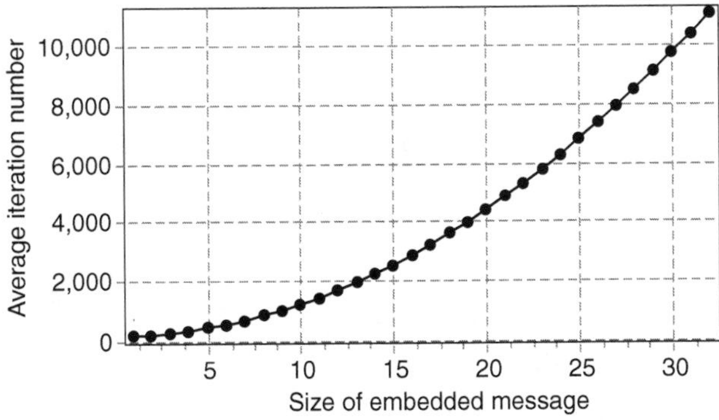

Figure 13.9 The relationship between the size of the embedded message and the required iterations.

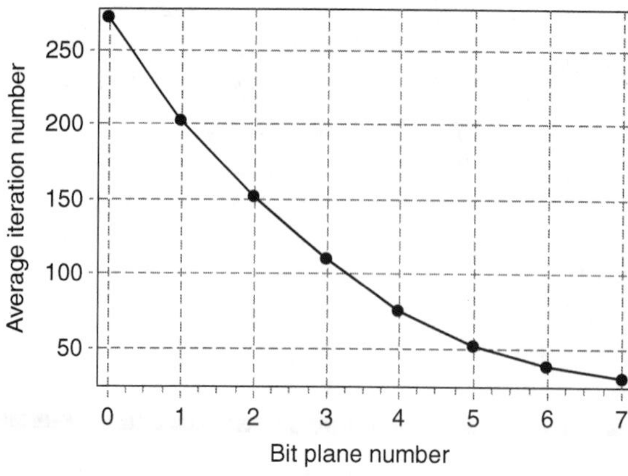

Figure 13.10 Embedding a 4-bit message into the different bit planes.

when we embed a 4-bit message into the different bit planes of DCT coefficients. We observe that the lower the bit plane used for embedding, the more the iterations required in our GA-based algorithm.

REFERENCES

Avcibas, I. and Sankur, B., "Statistical analysis of image quality measures," *J. Electron. Imaging*, vol. 11, no. 2, pp. 206–223, Apr. 2002.

Avcibas, I., Memon, N., and Sankur, B., "Steganalysis using image quality metrics," *IEEE Trans. Image Process.*, vol. 12, no. 2, pp. 221–229, Feb. 2003.

Brown, A., *S-Tools for Windows*, Shareware, 1994.

Chu, R., You, X., Kong, X., and Ba, X., "A DCT-based image steganographic method resisting statistical attacks," Proceedings of the International Conference on Acoustics, Speech, and Signal Processing, Montreal, Quebec, Canada, vol. 5, pp. 953–956, May 2004.

Cox, I. J., Kilian, J., Leighton, F. T., and Shamoon, T., "Secure spread spectrum watermarking for multimedia," *IEEE Trans. Image Process.*, vol. 6, no. 12, pp. 1673–1687, Dec. 1997.

Dumitrescu, S. and Wu, X., "A new framework of LSB steganalysis of digital media," *IEEE Trans. Signal Process.*, vol. 53, no. 10, Part 2, pp. 3936–3947, Oct. 2005.

Farid, H., "Detecting steganographic message in digital images," Technical Report, TR2001-412, Computer Science, Dartmouth College, 2001.

Fridrich, J., Goljan, M., and Hogea, D., "New methodology for breaking steganographic techniques for JPEGs," *Proceedings of EI SPIE*, Santa Clara, CA, pp. 143–155, Jan. 2003.

Holland, J. H., *Adaptation in Natural and Artificial Systems*, The University of Michigan Press, 1975.

Korejwa, J.,Jsteg shell 2.0, http://www.tiac.net/users/korejwa/steg.htm, 2006.

Martin, A., Sapiro, G., and Seroussi, G., "Is image steganography natural?" *IEEE Trans. Image Process.*, vol. 14, no. 12, pp. 2040–2050, Dec. 2005.

Marvel, L. M., Boncelet, C. G., and Retter, C. T., "Spread spectrum image steganography," *IEEE Trans. Image Process.*, vol. 8, no. 8, pp. 1075–1083, Aug. 1999.

PictureMarc, Embed Watermark, v 1.00.45, Digimarc Corporation, 2006.

Rencher, A. C., *Methods of Multivariate Analysis*, Wiley, New York, 1995.

Satish, K., Jayakar, T., Tobin, C., Madhavi, K., and Murali, K., "Chaos based spread spectrum image steganography," *IEEE Trans. Consum. Electron.*, vol. 50, no. 2, pp. 587–590, May 2004.

Shih, F. Y. and Wu, Y.-T., "Combinational image watermarking in the spatial and frequency domains," *Pattern Recogn.*, vol. 36, no. 4, pp. 969–975, Apr. 2003.

Shih, F. Y. and Wu, Y.-T., "Enhancement of image watermark retrieval based on genetic algorithm," *J. Visual Commun. Image Represent.*, vol. 16, no. 2, pp. 115–133, Apr. 2005.

Simmons, G. J., "Prisoners' problem and the subliminal channel," *Proceedings of the International Conference on Advances in Cryptology*, Santa Barbara, CA, pp. 51–67, Aug. 1984.

Wang, R.-Z. and Chen, Y.-S., "High-payload image steganography using two-way block matching," *IEEE Trans. Signal Process. Lett.*, vol. 13, no. 3, pp. 161–164, Mar. 2006.

Westfeld, A. and Pfitzmann, A., "Attacks on steganographic systems breaking the steganographic utilities EzStego, Jsteg, Steganos, and S-Tools and some lessons learned," *Proceedings of the International Workshop on Information Hiding*, Dresden, Germany, pp. 61–75, Sept. 1999.

Wolf, B.,StegoDos—Black Wolf's Picture Encoder, v0.90B, Public Domain, ftp://ftp.csua.berkeley.edu/pub/cypherpunks/steganography/stegodos.zip, 2006.

CHAPTER 14

SOLAR IMAGE PROCESSING AND ANALYSIS

We present the applications of image processing and machine learning techniques to automatically detect and classify solar events. The focus of automatic solar feature detection is on the development of efficient feature-based classifiers. We have conducted experiments using different classifiers for the detection of solar flare on the solar Hα (hydrogen alpha) images. These images were obtained from the Big Bear Solar Observatory (BBSO) in California. Series of Hα full-disk images are taken in regular time intervals to observe the changes of solar disk features. In addition, we use the classifier for the detection of coronal mass ejections (CMEs) in large angle spectrometric coronagraph (LASCO) C2 and C3 images that were obtained from the Solar and Heliospheric Observatory (SOHO), where we use filtering and enhancement techniques to preprocess the images and extract a set of critical features for event classification. The proposed method can detect and track solar events as well as measure their properties. This automatic process is valuable for forecasting and studying solar events since it dramatically improves efficiency and accuracy.

This chapter is organized as follows. Automatic extraction of filaments is presented in Section 14.1. Solar flare detection is described in Section 14.2. Solar corona mass ejections detection is presented in Section 14.3.

14.1 AUTOMATIC EXTRACTION OF FILAMENTS

A chromospheric filament usually marks a boundary between two opposite magnetic regions. Filaments may last for multiple days, changing their manifestation, but eventually they vanish. Their disappearance may end up with large CME associated with geomagnetic storms that affect the quality of terrestrial communication, reliability of power systems, and safety of space missions (Cleveland et al., 1992; Martens and Zwaan, 2001; Webb, 2000); therefore, filament tracking is an important task for solar observation. The existing filament detection systems use global thresholding and region growing (Gao et al., 2002) or local edge detection with linking edge vectors of similar angles in the TRACS project (*Toolkit for Recognition and Automatic Classification of Solar Features*, http://www.aai.com/AAI/NOAA/

Image Processing and Pattern Recognition by Frank Shih

NOAA.html). Localized median filter combined with least-square polynomial fit is used in limb darkening removal (Denker et al., 1999) as an initial step in solar image processing. Other techniques used in solar feature detection employ wavelet transforms (Irbah et al., 1999), multichannel blind deconvolution (Krejcí et al., 1998), Bayesian estimation (Molina et al., 2001), contrast and contiguity criteria (Preminger et al., 2001), and Markov chain Monte Carlo method (Turmon and Mukhtar, 1997). Additional elongated shape detection methods, such as the probabilistic saliency approach using deformable models (Orriols et al., 2000) and statistical snakes (Toledo et al., 2000) for tracking elongated structures, can also be applied for filament detection. Some approaches used in medical angiography for blood vessel segmentation (Walter et al., 2000; Zana and Klein, 1999; Zana and Klein, 2001) can be employed for filament detection. Other applicable techniques involve sequential morphological filtering (Shih and Pu, 1995; Shih and Puttagunta, 1995; Shih et al., 1995) and directional filtering (Soille et al., 1996).

The presented filament detection techniques were motivated by the solar research conducted at the Big Bear Solar Observatory in California, administered by the Department of Physics of the New Jersey Institute of Technology. From multiple methods of feature extraction, the work described focuses on the techniques using mathematical morphology. These techniques offer precise and fast discrimination of the required features and can be implemented in parallel. The often used method consists of two steps, image preprocessing and feature extraction. In the first step, the superfluous features are removed from the image as it is converted from grayscale to black and white (binary). The outcome is used for filament detection by multiple morphological operations involving directional structuring elements, that is, a specific application of disjunctive granulometric filters (Dougherty, 2001; Dougherty and Chen, 1999). The final result is a black-and-white image with the black features indicating the solar filaments.

In solar images, filaments appear as dark elongated objects of different shapes and sizes (Figs 14.1 and 14.2). Sometimes they may be broken into separate parts, as visible at the top of Figure 14.1. It is assumed that filaments of some importance have the area of more than 200 arcsec^2 (as observed in Hα full-disk pictures). Since the spatial CCD resolution used in the Big Bear Solar Observatory is 1 arcsec, the threshold area for filament detection is set to 200 pixels in 2000 $\times$ 2000 images. The most difficult part of filament tracking is its separation from the solar image background. Filaments are similar to other dark features in solar images, but may be defined as prominent, elongated, thin dark objects on the surface of the Sun.

14.1.1 Local Thresholding Based on Median Values

The goal of filament extraction is removing additional features of the solar chromosphere, such as flares, plages, faculae, and other differences in coronal brightness, leaving only the solar disk with dark spots (i.e., filaments) cleanly separated. In other words, filaments should be visible as black elongated shapes, with as few other black features (mostly solar surface granularity and sunspots) as possible.

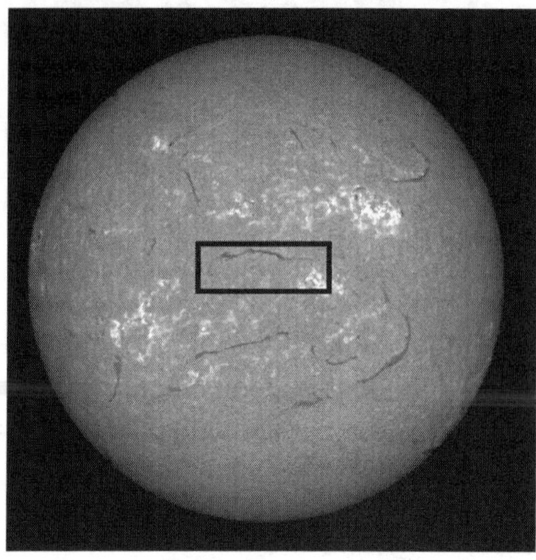

Figure 14.1 Image of solar disk observed at BBSO on 11 January 2002, 17:23:44 UT (a fragment with a filament used in further examples is marked with a black rectangle).

Thresholding is used for image segmentation. Thresholding with a single threshold level is an operation that involves a test against a function T of the form (Gonzales and Woods, 2008)

$$T = T[x, y, p(x, y), f(x, y)] \tag{14.1}$$

where $f(x, y)$ is the brightness level and $p(x, y)$ denotes the local property (such as the average brightness level of neighborhood centered at this pixel).

The thresholded image $g(x, y)$ is defined as

$$g(x, y) = \begin{cases} L-1 & \text{if } f(x, y) > T \\ 0 & \text{if } f(x, y) \leq T \end{cases} \tag{14.2}$$

where $L-1$ is the maximum brightness level of the resulting image.

The most intuitive method used to separate dark filaments from bright disk background is single-level thresholding applied to the whole image, called global thresholding. Unfortunately, due to a solar phenomenon called limb darkening, the brightness level of the disk background as well as of the solar features decreases toward the edge of the solar disk. Although the relationship between the dark and bright solar features is preserved, the absolute values of the same features are different in different parts of the solar disk. Therefore, global thresholding either correctly

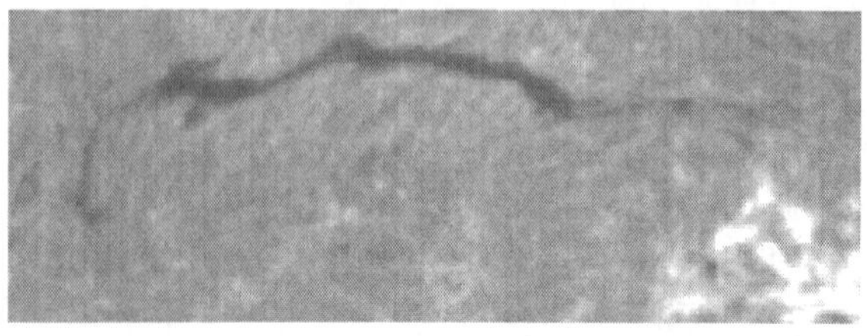

Figure 14.2 An example of a solar filament, unprocessed.

separates filaments in the disk center but merges them in the background at the disk edge, or discriminates the filaments correctly close to the disk edge but loses all details in the center.

To diminish the influence of limb darkening, local thresholding is considered. The function T depends on both $f(x, y)$ and $p(x, y)$, so this technique may change the threshold level based on the local differences between the background and feature values. To avoid creation of incorrect artificial filaments, the work window need be moved smoothly (pixel by pixel) throughout the whole image. We adopt a median filter as the local property function $p(x, y)$. Let the function T be described as

$$T = \text{med}_z(x, y) \tag{14.3}$$

where $\text{med}_z(x, y)$ is a median value of a $z \times z$ neighborhood centered at (x, y).

In the first step, the median value for the working window of all pixels in the image is computed and stored. In the next step, thresholding is applied to every pixel, with the threshold value being the appropriate median value retrieved for that pixel. Two methods are considered.

In a simple variant, the thresholding is based on the local median value. The results are promising, but in the very bright and very light areas some unimportant features are enhanced. This problem is particularly visible in the dark space outside the solar disk (Fig. 14.3).

In a modified method an additional preprocessing step is applied, in which very bright and very dark regions are directly transformed to black and white areas, respectively. The initial median calculation step remains unchanged, but the pixel brightness value is compared to the low and high cutoff values before the final thresholding. If the pixel brightness is lower than or equal to the lower cutoff margin value, then the output pixel brightness is assigned black; if the pixel brightness is higher than or equal to the higher cutoff margin value, the output pixel is set to white, otherwise the thresholding based on the local median value is applied (Fig. 14.4).

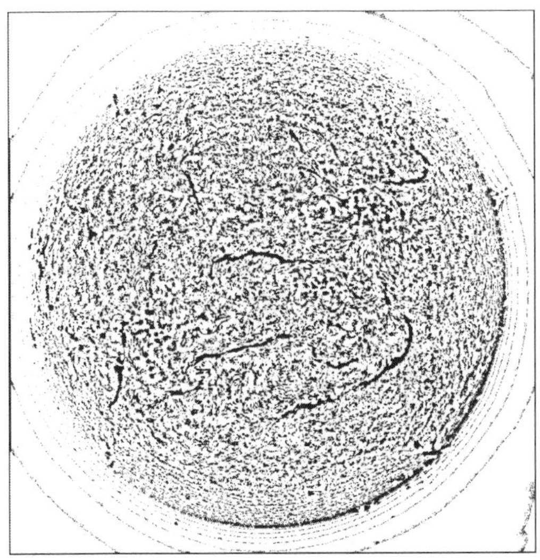

Figure 14.3 Solar disk, local thresholding.

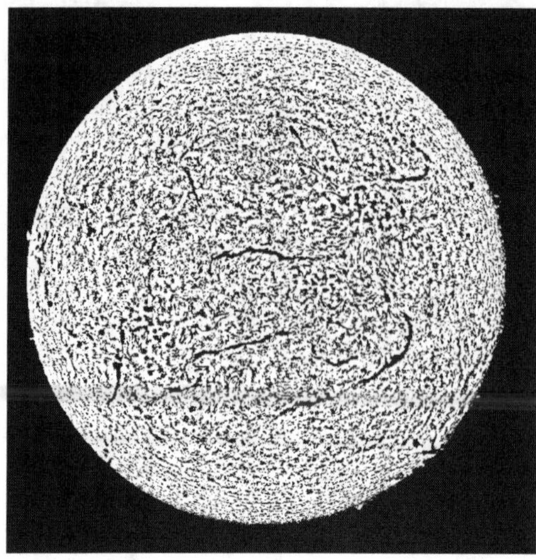

Figure 14.4 Solar disk, advanced local thresholding, with symmetrical cutoff values (20%).

In this case, the threshold function T can be described as

$$T = \begin{cases} c_{\text{inf}} & \text{if } \text{med}_z(x, y) \leq c_{\text{inf}} \\ \text{med}_z(x, y) & \text{if } c_{\text{inf}} < \text{med}_z(x, y) < c_{\text{sup}} \\ c_{\text{sup}} & \text{if } \text{med}_z(x, y) \geq c_{\text{sup}} \end{cases} \tag{14.4}$$

where c_{inf} and c_{sup} are lower and higher cutoff margin values, respectively (Fig. 14.5).

A few experiments using different cutoff margin values demonstrate that this method is suitable for separating dark solar features and does not create artificial features in very dark or bright regions. In some cases, artificial features appear in plage regions or in bright surroundings of sunspots because the local median method amplifies tiny brightness differences in regions with similar brightness. This problem can be eliminated using asymmetrical cutoff values.

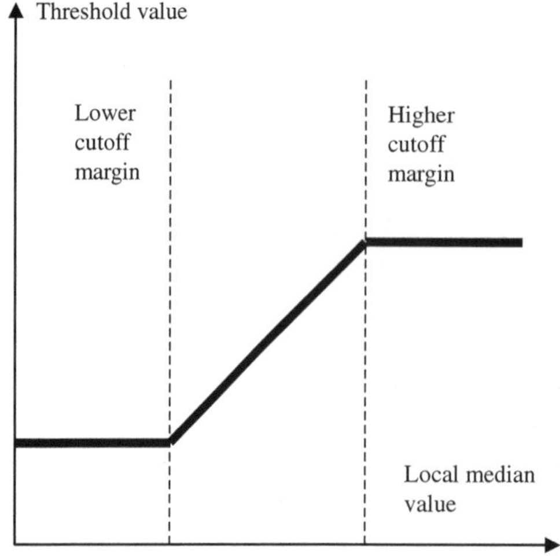

Figure 14.5 Threshold value for local thresholding with cutoff margins.

Since filaments are usually significantly colder than the gaps in solar chromosphere granularity and therefore darker, an additional modification is introduced for better separation of filaments from the surrounding background. If the thresholding value (the local median) is multiplied by a constant factor lower than 1, dark filaments are separated more precisely from slightly brighter granularity gaps. The threshold function T in this case can be described as

$$T = \begin{cases} c_{\text{inf}} & \text{if } \text{med}_z(x, y) \times m \le c_{\text{inf}} \\ \text{med}_z(x, y) \times m & \text{if } c_{\text{inf}} < \text{med}_z(x, y) \times m < c_{\text{sup}} \\ c_{\text{sup}} & \text{if } \text{med}_z(x, y) \times m \ge c_{\text{sup}} \end{cases} \tag{14.5}$$

where m is a local median modification factor and $0 < m < 1$.

After some experiments using different cutoff margin values, window sizes, and modification factors, it can be determined that the modified local thresholding technique by equation (14.5) is most suitable for further filament detection (Fig. 14.6).

14.1.2 Global Thresholding with Brightness and Area Normalization

Since local thresholding may enhance unnecessary local dark features, an alternative set of preprocessing methods was investigated. The solar disk image is normalized in such a way that the average brightness level is the same for every image and the limb darkening is removed. These operations allow for a simple sunspot removal and the separation of filaments through global thresholding using the same parameters for all preprocessed images. In addition, equal-area projection is investigated to simplify filament detection based on the area threshold.

Since the position and size of the solar disk in each image may vary, the first preprocessing step identifies the location of the disk center and the radius value. The proposed method detects disk edge pixels starting from the image center toward the

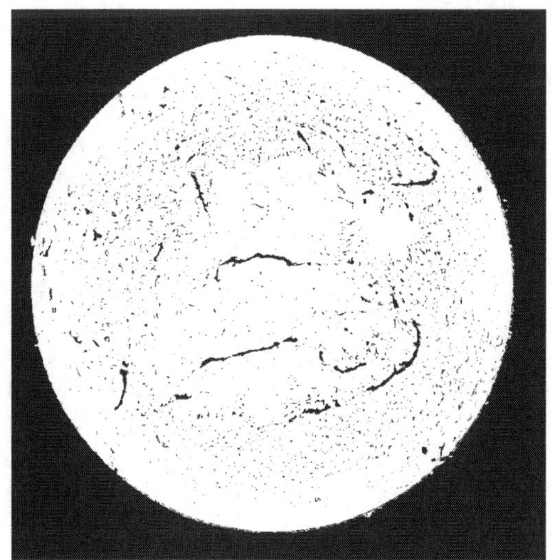

Figure 14.6 Solar disk, advanced local thresholding, with asymmetrical cutoff values ($c_{\text{inf}} = 20\%$, $c_{\text{sup}} = 40\%$) and modifying factor $m = 0.95$.

limb. The disk center is then calculated as the average of edge pixel coordinates, and the radius as the average distance of the edge from the center. Other methods can be used for disk center and radius determination, such as Denker et al. (1999) or Neckel and Labs (1984).

Solar images are not always taken under identical conditions, so the overall disk brightness may differ. An additional normalization step converts image brightness levels to a uniform space. The brightness of the image is scaled to achieve the luminosity value of 200 in the central disk region by calculating the average brightness of the central circular area within 100-pixel radius in the original image and scaling it to the average brightness of an "ideal" normalized image.

The brightness of filaments and other solar disk features diminishes from the disk center toward the edge due to limb darkening. This phenomenon can be approximated by a mathematical function. Finding what an exact transformation is, however, a difficult task. Denker et al. (1999) presented a method that relies on statistical properties of concentric rings. The average brightness in each ring is approximated by a separate polynomial function. We approximate brightness level of the whole disk by a single polynomial function of the radial distance from the solar disk center. This method is more suitable for further calculations than the usual function of angular distance (Neckel and Labs, 1994). Since the radius of the disk varies from image to image, the maximum radial distance is scaled to 1000. The data beyond the limb brightness level of 50 (after brightness normalization) are discarded to eliminate visual effect and to simplify further curve fitting. Based on the data gathered from 58 normalized solar images recorded in January and February 2002 at the BBSO, the average solar disk brightness level is calculated as a function of the scaled radial distance from the disk center. The brightness level function is approximated by a polynomial:

$$B(r) = c_0 + c_1 r + \cdots + c_n r^n \tag{14.6}$$

where r is the radial distance from the center, $B(r)$ is the brightness level at this distance, and $c_0, c_1, \ldots, c_n$ are the coefficients of the polynomial of the nth order.

The average brightness data for $r = 0, 1, \ldots, 1000$ are symmetrically duplicated for $r = -1000, -999, \ldots, -1$ to assure continuity of the approximation in the disk center, and the least-square curve fitting method is applied for polynomials of the 3rd to 10th order, yielding the results shown in Table 14.1 (Fig. 14.7).

The polynomial of the 10th order is used for further calculations. To remove limb darkening, new brightness values of every pixel in the solar disk are calculated by

$$B_n(x, y) = B_o(x, y) \cdot \frac{200}{B(r_a)} \tag{14.7}$$

where $B_n(x, y)$ is the new brightness value after limb darkening removal, $B_o(x, y)$ is the original brightness value of this pixel, and $B(r_a)$ is the average brightness level of the pixel with the adjusted radial distance r_a from the disk center, calculated according to equation (14.6). The value of r_a is given by the following formula:

$$r_a = \sqrt{(x_c - x)^2 + (y_c - y)^2} \cdot \frac{1000}{R} \tag{14.8}$$

TABLE 14.1 Polynomial Least-Square Curve Fitting Results

Polynomial order	3	4	5	6	7	8	9	10
Correlation	0.897128	0.97397	0.97397	0.991488	0.991488	0.99702	0.99702	0.999301
Average absolute error value	8.500201	4.209949	4.209949	2.447857	2.44786	1.5424	1.542395	0.746936
Maximum absolute error value	81.1135	49.6285	49.6285	31.2198	31.2198	19.4019	19.4019	11.0396
c_0	207.186	195.286	195.286	201.124	201.124	197.816	197.816	199.945
c_1	-5.90×10^{-17}	-3.84×10^{-16}	2.58×10^{-16}	4.88×10^{-16}	-5.24×10^{-16}	-8.34×10^{-16}	-5.93×10^{-16}	1.65×10^{-15}
c_2	-8.36×10^{-5}	3.50×10^{-5}	3.50×10^{-5}	-8.71×10^{-5}	-8.71×10^{-5}	3.14×10^{-5}	3.14×10^{-5}	-8.51×10^{-5}
c_3	-1.07×10^{-22}	3.49×10^{-22}	-9.24×10^{-22}	-1.02×10^{-21}	2.92×10^{-21}	2.81×10^{-21}	-1.70×10^{-21}	-3.33×10^{-21}
c_4	—	-1.38×10^{-10}	-1.38×10^{-10}	2.28×10^{-10}	2.28×10^{-10}	-4.23×10^{-10}	-4.23×10^{-10}	5.84×10^{-10}
c_5	—	—	8.23×10^{-28}	8.73×10^{-28}	-3.72×10^{-27}	-2.36×10^{-27}	1.88×10^{-26}	7.82×10^{-27}
c_6	—	—	—	-2.68×10^{-16}	-2.68×10^{-16}	8.59×10^{-16}	8.59×10^{-16}	-2.16×10^{-15}
c_7	—	—	—	—	8.87×10^{-34}	-6.24×10^{-34}	-3.43×10^{-32}	-1.36×10^{-32}
c_8	—	—	—	—	—	-6.03×10^{-22}	-6.03×10^{-22}	3.06×10^{-21}
c_9	—	—	—	—	—	—	1.74×10^{-38}	7.92×10^{-39}
c_{10}	—	—	—	—	—	—	—	-1.54×10^{-27}

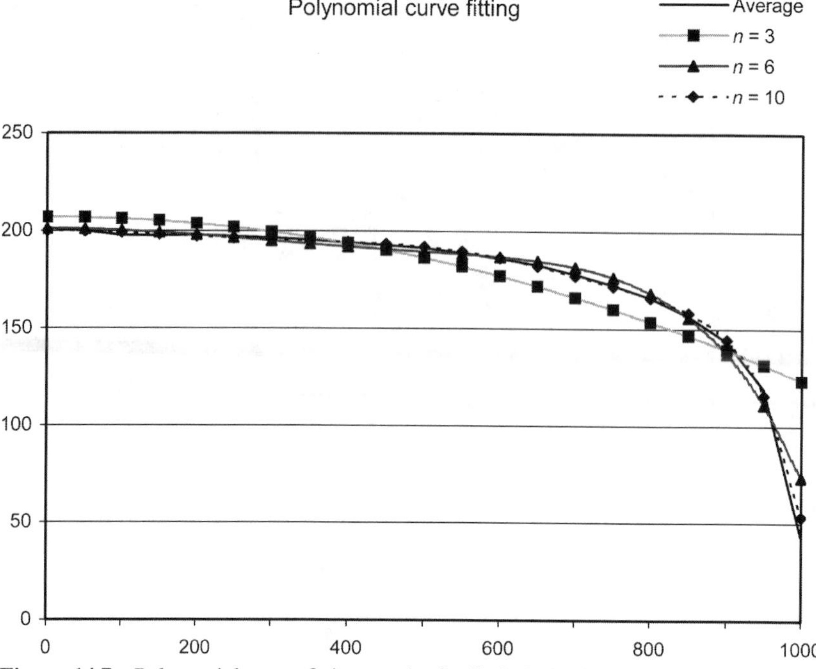

Figure 14.7 Polynomial curve fitting results for limb darkening approximation.

where (x_c, y_c) are coordinates of the disk center and R is the disk radius. The brightness level of the pixels outside the solar disk as well as of the outer 5% of the disk pixels (containing no valuable solar filament data) is set to the maximum brightness value to facilitate further processing (Fig. 14.8).

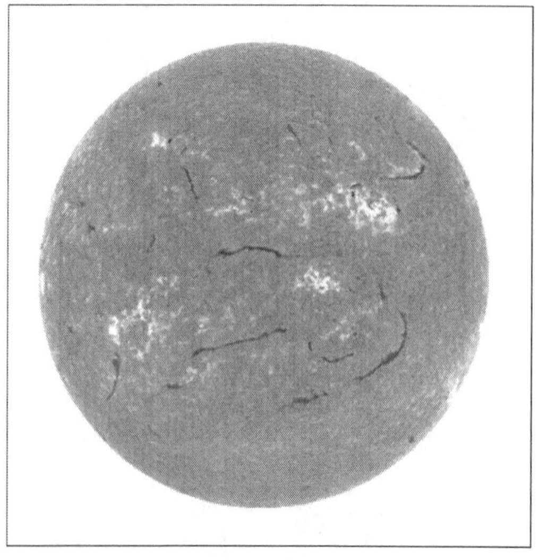

Figure 14.8 Disk brightness normalization (limb darkening removal).

Sunspots, as dark solar features, can be misrecognized as solar filaments in further solar image processing steps. Since sunspots are usually darker than filaments, they can be easily removed from the normalized solar image by applying global thresholding with threshold level below the usual filament brightness range. Using the threshold value of 140, the extracted sunspot areas are used as seeds for region growing in the normalized image above the filament threshold level. The brightness value of 186 is used as the region growing limit. The resulting sunspot regions are removed from the normalized image using region image inversion and logical union with the normalized image before further processing.

The disk is transformed to achieve uniform area of solar features throughout the disk using the calculated disk center and radius. The transformation applied is an equal-area projection with 90% of the visible hemisphere transformed and scaled back to the original disk size. After this transformation the features have a uniform area, but their shapes are distorted toward the edges and the visual data are more reliable in the disk center.

A sphere of a radius R can be described by

$$x^2 + y^2 + z^2 = R^2 \tag{14.9}$$

The equal-area projection can be defined as

$$r' = \sqrt{x^2 + y^2 + (R-z)^2} \tag{14.10}$$

where r' is the radial distance of the projected point from the center of projection. By substituting

$$x^2 + y^2 = r^2 \tag{14.11}$$

in equation (14.10), where r is the radial distance of the original point from the center of projection, we obtain

$$r' = \sqrt{r^2 + (R-z)^2} \tag{14.12}$$

Since the solar disk image can be considered as a scaled orthographic projection of an actual solar sphere, the value of z is unknown, but from equations (14.9) and (14.11), we derive

$$z = \sqrt{R^2 - r^2} \tag{14.13}$$

By combining equations (14.12) and (14.13), we obtain

$$r' = \sqrt{r^2 + R^2 - 2R\sqrt{R^2 - r^2} + R^2 - r^2} \tag{14.14}$$

$$r' = \sqrt{2R^2 - 2R\sqrt{R^2 - r^2}} \tag{14.15}$$

$$r' = R\sqrt{2\left(1 - \sqrt{1 - \frac{r^2}{R^2}}\right)} \tag{14.16}$$

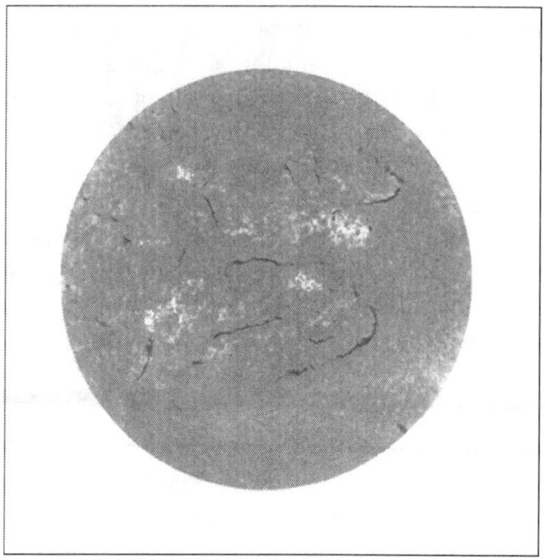

Figure 14.9 Equal-area projection with sunspots removed.

Since the diameter of the resulting equal-area projection is larger than the original image by a factor of $\sqrt{2}$, the projection is scaled back to the original size:

$$r'' = R\sqrt{1-\sqrt{1-\frac{r^2}{R^2}}} \tag{14.17}$$

where r'' is the scaled radial distance of the projected point from the center of projection. The inverse transformation to assure correct values for every pixel in the projected image can be obtained as

$$r = r''\sqrt{2-\frac{r''^2}{R^2}} \tag{14.18}$$

The resulting image is shown in Figure 14.9. Since the distortion of the projected features close to the disk edge is very high, the outer 10% of the disk is removed to prevent filament detection from extracting features in unreliable regions.

The final step of preprocessing is brightness thresholding. Using the threshold value of 185, most of the disk dark features are cleanly separated, including the majority of filaments as shown in Figures 14.10 and 14.11. A similar operation using single or multiple thresholding values may be used for the extraction of other solar features, such as sunspots, flares, plages, and so on.

14.1.3 Feature Extraction

The next step is the extraction of filaments from the black-and-white preprocessed solar image. After preprocessing, filaments are visible as large, elongated black gaps in a black-and-white solar image. All other features on the solar disk are also visible as black areas, but of small sizes with the exception of sunspots, which can be described as huge, roughly circular black regions of the disk. The objective of applying

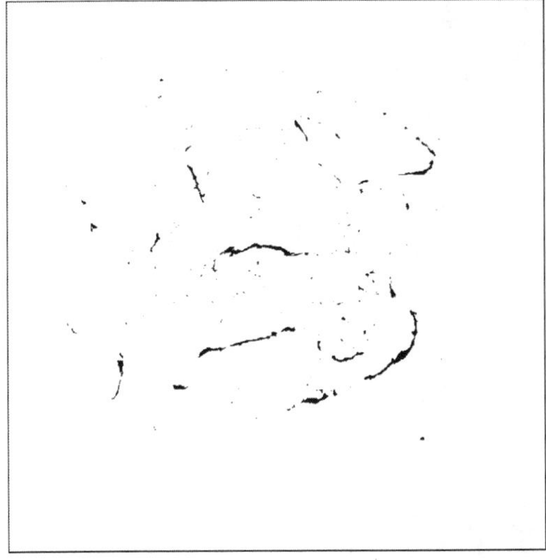

Figure 14.10 Thresholded image ($T = 185$) without equal-area transformation.

morphological operations to solar images is to eliminate all solar disk features except for filaments and, if possible, to merge separate sections of the filaments.

The simplest approach evaluated is a sequence of dilations followed by a sequence of erosions. A single closing operation (i.e., a dilation followed by an erosion) with a small structuring element is obviously insufficient for eliminating unnecessary features; therefore, a sequence of dilations followed by a sequence of erosions with a 3×3 circular structuring element is applied. Note that the effect of such sequences with a small structuring element is similar to a single application of closing with a large structuring element of the same shape. However, its speed is significantly improved. The result of a single closing operation with a 3×3 structuring element is shown in Figure 14.12. However, this method not only

Figure 14.11 Image thresholded ($T = 185$) after equal-area transformation.

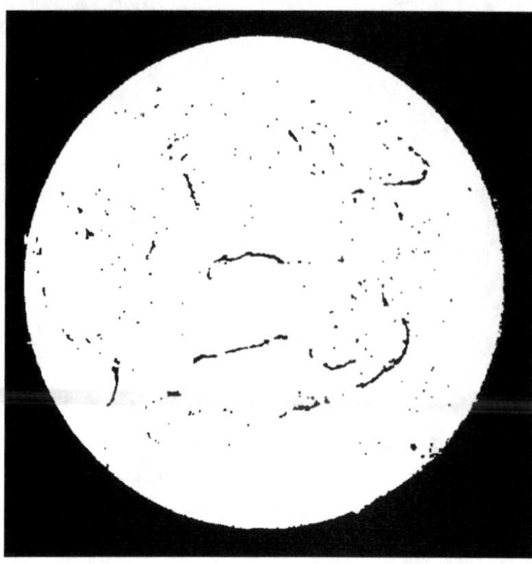

Figure 14.12 Solar disk, the first preprocessing technique, closing with 3×3 structuring element.

eliminates most of unnecessary solar features, but also removes some small filaments and disconnects parts of big ones. Therefore, it is deemed as inadequate for filament detection.

Since simple morphological operations cannot yield necessary results, more complex combinations of mathematical morphology methods are evaluated, as well as combinations of morphological methods with other image processing techniques.

Zana and Klein (1999, 2001) described a method for vessel recognition in a noisy retinal angiography employing mathematical morphology techniques. The retinal images they processed contain a different set of features from solar images, but the basic structure to be extracted, tiny blood vessels, can be described similarly to solar filaments—as thin, elongated objects. Since the preprocessed solar image contains binary information, some of their steps are irrelevant, but a revised technique can be used for filament detection. The method employed here is an application of disjunctive granulometric filters using directional linear structuring elements, with theory laid out by Dougherty and Chen (1999), Dougherty (2001), and the implementation enhancement by Soille et al. (1996).

The most important step is the separation of elongated shapes (i.e., filaments) from the granular background. Instead of a supremum of openings (Walter et al., 2000) used for the discrimination of bright vessels in medical images, a superposition (implemented as logical intersection) of closings is used for the separation of dark filaments in solar images. A set of directional linear structuring elements is used, with the best results for elements of size 11×11 (Fig. 14.13). The supremum of a set is defined as a quantity k, such that no member of the set exceeds k.

Applying the four directional elements (Fig. 14.13, elements a–d) renders promising results, but some filaments of nonvertical, nonhorizontal, or nondiagonal slope are missing and other filaments (especially the ones with high curvature) are separated into a few disconnected parts. Applying additional four directional elements (Fig. 14.13, elements e–h) improves the result significantly (Fig. 14.14).

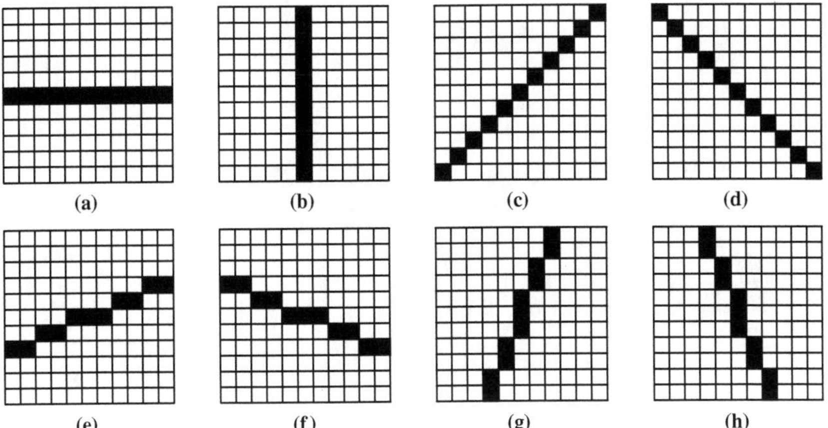

Figure 14.13 Eight directional linear 11 × 11 structuring elements with 90°, 0°, 45°, 135°, 67.5°, 112.5°, 22.5°, and 157.5° slopes, respectively.

The final superposition contains multiple short thin lines that may not be actual filaments, but granularity gaps or features in the sunspot vicinity. They may be eliminated by applying a closing operation with a 3 × 3 circular structuring element. Such an operation eliminates most of the spurious features, but applying this operation with a larger structuring element (or applying multiple dilations followed by the same number of erosions) also removes some minor filaments.

The dark features obtained by the superposition of directional closing operations (with or without additional "cleaning" closings) may be used as seeds for a region growing operation. The method checks the neighborhood of the detected features and compares it against the original preprocessed image. All connected black

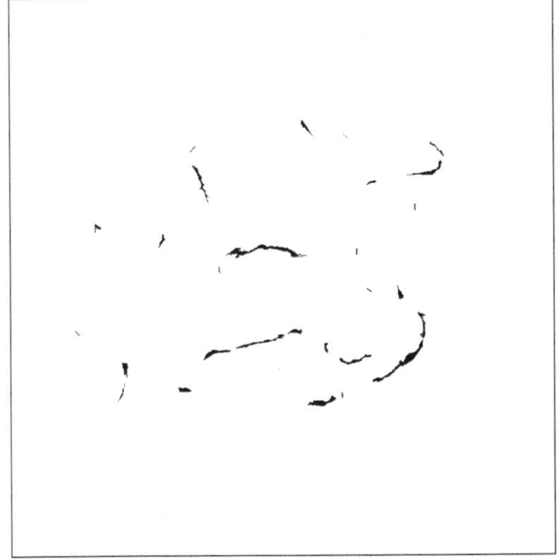

Figure 14.14 Solar disk, the second preprocessing technique, superposition of eight directional closing operations.

Figure 14.15 Solar disk, the second preprocessing technique, directional filtering, single closing, 8-neighborhood region growing.

points in the preprocessed image neighboring the detected filament are marked as belonging to the filament. Using 8-neighbor connectivity (see Fig. 14.15) yields better results than using 4-neighbor connectivity. Figure 14.16 shows the results of applying consecutive processing steps to a single filament.

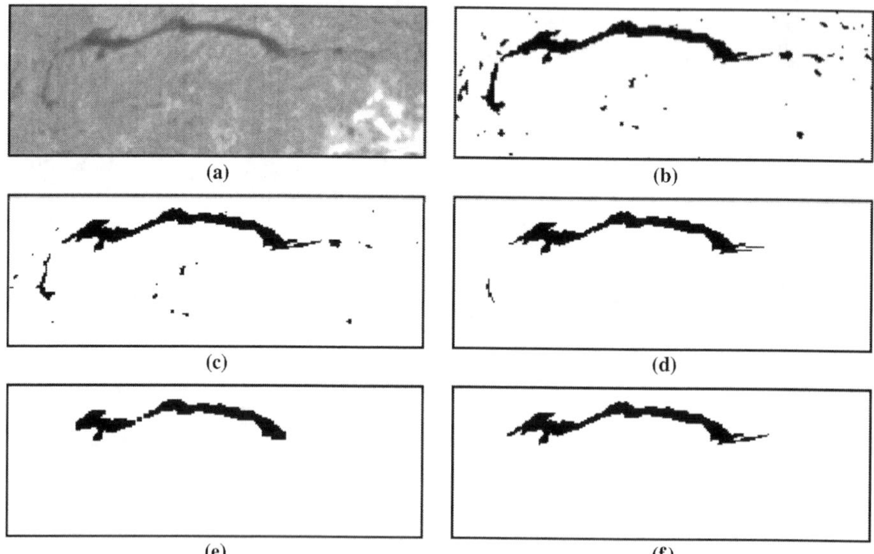

Figure 14.16 Selected filament after consecutive image processing steps: (a) original image, (b) preprocessing based on advanced local thresholding, (c) preprocessing based on image normalization, (d) the second preprocessing method and superposition of eight directional closing operations, (e) additional 3×3 closing, and (f) additional 8-neighborhood region growing.

14.1.4 Experimental Results

This section describes the experimental results of the proposed filament detection system. The filament detection rate and the processing time are based on the processing of 10 solar images gathered by BBSO in January 2002. Performance of the filament detection system is compared against the method presented by Gao et al. (2002). One of the original grayscale images of 1016×1016 used in the evaluation with all the dark features marked manually is shown in Figure 14.17. There are six large filaments (labeled LF), eight large filament segments (labeled T), six small filaments (labeled SF), and three sunspots (labeled SS) in the image.

Table 14.2 presents the results of filament detection techniques. The methods referred to in this table (and in the following tables) are described below.

- *Method I:* The method presented by Gao et al. and their result is shown in Figure 14.18.
- *Method II:* The first preprocessing method (i.e., local thresholding with cutoff margins and modifying factors) combined with morphological operations (i.e., superposition of eight recursive directional closings and region growing).
- *Method III:* The second preprocessing method (i.e., brightness normalization, limb darkening, sunspot removal, and global thresholding) with the superposition of eight recursive directional closings and region growing.
- *Method IV:* Similar to method III, with an additional equal-area normalization in the preprocessing stage.

The average preprocessing and feature extraction time and the average, minimum, and maximum of total processing time for 10 selected solar images are presented in Table 14.3. Experiments are implemented on a single-processor 1.2 MHz AMD Athlon PC with 512 MB of physical memory.

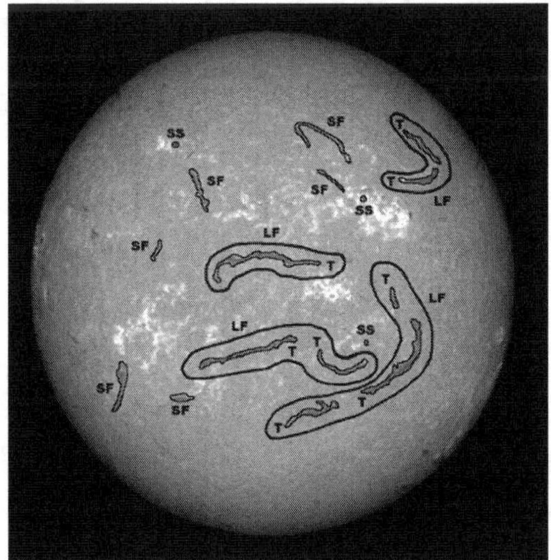

Figure 14.17 Solar disk with dark features marked.

TABLE 14.2 Filament Detection Results

Method	LF[a] (%)	LFS[b] (%)	SF[c] (%)	SMF[d]	AFMF[e]
Method I	50.2	33.3	5.1	0.0	0.2
Method II	100.0	98.1	90.4	0.7	10.5
Method III	96.7	85.4	74.2	0.0	5.5
Method IV	94.3	80.1	54.9	0.0	4.0

[a] LF: average detection rate of large filaments.
[b] LFS: average detection rate of large filament segments.
[c] SF: average detection rate of small filaments.
[d] SMF: average number of sunspots misrecognized as filaments.
[e] AFMF: average number of additional features misrecognized as filaments.

Gao et al. (2002) used the images that are already preprocessed (with dark current and flat field correction, removed limb darkening, and enhanced contrast) during the image acquisition process at BBSO; therefore, it is compared against the feature extraction stage of the other methods described. Their filament detection technique is based on binary thresholding with the half of the image median value as a threshold, followed by region growing. Pieces of filament segments are merged within 80-pixel neighborhood, and all segments within 40-pixel distance are merged into a single filament. Filaments with the total area smaller than a preset threshold (50 pixels) are ignored. The method detects on average 50.2% of large filaments, 33.3% of their segments, and only 5.1% of small filaments; however, there are a very less number of other dark features (including sunspots) recognized incorrectly as filaments. The low filament detection rate is mainly due to global thresholding based on the median value. The second stage, region growing, merges pieces of large filaments correctly and filters out all unnecessary dark features. Unfortunately, it also removes most of small filaments and some segments of large ones. The processing time

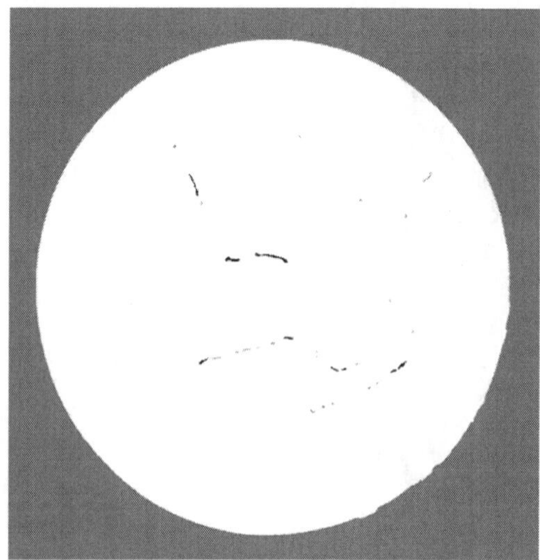

Figure 14.18 Solar disk with the method of Gao et al. applied. Black features are recognized filaments. Gray areas are rejected due to size limitation or closeness to the disk limb.

TABLE 14.3 Processing Time for Preprocessing and Feature Extraction Stages

Method	Average preprocessing time (ms)	Average feature extraction time (ms)	Average total time (ms)	Minimum total time (ms)	Maximum total time (ms)
Method I	N/A	1672.3	1672.3	851	2143
Method II	1193.8	6286.1	7479.9	7471	7501
Method III	1361.8	5507.0	6868.8	6850	6880
Method IV	1710.5	5515.0	7225.5	7201	7241

(the shortest among all the methods compared) highly depends on the number of the dark features in the image and cannot be significantly reduced by parallelization due to the nature of median and region growing algorithms.

The new filament detection technique based on directional morphological filtering achieves excellent results in detecting large filaments and good results for small filaments. The technique, however, misrecognizes some dark features as filaments. This deficiency could be improved by an additional step of rejecting features, which are too small or improperly shaped. The application of the first preprocessing method (modified local thresholding used in method II) results in a higher overall filament detection rate, but it is unable to reject all sunspots and incorrectly recognizes many dark features as filaments.

The second preprocessing method (based on image normalization used in method III) is slightly slower than the first one, but it is more suitable for separating solar features into a few categories due to their brightness level. Therefore, it is able to remove sunspots from the image. A large part of preprocessing techniques (limb darkening, sunspot removal, and thresholding) can be parallelized, leading to reduced processing time. The second preprocessing method with the additional equal-area projection step (used in method IV) achieves a lower filament detection rate and is

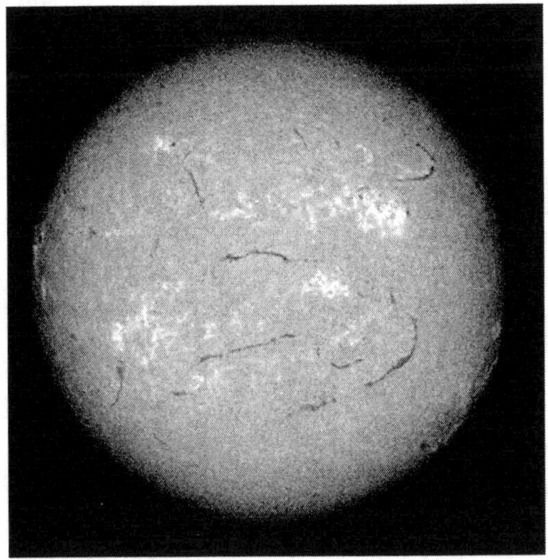

Figure 14.19 Solar disk, original image with Gaussian noise, 50% of the noise level and 50% noise distribution.

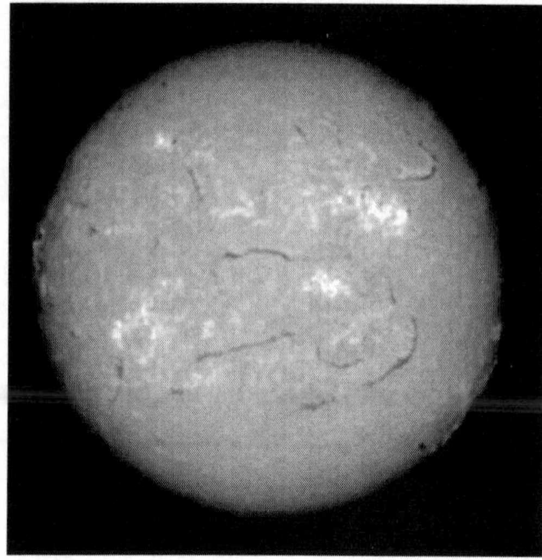

Figure 14.20 Solar disk, original image low-pass filtering, 5-pixel range.

slightly slower than other methods. It is more suitable for tracking filament size changes, as it extracts area-corrected filaments.

The proposed method is applied on a set of noisy images that were added noise of various nature. Gaussian noise (Fig. 14.19) is used to simulate random and electronic equipment-related interference, while smoothing (Gaussian blur and low-pass filtering, see Fig. 14.20) is used to simulate the effect of seeing and light high-altitude cloudiness, particularly common in solar observation.

The results in Table 14.4 and Figures 14.21 and 14.22 show that method II (with modified local thresholding) is suitable for processing noisy images. The detection

TABLE 14.4 Results of Filament Detection in Noisy Images (Method II)

Image noise	LF[a] (%)	LFS[b] (%)	SF[c] (%)	SMF[d]	AFMF[e]
Gaussian noise, 10% noise level, 100% distribution	100.0	100.0	83.3	0	9
Gaussian noise, 100% noise level, 10% distribution	100.0	100.0	83.3	1	9
Gaussian noise, 25% noise level, 25% distribution	100.0	100.0	83.3	0	10
Gaussian noise, 50% noise level, 50% distribution	100.0	87.5	66.7	0	3
Gaussian blur, 1 pixel	100.0	100.0	83.3	0	8
Gaussian blur, 3 pixels	100.0	87.5	66.7	0	2
Low pass, 1 pixel	100.0	100.0	83.3	0	11
Low pass, 5 pixels	100.0	87.5	66.7	0	3

[a] LF: average detection rate of large filaments.
[b] LFS: average detection rate of large filament segments.
[c] SF: average detection rate of small filaments.
[d] SMF: average number of sunspots misrecognized as filaments.
[e] AFMF: average number of additional features misrecognized as filaments.

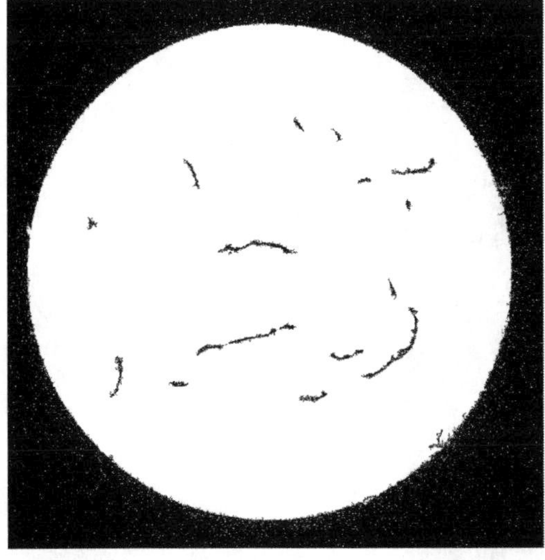

Figure 14.21 Solar disk, filaments detected in the image with Gaussian noise, 50% of the noise level and 50% noise distribution.

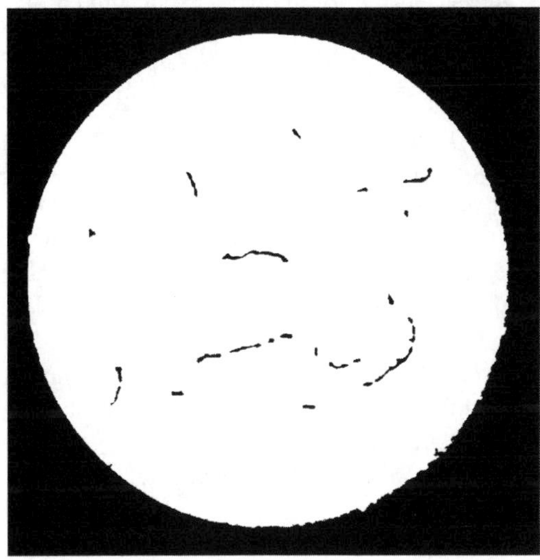

Figure 14.22 Solar disk, filaments detected in the image with low-pass filtering, 5-pixel range distribution.

rate for large filaments is excellent, with a satisfactory rate for smaller filaments. A slightly lower number of additional features are misrecognized as filaments due to smaller areas and discontinuities in the dark features.

14.2 SOLAR FLARE DETECTION

The sun is the source of space weather—the disturbances whose radiative, field, and particle energy directly impact earth. Solar activity changes the radiative and particle output of the sun, producing corresponding changes in the near-earth space

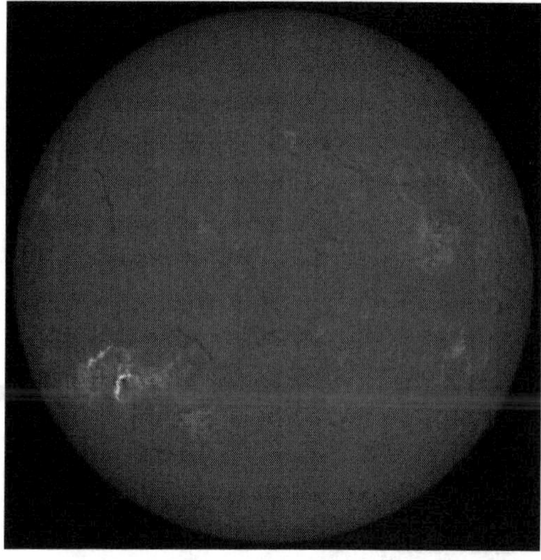

Figure 14.23 Solar flares on the left side of the image.

environment, as well as at the earth's surface. The most dramatic events on the sun, insofar as space weather effects are concerned, are solar flares, filament, and CME.

A flare is defined as a sudden, rapid, and intense variation in brightness. A solar flare occurs when magnetic energy that has built up in the solar atmosphere is suddenly released. Radiation is emitted across virtually the entire electromagnetic spectrum, from radio waves at the long-wavelength end, through optical emission, to X-rays and gamma rays at the short-wavelength end (Zirin, 1988). The amount of energy released is the equivalent of millions of 100-megaton hydrogen bombs exploding at the same time. Flares are often associated with CMEs. Figure 14.23 shows a full-disk Hα (hydrogen-alpha) image with a flare detected on the low-left corner.

A chromospheric filament usually marks a boundary between two opposite magnetic regions. Filaments may last for multiple days, changing their manifestation, but eventually they vanish. Their disappearance may end up with large CME associated with geomagnetic storms that affect the quality of terrestrial communication, reliability of power systems, and safety of space missions (Martens and Zwaan, 2001); therefore, the filament tracking is an important task of solar observation. In solar images, filaments appear as dark elongated objects of different shapes and sizes, and may be broken into separate parts, as shown in Figure 14.24. Qu et al. (2005) developed an automatic solar filament detection algorithm based on image enhancement, segmentation, pattern recognition, and mathematical morphology (Shih and Mitchell, 1989).

A CME is due to a large-scale rearrangement of the solar magnetic field that can induce a geoeffective solar wind structure. Each CME may carry away a mass of up to 1013 kg and release up to 1025 J of energy from coronal magnetic fields (Harrison, 1995). Following their discovery, CMEs were soon found to be correlated with the occurrence of geomagnetic storms. CMEs are easily observed at the solar limb, where they are seen against a dark background. However, the earth-directed, disk CMEs are much harder to detect, although they are the ones that have the most

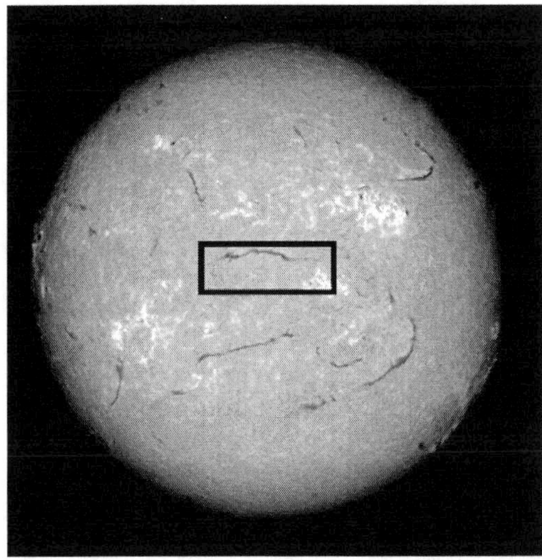

Figure 14.24 A fragment with a solar filament is marked by a rectangle.

important geomagnetic effects. Therefore, detecting the early manifestation of CMEs on the solar disk is essential to understanding earth-directed CMEs themselves. Establishing the correlation among the occurrence of CMEs and some other solar surface phenomena, such as filament eruptions and flares, is a long-standing problem in solar physics. After firm relationships are established, the correlated solar surface phenomena can provide an early warning of the occurrence of earth-directed CMEs and geoeffectiveness of CMEs. A key challenge to move this research area forward is to have real-time, automatic, and accurate detection and characterization of solar events and to use the results for real-time space weather forecasts.

CMEs have been observed from Large Angle and Spectrometric Coronagraph (LASCO) images by SOHO. LASCO images include three different coronagraphs: C1, C2, and C3, whose ranges are from 1.1 to 3, 1.5 to 6, and 3.7 to 30 solar radii, respectively. Coronagraph C1 has been inoperable since 1999. A detailed description of LASCO can be found in Brueckner et al. (1995).

There are many methods to detect solar flares. When a solar flare first appears on the Sun's surface, there is often a radiation covering the entire range of electromagnetic waves. For example, light curves of radio waves and X-rays would indicate on a set of flares. However, the location and properties of flares cannot be demonstrated in light curves. In the mean time, solar flares also emit high-velocity, charged particles that take 1 or 2 days to reach Earth. Our objective is to automatically detect and characterize solar flares in real time. Since most solar flares can be detected using Hα, we start with solar images observed in Hα using an 8 in. telescope in the BBSO in California, operated by New Jersey Institute of Technology.

Automatic solar flare detection is the key to space weather monitoring. It is very challenging since features of solar flares are complicated. Veronig et al. (2000) proposed an automatic flare detection method using the combination of region-based and edge-based segmentation approaches. Simple region-based methods are applied for a tentative assignment of flare activity, making use of one of the decisive flare characteristics: the high intensities. Borda et al. (2002) presented automatic detection of solar flares using the neural network technique. The network uses multilayer perceptron (MLP) with back-propagation training rule. They applied supervised learning that requires a lot of iterations to train the network.

Qu et al. (2003) compared three principal techniques, MLP, radial basis function (RBF), and support vector machines (SVM) classifiers, for solar flare detection. The preprocessing step obtains nine principal features of solar flares for input to the three classifiers. Experimental results show that by using SVM, the best classification rate of solar flares can be obtained. Qu et al. (2004) further extended the work to automatically detect flares by applying image segmentation and computing the properties. A solution is obtained for automatically tracking the apparent separation motion of two-ribbon flares and measuring their moving direction and speed in the magnetic fields. From these measurements under certain assumptions, the reconnection of the electric field as a measure of the rate of the magnetic reconnection in the corona can be derived. The automatic procedure is a valuable tool for real-time monitoring of flare evolution.

14.2.1 Feature Analysis and Preprocessing

Borda et al. (2002) used seven features for solar flare detection. After extensive investigations, we modify them into nine features as described below.

Feature 1: Mean brightness of the frame. When flares happen, their intensities become brighter. Let x_j denote the gray level of a pixel and N denote the number of pixels. The mean brightness $\bar{x}$ is represented as

$$\bar{x} = \sum_{j=0}^{N-1} x_j \tag{14.19}$$

Feature 2: Standard deviation of brightness. When flares happen, the standard deviation of their brightness becomes large. Let std denote the standard deviation and let V denote the variance.

$$\text{std} = \sqrt{V} \tag{14.20}$$

$$V = \sum_{j=0}^{N-1} (x_j - \bar{x})^2 \tag{14.21}$$

Feature 3: Variation of mean brightness between consecutive images. This is the difference of mean values between the current image and the previous image.

Feature 4: Absolute brightness of a key pixel. By comparing consecutive images, we can find the key pixel that has the maximum gray level difference. Usually, when flares happen, the absolute brightness of the key pixel becomes large. The position of the key pixel k is represented as

$$k = \text{where}\ (\max(\text{Im}_{\text{pre}} - \text{Im}_{\text{current}})) \tag{14.22}$$

where Im_{pre} denotes the vector of the previous image and $\text{Im}_{\text{current}}$ denotes the vector of the current image.

TABLE 14.5 Samples of the Nine Features and the Target Value

1	2	3	4	5	6	7	8	9	Target value
3994.25	2628.22	45	12474.00	431.23	3635.00	2011.43	1978.60	427.35	−1
3807.42	2458.64	187	16383.00	437.13	6484.00	9460.98	2430.92	7449.55	1

Feature 5: Radial position of the key pixel.
Features 1–5 focus on the global view of consecutive images. But these features are insufficient to obtain the information for small solar flares. Therefore, it is necessary to include local view of solar images.

Feature 6: Contrast between the key pixel and the minimum value of its neighbors in a 7×7 window. When flares occur, this contrast becomes large.

Feature 7: Mean brightness of a 50×50 window, where the key pixel is on the center. Since we assume that the key pixel is one of the pixels on the flare, the 50×50 window will include most of the flare. This region is the most effective for the solar flare detection.

Feature 8: Standard deviation of the pixels in the aforementioned 50×50 window.

Feature 9: Difference of the mean brightness of the 50×50 window between the current and the previous images.

See samples of the input nine features and the target value in Table 14.5.

14.2.2 Classification Rates

The solar Hα images of 2032×2032 pixels were obtained from BBSO. We selected the images of flare events starting from January 1, 2002 to December 31, 2002. The first step carefully divides the images into two classes: flare state and nonflare state. We try to include all the flare events that can be recognized by human eyes. We label the corresponding images as "flare state" and include the same number of "nonflare state" images. Because some microflares are difficult to be recognized even by human eyes, we regard them as "nonflare state." Second, we compute the nine features for a given solar image and put it into either the training or the testing data set. We use the training data set to train the networks and use the testing data set to obtain the classification rate. Finally, we obtain flare properties using image segmentation techniques since we are also interested in obtaining the properties of flares such as size, lifetime, and motion.

We developed the programs in Interactive Data Language (IDL) by Research Systems, Inc. The program runs on a DELL Dimension L733r with CPU time 733 Mhz and memory of 256 MB under Windows 2000. There are three steps in our program:

1. Preprocessing to obtain the nine features of solar flares
2. MLP, RBF, and SVM training and testing programs used for solar flare detection
3. Region growing and edge detection methods for obtaining the flare properties

For the experiments of solar flare detection, we capture one image per minute and process the current image with comparisons of previous images to obtain the nine features. First, we filter the noise of full-disk images using a Gaussian function and center these two images using the IDL FIT_LIMB function. Second, we compute the mean brightness and standard deviation for the two images. Third, we divide them to obtain the pixel with the maximum gray level difference. It is regarded as the key pixel to compute other features. After preprocessing steps, we use the nine features of the key pixel as the input to the neural network and SVM. Finally, we use region growing combined with Canny edge detector to analyze the region of the flares.

We test the MLP, RBF, and SVM methods based on the 240 events including preflare, main phase, and postflare. Half of them are used as the training set and the rest half as the testing set. For MLP, after about 1000 iterations for the training set of 120 events, we test this network with 120 events. The fraction of misclassified events is about 5%. The MLP uses supervised training and needs a lot of iterations to obtain the satisfied classification rate. For the RBF, we obtain a better classification rate and faster training speed. For the SVM, we obtain the best performance among the three methods, such that the misclassification rate is 3.3%. The results are shown in Table 14.6.

The MLP architecture is to perform a full nonlinear optimization of the network. Therefore, a lot of iterations to train the system to get the best result are needed. In Table 14.7, it is demonstrated that the result performs better with more iterations. The number of hidden nodes determines the complexities of neural networks. We can adjust the number of hidden nodes to obtain the best performance. In Table 14.7, we show the comparison of MLP with 11 hidden nodes and MLP with other number of hidden nodes. Test correctness of MLP with 11 hidden nodes is comparable to MLP with more hidden nodes. MLP with 11 hidden nodes has better performance than MLP with more or less hidden nodes.

RBF neural network, as compared with the MLP, is the possibility of choosing suitable parameters for each hidden unit without having to perform a full nonlinear optimization of the network (Bishop, 1995). Usually, the number of hidden nodes equals the number of classes. For our case, we want to use RBF classifier to separate two classes. We try RBF with different number of hidden nodes and find that RBF with two hidden nodes is good for our pattern classification. The parameters of RBF are center vector, the weight, and the radius. The training architecture can be quickly constructed by *K*-mean algorithm.

SVM is based on the idea of "minimization of the structural risk." To obtain the best classification rate, we consider the confidence interval and empirical risk. Complex system has high confidence interval and low empirical risk. We test our

TABLE 14.6 Classification Report Based on 120 Training Events and 120 Testing Events

Methods based on events	Classification rate	Training time	Testing time
MLP with 1000 iterations	94.2%	60.20 s	0.01 s
RBF	95%	0.27 s	0.01 s
SVM	96.7%	0.16 s	0.03 s

TABLE 14.7 Comparison of Different MLP Iterations and Hidden Nodes on 120 Training Events and 120 Testing Events

Methods based on events	Classification rate	Training time
MLP with 1000 iterations and 11 hidden nodes	94.2%	60.2 s
MLP with 100 iterations and 11 hidden nodes	69.2%	6.2 s
MLP with 3000 iterations and 11 hidden nodes	94.2%	179.3 s
MLP with 1000 iterations and 11 hidden nodes	94.2%	60.2 s
MLP with 1000 iterations and 6 hidden nodes	92.5%	46.7 s
MLP with 1000 iterations and 20 hidden nodes	94.2%	115.6 s

TABLE 14.8 Comparison of Different SVM Training Strategies on 120 Training Events and 120 Testing Events

Methods based on events	Classification rate	Training time	Testing time
Linear SVM	96.7%	0.16 s	0.03 s
SVM with polynomial kernel	95%	0.34 s	0.03 s
SVM with RBF kernel	90.83%	0.44 s	0.03 s

flare events using linear SVM and nonlinear SVMs, respectively, with polynomial kernel classifier and Gaussian RBF kernel classifier. Experimental results are shown in Table 14.8.

Table 14.8 shows that linear SVM is better than nonlinear kernel SVM for our events. Nonlinear SVM can reduce the empirical risk using complex systems, but it has higher confidence interval. Using linear SVM with the nine features of the solar flare, we have both low empirical risk and confidence interval to achieve the best total risk. Therefore, we prefer linear SVM to classify the flare patterns.

14.3 SOLAR CORONA MASS EJECTION DETECTION

The structures of CMEs can be grossly arranged into eight categories (Howard et al., 1985): halo, curved front, loop, spike, double spike, multiple spike, streamer blowout, diffuse fan, and complex. In an effort to get an intensity classification, Dai et al. (2002) analyzed the velocity, span, mass, and kinetic energy of CMEs and proposed three CME intensity categories, strong (including halo complex), middle (including double spike, multiple spike, and loop), and weak (including spike, streamer blowout, and diffuse fan). Among these categories, the halo CMEs, which appear as expanding, circular brightenings surrounding the coronagraph's occulter, are of great concern because they may carry away a mass of up to 10^{15} kg and are moving outward along the Sun–Earth line with speeds greater than 500 km/s (Dai et al., 2002). It appears that halo CMEs are an excellent indicator of increased geoactivity 3–5 days later (Brueckner et al., 1998), and therefore, the automatic and accurate detection and characterization of CMEs, especially halo CMEs, in real time can provide an early warning of the occurrence of potentially geoeffective disturbance. In Figure 14.25, examples of strong and weak CMEs in LASCO C2 images are pointed by arrows. Fu et al. (2007) proposed automatic detection of prominence eruption using consecutive solar images.

The traditional way of detecting CMEs is based on human observation. For example, the SOHO/LASCO group provided a CME catalog based on image enhancement and human detection. However, the human observation is subjective and slow, and decision rules may vary from one operator to another. During the last few years, image processing and pattern recognition techniques have been utilized in automatic CME detection. For example, Berghmans (2002) introduced an automatic

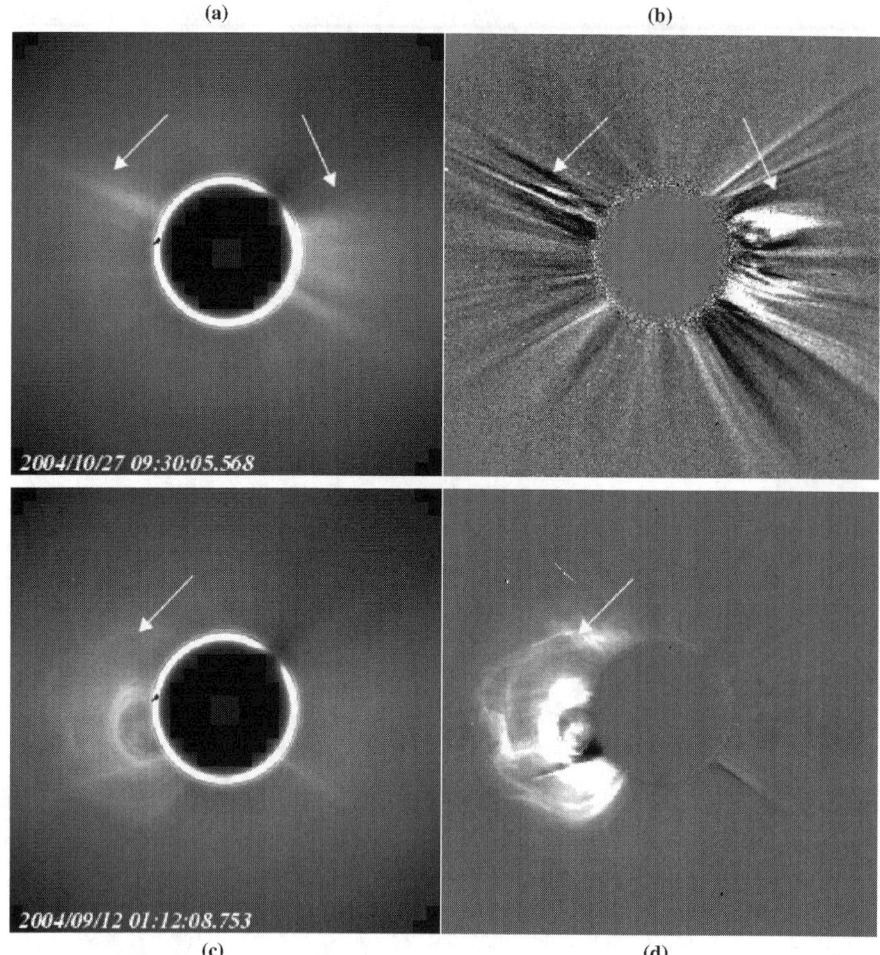

Figure 14.25 (a) Weak CMEs in a LASCO C2 image on October 27, 2004, (b) the same weak CMEs in the running difference image, (c) a strong CME in a LASCO C2 image on September 12, 2004, and (d) the same strong CME in the running difference image. CMEs are pointed by arrows.

CME detection procedure using image processing techniques such as image enhancement, thresholding, and motion tracking. Robbrecht and Berghmans (2004) developed an application, CACTus (Computer Aided CME Tracking), for automatic detection of CMEs in sequences of LASCO images that uses Hough transform to detect CMEs in the running differences. They also published their real-time CME detection results through the Internet.

We present an automatic three-step CME detection and classification algorithm. In brief, the three steps are (1) preprocessing, in which running difference images of LASCO C2 and C3 are automatically generated, (2) detection and characterization, in which the structural properties of CMEs (such as height, span angle) are characterized using image segmentation and morphological methods, and (3) classification, in which SVM classifiers are used to classify CMEs into three groups of strong, medium, and weak on the basis of characterization.

14.3.1 Preprocessing

SOHO/LASCO provides C2 and C3 images in a Flexible Image Transport System (FITS) format that can be downloaded from the SOHO/LASCO Web site. Image information such as size, center coordinates, and exposure time is written into the headers of FITS images. In our preprocessing step, LASCO images in different sizes are resized to the same size of 512 × 512 pixels and aligned according to coordinates of solar centers. Like most other CME detections, brightness normalization is applied according to the ratio of exposure time of LASCO images. Furthermore, if the gray level ranges of mean brightness between two consecutive images are different after the exposure time adjustment, the images are normalized using the mean brightness ratio of the current image to the previous image. In our preprocessing, when the ratio of their gray level is greater than 1.1 or less than 0.9, equation (14.23) is applied to correct the brightness of the current image.

$$M'_{\mathrm{c}} = M_{\mathrm{c}} \frac{\bar{x}_{\mathrm{p}}}{\bar{x}_{\mathrm{c}}} \tag{14.23}$$

where M_{c} and M'_{c} denote the current images before and after the normalization, and $\bar{x}_{\mathrm{p}}$ and $\bar{x}_{\mathrm{c}}$ denote the mean brightness of no-CME regions on the previous and current images, respectively. The mean brightness ratio is based on the regions with no CME because the mean brightness normalization may reduce CMEs' brightness. The pixel with a lower gray level than the median value of the image is considered to be the one located on the no-CME region.

In LASCO images, streamers are structured similarly to CMEs. The difference between streamers and CMEs are that streamers are steady bright structures, while CMEs are sudden brightness changes. To distinguish a CME from other stable bright structures (e.g., streamers) and stable bright noise, a running difference image, obtained by equation (14.24), is commonly used. The running difference image is also useful for the measurement of CMEs' moving front (i.e., the top moving edge) and foot point (i.e., the lower moving bottom).

$$D = M_{\mathrm{c}} - G \tag{14.24}$$

where G is the reference image and M_c is the current LASCO image. The reference image to use with the normal aspect is the image before the current image. This has the disadvantage that if some regions of the previous image are badly captured, the resulting running difference images are badly generated. Another issue is when a CME is evolved on a sequence of image frames, the running difference may succeed in detecting the moving front of the CME, but may fail in detecting its location of foot point because some parts of CME regions are overlapped on a sequence of images. In this book, a reference image G' is produced recursively by using the combination of the reference and previous images, as shown in equation (14.25).

$$G' = Gc_1 + M_c c_2 \tag{14.25}$$

where c_1 and c_2 denote the percentages of effectiveness on the reference and current images. The disadvantage of using a single image as the reference image is avoided. Based on our experiments, we set $c_1 = 90\%$ and $c_1 = 10\%$. However, if a strong CME is detected in the previous image frame, $c_1 = 100\%$ and $c_2 = 0\%$ are used.

In the next section, CMEs are segmented using a threshold based on high brightness regions. When a weak CME with small brightness enhancement is located in the dark region and a strong CME is located nearby, the weak CME may not be detected because the chosen threshold is focused on the strong one. To overcome this problem, a partial division image is used to segment CME regions based on the ratio of brightness between the current LASCO image and the reference image. The division image is obtained by dividing the LASCO image by the reference image. For the region with gray level close to zero, the division may enhance noise significantly and produce overflow errors. Therefore, only a partial region of the image is chosen. The partial division image V is shown in equation (14.26).

$$V = \frac{M_2}{G_2} \tag{14.26}$$

where G_2 denotes the region on the reference image with pixels brighter than the median value of the reference image and M_2 denotes the region corresponding to G_2 on the LASCO image. By using the relatively bright regions for G_2, the division over flow is avoided.

Some of LASCO images contain missing blocks that appear as dark or bright squares on image frames. Three criteria are proposed to find the missing blocks: (1) when a dark missing block first appears in the current image, its appearance on the current image and the running difference images are both dark; (2) its gray level of the current image is close to the minimum value of the image, and its gray level of the running difference image is less than a negative threshold t_1; (3) the number of pixels for an individual block region is set to be greater than a threshold t_2 because a missing block contains a large amount of pixels as compared to small noise. The thresholds t_1 and t_2, obtained from our experiments, are negative values of the standard deviation of the running difference in brightness and 50 pixels, respectively. A bright missing block can also be detected in the same way. Finally, those missing blocks in running difference images and division images are replaced by the minimum value of the images.

14.3.2 Automatic Detection of CMEs

Two consecutive LASCO images incorporated with the reference image produce two running difference images and two division images. The running difference images and division images are segmented into binary differences by applying a thresholding method. By tracking segmented CME regions in binary differences, the properties of CMEs are obtained.

14.3.2.1 Segmentation of CMEs Running difference images (D) and division images (V) are constructed by using consecutive LASCO images in the previous step. In this step, a thresholding method is adopted to D and V to produce binary differences in which CME regions are separated from the background. A fixed value of the threshold would produce unstable results because contrasts of running difference images vary from time to time. In our CME segmentation, the threshold is computed using the median and the standard deviation of D and V. Based on extensive experiments, the thresholds for D and V of C2 images are chosen as $m + s$, where m and s are the median and the standard deviation of C2's D and V images, respectively. Similarly, we obtain the thresholds for C3's D and V images, which is $m + 1.5s$. These automatic thresholds are robust with respect to different image contrasts. The final segmentation result is the summation of the two segmentation results from the difference and the division images.

After segmentation, morphological closing (Shih and Mitchell, 1989) is applied to the binary segmentation image to eliminate small gaps. A 5×5 structuring element is used to perform the binary closing. The closing method can smooth features and remove small noise. The 5×5 structuring element removes the noise whose width and length are less than 5 pixels and eliminates the gaps whose distance is less than 5 pixels.

In order to calculate the features of CMEs easily, binary segmentation images are reformed to $360 \times r$ binary angular images, where r is the radius to the edge of the C2 or C3 occulting disk in the LASCO image. The results of the segmentation, morphological closing, and the angular images are shown in Figure 14.26. The degree of angle [0, 359] is counted from north clockwise. The size of C2 angular images is 360 pixels $\times$ 156 pixels, and of C3 angular images is 360 pixels $\times$ 216 pixels.

14.3.2.2 Features of CMEs In the feature detection, all the segmented regions in a CME frame become CME candidate regions, but only continuously moving regions on consecutive images are treated as real CME regions. A CME candidate region is classified as a CME region if it occurs on the current image frame as well as on the previous image frame. The corresponding rules are given as follows: (1) the distance between the two centers of overlapping regions is less than t_3, where t_3 is the threshold for the maximum CME movement for an hour, and (2) the difference between the span widths of two overlapping regions is less than t_4, where t_4 is the threshold for the maximum degree span change for an hour. t_3 is chosen as 100 pixels for the LASCO C2 image, and as 40 pixels for the C3 image. t_4 is chosen as a half of the span width of the CME candidate. In addition, the speed and the new increasing area for CME regions are computed by comparing the corresponding CME regions on the two

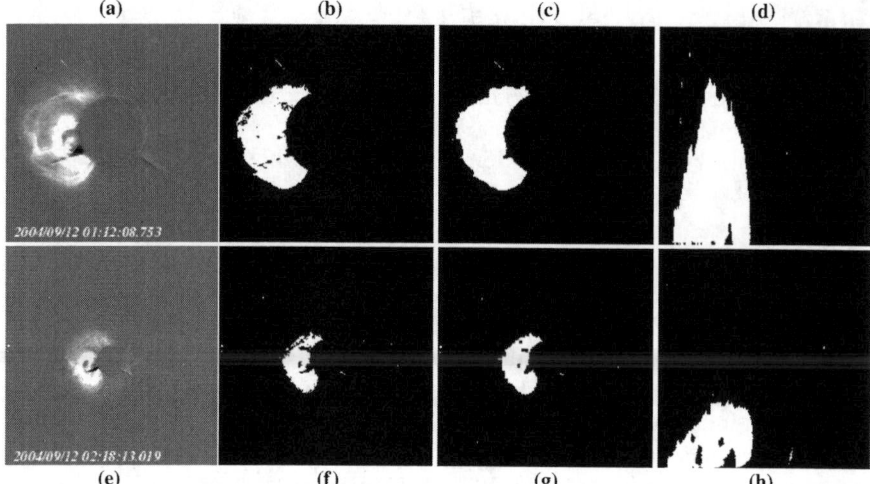

Figure 14.26 (a) The running difference of a LASCO C2 image, observed on September 12, 2004, (b) binary segmented result of the LASCO C2, (c) morphological closing result of the LASCO C2, and (d) angular image of the LASCO C2. From (e) to (h): the running difference of a LASCO C3 image, and its segmented, closing and angular, results, respectively.

binary angular images. Let us denote a CME region in the current image and its corresponding region in the previous image as A and A_{p}, respectively. The CME properties of A in an image frame are listed in Table 14.9.

The detection reports for a CME on August 7, 2005 are shown in Figures 14.27 and 14.28. The classification of strong, medium, and weak CMEs is presented in the next section. Figures 14.29 and 14.30 show the height and velocity for the CMEs on September 1, 2002 in C2 and C3 images, respectively. The speed of the CME is around 300 km/s, which is almost equal to the computed CME speed in the LASCO catalog.

14.3.3 Classification of Strong, Medium, and Weak CMEs

A strong CME consists of a large amount of fast moving mass. In our CME classification step, strong CMEs are selected from halo, curved front and complex CMEs according to Howard et al. (1985). In this work, the SVM classifier with a linear kernel is used for distinguishing the strong CME from others. For the present study, a CME can be represented by 22 features that are obtained in the previous step. The SVM training computes the corresponding weights for each input feature for the CME classification. Six features with significant weights are selected as the inputs for classification. The six input features are proven to be robust based on our comparisons on feature combination.

SVM is a powerful learning system for data classification. The idea of SVMs is to divide the fixed given input pattern vectors into two classes using a hyperplane with the maximal margin, which is the distance between the decision plane and the closest sample points (Vapnik, 1998). When support vectors (the points on the decision plane) are found, the problem is solved. Support vectors on the decision plane can be found

TABLE 14.9 The Properties of a CME Region

No.	Description of the CME properties
1	The exposure time of the LASCO image
2	The time interval between the current and the previous image
3	The pixel size of the LASCO image
4	The mean brightness value of the reference image
5	The mean brightness value of the current image
6	The mean brightness value of the running difference
7	The standard deviation of the running difference
8	The number of pixels for A
9	The threshold for segmenting A from the running difference
10	The maximum height (arcsecs from disk center) of A
11	The height of the center of A
12	The minimum height of A
13	The starting angle of A. The angle is calculated from North 0 clockwise
14	The angle of the center of A
15	The ending angle of A
16	The angular width of A
17	The height difference (h_1) between the maximum height of A and A_p
18	The height of the new moving region (h_2) that is obtained by subtracting A_p from A
19	The speed that is computed using h_1 divided by the interval time cadence
20	The speed that is computed using h_2 divided by the interval time cadence
21	The span width of the new moving region
22	The center angle of the new moving region

The features 17–22 are obtained using the corresponding regions on the two consecutive images.

by the classical method of Lagrange multipliers. Compared to other learning machine such as neural networks based on the empirical risk minimization (ERM), SVM is based on the structural risk minimization (SRM) inductive principle. The SRM principle is intended to minimize the risk functional with respect to the empirical risks in the ERM and the confidence interval (Vapnik, 1998). Because SVM considers both terms in the risk function, its classifier is better than neural networks that only consider the empirical risks. For the nonlinear case, SVM uses the kernel functions to transfer input data to feature space. For example, polynomial support vector classifier and Gaussian radial basis function kernel classifier are two popular nonlinear classifiers. In our study, a simple linear support vector machine is used. To increase the complexity of classification, the nonlinear kernel could be used in the future study. This is the first time that the SVM is applied to the CME classification. The comparisons between the linear and other kernels can be found in Qu et al. (2003). For further information regarding the SVM, readers can refer to Vapnik (1998).

The six inputs to our SVM classifier are the features with significant classification weights in Table 14.5 that are the mean brightness in the running difference image, the number of pixels in the running difference image, the angular width of span, the height of new moving region, the span width of the new moving region, and

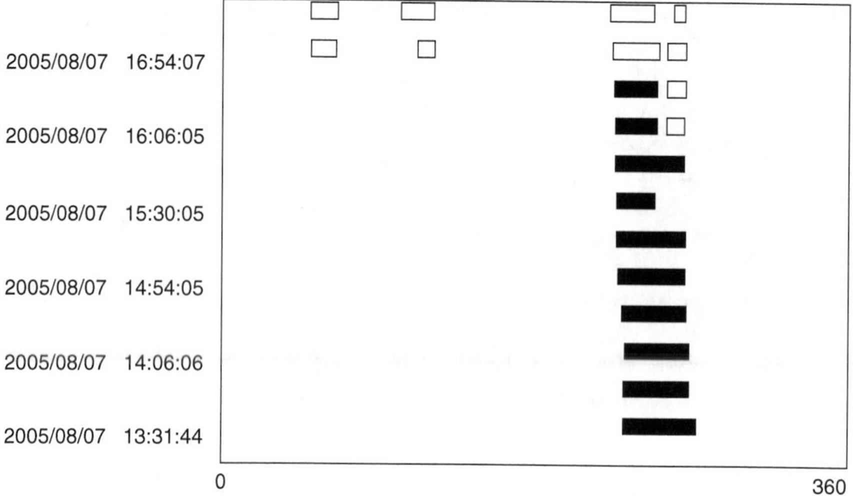

Figure 14.27 Detection and classification report for a detected CME using LASCO C2 images on August 7, 2005. Apparently, there is a strong CME on the West from 13:30 to 16:50 UT. Solid black and empty rectangles denote strong and weak CMEs, respectively.

the speed described in feature 20 of Table 14.9. A list of CMEs are randomly selected in 2004. Assuming that human classification for strong CMEs is 100% accurate, we can select 50 strong (halo, curved front and most complex) and 50 other (medium and weak) CMEs by searching through a sequence of running difference images. The SVM classifier is trained by human classification results with the 100 CMEs based on

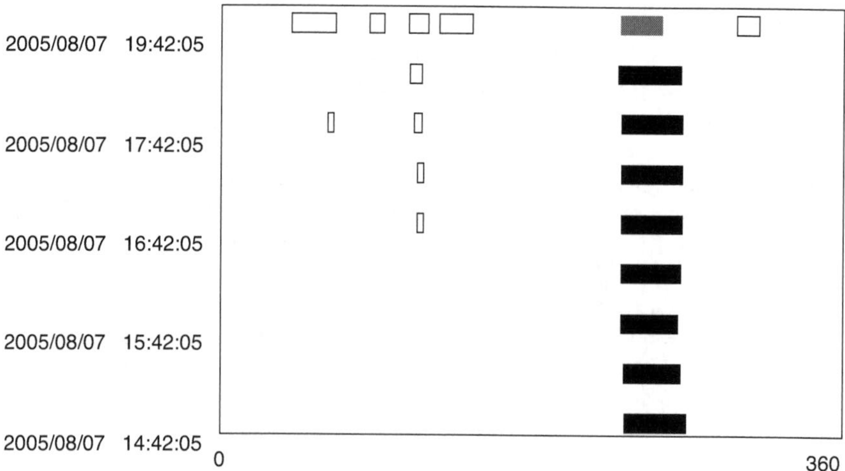

Figure 14.28 Detection and classification report for a detected CME using LASCO C3 images on August 7, 2005. There is a strong CME on the West from 14:40 to 19:40 UT. Solid black, solid gray, and empty rectangles denote strong, medium, and weak CMEs, respectively.

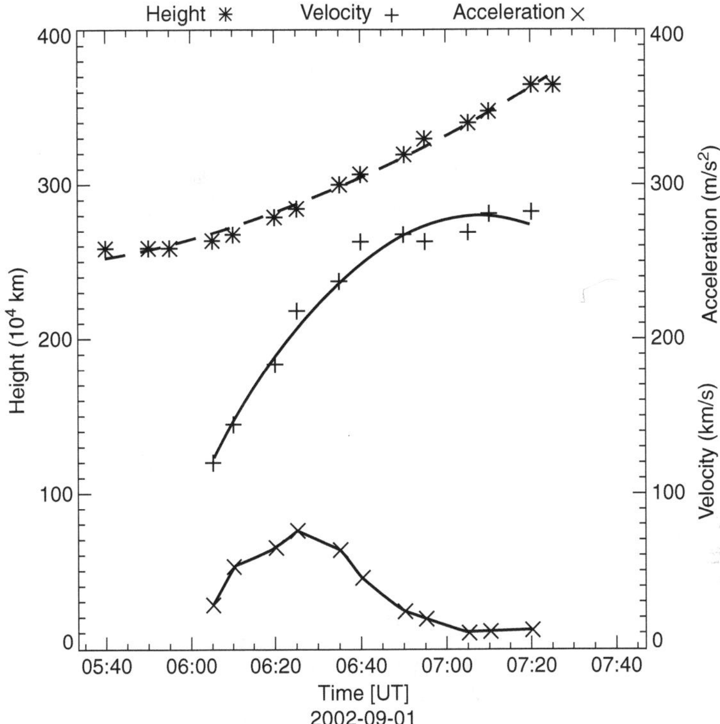

Figure 14.29 Height, velocity, and acceleration profile of a CME for LASCO C2 images on September 1, 2002.

the aforementioned six inputs. After training, the SVM classifier is able to classify strong CMEs from others automatically. The classification rate of the testing experiments is shown in Table 14.10.

After finding strong CMEs using our SVM classifier, further classification is applied to distinguish medium from weak CMEs using the rule proposed by Howard et al. (1985) and Dai et al. (2002). The speed of medium CMEs is greater than 300 km/s, while the speed of weak CMEs is less than 300 km/s.

14.3.4 Comparisons for CME Detections

We have developed the software to detect and characterize CMEs. The programs were developed in IDL by Research Systems, Inc., and run on a DELL Dimension 4600 PC with CPU time 2.8 GHz and memory of 512 MB under Linux. The computational time for detecting a CME using three LASCO images is about 5 s, which is far less than the observational interval. The results in our catalog include a list of CMEs, a sequence of CME image frames, the classification type of each CME frame, and the properties of each CME region, such as height, velocity, and angular width. The properties and classification results of CMEs are saved in our database available through our Web site.

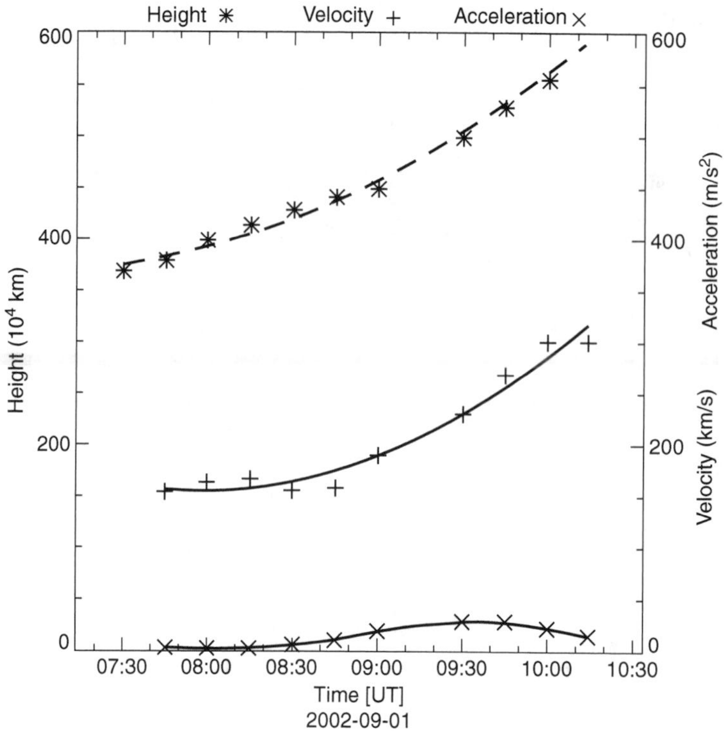

Figure 14.30 Height, velocity, and acceleration profile of a CME for LASCO C3 images on September 1, 2002.

There are two, previously developed, CME detection catalogs available to the public. From the LASCO Web site at http://cdaw.gsfc.nasa.gov/cme_list/, one can find the CME catalog created by visual inspection. Robbrecht and Berghmans presented their results at http://sidc.oma.be/cactus/. Our automatic detection results are currently shown at http://filament.njit.edu/detection/vso.html. It is difficult to compare the CME catalogs because there is no comprehensive catalog to be used as a reference to conduct the comparisons (Berghmans, 2002). A CME in one catalog may be considered as two CMEs in another catalog, and the beginning and ending time for a CME is hard to determine. The reasons are that the preprocessing methods and detection rules of three catalogs are different and human decision is subjective. The

TABLE 14.10 Success Classification Rates of the Strong and Nonstrong CMEs Based on 50 Strong CMEs and 50 Nonstrong CMEs

LASCO	Strong CMEs	Weak and medium CMEs
C2	94%	96%
C3	96%	96%

We assume the detection by human operators to be 100% accurate.

reference image and threshold selection may affect the final decision of a CME detection.

We select results between August 1 and 31, 2004 to perform comparisons among the three catalogs. The LASCO catalog, used as the reference, listed 65 CMEs in this period. Our catalog missed one weak CME that happened on the West at 316° on August 26, 2004 at 16:54 UT. CACTus missed three weak CMEs that happened on the East at 77° on August 2, 2004 at 23:06 UT, on the East at 94° on August 22, 2004 at 17:30 UT, and on the East at 111° on August 26, 2004 at 21:54 UT. The missed CMEs in our catalog and in CACTus are weak CMEs. On the other hand, some CMEs missed in LASCO can be detected in CACTus and our catalog. For example, the CMEs happened on the East on August 6, 2004 at 22:30 UT, on the East on August 7, 2004 at 18:54 UT, and on the Southeast on August 9, 2004 at 21:30 UT. By combining the results of CACTus and our catalog, the CME detection results are more complete and accurate than the LASCO catalog that is based on human eye detection. In the three catalogs, the center of principal angles, angular width, and speed of the strong CMEs are given as the important CME properties. The detected properties of CMEs in the different catalogs could be varied because the region merging criteria are different. As we said previously, a CME in one catalog may count as two in another. Overall, the center of principal angles, the angular width, and the speed, especially for the strong CMEs, are similar in all the three catalogs.

REFERENCES

Berghmans, D., "Automated detection of CMEs," The 10th European Solar Physics Meeting SP-506, Prague, Czech. Republic, pp. 85–89, Sep. 2002.

Bishop, C. M., *Neural Networks for Pattern Recognition*, Oxford University Press, Oxford, 1995.

Borda, R. A., Mininni, P. D., Mandrini, C. H., Gómez, D. O., Bauer, O. H., and Rovira, M. G., "Automatic solar flare detection using neural network techniques," *Sol. Phys.*, vol. 206, no. 2, pp. 347–357, Apr. 2002.

Brueckner, G. E., Howard, R. A., Koomen, M. J., Korendyke, C. M., Michels, D. J., Moses, J. D., Socker, D. G., Dere, K. P., Lamy, P. L., Llebaria, A., Bout, M. V., Schwenn, R., Simnett, G. M., Bedford, D. K., and Eyles, C. J., "The Large Angle Spectroscopic Coronagraph (LASCO)," *Sol. Phys.*, vol. 162, no. 1–2, pp. 357–402, Dec. 1995.

Brueckner, G. E., Delaboudiniere, J.-P., Howard, R. A., Paswaters, S. E., St. Cyr, O. C., Schwenn, R., Lamy, P., Simnett, G. M., Thompson, B., and Wang, D., "Geomagnetic storms caused by coronal mass ejections (CMEs): March 1996 through June 1997," *Geophys. Res. Lett.*, vol. 25, pp. 3019–3022, 1998.

Cleveland, F., Malcolm, W., Nordell, D.E., and Jack Zirker, J., "Solar effects on communications," *IEEE Trans. Power Delivery*, vol. 7, no. 2, pp. 460–468, Apr. 1992.

Dai, Y., Zong, W., and Tang, Y., "A quantitative research on the classification of coronal mass ejections," *Chin. Astron. Astrophys.*, vol. 26, no. 2, pp. 183–188, Apr. 2002.

Denker, C., Johannesson, A., Marquette, W., Goode, P.R., Wang, H., and Zirin, H., "Synoptic Hα full-disk observations of the sun from big bear solar observatory. I. Instrumentation, image processing, data products, and first results," *Sol. Phys.*, vol. 184, no. 1, pp. 87–102, Jan. 1999.

Dougherty, E.R., "Optimal conjunctive granulometric bandpass filters," *J. Math. Imaging Vis.*, vol. 14, no. 1, pp. 39–51, Feb. 2001.

Dougherty, E.R. and Chen, Y., "Granulometric filters," in *Nonlinear Filtering for Image Processing* (E. Dougherty and J. Astola,eds.), SPIE and IEEE Presses, Bellingham, WA, pp. 121–162, 1999.

Fu, G., Shih, F. Y., and Wang, H., "Automatic detection of prominence eruption using consecutive solar images," *IEEE Trans. Circuits Syst. Video Technol.*, vol. 17, no. 1, pp. 79–85, Jan. 2007.

Gao, J., Wang, H., and Zhou, M., "Development of an automatic filament disappearance detection system," *Sol. Phys.*, vol. 205, no. 1, pp. 93–103, Jan. 2002.

Gonzalez, R.C. and Woods, R.E., Digital Image Processing, 3rd edition, Prentice Hall, 2008.

Harrison, R. A., "The nature of solar flares associated with coronal mass ejection," *Astron. Astrophys.*, vol. 304, pp. 585–594, Dec. 1995.

Howard, R. A., Sheeley, N. R., Jr., Michels, D. J., and Koomen, M. J., "Coronal mass ejections: 1979–1981," *J. Geophys. Res.*, vol. 90, pp. 8173–8191, Sep. 1985.

Irbah, A., Bouzaria, M., Lakhal, L., Moussaoui, R., Borgnino, J., Laclare, F., and Delmas, C., "Feature extraction from solar images using wavelet transform: image cleaning for applications to solar astrolabe experiment," *Sol. Phys.*, vol. 185, no. 2, pp. 255–273, Apr. 1999.

Krejcí, R., Flusser J., and Šimberová, S., "A new multichannel blind deconvolution method and its application to solar images," Proceedings of the International IEEE Conference on Pattern Recognition, vol. 2, Brisbane, Qld., Australia, pp. 1765–1768, Aug. 1998.

Martens, P.C. and Zwaan, C., "Origin and evolution of filament-prominence systems," *Astrophys. J.*, vol. 558, no. 2, pp. 872–887, Sep. 2001.

Molina, R., Nunez J., Cortijo, F.J., and Mateos, J., "Image restoration in astronomy: a Bayesian perspective," *IEEE Signal Process. Mag.*, vol. 18, no. 2, pp. 11–29, Mar. 2001.

Neckel, H. and Labs, D., "The solar radiation between 3300 and 12500 Å," *Sol. Phys.*, vol. 90, no. 2, pp. 205–258, Feb. 1984.

Neckel, H. and Labs, D., "Solar limb darkening 1986–1990 ($\lambda\lambda$ 303 to 1099 nm)," *Sol. Phys.*, vol. 153, no. 1–2, pp. 91–114, Aug. 1994.

Orriols, X., Toledo, R., Binefa, X., Radeva, P., Vitria, J., and Villanueva, J. J., "Probabilistic saliency approach for elongated structure detection using deformable models," *Proceedings of the International Conference on Pattern Recognition*, vol. 3, Barcelona, Spain, IEEE Computer Society Press, pp. 1006–1009, Sep. 2000.

Preminger, D.G., Walton, S.R., and Chapman, G.A., "Solar feature identification using contrasts and contiguity," *Sol. Phys.*, vol. 202, no. 1, pp. 53–62, Aug. 2001.

Qu, M., Shih, F. Y., Jing, J., and Wang, H., "Automatic solar flare detection using MLP, RBF, and SVM," *Sol. Phys.*, vol. 217, no. 1, pp. 157–172, Oct. 2003.

Qu, M., Shih, F. Y., Jing, J., and Wang, H., "Automatic solar flare tracking using image processing techniques," *Sol. Phys.*, vol. 222, no. 1, pp. 137–149, July 2004.

Qu, M., Shih, F. Y., Jing, J., and Wang, H., "Automatic solar filament detection using image processing techniques," *Sol. Phys.*, vol. 228, no. 1, pp. 119–135, May 2005.

Robbrecht, E. and Berghmans, D., "Automated recognition of coronal mass ejections (CMEs) in near-real-time data," *Astron. Astrophys.*, vol. 425, pp. 1097–1106, 2004.

Shih, F. Y. and Mitchell, O. R., "Threshold decomposition of grayscale morphology into binary morphology," *IEEE Trans. Pattern Anal. Mach. Intell.*, vol. 11, no. 1, pp. 31–42, Jan. 1989.

Shih, F. Y. and Pu, C. C., "Analysis of the properties of soft morphological filtering using threshold decomposition," *IEEE Trans. Signal Processing*, vol. 43, no. 2, pp. 539–544, Feb. 1995.

Shih, F. Y. and Puttagunta, P., "Recursive soft morphological filters," *IEEE Trans. Image Processing*, vol. 4, no. 7, pp. 1027–1032, July 1995.

Shih, F. Y., King, C. T. and Pu, C. C., "Pipeline architectures for recursive morphological operations," *IEEE Trans. Image Processing*, vol. 4, no. 1, pp. 11–18, Jan. 1995.

Soille, P., Breen, E. J. and Jones, R, "Recursive implementation of erosions and dilations along discrete lines at arbitrary angles," *IEEE Trans. Pattern Anal. Mach. Intell.*, vol. 18, no. 5, pp. 562–567, May 1996.

Toledo, R., Orriols, X., Binefa, X., Radeva, P., Vitria, J., and Villanueva, J. J., "Tracking elongated structures using statistical snakes," *Proceedings of the IEEE Conference on Computer Vision and Pattern Recognition*, IEEE Computer Society Press, pp. 157–162, 2000.

Turmon, M.J. and Mukhtar, S., "Recognizing chromospheric objects via Markov chain Monte Carlo," *Proceedings of the International Conference on Image Processing*, vol. 3, IEEE Computer Society Press, pp. 320–323, Oct. 1997.

Vapnik, N. V., *Statistical Learning Theory*, John Wiley & Sons, Inc., New York, 1998.

Veronig, A., Steinegger, M., Otruba, W., Hanslmeier, A., Messerotti, M., Temmer, M., Brunner, G., and Gonzi, S., "Automated image segmentation and feature detection in solar full-disk images," *Solar and Space Weather Euroconference, European Space Agency (ESA)*, vol. 463, pp. 455–458, Sep. 2000.

Walter, T., Klein, J.-C., Massin, P. and Zana, F., "Automatic segmentation and registration of retinal fluorescein angiographies: application to diabetic retinopathy," *Proceedings of the First International Workshop on Computer Assisted Fundus Image Analysis (CAFIA)*, Copenhagen, Denmark, 2000.

Webb, D.F., "Coronal mass ejections: origins, evolution and role in space weather," *IEEE Trans. Plasma Sci.*, vol. 28, pp. 1795–1806, 2000.

Zana, F. and Klein, J.-C., "A multimodal registration algorithm of eye fundus images using vessels detection and Hough transform," *IEEE Trans. Med. Imaging*, vol. 18, no. 5, pp. 419–428, 1999.

Zana, F. and Klein, J.-C., "Segmentation of vessel-like patterns using mathematical morphology and curvature evaluation," *IEEE Trans. on Image Process.*, vol. 10, no. 7, pp. 1010–1019, 2001.

Zirin, H., *Astrophysics of the Sun*, Cambridge University Press, Cambridge, 1988.

INDEX

Image Processing and Pattern Recognition by Frank Shih